LEGAL ASPECTS OF CARBON TRADING

Legal Aspects of Carbon Trading

Kyoto, Copenhagen, and beyond

Edited by
DAVID FREESTONE
and
CHARLOTTE STRECK

OXFORD
UNIVERSITY PRESS

OXFORD
UNIVERSITY PRESS

Great Clarendon Street, Oxford OX2 6DP

Oxford University Press is a department of the University of Oxford.
It furthers the University's objective of excellence in research, scholarship,
and education by publishing worldwide in

Oxford New York

Auckland Cape Town Dar es Salaam Hong Kong Karachi Kuala Lumpur Madrid
Melbourne Mexico City Nairobi New Delhi Shanghai Taipei Toronto

With offices in

Argentina Austria Brazil Chile Czech Republic France Greece
Guatemala Hungary Italy Japan Poland Portugal Singapore
South Korea Switzerland Thailand Turkey Ukraine Vietnam

Oxford is a registered trade mark of Oxford University Press
in the UK and in certain other countries

Published in the United States
by Oxford University Press Inc., New York

First published 2009

British Library Cataloguing in Publication Data
Data available

Library of Congress Cataloging-in-Publication Data
Legal aspects of carbon trading : Kyoto, Copenhagen, and beyond / edited by David Freestone and Charlotte Streck.
p. cm.
Includes index.
ISBN 978-0-19-956593-1 (alk. paper)
1. Emissions trading–Law and legislation. 2. Carbon offsetting–Law and legislation. I. Freestone, David. II. Streck, Charlotte.
K3593.5C37 L44 2009
341.04 6342–dc22
2009028806

Typeset by Macmillan Publishing Solutions
Printed on acid-free paper by
Antony Rowe, Chippenham, Wiltshire

ISBN 978–0–19–956593–1

1 3 5 7 9 10 8 6 4 2

Acknowledgements

We are most grateful to the contributors to this volume for finding the time to meet our stretch targets for completion of their chapters.

We would like to thank The George Washington University Law School for providing research support to David Freestone during the Spring 2009 Semester and to Scott Nuzum and Natalia Roban for their outstanding research assistance during that period.

We are also grateful to John Louth, Merel Alstein and Bethan Cousins at OUP for helping us keep on schedule.

Finally, special thanks to Jenny Freestone and Michael Streck for their support and encouragement.

Preface

Climate change is the overriding environmental challenge of our time. The threats that it poses to global weather patterns and natural ecosystems are difficult to underestimate and if left unchecked the 'runaway' greenhouse effect could threaten the very future of human life on the planet. In the shorter term, the adverse effects of climate change, many of which are beginning to be felt already, risk rolling back many of the hard won advances made in human development.[1] In addition, the implementation of the measures necessary to reduce greenhouse gas emissions to the atmosphere will affect all sections of society as they will involve virtually all sectors of our economies.

In addressing climate change, we face tough choices. There are no simple solutions. Environment, economy and human development are closely interlinked; effective climate policy has to lead to a major shift in the way we live and do business that will ultimately influence everything and everybody. What is a challenge for developed, industrialized and sophisticated economies is even more daunting for developing countries, which will be asked to pioneer untested low-carbon development patterns. Moving to a low carbon economy has to become an integral part of our daily lives.

This shift to a low carbon development paradigm will require billions of dollars of new money to support, in turn, the redirection of trillions of dollars available from conventional funding and investment streams. The international community is urgently looking for means to promote investments in clean technology and systems worldwide, especially in developing countries. The 2007 Fourth Assessment Report of the IPCC, in particular Working Group III, signals clearly why this needs to happen:

> With current climate change mitigation policies and related sustainable development practices, global GHG emissions will continue to grow over the next few decades: CO_2 emissions between 2000 and 2030 from energy use are projected to grow 45 to 110% over that period. Two thirds to three quarters of this increase in CO_2 emissions is projected to come from non-Annex I regions, with their average per capita energy CO_2 emissions being projected to remain substantially lower than those in Annex I regions in 2030.

Estimates of the overall funding levels needed for adaptation, mitigation and technology transfer vary widely, but to return global greenhouse gas emissions to

[1] 'The Anatomy of a Silent Crisis' Global Humanitarian Forum Report, Geneva, 2009. Available at http://ghfgeneva.org/Portals/0/pdfs/human_impact_report.pdf accessed 29 July 2009.

2005 levels incremental costs will be in the order of US$ 200-300 billion per annum up to 2030.[2] As well as regulating existing public sector funds administered by mechanisms under the supervision of the UNFCCC, such as the Global Environment Facility (GEF), effective climate change financing arrangements must also take into account the role of private sector funding, of carbon market finance, and of bilateral and multilateral funding as well as of national budgets. In fact, it can be argued that the ability to leverage these other sources of funding, to encourage their growth and also to operate any international financial mechanisms in a complementary and compatible way, will rank among the most important features of a successful international climate change treaty regime.

More than a decade after the adoption of the Kyoto Protocol, it should be apparent that many of the elements of such a regime are already in place. Indeed it is the success of the carbon finance 'opportunity' that this volume explores. For example, the Clean Development Mechanism (CDM), born in the last hours of the negotiation of the Kyoto Protocol with modest expectations, is the corner stone of the international carbon market and offers a remarkable success story. The CDM initiated a paradigm shift in the modes of support of developing country action under a multilateral environmental treaty. When designing the CDM, negotiators no doubt relied heavily on the experience of the GEF and the Montreal Protocol Multilateral Fund; however, two essential innovations were introduced into the design of the CDM fundamentally altering the operational reality of the mechanism. First, project investment was linked to the creation of tradable emission certificates and, second, private entities were invited to participate in the mechanism, provided that such entities were authorized by state parties. These key features created the fundamentals of a truly global carbon market.

It is also clear that national policy makers have also been strongly influenced by the experiences gained in the implementation of the Kyoto Protocol flexible mechanisms, which include not only the CDM, but also Joint Implementation and international emission trading. Over the last few years they have increasingly opted for emissions trading as a preferred instrument of domestic climate policy. After the pioneering schemes in Denmark and the UK, the EU Emissions Trading Scheme is the largest such scheme ever developed. Since then subnational schemes have been established in New South Wales in Australia and in the North-east of the US. Further efforts led by California are underway in the Western US and also in Canada. In addition, at the time of writing, the US, Australia and New Zealand are all in the process of adopting national schemes.

[2] UNFCCC (2007). Investment and Financial Flows to Address Climate Change. The IEA cost estimates are higher than those of the UNFCCC. The EU Commission uses a Joint Research Council estimate of net global yearly incremental cost of EUR175 billion by 2020. The Project Catalyst using McKinsey's Abatement Cost Curve 2.0 calculates developing country abatement would require on average EUR210 billion per annum between 2010 and 2030.

But, even taken together, these substantial efforts are simply not enough. Scientific evidence suggests that the world has a reasonable chance of containing warming below 2°C if long-term greenhouse gas concentrations are stabilized below 450 parts per million (ppm). According to McKinsey, keeping to this 450 ppm pathway will require the world to abate 17 Giga-tonnes (Gt) of CO_2e-quivalent (CO_2e) in 2020 against a business as usual pathway. The McKinsey Global Abatement Cost Curve has identified 19 Gt of potential abatement, a major share of which will, of necessity, have to be in developing countries.

Carbon finance will therefore have to move to the next stage. Future regimes need to be more comprehensive in scope and in geographical coverage than the Kyoto Protocol to allow carbon finance to contribute effectively to the mitigation challenge. Future mechanisms need to build on experience and take into account lessons learned; they need to be bold and innovative in order to pioneer new and ambitious paths to provide incentives for the necessary mitigation measures. This volume sets out the most important lessons from a decade of carbon finance. It takes a legal and policy view, bringing together a group of the most renowned practitioners and scholars in the field of climate change law and policy. Their contributions are both academically rigorous and based on a wealth of practical experience. We are confident that the insights and accumulated expertise collected in this volume can make an important contribution to future institutional and policy choices, as well as to the successful implementation of those choices.

Charlotte Streck and David Freestone

(Brussels and Washington, DC)
1 August 2009.

Contents–Summary

Contents

II. GENERAL ISSUES

V. CARBON TRADING OUTSIDE KYOTO: NATIONAL AND SUB-NATIONAL SCHEMES

VI. VOLUNTARY MARKETS

VII. POST KYOTO: MOVING TOWARDS COPENHAGEN

List of Abbreviations

A21	Agenda 21
AAU	Assigned Amount Unit
AB	Appellate Body
AC	Abatement Certificate
ACCC	Australian Competition and Consumer Commission
AGBM	Ad Hoc Group on the Berlin Mandate
AGO	Australian Greenhouse Office
AICPA	American Institute of Certified Public Accountants
AIE	Accredited Independent Entity
AIJ	Activities Implemented Jointly
AOSIS	Alliance of Small Island States
AP	Accreditation Panel
APEC	Asia-Pacific Economic Co-operation
AWG–KP	Ad Hoc Working Group on Further Commitments for Annex 1 Parties under the Kyoto Protocol
AWG–LCA	Ad Hoc Working Group on Long-Term Cooperative Action under the Convention
A&R	Afforestation and Reforestation
BINGO	Business and Industry Non-Governmental Organization
BIT	Bilateral Investment Treaty
BOT	Build Operate Transfer
CAN	Climate Action Network
CBD	Convention on Biological Diversity
CCA	Climate Change Agreement
CCAR	California Climate Action Registry
CCS	Carbon Capture and Storage
CDM	Clean Development Mechanism
CER	Certified Emission Reduction
CFI	Court of First Instance (of the European Court of Justice)
CICA	Canadian Institute of Chartered Accountants
CISG	United Nations Convention on Contracts for the International Sale of Goods
CITL	Community Independent Transaction Log

CNCCP	China's National Climate Change Programme
COP	Conference of the Parties
COP/MOP	Conference of the Parties to the UNFCCC serving as the Meeting of the Parties to the Kyoto Protocol
CPC	United Nations Provisional Central Product Classification System
CPRS	Carbon Pollution Reduction Scheme (Australia)
CPV	Common Procurement Vocabulary
CSD	Commission on Sustainable Development
CSR	Corporate Social Responsibility
CTE	Committee on Trade and Environment (WTO)
CUP	Cambridge University Press
DEFRA	Department of Environment, Food and Rural Affairs (UK)
DNA	Designated National Authority
DOE	Designated Operational Entity
DoE	Department of Energy (US)
DSB	Dispute Settlement Body
DSU	Dispute Settlement Understanding
DTI	Department of Trade and Industry (UK)
EB	CDM Executive Board
EBRD	European Bank for Reconstruction and Development
ECCP	European Climate Change Programme
ECJ	European Court of Justice
EIA	Environmental Impact Assessment
EIs	Independent Entities
EIT	Economy in Transition
ENGO	Environmental Non-Governmental Organization
EPA	Environmental Protection Agency (US)
EPD	Environmental Protection Department (Hong Kong)
ER	Emission Reduction
ERPA	Emissions Reductions Purchase Agreement
ERU	Emission Reduction Unit
ETS	Emissions Trading Scheme
EUA	European Union Allowance
EU ETS	European Union Emissions Trading Scheme
FASB	Financial Accounting Standards Board

FDI	Foreign Direct Investment
FERC	Federal Energy Regulatory Commission
GAAP	Generally Accepted Accounting Principles
GATS	General Agreement on Trade in Services
GATT	General Agreement on Tariffs and Trade
GDP	Gross Domestic Product
GEC	Gas Electricity Certificates (Australia)
GEF	Global Environment Facility
GERT	Gas Emission Reduction Trading (Canada)
GHG	Greenhouse Gas
GIS	Green Investment Scheme
GMO	Genetically Modified Organisms
GST	Goods and Services Tax
HCA	Host Country Umbrella Agreement
IASB	International Accounting Standards Board
IBRD	International Bank for Reconstruction and Development
ICAO	International Civil Aviation Authority
ICC	International Chamber of Commerce
ICJ	International Court of Justice
ICSID	International Centre for the Settlement of Investment Disputes
IDA	International Development Association
IET	International Emissions Trading
IETA	International Emissions Trading Association
IFC	International Finance Corporation
IFI	International Finance Institution
IFRIC	International Financial Reporting Interpretations Committee
IFRS	International Financial Reporting Standards
IIFB	International Indigenous Forum on Biodiversity
IISD	International Institute for Sustainable Development
IPCC	Intergovernmental Panel on Climate Change
IPO	Indigenous Peoples Organization
IRR	Internal Rate of Return
IRS	Internal Revenue Service (US)
ISDA	International Swaps and Derivatives Association
ITL	Independent Transaction Log

JI	Joint Implementation
JI-SC	Joint Implementation Supervisory Committee
LCDZ	Low Carbon Development Zones (China)
lCER	long-term CER
LCIA	London Court of International Arbitration
LGMA	Local Government and Municipal Authorities
LoA	Letter of Approval
LULUCF	Land Use, Land-Use Change or Forestry
MDB	Multilateral Development Bank
MEA	Multilateral Environmental Agreements
MIGA	Multilateral Investment Guarantee Agency
MIT	Multilateral Investment Treaty
MOC	Modalities of Communication
MOU	Memorandum of Understanding
MRET	Mandatory Renewable Energy Target (Australia)
MS	Member States (of the EU)
NAO	National Audit Office (UK)
NAP	National Allocation Plan
NAPA	National Adaptation Programmes of Action
NDRC	National Development and Reform Commission (China)
NGO	Non-Governmental Organization
NJCL	National Journal of Constitutional Law (US)
ODA	Official Development Assistance
OECD	Organization for Economic Co-operation and Development
OPG	Ontario Power Generation (Canada)
ORER	Office of the Renewable Energy Regulator (Australia)
OTC	Ozone Transport Commission (Canada)
OUP	Oxford University Press
PCA	Permanent Court of Arbitration
PCF	Prototype Carbon Fund
PDD	Project Design Document
PIN	Project Information Note
PoA	Programme of Activities

PRC	Peoples Republic of China
PRE	Projects to Reduce Emissions programme (New Zealand)
QELRC	Quantified Emission Limitation and Reduction Commitments
REC	Renewable Energy Certificates (Australia)
RECLAIM	REgional CLean Air Incentives Market (US)
REDD	Reducing Emissions from Deforestation and Degradation
RGGI	Regional Greenhouse Gas Initiative (US)
RINGO	Research-Oriented and Independent Organizations
RIT	Registration and Issuance Team
RMU	Removal Unit
SBI	Subsidiary Body for Implementation
SBSTA	Subsidiary Body for Scientific and Technological Advice
SCM	Subsidies and Countervailing Measures
SEDA	Sustainable Energy Development Authority (Australia)
SERPA	Secondary Emissions Reduction Purchase Agreement
SGER	Specified Gas Emitters Regulation (Alberta, Canada)
SM	Standard Methods
SNLT	Sectoral No-Lose Targets
STO	Specific Trade Obligation
TBT	Technical Barriers for Trade
tCER	temporary CER
TRIMs	Trade-related Investment Measures
UN/ECE	United Nations Economic Commission for Europe
UNCED	United Nations Conference on Environment and Development
UNCITRAL	United Nations Commission on International Trade Law
UNDP	United Nations Development Programme
UNEP	United Nations Environment Programme
UNFCCC	United Nations Framework Convention on Climate Change 1992
UNGA	United Nations General Assembly
VCS	Voluntary Carbon Standard
VER	Verified Emission Reduction

VERPA	Verified Emissions Purchase Agreement
VVM	Validation and Verification Manual
WG	Working Group of the IPCC
WMO	World Meteorological Organization
WRI	World Resources Institute
WSSD	World Summit on Sustainable Development
WTO	World Trade Organization
WWF	World Wildlife Fund

List of Contributors

Michael Barrett is a corporate partner in the Bennett Jones LLP Climate Change and Emissions Trading practice group. Michael is based in Toronto, Canada where he advises Canadian and international clients on climate change issues. Michael speaks frequently to members of the Canadian business community on a variety of climate change related issues. He holds an LLB and B Comm from Queen's University and an MBA from York University's Schulich School of Business.

Thiago Chagas is a counsel at Climate Focus. He has five years experience working with international economic and environmental law, three of which dedicated to climate change law. As a legal counsel at Climate Focus, he has advised private and public entities in Europe and Latin America on legal issues related to business transactions in the carbon market as well as on institutional and regulatory aspects of the climate change regime. He holds a Bachelor of Law degree from Pontifícia Universidade Católica de São Paulo, Brazil and an LLM in International Law from the University of Edinburgh, UK.

Marie-Claire Cordonier Segger is Director of the Centre for International Sustainable Development Law (CISDL) in Canada and the UK. In this academic capacity, she has published widely and through the UN has provided legal advice in the Americas, Africa, and Asia on trade and investment agreements, the Cartagena Protocol and the Kyoto Protocol. She is a Fellow of the Lauterpacht Research Centre for International Law at Cambridge University, a Visiting Professor at the University of Victoria. She is also editor of a new series of books published by Cambridge University Press on Treaty Regimes in Practice. She also directs International Affairs at Canada's Ministry of Natural Resources, where she serves on the International Trade and Investment Centre, the Extractive Industries Transparency Initiative (EITI) Trust Fund, and the UN International Panel on Sustainable Resource Management. She holds an MEM from Yale, and civil and common law degrees from McGill University.

Claybourne Fox Clarke is a licensed US attorney currently based in Washington, DC. He has worked on numerous project due-diligence reports and carbon contracts in the carbon market. His qualifications also include specific experience consulting clients from a wide range of sectors on the requirements and market mechanisms of the Kyoto, EU ETS, and emerging US climate programs, along with over ten years of environmental law experience representing government and private entities in litigation and the drafting and implementation of environmental regulations. He received his LLM in international energy and environmental law from K U Leuven in Belgium where his focus was climate change law and policy. Other personal achievements include summitting several of the world's highest mountains, including Mount Everest.

Michael Coren is a specialist in carbon markets and forestry at Climate Focus. He has worked for the World Bank in Indonesia advising on the design of avoided deforestation policy with the Indonesia Forest Climate Alliance, and managed feasibility studies of forest carbon and ecosystem service projects across Southeast Asia. He has experience applying voluntary and compliance market standards, as well as financial and carbon modelling in tropical forests. Michael studied economics and forestry at the Yale School of Forestry and Environmental Studies focusing on economic factors affecting global distribution of forest carbon investments. Prior to arriving at Yale, Michael worked as a journalist writing and reporting for the *Globe and Mail*, *Outside* magazine, and *Newsweek*. He has served as the managing editor of the *Phnom Penh Post* in Cambodia and as the science and technology producer for CNN.com in Atlanta.

Allan Cook was from 2005 to 2007 Co-ordinator of the International Financial Reporting Interpretations Committee of the International Accounting Standards Board (IASB), having previously been for 13 years a Board member and Technical Director of the UK Accounting Standards Board. From 2001 to 2005 he was a member of the Technical Expert Group of the European Financial Reporting Advisory Group, the body which provides formal advice to the European Commission on the adoption of International Accounting Standards. His earlier career was divided between two Anglo Dutch multinationals, Unilever and Shell, separated by a secondment during 1979–81 as Secretary to the International Accounting Standards Committee, the predecessor body to the current IASB. While at Shell, he led for six years a working group providing the views of industry at OECD discussions on developments in financial reporting. He is a UK Chartered Accountant and a graduate of the London School of Economics and Political Science.

Jos Cozijnsen has been an independent consulting attorney on emissions trading based in the Netherlands since 1998. After working in private practice from 1985 to 1993, he worked for four years for the Netherlands' Environment Ministry as a member of the Dutch delegation to international climate meetings including Kyoto. He now consults for companies and institutions on national and international projects related to emissions trading. His core activities are: knowledge transfer and training; legal issues related to emissions trading; assisting companies regarding CO_2allocation procedures; identifying carbon market opportunities; assessing the carbon value of projects; and providing advice on CDM/JI procedures.

Kyle Danish is a partner in the law firm of Van Ness Feldman in Washington DC, and is coordinator of the Climate Change and Emissions Trading practice. His clients include electric power, oil and gas, and mineral resources companies; financials; offset providers; and think tanks. Mr. Danish has published extensively on issues of climate change law and policy. He is an Adjunct Faculty member at the American University Washington College of Law, and is an Associate Editor for the *Carbon & Climate Law Review*. Mr Danish holds a JD from Temple University School of Law, and a Master in Public Affairs from Princeton University's Woodrow Wilson School of Public and International Affairs.

Louisa Fitz-Gerald is a lawyer in the Sydney office of Baker & McKenzie's climate change team. She worked on the Australian Multi-State National Emissions trading Scheme taskforce and is currently on leave and working with the Cancer Council of New South Wales.

David Freestone is Lobingier Visiting Professor of Comparative Law and Jurisprudence at The George Washington University Law School in Washington DC, where he teaches International Environmental Law and International Climate Change Law. He is a former Deputy General Counsel at the World Bank, and a Visiting Professor at the UN University Institute of Advanced Studies. From 1996-2004 he was head of the World Bank's Environment and International Law Group, and, *inter alia*, legal adviser to the World Bank Prototype Carbon Fund team. Dr. Freestone is the General Editor of the *Legal Aspects of Sustainable Development* monograph series published by Martinus Nijhoff and is founder and Editor-in-Chief of the *International Journal of Marine and Coastal Law*. He previously collaborated with Charlotte Streck on *Legal Aspects of Implementing the Kyoto Protocol Mechanisms: Making Kyoto work* (ed., Oxford, 2005).

Markus Gehring is Lecturer in International and European Law at the University of Cambridge in the Centre of International Studies, and Fellow of Robinson College. He is also associate Professor of Law at Ottawa University Law Faculty. He practiced European competition and international trade law with Cleary Gottlieb in their Brussels office and is a member of the Frankfurt/Main Bar. Prior to moving to Cambridge, he was a tutor in Public International Law at University College, Oxford. He also serves as Lead Counsel of International Trade, Investment and Competition Law with the Centre of International Sustainable Development Law (CISDL) and is a member of the Sustainable Development Committee of the International Law Association (ILA). His recent books, with Dr Cordonier Segger, include *Sustainable Development in World Trade Law* (Kluwer Law International, 2005) and *World Trade Law in Practice* (Globe Publishing, 2007). He holds an LL.M from Yale and a Dr jur. from Hamburg.

Navraj Singh Ghaleigh has been a Lecturer in Public Law at Edinburgh since 2003. Previously a barrister in London and Lecturer at King's London, he undertook his graduate work at the University of Cambridge, the European University Institute (Florence) and the University of California, Berkeley (Fulbright Scholar). Navraj writes on issues relating to the law of climate change (both as a matter of PIL and EU law) and EU constitutional law. He has also advised various regulatory agencies, foreign governments and international organizations on issues in party funding law. Navraj is also a Research Associate at the University of Zurich's e-Democracy Centre.

Andrew Hedges is a partner with the law firm Norton Rose LLP. Previously head of the legal office at Scottish Power, he specializes in the areas of clean energy, carbon finance, and emissions trading. Andrew is particularly known for his comprehensive ability to straddle primary and secondary market carbon transactions. This has underpinned his work on the majority of the leading structured portfolio transactions that have either been marketed to selected secondary market buyers or privately negotiated since 2008. Clients regularly work closely with Andrew from the inception stage. Other aspects of his expertise include forestry projects and integrated supply chains for sustainably procured biomass. As REDD projects become more likely, clients value the expertise Andrew has gained through advising on actual projects.

Anthony Hobley is a banking lawyer with Norton Rose in London. He specializes in climate change and clean energy law as well as UK, EU, and international environmental law, and has spent more than 13 years with major law firms working on these issues.

Previously, he was General Counsel to the Climate Change Capital Carbon Fund and Director of Legal Policy for Climate Change Capital which he joined in September 1995. Currently the Chairman of London Climate Change Services (LCCS), he has played a key role in helping C4F to establish its key investment procedures and documentation. He is qualified as a solicitor of England, has a degree in Chemistry and Physics and recently completed an LLM in Environmental Law.

Lauren Hopkins is an Associate in the Washington, DC office of Beveridge and Diamond, PC with an environmental, litigation, and regulatory practice. She received her Juris Doctor, magna cum laude, from Vermont Law School and Master of Environmental Management from the Yale University School of Forestry and Environmental Studies, with emphasis on US and international environmental law, carbon finance and climate policy, and Kyoto Protocol implementation. Ms Hopkins received a Bachelor of Arts degree in Philosophy from the College of Charleston. She is a member of the Massachusetts Bar Association.

Jelmer Hoogzaad is an expert on the Kyoto flexible mechanisms at Climate Focus. His main expertise lies in JI and CDM project and portfolio risk assessment, programmatic CDM project development, Green Investment Schemes and JI track 1. Before joining Climate Focus, Jelmer was a project officer for the Dutch ERUPT program, where he was responsible for the development and implementation of risk management for all projects in the ERUPT portfolio. He graduated in 2004 in Natural Sciences and Innovation Management from Utrecht University in the Netherlands. Parallel to researching innovation towards using bio-energy systems in Ukraine as part of his doctoraal exam, Jelmer gained practical experience with renewable energy and energy efficiency in the context of an economy in transition while working for Cargill in Donetsk.

Matthias Krey has been Managing Director of Perspectives GmbH since 2007 and is also responsible for the risk assessment services of Perspectives. Matthias has worked in the field of climate policy since 2001. During his career Matthias has developed a number of approved CDM methodologies and has been managing various CDM projects in the UNFCCC registration process (eg energy efficiency, renewable energy, and N_2O abatement). Perspectives consults for the private sector, governmental and non-governmental clients in the international greenhouse gas market. In early 2009 it was acquired by Point Carbon AS to jointly form the new power house for independent offset project advice.

K Russell LaMotte is co-chair of the Climate Change and the International Environmental Practice Groups of Beveridge & Diamond, PC, where he advises clients in matters relating to US, regional, and global environmental regulatory regimes. He serves as co-counsel to the Carbon Offset Providers Coalition in the United States. He previously spent over ten years as a lawyer at the US Department of State, where he helped design, negotiate, or implement most of the major environmental and oceans agreements. He participated as legal adviser to the US delegation negotiating the market mechanisms provisions of the Marrakech Accords. He is a graduate of Harvard Law School.

Jolene Lin is Assistant Professor at the Faculty of Law, University of Hong Kong. She teaches International Environmental Law and a course on Law, Economics, Regulation and Development. Her research interests include climate change law, administrative law, and the use of market-based instruments in environmental policy. She is the International Environmental Law editor of the *Singapore Year Book of International Law* as well as an associate member of the Asia-Pacific Centre for Environmental Law (APCEL).

Michael Mehling is President of the Ecologic Institute in Washington DC, an environmental policy think tank and consultancy with offices in Berlin, Brussels and Vienna. In this position, he has led a range of research and advisory projects for government agencies, intergovernmental bodies as well as civil society institutions in North America, Europe and the developing world. In 2009, he contributed to the Lazarowicz report on Global Carbon Trading for the UK Prime Minister. Prior to joining Ecologic, he held academic positions at the Universities of Constance, Greifswald and Helsinki. He has published extensively on environmental law and policy, and is also the founding editor of the *Carbon & Climate Law Review*, a quarterly journal on climate regulation and the carbon market.

Axel Michaelowa is senior founding partner of the CDM consultancy, Perspectives, and teaches international climate policy at the University of Zurich. Axel is a member of the CDM Executive Board's Registration and Issuance Team and UNFCCC desk reviewer on baseline methodologies; he has contributed to the development of six approved CDM baseline methodologies. As one of the leading European experts on CDM institution building in developing countries, Axel has substantial experience in Asia and the Middle East/North Africa. Axel was a lead author on climate policy instruments in the 4th Assessment Report of the Intergovernmental Panel on Climate Change and is member of the board of the Swiss Climate Cent Foundation. He has published more than 100 articles on climate policy.

Maria Netto is a senior Climate Change Policy Advisor at the Environment and Energy Group of UNDP where she is responsible for overseeing the team integrating climate change considerations in UNDP operations. Over the last 14 years she has worked on financial and technical cooperation, mainstreaming of climate change considerations in development planning and policy and carbon finance. She was part of the UNFCCC secretariat team setting up the institutional framework for the implementation of the CDM and JI and acted as lead officer for procedural issues, methodologies for baseline and monitoring, small-scale project activities, afforestation and reforestation project activities and registration of project activities. In 2007, as team leader of the Financial Cooperation Unit of the UNFCCC Secretariat, she also led a study of investment and financial flows to address climate change.

Heike Santen is responsible for CER/ERU acquisitions at the Vattenfall Carbon Fund. Before joining Vattenfall, she worked at Perspectives Climate Change, where she was responsible for all CDM/JI Risk Management Services. Heike developed CDM methodologies, elaborated CDM feasibility studies and worked in policy advice. In previous positions, she led international cooperation projects in the field of environmental technologies, mainly in Latin America, where she gained extensive experience in negotiating and managing multi-stakeholder projects. Heike is a civil and environmental engineer and holds a PhD in engineering from Technische Universität Carolo-Wilhemina zu Braunschweig, Germany.

Robert O'Sullivan is the Executive Director of Climate Focus North America, based in Washington DC. Trained as a lawyer, he specializes in regulatory and policy aspects of the CDM and JI. His diverse experience covers afforestation and reforestation projects through to energy projects as well as management of CER purchase programs. He is an expert in carbon finance transactions, both on the buyers' and sellers' sides of the negotiation table and was one of the lead drafters of the CERSPA initiative. Prior to joining Climate Focus, Robert was a lawyer in the World Bank's Carbon Finance Business. He holds degrees in sciences, arts and law from the University of Queensland in Australia and an LLM from the American University in Washington, DC.

Michelle Passero is the Senior Climate Policy Advisor to The Nature Conservancy, California office. She has over 12 years of experience working in land conservation, environmental law, and policy with particular expertise in forest and climate policy and related conservation transactions. She has led and advised on processes to develop greenhouse gas (GHG) accounting standards for forest and land use projects, including the California Climate Action Registry Forest and Urban Forestry Protocols, as well as GHG project standards published by the World Resources Institute and World Business Council for Sustainable Development. Ms Passero has previously worked with Eco-Securities, the Pacific Forest Trust, The Natural Heritage Institute and Heller, Ehrman, White and McCauliffe. She holds an LLM in Sustainable International Development from the University of Washington, a JD from the University of San Francisco, and a BA in Political Science from the University of Vermont.

Markus Pohlmann is a legal counsel in the Co-financing and Project Finance unit of the World Bank legal department giving legal advice to the World Bank Carbon Finance Unit. In the World Bank, he provides legal support in managing existing carbon trust funds, setting up new carbon trust funds, and negotiating emission reduction purchase agreements. Furthermore, he is responsible for all issues involving the EU Emissions Trading regime and the World Bank's involvement in greening Assigned Amount Units in Eastern European countries. Prior to his work at the World Bank, Dr Pohlmann worked as legal counsel in the KfW legal department giving legal advice to the KfW Carbon Fund. He has a doctorate degree from the University of Hamburg in international carbon finance law and an LLM from the American University (Washington College of Law) in international business and environmental law.

Carly Roberts is an associate in the Carbon Finance and Climate Change team of Norton Rose in London. She has worked on primary and secondary market transactions including CDM Projects in China, Africa & India, JI projects in Russia, AAU/GIS transactions in Slovakia and voluntary projects (including forestry) in Indonesia and Africa. She has assisted major energy and resources companies and financial institutions create template carbon finance documentation. Prior to joining the Carbon Finance team her background was in energy and resources and corporate finance. She gained emerging markets experience while based in the Norton Rose Bahrain office. She originally joined Norton Rose from Freehills, one of Australia's top tier law firms. She is completing a Masters in Environmental Economics.

Kai-Uwe Schmidt is a senior programme officer at the Climate Change Support Team of the Secretary-General of the United Nations. For the last 14 years he has worked for the UNFCCC Secretariat in Bonn, leading a team supporting the negotiation, and subsequent implementation, of the concept of joint/cooperative implementation. He managed the work on the Clean Development Mechanism as secretary to the CDM Executive Board, including support to the Board's panel and working groups, related experts and the development and maintenance of the CDM information system. He was also in charge of the CDM registry. He holds a Masters Degree from the University of Geneva in Macroeconomics and Finance.

Sander Simonetti is a senior-associate at De Brauw Blackstone Westbroek NV and specialises in carbon trading and environmental finance (both under the Kyoto mechanisms of IET, CDM and JI, and the EU ETS, GIS and voluntary standards), as well as energy and natural resources law. He advises a broad range of international clients in carbon and energy related regulatory and contractual matters. In 2008, Sander was a visiting fellow at the Robert Schuman Centre for Advanced Studies of the European University Institute in Florence, carrying out research in the field of carbon and energy regulation.

Charlotte Streck is a founding partner and Director of Climate Focus and an international lawyer, expert in climate change and carbon finance law and policy. She advises on the regulatory framework of the United Nations Framework Convention on Climate Change, the Kyoto Protocol, and the European Union Emissions Trading Scheme. Before she founded Climate Focus in February 2005, Charlotte was counsel at the World Bank in Washington, DC for five years where she gained extensive experience in drafting and negotiating complex carbon finance agreements. She has a background in both law and sciences and holds a PhD in environmental law from Humboldt University in Berlin. She authored and edited five books and numerous articles on environmental law and policy, serves on several editorial boards, is a board member of the Global Public Policy Institute, an adjunct lecturer, at the University of Potsdam, Germany, and lead counsel for climate change of the Center of International Sustainable Development Law at McGill University, Montreal, Canada.

Gray Taylor is the leader of the Bennett Jones LLP Climate Change and Emission Trading practice group which focuses on climate change issues affecting businesses in Canada and around the world. Identified by Chambers as 'one of the fathers of the climate change industry' and qualified to practice in England, New York, and Ontario, Gray focuses on emissions trading transactions and advising on corporate governance and climate change business planning issues. Gray acts on complex transactions, such as representing the largest private sector participants in the World Bank US$1billion Umbrella Carbon Fund transaction and creating unique structures to facilitate participation by diverse entities in emission trading markets, including Natsource's Greenhouse Gas Credit Aggregation Buyers Pool, and innovative arrangements related to voluntary environmental credit creation and trading.

Christopher Tung is a partner in the Hong Kong office of Mallesons StephenJaques. He has been based in Hong Kong since 1992 and practices dispute resolution, construction, and environmental law. He specializes in climate change and clean energy matters and is a leading advocate of sustainable development law in Asia. Chris has also been involved in the development of cutting edge environmental and carbon risk management tools for aviation, banking, construction and energy sector clients.

Christina Voigt is postdoctoral research fellow and lecturer in international and environmental law at the University of Oslo, Department of Public and International Law, Norway. She obtained the First (Universität Passau, 1996) and Second (1999) German Legal State Examination, and holds an LLM in Environmental Law from the University of Auckland, New Zealand. She is the author of *Sustainable Development as a Principle of International Law: Resolving Conflicts between Climate Measures and WTO Law* (Nijhoff, 2008). Her current research focuses on issues relating to the safeguarding of the environmental integrity of the global carbon market.

Murray Ward is the principal of Global Climate Change Consultancy (GtripleC) which he founded in 2003. GtripleC provides high-level counsel to a range of international and New Zealand public and private sector clients. From 1996 to 2002 Murray led the New Zealand Ministry for the Environment's climate change team, steering the development of domestic climate change policy and as senior negotiator in NZ delegations. He is considered to be one of the key architects of the Kyoto framework, in particular for his international work on Land-Use, Land-Use Change and Forestry (sinks) and market trading mechanisms. GtripleC's current international focus is on the policy architecture of an enlarged multilateral climate change regime post-2012.

Matthieu Wemaëre is a Senior Lawyer with Climate Focus, and registered at the Paris and Brussels Bar Associations. Matthieu worked for six years for the European Commission, as a lawyer in the Waste Management Unit of DG ENV, and as a member of the European Commission Delegation for the UNFCCC negotiations on implementation measures of the Kyoto Protocol. Since 2003, he has provided assistance to some Member States' Governments in drafting national measures transposing the EU ETS and the Linking Directives and implementing the Monitoring Guidelines, as well as, the Registry Regulation. Currently, Matthieu also lectures on International and EU environmental law at the Aix-en-Provence Institute of Political Sciences (Masters course on environmental management and public affairs).

David M (Max) Williamson is counsel with Andrews Kurth, LLP in Washington, DC and New York City, where he manages a nationwide practice covering a range of litigation, transactional and environmental matters, including emerging ecosystem services trading markets and project development. Max currently represents the Carbon Offset Providers Coalition, an association of companies that finance and develop greenhouse gas reduction projects to generate carbon offsets for the voluntary and compliance markets. He did his undergraduate work at Yale College and received a JD and a Masters in environmental policy from Vermont Law School.

Rutger de Witt Wijnen is a partner at De Brauw Blackstone Westbroek NV, where he developed their international carbon finance and climate change practice. He has drafted and negotiated Emission Reductions Purchase Agreements for CDM, JI and VER projects in Latin America, Asia, and Eastern Europe. Rutger has also advised various states on carbon purchase agreements and the implementation of the Kyoto Protocol. Other clients include international and multilateral financial institutions, project developers and compliance buyers. Rutger is a member of the Technical Advisory Committee of the Gold Standard.

Martijn Wilder is a Partner and head of the Global Change practice at Baker & McKenzie in Sydney. He is an adjunct Professor in Climate Law, at the Australian National University in Canberra and Chairman of the New South Wales Climate Change Council. In addition, he is a member of Baker and McKenzie's projects and energy practice, where he specializes in domestic and international environmental law and assists on major projects. Martijn also advises key corporate clients on matters of Corporate Responsibility.

Table of Cases

EUROPEAN COURT OF JUSTICE

EUROPEAN COURT OF THE FIRST INSTANCE

ICSID

WTO

NATIONAL CASES

Canada

France

Hong Kong

UK

USA

Table of Treaties and Legislation

INTERNATIONAL TREATIES, CONVENTIONS AND RESOLUTIONS

UNFCCC DECISIONS

EUROPEAN UNION TREATIES, REGULATIONS, DIRECTIVES AND DECISIONS LEGISLATION

EC Treaty

Regulations

Directives

Decisions

NATIONAL JURISDICTIONS

PART I

INTRODUCTION

1

The International Climate Change Legal and Institutional Framework: An Overview

David Freestone

In December 2009 the Parties to the 1992 UN Framework Convention on Climate Change (UNFCCC) will meet at the 15th session of the Conference of the Parties (COP) in Copenhagen. The primary agenda item of the Copenhagen COP will be to agree the fundamental principles of the legal regime that will apply post 2012. With the new Obama administration now in power in Washington the United States is expected to be a major and positive participant at the conference for the first time for many years, but the challenges meeting the parties in addressing global climate change are enormous.

The UNFCCC was opened for signature in Rio de Janeiro in June 1992 and came into force on 21 March 1994.[1] It now has 192 Parties which have met at least annually since then. The details of the Convention and of the negotiating processes that it has initiated are discussed in more detail below, in order to set the scene for the rest of this volume. However it is important to start by emphasizing the importance of the UNFCCC and the process that it represents and to highlight its significant achievements as well as the major challenges that it faces. In the 17 years since the finalization of the Convention at the UN Conference on Environment and Development in Rio de Janeiro it has achieved virtual universality. In 1997 in Kyoto, its Parties negotiated an important Protocol that envisages reduction of Greenhouse Gases (GHG) by developed countries by an average of 5.2% from 1990 levels by 2008–2012.[2] In Marrakech in 2001, the parties also developed complex 'guidelines, modalities and rules' for the implementation of the innovative 'flexibility mechanisms' contained in the Protocol.[3]

[1] UN Doc Distr General A/AC.237/18 (Part II)/Add.1, 15 May 1992.

[2] FCCC/CP/1997/L.7/Add.1, Decision 1/CP.3 Adoption of the Kyoto Protocol to United Nations Framework Convention on Climate Change, Annex, reprinted in (1998) 37 International Legal Materials (ILM) 22.

[3] UNFCCC, Decisions 15–19/CP.7 and Annexes; reproduced in David Freestone and Charlotte Streck, *Legal Aspects of Implementing the Kyoto Protocol Mechanisms* (OUP, 2005) 566–622.

There are still outstanding issues: the fact that the US—until recently the largest single emitter of GHGs—is not a party; the fact that some states have expressed concerns about meeting their Kyoto emissions reduction targets and one, Canada, has openly stated that it will not try; and the fact that the reductions envisaged by the Kyoto Protocol will not by themselves solve the problem of climate change. But these, admittedly important, issues should not distract us from the significance of what has been achieved. The Kyoto Protocol was not designed to solve the problems of climate change; rather it was designed to set in motion a process by which the major economies of the developed world would begin to address in a meaningful way the means and modalities of radically reducing their carbon footprint. The model was the 1987 Montreal Protocol on Substances that Deplete the Ozone Layer. Although the basic text was negotiated in 1987 the Montreal Parties have successively agreed, and progressively moved, to ever more rigorous requirements for the elimination of these substances so that by 2006 the UN Secretary-General was able to announce that the use of ozone depleting substances had been virtually eliminated and the ozone hole is predicted to have closed by 2050.

Climate change poses an unprecedented and as yet still not fully understood threat to the global climate system upon which mankind depends; remedial action requires an extensive, and expensive, overhaul of industrial economies which are still heavily dependent on non-renewable carbon-based fossil fuels. For decades scientist have understood the chemical processes by which emissions of carbon dioxide and other gases might warm the planet through the so-called 'greenhouse effect', nevertheless it was not until the 1980s that international concern about anthropogenic impacts on the atmosphere through such emissions came to a head, and it was only in the last two decades of the 20th century that the United Nations took the first, somewhat faltering, steps towards recognizing and addressing the issue.

The aim of this chapter is to provide an overview of the complex regime established by the Convention, the processes leading up to the adoption of the Kyoto Protocol and key implementation challenges it has experienced. It introduces some of the key processes and institutions and describes some of the key concepts that will be further developed in later chapters. It will also discuss briefly the explosive growth of carbon trading and the challenges faced by the UNFCCC parties at Copenhagen. This is only intended as a scene setting, for the majority of these issues are dealt with in considerable detail in the following chapters.

1. The Road to Rio and the UN Framework Convention on Climate Change

The publicity surrounding the discovery and mapping of the growing 'ozone hole' over the Southern Polar region during the early 1980s, together with the swift and successful negotiation of the 1985 Vienna Convention for the Protection of the Ozone Layer[4] followed by its 1987 Montreal Protocol on Substances that Deplete the Ozone Layer[5] focused attention on the much wider and more complex issue of the impacts of greenhouse gases on the global atmospheric system. By the late 1980s climate change was already firmly on the international agenda, and in 1988 the UN General Assembly (UNGA) passed Resolution 43/53 on the Protection of the Global Climate for Present and Future Generations of Mankind.[6] It declared the issue to be one of 'common concern to mankind'[7] and endorsed the action of the UN Environment Programme (UNEP) and the World Meteorological Organization (WMO) in jointly establishing an Intergovernmental Panel on Climate Change (IPCC) to 'provide internationally co-ordinated scientific assessments of the magnitude, timing and potential environmental and socio-economic impacts of climate change and realistic response strategies.'[8] Two years later, in 1990, UNGA Resolution 45/212 established an Intergovernmental Negotiating Committee to develop a legal instrument on climate change. The outcome of a further two years of strenuous negotiations was the 1992 UN Framework Convention on Climate Change (UNFCCC).[9]

The UNFCCC was concluded in New York on 9 May 1992.[10] It was opened for signature in June 1992 as a part of the UN Conference on Environment and Development (UNCED) in Rio de Janeiro, Brazil. It now has 192 Parties. The basic objective of the Convention, set out in Article 2, is not to reverse greenhouse gas emissions but to stabilize them 'at a level that would prevent dangerous anthropogenic interference with the climate system.' Indeed, that article goes on to require that such stabilization should be achieved 'within a time frame sufficient to allow eco-systems to adapt naturally to climate change, to ensure that food production is not threatened and to enable economic development to proceed in a sustainable manner.'[11]

Article 3 then goes on to enumerate the principles by which the Parties should be guided in their actions to achieve this objective. These include a number of

[4] (1985) 26 ILM 1527. [5] (1987) 26 ILM 1550.

[6] UNGA Protection of Global Climate for Present and Future Generations of Mankind, UN Doc A/RES/43/53 (6 December 1988).

[7] See further Alan Boyle, 'International Law and the Protection of the Global Atmosphere: Concepts, Categories and Principles' in Robin Churchill and David Freestone (eds), *International Law and Global Climate Change* (Nijhoff: London, 1991) 17–20.

[8] The IPCC has three Working Groups (WGs): WGI assesses the physical scientific aspects of the climate system and climate change; WGII assesses the vulnerability of socio-economic and natural systems to climate change, negative and positive consequences of climate change, and options for adapting to it; and WGIII assesses options for mitigating climate change through limiting or preventing greenhouse gas emissions and enhancing activities that remove them from the atmosphere.

[9] UN Framework Convention on Climate Change (adopted 9 May 1992, entered into force 21 March 1994) 1771 UNTS 107 (UNFCCC).

[10] UN Doc Distr. General A/AC.237/18 (Part II)/Add.1.15 May 1992.

[11] UNFCCC, Art 2.

the innovative principles set out in the 1992 Rio Declaration, notably the precautionary principle,[12] and the principle of inter-generational equity.[13] Pre-eminent however is the new principle of 'common but differentiated responsibility'[14] discussed in more detail below. The Convention also recognizes a number of other principles such as the special needs of developing country Parties and of those 'that would have to bear a disproportionate or abnormal burden under the Convention',[15] as well as the right of all Parties to promote sustainable development[16] and the need to promote 'a supportive and open international economic system'.[17]

The generally hortatory obligations of all Parties are set out in Article 4(1), but the full significance of the common but differentiated obligations emerges in Article 4(2)[18] where the developed countries and countries with economies in transition (which are listed in Annex I) undertake to adopt policies and measures which will 'demonstrate that developed countries are taking the lead in modifying longer term trends in anthropogenic emissions consistent with the objective of the Convention.' While a number of particular factors are listed, the Article does recognize that 'a return by the end of the present decade to earlier levels of anthropogenic emissions ... would contribute to such modification.' The Convention does not however provide any modality by which this return to 'earlier' levels can be made—nor indeed does it indicate which particular earlier levels should be used as the benchmark. Nevertheless, Article 4(3) does provide

[12] UNFCCC, Art 3(3). See also Principle 15 of the Rio Declaration (UN Doc Distr. General A/Conf.151/5/Rev.1 13 June 1992). There is an extensive literature on the Precautionary Principle, see eg David Freestone, 'The Precautionary Principle' in Robin Churchill and David Freestone (eds), *International Law and Global Climate Change* (London/Dordrecht, 1991) 21. See also T O'Riordon and J Cameron (eds), *Interpreting the Precautionary Principle* (London, 1994); D Freestone and E Hey (eds), *The Precautionary Principle and International Law: The Challenge of Implementation,* (The Hague/London, 1996); Arie Trouwborst, *Evolution and Status of the Precautionary Principle in International Law* (The Hague, 2002).

[13] UNFCCC, Art 3(1) which provides that 'The Parties shall protect the climate system for the benefit of present and future generations of humankind, on the basis of equity and in accordance with their common but differentiated responsibilities and respective capabilities.' See also Principle 3 of the Rio Declaration (n 12 above). On the principle of intergenerational equity see Edith Brown Weiss, *In Fairness to Future Generations: International Law, Common Patrimony and Intergenerational Equity* (Hotei Publishing, 1989); Catherine Redgwell, 'Intergenerational Equity and Global Warming' in Churchill and Freestone (eds), *International Law and Climate Change* (London/Dordrecht, 1991) 41–56.

[14] UNFCCC, Art 3(1). The text of the article continues: 'Accordingly the developed country Parties should take the lead in combating climate change and the adverse impacts thereof'. See also Principle 7 of the Rio Declaration (n 12 above). For a wider discussion of the principle, see Christopher D Stone, 'Common but Differentiated Responsibilities in International Law' (2004) 98 American Journal of International Law 276–301. Also, more generally Lavanya Rajamani, *Differential Treatment in International Environmental Law* (Oxford, 2003).

[15] UNFCCC, Art 3(3). [16] UNFCCC, Art 3(4).

[17] UNFCCC, Art 3(5). For a discussion of the legal status of the 'guidance' under Art 3 see Birnie and Boyle, *International Law and the Environment* (2nd edn, Oxford, 2002) 525.

[18] As well as other articles of the Convention, see eg UNFCCC, Art 12.

an unequivocal obligation that the developed countries listed in Annex II[19] shall provide 'new and additional financial resources' to meet the 'agreed full costs' incurred by developing countries in meeting their communication obligations under Article 12 and to meet the 'agreed full incremental costs' in implementing the measures envisaged in Article 4(1) and which are agreed between the developing country Party and the 'entity or entities referred to in Article 11' (ie the Financial Mechanism).[20] The Annex II countries also undertake to assist countries particularly vulnerable to the impacts of climate change in meeting the costs of adaptation to those adverse effects[21] and to take all practicable steps to promote, facilitate and finance, as appropriate, the transfer of, or access to, environmentally sound technologies and know-how to other parties, particularly developing countries, to enable them to implement the provisions of the Convention.[22]

However Article 4(7) goes even further in imposing what might be called a 'blanket' implementation conditionality. It provides that: 'The extent to which developing country Parties will effectively implement their commitments under the Convention will depend on the effective implementation by developed Country Parties of their commitments under the Convention relating to financial resources and transfer of technology.'[23]

Articles 7–11 follow the pattern of the more recent multilateral environment agreements by establishing a fairly sophisticated institutional framework. The Conference of the Parties, as the 'supreme body', is charged with keeping the implementation of the Convention under regular review and is specifically granted the power to make decisions necessary to promote the effective implementation of the Convention.[24] In addition, the Convention establishes a Permanent Secretariat, and two 'Subsidiary Bodies.' The Subsidiary Body for Scientific and Technological Advice (SBSTA) is established, and its role defined, by Article 9 and the Subsidiary Body for Implementation (SBI) by Article 10. The Subsidiary Bodies meet in conjunction with the Conference of the Parties and also inter-sessionally.[25]

[19] Annex II differs from Annex I in that it does not contain the countries with economies in transition.

[20] Under UNFCCC, Art 21(3), the Global Environment Facility (see below) shall be 'the international entity entrusted with the operation of the financial mechanism referred to in Article 11 on an interim basis.' Note that Art 4(3) does qualify the obligation to provide 'agreed full incremental costs' by requiring that account be taken of the need for adequacy and predictability in the flow of funds and the importance of appropriate burden sharing among the developed country Parties.

[21] UNFCCC, Art 4(4).

[22] UNFCCC, Art 4(5).

[23] UNFCCC, Art 4(7) goes on to say that this implementation will also 'take fully into account that economic and social development and poverty eradication are the first and overriding priorities of the developing country Parties.'

[24] UNFCCC, Art 7, provided of course that these are within its mandate.

[25] The Subsidiary Bodies commonly meet in the week prior to the COP and inter-sessionally—usually in June at the seat of the Secretariat in Bonn.

The chief modality for the mobilization of the new financial resources required by the Convention is the Financial Mechanism defined by Article 11 that will provide financing on a grant or concessional basis. The Mechanism is to function under the guidance of, and be accountable to, the Conference of the Parties, but its operation shall be entrusted to one or more existing international entities. The Convention also specifically provides that the financial mechanism shall have an equitable and balanced representation of all Parties within a transparent system of governance. To this end the interim arrangements set out in Article 21(3) specifically designate the Global Environment Facility as the international entity entrusted with the operation of the financial mechanism but only 'on an interim basis' and require that it be 'appropriately restructured and its membership made universal to enable it to fulfil the requirements of Article 11'.[26]

2. The Global Environment Facility

The wording of Article 21(3)[27] demonstrates the considerable caution expressed at the UNCED Summit about the new and as yet unproven Global Environment Facility that had been set up by UNEP, UNDP, and the World Bank in 1991.[28] In the nearly two decades since then it has become an important part of the institutional landscape of the UNFCCC, so some brief appreciation of its origins and role, as well the resources it has available are appropriate.

Based on an idea from the 1987 Brundtland Commission,[29] in 1988 the French Prime Minister proposed the establishment of a grant-making fund devoted to addressing global environmental issues.[30] The outcome of a series of

[26] UNFCCC, Art 11 requires the Conference of the Parties at its first session to review the interim arrangements in Art 21(3) and to decide whether to maintain these arrangements. Within four years the financial mechanism shall be reviewed and 'appropriate measures' taken. These very cautious provisions indicate that the negotiators were not entirely happy with passing sole responsibility to the newly established GEF and it is equally clear that some fundamental changes in its governance structure would be necessary before the Parties would be willing to confirm its status as a [or the] financial mechanism for the Convention.

[27] And the equivalent similar provisions in the Convention on the Protection of Biological Diversity (Arts 21 and 39) and also Agenda 21, Chapter 33.16(a)(iii).

[28] For a detailed discussion see further, David Freestone, 'The Establishment, Role and Evolution of the Global Environment Facility: Operationalising Common but Differentiated Responsibility?' in Tafsir Ndlaye and Rüdiger Wolfrum (eds) *Liber Amicorum for Thomas A. Mensah: Law of the Sea, Protection of the Marine Environment and Settlement of Disputes* (Martinus Nijhoff, 2007) 1077–1107

[29] Report of the World Commission on Environment and Development, *Our Common Future* (Oxford, 1987) at 338, proposed the 'development of a special international banking programme or facility linked to the World Bank ... to finance investments in conservation projects and national strategies'.

[30] He also pledged 900 million French francs over a three-year period, joined by the Federal Republic of Germany. See Laurence Boisson de Chazournes, 'The Global Environment Facility Galaxy: On Linkages among Institutions' (1999) 3 Max Plank UN Yearbook 243–85, at 246 (citing '1989 Annual Meeting of the Board of Governors, Summary Proceedings,' at 79 and 81–2 respectively).

meetings was a proposal to the World Bank Board of Executive Directors for a Global Environment Trust Fund (GET) with a commitment from donors of some US$1.2 billion over a three-year pilot phase, approved on 14 March 1991.[31] In October 1991, UNDP, UNEP, and the World Bank concluded a unique tripartite arrangement to cooperate in the implementation of the Facility.[32] The World Bank was to act as Trustee of the Fund and with UNEP and UNDP it was also to act as an Implementing Agency.[33]

In part, the motive for the establishment of the GET in 1991 had been to acquire useful experience which could provide a useful input into the deliberations of the June 1992 Rio de Janeiro UNCED at which the UNFCCC was open for signature.[34] The 'restructuring process' required by UNCED involved negotiations among 73 states and was described by a participant as 'neither easy nor unimportant'.[35] After two years and seven meetings, it was completed in Geneva in March 1994. The text of a new 'Instrument for the Establishment of the Restructured Global Environment Facility'[36] was agreed and at the same time donors agreed to replenish the facility to the level of $US2 billion for the period 1994–1998.

The GEF was innovative in many ways. Its legal status may be unique and the collaborative mechanism between UN entities and a Bretton Woods institution that it represents is also highly novel.[37] The Instrument itself is a *sui generis* document which does not follow the format of an international treaty.[38]

The governance system for the restructured GEF draws on elements of both the UN and the Bretton Woods systems. It established a permanent Secretariat which, although housed within the World Bank, would be 'functionally independent' from the Bank.[39] This addressed concern voiced at UNCED that a

[31] A proposal for the establishment of a Global Environment Trust Fund was approved by the Bank's Board of Executive Directors by Resolution No 91–5 on 14 March 1991, (1991) 30 ILM 1735.

[32] 'Operational Cooperation under the Global Environment Facility'. Reproduced in (1991) 30 ILM 1766.

[33] The governing bodies of each of the agencies then approved the Arrangement.

[34] Preamble to the GEF concept paper (1991) 30 ILM 1739, 1740. For a comprehensive overview including all the major documentation see Stanley P Johnson (ed), *The Earth Summit* (Nijhoff, 1993).

[35] See Stephen A Silard, 'The Global Environment Facility: A New Development in International Law and Organization' (1994) 28 George Washington Journal of International Law and Economics 607–54, 633. The author is a former World Bank finance lawyer.

[36] For text of the Instrument for the Establishment of the Restructured Global Environment Facility, Report of the GEF Participants Meeting, Geneva 14–16 March 1994, see 33 ILM 1273 (1994) [Hereafter: 'the Instrument'].

[37] See J Werksman, 'Consolidating Governance of the Global Commons. Insights from the Global Environment Facility' (1995) 6 Yearbook of International Environmental Law 27–63.

[38] The Instrument is not registered with the UN. After the text had been negotiated, it was approved by three separate resolutions of the three Implementing Agencies. See further Freestone, (n 28 above); Boisson de Chazournes (1999) 3 Max Plank UN Yearbook at 255; and Boisson de Chazournes, 'The Global Environment Facility as a Pioneering Institution' (2003) GEF Working Paper.

[39] Instrument, para 21.

new organization should not result from the process, and also created a clear functional space between the activities of the GEF and those of the World Bank. A representative Council was established to meet twice a year, overseen by an Assembly with universal membership. [40]

At a functional level the GEF was given four primary financing focal areas: Ozone Depletion (to the extent not covered by the Montreal Protocol Multilateral Fund), Biological Diversity, International Waters and Climate Change.[41]

3. The Berlin Mandate

On 21 March 1994, the UNFCCC came into force. The First Session of the Conference of the Parties (COP1) was held in Berlin in April 1995. At the COP the major issue was the need for the Parties to address the vagueness of the commitments in the Convention. It was agreed to develop more concrete obligations in two respects: a more precise time frame for the reduction of emissions of GHGs restricted to the industrialized countries in Annex I. This became known as the Berlin Mandate.[42]

The Conference also endorsed the concept of a pilot phase for joint implementation activities. Some degree of co-operation had been envisaged by the 1992 Convention itself which in Article 4(2)(b) talks of the aims of 'returning [GHG emissions] individually, *or jointly*, to their 1990 levels . . .'[43] This pilot phase was to be called 'activities implemented jointly (AIJ) under the pilot phase'.[44] The objective was to allow investments by Annex I countries in a range of GHG mitigation projects outside their own borders to allow them to begin to understand more clearly the implications of developing such 'off shore' projects. The decision did not envisage Annex I countries being able to claim credit against future reduction targets for investments made under this AIJ pilot phase.

[40] The Instrument envisages the Assembly meeting every three years. It has the authority, inter alia, to 'consider for approval by consensus, amendments to the present Instrument on the basis of recommendations by the Council.' Instrument, para 14(d).

[41] Instrument, para 2.

[42] At the final plenary session of the Conference, the President, Angela Merkel introduced the review of the adequacy of commitments in Art 4.2(a) and (b) (FCCC/CP/1995/L.14, review of the adequacy of commitments in Art 4.2(a) and (b)). She said that the industrialized countries' agreement to specific measures and the developing countries' agreement to reaffirm and advance existing commitments meant that the process would advance. India for the 'G77 and China' said the decision to launch a process to strengthen the commitments made COP1 a success, noting that consultations were not easy. The EU stated its understanding that the wording 'developed countries/other Parties' in section II(2)(a) must be interpreted as 'developed countries and/or other Parties' and means that this sub-paragraph applies to Annex I Parties within the European Community, individually or jointly among themselves, in accordance with Art 4.2(a) and (b) of the Convention.

[43] The idea that the parties might seek to do this jointly was the basis for the AIJ pilot phase.

[44] FCCC/CP/1995/L.13,

Building on the mandate of COP1 an *ad hoc* group was established to advance the concepts of the adequacy of the commitments under Article 4 and the development of a new instrument to crystallize the inspirational wording of the Framework Convention. The Ad Hoc Group on the Berlin Mandate (AGBM) met regularly at COP2 and at the intervening Subsidiary Bodies meetings. The culmination of this process, which involved a series of submissions by the Parties of observations and of draft language, was the COP3 meeting in Kyoto.

4. The Kyoto Protocol

The Kyoto negotiations were not easy. On 11 December 1997, in the early morning of the day after that scheduled for the completion of COP3, after nearly 48 hours of continuous negotiation, the Parties to the 1992 UNFCCC concluded the long-awaited protocol. The 1997 Kyoto Protocol strengthens the commitments of the 1992 Convention—particularly Article 4(2) (a) and (b)—by setting out a firm schedule for reductions of GHG emissions by Annex I countries and firm targets to be met within an agreed commitment period (2008–12). In adopting the Protocol the COP recalled that it had decided in Berlin in 1995 at its first Conference to start a process by which it could take 'appropriate action' for the period beyond 2000 by the adoption of a protocol or another legal instrument at its third session; and recalled that one aim of that process was to strengthen the commitments in Article 4(2)(a) and (b) for developed country parties and other parties included in Annex I of the 1992 Convention both to require them to elaborate policies and measures and to set quantified limitation and reduction objectives (QELROs) within specified time frames for their anthropogenic emission be sources, and removal by sinks of greenhouse gases.[45]

The COP decision also recalled that according to the Berlin Mandate, the process would not introduce any new commitments for developing countries but would reaffirm existing commitments in Article 4(1) and continue to advance the implementation of these commitments in order to achieve sustainable development, taking into account Article 4(3), (5), and (7).

The specific commitments of Annex I parties are set out in Annex B to the Protocol. These envisage that during the five years of the 'commitment period' from 2008 to 2012, those parties will have reduced their emissions by an average of 5.2% from 1990 levels. The specific targets (or assigned amounts) however range from Iceland and Australia which were able to increase their emissions

[45] The UNFCCC and its Kyoto Protocol both purport to regulate greenhouse gases 'not controlled by the Montreal Protocol'. This form of words is used as some gases that contribute to global climate change also deplete the ozone layer and are already subject to the legal regulatory regime of the Montreal Protocol. To avoid undue repetition this important qualification is not used in the text of this chapter but should be borne in mind.

from the 1990 base levels (by 10% and 8% respectively) to the countries of the European Union which accepted an 8% reduction from 1990 levels.

Since 1990 many of the Annex I countries had substantially further increased their GHG emissions through growth in their economies. The commitments to reduce emissions below 1990 levels are therefore commensurably even more rigorous, and the potential economic impacts of these obligations are unparalleled in international environmental law.[46] However, the most innovative aspects of the Protocol may in fact not be these strict commitments but the introduction of so-called market mechanisms, initially called flexibility mechanisms (or flex-mechs) but now known as the Kyoto Mechanisms, into the process by which Annex I countries can meet their obligations. As indicated above, some degree of co-operation is envisaged by the 1992 Convention itself which in Article 4(2)(b) talks of the aims of 'returning [GHG emissions] individually, *or jointly*, to their 1990 levels . . .' The wording of the 1992 Convention is echoed in Article 3(1) of the Protocol which re-affirms that these commitments may be made individually or jointly, and has also been developed further in Article 4 in response to requests by the EU to allow it to be done within the EU group of nations as a whole. This mechanism, the so-called EU 'Bubble', requires that Parties in Annex I which reach an agreement to fulfil their commitments under Article 3 jointly shall be deemed to have met those commitments if their total combined aggregate emissions do not exceed their assigned amounts calculated in accordance with Annex B. Such an inter-state agreement must be formally concluded and notified to the Parties through the UNFCCC Secretariat; and in the event of failure of the group of states to meet the required reductions, each party becomes again individually liable for its own levels of emissions.[47]

5. The Kyoto Mechanisms

More radical than the joint responsibility agreements envisaged under UNFCCC Article 4 however are the provisions of Article 6 and Article 12 of the Kyoto Protocol—each of which envisages that in various ways emission reductions financed in other countries might be set off against the financiers' GHG reduction targets. The

[46] The potential impacts of the economic changes necessary to meet the commitments are different in kind from the very significant commitments made by developed countries under the 1987 Montreal Protocol (as amended and adjusted) to the 1985 Vienna Convention for the Protection of the Ozone Layer 26 ILM (1987) 1527. The text of the Montreal Protocol on Substances that Deplete the Ozone Layer; as changed by the London adjustments and the amendments of 1990, 1992, 1995 and 1997 can be found at <www.unep.org/ozone>, accessed 31 March 2009. For the trade and investment implications of these economic challenges, see Cordonnier-Segger and Gehring, chapter 4 below.

[47] See also Art 4(6) which applies the same principles to co-operation within a 'regional economic integration organization' (the standard treaty language for bodies such as the EU) which is itself a Party to the Protocol.

theory behind this approach is that the 'marginal abatement cost', ie the cost of financing an emission reduction, will usually be far higher in a relatively fuel efficient industrialized country than in a country such as an EIT or a developing country which may have less efficient fuel-use technology. As the global climate system benefits from these reductions wherever they are made, making reductions in an EIT or a developing country as part of a national strategy which also of course includes the introduction of domestic 'policies and measures' to reduce emissions at home, will make the cost of reaching these reduction target cheaper and increase the chances that they will actually be reached. The consequential advantages for the project host countries are new resource and technology transfers, including access to cleaner technology and contributions to sustainable development. In addition, Article 17 even contemplates a system of emission rights trading whereby one Annex I country might directly purchase from another Annex I country some of its rights to emit GHG, known as Assigned Amounts.

5.1 Article 6: Joint Implementation

Article 6 builds directly on the wording of UNFCCC, Article 4(2)(b) which envisages that countries listed in Annex I may act individually or jointly to meet their emission reduction objectives. Article 6 of the Kyoto Protocol allows any Annex I country to transfer to, or acquire from, another Annex I country, reductions of GHG emissions, described as Emission Reduction Units (ERUs), achieved by project activities.[48] These ERUs may be generated by any projects that reduce anthropogenic emissions of designated greenhouse gases or which enhance the anthropogenic removal by sinks[49] of such gases. The key feature of this mechanism is that all the emission reductions need to be brought about, and verified, by investments in specific projects.[50] It is also subject to a number of further requirements. First, these projects require the express approval of both the State Parties acting as transferor and transferee. The Protocol uses the term 'Parties', hence both States must be parties to the Protocol. Second, the reduction of GHGs that take place as a result of the project must also be 'additional' to any that would otherwise have occurred.[51] Third, no State Party can acquire ERUs if it is not itself in compliance with a number of other obligations under the Protocol.[52] Finally the acquisition of such ERUs is not a substitute for

[48] Once verified and reported in accordance with guidelines to be elaborated by the UNFCCC COP/MOP, Art 6(2).

[49] Such removal could include the sequestering of carbon by land use, land use change or forestry (LULUCF), in which case they are termed RMUs (Removal Units).

[50] In the case of Track 1 JI verification is the responsibility of the host country concerned; Track 2 JI projects are under the supervision of the JI Supervision Committee, discussed by Hoogzaad and Streck in chapter 8 and Hobley and Roberts in chapter 9.

[51] Sometimes called the 'but for' test: 'But for the project' . . . would not have happened.

[52] See Arts 5 and 7, Kyoto Protocol.

domestic action to reduce GHG emissions, for all such acquisitions must be 'supplemental to domestic actions.'[53]

In order to make it clear that the full risk of any non-compliance with other provisions of the Protocol rests entirely with the acquirer of the ERUs the Protocol goes on to provide that where an issue of non-compliance is raised, transfer and acquisition transactions may continue but entirely at the risk of the acquirer. As the usual modality for these transfers will be by investment or purchase, this is known as 'buyer risk.'[54]

The final issue of importance is that although the Protocol talks clearly about Party-to-Party transactions it does envisage others such as the private sector (or even international organizations, such as the World Bank) participating in the financing and organizing of such projects. Article 6(3) permits any party to authorize a legal entity to participate in 'action leading to the generation, transfer or acquisition' of ERUs, provided however that that authorization is under the responsibility of the authorizing State Party.

5.2 Article 12: the Clean Development Mechanism

This Article was the great surprise of the Kyoto Meeting.[55] The Brazilian delegation had proposed a form of Green Development Fund at an earlier meeting but the forms and modalities of the Clean Development Mechanism (CDM) defined in Article 12 were truly the results of the negotiations at Kyoto. Perhaps as a result of the fact that the negotiation of Article 12 took place in a different drafting group from that of Article 6, or perhaps deliberately, each article uses rather different language to describe concepts that are essentially very similar. The most important aspect of Article 12 is the fact it is to be used by Annex I countries to finance emission reductions in countries which have not made commitments under the Protocol to meet GHG emission reduction targets, ie developing countries which are not listed in Annex I. Article 12(2) sets out clearly the objective which is 'to assist Parties not included in Annex I to achieve sustainable development and in contributing to the ultimate objective of the Convention, and to assist Parties included in Annex I in achieving compliance with their ... commitments under Article 3.'

The particular appeal as well as the tension of the CDM however lies in the fact that States with commitments would be financing projects in countries

[53] This term is not defined, but the European Union has decided that this means at least 50% by domestic policies and measures.

[54] Article 6(4) reads: 'If a question of implementation by a Party included in Annex I of the requirements referred to in this Article is identified in accordance with the relevant provisions of Article 8, transfers and acquisitions of emission reduction units may continue to be made after the question has been identified, *provided that any such units may not be used by a Party to meet its commitments under Article 3 until any issue of compliance is resolved.*' (Emphasis added).

[55] Jacob Werksman, 'Unwrapping the Kyoto Surprise' (1998) 7 Review of European Community and International Environmental Law 147, 151.

without commitments. This should mean of course that developing countries share the benefits of project investments in clean technology within their economies, but also that a far wider pool of possible reductions are possible than would be available through reductions in Annex I countries alone. Even if the majority of the responsibility for the historical emissions leading to climate change can be laid at the doors of the developed world, a reduction of emissions anywhere in the world has an equally beneficial impact on the global climate system, and it is in the developing world—notably in India and China—where much of the increase in future emissions of GHG seems likely to take place in the decades to come. A mechanism therefore that encourages investments in climate friendly projects in developing countries was also thought to have a beneficial impact on the global climate system as well as a net beneficial impact on sustainable development. As a number of the later chapters in this book demonstrate,[56] there is however also a global community interest in ensuring that Annex I countries or entities only be permitted to claim credit for those projects in developing countries which demonstrably reduce GHG emissions. Hence the Protocol envisaged a supervisory structure—the CDM Executive Board acting under the guidance of the COP/MOP[57]—including an emission reduction verification and certification system. CDM projects need to be validated and registered as CDM projects by 'operational entities' designated by the COP/MOP who are also responsible for verifying and certifying that emission reductions actually occurred. Validation of the project through the designated operational entity (DOE) is the modality for the certification of the project as contemplated by Article 12(5). DOEs also certify the emission reductions generated by the project. Emission reductions so certified are therefore termed Certified Emission Reductions (CERs), to distinguish them from the products of Article 6 projects, which are termed Emission Reduction Units (ERUs).

The administrative expenses of the Convention servicing these bodies are covered inter alia by a levy of 'a share of the proceeds' from certified project activities. This share is used to 'assist developing countries that are particularly vulnerable to the adverse effects of climate change to meet the cost of adaptation.'[58] In a number of other respects CDM projects resemble Article 6 projects. Participation in CDM projects must be voluntary; projects must manifest real measurable and long-term benefits relating to mitigation of climate change; and

[56] See chapters on the CDM below: Netto and Schmidt, chapter 10; Krey and Santen, chapter 11; Michaelowa, chapter 12; and Voigt, chapter 13.

[57] Art 12(4). See section 6 below.

[58] The share of proceeds for adaptation measures is fixed at 2%. In addition, the participants in a CDM project have to pay an administrative fee of US$0.10 for the first 15,000 CERs issued for each project. An additional US$0.20 has to be paid for all subsequent CERs. FCCC/KP/2005/CMP/8/Add.1, Decision 3/CMP 1, Modalities and procedures for the clean development mechanism as defined in Article 12 of the Kyoto Protocol, para 15(a) and the relevant decision of the Executive Board at its 23rd meeting (Annex 35 to the report of the Executive Board).

a project activity generating CERs must be 'additional' to that which would have occurred in its absence.[59] Participation in the CDM is also open to the involvement of private and/or public entities,[60] subject to the guidance of the Executive Board. More significant however is the absence from Article 12 of the express distinction (made in Article 6) between emission reduction activities and those designed to enhance anthropogenic removal by sinks of GHGs. The COP/MOP has overarching responsibility to provide 'authority and guidance' and is responsible for elaborating 'modalities and procedures'[61] for the operation of the CDM. At COP7 it was decided that 'afforestation and reforestation' projects were CDM eligible[62] and the modalities for such projects were finalized at COP9 in Milan in December 2003,[63] although the COP has yet to reach final decisions on whether other sink projects will be CDM eligible in the longer term.[64]

5.3 Article 17: Assigned Amount Trading

The third mechanism envisaged by the Kyoto Protocol allows the trading of parts of Assigned Amounts (or Assigned Amounts Units (AAUs)) among Annex I Countries (Article 17 of the Protocol).[65] This is discussed in detail in chapter 7.

The rules for IET have now been laid down.[66] 'Everybody, whether a state or not, whether a Party to the Kyoto Protocol or not, can trade Kyoto emission rights. But if such trade leads to a transfer, such transfer can, as a general rule, only be effected by Parties to the Kyoto Protocol.'[67] For this reason, the trading of AAUs between governments not linked to any emission reducing activity is still not, or not yet, widely supported by Annex I countries in need of AAUs.

[59] See Art 12(5)(c), and Michaelowa, chapter 12 below.

[60] cf. Art 6(3) which says legal entities may participate in 'actions *leading to* the generation, transfer or acquisition . . . of emission reduction units. This could be interpreted as saying that they may not be transferred or acquired (and therefore perhaps not owned?) by such entities. By contrast Art 12 envisages 'participation . . . in the acquisition of certified emission reductions' suggesting that private/public entities might own these CERs.

[61] Article 12(7) empowers the COP/MOP at its first meeting, to 'elaborate modalities and procedures with the objective of ensuring transparency, efficiency, and accountability through independent auditing and verification of project activities.'

[62] FCCC/CP/2001/13/Add.2 Decision 17/CP.7, para 7, Modalities and procedures for the clean development mechanism as defined in Article 12 of the Kyoto Protocol.

[63] FCCC/KP/2005/CMP/8/Add.1, Decision 5/CMP1.Modalities and procedures for afforestation and reforestation project activities under the clean development mechanism in the first commitment period of the Kyoto Protocol.

[64] The treatment of land use, land use change and forestry project activities under the CDM in future commitment periods will be decided as part of the negotiations of the second commitment period, FCCC/KP/2005/CMP/8/Add.1, Decision 5/CMP1 recital 3.

[65] Assigned Amounts' refers to the quantity of GHGs a Party to the Kyoto Protocol is allowed to release in the global atmosphere as calculated on a yearly basis in Annex B of the Protocol (Article 3 of the Protocol).

[66] Annex to Decision 11/CMP.1.

[67] See Simonetti and de Witt Wijnen, chapter 7, below.

Most governments are reluctant to use taxpayers' money for the purchase of abstract paper values and have expressed a clear preference for 'greened' AAUs. Green Investment Schemes (GISs) offer a way to earmark AAU revenues for environment-related activities. Selling countries can choose to reinvest their proceeds into measures that further reduce their emissions, thereby freeing up additional AAUs, or to support measures with other environmental benefits. A credible GIS increases the value of the traded AAUs and is in the interest of both the selling and the buying country. The World Bank has been pioneering the design of GISs.[68]

This Mechanism is also referred to as 'emission trading', a term commonly used to describe the trading in emission rights. Article 17 of the Kyoto Protocol forms the basis for a global emissions trading system among Annex B Parties, which may be used by Annex B Parties for compliance purposes. Since international emission trading involves the transfer of AAUs, actual trades could only be settled once the first commitment period of the Kyoto Protocol started in 2008. Even before the start of 2008 however emissions trading at a national and regional level has gone through a period of unprecedented popularity. Within the last few years, emission trading systems in Australia, the UK and Denmark, followed by an EU-wide emission trading system, have been implemented and a voluntary market in emission reductions has also flourished, in which private actors have been the most important participants.[69] Legislative conditions for State level emission trading are currently being created in a number of US states, and even before the election of Barak Obama in November several bills proposing the establishment of a federal emission trading system were being considered by the US Congress. All these initiatives are discussed in detail later in this volume.[70]

6. Implementation of the Kyoto Mechanisms

Immediately after the finalization of the Kyoto Protocol, the Parties to the UNFCCC began negotiating the ways in which its mechanisms would operate. After a protracted negotiating process which included an extra full COP Meeting (COP6.5 or COP6bis), by the time that the previously scheduled Seventh

[68] See Jane Ebinger and Jari Vayrynen, 'The Greening of AAUs and the Interface with Joint Implementation,' in David Freestone and Charlotte Streck, (eds) Special Issue: *The Kyoto Protocol: Current Legal Status of Carbon Finance and the Flexible Mechanisms* (2007) 15–2 Environmental Liability 73–80.

[69] On the voluntary carbon market, see Passero, chapter 23 below, and on the role of private actors see Lin, chapter 6.

[70] On the EU ETS see Pohlmann, chapter 16 and Ghaleigh, chapter 17; on Australia, Wilder and Fitz-Gerald, chapter 20; on Canada, Taylor and Barrett, chapter 21; and on the US, LaMotte, Williamson and Hopkins, chapter 18 and Danish, chapter 19 below. On the linking of schemes see Mehling, chapter 5.

Session of the Conference (COP7) took place in Marrakech in November 2001 the parties were in a position to agree a major package of measures to enable the Kyoto mechanisms to start operations. This bundle of some 262 pages and 39 decisions was called the Marrakech Accords. As will be seen in the following pages of this book, the Accords provide a much needed framework of 'guidelines, modalities and rules' for moving forward with implementation of the Protocol; however a number of key issues remained still to be settled either through practice or by further COP decisions.

In 2000 there had been other setbacks to the Kyoto process. Although the US under the Clinton administration had signed the Protocol, after the November 2000 US Presidential election the incoming administration of George W Bush announced its intention to withdraw from Kyoto—indeed to 'unsign' the Protocol. To enter into force the Kyoto Protocol required ratification by 55 Parties to the Convention including ratification by Annex I Parties 'which accounted in total for at least 55 per cent of the total carbon dioxide emissions for 1990 of the Parties included in Annex I'. [71] The result of the US decision was therefore that virtually all the other Annex I parties—including the Russian Federation—had to ratify the Protocol in order to bring it into force. Paradoxically, the decision of the world's largest GHG emitter to withdraw from the process seems to have galvanized the remaining UNFCCC Parties to move forward with the finalization of the Marrakech Accords and to proceed with the establishment of the bodies—such as the CDM Executive Board—and of the procedures to enable an immediate start of the Kyoto Mechanisms once the Protocol did enter into force.[72] On 16 February 2005, after the Russian Federation had ratified the Protocol it came into force, and the First Meeting of the Parties to the Protocol (MOP) meeting with the 11th Session of the Conference of the UNFCCC Parties (jointly COP11/MOP1) took place in Montreal in 2005.

The Kyoto Mechanisms had put a market value on GHG emission reductions even before the Kyoto Protocol entered into force. Starting from modest beginnings there has been an explosion in the size of the carbon market created by the Kyoto Protocol but supplemented by a number of regional and national emission trading schemes, of which the largest by far is the EU Emission Trading Scheme (EU ETS). In 2008 the World Bank estimated that the total size of the carbon market in 2007 was some US$67 billion—of which some US$50 billion was from the EU ETS.[73] Recent figures for 2008 suggest that

[71] Kyoto Protocol, Art 25(1).

[72] Indeed many of the preliminary procedures such as validation of project baseline methodologies by the Executive Board and it Methodology panel had already begun before entry into force.

[73] K Capoor and P Ambrosi, 'State and Trends of the Carbon Market 2008' (Washington, DC: The World Bank, 2008) <siteresources.worldbank.org/NEWS/Resources/State&Trendsformatted 06May10pm.pdf>, accessed 9 April 2009.

the total size of the market in 2008 at a time of an incipient economic recession was over US$120 billion.[74]

Meanwhile the 2007 publication of the Fourth Assessment Report of the Intergovernmental Panel on Climate Change (IPCC), and the publicity surrounding their award of the Nobel Prize,[75] has sharply focussed public and political attention on the very real risk of dangerous climate change of which they warn. Indeed the IPCC Fourth Assessment showed that the current situation and prognosis for dangerous climate change is already far worse than had previously been envisaged.[76] In addition, the 2007 *Stern Review of the Economics of Climate Change* suggests that if serious action is not taken within the next 15–20 years then the costs of coping with climate change could be in excess of 20% of total global income, annually.[77] The World Bank has estimated total global income currently at some US$35 trillion per annum, rising by 2050 to perhaps US$135 trillion, when the global population is estimated to be some 9 billion, with major relocations of population in developing countries and a huge demand for new infrastructure and power sources.[78] 20% of US$135 trillion in 2050 is some US$27 trillion per annum. These are enormous sums of money that put the current need for serious investments in new technology and innovative approaches to tackling both mitigation and adaptation into proper perspective.

Against this background, and although it is only a modest first step in tackling the problem, the Kyoto Protocol and various regional emission trading schemes have already had an important impact in making companies internalize and integrate a price for GHG emissions into their operations. Trading in carbon credits, allowances and emission rights now forms part of the GHG emissions management of most major industrial GHG emitters and very few corporate players today have not yet engaged in this fast growing market, which has been expanding exponentially for more than five years. The trading of carbon credits has lost much of its exotic flavour and is considered a serious factor in the closely related oil, gas and electricity markets.

In this rapidly evolving market, the Kyoto Mechanisms provide the reference point for a growing number of regulations, systems and standards, and the Clean Development Mechanism (CDM) remains the only mechanism that links

[74] Point Carbon, 'Carbon 2009: Emissions Trading Coming Home' (Point Carbon, March 2009) <http://www.pointcarbon.com/1.1083366>, accessed 9 March 2009; and K Capoor and P Ambrosi, 'State and Trends of the Carbon Market 2009' (The World Bank, Washington, DC, 2009) <http://siteresources.worldbank.org/INTCARBONFINANCE/Resources/State___Trends_of_the_Carbon_Market_2009-FINAL_26_May09.pdf>, accessed 1 June 2009.

[75] Together with former US Vice President Al Gore.

[76] Intergovernmental Panel on Climate Change, 'IPCC Forth Assessment Report' (2007) <http://www.ipcc.ch/ipccreports/ar4-syr.htm> accessed 9 April 2009.

[77] Commissioned by the then UK Chancellor of the Exchequer, Gordon Brown: 'Stern Review on the Economics of Climate Change' (2007) <http://www.hm-treasury.gov.uk/sternreview_index.htm>, accessed 9 April 2009.

[78] Kirk Hamilton and Ian Johnson, *Responsible Growth to 2050* (The World Bank, 2004).

emission reductions achieved in developing countries to the global carbon markets. The link between the Kyoto Protocol mechanisms and the EU Emission Trading Scheme through the adoption of the 'Linking Directive'[79] has increased the attractiveness of credits generated by CDM and JI projects for private sector participants. Of the two mechanisms, to date it has been the CDM which has attracted most attention (and the majority of funds—more than US$ 12 billion in 2007). JI has been put at a disadvantage compared with the CDM, not only because the countries holding the biggest JI potential (Russia and Ukraine) have been slow to develop an approval framework, but also because JI does not benefit from the 'prompt start' provisions of the CDM. The prompt start of the CDM is based on Article 12(10) of the Kyoto Protocol and allows emission reductions obtained between 2000 and the beginning of the Protocol's first commitment period to be used to meet Annex I commitments. Consequently, the CDM started operating shortly after the adoption of the Marrakech Accords in 2001 and the Parties authorized the retroactive crediting of emission reductions from 2000. As the first international mechanism to start operating, the CDM gave project developers time to design and develop emission reduction projects. The establishment of the basic JI infrastructure was dependent on the Kyoto Protocol entering into force, which in turn delayed the establishment of the JI Supervisory Committee (JISC, the equivalent to the CDM Executive Board) until December 2005. However, in parallel to the activities of the JISC an increasing number of JI projects were being designed and as we will see,[80] with the beginning of the Kyoto commitment period in 2008, the market in ERUs has gained momentum.

At the time of writing, more than 1500 CDM projects have been internationally approved and registered. As the later chapters in this book show clearly, most of the experience in the development of emission reduction projects, the development of baseline and monitoring methodologies and the financial structuring of carbon finance can thus be attributed to the CDM.[81] This enthusiasm for the CDM is surprising, bearing in mind that the CDM is the most complex of the mechanisms established under the Kyoto Protocol. It is governed by a complicated (unsystematic and not readily available) set of rules, decisions and guidance documentation established by the COP/MOP, but drafted and interpreted by the Executive Board, its panels and working groups as well as the UNFCCC Secretariat. The process is complex and not always transparent and there is ample room for criticism. Some see the practice of the CDM leaning too much toward the objective of generating cost-efficient emission reductions—mostly through projects involving the simple destruction of

[79] Directive 2004/101/EC of the European Parliament and of the Council of 27 October 2004 amending Directive 2003/87/EC establishing a scheme for greenhouse gas emission allowance trading within the Community, in respect of the Kyoto Protocol's project mechanisms.

[80] See Hoogzaad and Streck, chapter 8 and Hobley and Roberts, chapter 9

[81] See chapters 10–13 below.

high potential GHG gases—losing sight of the impact of the CDM on sustainable development, the other objective of the mechanism. Others complain about the long and costly procedures, the dubious added value of the additionality test and the lack of transparency of the process.[82] While it is obviously necessary to ensure the environmental integrity of the process through rigorous standards and evaluation mechanisms, the CDM institutions are currently being challenged by the very popularity of the mechanism. The increasing number of projects processed by the Executive Board and its panels are demonstrating the obvious constraints and limitations of using UN procedures and committees to regulate a private market. Despite these valid points of criticism, the CDM must be considered a success. The measure of its success lies not only in the reduction of GHG emissions that it has facilitated—for so far these are still too small to change the emission trends on a global scale—but more importantly in the fact that the CDM has helped to create a global partnership between countless actors united in their efforts to finance emission reducing projects and create emission reductions. The CDM has proven to be a global market experience of unprecedented scale. It has brought the idea of market based approaches to parts of the world new to the notion of a global market. It has helped to leverage funds for renewable energy technologies and other emission reducing activities. And, last but not least, it has helped to test methods to calculate emissions and emissions reductions, develop monitoring protocols, and an essential infrastructure of emission registers. All these issues are explored fully in the chapters that follow.

By contrast with the CDM, in JI, where the host country is ultimately responsible for all project activities, the host country has the right, under the so-called Track I procedure, to verify emission reductions as being additional as soon as it has the proper monitoring and accounting framework in place. Even if the Annex I country hosting a JI activity does not meet the full set of accounting criteria, it can nevertheless participate in JI, in which case the rules and procedures are modelled after the CDM (Track II JI). But even then the country needs to meet the reporting requirement of the Protocol in order to issue and transfer ERUs.[83] Consequently, Article 12(7) states that the COP/MOP 'shall' at its first session elaborate modalities and procedures with the objective of ensuring transparency, efficiency and accountability through independent auditing and verification of project activities.[84] COP/MOP1 complied with this mandate when it established the JISC and elaborated the basic procedures governing Track II JI. Since its establishment the JISC has completed an impressive amount of work, generally in an efficient and cooperative manner. The Committee relies wherever possible on the model documents and procedures of the CDM Executive Board without uncritically copying them. It

[82] And the inability to challenge the decisions—see particularly Voigt, chapter 13 below.

[83] FCCC/CP/2001/13/Add.2 Draft Decision-/CMP/1 (art 6), Annex, para 25.

[84] In 2001, the COP adopted the Marrakech Accords that set out the necessary guidelines and modalities. These were then submitted to and approved by the COP/MOP at its first session.

evaluates, interprets and modifies the CDM rules wherever it considers it inadequate, bearing in mind that JI is a mechanism implemented between two countries with quantified emission reduction targets where ultimate liability for any overestimation of emission reductions lies with the host country.

7. Carbon Contracting

Any type of transaction involving the transfer of carbon credits needs to rely on the proper set of contracts. Once issued, CERs and ERUs as well as allocated allowances under various emission trading schemes are traded on the so-called secondary market. This is an area of law which, as we will see below, is becoming extremely sophisticated and complicated.[85] Most of these secondary transactions are governed by standard and relatively simple spot contracts. However, most JI and CDM credits are still traded under forward contracts (commonly referred to as Emission Reduction Purchase Agreements or ERPAs).[86] The implementation of JI/CDM projects leads to the establishment of legal links between the different sovereign Parties and the private or public project participants involved. Carbon contracts governing the implementation of the Kyoto Mechanisms include the following type of contractual instruments:

- Unilateral instruments of Non Annex I and Annex I Parties, such as Approval Letters for JI and CDM project activities;[87]
- Bilateral instruments between Annex I and non-Annex I Parties, such as Host Country Agreements or Memoranda of Understanding that govern the transfer of AAUs [88] or ERUs under Article 6 or Article 17 of the Kyoto Protocol; or

[85] On the secondary market see Hedges, chapter 15.

[86] In the course of negotiating the very first project of the Prototype Carbon Fund in Latvia, the World Bank transformed its traditional project investment finance agreement into a totally new instrument—an Emissions Reduction Purchase Agreement. This is discussed in detail in David Freestone, 'The World Bank's Prototype Carbon Fund: Mobilising new Resources for Sustainable Development' in S Schemmer-Schulte and Ko-Yung Tung (eds) *Liber Amicorum for Ibrahim S I Shihata* (Kluwer Law International: The Hague, 2001) 265–341. The basic structure was that in return for the project generating emission reductions verified by an independent certifier, the PCF would pay agreed amounts. This model, with modifications, appears to have been the one that the market has followed and recent developments are discussed in this volume: see for JI transactions Hobley and Roberts, chapter 9; and generally Wilder and Fitz-Gerald, chapter 14.

[87] Letters of Approval have an interesting double nature as administrative instruments under the domestic law of the host country and at the same time as unilateral statements under international public law generally directed towards an Annex I Party and authorizing private entities to participate in the project

[88] For the CDM, the Executive Board has clarified that a project participant is either a Party involved or a private and/or public entity authorized by a Party to participate, under the Party's responsibility, in CDM project activities which takes decisions on the allocation of CERs from the CDM project activity under consideration. See UNFCCC, 'Glossary of CDM terms' <http://cdm.unfccc.int/Reference/glossary.html>, accessed 9 April 2009.

- Bilateral or multilateral instruments between project participants, which may include Annex I and non-Annex I Parties, private and public sector entities, and which govern the implementation of the emission reduction project, risk allocation and the transfer of funds and carbon credits.

The first two categories of agreements focus on the authorization of the project by the Parties involved and the transfer of the internationally defined carbon credits to the extent that sovereign action is necessary. JI project activities rely on the confirmation of the host government to convert AAUs into ERUs once they have been verified. In the case of the CDM a simple host country approval is sufficient to allow the issuance of credits. In this volume, we focus in our analysis on the last of the three categories, which includes the commercially structured contracts that govern the purchase and sale of rights to the GHG emission reductions (or removals) generated by CDM or JI projects. CDM and JI transactions need to be based on innovative and robust contracts which enable the parties successfully to implement the project long after the attention of the initial project developers and consultants has moved on to other tasks. Carbon contracts define the relationship between parties in an emerging market, which is characterized by a wide variety of uncertainties and risks. The contracts need to record the agreement between the parties, identify responsibilities, allocate risks, establish rights and create clear and enforceable obligations. There have been various efforts to standardize ERPAs and other carbon related documents. Experience shows, however, that carbon projects come in all kind of forms and varieties. Regional context, project type, project size and financial status of the project and project owner determine the conditions of and the format of forward CDM/JI transactions. Still, publicly available contract models assist buyers and sellers of emission reductions in the definition of the key terms, in the proposal of model formulation, and in providing a starting point for negotiations. The World Bank and the International Emission Trading Association have developed standardized and publicly available ERPA templates.[89]

8. Public Sector Financing for Climate Change

This section will look at the ways in which the donor community has been able to respond to the challenge of climate change and the funding requirement provisions of the UNFCCC. It should be pointed out at the beginning that despite the fact that although a number of different funds have been established,

[89] See, eg, The World Bank Carbon Finance Unit, 'Document Library' <http://wbcarbonfinance.org/Router.cfm?Page=DocLib&ItemID=24703>, accessed 9 April 2009; International Emissions Trading Association, 'Emissions Reduction Purchase Agreement version 3' <http://www.ieta.org/ieta/www/pages/download.php?docID=1793>, accessed 9 April 2009.

the scale of funding has not—until perhaps recently—been commensurate with the needs of the tasks in hand.

Like the UNFCCC itself, the Instrument of the Restructured GEF envisages 'agreed incremental financing' to be provided by the Facility. However in addition to climate change the Instrument also identifies three other focal areas: ozone depletion (to the extent not covered by the Montreal Protocol Multilateral Fund), international waters, and conservation of biodiversity. Moreover, in 2002 the text of the 1994 Instrument was amended and a further two focal areas were added to mandate support for land degradation, particularly desertification and deforestation, and persistent organic pollutants.[90] In addition, paragraph 3 of the Amendment recognizes that 'the agreed incremental costs of activities to achieve global environmental benefits concerning chemicals management as they relate to the [existing six] focal areas shall be eligible for funding.' The outcome is that there are currently 15 GEF Operational Programmes (OPs) approved by the Council that map out strategy in each focal area and provide synergies and crosscutting themes between the focal areas.[91] Of these 15 OPs, four relate to climate change.[92]

The GEF only finances 'incremental costs'—the cost of those activities and investments that represent a gain to the international as opposed to the national environment. In relation to climate change this concept may be relatively easy to calculate—or at least to understand. Here the GEF might, for example, finance the extra cost of converting a 'least-cost' but climate damaging project (such as a coal-fired power station) to a climate friendly project (such as a wind farm or a solar thermal plant). In such a situation, the country contributes the amount it would have contributed to the 'least-cost' project and the GEF finances the additional (incremental) costs of the climate friendly technology. However, these incremental cost calculations are more difficult in relation to other focal areas.[93]

Since 1994 the GEF has been replenished four times: the 'restructuring' involved pledges of US$2 billion to this new Trust Fund, for the four-year period 1994–1998.[94] In March 1998 the second replenishment (GEF-2) was

[90] Instrument as amended, Para 2.

[91] The full texts of the OPs are available on the GEF website. See <http://www.gefweb.org>, accessed 9 April 2009.

[92] Climate Change:

OP5. Removal of Barriers to Energy Efficiency and Energy Conservation;

OP6. Promoting the Adoption of Renewable Energy by Removing Barriers and Reducing Implementation Costs;

OP7. Reducing the Long-Term Costs of Low Greenhouse Gas Emitting Energy Technologies;

OP11. Promoting Environmentally Sustainable Transport.

[93] The Council considered this issues specifically in relation to the new land degradation focal area, see 'Legal, Operational and Financial implications of an Amendment to the GEF Instrument to reflect the designation of the GEF as a financial mechanism of the UNCCD' (GEF/C.30/7. December 2006), at para 37ff.

[94] And the first of the Overall performance assessments which have preceded replenishments: GEF Secretariat, *Study of GEF's Overall Performance* (1998, GEF).

completed with commitments totalling US$2.75 billion for the four-year period to 2002.[95] In August 2002 after a long negotiating process the GEF was replenished for a third time to a total sum of US$2.92 billion for the period 2002–2006. After another evaluation, [96] and another protracted negotiation, and despite major reductions in the pledges of two major donors (the US and Japan) the GEF was replenished to a sum of US$3.12 billion.[97] The Third OPS found that the impact of the GEF was not restricted to the $6.2 billion dollars that it had itself committed since 1991—of which some US$1.75 billion was to the climate change portfolio[98]—but also included the more than $20 billion that had been mobilized in co-financing by a range of international and bilateral agencies as well as the private sector. In the area of climate change while its role in mitigating changes is obviously minor, the study found that it had 'played an important catalytic role in developing and transforming markets for energy and mobility in developing countries.'[99] Looking at the replenishment figures however what is most noteworthy is that since 1998 the GEF has effectively been functioning with level, maybe even decreasing, funding while at the same time it has been given new focal areas, wider ranges of interlocutors, and extra tasks and responsibilities have been imposed by the donors. The contrast with the size of the resources mobilized by carbon finance discussed above could not be starker.

8.1 The new Climate Funds

Another important outcome of COP7 in Marrakech in November 2001 was that the COP invited the GEF as the financial mechanism of the Convention to establish and operate two new funds related to the UN Framework Convention on Climate Change. By decision 7/CP.7 it decided to establish a Special Climate Change Fund[100] and a Least Developed Countries Fund.[101] By decision 10/CP.7 it also established a new fund related to the Kyoto Protocol (the Adaptation Fund) which would be financed by a 'share of the proceeds' of Clean Development Mechanism project activities, as envisaged by Article 12 of the Kyoto Protocol and by additional funding invited from those Annex 1 Parties intending to ratify the Protocol.[102] While COP7 did not provide full guidance on these funds, the COP Decision did lay down certain types of activities which might be financed by these funds.

[95] And see GEF Secretariat, The First Decade of the GEF: The Second Overall Performance Evaluation (25 January 2002, GEF).

[96] OPS3: Progressing towards Environmental Results (June 2005, GEF).

[97] *Summary of the negotiations on the Fourth Replenishment of the GEF Trust Fund.* Paper for the Third GEF Assembly, Cape Town, South Africa, August 2006 (GEF/A.3/6). A major aspect of the replenishment was the approval of a Resource Allocation Framework (RAF) that had been first raised at the second replenishment discussions, see further Freestone, n 28, 1099–1102.

[98] OPS 3 p. 30. [99] Ibid, p. 4. [100] 7/CP.7, Para 2. [101] Ibid, Para 6.

[102] 10/CP.7, Para 2.

8.1.1 The Special Climate Change Fund

This Fund will be available to finance activities, programmes and measures relating to climate change, in accordance with paragraph 2 of decision 7/CP.7, that are complementary to those funded by the resources allocated to the climate change focal area of the GEF and by bilateral and multilateral funding, in the following areas: (a) adaptation, in accordance with paragraph 8 of decision 5/CP.7; (b) transfer of technologies, in accordance with decision 4/CP.7; (c) energy, transport, industry, agriculture, forestry and waste management; (d) activities to assist developing country Parties referred to under Article 4, paragraph 8(h) (ie countries whose economies are highly dependent on income generated from the production, processing and export, and/or on consumption of fossil fuels and associated energy-intensive products) in diversifying their economies.[103]

8.1.2 Least Developed Countries Fund

Initial guidance on the Least Developed Countries Fund was provided by decision 7/CP.7: (a) as a first step, to provide funding from the LDC Fund to meet the agreed full cost of preparing the National Adaptation Plans of Action (NAPAs), given that the preparation of NAPAs will help to build capacity for the preparation of national communications under Article 12, paragraph 1, of the Convention; (b) to ensure complementarily of funding between the LDC Fund and other funds with which the operating entity is entrusted; (c) to ensure separation of the LDC Fund from other funds with which the operating entity is entrusted; (d) to adopt simplified procedures and arrange for expedited access to the Fund by the least developed countries, while ensuring sound financial management; (e) to ensure transparency in all steps relating to the operation of the Fund; (f) to encourage the use of national and, where appropriate, regional experts; (g) to adopt streamlined procedures for the operation of the Fund.

8.1.3 Kyoto Protocol Adaptation Fund

In accordance with paragraph 1 of Decision 10/CP.7, the Kyoto Protocol Adaptation Fund is to finance concrete adaptation projects and programmes as well as the following activities identified in paragraph 8 of decision 5/CP.7: (a) starting to implement adaptation activities promptly where sufficient information is available to warrant such activities, *inter alia*, in the areas of water resources management, land management, agriculture, health, infrastructure development, fragile ecosystems, including mountainous ecosystems, and integrated coastal zone management; (b) improving the monitoring of diseases and vectors affected by climate change, and related forecasting and early-warning

[103] In accordance with Decision 5/CP.7.

systems, and in this context improving disease control and prevention; (c) supporting capacity building, including institutional capacity, for preventive measures, planning, preparedness and management of disasters relating to climate change, including contingency planning, in particular, for droughts and floods in areas prone to extreme weather events; (d) strengthening existing and, where needed, establishing national and regional centers and information networks for rapid response to extreme weather events, utilizing information technology as much as possible.

8.2 Governance of the Climate Change Funds

However the main issues have not been with the mandate of these new funds but with their governance. After COP7 the Trustee and the Secretariat of the GEF proposed arrangements for the establishment and operation of these new funds.[104] These envisaged that the World Bank would act as Trustee for the three funds but that they would each be managed separately from the GEF Trust Fund. Mobilization of resources for these funds would also not be through the four-yearly GEF replenishment process but through a separate process managed by the Secretariat. It was also agreed that the operational policies and procedures and governance structure of the GEF would apply to these voluntary funds. These arrangements were reported to the 8th session of the Conference of the Parties (COP8) in New Delhi. Further guidance was provided by COP8 on the LDC Fund as a result of which the LDC Trust Fund was made operational and by spring of 2003 pledges in the order of US$15 million were received from nine donors.[105]

8.2.1 LDCF and SCCF

Since that time procedures have been established for the governance of the LDCF and the SCCF. The GEF Council meets as the LDCF/SCCF Council.[106] Any Council member is eligible to take part and participate or attend as an observer. Formal voting is, as normal, by consensus but in the event of a vote the normal GEF rule would be amended to require both 60% majority of GEF Participants represented at the LDCF/SCCF Council and 60% majority of the total contribution to such fund. [107]

[104] GEF Sec prepared a paper for the May 2002 GEF Council Meeting, GEF/C.19/6, *Arrangements for the Establishment of the New Climate Change Funds*.

[105] For further details see: GEF/C.21/5/Rev.1. Operation of the LDC Trust Fund for Climate Change.

[106] The first such meeting took place in December 2006. GEF, 'LDCF/SCCF Council Meeting, December 8, 2006' <http://www.gefweb.org/interior.aspx?id=16718> accessed 1 June 2009.

[107] As of November 2006 pledges to the LDCF stood at US$115 million and to the SCCF at US$ 37.7 million. *Status Report on the Climate Change Funds* (GEF/LDCF/SCCF.1/Inf.2/Rev. 1). Average portfolio size for the LDCF in 2009 is estimated at US$114 million and US$87 million for the SCCF <http://www.gefweb.org/uploadedFiles/Documents/LDCFSCCF_Council_Documents/LDCFSCCF6_June_2009/LDCF.SCCF.6.5.pdf>, accessed 1 June 2009.

8.2.2 The Adaptation Fund

In the COP decision cited above,[108] the Conference of the Parties had decided that the Adaptation Fund should be operated by an entity entrusted with the financial mechanism of the Convention. It will be recalled that under the Kyoto Protocol Article 12(8) a 'share of the proceeds from certified emission reductions' under the Clean Development Mechanism shall be made available 'to assist developing country Parties that are particularly vulnerable to the adverse effects of climate change to meet the costs of adaptation.' Hence this Fund is not intended to be primarily a donor fund—although donors may contribute to it. The COP felt therefore that a different governance structure was appropriate than that used by the GEF.

The GEF Council did make a submission to the UNFCCC COP/MOP regarding possible Arrangements for the Management of the Adaptation Fund (AF).[109] However in Bali it was decided to establish a separate governance structure with an Adaptation Fund Board chosen by the COP itself from among its members.[110] The Board met for the first time in March 2008 in Bonn. In the meantime the COP invited the GEF Secretariat to act as the secretariat 'on an interim basis' and the Word Bank as Trustee—also on the same interim basis. A key role of the Trustee is the 'monetization' of the CERs allocated as the share of the proceeds of CDM projects. The World Bank made detailed proposals as to the way that it would handle this task, clearly defining its responsibilities and liabilities—these were accepted at COP14 in Poznan in December 2008.[111]

8.3 The World Bank Climate Investment Funds

After the World Bank/IMF Annual Meeting in the autumn of 2007, in the follow up to proposals from the Gleneagles G8 Summit in 2005, a number of donors, led by the United Kingdom began to talk to the World Bank about establishing a Climate Investment Fund. These proposals were formally approved by the World Bank's Board of Executive Directors on 1 July 2008 and in the following September pledges totaling nearly US$6.2 billion were received from 10 donors.[112]

[108] Decision 10/CP.7.

[109] Submission of the Global Environment Facility (GEF) on Possible Arrangements for the Management of the Adaptation Fund. GEF/C.29/Inf.2. 8 August 2006.

[110] Decision 5/CMP.2. The Board's composition would be: two from each of five UN regions; one from the Small Island Developing States (SIDS); one from the LDCs; two from Annex 1 Parties; and two from non-Annex I Parties. See also Adaptation Fund, 'Meetings' <http://www.adaptation-fund.org/meetings.html>, accessed 9 April 2009.

[111] See Decision 1/CMP.4. Adaptation Fund (Addendum) Annex III at p. 14 see <http://unfccc.int/resource/docs/2008/cmp4/eng/11a02.pdf>, accessed 20 July 2009.

[112] Australia, France, Germany, Japan, The Netherlands, Norway, Sweden, Switzerland, the United Kingdom, and the United States. For details see The World Bank, 'Donor Nations Pledge Over $6.1 Billion to Climate Investment Funds' (26 September 2008) <http://web.worldbank.org/WBSITE/EXTERNAL/NEWS/0,,contentMDK:21916602~pagePK:34370~piPK:34424~theSitePK:4607,00.html>, accessed 9 April 2009.

The funds, to be disbursed as grants, highly concessional loans, and/or risk mitigation instruments, are administered through all the multilateral development banks and the World Bank Group. The World Bank is the Trustee. There are two funds. The Clean Technology Fund will invest in projects and programs in developing countries that contribute to the demonstration, deployment, and transfer of low-carbon technologies. The projects or programs must have a significant potential for long-term greenhouse gas savings. The second fund, the Strategic Climate Fund, is broader and more flexible in scope. It will serve as an overarching fund for various programs to test innovative approaches to climate change. The first program under this fund is a pilot aimed at increasing climate resilience in developing countries. A Forest Investment Program and a Scaling-Up Renewable Energy Program are also expected to be created during 2009.

In creating the funds, and in support of the Bali Action Plan, participants took care to recognize the primacy of the UNFCCC global climate negotiations, and to support those negotiations. All funds and programs under the CIF have a sunset clause in order not to prejudice UNFCCC deliberations regarding the future of the climate change regime. The design of the funds ensures that developing countries have an equal voice in the governance structures of the funds, and decisions on the use of funds are to be made by consensus. An annual Partnership Forum will be held to provide a venue for talks on the strategic directions, results, and impacts of the CIF, as well as provide a platform to share lessons learned as widely as possible.

9. The Bali Roadmap

The most important outcome of the Bali COP in December 2007 was the approval of the Bali Action Plan.[113] This Action Plan includes the so-called 'roadmap' for the forthcoming negotiations which will aim to develop by 2009 a legal instrument to replace the obligations of the 1997 Kyoto Protocol when its commitment period expires in 2012, and which are due to culminate at the COP15 meeting in Copenhagen in December 2009. The 'roadmap' is of particular importance in that the UNFCCC Parties agreed to consider 'measurable, reportable and verifiable nationally appropriate mitigation actions' for *all* parties (including developing country parties), although developed country parties also agree to consider 'commitments, ... including quantified emission limitation and reduction objectives'.

Unlike the Berlin Mandate—which only referred to Annex I countries and only considered quantified emission limitation and reduction objectives (so-called QELROs)—the agenda mapped out by Bali is potentially much wider. A whole range of issues must be on the table. Since Kyoto, the emission of a number of

[113] Decision 1/CP.13.

middle income non-Annex I countries have increased steeply—notably China (which some reports suggest has overtaken the US as the biggest GHG emitter), India, and Brazil. While there is strong resistance among such countries to taking on the same sort of targets as Annex I countries assumed at Kyoto, there are a range of other options which are discussed in a number of the following chapters. Rather than economy-wide targets, for example, which do seem set to continue for Annex I countries, it might be possible that middle income countries agree to adopt sectoral 'no-lose' targets. These would not necessarily carry sanctions for non-compliance but would facilitate the inflow of investment and technology. These might relate to specific sectors such as, for example, the electricity generation or transmission sector, or for emission intensive industries such as cement or aluminium, or even for transportation. These agreements could be part of a global package or be linked perhaps to separate technology transfer agreements 'off-line' from the main negotiation.[114] In addition as the agenda moves towards a more comprehensive emissions framework, significant non-covered sectors—such as aviation or maritime bunker fuels—need to be brought somehow within the negotiations.[115]

Bali makes it clear that for Annex I countries increased quantified limitation and reduction objectives (QELROs) are still definitely on the table. In order to prepare for these obligations a number of Annex 1 countries are implementing or (as with the case of the EU) enhancing existing, carbon trading schemes. The growth of regional and national emission trading schemes is dealt with in the following chapters covering the EU,[116] Australia[117] Canada,[118] and China,[119] as is the preparatory work which has started to bring the US into the global system. The incoming Obama administration has signalled that the US will adopt a cap-and-trade system, but important aspects of this are being worked out.[120]

At the same time the future of the Kyoto Mechanisms is also under discussion. What would be the future for the Clean Development Mechanism in a world where more countries have targets—even 'no-lose' ones? Given the substantial achievement of the CDM in mobilizing resources for sustainable development, it is possible to have a vision for the CDM after 2012[121]—even if as a minimum if were to be only for the poorest and smallest countries, as the European Commission has suggested. Joint Implementation, which has until recently been the 'wallflower' of the Kyoto Mechanisms, is beginning to take off.[122] In a post-2012 world project-based investments between countries with

114 As Ward explains in chapter 24
115 Aviation is discussed by Chagas and Clarke in chapter 28.
116 See Pohlmann, chapter 16 and Ghaleigh, chapter 17.
117 Wilder and Fitz-Gerald, chapter 20. 118 Taylor and Barrett, chapter 21.
119 Tung, chapter 22.
120 Explored by LaMotte, Williamson and Hopkins, chapter 18 and Danish, chapter 19.
121 As Figueres and Streck point out in chapter 26.
122 See Hoogzaad and Streck, chapter 8 and Hobley and Roberts, chapter 9,

economy-wide or even sectoral targets may have an important role to play in the future. All these issues are covered in detail in the following chapters.

Another new agenda for Copenhagen is also set out in the Bali Action Plan which commits the parties to consider:

(iii) Policy approaches and positive incentives on issues relating to reducing emissions from deforestation and forest degradation in developing countries; and the role of conservation, sustainable management of forests and enhancement of forest carbon stocks in developing countries;[123]

Although agriculture, forestry and other land use (AFOLU) accounts for some 20% of the total amount of carbon which exists on the planet, it was nevertheless decided at Marrakech that only reforestation and afforestation projects would be eligible for consideration under the CDM—and indeed to date only two such projects have been approved by the CDM Executive Board.[124] In 2005 at COP11, Papua New Guinea and Costa Rica—with support from a number of important forested countries—first put forward a formal proposal to include the crediting of benefits from avoiding further deforestation. The 2007 Stern Review identified avoided deforestation as the cheapest option to mitigate increases in emissions of greenhouse gases. Sophisticated monitoring mechanisms—often involving satellite surveillance—need to be put into place and effective compensation systems devised to encourage governments and their nationals to stop cutting down trees.[125]

10. Conclusions

This chapter has sought to provide an introduction to the UNFCCC system and the key institutions as well as to the processes leading up to, and from, the adoption of the Kyoto Protocol. It has aimed to introduce some of the key concepts that will be further developed in later chapters. With the release of the IPCC Fourth Assessment Report it is clear that radical action needs to be taken to avoid dangerous climate change. The Stern Review forcefully reminds us that early, radical action in reducing emissions of GHGs is also the most economically sensible approach. At the Copenhagen summit UNFCCC parties will be negotiating the sorts of major commitments that need to be made to address seriously the challenge of climate change. Stern suggests that action within the next 10–15 years to keep carbon concentrations under that critical level could cost 1% of global product—ie some US$3.5 trillion a year. Although discussion of the new Global Climate Fund is ongoing, it is unfortunately clear from the experience of the GEF that public sector financing—through government

[123] n 113 above, para 1(b)(iii).

[124] As we see from Pohlmann, chapter 16, the EU ETS does not currently count ERs from sequestration projects at all.

[125] For a full discussion, see O'Sullivan and Saines, chapter 27.

donors—is unlikely by itself to generate the level of resources that are necessary. Among the alternative or additional approaches, by contrast, carbon trading—worth more than US$120 billion in 2008—can, if well managed, provide access to huge resources from the private sector that will be necessary to meet this challenge. The objective of the rest of this volume is explain the way that these markets can be put to work in the global effort to lower GHG emissions, decarbonize our economies, and to achieve truly sustainable development.

PART II

GENERAL ISSUES

2

Legal Ownership and Nature of Kyoto Units and EU Allowances

Matthieu Wemaere, Charlotte Streck and Thiago Chagas

1. Introduction

The Kyoto Protocol is based on the principle of achieving quantitative reductions in greenhouse gas (GHG) emissions by establishing binding emission targets on individual parties to the Protocol. While the overall goal is the reduction of GHG emissions, the Protocol allows for a degree of flexibility as to where the emission reductions can be generated. The flexibility is created through a mechanism for trading emission rights (Article 17), and two mechanisms for creating project-based emission credits which are measured and verified against baseline emission levels (Articles 6 and 12).[1] The base units of the Kyoto Protocol are Assigned Amount Units (AAUs) where each unit is equivalent to one metric tonne of carbon dioxide emissions equivalents assigned to an industrialized (Annex B) country pursuant to Article 3 of the Kyoto Protocol. Assigned Amount Units can be acquired and transferred in accordance with Article 17 of the Kyoto Protocol. Emission rights associated with Article 12 of the Kyoto Protocol are Certified Emission Reductions (CERs) generated by project activities approved under the Clean Development Mechanism (CDM). Certified Emission Reductions are issued by the Executive Board of the CDM on the basis of an independent report that verifies and certifies reductions in GHG emissions. Joint Implementation (JI) projects pursuant to Article 6 of the Kyoto Protocol generate Emission Reduction Units (ERUs) which come into existence through a conversion of an AAU or Removal Unit (RMU) from the accounts of the host country. Removal Units result from eligible sink activities in accordance with Articles 3.3 and 3.4 of the Kyoto Protocol.[2] The most recent

[1] Decision 2/CMP.1 sets out the principles, nature and scope of all three mechanisms. Decisions 3/CMP.1, 4/CMP.1 (CDM), 9/CMP.1, 10/CMP.1 (JI), and 11/CMP.1 (ET) contain operational rules for Joint Implementation, the CDM and emissions trading.

[2] For the rules governing the issuance of RMUs, start at UNFCCC/CMP/2005/16/CMP.1 (LULUCF) and 13/CMP.1 (*Modalities for the accounting of assigned amounts*): Modalities for the accounting of assigned amounts under Art 7(4) of the Kyoto Protocol, Annex, § 25.

addition to the family of Kyoto units are tCERs and lCERs. These were defined by the 9th Session of the Conference of the Parties to the UN Framework Convention on Climate Change (UNFCCC) in 2003. Temporary CERs (tCERs) or long-term CERs (lCERs)[3] are associated with land use change and forestry activities which create removals in GHG under the CDM.

This plethora of units, which determine the currencies of the international carbon market, have been complemented by domestic GHG emission rights. Several national and regional emissions trading schemes have already been established or are currently being designed. Most prominently, the 'Directive establishing a scheme for greenhouse gas emission allowance trading within the European Community' (EU Emissions Trading Directive) defined the tradable item as a transferable allowance to emit one tonne of carbon dioxide equivalent during a specified period.[4] The EU Emissions Allowances Trading Scheme (EU ETS) is intended to help the European Community[5] meet its Kyoto target, of an 8% reduction in emissions relative to 1990 levels, during the Kyoto Protocol's first commitment period of 2008–2012. The EU Emissions Trading Directive was transposed in 27[6] emission trading systems of EU member states, whereby tradable allowances are allocated to covered installations.[7] Each member state had to integrate the concept of emission allowances (EU Allowance) into the domestic legal system.

This chapter aims to analyse the legal nature and characteristics of the aforementioned different units. Clear decisions on the nature and the treatment of these units are important to give legal security and certainty to both Governments and private (or public) entities, and provide security and confidence to the trading system. Where legislation defines tradable units differently, but where the trading systems are linked, such as in the EU ETS, difference in treatment can make one sub-market of the overall system more attractive than others. Different treatment can thus lead to, intended and—worse—unintended, distortions of the market.

The analysis in this chapter is ambitious in its scope as the different units pertain to markedly different legal frameworks including public international law, both civil and common law systems, and rights based on private contracts. They are also embedded in different legal traditions: civil law and common law

3 Decision 4/CMP.1. Modalities and procedures for afforestation and reforestation project activities under the clean development mechanism in the first commitment period of the Kyoto Protocol, Annex, §1(g), (h).

4 Directive 2003/87/EC of the European Parliament and of the Council of 13 October 2003 establishing a scheme for greenhouse gas emission allowance trading within the Community and amending Council Directive 96/61/EC.

5 The European Community ratified the Kyoto Protocol on 31 May 2002.

6 With the accession of Romania and Bulgaria to the EU in 2007, the EU ETS is now applicable to the 27 EU member states. In addition, Norway, Iceland, and Liechtenstein also joined the EU ETS in 2008.

7 The first trading period, also referred to as the 'trial period', ran through 2005 to 2007. The second trading period started in 2008 and will be effective until to the end of 2012. The EU ETS second trading period coincides with the first commitment period of the Kyoto Protocol.

systems traditionally define property rights differently—but, however, applying the same basic criteria based on the maxim '*usus, abusus, fructus*'[8] to characterize property. We are dealing with a new set of rights which carries a number of features that justify the classification of emission rights as instruments '*sui generis*'. These rights may not be assimilated easily by any existing legal system, requiring the nature and ownership to be newly determined by domestic law.

2. Carbon Units

All of the carbon units, AAUs, ERUs, CERs, RMUs, and EU Allowances are all equal to one metric tonne of carbon dioxide equivalent and each have a unique serial number. They are first and foremost accounting units, which are tracked and recorded through national registries that are established and maintained by Annex B parties. At the same time they are also tradable instruments which share two important determining features:

(1) they represent an entitlement to release a certain quantity of GHG emissions into the atmosphere; and
(2) they are transferable under certain established conditions.

The creation of transferable rights in the context of emissions trading schemes also represents the main difference between these schemes and instruments rooted in the traditional command and control approach of environmental policy. While in command and control instruments the entitlement to emit is granted on an individual basis and is conditional upon operating an installation to certain standards, it is not, however, a transferable instrument.

Despite the general similarity of the listed carbon units, they differ between each other in significant aspects: They are either created or allocated through a regulating body (international or domestic), and are subject to different legal regimes (mandatory/voluntary), which implies that they differ in the scope of their use. This ranges from compliance with international or national legally binding reduction obligations to the fulfillment of voluntary commitments.

3. The Cornerstones of Emissions Trading

3.1 The ethical dimension

Emissions trading is based on the creation of tradable entitlements to emit a certain quantity of a pollutant. This concept, which relies on the privatization of

[8] This maxim refers to the right of an owner to use (*usus*), discharge or transfer (*abusus*), and benefit from (*fructus*) a thing or right he or she possesses.

formerly publicly held rights, has drawn significant criticism.[9] The idea of dividing the atmosphere in to small pieces and distributing it to polluting industries has been interpreted as a form of 'privatization of the atmosphere' leading thus to an abuse of the global commons (that is, the responsibility of the State to protect, not to give away).[10] It has been argued that by looking at the sky as though it were a common resource free of any substantial public interest other than the protection of its quality, Governments have taken a too narrow view and neglected their capacity as trustee of public resources by transferring a significant public resource to private hands, especially if this happens free of charge.[11] These critical voices are complemented by those of environmentalists who reject the very notion of 'rights' to pollute, let alone a profitable market based on these rights.[12]

Responding to these concerns requires us to take a step back and reconsider the nature of the right which has been created for trade among private entities. The question is two-fold: first, has the State the necessary power to establish an emissions trading system, and second, what type of right does the Government bestow on private entities which participate in such schemes.

With respect to the first question, it is important to remember that the management of global commons is largely based on property rights. This holds true for command-and-control as well as market-based approaches. In the process of formulating and implementing environmental policy, the State has a choice of instruments which represent different types of management systems. The difference between the instruments often lies only in the nature and extent of property rights imposed. The State is entrusted to manage, maintain and conserve public goods for the benefit of its people. Effective as well as cost-efficient protection of the good should be the determining factor in the choice of the policy instrument, which will eventually depend on the type of good to be protected and the transaction costs related to a certain policy instrument. Regardless of the instrument chosen, however, the State's environmental protection efforts are inevitably property-based in that all solutions to the 'tragedy of the commons'[13] involve the imposition of property rights (private/individual, common, or state/public) on formerly unowned (nonproperty or open-access) resources.[14] When turning to command-and-control regulations the State asserts

[9] Peter S Menell, 'Institutional Fantasylands: From Scientific Management to Free Market Environmentalism' (1992)15 Harv JL & Pub Pol'y, 489; Michael C Blumm, The Fallacies of Free Market Environmentalism, (1992) 15 Harv JL & Pub Pol'y, 371; Gerald Torres, Seventh Annual Lloyd K. Garrison Lecture On Environmental Law 'Who Owns the Sky?' (2001) 18 Pace Envtl L Rev 227.

[10] See Torres, n 9 above.

[11] See Torres, n 9 above, 286.

[12] Robert W Hahn and Gordon L Hester, 'Where Did All the Markets Go? An Analysis of EPA's Emissions Trading Program' (1989) 6 Yale J on Reg 109, 142–3.

[13] Garrett Hardin, 'The Tragedy of the Commons' (1968) 162 Science 1243. The 'tragedy of the commons' relates to the problem that free access to the commons leads to its overuse and ruination.

[14] Daniel H Cole, 'From Local to Global Property: Privatizing The Global Environment?: Clearing The Air: Four Propositions About Property Rights And Environmental Protection' (1999) 10 Duke Env L & Pol'y F 103.

public property rights. When creating transferable emission rights, the State imposes a combination of public and private property rights on the environmental goods: it first converts non-property into public property before it converts the same property into private property.[15] The most prominent distinction between the two approaches is that the creation of non-tradable permits determines the extent of pollution-reduction at the installation level, while the creation of tradable permits enables the operators of covered installations to allocate the costs of achieving pollution reductions most efficiently via the market.[16]

This leads to consideration of what sort of right it is that States create in tradable permits. Generally, the assignment of allowances does not lead to the 'privatization of the atmosphere' itself. It is not the 'air' or the 'atmosphere', nor even the 'emissions' which are subject to distribution. Instead, the assignment of quantified emission reduction targets or the allocation of allowances for a given period of time aims at regulating the use of the atmosphere (not the access to, or conditions to share, the 'resource'). It does not create any right or entitlement for the benefit of allowance holders in the atmosphere itself. The State does not transfer ownership in a parcel of air, but instead creates the right to release a certain amount of GHGs into the atmosphere. The right that is created through this action is of regulatory nature; the market a 'permit' market.[17] Other than public property, for which government agents manage the resource in question, or private property for which private parties allocate, manage, and transfer rights in the benign light of government, regulatory property is characterized by government agents allocating and managing inalienable rights.[18]

Consequently, the question of where exactly the boundaries of the State's sovereign right to the atmosphere lie does not need to be answered in this context. The State, often mandated through a constitutional obligation, has the right and the obligation to protect and regulate the sustainable use of the atmosphere. If the legal nature of the atmosphere as *res communis* or common good were to be recognized, it would not limit the State in fulfilling its mandate in regulating use and access to the resource. When creating international law to ensure the global access and management of public goods, states agree to regimes of common management and regulation in the understanding that these

[15] This sequence, however, does not give the answer on how modern property rights evolve. In this context Robert Nelson argues that, indeed, it is the transformation of private property to public property rights is how these rights evolve. The benefits that flow from regulation are seen as being no different than the stream of benefits that comes from any other form of property. See Robert H Nelson, 'Private Rights to Government Actions: How Modern Property Rights Evolve' [1986] U Ill L Rev 361, 364.

[16] See Cole, n 14 above.

[17] Most of the markets that exist today do not trade in private property, rather they trade in regulatory property. See Bruce Yandle, 'From Local to Global Commons: Private Property, Common Property, and Hybrid Property Regimes: Grasping For The Heavens: 3-D Property Rights And The Global Commons' (1999) 10 Duke Env L & Pol'y F 13.

[18] See Yandle, n 17 above at 6.

common spaces, such as oceans,[19] the common airspace, and the atmosphere[20] form part of the common property or common concern of humankind.[21] The assignment of AAUs in the context of the Kyoto Protocol does not contradict with this general principle. It neither leads to a regulation of the climatic system for the benefit of a limited number of countries (with targets), nor does the allocation of AAUs create any right or entitlement to modify the climatic system. Instead it sets an obligation to preserve the beneficial effects of the climate for future generations.

3.2 The basic concepts

Before engaging in a more detailed analysis of the units defined under the Kyoto Protocol, it is valuable to briefly describe the main underlying concepts which determine the features of emissions trading schemes: (i) emissions; (ii) allowances; (iii) allocation of authorizations to emit in the context of cap and trade schemes; and (iv) emission credits generated in baseline and credit systems.[22]

3.2.1 Anthropogenic GHG emissions

Anthropogenic GHG emissions are the human-induced release of substances into the atmosphere that contribute to the phenomenon commonly referred to as 'global climate change'. While the natural cycle of carbon dioxide (CO_2) and other gases also results in the release of GHG into the atmosphere, it is the increase in the release of such gases as a result of human activities which causes the problem of global climate change. Since our control over the natural cycle of

[19] United Nations Convention on the Law of the Sea, UN Doc A/Conf. 62/122, reprinted in (1982) 21 ILM 1261.

[20] Because of the opposition of some states' parties to the Part XI of the 1982 UN Convention on the Law of the Sea, it was preferred to refer to the concept of 'common concern of mankind" instead of 'common property of mankind' in order to define legally the climatic system in the Preamble of the UNFCCC, further to UN General Assembly Resolution 43/53 of 6 December 1988.

[21] Under international law the concept of common property is applied to refer to spaces beyond the territorial sovereignty of States, such as the high seas and the superjacent airspace, in which no single State enjoys exclusive right to use and exploit the natural resource. In turn, the atmosphere, including the climate system and the ozone layer, is often attributed the status of 'common concern of human kind'. This concept aims at stressing the totality and unit of the resource, which overlaps both national jurisdictions and common spaces and gives any State (whether directly injured or not) a legitimate claim to protect the interest of the international community as a whole. In this respect, see Alan Boyle, 'International Law and the Protection of the Global Atmosphere: Concepts, Categories and Principles' in Robin Churchill and David Freestone (eds), *International Law and Global Climate Change* (Nijhoff, 1991) and P Birnie and A Boyle, *International Environmental Law & the Environment* (2nd edn, Oxford, 2002) 500–2.

[22] The present discussion is focussing on the legal ownership and nature of GHG emissions; however, interesting lessons can be drawn from other emission trading programs covering other air pollutants, such as sulphur dioxides and nitrogen oxides (schemes are in place in the USA, Canada, and the Netherlands, among others).

GHGs is limited, efforts to curb GHG emissions focus on anthropogenic emissions.

GHG emissions result from a wide range of activities and often represent the unintended by-product of an activity. They are therefore residues of combustion, industrial, and agricultural processes. An interesting parallel can be drawn between the problem of waste, which like GHG emissions, is not produced voluntarily. An object, a material, or a substance becomes waste whenever the holder decides so or is obliged to dispose of it. By similar analogy, GHG emissions could be regarded as gaseous wastes at the point of time when they are released into the atmosphere. Discharges of hazardous substances into the soil or watercourses represent, under certain circumstances, a disposal operation which may be prohibited or subject to an administrative authorization. However, air pollutants or emissions are not usually defined as waste because they are intangible. Waste may have an economic value which can be either positive or negative for the holder, who may also be the owner. The economic value of waste is a direct consequence of the evolution of legislation requiring a reduction of waste (prevention) or the ecologically sound management or disposal of waste, which comes at a certain cost. Again, similar reasoning could be applied to GHG emissions which alone have no economic value (gases are usually not reused, with few exceptions such as the use of furnace gases for energy production or liquefaction of CO_2). Any economic value attached to GHG emissions arises from the right to transfer defined units, which represents the authorization to emit a certain quantity of GHGs.

3.2.2 Emissions trading systems

There are two basic forms of emissions trading systems: cap-and-trade systems that are based on an initial allocation of emission rights through a government or other regulatory body, and baseline-and-credit systems where emission credits are generated through the implementation of an activity which reduces emissions against a baseline. In theory, both the cap-and-trade approach and baseline-and-credit systems may be subject to absolute or relative emissions caps. Any emissions trading system may also combine features and credits from both approaches.

A key element in cap-and-trade systems is the procedure in which allowances are allocated. Governments establishing this form of emissions trading scheme are required to set the total 'fringe' amount of the resource that may be safely and sustainably used, before, in a second step, dividing this amount up and allocating that usable total in the form of individual tradable allowances.[23]

[23] Bruce A Ackerman and Richard B Stewart, 'Reforming Environmental Law: The Democratic Case for Market Incentives' (1988) 13 Colum J Envtl L 171, 184. Ackerman and Steward add another four bureaucratic measures that are required to ensure a functioning trading scheme: a system for monitoring; a record-keeping system; a reporting procedure; and the enforcement of the limits both externally (to exclude interlopers) and internally (to constrain insiders' overuse).

Under Article 3.1 of the Kyoto Protocol, allowances are assigned only to parties by virtue of international law. Governments on the other hand can translate the international law targets applying to their country through the allocation of allowances to private and public entities under national emissions trading schemes. In both instances, the allocation fixes the overall quantity of emissions that can be emitted in a given jurisdiction and therefore puts a cap on emissions. It is the starting point before the trading of allowances can happen and is also the pre-determination of the overall environmental benefits that can be expected from the system. Only voluntary trading schemes operate without a cap relying on non-binding initiatives to generate emission reductions and participate in the trade. Frameworks based on mandatory schemes regulate, through the allocation process, the use of the atmosphere to discharge certain substances from a quantitative point of view. Emission rights in mandatory cap-and-trade systems are of governmental origin but once allowances are allocated they can be traded without any restriction other than those foreseen by the applicable regime.

3.2.3 Allowances

Allowances generally represent the authorization or entitlement to emit a certain amount of GHGs. In the allocation process the regulatory authority decides how many allowances an emitter receives. When the Kyoto Protocol allocates Assigned Amounts to state parties it distributes allowances, as does a national emissions trading scheme when it develops an allocation plan. The entitlement to emit a certain amount of specified substances is generally not different from the traditional command and control approach under which an operator of an installation needs a permit to release up to a certain quantity of emissions for specified period of time. However, the transferability, and when different schemes are linked, the fungibility among units, makes the holding of allowances fundamentally different from the holding of a traditional command and control permit. An allowance not only carries the authorization to emit a certain substance, it also embodies a right to be sold and transferred. While the allowance does not create a right in the emissions itself, it creates a transferable entitlement to emit a certain substance. This right is different depending under which regime the right has been created.

Article 17 of the Kyoto Protocol establishes the right of Annex B parties to trade Kyoto units under the framework of the Protocol. The right can be exercised subject to certain conditions and limitations. Such limitations include the necessity to fulfill a set of eligibility criteria and to maintain a certain minimum amount on the registries of the party (commitment period reserve).[24] The right

[24]The Marrakech Accords as adopted at the COP/MOP 1 require Annex B parties to hold a minimum level of AAUs, CERs, ERUs, and/or RMUs in a 'commitment period reserve' which cannot be traded. This is calculated as 90% of the assigned amount or the amount of emissions reported in the Party's most recent emission inventory, whichever is the lower figure. Decision 11/CMP.1, Annex (6).

to participate in the trade with AAUs may be delegated to private and public legal entities by virtue of government authorization. The AAUs allocated through the Kyoto Protocol could be thus regarded as a mixture of a sovereign right (to be used to fulfill an international obligation) and a public property right of an Annex B government (to make use of the value of its assigned units, for instance, by selling them to another Annex B party).

Allowances can also create property rights or quasi property rights of private entities holding allowances allocated under a domestic scheme. The exact legal nature of the allowances depends on the legal system under which the allowance is held and may include an authorization or permit to emit, an administrative or public right, a private property right, a security, a financial instrument, a good, or a commodity. More recently the argument has been made that emissions rights also present clear currency-like characteristics, similar to a monetary instrument.[25] Indeed, in most cases the allowance will show features of a number of these categories and have a mixed nature.

Generally, allowances are designed for simplicity, transparency, and security. The strengths of the concept of an allowance derives from their impersonality, their simplicity and explicitness, their usefulness for diffuse and wide-ranging resources, their openness to new entry, and their adaptability to changes in human demand.[26] In this respect they differ from other environmental management systems which rely on the distribution of responsibilities with respect to a particular source or problem by a difficult set of (negotiated) rights between different affected parties. An interesting hybrid between such more complex and multidimensional rights are verified emission credits.

3.2.4 Emission credits

In contrast to allowances, emission credits are generated privately. Governments may recognize these credits under a national emissions trading scheme, if the credits and the process of generating the credits fulfill certain defined criteria. However, governments do not allocate these rights. Emission credits represent a reduction of GHG emissions resulting from a defined project activity, calculated on the basis of the comparison between the level of verified actual emissions and a counterfactual scenario (defined as the baseline scenario). They are of private 'origin' and the credits are generated by entities as a direct consequence of a certain investment in a project. In order to ensure that

[25] The argument that that emissions rights present clear currency-like features has been pertinently put forward by Jillian Button in 'Carbon: Commodity or Currency? The Case for an International Carbon Market Based on The Currency Model' (2008) 32 Harv Envtl L Rev 571, 574.

[26] Carol M. Rose, 'From Local to Global Commons: Private Property, Common Property, and Hybrid Property Regimes: Expanding The Choices For The Global Commons: Comparing New Tradable Allowance Schemes to Old-Fashioned Common Property Regimes' (1999) 10 Duke Env L & Pol'y F 45, 70.

the emission reductions are real and measurable, most regimes when defining emission credits require an independent verification. Under the Kyoto Protocol, verified emission reductions constitute the basis for both JI and CDM credits. They are also the unit which is transferred and traded in voluntary schemes; either as a right defined by the underlying contract or as a unit defined by a voluntary carbon standard.

CERs come into existence through issuance by the Executive Board of the CDM, in a number which equals the verified emission reductions stated in an independent verification report. The Joint Implementation ERUs represent a hybrid between an allowance and an emissions credit. Based on verified emission reductions, they are converted and re-labelled AAUs or RMUs included in the cap of the host country. To avoid double-counting of emissions reductions within the Annex B cap of AAUs, for each ERU it issues the host country is required to cancel a corresponding amount of RMUs or AAUs from its assigned amount. The host country is ultimately liable for achieving the target assumed under the Kyoto Protocol and thus also for emission reductions generated by a JI project implemented in its territory.

4. Nature and Legal Ownership of Emission Rights

Environmental markets depend on rights that are defined, enforced, and transferable. An allowance carries these features and represents an asset or (quasi) property right for the owner. The limits of the property may be defined by the State. An understanding of the underlying right is however essential to establish the limits to the right. Additionally, the right needs to be transferable. This is because although well defined and enforced rights allow the owner to enjoy the benefits of using his property, legal restrictions on the sale of that property can preclude the potential for gains from trade.[27]

Allowances to emit a certain amount of a pollutant are established by law and the scope of the authorization is generally defined in the legal act establishing the system. Instead of providing clear definitions about what allowances or emission rights are, they normally describe what they entitle the holder to do.[28] The resulting right represents a hybrid between a purely public and a purely private right, which has been described as a 'regulatory' right. As a result, emissions rights are somewhere between an administrative grant and private property. While there are benefits to this dualism, the distinction between the public law status and the private law status becomes important when it comes to the treatment of allowances under taxation of accounting schemes.

[27] See Yandle, n 17 above, at 3. [28] See Button, n 25 above.

4.1 Public international law

4.1.1 The Kyoto Protocol

All tradable units defined by the Kyoto Protocol have the same definition: they are defined as a unit of 'one metric tonne of carbon dioxide equivalent, calculated using global warming potentials defined by decision 2/CP.3 or as subsequently revised in accordance with Article 5'.[29] These units are all created by an act of international law, which puts obligations upon its member states. It does not oblige any private or public entity, other than the signing and ratifying state. Therefore all emission rights under the Kyoto Protocol are created, owned and held by states.

These rights need to be reconciled with the statement in the Marrakech Accords that 'the Kyoto Protocol has not created or bestowed any right, title or entitlement to emissions of any kind on parties included in Annex I'.[30] The meaning of this sentence needs to be understood in the political context of the Kyoto Protocol:

- As elaborated above, the Kyoto Protocol does not create any rights to emissions or the atmosphere, it only creates the right to a defined pollution in a defined timeframe.
- The allocation of AAUs does not grandfather any future authorizations to emit a certain quantity of pollutants. It does not bestow Annex B parties with any entitlements beyond the first commitment period (2008–2012) of the Kyoto Protocol.

Another concern is that the Kyoto Protocol also clearly envisages the participation of entities other than states in the Kyoto mechanisms.[31] The Treaty itself, however, does not transfer any direct rights or obligations to entities other than its parties. Since the Kyoto Protocol does not bestow any rights and obligations to private sector entities, such rights need to be created either through implementation legislation or through single government acts, such as an individual authorization. In order to allow private entities to hold, own and trade units defined under the Kyoto Protocol, they need to be authorized to do so.[32] A government's approval of a CDM and JI project normally includes the authorization of the relevant non-state entities to participate in the project, and

[29] Decisions 9/CMP.1, 3/CMP.1 and 11/CMP.1 of COP/MOP 1.

[30] FCCC/CMP/2005/8/Add.1, Decision 2/CMP.1 Principles, nature and scope of the mechanisms pursuant to Articles 6, 12 and 17 of the Kyoto Protocol, preamble.

[31] Article 6.3 foresees the authorization of private sector entities to participate in JI. Article 12.9 of the CDM refers to the involvement of private or/public sector entities in the mechanism.

[32] Decision 9/CMP.1 Guidelines for the implementation of Article 6 of the Kyoto Protocol, Annex, § 29; Decision 3/CMP.1 Modalities and Procedures for a Clean Development Mechanism Annex, § 33; Decision 11/CMP.1 Modalities, rules and guidelines for emission trading under Article 17 of the Kyoto Protocol, Annex, § 5.

authorize the participants to own and hold CERs or ERUs. Regardless of whether it is created through a general act (law) or through a governmental decision, with the authorization to participate in a project and to hold and transfer CERs or ERUs, a government transfers its right to an ERU or CERs to the authorized entity.

4.1.2 Treatment of Kyoto Units under the WTO

The interface between the Kyoto Protocol and the WTO is part of a broader debate on the relationship between WTO and multilateral environmental agreements (MEAs), such as the UNFCCC or the Kyoto Protocol, which should be mutually supportive.[33] MEAs and international trade law however can come into conflict, for instance when MEAs' trade measures are used for compliance or enforcement purposes. A preliminary question is whether the relationship between the Kyoto Protocol and the WTO is covered by the Doha Agenda, and this depends on specific trade obligations that MEAs may contain. The Kyoto Protocol, however, does not contain provisions that would affect trade in a restrictive manner. The trade related provisions contained in the Kyoto Protocol are not specific enough to be in conflict with the WTO regime. Actually, some interactions of the Kyoto Protocol and the WTO regime could lead to some mutually beneficial situations. These include, for example, reduction of trade barriers concerning environmental goods and services related to climate change mitigation, or reduction of fuel subsidies as part of a discussion on mainstreaming environment and development.

If there is no direct conflict between the Kyoto Protocol itself and international trade law, some tension could still arise concerning measures adopted by parties for the implementation of the Kyoto Protocol. For instance, implementing measures for climate mitigation that are technical regulations such as the EU measures on labelling and certification of energy efficiency may raise issues associated with non-tariff (or technical) barriers to international trade.[34] Similarly, national policies (including, for instance, carbon taxes and cap-and-trade schemes) establishing a carbon constrained regime for local industry generate competitiveness concerns and are therefore prone to the creation of trade restrictive measures by governments. These measures aim at counter-balancing negative economic impacts on local production *vis-à-vis* foreign competitors operating in an unregulated emissions environment. Both the EU (in laying down the provisions for the review of the third phase of its EU-wide

[33] See in particular Art 2.3 of the Kyoto Protocol and Art 3.5 of the UNFCCC stating clearly that the adverse effect on international trade should be minimized. See further chapter 4 below.

[34] The international rules applicable to technical barriers to trade are set out in the Agreement on Technical Barriers to Trade. The rules apply in general to technical regulations, standards and conformity procedures for trade in products.

emissions trading scheme[35]) and the United States (in some recent proposals for a federal cap-and-trade system[36]) have considered possible border adjustment measures to protect their trade-exposed sectors and to further stimulate the adoption of equivalent emissions control by other exporting countries.

As far as international emissions trading under Article 17 of the Kyoto Protocol is concerned, a key question is whether Kyoto Units are 'goods' under the General Agreement on Tariffs and Trade (GATT 1994) (including Trade-related Investment Measures [TRIMs] and Subsidies and Countervailing Measures [SCM]) or 'services' under the General Agreement on Trade in Services (GATS) (including its fifth protocol on financial services).

Government-to-government trading of Kyoto units does not create a market in goods or services, rather it corresponds to a sovereign to sovereign exchange of commitments: the sovereign exchange of AAUs could be viewed as the reallocation of the overall assigned amount established by the 'cap' rather than the creation of a market.[37] In that case, there is no interaction with WTO rules.

The situation becomes more complex, however, when there is trade in emission allowances between private entities. Due to their ambiguous legal nature, it is not evident which part of international trade law would be applicable for the trade in emission rights. Allowances do not seem to fall under the definition of 'goods' or 'products' as defined by the GATT. Thus, the transfer and mutual recognition of allowances are not covered by the WTO disciplines.[38] This is confirmed by WTO case law where goods are defined as material things with an intrinsic value. Additionally, the Harmonized Commodity Description and Coding System elaborated by the World Custom Organization do not cover the category of certificates, if one can accept that allowances are above all certificates irrespective of their legal nature under various legal systems. Kyoto units could, however, be considered as a 'service' under the GATS. The GATS does not provide a general definition of services, it rather treats services as those things that do not fall under the category of goods. However, neither the WTO Services Sectoral Classification List[39] nor the

[35] See Directive of the European Parliament and of the Council amending Directive 2003/87/EC so as to improve and extend the greenhouse gas emission allowance trading system of the Community: COM (2008) 16 Final, Brussels 23.1.2008. Recital (26) stipulates that the Commission will review the situation of trade-exposed industries under the EU ETS by 30 June 2010 and, if necessary, consider the introduction of a 'carbon equalization system' to put EU and third-country industries on a comparable footing. Obliging importers to surrender allowances for products imported from countries with no equivalent emissions control is one of the options that can be considered by the Commission in its review process.

[36] See, for example, the Lieberman-Warner Bill (which was blocked in the US Senate in June 2008) which would have established a governmental agency to determine which countries have failed to implement similar greenhouse gas reduction efforts and require US importers of products from those countries to purchase allowances from an international reserve allowance program.

[37] Jake Werksman, 'Greenhouse Gas Emissions Trading And the WTO' (1999) 8 RECIEL 1, 3.

[38] See Werksman, n 37 above 1, 2.

[39] <http://www.wto.org/english/tratop_e/serv_e/serv_e.htm>, accessed 2 June 2009.

United Nations Provisional Central Product Classification system list Kyoto or other emission units as services.[40] Trading of Kyoto units involves services, such as exchange services by brokers or trustee services by Banks and financial managers, which would fall under the Annex on Financial Services of the GATS.[41] Therefore, remaining concerns with respect to the applicability of WTO agreements relate more to the domestic implementation of measures concerned with the Kyoto Protocol and secondary markets on derivatives (financial services) than to the actual transfer of Kyoto units or allowance.

It is commonly accepted that a cap-and-trade regime is 'inherently more compatible with rule-based international trading systems than other environmental policy instruments are'.[42] However, the allocation process is likely to affect the competitive relationship between products and services that are governed by international trade agreements. Countries can be seen as subsidizing industries when they waive revenue otherwise due. They could also be seen to be directly subsidizing renewable energies or other clean products. If these preferences undermined the competitive relationship between products in similar or domestic markets, they may be regarded as countervailing subsidies as defined in the SCM and subject to a challenge or direct countervailing measures.[43] However, a closer look at the SCM's provisions does not support this assessment: the allocation of emission rights can neither be regarded as a 'financial contribution' nor as an 'income or price support'. It is also questionable whether the transfer of an allowance to private entities does confer a 'benefit' in the sense of Article 1 of the SCM agreement, since the allocation of an allowance to private entities is accompanied by an obligation limit the emission of GHG to the number allowances held.[44]

4.2 Domestic legal systems

4.2.1 The EU ETS

The EU ETS Directive (Directive 2003/87/EC), Article 3(a), defines an 'allowance' as '. . . an allowance to emit one tonne of carbon dioxide equivalent during a specified period, which shall be valid only for the purposes of meeting

[40] <http://unstats.un.org/unsd/class/family/family2.asp?Cl=9>, accessed 2 June 2009.

[41] Paragraph 5 (definitions) of the GATS Agreement states that negotiable instruments are included within the notion of financial services.

[42] Annie Petsonk, Environmental Defence, presentation at the transatlantic dialogue on climate change, Washington DC (11 June 2003).

[43] This question is particularly interesting taking into account the recent OECD decision to subtract the corresponding financial value of CERs from ODA reported by Annex I countries: ODA Eligibility Issues For Expenditures Under the Clean Development Mechanism (CDM), 16 March 2004, DCD/DAC (2004)17.

[44] However, allowances have been classified as subsidies by a number of authors. For example see, Christian Koenig, Jens-Daniel Braun, René Fromm, 'Beihilferechtliche Probleme des EG-Emissionsrechtehandels', ZweR (2003), 2/2003, 153.

the requirements of this Directive and shall be transferable in accordance with the provisions of this Directive . . .'. It also prescribes the main rights embodied in an allowance:

- Any person may hold allowances (Article 19.1).
- It can be transferred between persons in the EU and between persons in third countries where allowances are recognized (Article 12.1).
- It can be used for compliance with the Permit (Article 12.3).
- It can be 'destroyed' through cancellation (*abusus*) (Article 12.4).

However, this definition does not define the legal nature and ownership of allowance beyond the main characteristics of a tradable instrument. It is also silent on the legal nature of the allowances, especially on whether the allowance is to be regulated as a security or a financial instrument under the Investment Services or Market Abuse Directives[45] or whether it ought to be treated as a good or as a service. The initial draft prepared by the European Commission proposed to define allowances as administrative authorizations, making a clear analogy with SO_2 allowances referred to in §403(f) of the 1990 amendments to the US Clean Air Act (see for further detail below). However, the Legal Service of the Commission opposed this inclusion into the Commission proposal before it was tabled before the Council and the Parliament. It was argued there was a need to preserve the subsidiarity principle as well as member states' national legal systems' particuliarities. Consequently, it is up to the member states to define the legal nature of EU Allowances, which is dependent on their national legal systems and traditions.

The EU emissions trading system has been adopted in the form of a Directive that is legally based on Article 175 §1 of the EC Treaty (environmental protection). The Directive prescribes a defined objective (to promote GHG emission reductions in an economically efficient manner) but leaves member states the choice of the means and form to achieve it. The intervention of the Community in establishing a European emissions trading system, as opposed to allowing it to the discretion of each member state, has been mainly justified by the necessity to ensure the integrity of the internal market and the need to avoid distortion of competition.[46] In accordance with the principle of subsidiarity, Community Law only regulates and defines the EU ETS to the extent necessary to ensure the smooth operation of the EU ETS. The further refinement of the

[45] Directive 2004/39/EC of the European Parliament and of the Council of 21 April 2004 on markets in financial instruments amending Council Directives 85/611/EEC and 93/6/EEC and Directive 2000/12/EC of the European Parliament and of the Council and repealing Council Directive 93/22/EEC (Investment Services Directives) OJ L145, 30/04/2004; Commission Directive 2004/72/EC of 29 April 2004 implementing Directive 2003/6/EC of the European Parliament and of the Council as regards accepted market practices, the definition of inside information in relation to derivatives on commodities, the drawing up of lists of insiders, the notification of managers' transactions and the notification of suspicious transactions (Market Abuse Directive) OJ L162, 30/04/2004.

[46] Preamble 7 of the EU ETS.

definition and the treatment of allowances under contractual and tax law remains the responsibility of the member states. It is interesting to note that the private sector has expressed discomfort with the fact that the European Commission has used a Directive, rather than a Regulation, to implement the ETS.[47] The differences in legal treatment (as a permit, good, right, or financial instrument) and under domestic tax and accounting systems are perceived as creating major uncertainties and thus as barriers to the emergence of a liquid market. The less defined and secure the allowances are, the less likely potential buyers will be willing to invest in them. It is worth noting that in the recently adopted text amending the EU ETS Directive to improve and expand the scheme for a third trading period, an EU centralized procedure was established for the definition of a cap and allocation of allowances. However, no reference was made in the new text in relation to harmonizing the legal treatment of allowances.[48]

4.2.2 Under EU domestic schemes

EU member states gave effect to the EU ETS Directive by transposing it into their domestic legal systems. As feared by the private sector, the lack of a central harmonization of the legal status of allowances resulted in an array of different legal classifications by individual member states.[49] The legal treatment of an allowance raises the following questions:

- Will the trading be supervised by a regulatory authority; as it is the case for financial instruments? This would make the trade for appraised entities easier, however, it would exclude many operators of covered installations since they do not meet the capital retaining requirements of banks or authorized traders.
- Can the allowance be revoked and cancelled (with or without compensation)?
- Can the allowance be encumbered and used as credit security?
- What is the treatment of allowances in case of insolvency and bankruptcy?
- Will the use and trade of allowances attract value added tax (VAT)? If so and in the case of cross-border trades, will it accrue within the acquiring or transferring country?

[47] This position has been expressed by many representatives of private banks and trading houses lastly at the occasion of a workshop held in London and co-sponsored by the Foundation of International Environmental Law and Baker McKenzie (London, King's College, 24–25 May 2004).

[48] A Directive is an obligation of result—leaving EU Member States to chose the form and methods for implementation, whereas a Regulation is law per se and directly applicable in all the legal systems of the Member States. The text for the revised EU ETS for a third trading period was approved by the EU Parliament on December 2008 and the amended EU ETS Directive is at the time of writing awaiting publication. See n 35 above. On the ETS, see also chapter 16 below.

[49] For a more in-depth overview of the legal issues arising as a result of the different approaches to the classification of allowances by EU member states, see also K Anttonen, M Mehling and K Upston-Hooper, 'Breathing Life into the Carbon Market: Legal Framework of Emissions Trading in Europe' (2007) 16 Eur Env Law Rev 96.

The main difference in the legal treatment afforded to allowances centres on whether the allowances are classified as financial instruments (subject thus to national financial regulations) or as commodities. For instance, while in Sweden allowances are deemed financial instruments subject to regulation by the financial services authorities, in Austria, Germany, France, Italy, Poland, Portugal, and Spain the same allowances are currently treated as tradable commodities for the purposes of financial regulation. Furthermore, in most member states a differentiation in treatment is made depending on whether an allowance is traded on the spot market (commodity-like) or on the forward market (derivative-like, depending on the underlying contractual arrangement).[50]

The accounting treatment of allowances also varies considerably among EU member states. Several countries such as Spain, Finland, Italy, Malta, and Portugal treat allowances either as intangible assets or as financial instruments. Others, such as France, Netherlands, and Germany stipulate that allowances must be recorded as tangible assets or inventory. Some countries such as Greece and the United Kingdom have not yet defined any specific accounting treatment for allowances.[51] In this respect, it is important to highlight the attempt made by the International Financial Reporting Interpretations Committee (IFRIC) which issued guidance on the accounting to be adopted under emissions trading schemes in 2004. The IFRIC clarified that allowances should be recorded as intangible assets and the actual emissions as liabilities.[52] However, this guidance was withdrawn by the International Accounting Standards Board (IASB) at its meeting of June 2005.[53]

With respect to the tax treatment, the applicability of VAT in the sale and purchase of emission allowances has also raised a number of issues. Although the 6th VAT Directive provides a harmonized VAT system for the EU, there is still significant variation in application of this Directive, as its provisions are subject to interpretation by member state case law and taxation practices.

When emission allowances are transferred they are treated as a supply of a service under Article 6 of the Directive. When emission trading occurs within a member state the transaction is subject to the VAT of that state. However, if there is a transaction between member states then there is some uncertainty as to which country the VAT is to be applied in. This has resulted in situations of double or zero taxation being applied.[54] The EC Advisory Committee on VAT

[50] For a summary of the legal treatment afforded to allowances within EU Member States see Section 14 of the 'Application of the Emissions Trading Directive by EU Member States—Reporting Year 2008', published on 5 January 2009 and available at <http://www.eea.europa.eu/publications/technical_report_2008_13>, accessed 2 June 2009.

[51] Ibid, at 75.

[52] For a summary of the discussions and interpretations of the IFRIC on emissions rights see <http://www.iasplus.com/interps/ifric003.htm#withdraw>, accessed 11 January 2009.

[53] See IASB public statement of withdrawal at <http://www.iasplus.com/pressrel/0507withdrawifric3.pdf>, accessed 2 June 2009.

[54] K Anttonen, M Mehling and K Upston-Hooper, 'Breathing Life into the Carbon Market: Legal Framework of Emissions Trading in Europe' (2007) 16 Eur Env Law Rev 96, 100.

issued a non-binding statement on this in 2004 which helps clarify the issue.[55] This stated that there is a 'reverse charging rule', meaning that the transferee accounts for VAT in its own country, rather than being liable for the tax in the country where the emission transfer initiated.

It should be noted that these VAT rules only apply to emissions transfer, and other aspects such as auctioning, derivative contracts arising out of the transaction, and advisory services related to the transaction will be subject to separate VAT rules. In general the initial grant of allowances by the Government is unlikely to require VAT as it is a public function under Article 4(5) of the Directive. If the allowances are auctioned the situation is a little more unclear. Although it is also a public function, Article 4(5) provides that such transactions will be subject to VAT if to do otherwise would result in 'significant distortions of competition'.

In sum, the EU Allowance does not fit easily in any legal system of the EU Members. It can be deemed as a right '*sui generis*' in many jurisdictions, carrying the following features: (i) transferable permit; (ii) an administrative public right; (iii) an intangible good or a commodity; and (iv) a security or a financial instrument. It will also depend on the legislative purpose for how the property rights (and obligations) to the allowance will be defined.

4.2.3 Emissions trading experiences in the US

SOx/NOx trading

While climate regulation is still being prepared by the US congress, it may be worthwhile to examine the past experiences with the implementation of emissions trading schemes in the US. The first emissions trading scheme implemented in the US came to life with the 'acid deposition control' program established in Title IV of the 1990 Clean Air Act Amendments.[56] The Act sought to halve sulphur dioxide (SO_2) emissions by the year 2000 and created a transferable emissions 'allowance' scheme to accomplish this goal. The regulating authority has set emission quotas for the covered installations. The Act does not specify how sources are to meet their emissions reduction targets, and the law does not even require sources to reduce emissions to the levels set by Congress but only to possess allowances equal to their actual emissions.[57]

The authorization to emit then gains a quality which confers a quasi-property right upon the holder of the unit. The holder is authorized to hold, sell or use the unit. However, whereas these rights are enforceable against third parties, they enjoy only limited enforceability against governments. The most elaborate restriction with respect to the use of allocated emission rights can be found in § 403(f) of the 1990 amendments to the Clean Air Act which carefully defines the legal nature of emission allowances:

[55] Document TAXUD/1625/04Rev.1-Working Paper No 443 Rev 1.
[56] Pub L No 101–549, 104 Stat 2399 (codified as amended at (1994) 42 USC 7401–671).
[57] See Cole, n 14 above.

(f) Nature of allowances

An allowance allocated under this subchapter is a limited authorization to emit sulfur dioxide in accordance with the provisions of this subchapter. Such allowance does not constitute a property right. Nothing in this subchapter or in any other provision of law shall be construed to limit the authority of the United States to terminate or limit such authorization. Nothing in this section relating to allowances shall be construed as affecting the application of, or compliance with, any other provision of this chapter to an affected unit or source, including the provisions related to applicable National Ambient Air Quality Standards and State implementation plans. Nothing in this section shall be construed as requiring a change of any kind in any State law regulating electric utility rates and charges or affecting any State law regarding such State regulation or as limiting State regulation (including any prudency review) under such a State law.

Nothing in this section shall be construed as modifying the Federal Power Act (16 U.S.C. 791a et seq.) or as affecting the authority of the Federal Energy Regulatory Commission under that Act. Nothing in this subchapter shall be construed to interfere with or impair any program for competitive bidding for power supply in a State in which such program is established. Allowances, once allocated to a person by the Administrator, may be received, held, and temporarily or permanently transferred in accordance with this subchapter and the regulations of the Administrator without regard to whether or not a permit is in effect under subchapter V of this chapter or section 7651g of this title with respect to the unit for which such allowance was originally allocated and recorded. Each permit under this subchapter and each permit issued under subchapter V of this chapter for any affected unit shall provide that the affected unit may not emit an annual tonnage of sulfur dioxide in excess of the allowances held for that unit.

Although the Act clearly states that an allowance does not create a property right, it seems that, between the contracting parties, all normal property rights (*usus*, *fructus* and *abusus*) are available.[58] However, the US legislator did not want to create property in order to avoid being held liable for compensation should the allowance be devaluated or withdrawn from the market. This has led some to describe § 403(f) as 'premised on the confusion between property rights in something and the thing itself.'[59] Indeed, it does expressly define an allowance as not creating a property right, yet allows free transfer and thus recognizes the property rights *in* the emission allowance, because utilities can receive, hold

[58] Restatement of the Law, Property 1936, The American Law Institute, §5 (e) Complete property. The totality of these rights, privileges, powers, and immunities which it is legally possible for a person to have with regard to a given piece of land, or with regard to a thing other than land, that are other than those which all other members of society have as such, constitutes complete property in such land or thing other than land. This totality varies from time to time, and from place to place, either because of changes in the common law, or because of alterations by statute. Thus if the law should come to be that no person could build a five-story building on his land, the totality of privileges that every person has who owns land would be correspondingly diminished. So if a zoning ordinance were passed, the totality of interests would be affected, to the extent of the ordinance, for persons owning land within the district to which the ordinance applied. At any one time and place, however, there is a maximum combination of rights, privileges, powers and immunities in the land that is legally possible, and which constitutes complete property in the land, or thing other than land.

[59] See Cole, n 14 above.

(ie possess), and transfer (ie alienate) allowances. Utilities and all other allowance holders (such as private NGOs) can exclude all others, besides the government, from interfering with their possession, use, and disposition of allowances.[60]

GHG trading

More recent developments in the US relate to GHG cap-and-trade initiatives both at the regional as well as the federal level. The Regional Greenhouse Gas Initiative (RGGI) is the first US mandatory GHG emissions trading scheme. It has become operational in 2009 and sets binding emissions caps for electricity producers in 10 eastern American states. The RGGI was agreed in 2005 through a memorandum of understanding in which the states of Connecticut, Delaware, Maine, New Hampshire, New Jersey, New York, and Vermont agreed to adopt a common 'CO_2 Budget Trading Program'.[61]

In August 2006, the RGGI states issued a Model Rule detailing the rules for the CO_2 Budget Trading Program. The Model Rule provides for a central design to each individual state emissions reductions trading scheme, specifying the general rules which apply to all participant states, including provisions on applicability of the scheme, compliance requirements, overall state allowances budget, allocation process and penalties in the case of non-compliance by regulated entities.[62]

In defining the legal status of emission allowances, the RGGI Model Rule adopted the same cautious and restrictive approach used by the Clean Air Act with respect to SO_2 allowances. Section 1.2 of the Model Rule defines allowances as 'A limited authorization by the [Regulatory Agency] under the CO_2 Budget Trading Program to emit up to one ton of CO_2, subject to all applicable limitations contained in this Part'.[63] The Model Rule then clarifies that 'No provision of this regulation shall be construed to limit the authority of the [Regulatory Agency] to terminate or limit such authorization to emit. This limited authorization does not constitute a property right'.[64]

The exclusion of property rights features to the RGGI allowance is reinforced by Section 1.5(7), which states that 'A CO_2 allowance allocated by the [Regulatory Agency] under the CO_2 Budget Trading Program does not constitute a property right'.[65]

At the federal level, all recent proposals, such as the defeated Climate Security Act (also known as the Lieberman Warner Bill) or the American Clean Energy and Security Act of 2009 (also known as the Waxman Markey Bill), the most

[60] Markus Gehring, Charlotte Streck, Emissions Trading: Lessons from SO_2 and NO_x Emissions Allowance and Credit Systems, IN PRESS.

[61] The RGGI memorandum of understanding is available at <http://www.rggi.org/about/history/mou>, accessed 17 February 2009. Currently the following US states are members of the RGGI: Connecticut, Delaware, Maine, Maryland, Massachusetts, New Hampshire, New Jersey, New York, Rhode Island, and Vermont.

[62] See RGGI Model Rule available at <http://www.rggi.org/about/history/model_rule>, accessed 17 February 2009.

[63] Ibid, Section XX–1.2 (k).

[64] Ibid.

[65] Ibid.

prominent proposals for GHG emissions trading schemes in the US, seek to establish an economy-wide cap-and-trade scheme which would regulate GHG emissions as of 2012. The Lieberman Warner Bill was first introduced in the Congress in October 2007 but its passage was blocked by the Senate in June 2008.[66] The Waxman Markey Bill is still under consideration. Similar to the SO_2trading scheme and RGGI, both federal emission trading regulations also make clear that regulated entities would not be endowed with any property rights over allowances. Section 201 of the Lieberman Warner Bill and Section 721(c) of the Waxman Markey Bill expressly state that 'an emission allowance shall not be a property right'.[67]

Overall, it can be concluded that although not being a property right itself, allocated emission rights usually carry characteristics of 'property rights' such as transferability or the protection of allowances against arbitrary confiscation. While the state or federal government protects its flexibility to create or cancel units in order to manage a cap and trade system, or to withdraw or suspend allowances found to have been improperly allocated, it allocates to the private sector 'quasi' property rights which allow free use and transfer of these units. In this way the state creates what has been described as a regulatory right.

4.2.4 Emerging emissions trading schemes—Australia and New Zealand

In Australia and New Zealand the desire to create a harmonized carbon market seems to have prevailed over fears of narrower state regulatory powers on emissions by regulated entities. In New Zealand the Climate Change Response (Emissions Trading) Amendment Act 2008 (Emissions Trading Act) introduced an emissions trading scheme which covers GHG emissions from several sectors of the economy, including forestry, transport, industrial processes, stationary sources, agriculture, and waste.[68]

In contrast to the approach adopted so far by the EU and the US, the New Zealand ETS classifies an emission allowance in the property category of investment securities under the Personal Property Securities Act 1999. It thus extends the protection given by this Act to purchasers of emissions units.[69]

[66] The Lieberman Warner Bill was passed out of the Environment and Public Works Committee and an amended version was brought to the Senate floor in June 2008. The Bill was eventually rejected.

[67] See section 1201(c) of the Lieberman Warner Bill (S.3036), 110th Congress, 2nd Session; section 721(c) of the Waxman Markey Bill (p 402), 111th Congress, 1st Session HR 2554.

[68] The Climate Change Response (Emissions Trading) Amendment Act 2008 became law in September 2008. However, late in 2008 the newly elected New Zealand Government established a review committee to conduct a review of the rules of the New Zealand ETS. It is expected that any subsequent amendments to the New Zealand ETS will be passed in the course of 2009.

[69] In particular, Section 97 of the Personal Property Securities Act 1999 sets out the rules governing priority between the interests of different purchasers of an investment security. See <http://www.med.govt.nz/templates/MultipageDocumentPage____33790.aspx>, accessed 6 February 2009.

Section 65 of the Emissions Trading Act amends the definition of an investment security pursuant to section 16(1) of the Personal Property Securities Act 1999 to include emissions units along with instruments 'evidencing a futures contract, or a warrant or option or share, right to participate, or other interest in property or an enterprise'.[70]

In addition, the Draft Settlement Systems, Futures, and Emissions Units Bill, which is expected to go through the legislative process during 2009, proposes additional amendments to Personal Property Securities Act 1999 in order to further regulate and clarify the legal treatment of emission allowances. It comprehensively defines emissions unit as:

(b) personal property that:

(i) is created by, or in accordance with, any enactment (whether of New Zealand or of another country), rule of law, or contractual provision as:
(A) one of a fixed number of units issued by reference to a specified amount of greenhouse gas? or
(B) evidence of a specified amount of reductions, removals, avoidance, or sequestration of greenhouse gas emissions? and

(ii) can be surrendered, retired, cancelled, or otherwise used to:
(A) offset greenhouse gas emissions under any enactment (whether of New Zealand or of another country), rule of law, or contractual provision? or
(B) enable a person who surrenders, retires, cancels, or otherwise uses it to claim an environmental benefit.[71]

The Draft Settlement Systems, Futures, and Emissions Units Bill also further clarifies the situations in which a person takes possession of a security interest that is an emissions unit and thus may have his or her security interest perfected for the purposes of the Personal Property Securities Act 1999.[72]

The Australian Carbon Pollution Reduction Scheme (CPRS), in turn, is expected to begin operations in July 2011 and will cap greenhouse gas emissions from stationary energy, transport, fugitive, industrial processes, waste, and forestry sectors across the country.[73] Similarly, the Australian Government

[70] See Climate Change Response Emissions Trading Act 2008, No 85, of 25 September 2008 available at <http://www.legislation.govt.nz/act/public/2008/0085/latest/DLM1130932.html>, accessed 6 February 2009.

[71] See Clause 33 of the Draft Emission Units, Settlement Systems, and Futures Bill 252–1 (2008) available at <http://www.legislation.govt.nz/bill/government/2008/0252-1/latest/versions.aspx>, accessed 3 February 2009.

[72] Ibid, at Clause 34.

[73] The Australian Carbon Reduction Pollution Scheme is still in its design phase.

proposes that emission permits under the CPRS will be granted the status of a personal property right. However, the CPRS goes a step further in the regulatory treatment of emission rights in that it explicitly refers to the allowance holder's right to compensation in the event of expropriation of its emission rights by the Government.

According to the White Paper published in December 2008, the Australian Government had to consider two regulatory options for the treatment of the so-called 'carbon pollution permits': (i) to define permits as a limited compliance instrument or license that could be readily extinguished by the government without compensation; or (ii) to attribute these pollution permits the character of personal property, entitling the holder to surrender or transfer the permit and to be compensated in the event of expropriation. The Australian Government, highlighting the need for market certainty and legal security, decided to pursue the second option. According to the latter alternative, emission permits will be treated as personal property and legal title will pass on at moment of entry into the registry. The State will not be able to extinguish or relinquish permits without compensation, unless such permits have been obtained through misrepresentation or fraud.[74]

In sum, the New Zealand and the Australian experience have so far put considerably greater emphasis on creating regulatory certainty for the development of a domestic carbon market than have their counterparts in the EU and the US. These emerging emissions trading scheme are an exception in that they expressly define the property rights nature of the emissions rights created within their regulatory framework.

5. Conclusions

The Kyoto Protocol creates sovereign rights for parties trading Kyoto units. Kyoto parties can then authorize private entities to participate in trading under the international instrument. Therefore, it is important that rights to the units are recognized on the domestic level and that they relate to the right to transfer something of economic value which can be used for compliance or other purposes.

The discussion of the different features of emission allowances has made it obvious that, although it is in the best interest of a liquid and functioning market, that governments are often reluctant to attach a property right per se to an emission allowance. Instead of clearly defining the boundaries of the rights they create, they often leave it up to the market and the private sector to explore

[74] See Chapter 8, section 8.2, of the Carbon Pollution Reduction Scheme: Australia's Low Pollution Future, White Paper, 15 December 2008, available at <http://www.climatechange.gov.au/whitepaper/report/index.html>, accessed 6 February 2009.

the limits of the rights they hold in 'learning by doing'. Apart from the recent developments in New Zealand and Australia, the EU ETS, the RGGI, as well as one of the most prominent bills for a federal a cap-and-trade program in the US, have established the allowance as an 'authorization' to emit, without giving further detail on the legal nature of these rights.

When creating emission rights under national law, from the government's viewpoint the main advantage of an ambiguous right is the cautionary approach that it represents. The creation of narrow and concretely defined rights would pose significant difficulties for regulators who need the flexibility to create or cancel such rights. However, at the same time, there is also keen awareness that certainty, for example that emission allowances should be protected against arbitrary confiscation, is vital for the creation of an active trading market. It is the hybrid nature of emission rights, which does make it difficult to clearly categorize such rights as public or private. They are a hybrid, which is best described as a regulatory right. The right is transferable and enforceable against other private parties; however, it enjoys limited enforceability against the State, which wishes to retain the overall control over the allowances it has created. As a result, holders of tradable emission rights generally have the exclusive right to hold, transfer and use these rights and the administrative agency should refrain from arbitrary confiscation or discounting these rights.

While the exact nature of emissions rights still differs considerably between jurisdictions, it is possible that, as the carbon market matures and linkages between different emission trading schemes are created, States begin to attribute a greater emphasis to legal certainty and market harmonization in the trading of emission rights. Such process may lead to a gradual consensus of the need to formally endorse the property features of emissions rights as a tool to a more effective and stable international carbon market.

3

Accounting for Emissions: From Costless Activity to Market Operations*

*Allan Cook***

1. Introduction

The economic transformation of greenhouse gas emissions from a costless by-product to a significant market cost poses challenges that are not confined to scientists and politicians. The 'cap and trade' model which is now the most favoured method of governments for motivating industry to reduce emissions has also raised new and quite difficult problems in the field of financial reporting. This chapter examines, in the context of the European Union's new Emissions Trading Scheme under the Kyoto Protocol, the issues surrounding the aborted attempt by the International Accounting Standards Board (IASB) in early 2005 to regulate the accounting for 'cap and trade' schemes. The chapter argues that the features that made this model attractive to governments were precisely the ones that accountants found difficult to capture under existing standards. After showing why the challenge has to be faced, the chapter suggests a possible way forward that the IASB might consider when it revisits the subject, as it is now doing.

The problems surfaced in a particularly acute form when the IASB attempted through its International Financial Reporting Interpretations Committee (IFRIC) to develop an accounting régime in time for the start of Phase I of the EU Emissions Trading Scheme (ETS) in January 2005. Though the ETS was not the earliest example of a 'cap and trade' model, it operated on a far larger scale than any of its predecessors. Furthermore, by the time it came into effect,

* This chapter originally appeared as an article in Accounting, Organizations and Society (AOS) Vol. 34 pp 456–468 copyright Elsevier 2009. The editors are grateful to Anthony Hopwood, editor of AOS, and to Elsevier for permission to include the article in this book.

** Allan Cook was the Co-ordinator for the International Financial Reporting Interpretations Committee (IFRIC) of the International Accounting Standards Board (IASB) at the time that the issues described in the paper came to a head. He is particularly grateful to Henry Rees, the IASB senior project manager responsible for the topic, for permission to draw on his work for the IFRIC and the IASB and to Professor Donald Mackenzie for his advice on the relevance of existing papers dealing with the economic implications of various possible mechanisms for reducing emissions. However, the views expressed are the author's own and do not necessarily reflect those of either of these gentlemen or the members of the IASB or the IFRIC.

financial reporting principles had become far more rigorous than previously, with the result that the IFRIC did not feel able to accept some of the easy compromises that had served in the past.

Although this chapter analyses the accounting problems in the context of the EU ETS, it will be seen that similar difficulties are inherent in any 'cap and trade' scheme that does not involve auctioning or other forms of payment for the entire allocation of allowances by a government or supra-national authority.

2. Background—an Aborted Interpretation

While Europe was preparing for the adoption of International Accounting Standards (IASs) by all listed companies from January 2005, the IASB was urged to develop mandatory guidance for the financial reporting of Emission Rights, the basis for the EU's new scheme designed to reduce emissions by major polluters.

In view of the specialized nature of the subject, the IASB asked the IFRIC to undertake the task. The IFRIC published a draft Interpretation in May 2003 and, after considering public comments, developed an Interpretation, which the IASB issued in December 2004, just within the European Commission's deadline for the adoption of IASs.[1]

The result was a public outcry. Companies complained that application of the Interpretation would force them into showing a completely distorted picture of their performance in their annual and interim financial statements. The IASB, while recognizing that the IFRIC had made a valid interpretation of the relevant IASs, accepted that the end result was confusing in certain respects. Perhaps fortunately, the expected market for Emission Rights was slow to develop and the IASB took the opportunity to withdraw the Interpretation in June 2005, only six months after it had been issued.

What was the cause of this reverse?

2.1 The problem

The Emission Rights affair is worth studying because it illustrates the problems faced by standard setters as they explore the frontiers of accounting. Just as the science of medicine sometimes advances by examination of its failures to deal with newly presenting diseases, so, in other fields, an understanding of the roots of failures can lead to a fuller understanding of the discipline involved, its limitations and its possibilities.

[1] International Accounting Standards Board, IFRIC 3 'Emission Rights' (Withdrawn June 2005). See also IASB, 'IASB withdraws IFRIC Interpretation on Emission Rights' (Press Release) (June 2005) London <http://www.iasplus.com/pressrel/0507withdrawifric3.pdf>, accessed 16 March 2009.

Three features in particular lay at the heart of the challenge posed to standard setters as they sought to deal with Emission Rights:

(1) a previously costless activity had become costly;
(2) governments mitigated the cost;
(3) by means of marketable allowances.

It is the combination of these features that has so far proved an insuperable problem for accounting to reflect. Before examining the accountants' problem, we need to understand what led governments to adopt such a scheme in their attempts to reduce emissions.

3. Controlling Emissions—Alternative Approaches

There is a voluminous literature, dating back over 30 years, discussing the relative merits of different forms of economic instrument for achieving government policies with respect to the use of a scarce resource or service. The choice in a given situation depends on many factors, including the degree of uncertainty regarding the marginal costs and benefits of a particular course.

Economic instruments divide broadly into those that set a price for the good or service, leaving the quantity demanded to adjust accordingly, and those that set a total quantity, leaving the price to be determined by market forces. In the field of emissions an example of a price instrument is a carbon tax: a previously costless, or less costly, activity becomes costly and the cost is borne in proportion to the level of emissions involved. An example of a quantitative instrument is a system under which government issues a predetermined total of licences to emit. If the licences are auctioned rather than issued free and emitters are allowed to trade them, it can be shown that under idealized conditions either instrument will result in the same optimal price and quantity. Note that an ability to trade licences is necessary for a quantitative system to achieve equivalence with a price instrument.

Of course, conditions are rarely idealized nor is it only economic factors that determine political decisions by governments or inter-governmental agreements. This paper does not attempt to evaluate the relative merits of price versus quantitative instruments in the control of emissions.[2] Rather, the paper examines the accounting implications of the quantitative systems that have been put in place by governments. Although price instruments also exist, they are not currently as widespread as quantitative systems and, in any event, do not raise any new accounting issues.

[2] An excellent, recent analysis of those issues is given by Cameron Hepburn, 'Regulation by Prices, Quantities or Both: A Review of Instrument Choice' (2006) 22 Oxford Review of Economic Policy 226–47. That article explains why some economists believe that the relationships between the marginal costs and benefits of alternative instruments to reduce emissions suggest that, in theory, a price instrument may be more efficient (ie arrive closer to the optimum price and quantity) than a quantitative instrument.

The principal quantitative instruments in use for controlling emissions are 'baseline and credit' and 'cap and trade'.

3.1 Baseline and credit

Under this system a government allocates allowances based on some assessment of a normal rate of emissions. For each entity a baseline is set below which no charge for emissions will be made. Entities that manage to operate below the baseline for a period are given credits, which can be sold through a market to those which exceed their baseline. At the end of the period any entity that has exceeded its baseline must deliver to the government sufficient credits, bought in the market, to cover the excess.

A weakness of this system is that allowances that are traded may be insufficient to sustain a market. Moreover, the government may lay itself open to charges of unfair discrimination in its determination of the baselines for different entities.

3.2 Cap and trade

Under a cap-and-trade system the government mitigates the cost for entities, not by setting a baseline, but by allocating tradable allowances for the period. After the end of the period each entity must pay for its emissions by surrendering allowances granted or buying in from the market any shortfall. Since the allowances are allocated from the beginning of a period, in theory, an entity that perceives a market opportunity can sell more than the amount of its expected savings in the hope of buying back later at a more advantageous price.

This system places greater reliance on the market mechanism than does the baseline and credit method. Although initially all allowances are allocated by governments for nil consideration, it is intended, under the EU scheme, that a gradually increasing proportion should be auctioned rather than granted free. Some believe that an auction of all allowances would be the most effective means of reducing total emissions by ensuring that the savings that were easiest to make were addressed the first. However, too rapid a move to the auction of all allowances might be counterproductive by obliging entities to pass on the cost to the consumer instead of encouraging them to seek savings.

3.3 The EU scheme under the Kyoto Protocol

The EU scheme covers all its Member States and was adopted as the principal means for the EU to meet its emission reduction commitments under the Kyoto Protocol. The Protocol set reduction targets in the form of Allowed Amount Units (AAUs) of emissions by country, but allowed for countries to group together to redistribute their total allowances among themselves. Under the EU scheme, the Union's overall target of 8% reduction was redistributed by giving the UK, for example, a target of 12.5% reduction but allowing Portugal an increase of 27%. Under the EU scheme entities may participate in the Kyoto Clean Development Mechanism, which enables them to acquire further allowances by undertaking a development to reduce emissions in another country.[3]

Phase I of the EU scheme was for an initial period of three years and was succeeded by the first full five-year phase under the Kyoto Protocol running from 1 January 2008 to 31 December 2012. For the initial phase, 95% of allowances had to be allocated free to entities. During the following phase, only 90% must be free. The remaining 10% may be auctioned, thus obliging most entities to have recourse to the market, either at the time of the auction or subsequently.

Allowances are allocated in January to cover emissions up to the end of December; settlement is required by the end of the following April. Entities that during the initial phase failed to deliver sufficient allowances to cover their level of emissions are fined €40 per tonne of shortfall. During the second phase this is expected to rise to €100 per tonne. Payment of the fine is not a substitute for delivery of the required allowances. The shortfall must be made up from the subsequent year's allocation plus market purchases.

Allowances are not themselves a licence to emit: each entity covered by the scheme must obtain from the government an emissions permit. If the entity ceases production during the year, it must surrender its emissions permit, but may sell any surplus allowances on the market.

All features of the scheme are focused on the objective that somewhere in the world emissions get reduced, even if not necessarily by the entity whose expected emissions exceed its allocation of allowances, nor even within the jurisdiction of the government determining that allocation. By allowing an entity that ceases production to sell its allowances, the scheme maintains the originally assessed level of allowances for the economy as a whole. It is not necessary to claw them back from the entity to which they had been granted. Moreover, an efficient producer may judge that it can reduce more emissions and earn more credits by undertaking a project to enable a less efficient entity to reduce its emissions than it could by investing the same level of resources in reducing its own emissions. The design of the scheme encourages it to take the former, more effective course.

[3] Donald MacKenzie gives a good illustration of the scientific imponderables that underlie 'making things the same' by using allowances received in one industry to offset the cost of emissions in another. The paper stresses the important role of accounting if market-based schemes are to motivate successfully by making carbon costs visible. Donald MacKenzie, 'Making Things the Same: Gases, Emission Rights and the Politics of Carbon Markets' (2009) 34 Accounting, Organizations and Society 440–55.

4. The Accountants' Problem Rooted in Government Objectives

From the above description we can see why the three features causing the most problems for accountants are particularly attractive from the point of view of governments.

- Emission reduction schemes are designed to give a cost to a previously costless activity, in order to motivate producers to regard emissions as an input cost that must be monitored and controlled like any other.
- However, unlike licences for oil and gas exploration or 3G development, emissions costs are being introduced only slowly, by gradual reduction of the total level of allowances or the proportion of the total that is allocated free. This strategy avoids a shock to existing systems for the production, supply, and use of energy, which might result in the new cost simply being passed on to the consumer instead of being controlled at the point where savings were possible.
- The cap-and-trade system responds well to the previous two objectives while also making most effective use of the market. Use of the market both simulates other real costs and is the most efficient means of passing the incentives to save emissions to those best able to make such savings at an early date.

The remainder of this paper analyses the accounting problems created by these very reasonable governmental objectives.

5. One Accounting Solution: Maintain the Status Quo

Many producers and others commenting on this subject to the IASB asserted that the key to the accounting solution lay in dealing only with the marginal effect on cost. They argued that, as long as an entity emitted no more than the amount covered by its emissions allowances, no new cost emerged. The only cost would be that of acquiring additional allowances from the market: the only credit, the proceeds of a sale of surplus allowances to the market.

This solution relied on netting the benefit of tradable allowances received against the newly created cost of emissions, as though it were dealing with a baseline and credit scheme. It sought to simulate as far as possible the status quo ante. The question for accountants was whether it was true that the status quo had not changed: whether in fact something was being overlooked if income from the grant of allowances was netted against the cost of emissions. The circumstances in which costs may be netted against income are, of course, severely limited both under IASs and under the EU Fourth Directive.

The weakness of the netting solution is that it ignores the market-oriented thrust of cap-and-trade schemes. Although beginning slowly, the market for allowances seems likely to play an increasingly important part in governments' attempts to reduce emissions both in the EU and internationally. The netting solution attributes a cost to purchased allowances, but not to those granted to an entity by government. Yet the allowances themselves are fungible: neither the government, nor the market distinguishes allowances according to their source. Allowances are assets in their own right, as demonstrated by their ability to be sold for cash and even to be transferred between different schemes under the Clean Development Mechanism of the Kyoto Protocol.

The netting solution also ignores the reality of the liability for emissions, a liability that exists and behaves quite independently from the allowances held by the emitting entity. It is an understatement of that liability to net it against allowances held. The netting of assets and liabilities is restricted, if anything, even more severely than the netting of income and costs.

Even as a practical expedient, the netting solution breaks down as soon as an entity begins to trade its allowances. From that point the significance of the initial grant disappears as granted allowances are replaced by others bought in from the market. At the extreme, the whole of the initial grant may have been sold and replaced; on what basis is the amount of offset then to be calculated? If the reply is that no offset should be allowed in such a case, there would be no comparability between one entity that had retained its original grant of allowances and one that had sold but replaced them, even if there had been no significant price movements in the interim. If the reply is that offset is to be allowed, some method would have to be devised to limit the amount of the offset to the volume and unit price of allowances at the time the grant was made. The reason is that any variation of the current value of allowances from the amount of the grant would be the result of price movements or market transactions rather than an adjustment of the original grant.

6. The Issues for the IFRIC

The IFRIC approached the problem by analysing separately the nature of the allowances granted and the obligation to pay for emissions by delivering allowances.[4] It asked:

- Are the allowances an asset:
  - if purchased?
  - if allocated by government?

[4] This section summarizes the requirements and rationale contained in IFRIC 3 *Emission Rights*. Robert Casamento gives a fuller description of the debate surrounding the finalization of IFRIC 3; Robert Casamento, 'Accounting for and Taxation of Emission Allowances and Credits' in David Freestone and Charlotte Streck (eds), *Legal Aspects of Implementing the Kyoto Protocol Mechanisms: Making Kyoto Work* (OUP: Oxford, 2005) 55–70.

- If so, what kind of asset?
  - Is it a financial instrument? or
  - Is it an intangible?
- If the allowances are an asset, is the credit (the grant) on receipt:
  - income? or
  - a liability?

6.1 Allowances

Most people would instinctively agree that purchased allowances are an asset, probably because they have a cost. However, the IASB Framework for the Preparation and Presentation of Financial Statements ('the Framework') does not require a cost. It defines an asset as 'a resource controlled by the entity as a result of past events and from which future economic benefits are expected to flow to the entity'. The IFRIC found that allowances allocated by government, no less than those purchased, met this definition. They are controlled by the entity from the moment of the past event—allocation—by which they were acquired. They are expected to generate future benefits for the entity either when tendered in settlement of an emissions liability or through sale into the market. As noted above, the rights attaching to them are indistinguishable from those of purchased allowances, a point that is likely to assume practical, as well as conceptual importance as markets develop and entities begin to trade significant portions of their allocations.

Although, as we shall see later, this conclusion gave rise to uncomfortable consequences, it is difficult to challenge it on conceptual grounds. However, in determining whether allowances were a financial instrument or an intangible asset, the IFRIC encountered a more artificial difficulty of the IASB's own making. It would have been convenient if the allowances could have been classified as one of the classes of assets that IASs require to be re-measured to fair value at each reporting date, with gains and losses reported in profit or loss. The reason is that under IAS 37 *Provisions, Contingent Liabilities and Contingent Assets* the liability for emissions must be measured and reported in that way. IASs require financial instruments held for trading to be on this basis but the IFRIC decided that the allowances did not meet the, somewhat detailed, definition of a financial instrument. Nor did they fall within the limited class of other assets that could be measured like financial instruments.

Instead, the IFRIC decided that the allowances fell within the definition of an intangible asset. Although such assets may be re-measured to fair value if they can be traded in an active market, IASs currently require the resulting gains and losses to be reported outside profit or loss. Symmetry of treatment with that of the liability, therefore, eluded the IFRIC.

6.2 Emissions liability

The IFRIC decided that the obligation to pay for emissions by delivering allowances to the government after the end of the period was a liability. It clearly fell within the Framework's definition, 'a present obligation of the entity arising from past events, the settlement of which is expected to result in an outflow from the entity of resources embodying economic benefits'. An obligation arose as emissions were made (the past event), which would require the entity to part in the future with resources (the allowances) embodying economic benefits: the ability to be used to settle further obligations or be sold in the market.

Some argued that, even if the asset and liability were distinct, they should be permitted to be offset, since the ultimate use of the allowances was to settle the liability. However, as noted above, the allowances would not necessarily be used to settle that entity's liability. Furthermore, the allowances were not a debt owed by the government, which might have qualified as offsettable against a liability to the government.

Others argued that hedge accounting should be permitted between the allowances and the liability. Certainly, the existing rules for hedge accounting would not permit that but, since hedge accounting is essentially an artificial convention, this could have been a route worth exploring. As with all forms of hedge accounting, however, such a solution would enable one entity to portray itself differently from another entity in the same or a very similar economic position. Some of the practical complexities that would be involved in applying rules for such hedge accounting have been referred to above.

6.3 Grant of allowances

If the allowances are recognized as an asset, it is necessary to determine how the corresponding credit for free allocation of allowances (the grant) should be treated. The IFRIC decided that any allocation below fair value fell within the definition of government grants, essentially a government's 'transfers of resources to an entity in return for past or future compliance with certain conditions relating to the operating activities of the entity'. It argued that the obligation under an emission rights scheme to reduce emissions or deliver allowances was a condition 'relating to the operating activities of the entity'.

At the outset, the grant and the allowances are measured at the fair value of the allowances at the time the grant is made. As noted above, under IASs the allowances may, but do not have to be, re-measured to fair value; the grant, however, is not re-measured in like manner.

Under the existing version of IAS 20 *Government Grants* income from the grant is recognized gradually in line with recognition of the activity to which the grant relates. For example, a grant towards the purchase of a machine is recognized in the income statement over the same period as depreciation of the machine. In a paragraph of the draft Interpretation that attracted virtually no comment, the IFRIC noted the IASB's long-standing intention to propose amendment of IAS 20: the IASB had already discussed publicly its proposal to require income from a government grant to be recognized over the period to the point at which the grant could no longer be recalled. The rationale for the change was that beyond that point no liability existed: therefore, the whole grant needed to be recognized in income. The effect of such a change would be that a grant of emissions allowances would have to be recognized in income immediately on its receipt at the start of the annual cycle.

7. Effects of the IFRIC Decisions

While each of the IFRIC's decisions outlined above was based soundly on existing IASs, which in their turn, had stood the test of time, the combined effect when applied to emission rights produced some very strange results. The various measurement bases and recognition points in the income statement are summarized in Table 3.1 and the effect on the balance sheet is illustrated graphically in Figure 3.1. For simplicity, the analysis that follows ignores the effects of market transactions. Moreover, no attempt has been made to portray the complications that arise when an entity's financial year does not correspond with the annual cycle for the allocation and settlement of allowances.

When allowances are received at the start of the annual cycle, they are recognized at fair value, which is the market price for allowances at that date. Since the entity does not pay for allowances received from the government, it recognizes the grant at the same amount as the allowances. Unless and until IAS 20 is amended, the grant is recognized initially as a liability. Profit or loss at that point is therefore not affected.

Table 3.1. Measurement bases and recognition points

IFRIC Requirements	Measurement basis	Recognition point in income statement	Re-priced
Grant	Fair value of allowances at start of annual cycle	In line with emissions *but could change to* start of annual cycle.	Not re-priced
Allowances	Historical cost *or* Fair value		If historical cost, on disposal. If fair value, continuously in equity.
Emissions	Fair value of allowances	As emitted	Cumulative continuously in profit or loss.

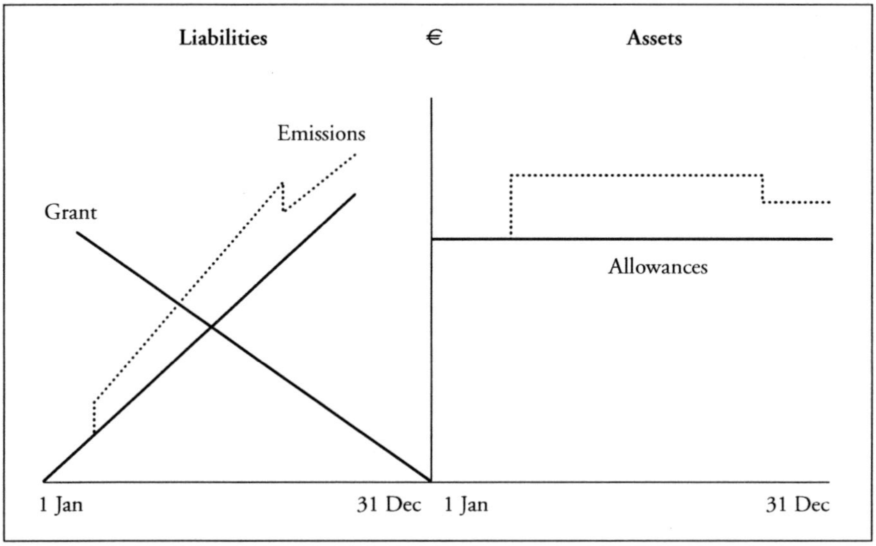

Figure 3.1. Effect on balance sheet

As emissions are produced, a new liability emerges for the amount that will have to be paid to the government in the form of allowances after the end of the annual cycle. In common with most other liabilities, this emissions liability is re-measured to 'current settlement value', in this case essentially fair value at each reporting date. The grant is gradually transferred from liability status to income to compensate for the emissions cost incurred. However, unlike most other liabilities, the liability for the grant is not re-measured for price movements. The reason is that, as explained above, it is not a true liability but in reality no more than deferred income, which is being recognized for convenience sake in line with the cost it is designed to subsidize.

Various imbalances now begin to emerge. Even if the entity continues throughout the annual cycle to expect that its cumulative emissions will be exactly covered by the grant of allowances, the effect of price changes will generally prevent the diminishing liability for the grant from exactly compensating for the growing liability for the cost of emissions. Furthermore, the effect of price changes on the asset side of the balance sheet differs from that on the liabilities side. If the entity chooses not to measure the allowances at its fair value, as it is permitted to do, there will naturally be no equivalent price movements on the assets side. However, even if the allowances are re-measured, the effect of price changes during the year differs from that affecting the liability. The reason is that a full year's allowances are held by the entity from the start of the annual cycle, whereas the liability for emissions builds up only gradually. It is therefore only at the end of the annual cycle that an entity that fair values its allowances will find that the price movements on the allowances broadly compensate for those on the liability. If the entity measures its allowances at historical cost, they will only be adjusted for cumulative price changes at the time of settlement, four months into the subsequent annual cycle.

In the graph in Figure 3.1 the dotted line curves show the effects of one upward and one downward price movement during the year. Note that the price change that occurs early in the annual cycle has nearly twice the effect on the allowances as on the liability for emissions. When the second price change occurs, towards the end of the annual cycle, the two effects are nearly equal.

The third column of Table 3.1 summarizes the effects of price changes on the income statement. As in the balance sheet, the effect of price changes on the cost of emissions increases as the emissions liability builds up. However, the imbalances noted in relation to the balance sheet are aggravated by the requirement under existing IASs that, if an intangible asset is revalued, the gain or loss should generally be recognized in equity rather than in profit or loss. Even by the end of the annual cycle, therefore, an entity with allowances that broadly match the total of its annual emissions and which re-values those allowances will report price change effects in profit or loss without the compensating effects of those same price changes on its allowances.

8. Evaluation

How far was the widespread criticism of IFRIC 3 justified and how far was it essentially a rejection of the real implications of a new phenomenon? The critics concentrated on the volatility generated in the income statement over the course of the annual cycle, at the end and even in some cases into the next period. To what extent was that volatility artificial and what could reasonably be done to reduce the artificial element?

The least justifiable aspect of the Interpretation and probably the easiest, at least in principle, to remedy was the mismatch arising from recognition of gains and losses on the emissions liability in profit or loss and on the allowances in equity. IASs apply the same revaluation rules to intangible assets as to property, plant, and equipment, effectively treating them all as long-term assets. Emission allowances, however, are current rather than long-term in nature and it would be appropriate for gains and losses on them to be recognized in profit or loss along with other trading gains and losses.

A simple, though unchanged, answer can also be given to the fundamental question identified at the start of this paper, whether allowances granted by government should be recognized separately from the emissions to which they relate. For the reasons discussed already, these two items must be recognized separately, in order to reflect the various ways in which the allowances can be used.

There are then two questions that need to be considered together:

- When should the grant be recognized in income? and
- Should any balance of the grant that is recognized as a liability be re-measured for price changes?

We have seen that under existing IASs a government grant is recognized in income in line with the costs it is designed to subsidize. However, the IASB has made no secret of its desire to revise this requirement to one that would recognize a grant as a liability only so long as it is repayable if its conditions cease to be met.

Emission allowances are in principle current rather than long-term, depreciable assets. Unlike other assets that are the subject of grants, emission allowances are designed to be ultimately payable to the government in settlement of the newly created environmental liability. Some would say that such payment is not the same as repayment of a grant because its conditions cease to be met. They point out that, if the entity reduces its emissions below the level of its allowances for whatever reason, even if it ceases that line of business, it does not have to pay back the allowances to the government but may sell them on the market. To the extent that emissions have occurred, IFRIC 3 recognizes a liability and re-measures it continuously at current market prices. The question is whether the same recognition can be given to the unused portion of the grant, which will be available to cover future emissions. If it can, that portion of the grant would be recognized as a liability, and in principle, should be re-measured continuously to current prices; if it cannot, the whole grant should be recognized in income upon receipt. IFRIC 3's treatment—which is to recognize a liability for the grant but not re-measure it—is unlikely to survive a revision of the government grants standard.

Although treating the unused portion of the grant in the same way as the emerging liability for past emissions would satisfy most critics by removing what they perceive as artificial volatility from the income statement, there is at present no obvious principle within the Framework by which it could be justified. If one is to be found or an exception made, it is necessary to examine the reasons for the disquiet shown by so many people for IFRIC 3's strict application of existing principles.

The reason is essentially that the government is giving at the start of the year what it will in all probability take away at the end. Managers resent recognizing a profit early in their reporting period on an asset that will have to be surrendered at the end of the period in settlement of a liability that the asset was designed to meet. The same logic is at work as leads managers to seek hedge accounting for transactions deemed to be incomplete until a further transaction is completed in a later period. An important difference from hedge accounting, however, is that emission allowances are given by and repaid to the same party. In that sense, the initial gain is more closely related to the eventual cost than are the gains and losses on hedged and hedging items, which are unrelated except in the mind of the manager.

It is true that for the duration of the annual cycle the allowances have an independent existence from the emissions cost that they will be required to settle. That requires their separate recognition from the moment of receipt. However, it does not necessarily require immediate recognition in income of the grant.

Although one *can* view the grant as without conditions once it has been made, that view ignores one of the two basic purposes of the grant. The first purpose is to provide the entity with the resource to settle a predetermined portion of its expected level of emissions. The second is to create a market for the allowances that are used in settlement. Those who view the grant as without conditions focus on the second purpose, noting that the allowances granted may be sold for the entity's own account. However, the first purpose is the more fundamental. The first purpose suggests that the grant of allowances is akin to the granting of a loan, which will be called in at the end of a period, except insofar as the entity is able to reduce its emissions below the level of the allowances. At intermediate reporting dates the cumulative level of emissions is unlikely to have reached the level of the allowances. The unused portion of the grant is therefore a kind of conditional liability, from which the entity can only escape if it can reduce its total emissions for the annual cycle below the level of its allowances. It is conditional, not because the event to which it relates has not yet occurred (which would not justify recognition as a liability), but because the allowances received have to be repaid unless savings in emissions can be made. In other words, the past event required by the Framework for recognition of a liability is the receipt of the allowances rather than the production of emissions, which still lies in the future.

If the grant can be recognized as a conditional liability, the question arises how to avoid double counting in respect of both the emissions liability itself and the liability to repay the allowances. The previous paragraphs refer to the 'unused' portion of the grant, ie only that portion that has not yet been credited to income to cover a proportion of the emissions produced to date. But, if the condition attached to the grant is that annual emissions should be reduced below the level of allowances, why should not the whole grant remain a liability unless or until such an event occurs?

The answer lies in the explicit linkage of the grant to the expected future liability. We have earlier likened the grant to a loan because it is in most cases repayable to the grantor after the end of the annual cycle. However, it is instructive to contrast the different effects of a simple loan of cash and a grant of emissions allowances. The emissions liability itself is assumed to be the same in both cases, building up day by day as emissions are produced. In the first scenario, a loan, the accounting entries would be:

Dr	Cash
Cr	Loan liability
Dr	Income statement
Cr	Emissions liability
Dr	Emissions liability
Cr	Cash
Dr	Loan liability
Cr	Cash

In other words, two cash payments are required: one to settle the emissions liability and the other to repay the loan.

By contrast, in the second scenario, the grant of emissions allowances, the recipient of the grant makes only one payment. The entries are:

Dr	Allowances
Cr	Grant liability
Dr	Income statement
Cr	Emissions liability
Dr	Grant liability
Cr	Income statement
Dr	Emissions liability
Cr	Allowances

Unlike in the first scenario, the emissions liability is extinguished by the repayment of the asset granted, the allowances. Insofar as a proportion of the emissions liability can be covered by allowances, there are not two liabilities but only one. Accordingly, as the emissions liability, which is the fundamental one, builds up, the grant liability reduces.

The above analysis suggests that the conditionality of a grant of emissions allowances should be expressed, not simply in terms of the obligation to repay the allowances unless emissions can be reduced, but in terms of the use to which they must then be put, ie settlement of the emissions liability.

9. Conclusion

A finding along the above lines would resolve most of the difficulties encountered by IFRIC 3. The question remains whether the IASB would be faithful to its mission if it interpreted the Framework in such a way. We noted at the start of this paper that the issue of emission rights illustrates the difficulties of standard setting at the frontiers of accounting, of being able to discern the limitations but also the possibilities of financial reporting. The IASB knows that it must resist the temptation, so often urged on it by its critics, to resolve every problem by a 'quick fix' that relies on a standalone rule rather than clear application of a principle in the Framework. That way lie increasing complexity and inconsistencies.

To assist in the search for a principled solution, this chapter has drawn attention to certain features of the cap-and-trade scheme that, in combination, are probably unique: a grant of assets having two functions: an immediately marketable security, and a 'currency' which will be the sole means of satisfying a highly probable future liability *to the grantor*.

The Framework necessarily focuses on the financial position at the reporting date. However, the implications of expected future events are taken into account to the extent that they throw light on the existence of assets and liabilities at that date. This paper suggests that a grant of assets that are extremely likely to have to be repaid is a liability, notwithstanding that the assets can be used independently in the meantime in market transactions.

Annex to Chapter 3

Summary of IFRIC 3's Illustrative Example

The following schedules summarize the Illustrative Example in IFRIC 3, first on the assumption that the allowances are accounted for at cost until disposed of and, secondly, on the assumption that they are continuously re-measured to fair value.

Other assumptions are:

- The annual cycle for the allocation of allowances and assessment of emissions, January – December, coincides with the entity's financial year.
- The entity receives a grant of allowances covering 12,000 tonnes of emissions.
- The fair value of allowances is:
  - 1 January CU10 (ie 10 currency units) per tonne
  - 30 June CU12 per tonne
  - 31 December till settlement CU11 per tonne
- Emissions
  - January – June 5,500 tonnes
  - July – December 7,000 tonnes
- At 30 June the entity still expects emissions of 12,000 tonnes
- On 31 December the entity buys at CU11 per tonne additional allowances to cover 500 tonnes
- Annual emissions are settled four months after the year end.

Emission Rights: Summarized Example of IFRIC 3 Requirements: Cost Model

	1 Jan	Jan–June	30 June	June–Dec	Jan–Dec	31 Dec	Jan–April	30 April
INCOME STATEMENT								
Govt grant 12,000 tonnes @ CU10		(55,000)		(65,000)	(120,000)			
Gain on disposal 12,000×CU1							(12,000)	
EXPENSE								
Emissions 5,500 tonnes @ CU12		66,000						
Cumulative 12,500 @ CU11					137,500			
CU137,500 – CU66,000				71,500				
(Profit) loss		11,000		6,500	17,500		(12,000)	
BALANCE SHEET								
EQUITY								
Retained earnings [neg]			11,000			17,500		5,500
(Gain)/loss on allwnces revaln								
Total equity [neg]			11,000			17,500		5,500
ASSETS								
Allwnces at initial FV & cost	120,000		120,000	5,500	5,500	125,500	(137,500)	
Repriced on disposal to CU11							12,000	
Cash 500 tonnes bought @ CU11				(5,500)	(5,500)	(5,500)		(5,500)
Total assets	120,000		120,000			120,000		(5,500)
LIABILITIES								
Govt grant Deferred Inc	(120,000)		(65,000)					
Govt grant offsetting						–		
emissions in period								
5,500/12,000 × CU120,000		55,000			120,000			
120,000 – 55,000 = 65,000				65,000				
Cum liability to deliver allwnces								
5,500 @ CU12		(66,000)	(66,000)					
12,500 @ CU11 = 137,500				(71,500)	(137,500)	(137,500)	137,500	–
Total liabilities	(120,000)		(131,000)			(137,500)		–
Total assets/(liabilities)	–		(11,000)			(17,500)		(5,500)

Emission Rights: Summarized Example of IFRIC 3 Requirements: FV model

	1 Jan	Jan–June	30 June	June–Dec	Jan–Dec	31 Dec	Jan–April	30 April
INCOME								
Govt grant 12,000 tonnes @ CU10		(55,000)		(65,000)	(120,000)			
EXPENSE								
Emissions 5,500 tonnes @ CU12		66,000						
Cumulative 12,500 @ CU11					137,500			
CU137,500 – CU66,000				71,500				
(Profit) loss		11,000		6,500	17,500			
BALANCE SHEET								
EQUITY								
Retained earnings [neg]			11,000			17,500		
(Gain)/loss on allwnces revaln		(24,000)	(24,000)	12,000	(12,000)	(12,000)		
Total equity			(13,000)			5,500		5,500
ASSETS								
Allwnces at current FV	120,000	24,000	144,000	(12,000)	12,000	132,000	(137,500)	–
500 tonnes bought @ CU11				5,500	5,500	5,500		
Cash				(5,500)	(5,500)	(5,500)		(5,500)
Total assets	120,000		144,000			132,000		(5,500)
LIABILITIES								
Govt grant								
Deferred Inc	(120,000)		(65,000)					
Offsetting emissions						–		
in period								
(5,500/12,000) × CU120,000		55,000			120,000			
120,000 – 55,000 = 65,000				65,000				
Liability to deliver allwnces								
5,500 @ CU12 = CU66,000		(66,000)	(66,000)					
12,500 @ CU11 = CU137,500				(71,500)	(137,500)	(137,500)	137,500	–
Total liabilities	(120,000)		(131,000)			(137,500)		–
Total assets/(liabilities)	–		13,000			(5,500)		(5,500)

4

Trade and Investment Implications of Carbon Trading for Sustainable Development

Marie-Claire Cordonier Segger and Markus Gehring[*]

1. Introduction

A complex matrix of international rules has emerged to guide policies to prevent climate change, encouraging the development of more sustainable global energy, forestry, transportation, agriculture, housing, and other systems. However, a great deal remains to be done.[1] Both developed and developing countries are considering new regulatory measures to reduce the greenhouse gas emissions which cause climate change, as part of a broader commitment to adopt new laws to promote more sustainable development.[2]

Evolving international economic regimes are an important piece of this puzzle. To date, multilateral efforts to liberalize trade and investment and to reduce global greenhouse gas emissions have proceeded largely along separate tracks.[3] One system is defined by the treaties establishing the World Trade Organization (WTO) and their annexes, along with over 3,000 regional and bi-lateral trade and investment treaties. The other is shaped by the UN Framework Convention on Climate Change, its Kyoto Protocol, and many related international arrangements. Though both systems of international rules share a common sustainable development objective, the relationship between the two is complex. As many have noted, these regimes 'are likely to come into closer contact as

* With thanks and acknowledgements to CISDL Associate Fellow Gareth Price LLB (UEA & Leuven), LLM (McGill), and CISDL Legal Research Fellow Sebastien Jodoin, BCL/LLB (McGill), LLM (LSE) for their excellent legal research and substantive contributions. Thanks also to CISDL Lead Counsel Dr Charlotte Streck, CISDL Legal Research Fellows Dr Christina Voigt and Prof Kate Miles, and to Richard Benwell, for their insights and advice. The views expressed herein are personal academic opinions of the authors, not meant to reflect positions of any government.

1 United Nations Environment Programme, *10 Years After Rio: The UNEP Assessment* (UNEP: New York, 2002).

2 N Stern, *Stern Review on the Economics of Climate Change* (Cambridge University Press: Cambridge, 2007).

3 S Charnovitz, 'Trade and Climate: Potential Conflicts and Synergies' in *Beyond Kyoto: Advancing the International Effort Against Climate Change* (Pew Center: Arlington, 2003) at 141.

climate policies lead to significant economic effects.'[4] International trade and investment treaty rules may also affect the viability and effectiveness of new regulations to address climate change.[5] Essentially, just as trade and investment rules can and should shape climate policy, so climate rules can and do shape trade and investment policy. While climate laws and policies might restrict or constrain certain kinds of economic development, they can also provide incentives for new kinds of development. Indeed, many climate change measures can even be characterized as trade and investment rules themselves.

New policies and regulations are being set in place to establish Emissions Trading Schemes (ETS) which aim to reduce greenhouse gas emissions, and to stimulate investment in more sustainable, low-carbon development. There are important differences between international disciplines that govern the use of trade and investment related barriers by States, international rules that encourage States to trade emission certificates and investment in low-carbon development, and domestic measures that encourage firms to trade Emission Reduction Units (ERUs) or develop renewable energy technologies. But the legal rules that govern international emissions trading and investment in clean development can—and should—be considered part of the wider body of evolving trade and investment law. Just as many trade provisions in the WTO Agreements contribute to the WTO's sustainable development objective, so do the *lex specialis* trade and investment provisions in the 1992 UN Framework Convention on Climate Change (UNFCCC) contribute to the sustainable development objective of the global climate regime. For instance, an ETS can provide financial incentives for firms and others to develop and adopt innovative new technologies by capping emissions and creating a market for emission reduction units.[6] In certain circumstances, an ETS may even provide incentives for the transfer of new sustainable development technologies to developing countries, for example when an Emission Rights Purchase Agreement (ERPA) provides for such transfer in relation to Clean Development Mechanism (CDM) certificates. More generally, the innovative and coherent design of future trade, investment, and climate change rules—building international cooperation on careful legal analysis—can make an important contribution to sustainable development.

This chapter briefly examines, based on recent legal analysis and scholarship in the field of sustainable development, the trade and investment implications of carbon trading. First, we consider how trade regimes, investment rules and emissions trading schemes can serve as tools for sustainable development. Then, we examine the trade and investment implications of new emissions trading

[4] *Ibid* at 141.

[5] C Voigt, *Sustainable Development as a Principle of International Law–Resolving Conflicts between Climate Measures and WTO Law* (Martinus Nijhoff: Leiden, 2009).

[6] D Freestone and C Streck (eds), *Legal Aspects of Implementing the Kyoto Protocol Mechanisms: Making Kyoto Work* (OUP: Oxford, 2005) Section VII.

regulations, suggesting ways to minimize overlaps with trade and investment disciplines when drafting carbon-related regulations, and addressing issues related to technology transfer. Next, we consider how trade and investment regimes might be refined to complement and support climate change related measures. We conclude with a call for integration and, at a minimum, coherence among trade, investment, and climate change regimes that share common sustainable development objectives.

2. Trade Regimes, Investment Rules and Emissions Trading Schemes: Tools for Sustainable Development?

2.1 International trade law obligations

Benefits of trade liberalization are not available to States and their citizens by right—governments negotiate treaties to secure them. In the WTO and many regional trade agreements, States have committed to lower tariffs and to secure non-discrimination and most favoured nation (MFN) status for trading partners. Briefly, four main obligations provide the foundations of world trade law.[7] These treaty rules aim to secure stability and lower costs for firms from all State Parties to trade agreements. They may also affect the design of new laws to reduce greenhouse gas (GHG) emissions and encourage low carbon investments.

First, States have committed to convert many non-tariff barriers to trade, such as quotas, into more easily measured tariffs. They have reciprocally agreed to be bound by schedules of specific tariff rates, and to continue negotiating reductions of these rates. The main commitments of WTO members are contained in individual detailed Country Schedules attached to the General Agreement on Tariffs and Trade (GATT)[8] for goods, and the General Agreement on Trade in Services (GATS)[9] which lists services. GATT rules apply to all goods unless they are reserved or excepted. GATS rules use a 'positive-list' approach whereby only services listed by WTO Members are covered. These schedules provide comprehensive listings of all the products for which a WTO member has accepted a commitment to a binding tariff at a particular level. Each WTO member maintains a 'schedule of concessions', which is essentially a list of the tariff rates that it imposes on imported goods. When members 'bind' a certain tariff in their schedule, they guarantee that their customs tariff on the product in question will not be higher than the bound rate. Members have no obligation under WTO law to lower their bound tariff rates, but they do commit to continue negotiations to

[7] M Gehring, J Hepburn and MC Cordonier Segger, *World Trade Law in Practice* (Globe Law & Business: New York, 2006) 17.

[8] General Agreement on Tariffs and Trade (adopted 30 October 1947, provisionally entered into force 1 January 1948) 55 UNTS 194, CTS No 31 (1948) (GATT).

[9] General Agreement on Trade in Services (adopted 15 April 1994, entered into force 1 January 1995) 1869 UNTS 183.

bind more product lines and to lower the tariffs on product lines already bound.[10] The WTO Committee on Trade in Goods and Committee on Trade in Services monitors implementation of these rules. This commitment to provide market access on a non-discriminatory basis can curtail the type of rules that States adopt, affecting a State's ability to selectively restrict certain imports.[11] Article XI:1 of the GATT, the provision that prohibits quantitative restrictions, has been used to evaluate the GATT-consistency of natural resource and environment related bans, for example in the *US – Tuna Case*[12] and the *US – Shrimp Case*.[13]

Second, Members commit to extend the MFN status[14] to each other, precluding discrimination between the ways goods, services, and investments from different economic partners are treated. Third, Members also commit to extend 'national treatment' to other Members for goods, certain services and other aspects of their trade policy.[15] This prevents WTO members from treating the 'like products' of other members differently,[16] and prevents like products from being distinguished on the basis of the process and production methods (PPMs) used to make them. These rules discipline the types of health, environmental, natural resource management, consumer safety and other standards that WTO members can apply to products if exceptions are not secured.[17] MFN and national treatment commitments are also implemented through the WTO Agreement on Technical Barriers to Trade (TBT Agreement)[18] which addresses technical regulations and standards and the WTO Agreement on the Application of Sanitary and PhytoSanitary Measures (SPS Agreement)[19] which addresses health and plant safety regulations and standards. The TBT Agreement and the SPS Agreement define when certain restrictions on trade are allowed, seeking to limit

[10] Gehring, Hepburn and Cordonier Segger (n 7 above) 97.

[11] Ibid; M Trebilcock and R Howse, *Regulation of International Trade* (Routledge: New York, 2005) 336.

[12] *United States – Restrictions on the Import of Tuna* (1991) GATT BISD 39S/155, (1991) 30 ILM 1594.

[13] WTO, *United States: Import Prohibition of Certain Shrimp and Shrimp Products – Report of the Panel* (15 May 1998) WT/DS58/R; See also WTO, *United States: Import Prohibition of Certain Shrimp and Shrimp Products-Report of the Appellate Body* (12 October 1998) Doc WT/DS58/AB/R.

[14] GATT Art I:1, General Agreement on Trade in Services (adopted 15 April 1994, entered into force 1 January 1995) 1869 UNTS 183, 44 art II:2, Agreement on Trade-Related Aspects of Intellectual Property Rights (adopted 15 April 1994, entered into force 1 January 1995) 1869 UNTS 299, art 4.

[15] GATT Art III, GATS (n 14 above) Art XVII, TRIPs (n 14 above) Art 3.

[16] Likeness is determined by the dispute settlement mechanism in the WTO on a case-by-case basis, taking into account criteria such as 'the product's end-uses in a given market; consumers' tastes and habits, which change from country to country; the product's properties, nature and quality' (Report of the Working Party on Border Tax Adjustments, BISD 18S/97, para 18).

[17] There is a growing opinion which argues that in the GATT the same rules as in the TBT should be in force permitting PPMs to be taken into account under certain conditions.

[18] Agreement on Technical Barriers to Trade (adopted 15 April 1994, entered into force 1 January 1995) 1868 UNTS 120.

[19] Agreement on the Application of Sanitary and Phytosanitary Measures (adopted 15 April 1994, entered into force 1 January 1995) 1867 UNTS 493.

protectionism (the use of regulations to unfairly privilege domestic firms *vis-à-vis* the firms of trading partners).[20] As such, for instance, the SPS Agreement essentially provides specific restrictions on the types of phyto-sanitary standards governments should adopt, conditioning the relevant GATT rules and exceptions.[21] WTO TBT and WTO SPS Committees study and debate these issues, and can grant time-limited exceptions to developing countries in light of their particular financial, trade, and development needs.[22] WTO Members also commit to protect intellectual property rights through the WTO Agreement on Trade Related Aspects of Intellectual Property Rights[23] (TRIPs Agreement), to discipline subsidies in the WTO Agreement on Subsidies and Countervailing Measures[24] (Subsidies Agreement), to discipline government procurement through the WTO Agreement on Government Procurement,[25] and to discipline investment measures related to trade in goods in the Agreement on Trade-Related Investment Measures[26] (TRIMs). Such obligations might affect government attempts to regulate in relation to climate change. The TRIPs Agreement obliges WTO Members to set laws in place to protect intellectual property rights, potentially affecting technology transfer. The Subsidies Agreement disciplines the types of subsidies WTO members can provide, potentially affecting emission reduction incentives. The WTO Government Procurement Agreement and the TRIMs are minimalist accords, as governments were unwilling take on significant restrictions in these areas.[27] For instance, the TRIMs disciplines apply only to measures that affect trade in goods, committing to notify certain specific trade-related investment measures that discriminate against foreigners or foreign products.[28] However, if more stringent disciplines are adopted on government procurement or investment, they might constrain schemes for public purchasing of lower-carbon products, or climate regulations affecting foreign investors.

Fourth, Members agreed to transparency and notification obligations, as well as binding, peaceful settlement of trade disputes.[29] The WTO's Dispute Settlement Understanding (DSU) aims to provide 'security and predictability to

[20] M Echols, *Food Safety and the WTO: The Interplay of Science, Culture and Technology* (Kluwer Law Int: The Hague, 2001), C Button, *The Power to Protect: Trade, Health and Uncertainty in the WTO* (Hart Publishing: Oxford, 2004) 43–90.

[21] Button (n 20 above) 10–11.

[22] See WTO's Documents Online database <http://docsonline.wto.org> using document symbol G/SPS/GEN for all documents of the SPS Committee, including those related to exceptions for developing countries.

[23] TRIPs (n 14 above).

[24] Agreement on Subsidies and Countervailing Measures (adopted 15 April 1994, entered into force 1 January 1995) 1867 UNTS 14.

[25] Agreement on Government Procurement (adopted 15 April 1994, entered into force 1 January 1995) 1867 UNTS 194.

[26] Agreement on Trade-Related Investment Measures (adopted 15 April 1994, entered into force 1 January 1995) 1868 UNTS 186.

[27] Gehring, Hepburn and Cordonier Segger (n 7 above) 139.

[28] Ibid.

[29] GATT, Art X; GATS (n 14 above), Art III; TRIPs (n 14 above) Art 63; Agreement on Technical Barriers to Trade (adopted 15 April 1994, entered into force 1 January 1995) 1868 UNTS 120, Art 10; Agreement on the Application of Sanitary and Phytosanitary Measures (adopted 15 April 1994, entered into force 1 January 1995) 1867 UNTS 493 Agreement, Art 7.

the multilateral trading system'.[30] WTO members settle disputes through consultation and other mechanisms, but if attempts fail to defuse a trade-related conflict, members can submit the disagreement to adjudication by a Panel, with appeals to an Appellate Body. If a decision is not implemented by the losing party, the Dispute Settlement Body can authorize retaliation (eg suspension of trade concessions such as preferential tariffs).

The WTO Agreements are not without exceptions for measures related to sustainable development. First, Article XX of the GATT allows WTO members to violate WTO disciplines in certain circumstances, such as for the protection of health, the environment or conservation of exhaustible natural resources. Article XX reads, in relevant part:

> Subject to the requirement that such measures are not applied in a manner which would constitute a means of arbitrary or unjustifiable discrimination between countries where the same conditions prevail, or a disguised restriction on international trade, nothing in this Agreement shall be construed to prevent the adoption or enforcement by any contracting party of measures: . . .
>
> (b) necessary to protect human, animal or plant life or health; . . .
>
> (g) relating to the conservation of exhaustible natural resources if such measures are made effective in conjunction with restrictions on domestic production or consumption; . . .[31]

Similar exceptions were agreed in the GATS and the TRIPs Agreement. As noted by the Retrospective Analysis of the 1994 Canadian Environmental Review of the WTO, GATT Article XX is an important safeguard for a State's ability to regulate for sustainable development.[32] However, once a violation of trade law obligations has been established, the burden to defend environmental and social measures falls upon the WTO Member State invoking the exception.[33] Article XX exceptions have been tested in WTO disputes related to several topics highlighted in Agenda 21 and the Johannesburg Plan of Implementation.[34] For instance, States have made claims related to the use of genetically modified organisms (*EC – Approval and Marketing of Biotech Products*),[35] the enforcement of domestic IP laws (*Denmark – Measures Affecting the Enforcement of Intellectual Property Rights*),[36] marine animal protection laws

[30] Understanding on Rules and Procedures Governing the Settlement of Disputes (adopted 15 April 1994, entered into force 1 January 1995) 1869 UNTS 401, (DSU) Art 3.2.

[31] GATT, Art XX.

[32] DFAIT, *Retrospective Analysis of the 1994 Canadian Environmental Review – Uruguay Round of Multilateral Trade Negotiations* (DFAIT: Ottawa 1999).

[33] V Lowe, *International Law* (OUP: Oxford, 2007) 219–20.

[34] Agenda 21; online: <www.un.org/esa/sustdev/agenda21.htm>, accessed 31 March 2009; Johannesburg Plan of Implementation (JPOI), Report of the World Summit on Sustainable Development, Johannesburg UN Doc A/CONF.199/20.

[35] WTO, *EC–Approval and Marketing of Biotech Products* (29 September 2006) WT/DS291/R.

[36] WTO, *Denmark – Measures Affecting the Enforcement of Intellectual Property Rights* (21 May 1997) WT/DS83/1.

(*US – Shrimp/Turtle*),[37] domestic legislation (*US – Section 211 Appropriations Act*),[38] the regulation of carcinogenic asbestos (*European Communities – Measures Affecting Asbestos and Products Containing Asbestos*),[39] and waste management (*Brazil – Measures Affecting Imports of Re-treaded Tyres*).[40] Such cases have been inconclusive, as unsurprisingly, the trade dispute settlement body appears to place highest priority on trade law obligations. A further systemic exception involves the recognition of non-reciprocal special and differential treatment for developing countries,[41] and at Article XXIV:5 of the GATT, WTO members also exclude customs unions and bilateral or regional free trade areas from compliance with WTO disciplines in certain circumstances.[42] These regional agreements are important, establishing both disciplines which might affect the adoption of domestic and international carbon rules, and measures to promote sustainable development and environmental cooperation.

2.2 Regional trade and investment law obligations

Nearly all States today participate in bilateral or regional free trade arrangements (RTAs). These treaties are discriminatory by nature, designed to grant certain trade preferences to the Parties. They can be agreed between developed States, developing States, or among both a developed and a developing State.[43] As of July 2007, 380 regional trade agreements had been notified to the WTO, with over 400 expected to be in operation by 2010.[44] Concerns have been raised, including by the WTO, about this proliferation of regional accords.[45] Furthermore, depending on their specific provisions, regional trade agreements might encourage growth in unsustainable industries or unsustainable levels of resource exploitation leading to related physical impacts, and might also constrain the

[37] WTO, *United States – Import Prohibition of Certain Shrimp and Shrimp Products* (12 October 1998) WT/DS58/AB/R.

[38] WTO, *United States – Section 211 Omnibus Appropriations Act of 1998* (1 February 2002) WT/DS17/AB/R.

[39] *EC – Measures Affecting Asbestos and Asbestos – Containing Products* (12 March 2001) WT/DS135/AB/R.

[40] WTO, *Brazil: Measures affecting Imports of Retreaded Tyres – Report of the Panel* (12 June 2007) WT/DS332/R.

[41] 1979 'Enabling Clause' decision of the GATT Contracting Parties, see *European Communities - Conditions for the Granting of Tariff Preferences to Developing Countries* (7 April 2004) WT/DS246/AB/R.

[42] Bartels and Ortino, *Regional Trade Agreements and the WTO Legal System* (OUP: Oxford, 2007) 3.

[43] CA Cooper and BF Massell, 'Toward a General Theory of Customs Unions for Developing Countries' (1965) The Journal of Political Economy 461.

[44] WTO, 'Regional Trade Agreements Gateway' <http://www.wto.org/english/tratop_e/region_e/region_e.htm>; WTO, 'WTO launches new database on regional trade agreements' WTO Doc Press/548, 14 January 2009.

[45] WTO, 'Regional Trade Agreements: Scope of RTAs' <http://www.wto.org/english/tratop_e/region_e/scope_rta_e.htm>, accessed 31 March 2009.

adoption or implementation of new standards. However, small-scale regional trade agreements may also provide a useful testing ground for trade policy innovations, including for the sustainable development of low-carbon economies.[46]

Furthermore, and of perhaps greatest relevance to new laws related to emissions trading schemes and more sustainable, low-carbon economic development, in recent decades more than three thousand International Investment Agreements (IIAs) have been negotiated,[47] supplemented internationally by rules and dispute settlement procedures developed through the United Nations Commission on International Trade Law (UNCITRAL) and the International Centre for Settlement of Investment Disputes (ICSID). Private sector investment could help to finance the adoption of low-carbon technologies.[48] Some even argue that 'private finance [is] now the biggest show in town.'[49] These IIAs seek to create favourable conditions and stable frameworks for the treatment of foreign investors and investments, encouraging private sector investment in developing countries. The IIAs obligations normally guarantee a minimum standard of treatment, or 'fair and equitable treatment' toward the foreign investor in 'like circumstances'. Some IIAs commit to 'stabilization clauses' which can exclude IIA-covered investments from changes in the law of host States. Such clauses may be important to future attempts to develop domestic climate rules. The 'legitimate expectations' of the investor regarding a regulatory framework may become grounds for a potential challenge by a foreign investor toward an (unfavourable) change in circumstances due to new climate change regulations, including emissions trading schemes that impose significant new costs on private firms.[50] By December 2007, there were 280 known IIA arbitrations, in which foreign investors challenged governments in confidential 'investor-State' dispute settlement proceedings by invoking investment agreement clauses.[51] Potentially, these IIAs, measures and rule-making bodies are more likely than trade law to constrain carbon trading and related regulatory measures. However, it may be possible to design emissions trading systems carefully, to avoid becoming embroiled in

[46] MC Cordonier Segger, 'The WTO, Regional Trade Agreements and Sustainable Development' in Bartels and Ortino (n 42 above).

[47] A Newcombe and L Paradell, *Law and Practice of Investment Treaties: Standards of Treatment* (Kluwer Law International: Leiden, 2009) 57–64; UNCTAD Secretariat, 'International Investment Rulemaking' 22 May 2007 TD/B/COM.2/EM.21/2; UNCTAD, Latest Developments in Investor-State Dispute Settlement, IIAS Monitor No 1 (2008) UNCTAD/WEB/ITE.IIA/2008/3.

[48] D Murphy, 'Technology in a Post 2012 Climate Regime', paper presented to 'A Way Forward: Canadian & International Perspectives on Post 2012 Climate Policy', 4 March 2008, Ottawa, Canada.

[49] M Klein and T Harford, 'The Market for Aid' (World Bank Publications: Washington DC, 2005) at 51.

[50] K Miles, 'International Investment Law and Climate Change: Issues in the Transition to a Low Carbon World' (2008), Society of International Economic Law, Inaugural Conference, June 2008, Geneva Switzerland 19.

[51] A Newcombe and L Paradell, *Law and Practice of Investment Treaties* (Kluwer Law Int: Alphen, 2009) 59.

disputes of this kind. It is also possible to design international investment agreements and trade agreements to ensure that legitimate new energy, transportation, forestry, waste management and other measures are—at a minimum—not frustrated, and perhaps even promoted.

3. Designing New GHG Regulatory Measures: Trade and Investment Considerations

3.1 Sustainable development objectives of the global climate regime

In the 1992 UNFCCC and its 1997 Kyoto Protocol, States commit to both environmental protection *and* sustainable development objectives. The treaties repeatedly emphasize the importance of sustainable economic development, especially for over 140 developing country Parties. As noted at Article 3(4) of the UNFCCC, each Party, when considering which policies would be appropriate for protecting the climate system, should 'tak[e] into account that economic development is essential for adopting measures to address climate change'.[52] Trade and investment liberalization is significant in this regard, providing both constraints and opportunities for developing countries. In the UNFCCC, Parties optimistically agreed to 'cooperate to promote a supportive and open international economic system' which, it was believed, would 'lead to sustainable economic growth and development in all Parties, particularly developing country Parties, thus enabling them better to address the problems of climate change'.[53] It was essentially hoped that in accordance with the now-discredited 'Environmental Kuznets Curve' theory,[54] trade liberalization would lead to higher incomes which would consequently lead to sustainable development in developing countries. However, by 1997, States had recognized the more complex series of linkages between climate change, policies to address it, and international trade.[55] In Article 2(3), Annex I Parties agreed to 'strive to implement policies and measures ... in such a way as to minimize adverse effects, including the adverse effects of climate change, effects on international trade, and social, environmental and economic impacts on other Parties, especially developing

[52] United Nations Framework Convention on Climate Change (adopted 9 May 1992, entered into force 21 March 1994) 1771 UNTS 107 (UNFCCC), Art 3(4).

[53] UNFCCC, Art 3(5).

[54] This theory proposed that as economic wealth increases, populations will demand higher environmental standards and environmental quality will automatically improve, without taking into account that certain environmental damage is irreversible, that many environmental problems are cumulative and global, or that the beneficiaries of economic growth do not necessarily live or vote where damage is done. See H Nordstrom and S Vaughan, 'Trade and Environment' (1999) WTO Special Studies 4, Geneva.

[55] M Dowden, *Climate Change & Sustainable Development; Law, Policy and Practice* (EG Books: London, 2008) at 56.

country Parties'.[56] This recognized that States may need to balance adverse effects of climate change with the adverse effects of measures on international trade, in order to minimize both.

3.2 Trading for GHG emissions reductions

There are important differences between world trade rules and international emissions trading rules, though both function by encouraging exchanges. Most trade and investment is carried out by private firms exchanging goods and services, with international trade and investment treaties to discipline the types of measures States may adopt, *inter alia* through schedules of 'bound' tariff rates and investor protection commitments. In contrast, international emissions trading is carried out by States themselves, with climate change treaties to facilitate exchanges among States of their 'rights to emit GHGs' and encourage investment in low-carbon development. According to Article 2 of the UNFCCC, the Parties seek to 'achieve ... stabilization of greenhouse gas concentrations in the atmosphere at a level that would prevent dangerous anthropogenic interference with the climate system'.[57] The Kyoto Protocol, in Articles 2 and 3, provides flexibility for Annex I Parties (from developed countries and economies in transition) to achieve their assigned emissions limitations.[58] Emissions trading, governed by Article 17 of the Protocol, is seen as an important instrument to achieve Annex 1 Parties' GHG emissions reduction commitments. As noted by Robinson et al, for the regulator, the 'cost of complying with an emissions trading scheme as a whole are typically less than under traditional command and control regimes, as the trading aspect ensures that reductions take place where they can be achieved most cost-effectively'.[59] The trading may also play a useful role by providing incentives for firms to innovate, advancing the development and deployment of more environmentally sound technologies. If the carbon-trading regime established by the Kyoto Protocol did overlap with other international trade and investment liberalization commitments, a trade tribunal could conclude that the emissions trading among States is governed by the *lex specialis* of the climate treaties.

Given the Kyoto Protocol's general openness to 'sovereign' emission trading, many countries have adopted climate change strategies that establish domestic emissions trading systems, often making use of the Protocol's CDM. Several leading developed countries have moved forward to pilot domestic emissions

[56] Kyoto Protocol to the UNFCCC (adopted 10 December 1997, entered into force 16 February 2005) 37 ILM 22 (Kyoto Protocol), Art 2(2).
[57] UNFCCC, Art 2.
[58] Kyoto Protocol, Arts 2 and 3.
[59] J Robinson, J Barton, C Dodwell, M Heydon, and L Milton, *Climate Change Law: Emissions Trading in the EU and the UK* (Cameron May: London, 2007) at 44.

trading schemes, both for greenhouse gases and other significant pollutants.[60] Some developing countries, constrained by capacity and resources, are already considering design and use of these instruments. Such domestic carbon trading schemes raise further trade and investment issues.

Current domestic emissions trading schemes are built on a variety of different legal rules and frameworks. At the basic level, there is a distinction between a cap-and-trade system and a baseline-and-credit system.[61] For instance, like the EU's emission trading scheme, Australia's proposed scheme (to be introduced from 2010 under the Rudd administration) adopts a cap-and-trade approach, which is promoted as an effective way to limit carbon emissions while continuing to provide business incentives in an economy focused on natural resource development.[62] The United States' trading scheme for NOx emissions imposed a cap of 143,000 tonnes in 2003, with the baseline year ranging from 1990 to 1995. As of 2007, this approach brought emissions to 5% below the cap for that year.[63] Alternatively, the Canadian province of Ontario operates a mixed emissions reductions scheme for nitrous oxide and sulphur dioxide (NO and SO_2),[64] attempting to combine cap-and-trade features with those of a baseline-and-credit system.[65] The innovative linkage of allowances and emissions reduction credits seeks to ensure dynamic interplay in a restricted and fragile market.

As such, domestic sustainable development objectives, as well as a State's trade and investment profile, can and do shape the fundamental design of any emissions trading system.[66]

3.3 Trade and investment implications of carbon trading

Depending on regulatory instrument choices, and how they are designed and implemented, carbon-trading schemes can have important implications for trade

[60] M Gehring and K Price 'Implementing the Kyoto Protocol in Canada and the UK: A Discussion of the Economic Instruments Employed' in CPM Waters (ed), *British and Canadian Perspectives on International Law* (Martinus Nijhoff: Leiden, 2006) 255, 262.

[61] UNFCCC National Reports; online <www.unfccc.int/national_reports/items/1408.php>, accessed 31 March 2009.

[62] See Carbon Pollution Reduction Scheme Green Paper, Australian Government, Department of Climate Change (July 2008), Summary Report, online: <http://www.climatechange.gov.au/greenpaper/index.html>, accessed 31 March 2009.

[63] US Environmental Protection Agency: Clean Air Markets; online: <www.epa.gov/airmarkets>, accessed 31 March 2009.

[64] Environmental Protection Act 1999, Ontario Regulation 397/01. See also Ontario Emission Trading Registry Introduction, online: <www.ene.gov.on.ca/envision/air/etr/index.htm>, accessed 31 March 2009.

[65] See especially, M Gehring and C Streck, 'Emissions Trading: Lessons from SOx and NOx emissions Allowance and Credit Systems Legal Nature, Title, Transfer and Taxation of Emission Allowances and Credits' (2005) 35 Environmental Law Reporter 10219 at 10227.

[66] See eg D Hirsch, A Bergman and M Heintz, 'Emissions Trading – Practical Aspects' in M Gerrard (ed), *Global Climate Change and U.S. Law* (American Bar Association: Chicago, 2007).

and investment law. Complementarity and mutual supportiveness is needed. A foreign trading partner or investor that perceives itself to be economically disadvantaged by an emissions trading scheme could launch a claim under the WTO, a regional trade tribunal, or an investment treaty investor-State tribunal, and the regulators would be required to spend significant resources defending their embryonic regime. For domestic carbon trading, both basic elements and accompanying measures, if not carefully designed, might discriminate against firms from foreign trading partners or foreign investors. The issues can be divided into two general categories. First, specific trade disciplines agreed in the WTO and other trade agreements might, if not properly analysed and characterized, appear to disallow the use of certain carbon trading measures. Second, and perhaps most problematic, recently agreed investment treaty commitments may, if not carefully interpreted, be used to challenge carbon trading measures, constraining their adoption or limiting their effective implementation.

3.3.1 Trade law implications of carbon trading measures

Disciplines agreed in the WTO might inadvertently or otherwise be interpreted by regulators or dispute settlement bodies in a way that unduly constrains the effectiveness of emissions trading schemes. There may be direct regulatory effects, where a trade authority or dispute settlement mechanism questions or disallows certain measures, or indirect regulatory effects, where environment and development authorities could be dissuaded from using carbon reduction measures with threats of trade challenges (so-called 'regulatory chill').

One trade issue that is often raised, for instance, involves the use of government subsidies. If an allowance, credit or unit that grants a right to produce carbon emissions were characterized as an unfair government subsidy for the purposes of a regional trade agreement or, more generally, WTO rules on subsidies and countervailing measures, allocations of emissions trading systems might be challenged in trade law. Or, parallel to a domestic ETS, a regulator may provide incentives for firms to reduce carbon emissions or adopt new technologies. Such measures could be characterized as an inappropriate border measure, likely to be inconsistent with WTO rules.[67] For instance, a blanket exemption from tax payments has been judged to be a subsidy.[68] However, in many cases, trade rules are structured to accommodate such situations. Most trade rules on subsidies, including in the WTO, initially provided for 'windows' or reservations for environmental measures, especially for subsidies meant to encourage the adoption of new technologies. Of course, as trade liberalization

[67] J Frankel, 'Environmental Effects of International Trade' Harvard Kennedy School Working Paper Series (2009) 39; online: <http://web.hks.harvard.edu/publications/getfile.aspx?id=335>, accessed 15 June 2009 (with thanks to Christina Voigt for drawing this article to the authors' attention).

[68] See especially, *US-Canada WTO Corn Trade Dispute* WTO Doc WT/DS357/11; discontinued.

continues and rules are refined through dispute settlement, such windows might become more limited. In WTO negotiations, some interests have proposed to set limits on the 'green box' subsides, the WTO-recognized category of subsidies which are permitted due to their environmental objectives, so that subsidies authorized in one State may not be recognized as legitimate by others.[69] In any case, many carbon reduction subsidies could still conform due to their lack of significant trade impact. In the case of ETS permit allocations, such a trade impact might be assessed by a comparison with any previous, less effective rules. And in most instances, GHGs caps place an additional burden on the company, as they find themselves generally at a disadvantage *vis-à-vis* non-regulated competitors. Viewed in this light, it would be difficult to challenge an ETS using trade rules on subsidies. Furthermore, trade and investment issues affect the political feasibility of new laws and policies to address climate change. Were a State to attempt to introduce a carbon tax for a carbon-intensive project, this could jeopardize the international competitiveness of its domestic companies.[70] To address these concerns, States may seek to implement border tax adjustments (BTAs).[71] The use of BTAs has been proposed as a solution to the potential distortions created by an ETS emission credit requirement:

> For legal purposes ... border tax adjustments ... amount to two different measures which follow a distinct regime: The first measure, refunds for exports, has to stand the test whether it constitutes an outlawed subsidy. The second measure, taxes charged on imports, has to fend off the suspicion that it represents an illegal discrimination.[72]

As suggested by Joost Pauwelyn, a State seeking to implement carbon trading provisions could utilize BTAs to ensure continued competitiveness, and to avoid challenges of discrimination, the proponents could argue that imports are being required to pay a carbon tax at the border to equalize competition between actors, where 'the tax is then simply the extension to imported products of the tax or cost of holding emission allowances imposed on domestic producers'.[73] The opposing argument would highlight that, in the context of emissions trading, the allowances (which are levied on imported products to mirror their

[69] Canadian Federation of Agriculture, Trade Policy Statement; online: <http://www.cfa-fca.ca/pages/index.php?main_id=61>, accessed 31 March 2009.

[70] It is also something which may raise questions of the 'regressiveness' of any carbon tax, meaning that already poorer actors from developing countries would be penalized in the short term; though question whether the long term impact on the poor in the event of no carbon taxing would not be worse. See Stephen Tindale and Chris Hewitt, Must the Poor Pay More? Sustainable Development, Social Justice and Environmental Taxation' in Andrew Dobson (ed), *Fairness and Futurity: Essays on Environmental Sustainability and Social Justice* (OUP: Oxford, 1999).

[71] Report of the Working Party on Border Tax Adjustments, BISD 18S/97, 2 December 1970.

[72] R Ismer and K Neuhoff, 'Border Tax Adjustments: A feasible way to address nonparticipation in emissions trading' Cambridge Working Paper Series CWPE (Cambridge University Press: Cambridge, 2004) at 9.

[73] Joost Pauwelyn 'U.S. Federal Climate Policy and Competitiveness Concerns: The Limits and Options of International Trade Law' Working Paper (Duke University Press: New York, 2007) at 41.

carbon costs of production in a 'non-regulated' State) are often allocated free of charge to domestic actors, raising claims of 'national treatment' violations.[74] Furthermore, it is unclear whether such BTAs would avoid challenges where the tax concerned an input such as energy, which is fully consumed and not present in the final product itself. The *US – Superfund* dispute offers some guidance,[75] where the WTO Panel permitted BTAs for chemicals used during production—though these chemicals were also still present in the final product.[76] Just as an ETS could be seen in subsidies terms as a tax, an ETS could be characterized as having the effect of a tax, permitting equalizations. A scheme characterized as a unilateral 'carrot or stick' BTA could be a promising avenue for emissions trading regimes within the framework of global trade rules.[77] It would be important to calculate the ETS equivalent BTA conservatively, and to be prepared to address challenges in trade or investment tribunals.[78]

In another example, as mentioned above, the WTO's Agreement on Technical Barriers to Trade (TBT) seeks to ensure that technical regulations are not 'more trade restrictive than necessary to fulfil a legitimate objective' and do not discriminate between 'like products.' [79] Regulations that take into account process and production measures (PPMs), restricting the entry of goods which have been produced in a carbon-intensive way, could be open to trade law challenges based on non-discrimination and prohibitions on quantitative restrictions. As noted above, the product itself is not different from its competing product—it is only the production method that differs. As such, the products *simplicter* remain 'like' and differing treatment would be caught by national treatment or non-discriminatory provisions. For the ETS, this distinction means that States could have trouble giving significant trade advantages to products produced under the application of a national or even international ETS. For instance, the EU has a firm commitment to promote climate protection internationally,[80] and its scheme allows covered emitters to benefit from

[74] With thanks to Christina Voigt for her input on this argument.

[75] 'United States – Taxes on Petroleum and Certain Imported Substances', Report of the Panel, Doc L/6175 – 34S/136, 1987.

[76] R Tarasofsky, 'Heating Up International Trade Law: Challenges and Opportunities Posed by Efforts to Combat Climate Change' (2008) 1 CCLR 7 at 11.

[77] Z Zhang, 'Multilateral Trade Measures in a Post 2012 Climate Change Regime?: What can be taken from the Montreal Protocol and the WTO?' MPRA Paper No 12782; online: <www.mpra.ub.uni-muenchen.de/12782>, accessed 31 March 2009.

[78] *Ethyl Corporation v Canada*, Award on Jurisdiction, 24 June 1998, (1999) 38 ILM 708.

[79] Agreement on Technical Barriers to Trade (adopted 15 April 1994, entered into force 1 January 1998) 1186 UNTS 276, Art 2.2.

[80] This will become stronger after the Lisbon Treaty enters into force as it explicitly commits the Union to in the new Art 191 (ex 174) of the Treaty on the Functioning of the European Union '1. Union policy on the environment shall contribute to pursuit of the following objectives: [...] — promoting measures at international level to deal with regional or worldwide environmental problems, and in particular combating climate change.'

CDM and Joint Implementation (JI) credits, though only up to a specified limit.[81] If standards were perceived as being based on PPMs and appeared to discriminate between products from different countries, any preferential treatment in terms of tariffs for those products could be challenged in the WTO and other regional trade dispute settlement fora.[82] However, few trade rules prevent general use of labels or certification schemes. Such 'eco-labelling' allows the consumer to know that certain goods were produced in a more environmentally friendly (or, at least, less environmentally harmful) manner than the competing product.[83] As noted by Simon Baughen:

> Caution as regards PPMs is perhaps understandable, in that they can be seen as one [WTO] member's attempt to impose its environmental standards on other members. However, the issue of PPMs may, in future, come up in the rather different context of transboundary spill-overs, where the objection to the way in which a product is manufactured is based on adverse environmental consequences felt in the member state imposing the measure. This could well occur in the context of the contribution to global warming made by the carbon emissions produced from a particular mode of production adopted by a member.[84]

Taking this proposition one step further, in emissions trading schemes, the practice of 'climate compliant' self-labelling could in theory fall within prohibition on PPM-based measures, should a State use such voluntary declarations or self-labels to assign legal consequences. However, where the impact on sustainable development and the environment is 'transboundary' in nature, such as climate change and carbon emissions, then the measure could no longer be seen as extra-territorial but rather becomes one which WTO Panels, in the *US – Shrimp* dispute and others, have recognized as being within the competence of States.[85] Emissions trading could be characterized as addressing such 'transboundary' issues.

Sustainable development has been recognized as an objective of world trade law under the WTO, both by the Panel and Appellate Body in the *US – Shrimp* and subsequent disputes, and in the 2001 Doha Declaration.[86] Two further world trade law arguments could, as such, be used generally by regulators and trade dispute settlement bodies to address any such areas of overlap.

[81] S Gardner, *EU ETS: The Winners and Losers of EU Emissions Trading* (Climate Change Corp: London, 2008). As a group, ETS participants are allowed to buy up to 1.4 billion CDM credits during the 2008 – 12 trading phase.

[82] A Cosbey, *Environment and Trade: A Handbook* (UNEP/IISD: Winnipeg, 2000).

[83] An interesting problem would be whether a carbon market 'seller'—habitually selling credits rather than purchasing them-could voluntarily eco-label itself or its product.

[84] See eg Simon Baughen, *International Trade and the Protection of the Environment* (Cavendish: New York 2007) 4.

[85] WTO, *United States: Import Prohibition of Certain Shrimp and Shrimp Product Report of the Appellate Body* (12 October 1998) Doc. WT/DS58/AB/R.

[86] Ibid, Doha Ministerial Declaration (14 November 2001) WT/MIN(01)/DEC/1 <http://www.wto.org/english/thewto_e/minist_e/min01_e/mindecl_e.htm>, accessed 31 March 2009, M Gehring and MC Cordonier Segger (eds), *Sustainable Development in World Trade Law* (Kluwer Law International: The Hague, 2005) at Introduction.

First, trade dispute settlement bodies may carefully consider the definition of 'like products' when assessing the legality of measures designed to combat carbon emission consequences. It is possible that States would be permitted, under trade law, to take a product's GHG emissions into account in determining its 'likeness' with another product.[87] The *EC – Asbestos* dispute remains indicative of the current stance on discrimination of like products, demonstrating that in certain instances, such as when a carcinogen is being substituted in a marketplace with potentially less carcinogenic alternatives, the WTO Appellate Body will take minute physical differences into consideration, shifting the burden of proof onto the challenger to demonstrate that their goods are indeed 'like' a less harmful substitute.[88]

In the alternative, any measures might still be justified under GATT XX(g) exceptions for measures to conserve exhaustible natural resources, or GATT XX (b) exceptions for measures necessary to protect human, plant or animal life and health. The Panel in the *US – Shrimp* dispute explicitly reserved the right for environmental measures to be excused from WTO obligations through Article XX(g), provided that similar products from other States were not given preferential treatment through special side agreements. It would be hard to describe the Kyoto Protocol, as an international agreement open to all WTO Member States, as setting discriminatory or exclusive standards.[89] To prove that a measure is 'necessary' to protect health or the environment, as noted by the Panel in the *Brazil – Retreaded Tyres* dispute, it may be sufficient to demonstrate, on the balance of qualitative evidence, that it is *likely* to contribute to achieving the legitimate health or environmental objectives.[90]

3.3.2 Investment law implications of carbon trading

While emissions trading can raise certain specific trade issues depending on the design of the regime, the most significant challenges are likely presented by recent developments in international investment law. Of course, international investment agreements (IIAs) could potentially promote sustainable development, by ensuring more stable investment climates and thus encouraging investors to provide private funds for CDM projects and for JI. However, in these accords, States have also agreed on disciplines that have been used recently to challenge regulatory measures related to sustainable development.[91] As noted

[87] Tarasofsky (n 76 above) at 7; Miles (n 50 above).

[88] M C Cordonier Segger and M Gehring, 'Precaution, Health and the World Trade Organization: Moving Toward Sustainable Development' (2003) 29 Queens Law Journal 133.

[89] Committee on Trade and Environment, Committee on Trade and Environment Special Session, *Matrix on Trade Measures Pursuant to Selected Multilateral Environmental Agreements* (14 March 2007) WT/CTE/W/160/Rev.4.

[90] *Brazil – Retreaded Tyres* (n 40 above)

[91] Miles (n 50 above) at 26. See also M Gehring, MC Cordonier Segger and A Newcombe, *Sustainable Development in World Investment Law* (Kluwer Law International: The Hague, 2009) (forthcoming).

above, in IIAs, Parties often grant foreign investors the right to challenge host States in investor-State arbitral tribunals under UNCITRAL or ICSID rules, particularly on claims related to performance requirements, fair and reasonable treatment, expropriation, and transparency. These privileges may be used to challenge carbon trading measures, depending on how new domestic schemes are designed, and how the IIAs are interpreted.[92] As with trade law, these challenges have both direct effects, where a State is asked to compensate an investor or group of investors for the economic impact of new carbon regulations, or indirect regulatory effects, where environment and development regulators are 'chilled' from adopting or implementing carbon reduction measures due to threats of investor-State litigation. Several examples can be provided to illustrate these implications.

First, on an almost theoretical level, emissions trading schemes may not always be classified as pure market-based instruments,[93] absent of 'command and control' origins. The very existence of a government-imposed cap on the amount of carbon that can be emitted by a given sector is evidently 'control' in nature. That cap establishes a performance requirement, but allows for the market to set the price of carbon emissions and for firms to choose abatement technologies to meet the standard. However, even the introduction of new performance standards could pose questions under certain investment treaties. In such treaties, the US and Canada have sought to prevent or constrain the use of performance requirements or standards that were once popular in developing countries, as a way to enhance the value of an investment by mandating a certain way of producing a product such as sourcing local services, labour or content (local content requirement), or earning foreign exchange through export requirements.

Second, the scope and nature of the emissions trading rules are important. The design and legal consequences of ETS implementation potentially raises claims of 'indirect expropriation'.[94] States choose whether their scheme will be limited to a particular sector or economy-wide in application, and which jurisdictions will be subjected to (or allowed entry into) the scheme. Emitters targeted by such systems can include both direct emitters of carbon, such as power plants or even car owners, and also those 'further upstream' in the chain of carbon use, such as oil companies or petroleum refiners. For instance, the European Union (EU) ETS covers power and industry sectors only,[95] and focuses

[92] J Werksman, KA Baumert and DK Navroz, 'Will International Investment Rules Obstruct Climate Protection Policies' Climate Notes (World Resources Institute: Washington DC, 2001).

[93] Robinson et al (n 59 above) at 44.

[94] S Huq and H Reid, 'Benefit Sharing under the Clean Development Mechanism' in D Freestone and C Streck (eds), *Legal Aspects of Implementing the Kyoto Protocol Mechanisms: Making Kyoto Work* (OUP: Oxford, 2005) at 229.

[95] Although extensions to the scheme are continually under review. Most recently, the Aviation industry is intended to be subjected to the system.

simply on addressing CO_2 emissions.[96] Depending on the data collected and economic impact assessments, States decide whether their schemes will be comprehensive, or simply sectoral, and whether partial coverage can achieve their objectives. Such choices may affect the competitiveness of national companies against each other and against foreign companies. If investments in foreign investor dominated sectors were seen as being unjustly targeted by stringent and costly requirements 'tantamount to expropriation' while other domestic-investor dominated heavy emitters were excluded from the scheme, challenges might be issued under IIAs. If a regulator places a cap on the use of carbon in some sectors and not others, there is the potential for such measures to be characterized as indirect expropriation of that company's investment. Similarly, if the cost of carbon certificates becomes high enough to threaten the economic viability of certain investments (for instance fossil fuel exploration and development, or a coal-fired power plant), the carbon measure could also be deemed 'tantamount to expropriation'. The core debate focuses on who bears the risks of private investments into 'high carbon' sectors—host governments or investors? Essentially:

> ... if a government measure is undertaken for a clear public welfare purpose (such as health and safety, environment, public morals or order, etc.), and is non-discriminatory, but has the effect of harming a ... foreign investor, under what circumstances can that measure be held to be an indirect expropriation, for which the government must pay compensation?[97]

As demonstrated in the *Ethyl v Canada Case*, [98] claims of indirect expropriation can be made when new government measures affect the value of a foreign investment in a specific or unique industry. Such issues could arise for governments implementing climate change measures, as in some countries, carbon-intensive industries can be dominated by multinational extractive enterprises (foreign investors) with the necessary know-how and capital for exploration and development. Moreover, the resulting repeal by Canada of the MMT ban clearly demonstrated the indirect effect of a foreign investor challenge on government policy directions.[99] Indeed, if a developed State such as Canada could be perceived to have 'chilled' its regulatory decisions due to international investment law obligations, IIAs, and expropriation challenges, it seems possible that a developing country might face even higher pressure to avoid necessary regulatory changes. Whether or not the developing country could in fact afford to

[96] AD Ellerman and P Joskow, 'The European Union's Emission Trading System in Perspective' (Pew Center: Arlington, 2008).

[97] A Cosbey, 'NAFTA's Chapter 11 and the Environment', Discussion Paper for CEC Public Workshop, Mexico City, 24 March 2003, at 3.

[98] *Ethyl* (n 78 above).

[99] A Newcombe, 'Regulatory Expropriation, Investment Protection and International Law: When is Government Regulation Expropriatory and When Should Compensation be Paid?' (LLM Thesis, University of Toronto 1999) (unpublished); online: <http://ita.law.uvic.ca/documents/regulatoryexpropriation.pdf>, accessed 31 March 2009.

compensate for the expropriation is a particularly pressing issue in the case of climate change measures, including ETSs.

Third, the actual legal nature of an allowance has been flagged as a concern in the design of various US emissions trading systems. For instance, the SO_2 trading system under the Acid Rain program of the 1990 Clean Air Act raised the issue of property rights. The possibility that an allowance would constitute a property right potentially raised arguments in the US, based on 'taking of property' under the Fifth Amendment of the US Constitution. This was resolved by classifying the allowances as tradable goods which were, nevertheless, not property rights as such.[100] However, legal concerns remain that contractual or property rights might be subject to claims by those holding the credits, in the situation where regulations are introduced which alter the value of the allowance or credit in question.[101] The decision to imbue allowances with property-like status could potentially open governments to allegations of expropriation under investment treaties, should the value or quantity of these allowances be reduced in the future. The potential for such an approach to conflict with international investment law is evident, and could lead to investor-state disputes. Having said this, the 'quasi-property rights' character of emission certificates is now widely recognized and seen as a necessary condition of many emission trading schemes.

Fourth, it is important to consider how allowances or emission reduction credits (ERCs) are allocated amongst the participants in an emissions trading scheme. The allocation of allowances by the government to the actors—whether these are particular industry-specific actors, or 'carbon-intensive parties' or any other pre-determined set of actors—can be problematic in trade and investment terms. Allocations indicate the degree to which carbon can be emitted, within a system, and thereby 'pre-determines the overall environmental benefits that can be expected from the system'.[102] Allocation, ie assigning certain certificates, is an intensely political process, and compromises are often necessary. Both trade and investment concerns can be triggered by allocations of credits which are discriminatory, or not 'fair and reasonable'. Not only could this process raise concerns for the competitiveness of firms and operators within the domestic and international markets, but it also raises concerns as to discrimination toward non-national actors which compete in the targeted market. Among various options open to designers of ETSs, free allocation, allocation based on 'grandfathered rights', allocation based on more modern 'baseline', and partial

[100] Gehring and Streck (n 65 above).

[101] See especially, N Fichthorn and A Wood, 'Preserving the SO_2 Market' Environmental Finance (September 2002); online: <www.emissions.org/publications/member_articles/ef9ema2002.pdf>, accessed 31 March 2009.

[102] M Wemaere and C Streck, 'Legal Ownership and Nature of Kyoto Units and EU Allowances' in David Freestone and C Streck (eds), *Legal Aspects of Implementing the Kyoto Protocol Mechanisms: Making Kyoto Work* (OUP: Oxford, 2005) at 41.

auctioning pose similar problems. If any non-national actors within the territory do not receive precisely the same allowances as comparable national actors, these firms can argue that they have been prejudiced in the market if they incur higher costs to reduce their carbon output or find the resources to pay for their continued output. This could be held to violate trade obligations of non-discrimination and national treatment, but more importantly, it could also be characterized as going beyond the 'fair and equitable treatment' standard promised to foreign investors in most IIAs. For instance, in light of the EU's design choices, windfalls may have been received by certain Parties through the free allocation process.[103] In another example, the New Zealand Climate Change Response (Emissions Trading) Amendment Act 2008, which came into force in September 2008, will gradually incorporate sectors of the New Zealand economy until 2013, and permits some allocations free of charge.[104] Still, as the proposed scheme is designed to be much broader in scope than for example the EU ETS or RGGI, distortions may be less relevant. The further option of 100% auctioning resolves many such concerns, though this can still entail competitive consequences depending on the frequency, size, and accessibility of auctions, should it be shown that *in effect,* regulations made participation more challenging for foreign firms. This problem is difficult but not impossible to address. For instance, many ETS regulations have incorporated provisions specifically ensuring no distinction between national or foreign-owned companies.

Another basic design element that triggers trade and investment issues involves the commitment to regulatory transparency, which may well support the designers of ETSs. It has been argued that emissions trading 'may also be more transparent and accessible than traditional command and control schemes: anyone wishing to challenge the environmental effectiveness of the trading regime can question directly the level of the overall cap rather than having to unravel the, often complex, relationship between specific controls applied to an individual plant and an ambient environmental quality standard'.[105] The transparency of domestic law and policy-making process is important to any potential investor.[106] However, as mentioned above, investors can seek investment treaty guarantees against changes in government policy (not just fiscal or

[103] Ellerman and Joskow (n 96 above).

[104] Climate Change Response (Emissions Trading) Amendment Act 2008, 25 September at s 73. See Emissions Trading Scheme for New Zealand, Report to Business New Zealand (NZIER Publishing: Wellington, 2007).

[105] Robinson et al (n 59 above) 45. See also, RA Stewart, 'Economic Incentives for Environmental Protection: Opportunities and Obstacles' in RL Revesz, P Sands and RA Stewart (eds), *Environmental Law, the Economy and Sustainable Development* (Cambridge University Press: Cambridge, 2000).

[106] The Preamble to the Aarhus Convention expressly calls for 'transparency in all branches of government' when implementing provisions related to the Aarhus Convention's aims: Convention on Access to Information, Public Participation in Decision-Making and Access to Justice in Environmental Matters, adopted 25 June 1998, entered into force 30 October 2001) 38 ILM 517.

tax policy, but also environment and development policy),[107] hoping to stabilize regulations for the lifetime of an investment.[108] A regulator may need to make it clear to potential investors that post-establishment decisions, ie governmental decisions influencing the investment after it has been made, will take State commitments under the UNFCCC and the Kyoto Protocol into account. Moreover, transparency works in many directions. As discussed elsewhere, impact assessments and other such requirements can contribute to stability rather than detracting from it, by generating valuable investment intelligence and creating a more level playing field for investors.[109] To that end, by securing transparent policy decisions, States might insulate their new climate policies from formal investor-State challenges, while also contributing to their clarity and legitimacy.[110]

Each of these aspects can influence the effectiveness of an emissions trading scheme. Additionally, these basic design questions are accompanied by myriad technical and practical rules, and trade and investment rules apply to the *de facto* effect of measures, not just their letter or intention. These include, but are not limited to, measurement and monitoring of emissions, calculation of individual baselines, and enforcement of sanctions for non-compliance. For instance, in addition to possible civil or criminal pollution-related penalties, the NOx trading scheme requires the surrendering of three allowances for every tonne not accounted for, for the next trading period. But many investment accords seek to prevent or constrain the use of performance requirements or standards that were once popular in developing countries as a way to enhance the value of an investment by mandating a certain way of producing a product such as sourcing local services, labour or content (local content requirement), or earning foreign exchange through export requirements.

A further concern was also mentioned above. Certain IIAs contain clauses agreeing to freeze the laws of country to the time the investment was made, or not to apply new laws to the investment, or to bear the costs of all regulatory changes affecting an investment: so-called stabilization clauses. These commitments could be problematic from the standpoint of ETS regulations. The principle difficulty posed is that States are bound to continue treating the investment in a certain way which may become no longer viable in light of the UNFCCC objectives, and the developing scientific discoveries that have driven the evolution of the climate regimes.

[107] Pauwelyn (n 73 above); A Ross-Robertson, 'Is the Environment Getting Squeezed out of Sustainable Development?' [2003] Public Law 249.

[108] Baughen (n 84 above) Chapter 7.

[109] M Gehring, MC Cordonier Segger and A Newcombe (eds) *Sustainable Development in World Investment Law* (Kluwer Law Int.: The Hague, 2009) (forthcoming).

[110] Tarasofsky (n 76 above).

As seen in the *Methanex v US* dispute,[111] not all investor-State tribunals are willing to interpret their jurisdiction over regulators so broadly. While some IIAs can be used to question environmental protection and sustainable development regulations, others clearly cannot be extended so far.[112] Still, where States enact measures, including ETSs, to favour low-carbon development over carbon-intensive projects, especially in developing countries where the extractives sectors are dominated by a few foreign investors, these interests could well try to frame investor-State challenges in terms of performance requirements, discrimination, fair and equitable treatment, or expropriation. Where such allegations are raised, it will be important to secure an appropriate interpretation of the concept of 'like circumstances'. It is only where two firms are in 'like circumstances' and receive different treatment that tribunals could find discrimination. For future climate change measures, the inclusion of public interests and carbon emissions in one proposed project as compared to another which does not take such issues into account might serve to distinguish hitherto 'like' parties from one another. The recent ICSID arbitration between *Parkerings* and Lithuania demonstrates this principle.[113] The case concerned two competing firms from Norway and the Netherlands in their tenders to construct and operate traffic facilities in the Lithuanian capital Vilnius. The Norwegian company (FET) argued discriminatory treatment in favour of the Dutch firm and the case turned on the concept of 'likeness.' The Tribunal ultimately found that the State could take into account such public policy issues when deciding how to treat different projects: 'Underscoring that each State has an undeniable right to exercise its sovereign legislative power—albeit in a reasonable and fair manner—and that an investor must anticipate a possible change of circumstances, and thus structure its investment in order to adapt it to the new legal environment (particularly of a country in transition), the tribunal rejected the FET claim, concluding that the Republic of Lithuania had not given any explicit or implicit promise that the legal framework of the investment would remain unchanged'.[114] The tribunal's approach has been welcomed in leading legal scholarship on these issues:

This decision points to the ecological impact of an investor's project as a determinative factor in the like circumstances test. If this approach is followed in future investor-state disputes, then the potential for non-discrimination requirements in international

[111] *Methanex v USA*, Award on Jurisdiction, 28 August 2002.

[112] J Lawrence 'Chicken Little Revisited: NAFTA Regulatory Expropriations After *Methanex*' (2006) 41 Georgia Law Review 261.

[113] *Parkerings-Compagniet AS v Republic of Lithuania*, ICSID Arbitration Case No ARB/05/8, Award, 11 September 2007, para 375, 392 at 2 October 2007; online: <http://ita.law.uvic.ca/documents/pakerings.pdf>, accessed 15 June 2009.

[114] UNCTAD, Latest Developments in Investor-State Dispute Settlements, (UNCTAD: Geneva, 2008) at 5; online: <www.unctad.org/en/docs/iteiia20083_en.pdf>.

investment agreements to frustrate climate change mitigation regulation will be significantly reduced.[115]

Learning from this experience, negotiators may need to recognize the importance of maintaining flexibility for climate change measures in investment treaties, while regulators take care to design the rules for carbon trading, and clean technology investments, to avoid discrimination between industries in 'like circumstances.' This approach can extend to implementation of emissions trading systems. For instance, in s 60 of the New Zealand Climate Change Response (Emissions Trading) Amendment Act 2008, a State authority can exempt otherwise regulated participants under the Act from complying with the emissions trading provisions. For firms with which the Crown signed a negotiated greenhouse gas agreement before 31 December 2005, such an exception may be granted, providing both stability for existing agreements and flexibility for government authorities.[116] The flexibility provided by the New Zealand Act may be beneficial when addressing expropriation or other investment related claims.

Further potential refinements to ensure that trade and investment laws, do not unduly constrain domestic regulatory flexibility to address climate change, and might even promote more sustainable development, are also being designed and tested particularly in regional treaties which advance beyond the globally agreed disciplines.

4. Improving Trade and Investment Regimes for Sustainable Development

The obligations of States under international trade and investment law might intersect with certain elements of climate change regulations, requiring careful work to design compatible measures to establish emissions trading schemes, and may potentially lead to constraints on policy and law-making flexibility.[117] However, as noted in the 1987 Report of the World Commission on Environment and Development,[118] the 1992 Rio Declaration and Agenda 21,[119] and the 2002 Johannesburg Declaration and Plan of Implementation,[120] trade and investment could also provide important contributions to sustainable development. Just as in the international climate regime, in many trade agreements and international investment agreements, Parties explicitly highlight their shared commitment to sustainable development as part of the object or purpose of the treaty. For instance, the North American Free Trade Agreement (NAFTA)

[115] Miles (n 50 above) at 32.

[116] NZ Climate Change Response (Emissions Trading) Amendment Act 2008; Richard Benwell, correspondence with authors, with thanks.

[117] See eg Miles (n 50 above).

[118] World Commission on Environment and Development, *Our Common Future* (OUP: Oxford, 1987).

[119] Rio Declaration on Environment and Development (1992) 31 ILM 874; Agenda 21 (n 34 above).

[120] JPOI (n 34 above).

includes a reference to the need to 'promote sustainable development' within its Preamble.[121] Both the Canada–Chile Free Trade Agreement and Chile–US Free Trade Agreement also recognize the importance of strengthening capacity to protect the environment and promote sustainable development.[122] The Canada–Peru Free Trade Agreement makes explicit reference, in the chapter entitled 'Investment', to corporate social responsibility and the need for Parties to encourage enterprises to incorporate such standards into their internal policies.[123] The EU–Chile Association Agreement goes further, committing to implement their accord in accordance with the 'principle of sustainable development',[124] and EU economic negotiations with Central America seek to 'harness globalisation in support of sustainable development' and 'ensure an appropriate balance between economic, social and environmental components in a sustainable development context'.[125] Sustainable development is a key objective of the world community, not only in the abstract, but in the very arena that has most sought to encourage economic growth—investment and trade policy and law.

In regional trade and investment agreements, States have gone further in certain instances, seeking to promote sustainable development through the inclusion of innovative yet practical international instruments. Several preventive provisions, cooperative mechanisms and new trade liberalization enhancement initiatives can be identified. There are also important procedural innovations which can support sustainable development. Many legal options are available to States seeking to deliver on a commitment to sustainable development in a regional trade and investment regime, either as a principle or an objective.[126]

First, States can include introductory provisions which signal the Parties' commitment to sustainable development, such as preambular commitments to 'promote sustainable development' as a 'joint resolution' of the Parties to the accord, or other initial provisions which commit to engage in the accord in accordance with a 'principle of sustainable development'.

Second, States can include provisions which create 'windows' or exemptions from trade rules, where trade obligations might otherwise constrain regulators

[121] North American Free Trade Agreement, 32 ILM 289 (1993).

[122] Canada – Chile Free Trade Agreement, 36 ILM 1079; US – Chile Free Trade Agreement, 42 ILM 1026.

[123] Canada – Peru FTA, signed 29 May 2008, Chapter 8 'Investment', Article 810. See also Angela DelfinoAngela Delfino et al, *Corporate Social Responsibility and Climate Change, in Climate Change: A Guide to Carbon Law and Practice* (Globe Business Publishing: London, 2008) at 157.

[124] EU – Chile Association Agreement, 30.12.2002; online: <http://ec.europa.eu/trade/issues/bilateral/countries/chile/euchlagr_en.htm>, accessed 15 June 2009.

[125] Draft EU – Central America Negotiating Directive (2007) at para(s) 3.4 and 3.7; the States involved are Costa Rica, El Salvador, Guatemala, Honduras, Nicaragua, Panama; online: <www.bilaterals.org/article.php3?id_article=8336>, accessed 31 March 2009.

[126] M Gehring and MC Cordonier Segger (eds), *Sustainable Development in World Trade Law* (Kluwer Law Int: The Hague, 2005) at Introduction. See also MC Cordonier Segger, 'Sustainable Development in Regional Trade Agreements,' in *Regional Trade Agreements and the WTO Legal System*, L Bartels & Fortino (eds.) (OUP: Oxford, 2006) at 319–339.

and policy-makers, mitigating their effects. For instance, in trade and investment agreements, many States adopt general exceptions for measures related to the conservation of exhaustible living and non-living natural resources, and the use of measures, including environmental measures, necessary to protect human, animal, or plant life or health.

States may further adopt specific exceptions in sections of the trade and investment treaty where it is clear that trade rules on *inter alia* sanitary and phytosanitary standards, technical barriers to trade, intellectual property rights, public procurement, services, or investment, might constrain the use of environmental and social measures. States may insert explicit reservations by the Parties of socially or environmentally sensitive sectors (such as parks, land use planning, energy policy, and other natural resources reserved from investment provisions, or health and education sectors from services disciplines), often linking these reservations to the findings of sustainability impact assessments or environmental assessments of the trade agreements. States can also include general interpretive statements to guide potential areas where trade rules could otherwise constrain the use of measures agreed in other international (or regional) agreements.

Third, States can negotiate mechanisms for 'value-added' but parallel (non-integrated) social and environmental cooperation strategies. These may include parallel agreements (or chapters, or both) for cooperation on environmental and social matters; the development of institutions for social and environmental cooperation including carbon trading; the agreement to adopt and implement common work programmes on specific environmental or social projects such as emissions monitoring and registration, particularly when accompanied by reliable capacity-building, technology transfer and financing commitments; and even factual report or complaints mechanisms to provide recourse when it appears that environmental or social rules are being violated in order to gain trade or investment related advantages.

Fourth, States many include constructive 'sustainable development' oriented trade and investment rule enhancement initiatives, where a positive 'triple-win' might be achieved within the trade agreement. These may include, for instance, sanitary and phytosanitary provisions which promote scientific cooperation and risk assessment to improve levels of health or environment protection; government procurement provisions which make public purchasing of low carbon goods or services more affordable; technical barriers to trade provisions to implement non-discriminatory certification processes and promote mutual recognition; intellectual property rights provisions which support low carbon technology transfer or respect for traditional knowledge; investment provisions which privilege socially responsible corporations and low-carbon investments; measures to promote reductions in illegal trade in forestry products; measures to secure additional liberalization of environmental goods and services including waste management or low-carbon transportation; or measures to secure reductions in unsustainable fossil fuel development subsidies.

Finally, certain procedural innovations can be undertaken by the Parties during the trade negotiations to promote sustainable development, and secure the integration of environmental and social concerns into a trade and investment treaty. Such process changes may also assist Parties and others in identifying useful innovations that might be included in a trade or investment agreement. For instance, States may undertake *ex-ante* (or ongoing) environment, development, human rights or sustainability impact assessments and reviews of trade liberalization policies and draft treaties. The outcomes of these assessments, as mentioned above, may be used to identify the areas where preventive, cooperative or enhancement initiatives could be useful in a trade or investment treaty. States may also host consultations between economic, environment and development authorities. They may agree upon, or strengthen, diverse mechanisms to ensure transparency and public participation in trade negotiations, and they may also establish new mechanisms to inform tribunals about sustainable, low-carbon development issues, including amicus curiae, public participation and expert consultation measures.

It is not yet clear which strategies or instruments will have the most success in helping to integrate social and economic development and climate change policy. It is likely that no one single measure provides the solution to all sustainable development challenges. Rather, many different provisions may be needed throughout the treaty. Certain instruments, such as the normative or regulatory evaluation elements of *ex-ante* sustainability impact assessments, are still underdeveloped. Others, such as the new system of certification to ensure that forestry products traded from Peru to the United States are not obtained through illegal logging which might provide a future model for low-carbon certification in some sectors, are simply very new.[127] Such provisions alone will not necessarily ensure that emission reduction priorities are given more weight by the Parties in complying with their obligations, or by dispute settlement bodies in interpreting Agreements, as compared to the other relevant objectives of Agreements. However, they appear likely to contribute to the achievement of a greater degree of integration in the trade agreements. This is an important first step toward sustainable development.

5. Technology Transfer Towards Low-Carbon Economies

For sustainable development, the need for 'mutually supportive' trade, investment, and carbon trading rules extends beyond reductions in potential overlaps or conflicts between regimes. The development and transfer of new technologies and practices to reduce greenhouse gas emissions is strongly encouraged in

[127] United States – Peru Trade Promotion Agreement, signed 12 April 2006 <http://www.ustr.gov/Trade_Agreements/Bilateral/Peru_TPA/Final_Texts/Section_Index.html>, accessed 31 March 2009; ICTSD 'US – Peru Bilateral to Address Illegal Logging, Boost MEA Implementation' Bridges Trade BioRes, 22 June 2007 <http://ictsd.net/i/news/biores/9122>, accessed 31 March 2009.

the UNFCCC and the Kyoto Protocol. In Article 4 of the UNFCCC, headed 'Commitments', Parties agree to 'promote and cooperate in the development, application and diffusion, including transfer, of technologies, practices and processes that control, reduce or prevent anthropogenic emissions of greenhouse gases not controlled by the Montreal Protocol in all relevant sectors, including the energy, transport, industry, agriculture, forestry and waste management sectors'.[128] In the Kyoto Protocol, similarly, Parties commit to ensure the transfer of technology to developing country Parties when implementing climate change measures through 'research on, and promotion, development and increased use of new and renewable forms of energy, of carbon dioxide sequestration technologies and of advanced and innovative environmentally sound technologies'.[129] At the 8th session of the Conference of the Parties to the Convention, Ministers issued a Declaration calling for technology transfer to be strengthened, including 'concrete projects and capacity-building in all relevant sectors such as energy, transport, industry, health, agriculture, biodiversity, forestry and waste management.' This Declaration further recognized that 'technological advances should be promoted through research and development, economic diversification and strengthening of relevant regional, national and local institutions for sustainable development'.[130]

Future carbon trading measures may need to consider technology transfer needs, within the context of trade and investment rules. It is important to consider how emission trading schemes can create the economic incentives to advance development *and transfer* of more sustainable technologies, especially from developed to developing countries as prioritized in the Convention and the Protocol. Emissions trading schemes provide incentives for firms to develop and commercialize new GHG reduction or elimination technologies. But can these technologies, through trade and investment links, also be transferred to developing country Parties to the UNFCCC regime? Do new trade and investment rules provide constraints or opportunities for national, regional, and global efforts to achieve these goals?

At present, a great deal of technology transfer still takes place through official development assistance (ODA). ODA remains critical for developing country Parties 'as a source of support for sector reforms necessary to attract capital to energy markets, and also help commercialize investments in environmentally promising but initially higher-cost technologies'.[131] However, the degree of significant technology transfer has been far from sufficient to achieve the objectives enshrined in the UNFCCC.[132] The contribution of trade and

[128] UNFCCC, Art 4(c). [129] Kyoto Protocol, Arts 2(1)(a)(iv).

[130] Delhi Declaration on Climate Change and Sustainable Development, Decision 1/CP.8 para (i) (1 November 2002).

[131] Alan S Miller, 'International Trade and Development' in *Global Climate Change and U.S. Law*, Michael B Gerrard (ed) (American Bar Association Publishing: Chicago, 2007) at 287.

[132] Expert Group on Technology Transfer; online: <http://unfccc.int/essential_background/convention/convention_bodies/constituted_bodies/items/2581.php>, accessed 31 March 2009.

investment law to this challenge has been mixed. In some ways, it can be positive. For instance, increased stability and predictability in investment flows could be central to ensuring that Clean Development Mechanism (CDM) opportunities are realized and that developing countries can begin to gather the resources necessary to move toward a low-carbon economy. As another example, increased trade liberalization could provide access to new technologies, services, and products. Indeed, Paragraph 31(3) of the Doha Development Agenda calls for liberalization of environmental goods and services,[133] and reductions in trade barriers for environmentally sound technologies might permit more viable economies of scale, or improve access.[134]

However, as some have noted, high licensing fees, protected by trade obligations related to intellectual property rights (IPRs), may also be contributing to the insufficient transfer of technology.[135] Expensive licensing fees can threaten the possibility of effective technology transfer by making the use of new carbon-reduction technologies, carefully stimulated by relevant emissions trading schemes, impossible for developing country investors and policy-makers in the majority of the world to contemplate. For developing countries and developed country investors, the Protocol's CDM provides one valuable means to stimulate investment in clean energy technologies and greenhouse gas mitigation from governments and the private sector as well. It provides opportunities for developed Annex I Parties to satisfy their obligations under the Convention and the Protocol whilst also allowing the host project States to benefit from improved local sustainability standards and the transfer of more sustainable low-carbon technology.[136] The CDM provides much-needed funds to develop and apply new technologies, and to capture carbon emissions, returning certified emission reduction credits to the investor. Technology transfer requirements can be included in CDM emission reduction purchase agreements, ensuring that a balance is found between the needs of developing countries for new technologies and practices, and the needs of investors and others who develop clean energy and low carbon technologies in different sectors. The vast majority of CDM projects, to date, are taking place in China, with India and Brazil also hosting a significant number. Of these projects, renewable energy continues to be the predominant type of project undertaken, although GHG elimination is also prevalent, especially in China.[137] Expanded CDM, including programmatic projects, will be central to technology transfer in the future.[138] Effective transfer

[133] Tarasofsky (n 76 above) at 12.

[134] World Bank, International Trade and Climate Change: Economic, Legal and Institutional Perspectives (World Bank: Washington DC, 2007) at 70.

[135] Miller (n 131 above) at 294.

[136] Chapter 10 below and M Netto and K Barani Schmidt, 'CDM Project Cycle and the Role of the UNFCCC Secretariat' in D Freestone and C Streck (eds), *Legal Aspects of Implementing the Kyoto Protocol Mechanisms: Making Kyoto Work* (OUP: Oxford, 2005) at 192.

[137] Pew Center on Global Climate Change, *Clean Development Mechanism Status Report*, 2008, online: <http://www.pewclimate.org/docuploads/cdm-backgrounder.pdf>, accessed 31 March 2009.

[138] Murphy (n 48 above).

of technology is essential if more coherent trade, investment and climate change cooperation is to achieve global sustainable development objectives.

6. Strengthening International Legal Coherence to Resist Climate Chaos

Current intersections between global and regional rules on trade, investment, and climate change are far from being well understood. How ought countries proceed, especially developing countries, faced with these dilemmas and overlapping legal regimes? On a global scale, there is a need for fast but effective regulatory action in the fields of trade and investment law, as well as in climate law. Not only must carbon reductions be achieved at a much faster pace, but trade and investment in sustainable low-carbon alternatives must also increase. Of equal importance is the improvement of technology transfer to ensure developing countries are actually able to develop more sustainable systems for transportation, energy, housing, agriculture, forestry and other sectors. In this light, it may well be that mechanisms such as effective domestic and international carbon-trading schemes and detailed sectoral agreements will be part of the way forward beyond the terms of the Kyoto Protocol, particularly for developing country Parties where economy-wide participation is impractical.

The UNFCCC and the Kyoto Protocol establish a comprehensive system in which economy-wide emission targets are set, but this does not prevent future more detailed progress in specific sectors (or ETS) to ensure mitigation. If sectoral agreements are undertaken, this could increase the contribution of those States currently unwilling to commit to economy-wide agreements. From a developing countries' perspective, the feasibility of entering into either an isolated or even linked framework of sector by sector agreements is significantly increased. Success in a few well defined and monitored areas could well lead to greater foreign investment in carbon reduction technologies, and could also focus private sector innovation efforts. Also, certain sectors such as steel and aluminium production and distribution have more concentrated global actors than do other sectors such as transport. Therefore, any potential negotiations would be simplified with the interests of these actors most carefully recognized.

But perhaps one the biggest advantages of sector-by-sector progress would be the greater ease in addressing competition concerns of the relevant sector actors. The economy-wide approach of the UNFCCC and the Kyoto Protocol (to some degree reflected in the more localized efforts, such as the EU ETS) opens room for actor favouritism by governments. If regulatory space is then used to introduce a national ETS, for instance with free allocation of certificates, the resulting competitive advantage for one actor over its foreign rivals can lead to real legal questions under trade and investment rules—most particularly non-discrimination / fair and reasonable treatment arguments. 'In contrast, an international sectoral approach could ensure that all global competitors in the given sector undertake mitigation efforts, whether fully comparable or differentiated to reflect equity considerations'.[139] Though this does not yet ensure that distortions are not maintained via the national implementation/assignment of rights/removal of rights, a main advantage of sectoral agreements is that they remove the distortion between current Annex I and non-Annex I countries.

From a developing country perspective, there are a number of options in terms of design and implementation of sectoral agreements. At the basic level, States may create independent sector agreements, or may seek to link sectors and allow trading between them, such as in the EU ETS. The EU ETS is an installation specific scheme, so is the CDM. Sectoral agreements could be government 'owned' and administrated schemes, which would have to be broken down to commitments on the installation level (similar to Annex I commitments which, however, span a whole economy). Unified or harmonized policies could be adopted in a variety of sectors, such as technology-based standards, basic taxes or best-practice standards, between States. These States need not necessarily be identical groups of countries for each participating sector.[140] Which sectors would most appropriately be open to progress in terms of climate benefit depends upon a number of varying factors. For instance, a sectors propensity to decrease its rate contribution toward global emissions from 2012 as 'year 0' could be a principal indicator of suitability for international action. From this standpoint, 'emissions from the transportation sectors are the fastest growing source in most countries and pose perhaps the greatest long-term challenge'.[141] Equally, the economic benefits or options present in a given sector might well prove a useful indicator of its availability for international agreement.

Could emissions trading be an option for international progress *within selected sectors*? The viability of this approach could depend not only on the industry-sector selected but also the substantive content of the agreement. If national allocation of credits were to be distributed by national governments, as is most likely to be the case, then agreed principles for such allocation could be based on international guidelines to create a level economic playing field, and pre-empting potential trade or investment claims relating to discrimination.

The interpretation of international investment treaty law will become central in ensuring that there is coherence with sustainable development policies and thereby avoiding litigation and conflict.[142] To that end, IIAs can include express conflict clauses where matters of sustainable development are at issue. Policies aimed at combating climate change might be also be introduced into the treaties and/or

[139] D Bodansky, 'International Sectoral Agreements in a Post 2012 Climate Framework', Pew Center on Global Climate Change Working Paper (Arlington: Pew Center, 2007) at 6.

[140] Ibid. [141] Ibid at 9.

[142] SE Gaines, 'International Trade, Environmental Protection and Development as a Sustainable Development Triangle' (2002) 11 Rev EC & Int Env Law 259.

parallel sustainable development cooperation mechanisms by States.[143] Running parallel to this, investment tribunals can recognize that protection and/or promotion of sustainable development values within a given project does affect the circumstances surrounding the investor, often placing superficially similar investments in different circumstances. In investment law, judicious interpretive statements would offer guidance to investors and regulators, helping States to understand under what conditions different treatment would not be found to be discriminatory. In that regard, the *Parkering* decision is to be welcomed and it is hoped that that tribunal's reasoning is carried forward in other disputes.

At the same time, certain basic steps can be taken by regulators to minimize the likelihood of challenges based on indirect expropriation and (un)fair and (in) equitable treatment clauses, in the context of emissions trading schemes. Allocation of allowances and the scope of actors and sectors covered by a domestic scheme are important design features which will affect the coherence of State's trade, investment, and carbon trading initiatives.

In terms of ensuring coherence between future climate change policies and trade rules, a number of practical factors have been identified. There clearly are win-win-win trade, investment and climate change approaches, such as cooperation programmes for monitoring and reduction of emissions or the transfer of low-carbon technologies, the liberalization of trade in environmental goods and services related to carbon trading, or the careful design of government procurement rules to exempt green procurement schemes. Future regional trade agreements can seek to explicitly incorporate provisions on climate change and other sustainable development objectives, as well as ensuring transparency for future investors.

But guidance for States is needed in this area, and lessons must be taken from actions taken to date, given the need to both implement measures to conserve exhaustible natural resources and still not unjustifiably restrict international trade or investment.[144] As stated, the Protocol expressly notes that in seeking to achieve a global reduction in carbon emissions, it must also implement measures to minimize adverse effects on both he world's climate and international trade.[145] Cooperation between the UNFCCC and the WTO is therefore not merely 'likely',[146] but essential for future coherence in this area. At the forefront of this must be a committed response to the UNFCCC's call to 'accelerate innovation in the development, deployment, adoption, diffusion and transfer of environmentally sound technologies among all Parties, and particularly from developed to developing countries, for both mitigation and adaptation'.[147]

[143] Baughen (n 84 above).
[144] UNFCCC, Art 3(5); WTO, Art XX(g).
[145] Kyoto Protocol, Art 2(3).
[146] Charnovitz (n 3 above).
[147] UNFCCC Decision 3/CP.13, 2007.

5

Linking of Emissions Trading Schemes

*Michael Mehling**

1. Introduction

As carbon markets emerge in various jurisdictions across the globe, their integration by way of a trading link is attracting interest among scholars and practitioners alike. By expanding the scope and range of available mitigation options, linking promises greater diversity of abatement costs in the market, and thus more efficient achievement of greenhouse gas mitigation objectives.[1] A trading link also creates larger, more liquid carbon markets, reducing volatility and limiting the potential for market abuse. And as an additional benefit, linked trading schemes may result in fewer competitive distortions for participants and thus a reduced threat of leakage.[2] Yet while lowered compliance costs and greater liquidity would seem valuable advantages in their own right, linking of carbon markets may even have a bearing on the global scale, affecting the dynamic of international efforts to mitigate climate change.

Indeed, linking has been described as the 'the de jure or de facto post-2012 international architecture' for climate mitigation.[3] With negotiations on a global climate regime perpetually threatened by diplomatic stalemate, linking

* Acknowledgment: I am indebted to Harro van Asselt, David Freestone, Erik Haites, Ashley Stafford, Charlotte Streck and Andreas Tuerk for valuable comments and suggestions; any remaining errors or inaccuracies are mine alone.

[1] R Baron and C Philibert, *Act Locally, Trade Globally: Emissions Trading for Climate Policy* (OECD: Paris, 2005) 123: 'The economic case for linking is clear. Linking various systems and emissions targets under a single emissions trading umbrella would help deliver a common environmental goal at least-cost, as each participant would now have access to a broader range of mitigation options.' Essentially, linking allows mitigation efforts to shift across systems towards the location where abatement costs are lowest.

[2] O Edenhofer, C Flachsland and R Marschinski, 'Towards a Global CO_2 Market: An Economic Analysis' (2007) Expertise for the Policy Planning Staff in the Federal Foreign Office, 7 <http://www.pik-potsdam.de/news/press-releases/files/PIK_Linking_ETS_2007_engl.pdf>, accessed 30 November 2008.

[3] RN Stavins, 'Linking Tradable Permit Systems: Opportunities, Challenges, and Implications' (7th Annual Workshop on Greenhouse Gas Emission Trading, 9 October 2007, Paris, France) <http://www.iea.org/Textbase/work/2007/ghget/Stavins.pdf>, accessed 30 November 2008.

provides an optional 'fallback in case of multilateral collapse'[4] driving a bottom-up process in which 'various domestic emissions trading schemes ... transform themselves into a global emissions market in a progressive and organic manner'.[5] In such a scenario, linking might arguably help avoid many of the constraints associated with international lawmaking, thereby offering a flexible and potentially more viable option for harnessing the economic benefits of emissions trading across geographic and political boundaries.[6] Whether a mere complement to international trading or the foundation of a global carbon market,[7] in other words, linking is certain to play a part in the future of greenhouse gas regulation, and may even be key to achieving a global reference price for carbon in what could become a market valued at several trillion US$ annually.[8]

As might be expected from any endeavour of this scale, however, linking of carbon markets will invariably give rise to a number of broader implications and possibly unintended consequences. For instance, price convergence across trading schemes, in itself a desired outcome of linking, can also have less welcome

[4] K Tangen and H Hasselknippe, 'Converging Markets' (2005) 5 International Environmental Agreements: Policy, Law and Economics 47, 52; the possibility of an international architecture arising from coordination 'on key elements necessary for successful international trading markets in order to foster the emergence of a plurilateral and, eventually, a fully integrated international trading system' was already foreseen by RB Stewart and P Sands, 'The Legal and Institutional Framework for a Plurilateral Greenhouse Gas Emissions Trading System' in UNCTAD (ed), *Greenhouse Gas Market Perspectives: Trade and Investment Implications of the Climate Change Regime* (UNCTAD: Geneva, 2001) 10; as described more recently by C Flachsland and others, *Developing the International Carbon Market: Linking Options for the EU ETS* (Potsdam Institute for Climate Impact Research (PIK), Potsdam 2008) 8: 'if no agreement on a global trading system is achieved within UNFCCC negotiations by 2009, linking offers an opportunity to keep and build political momentum for constructing a global carbon market in the mid- to long term.'

[5] C Egenhofer and N Fujiwara, 'The Contribution of Linking Emissions Markets to a Global Climate Change Agreement: Feasibility and Political Acceptability' (2006) Final Report of a study prepared for the Economic and Social Research Institute, Cabinet Office, Government of Japan <http://www.esri.go.jp/jp/prj-2004_2005/kankyou/kankyou17/02-1-R.pdf>, accessed 15 June 2009.

[6] Importantly, linking offers the opportunity to define 'common but differentiated responsibilities' in accordance with the United Nations Framework Convention on Climate Change (adopted 29 May 1992, entered into force 24 March 1994) 1771 UNTS 107 (UNFCCC), as it allows specifying a collective mitigation objective regardless of how emission reduction responsibilities are initially allocated across nations, see J Jaffe and RN Stavins, 'Linkage of Tradable Permit Systems in International Climate Policy Architecture' (2008) Harvard Project on International Climate Agreements Discussion Paper 08–0710, 10 <http://environment.tufts.edu/downloads/LinkageOfTradablePermitSystems.pdf>, accessed 15 June 2009. Unlike top-down emissions trading at the international level, moreover, a link between existing carbon markets would build on several years of experience with this policy instrument at the national and regional scale.

[7] For a detailed description of alternative scenarios involving carbon market links, see Flachsland and others (n 4 above), 7.

[8] Point Carbon, *Carbon 2008—Post-2012 Is Now* (Point Carbon: Oslo, 2008), 17, forecasting a global carbon market worth €2 trillion by 2020, assuming a market volume of 38 Gt and a carbon price of €50 in 2020; moreover, according to a recent survey among stakeholders and observers in the carbon market, 73% of respondents expect a global reference carbon price by 2020 due in part to linking, see Point Carbon, *Carbon Market Transactions in 2020: Dominated by Financials?* (Point Carbon: Oslo, 2008), 48.

distributional impacts: market participants in schemes with rising prices will face increased compliance costs,[9] which may, in turn, affect energy prices in the broader economy and have a detrimental effect on industrial competitiveness as well as vulnerable segments of society.[10] Once implemented, moreover, a link will not only influence carbon prices in the markets it connects; depending on its design and that of the linked trading schemes, it can also compromise the effectiveness of carbon markets as an instrument to mitigate greenhouse gases: by virtue of the flexibility afforded to participants, certain features of each scheme will be perpetuated throughout the linked market, including price ceilings, generous offset provisions, and weak enforcement rules. As a result, linked trading schemes may achieve fewer emissions reductions than they would have achieved in aggregate if each had operated separately.

As mentioned earlier, an intended benefit of linking is the ability to explore differences in abatement cost across jurisdictions. Depending on the stringency of the reduction objectives, however, such differences will also translate into heterogeneous carbon prices in each market; creation of a link is therefore likely to result in substantial capital flows across national borders, influencing the balance of payments of participating states.[11] Certain types of links, moreover, can limit domestic control over trading schemes and their operation.[12] And finally, if linking offers an alternative to more unwieldy formal processes of international lawmaking, it also raises the question of whether this increased flexibility entails risks, such as a loss of legitimacy and accountability.[13]

Given these ramifications, integration of carbon markets will require a careful assessment of its political and economic viability, which, in turn, largely depends on the respective compatibility of the underlying emissions trading schemes. Because design features in each scheme can extend to all other linked schemes, and differences—if too pronounced—can affect the integrity of the overall market as well as the direction of trade flows, a certain degree of harmonization is essential for linking. A growing body of research has therefore focused on comparing schemes by looking at the stringency of targets and timeframes, the equivalence of requirements for participating sectors, the definition of trading

[9] Once linked, schemes with a lower carbon price relative to other schemes will see an increase in demand, creating upward pressure on their domestic carbon pricing; although this may give rise to unwanted distributional impacts, the net effect of linking will usually remain positive for all linked schemes.

[10] Baron and Philibert (n 1 above) 137 give a poignant example when stating that 'actors in the high-cost system may feel that they are financing emissions reductions—including, in some cases, industrial modernization—in the low-cost system' as a result of linking.

[11] J Kruger, WE Oates, and WA Pizer, 'Decentralization in the EU Emissions Trading Scheme and Lessons for Global Policy' (2007) 1 Review of Environmental Economics and Policy 112, 119.

[12] J Jaffe and RN Stavins, 'Linking Tradable Permit Systems for Greenhouse Gas Emissions: Opportunities, Implications, and Challenges' (2007) Report prepared for the International Emissions Trading Association (IETA), 18–20 <http://www.ieta.org/ieta/www/pages/getfile.php?docID=2733>, accessed 30 November 2008.

[13] See below, Section 4.5.

units, and procedural aspects such as monitoring, reporting, verification, and enforcement.[14]

Less attention has been devoted to the legal and normative implications of linking, despite their obvious importance for emissions trading on a broader level. Generally speaking, carbon markets are sophisticated entities, born out of a political decision to constrain certain forms of polluting behaviour; an artifice by nature, thus, their very existence is contingent on the adoption of a framework of defined objectives, procedures and substantive rules and principles.[15] Moreover, these markets are themselves embedded in regulatory traditions and conditioned by fundamental determinations on the boundaries of individual and collective freedom, the nature and scope of public authority, and the general parameters of economic activity. Accordingly, neither the creation nor the subsequent operation of carbon markets can be fully understood without an appreciation of the legal and normative dimensions of emissions trading.

If carbon markets are inherently embedded in the realm of norms, it stands to reason that their linkage will also raise important questions of law. Yet legal and normative considerations have featured only marginally in existing research on the conditions and implications of a trading link.[16] Providing a systematic overview of such considerations is the main objective of this chapter, which sets out by drawing a conceptual background for the subsequent analysis, defining central characteristics and different forms of linking in the second section. It then proceeds to explore the specifically normative and jurisprudential implications of linking in the third section, and distinguishes these from broader questions of political expedience and economic rationality.

Legal categories become relevant in the establishment of a link between emissions trading schemes, notably as regards the formal classification, legal nature, and procedural requirements of the instruments available to create the link and govern its ensuing operation; a typology of such instruments is therefore offered in the fourth section. Moreover, a trading link will not emerge in a normative vacuum, and can subsequently interact with substantive rules and principles at the level of international, regional, or domestic law. Such interactions, which may range from tensions and conflicts to mutual synergies, are briefly highlighted in the fifth section, which also draws on case studies to illustrate the connections of any link with the wider body of norms and principles underlying each legal system.

[14] For an overview of relevant studies, see M Mehling and E Haites, 'Mechanisms for Linking Emissions Trading Schemes' (2009) 9 Climate Policy 169–184.

[15] For an in-depth analysis, see K Anttonen, M Michael and K Upston-Hooper, 'Breathing Life into the Carbon Market: Legal Frameworks of Emissions Trading in Europe' (2007) 16 European Environmental Law Review 96–115.

[16] Of more than 50 studies and published reports focused on the issue of carbon market linkages (individual citations on file with author), fewer than five address legal and normative considerations.

2. Linking: A Conceptual Framework

2.1 Definition and concept

Numerous definitions of linking have been brought forward to date, and they tend to share a set of common features. On a general level, emissions trading schemes are linked if a participant in one scheme can use a carbon unit issued under another scheme to meet compliance obligations.[17] Thus, as a result of linking, units are considered equivalent for compliance purposes without requiring some form of individual review and approval prior to each transaction.[18] In its most basic rendition, a link can consist of a simple provision stating the equivalence of foreign units;[19] yet recognition alone will not yet facilitate actual trading.

For transactions to occur between linked schemes, a link also needs to allow the flow of units across schemes.[20] Different approaches are conceivable to channel such a flow of units: a comprehensive link with a high degree of market integration, for instance, will be accompanied by an electronic connection of registries across schemes, permitting the direct transfer from accounts in one scheme to the accounts of participants in other schemes. Alternately, the flow can also be accomplished by cancelling units in one scheme and counting them towards holdings in another scheme, which is simplified if foreign entities are able to open and maintain accounts in the linked scheme.[21] Finally, participants

[17] E Haites, 'Harmonisation between National and International Tradeable Permit Schemes' (2003) CATEP Synthesis Paper, OECD Doc. CCNM/GF/SD/ENV(2003)2/FINAL, 5 <http://www.oecd.org/dataoecd/11/63/2957623.pdf>, accessed 30 November 2008; throughout this chapter, 'carbon units' will be used as a collective reference to allowances or credits issued under a carbon trading or offset scheme, regardless of whether that scheme is mandatory or voluntary, and whether it is premised on a 'cap-and-trade' or a 'baseline-and-credit' approach, see Jaffe and Stavins (n 12 above), 3–5; examples of carbon units are the unit types defined by the Kyoto Protocol, notably: Assigned Amount Units (AAUs), Emission Reduction Units (ERUs) for Joint Implementation (JI) projects, CERs for Clean Development Mechanism (CDM) projects, and Removal Units (RMUs) for net removals by certain sinks, see Decision 13/CMP.1, 'Modalities for the Accounting of Assigned Amounts under Article 7, Paragraph 4, of the Kyoto Protocol' (2006) UN Doc FCCC/KP/CMP/2005/8/Add.2 Annex, No 11.

[18] Flachsland and others (n 4 above), 7; Stewart and Sands (n 4 above) 9.

[19] Typically, however, a link will also specify operational details aside from the units that will be accepted, for instance individual or aggregate limits on imported units, and possible adjustments ('exchange rates') to the units, see Mehling and Haites (n 14 above), 176.,

[20] MJ Mace and others, 'Analysis of Legal and Organisational Issues Arising in Linking the EU Emissions Trading Scheme to other Existing and Emerging Emissions Trading Schemes. Study Commissioned by the European Commission, DG Environment, Climate Change and Air' (2008) 3 <http://www.field.org.uk/files/Linking%20emission%20trading%20schemes_0.pdf>, accessed 30 November 2008.

[21] For instance, the administrator of a scheme could establish an account in the registry of a linked scheme to receive foreign units; participants in that scheme wishing to use its units for domestic compliance could also open an account in said registry, acquire the units, and have them transferred to their account. Once units are to be used for compliance purposes, participants could transfer the units to the foreign account of the administrator, which would cancel or otherwise retire the units and issue a corresponding amount to participants in their domestic scheme, see Mehling and Haites (n 14 above) 171.

might also engage in arbitrage activities or commercial mechanisms operating across schemes; such transactions would be limited in scope, however, and will not be further discussed here. [22]

2.2 A typology of linkages

A link between carbon markets can assume various manifestations, with differences in degree, scope, and the direction of trading flows. Certain legal implications will depend on what type of link is implemented, highlighting the importance of a careful distinction. Conceptually, a link can be either direct or indirect, with direct linking conditional on an explicit decision by at least one of the linked jurisdictions.[23] Direct links allow trade both within and between different schemes,[24] moreover, and can be further distinguished on account of trading flows and whether they allow trading in one or more directions.[25] In theory, the greatest economic benefits will follow from a direct bilateral or multilateral link, in which two or more jurisdictions agree on the mutual recognition of units and trade can occur in various directions across schemes,[26] be it by way of connected registries, a gateway,[27] or a procedure to credit units acquired and cancelled in another scheme.

Direct linking does not necessarily have to operate in both directions, however, and indeed most links currently in place are unilateral;[28] such links emerge when only one jurisdiction is willing to recognize units from another scheme, taking advantage of the fact that trading schemes commonly allow foreign

[22] E Haites and F Mullins, 'Linking Domestic and Industry Greenhouse Gas Emission Trading Systems. Report prepared for the Electric Power Research Institute (EPRI), International Energy Agency (IEA) and the International Emissions Trading Association (IETA)' (2001) <http://www.iea.org/textbase/papers/2001/epri.pdf>, accessed 30 November 2008, 72; such indirect flows are possible due to the conceptual distinction between a transfer of units and unit trading: while a transfer is only possible within or between connected registries and eligible participants, trading-buying and selling-can occur between any entity, provided it is not explicitly ruled out, see RW Wijnen, 'Emissions Trading under Article 17' in D Freestone and C Streck (eds), *Legal Aspects of Implementing the Kyoto Protocol Mechanism: Making Kyoto Work* (OUP: Oxford, 2005) 410.

[23] Jaffe and Stavins (n 12 above) 11.

[24] J Ellis and D Tirpak, 'Linking GHG Emission Trading Schemes and Markets' (2006) OECD Doc COM/ENV/EPOC/IEA/SLT(2006)6, <http://www.oecd.org/dataoecd/45/35/37672298.pdf>, accessed 30 November 2008, 8.

[25] For the purposes of this assessment, multilateral links will be treated no differently from parallel bilateral links, obviating a further distinction between bilateral and multilateral linking.

[26] Ellis and Tirpak (n 24 above) 9; Flachsland and others (n 4 above) 9; Mehling and Haites (n 14 above) 181.

[27] A gateway, or clearing house, may help control trading flows or account for differences between schemes and tradable units, see A Roßnagel, 'Evaluating Links between Emissions Trading Schemes: An Analytical Framework' (2008) 2 Carbon and Climate Law Review 394, 397.

[28] See, for instance, Regional Greenhouse Gas Initiative Model Rule of 15 August 2006 sec. XX-10.3(b)(1); for a broader overview, see the overview in Mace and others (n 20 above) 77; Mehling and Haites (n 14 above) 172–3.

natural or legal persons to acquire and hold units,[29] or when the link extends to a credit system that generates units for greenhouse gas mitigation projects.[30] Unlike their direct counterparts, indirect links can evolve between systems even absent an explicit linking decision. A scheme that establishes a unilateral or bilateral link to another scheme also establishes indirect links with any other schemes to which the latter is linked,[31] and developments in one scheme will subsequently affect the supply and demand for units in each other scheme.[32]

3. Identifying Legal and Normative Questions

By and large, the issues discussed in the context of linking have been conceptual in nature, focusing on the mutual compatibility of different emissions trading schemes and the potential effects of a link on their economic and environmental performance. Studies to date collectively infer that variations in the design of emissions trading schemes can hamper the prospects for a market linkage; and while technical solutions may help overcome such differences, they tend to lessen the benefits of linking or affect the environmental integrity of underlying markets.[33] Needless to say, conclusions such as these are of vital importance when assessing the expediency of a trading link. Yet as most studies will also affirm, linking remains possible even when central design elements of the affected carbon markets differ.[34]

While essential to the operation of a trading link, for instance, the mutual recognition of units has been ultimately declared a 'political issue', with operational details left to the volition of decision makers;[35] likewise, monitoring, reporting, and verification procedures are vital to confidence in the market and the value of traded units, yet there, again, significant differences do not preclude linkage.[36] In other words, a high degree of harmonization between connected markets may be desirable, but is by no means a necessity. Certain design differences can be

[29] Given the potential effect on carbon pricing, schemes may want to restrict the ability of other schemes to create a unilateral link; this can occur, for instance, by limiting eligibility to open an account and hold units to entities subject to domestic compliance obligations as well as additional safeguards, such as penalties for transactions benefitting purchasers in other schemes.

[30] An example of such a credit system is the Clean Development Mechanism (CDM) defined by the Kyoto Protocol, as well as different national offsetting schemes.

[31] Mehling and Haites (n 14 above) 171.

[32] For instance, two systems can become linked indirectly through direct links to a common third system, such as the Clean Development Mechanism (CDM), see Jaffe and Stavins (n 12 above) 8–9.

[33] Haites and Mullins (n 22 above), 67; Kruger and others (n 11 above) 122–5.

[34] See, for instance, Mace and others (n 20 above) 51: 'linking emissions trading systems is possible even between quite different systems.' Jaffe and Stavins (n 12 above) ES-5: 'differences between systems can remain without undermining the case for linking'

[35] W Blyth and M Bosi, 'Linking Non-EU Domestic Emissions Trading Schemes with the EU Emissions Trading Scheme', OECD Doc. COM/ENV/EPOC/IEA/SLT35, 33 <http://www.iea.org/textbase/papers/2004/non_eu.pdf>, accessed 30 November 2008.

[36] S Bode, 'Implications of Linking National Emission Trading Schemes Prior to the Start of the First Commitment Period of the Kyoto Protocol', HWWA Discussion Paper 214, 3 <http://papers.ssrn.com/sol3/papers.cfm?abstract_id=376202>, accessed 30 November 2008.

overcome by technical solutions and relatively simple adjustments, such as conversion rates for different units or a software interface to connect separate registries. Other differences may well preclude decision makers from agreeing to a link, such as a price containment mechanism that is extended to other schemes by virtue of the link; a brief overview of potential barriers is provided in Table 5.1 below.

Yet questions of compatibility ultimately reside in the political domain:[37] their solution depends on deliberation, reconciliation, and, to some extent, mutual concession, with account given to the interaction of different design features and the broader context of domestic climate and energy policies. Establishing a link between emissions trading schemes is thus largely a matter of whether the benefits of linking outweigh its potential risks, and whether remaining differences and risks can be tolerated.

Table 5.1. Design features of emissions trading schemes and their relevance for linking[38].

Design Feature		Mutual Compatibility
Scope and timeline	Covered sectors	Desirable
	Commitment periods	Desirable
	Continuity of scheme	Essential
	New entrants and closures	Optional
	Ex-post adjustments	Essential
Tradable units	Fungibility of units	Desirable
Reduction obligation ('Cap')	Absolute v relative	Essential
	Environmental stringency	Desirable
Allocation	Allocation method	Optional
Banking	Possibility of banking units	Desirable
Cost containment	Price ceilings	Essential
	Borrowing	Essential
	Offset crediting	Desirable
Competition and leakage	Leakage control mechanisms	Optional
Governance	Monitoring and verification	Optional
	Enforcement rules	Desirable
	Registries	Optional

[37] Only in rare cases, essentially political aspects may also acquire legal relevance if they affect issues beyond the scope of legislative powers, that is, subject matters where the competence of legislators is restricted by fundamental doctrines and rights; for instance, if the linking of two separate carbon markets results in such a loss of environmental integrity that the—legally enshrined—policy aims of either trading scheme are clearly undermined, they may compromise its proportionality as a restriction on economic freedom and thus infringe on vested rights under national constitutions or other legal precepts.

[38] For an in-depth analysis, see A Tuerk, M Mehling, C Flachsland and W Sterk, 'Linking Carbon Markets: Concepts, Case Studies and Pathways' 9 (2009) Climate Policy 341–57.

It would be mistaken, however, to assume that linking can occur based on political motives alone, that is: free of any legal or normative constraints. Adopted by regulatory decision, trading schemes do not come into existence in a normative vacuum. Rather, they operate in a multilayered framework of established rules, principles, and procedures constituting the global legal order. Under domestic law, each trade among market participants is subject to regulatory choices regarding contractual relations, accounting, taxation, and financial services; each public intervention, in turn, governed by constitutional doctrines and institutional mandates which are themselves constrained by individual and collective rights of the participants.[39]

Cooperation between states, meanwhile, is governed by the broad and less determinate prescriptions of international law, as set out in treaties, general principles and customary practice—a system lacking clear hierarchy and centralized authority, yet nonetheless distinct from the larger arena of international diplomacy and cooperation.[40]

Carbon markets are highly regulated; and this relevance of norms also extends to any link between such markets. Just as the trading schemes it aspires to connect, a link will emerge within a legal and normative context from which derogation is not a matter of mere political expedience.[41]

Still, not every rule applicable to carbon trading in linked markets will be related to the link as such. When ascertaining the legal and normative implications of a link, thus, a distinction needs to be made between the legal implications of linking and the countless regulatory issues arising in the general context of each carbon market. Treatment of carbon units under property, accounting, and taxation rules, for instance, is an issue typically governed by the domestic legislation of participating jurisdictions, regardless of whether the units have been traded across a market link.[42] As a rule, carbon units will be subject to the

[39] Anttonen, Mehling and Upston-Hooper (n 15 above) 96.

[40] On the particular function and nature of international law, the classical treatise remains H Lauterpacht, *The Function of Law in the International Community* (Clarendon Press: Oxford, 1933). Later, of course, the realist movement denied international law a role separate from international relations and the power dynamics among states, see HJ Morgenthau, *Politics Among Nations: The Struggle for Power and Peace* (5th edn, Alfred A. Knopf: New York, 1978); more recent scholarship, however, has affirmed the flowing boundaries between international law and politics while still maintaining its specific nature as a normative order and formal argumentative practice, see, for instance, M Koskenniemi, *From Apology to Utopia: The Structure of International Legal Argument* (CUP: Cambridge, 2006).

[41] This is, of course, a quintessential feature of the law; yet generations of legal theorists and social philosophers have debated over the most accurate definition of the law and its domain. Defined *per genus et differentiam*, there is a wavering consensus that law belongs to the wider category of rules of behavior, but is distinct from other social norms, such as moral or religious precepts; and while explanations of this distinct nature diverge, they commonly focus on the authority and legitimacy imbued through specified conditions of validity, such as a legislative procedure and adoption through an elected legislature, and the possibility of enforcement, see generally HLA Hart, *The Concept of Law* (Clarendon Press: Oxford, 1961), 18–96.

[42] M J Mace, 'The Legal Nature of Emission Reductions and EU Allowances: Issues Addressed in an International Workshop' (2005) 2 Journal of European Environmental and Planning Law 123, 125.

regulatory framework of the jurisdiction in which they are held;[43] by virtue of the link, the primary benefit conferred with every unit, namely the ability to comply with greenhouse gas mitigation objectives,[44] will also be recognized in the jurisdiction of the recipient. But such recognition will not extend to any secondary rights or obligations that may have prevailed in the jurisdiction of origin, such as fiscal duties or privileges, unless also enacted in the receiving jurisdiction or expressly specified by the linking arrangement.[45]

Occasionally, differences across linked schemes may have legal consequences as trading occurs through the link: when units afford their holders in one jurisdiction a more favourable status under contractual or property rules than their counterparts in other jurisdictions, such preferential treatment may prompt a strategic shift of units and 'forum shopping'.[46] Likewise, legal remedies available in one jurisdiction may still be applicable to units after they have been transferred across a trading link, affording, for instance, compensation rights in the event of breach of contract.[47] Strictly speaking, however, these implications are

[43] An analogy can be drawn to international trade with products; for instance, a vehicle may be subject to strict environmental standards and high ownership taxes in its country of origin, but, once exported to the country of destination, will be subject to the less onerous rules applied there.

[44] The legal interest embodied in a carbon unit can vary from jurisdiction to jurisdiction, and range from a property right with all the ensuing privileges to a mere exemption from a statutory prohibition; even within a trading system, such as the scheme established in Europe, units are often accorded a different status under participating jurisdictions, for instance as intangible rights (Finland), financial instruments (Sweden), regulatory property (United Kingdom), or an immaterial good (France), see Anttonen, Mehling and Upston-Hooper (n 15 above), 98; M Wemaere and C Streck, 'Legal Ownership and Nature of Kyoto Units and EU Allowances' in D Freestone and C Streck (eds), *Legal Aspects of Implementing the Kyoto Protocol Mechanisms* (OUP: Oxford, 2005) 50; and see above chapter 2. At the international level, parties to the Kyoto Protocol decided that units have 'not created or bestowed any right, title or entitlement to emissions of any kind', FCCC/KP/CMP/2005/8/Add.1, Decision 2/CMP.1 Principles, nature and scope of the mechanisms pursuant to Articles 6, 12 and 17 of the Kyoto Protocol.

[45] In part, this follows from the doctrine of territorial sovereignty, which affirms that a state exercises the supreme, and normally exclusive, authority within its territory, R Jennings and A Watts (eds), *Oppenheim's International Law, Volume I: Peace* (9th edn, Longman: London, 1992), 564; subject to subsequent commitments entered under international law, thus, states have unconditional jurisdiction within their territory, including judicial, legislative, and administrative competence, a corollary of which is the general prohibition of measures taken in the territory of another state without their consent (also known as the principle of non-intervention in the internal affairs of other states), I Brownlie, *Principles of Public International Law* (7th edn, OUP: Oxford, 2008) 299–322.

[46] R Betz and A Stafford, 'The Policy Issues arising with the Linking of International Emissions Trading Schemes' 27 (2008) Australian Resources and Energy Law Journal 86, 95–6, list relevant examples, for instance pooling of units in the scheme affording the strongest proprietary rights to holders; given the administration and transaction costs of unit transfers and the recurring need to return units for surrender at the end of compliance periods, however, a large-scale shift would not seem practical. Moreover, a clearer distinction is needed between legal consequences attached to units, and rights and duties accruing to owners or market participants irrespective of where the taxable object is held: even if units are shifted to the jurisdiction with more favourable taxation rules, for instance, fiscal liability will not automatically follow the object of taxation, but may remain with the owner at the registered seat of business.

[47] Betz and Stafford (n 46 above), 96, outlining a scenario in which units are proprietary rights, potentially entitling the buyer to an equitable interest in the units or a declaration that the third party who received the units holds them on trust for the buyer.

neither a condition nor consequence of the link itself, but a result of the general differences between the underlying legal systems. Within limits, a link may account for such differences; yet circumstances will differ in each individual case.

Summing up, the challenges raised by linking are largely political in nature; and of the legal and regulatory issues faced when trading across a link, most will not follow from the link as such, but from the general operation of the underlying carbon markets within general law, as described throughout the various other chapters of this book. Invariably, then, one will be prompted to ask: what remaining legal and normative questions are a genuine condition or consequence of the link?

On a general level, such questions are likely to emerge in two respects: first, during the actual process of establishing the link, which necessitates recourse to recognized sources of law and legal procedures; and, second, in the event of a conflict between the link as such with substantive legal norms and principles, whether these originate in international, regional or domestic law. The first category relates to the different instruments available under the law to set up a link and govern its continued operation, including procedures required for their adoption; the second category relates to the subsequent interactions of the link—notably the mechanisms and procedures used for its operationalization—with substantive rules and principles at the level of international or domestic law, which may range from tensions and conflicts to mutual synergies.[48] Both categories are mutually contingent in that a failure to observe requirements under one category may also result in a violation of requirements under the other. In the remainder of this chapter, each category will be further illustrated through examples and case studies.

In the end, it bears noting that the boundaries between the legal, political and normative dimensions of carbon trading are in a constant state of flux. Legal reasoning—and hence the development of law and its interpretation—are perpetually caught in a dynamic tension between formal arguments based on legal doctrine and positive fact, and normative prescriptions invoked to change or censure extant realities. While law will remain the essential framework of all mandatory carbon markets, instilling confidence and guiding or constraining behaviour along predictable channels, it will thus also create a context for debate and a shared horizon on which interests and expectations of market participants, decision makers and other stakeholders converge. Seeing how carbon markets evolve around sophisticated networks of interests and political objectives, reflecting an innovative and highly debated paradigm shift in the approach to governance, this creative influence of law also merits consideration, although it is altogether too elusive for further analysis here.

[48] This second category also comprises normative considerations which may not form part of the positive legal order, but nonetheless spell out important general tenets such as legitimacy, fairness, and due process.

4. Designing a Market Link: Legal Nature and Process

4.1 Unilateral linking

A link between emissions trading schemes can assume varying degrees of formality, with implications for its legal nature and the procedural requirements of its adoption.[49] As outlined above, a unilateral link can be established through inclusion of a clause in the architecture of each trading scheme, specifying the conditions for recognition of foreign units.[50] Unless otherwise specified, the procedures for adoption and its legal nature will follow the instrument constituting the scheme, which is, in most cases, formal statutory legislation.[51] Because the clause establishing the link remains within the remit of national jurisdiction, the link can also be unilaterally altered or terminated at any point in time. Absent some form of international commitment, the implementing entity will not be bound by its decision to create a link.[52]

4.2 Bilateral and multilateral linking

At the core of any link between emissions trading schemes lies the recognition of units issued under a foreign trading scheme; this applies to bilateral or multilateral integration of carbon markets as much as it applies to unilateral linking, with one exception: in a bilateral or multilateral link, recognition must be mutual so as to allow trading flows in more than one direction.[53] As a result, these links will generally necessitate some form of coordination between schemes to synchronize the required adjustments, ranging from the mere decision to simultaneously accept foreign units for compliance purposes to more ambitious

[49] For a more detailed assessment, see M Mehling, 'Bridging The Transatlantic Divide: Legal Aspects of a Link between Regional Carbon Markets in Europe and the United States' (2007) 7 Sustainable Development Law and Policy 46, 47; MJ Mace and J Anderson, 'Transnational Aspects of a Linked Carbon Market' (2008) 2 Carbon and Climate Law Review 190, 193–4.

[50] Recognition can be provided by legislation that delegates to administrative or executive authorities the power, subject to criteria and procedures established in the legislation, to determine whether units from other schemes should be accorded recognition for purposes of domestic regulatory compliance, see Stewart and Sands (n 2 above), 26.

[51] Still, different options are conceivable; depending on the regulatory context, a link may require a statutory amendment, a substatutory regulation, or simply a decision by the scheme administrator, see Mehling and Haites (n 14 above), 177.

[52] An illustration is Directive 2004/101/EC of 27 October 2004 amending Directive 2003/87/EC establishing a Scheme for Greenhouse Gas Emission Allowance Trading within the Community, in respect of the Kyoto Protocol's Project Mechanisms [2004] OJ L338/18, which links the European emissions trading scheme to the Clean Development Mechanism and Joint Implementation; it is a secondary act of legislation adopted by the European Community, and can be repealed or amended at any time by the Community legislator.

[53] For a detailed description of the possible elements of a mutual recognition arrangement, including a draft proposal, see Mace and others (n 29 above), 80–4.

levels of integration, such as an agreement upon the trajectory of reduction obligations in each scheme.[54] Once again, different modalities are available to implement such a decision. Pursuant to the general doctrine of sources of law, such a negotiated understanding will only be binding on the linked jurisdictions if it meets the formal requirements of an international treaty. As one of the recognized sources of international law,[55] a treaty is an expression of state sovereignty bounded by voluntary consent,[56] and is governed by international law in relation to its validity, application, interpretation and enforceability. Failure to observe the terms set out under the treaty counts as a breach of international law, incurring consequences under customary international law and the possibility of countermeasures.[57]

Due to their formal nature, treaties offer a transparent and predictable framework for transactions across linked trading schemes; yet they are also subject to a number of restrictions. Only formal subjects of international law may enter into treaties, notably sovereign states and international organizations; by default, regional and local entities, such as the constituent units of a federation, will be excluded unless otherwise specified in the national constitution.[58] In a majority of cases, the adoption of international treaties—and especially multilateral treaties—also entails a cumbersome and often challenging ratification process,[59] with restrictions set out both under international law and in domestic constitutions or organizational mandates. Likewise, a withdrawal from the treaty

[54] Jaffe and Stavins (n 12 above), 51; such an agreement would pre-empt any incentive for linked markets to weaken their respective reduction targets and selling more units into the foreign scheme.

[55] See Statute of the International Court of Justice (adopted 26 June 1945, entered into force 24 October 1945) (1945) 39 AJIL Supp 215, art 38(1); treaties are defined as any 'international agreement concluded between States in written form and governed by international law, whether embodied in a single instrument or in two or more related instruments and whatever its particular designation', see Vienna Convention on the Law of Treaties (22 May 1969, entered into force 27 January 1980) 1155 UNTS 331 (VCLT) art 2(1)(a); accordingly, it is irrelevant whether a treaty is designated as a 'convention', 'charter', 'covenant', 'protocol', 'pact' or 'agreement', it will still be considered an international treaty for legal purposes, see A Aust, *Modern Treaty Law and Practice* (2nd edn, CUP: Cambridge, 2007), 23; once in force, a treaty is binding upon its parties and must be performed in good faith, as stipulated in the customary maxim of *pacta sunt servanda*, see VCLT, art 26.

[56] *S.S. Wimbledon* (*France, Italy, Japan and UK v Germany*) (Merits) PCIJ Rep Series A No 1; the consent to be bound by a treaty may be expressed by signature, exchange of instruments constituting a treaty, ratification, acceptance, approval or accession, or by any other means if so agreed, see VCLT, art 11.

[57] A breach of treaty constitutes an international wrong that entitles other parties to terminate, or withdraw from, the treaty or suspend its operation, see VCLT, art 60; where damages result from the breach, injured parties may seek cessation of the wrongful conduct, an assurance of non-repetition, and reparations, or even resort to (proportionate) countermeasures; often, sanctions and the conditions for their application will be defined in the treaty itself as part of a negotiated compliance or dispute settlement mechanism.

[58] See Aust (n 55 above), 63–4.

[59] While ratification is only one of many expressions of consent to be bound by a treaty, it remains common practice for the adoption of international environmental treaties; it requires the execution of an instrument of ratification by the executive and, under many national constitutions, legislative approval, Aust (n 55 above), 103–4.

and subsequent amendments are subject to formal constraints,[60] implying that any provision for adjustment or suspension of the link, for instance to account for changing circumstances, should already be included in the treaty from the outset.[61] Still, only a treaty is likely to achieve the greatest degree of market integration: beyond affording the option of carbon trading across market boundaries, it also facilitates creation of institutional structures to administer the link, with a mandate and conferred powers that may range from market supervision to the adoption of uniform standards and procedures.[62]

4.3 Reciprocal unilateral linking

A bilateral or multilateral link can also be created by way of a political understanding on the mutual recognition of carbon units, coupled with domestic legislation adjusting each scheme. In legal terms, this alternative will be similar to the unilateral link described in the preceding section, albeit with the difference that affected jurisdictions will establish unilateral links on a reciprocal basis. At the level of implementation, in other words, the link will still be based on an adjustment to the instruments establishing each trading scheme. Such reciprocal links have the benefit of obviating lengthy ratification procedures and avoiding other restrictions imposed by international law, such as the exclusion of jurisdictions other than Federal States; moreover, they leave each scheme with the flexibility to terminate the link or adapt it to changing circumstances as needed.[63]

Coordination between markets can be achieved through informal negotiations, or—at a slightly more formal level—by way of a Memorandum of Understanding[64] and technical standards. Still, while they document a common

[60] Pursuant to VCLT, art 39, a treaty may only be amended 'by agreement between the parties', meaning consensus, unless otherwise specified; and likewise, under VCLT, art 54, termination or withdrawal may only take place in conformity with the provisions of the treaty, or by consent of all the parties after consultation with the other contracting states. Unilateral amendment and withdrawal are thus, as a rule, no longer possible.

[61] For details on the likely substance of such a provision, see Mehling and Haites (n 14 above), 179; E Haites and X Wang, 'Ensuring the Environmental Effectiveness of Linked Emissions Trading Schemes Over Time' (2009) 14 Mitigation and Adaptation Strategies to Global Change (MITI) (forthcoming).

[62] Flachsland and others (n 4 above), 29–31 propose establishing an 'international clearinghouse' to provide a forum for coordination of regulatory issues and create an infrastructure and information basis for management of the combined market; such a clearinghouse could feature, *inter alia*, a common registry, guidance on the definition of reduction targets, adoption of monitoring, verification and enforcement standards, administration of exchange rates and gateway reserves, and a formalized, periodic review of the market linkages; Stewart and Sands (n 2 above), above, 28 propose a multilateral non-compliance procedure, moreover, allowing participants to raise claims of non-enforcement in a linked jurisdiction and setting out verification procedures.

[63] Mehling and Haites (n 14 above), 180.

[64] A Memorandum of Understanding (MoU) is 'an instrument concluded between states which they do not intend to be governed by international law (or any other law) and, consequently, is not legally binding', Aust (n 55 above), 32; aside from greater flexibility and convenience, an MoU also offers other features that may be considered advantageous, such as confidentiality: unlike treaties, there is no requirement to publish or deposit MoUs.

intent and desired outcome, each of these options lacks the binding power of a formal commitment; consequently, the link will only remain operational as long as parties find it expedient, and may experience sudden adjustments or even suspension from one of the participating jurisdictions. Without prior notice, such abrupt changes can have a significant impact on the market for carbon units and the broader economies of participating jurisdictions.[65] Where the legal certainty and transparency offered by reciprocal legislation is insufficient, yet parties still want to retain flexibility in the subsequent administration of the market link, an 'umbrella treaty' may be considered, setting out the essential elements of the link and complemented by more detailed technical and operational provisions in one or more informal Memoranda of Understanding or common standards.[66]

4.4 The linking process: institutional considerations

As carbon markets become more integrated over time, the discussion on linking of emissions trading schemes will also invariably extend to considerations of governance through institutional arrangements that go beyond the link as such.[67] Such arrangements may range from loose cooperation between linked jurisdictions to an international organization endowed with formal legislative and enforcement powers, and are likely to unfold in a dynamic and evolutionary process.[68] At an early stage of integration, cooperation may occur through informal networks geared towards an exchange of information, promotion of uniform approaches and standards, stakeholder involvement, and outreach activities. Rather than adopt binding standards or recommendations, these networks would be largely limited to issuing recommendations and providing advice on the implementation and harmonization of trading schemes.[69]

As integration becomes more aligned with domestic political priorities, however, participating jurisdictions may be willing to consider more formal arrangements to promote further market integration, such as an umbrella agreement harmonizing certain features of the domestic trading schemes and specifying mandatory procedures. Such an umbrella agreement could outline minimum standards for monitoring, reporting and verification, or seek to harmonize technical details, such as the registry software used by participating jurisdictions. Procedures could include mutual notification and information duties, external review or reciprocal monitoring of the trading schemes, and periodic meetings of representatives from each trading scheme to discuss items

[65] Mace and others (n 20 above), 75–6; parties to the link might agree upon information and early notification duties as a way of pre-empting abrupt developments.

[66] Aust (n 55 above), 43; see also below, Section 4.4.

[67] Stewart and Sands (n 2 above), 11.

[68] See Tuerk and others (n 38 above).

[69] An example for such cooperation already exists with the International Carbon Action Partnership (ICAP) launched on 29 October 2007 by more than 15 national and regional governments, expressly aimed at creating a 'forum to discuss relevant questions on the design, compatibility and potential linkage of regional carbon markets', see ICAP Political Declaration, <www.icapcarbonaction.com/docs/icap_declaration.pdf>, accessed 30 November 2008.

for harmonization, such as cost containment mechanisms. An umbrella agreement may also be used to create an institution with limited powers, such as a treaty secretariat or clearinghouse facilitating trading and continued operation of the market links through coordination of meetings, collection and circulation of information, and general logistical functions such as registry maintenance.

At a later stage of integration, participating states may even opt for the establishment of an international or supranational organization, with independent legal personality, a constitutive mandate, and defined governance structures. Aside from individually defined responsibilities, such an organization could also be afforded genuine powers to adopt and enforce rules for market participants and linked jurisdictions in pursuit of a more broadly conferred mandate. Its objectives could include:

- *facilitating market integration and convergence* by ensuring comparability of reduction pathways, sustaining compatibility of central design features, and avoiding fragmentation of national markets over time
- *upholding environmental performance and integrity* by implementing robust monitoring and enforcement structures, ensuring adequate administrative and regulatory capacities, and securing additionality of offset credits
- *safeguarding market efficiency and integrity* by managing volatility and price extremes in the market, avoiding manipulation by dominant market actors, ensuring market transparency in spot and derivative trading, regulating exchanges and over-the-counter trading, and regulating speculation and risk management with derivatives

Aside from the ability to collect and, where necessary, demand information on different aspects of market operation,[70] such an institution could, for instance, be given control over market access and accountability,[71] oversight of speculation and market abuse,[72] and management of carbon prices.[73] Ultimately, a

[70] Market transparency in spot and derivative trading can be upheld by collecting relevant market information (settlement prices, volumes, open interest, and opening and closing ranges) from exchanges, market intermediaries, and over-the-counter market participants, affording access to transaction records, publishing non-confidential data, and adopting uniform financial reporting standards for emission allowances and credits, see, M Mehling 'Global Carbon Market Institutions: An Assessment of Governance Challenges and Functions in the Carbon Market' (2009) Study prepared for the Office of Climate Change, London <http://www.decc.gov.uk/Media/viewfile.ashx?FilePath=What%20we%20do\Global%20climate%20change%20and%20energy\Tackling%20Climate%20Change\Emissions%20Trading\Lazarowicz%20report\1_20090720095853_e_@@_ClimateStrategiesGovernanceBackgroundPaperJuly2009.pdf&filetype=4>, accessed 20 July 2009.

[71] Usually, this will involve requiring licensing and registration of brokers, traders, and exchanges subject to defined criteria on training requirements, reporting and record keeping, and ability to maintain market integrity and publicize transaction data; it can also extend to participation in market monitoring and imposition of sanctions, and operation of a central clearing facility, see Mehling (n 70 above).

[72] Here, the imposition of bidding and position limits, minimum margin requirements, and sanctions are conceivable functions, including the ability to initiate judicial proceedings against violations, see Mehling (n 70 above).

[73] Such control can include the definition of conditions for banking and borrowing, setting reserve auction prices and trigger prices for increased supply, and defining a target price path with periodic adjustments, see Mehling (n 70 above).

central institution mandated with broad governance functions may even assume powers akin to those presently exercised by central banks, including the creation of a unit reserve for strategic intervention in the unit supply, as well as powers currently limited to national or regional entities, such as administration of the allocation and auctioning process for units.

Establishing such an institution would, again, be a matter of adopting a binding international treaty between participating jurisdictions. Aside from defining objectives and responsibilities, such a treaty—often termed an organizational 'charter' or 'establishing treaty'—will also have to specify its own governance structures and procedures. Typically, this will necessitate creation of a body of appointed officials with specified terms and conditions of appointment, and with rules on geographical and professional diversity.

4.5 The linking process: normative considerations

Linking emissions trading schemes has the potential to facilitate the transfer of significant revenue streams across national borders. Irrespective of its design, thus, the creation of a trading link will have clear economic relevance for market participants and, on a broader level, for civil societies in the affected jurisdictions. Any such measure adopted and implemented by a sovereign power needs to meet certain criteria to minimize later challenges, be it through judicial channels or in the arena of political debate. In particular, a trading link and institutional arrangements adopted for its implementation should raise no doubts as to their legitimacy, which is the main justification of political authority and a vital condition for institutional stability and adequate compliance by their addressees. Legitimacy can derive from various sources, notably from popular consent expressed through a system of elected representatives or, at the level of States, through treaties that are, again, endorsed by representatives.[74] A linking arrangement based purely on administrative or technical regulations, without a corresponding mandate in formal law, thus risks being perceived as deficient in legitimacy; with rising trade flows and ensuing impacts on the balance of payments of participating jurisdictions, such a linking arrangement will likely be challenged, either on legal grounds or at the level of public opinion. A formal treaty and clear statutory mandates can help forestall any such concerns.

Another important condition for the social acceptability of a linking arrangement can be expressed in procedural terms during the process of establishment, but also in its subsequent operation. At all stages of development, a linking arrangement should seek to ensure transparency;[75] provisions on linking should be

[74] D Bodansky, 'The Legitimacy of International Governance: A Coming Challenge for International Environmental Law?' (1999) 93 American Journal of International Law 596; it bears noting, however, that criteria for the legitimacy of public authority may vary across political and cultural boundaries.

[75] N Stern, *The Economics of Climate Change: The Stern Review* (CUP: Cambridge, 2006) 481.

clearly worded and precise, the processes leading to their adoption open to public scrutiny. Involvement by affected stakeholders and the public when designing the link, for instance when deciding on any restrictions that may apply to trade between linked schemes, can further help improve acceptance and confidence in the market.[76] Once the linking arrangement enters into force, disputes and irregularities may arise across the link between participants in each emissions trading scheme, necessitating adequate dispute settlement mechanisms, but also raising the question of accountability, both with regard to market participants and supervising institutions. Likewise, any activities relating to the trading link should generally be governed by principles of good faith, equity and fairness.

Moreover, both at the level of national and international law, a number of principles have emerged that govern both public and private activities relevant to the environment. Most importantly, these include the precautionary principle and the principle that the polluter should pay; while these principles may generally prove difficult to enforce in court and through other binding channels, they still have autonomous normative value.[77] Accordingly, their material stipulations should be, where possible, observed when designing a link across emissions trading schemes, for instance when considering possible impacts on environmental integrity resulting from the link.

5. Implementing a Market Link: Compliance with the Regulatory Framework

As trading schemes evolve from the early stages of conceptual debate to more advanced stages of implementation, emerging details of their operational design also allow an assessment of their interaction with substantive rules and principles at the level of international or municipal law. Likewise, the mechanisms and procedures used to implement a link can be measured against the framework of existing norms, both codified and customary.

Over time, the sustained viability and political acceptance of a trading link will depend on its ability to secure consistency with a number of unwritten norms and principles recognized across jurisdictions, guiding any form of public interference with economic activities and individual freedom.[78] For instance, the ability of measures adopted by public authority to promote and sustain a desired outcome may determine their necessity, and by extension their proportionality.

[76] J Ebbesson, 'Public Participation' in D Bodansky, J Brunnée and Ellen Hey (eds), *The Oxford Handbook of International Environmental Law* (OUP: Oxford, 2007) 681.

[77] N de Sadeleer, *Environmental Principles: From Political Slogans to Legal Rules* (OUP: Oxford, 2002), 305–40.

[78] Such norms can be rooted in customary law, constitutional doctrine, or have the status of common legal doctrines applied in all municipal legal systems, particularly rules of procedure, evidence, and jurisdiction; as general principles intrinsic to the idea of law and basic to all legal systems, they have been described as the common 'sediment' of the law, see K Tuori, *Critical Legal Positivism* (Ashgate: Dartmouth, 2002), 197–216.

In the case of linking, this may become relevant if the failures of integrity in one emissions trading scheme negatively influence the environmental effectiveness of other linked emissions trading schemes.

Yet while such norms and principles are undoubtedly important for the evolution of linking over the longer term, their immediate effect will remain more limited. By contrast, compliance with material provisions of international, regional and municipal law will be peremptory from the outset: a link must observe all constraints imposed by positive law, whether the latter originates in legislative statutes or substatutory ordinances and decrees, an international treaty, or customary practices accepted as law. Otherwise, it not only risks being annulled through a judicial challenge, but also may undermine the formal basis of any transactions carried out under the link. Generalizations are difficult when assessing the relationship of a link to positive law, as any such assessment will, by necessity, depend on the particular circumstances of each individual case. Some broad inferences can nonetheless be drawn from international law and domestic case studies; these are summarized below.

5.1 Compliance with international law

Under public international law, which is the body of rules governing the conduct of nation states and international organizations, as well as their relationship with each other and with natural and juridical persons,[79] a number of issue areas can acquire relevance for links between emissions trading systems. By default, any linking arrangement created and operating within the realm of international law will be bound by the stipulations of general international law and the international law of treaties, both of which set out doctrines and procedures necessary to any legal system. A link based on an international treaty, for instance, will be governed by the customary doctrines reflected in the Vienna Convention on the Law of Treaties as regards its conclusion, interpretation, amendment, and termination.[80]

An important aspect to consider under international law, moreover, is the Kyoto Protocol to the United Nations Framework Convention on Climate Change,[81] which has created tradable units recognized for compliance with the quantified emissions limitation and reduction objectives that it sets out for certain industrialized countries during the first commitment period.[82] In order to avoid a breach of their international obligations, parties that have entered

[79] V Lowe, *International Law* (OUP: Oxford, 2007), 5–6.

[80] VCLT, arts 6–18, 31–33, 39–41, and 42–45, respectively.

[81] Kyoto Protocol to the United Nations Framework Convention on Climate Change (10 December 1997, entered into force 16 February 2005) (1998) 37 ILM 22 (Kyoto Protocol).

[82] See Kyoto Protocol, Art 3(1), which specified that such parties 'shall, individually or jointly, ensure that their aggregate anthropogenic carbon dioxide equivalent emissions' of specified greenhouse gases 'do not exceed their assigned amounts, calculated pursuant to their quantified emission limitation and reduction commitments inscribed in Annex B'.

such obligations need to avoid a disparity between actual emissions and the number of units assigned under the Protocol.[83] If two parties to the Kyoto Protocol link their national trading schemes, they can ensure congruence between Kyoto units retired at the end of the compliance period and real emissions by either basing their domestic units on units recognized under the Kyoto Protocol,[84] or by 'shadowing' each trade of domestic units across schemes with a concurrent transfer of Kyoto units between jurisdictions.[85] Alternately, they may create an exchange mechanism for domestic units, with trade across jurisdictions occurring only in Kyoto units.[86]

Carbon units generated in states that are not a party to the Kyoto Protocol cannot be used for compliance with the Kyoto Protocol, even if the emissions reduction they reflect is genuine and additional.[87] As a result, parties with quantified emission reduction and limitation obligations are unlikely to link their national emissions trading schemes with schemes in countries which have not ratified the Kyoto Protocol, given that units purchased from the latter will not be accompanied by Kyoto units; complex gateway arrangements can be implemented to partially overcome this obstacle, although they will increase transaction costs and thus prove less attractive.[88] Under the current regime, therefore, linkages between Kyoto parties subject to quantified emissions targets and all other jurisdictions are unlikely to emerge. A similar assessment will also apply to any international agreement setting out mitigation commitments for the period beyond 2012, when the first commitment period of the Kyoto Protocol expires, provided such an agreement retains the concept of proprietary and tradable compliance units.

[83] J Bazelmans, 'Linking the EU ETS to other Emissions Trading Schemes' in M Faure and M Peeters (eds), *Climate Change And European Emissions Trading Lessons for Theory and Practice* (Edward Elgar: Cheltenham, 2008), 301.

[84] Pursuant to Commission Regulation (EC) No 2216/2004 of 21 December 2004 for a standardized and secured system of registries pursuant to Directive 2003/87/EC of the European Parliament and of the Council and Decision No 280/2004/EC of the European Parliament and of the Council (Registries Regulation) [2004] OJ L386/1 art 45, allowances issued within the European emissions trading scheme are converted Assigned Amount Units (AAUs) to which a unique unit identification code has been added.

[85] Haites (n 17 above), 12.

[86] Such a mechanism would allow sellers to exchange domestic units in their jurisdiction for Kyoto units, have the latter transferred to the jurisdiction of the purchaser, where domestic units are then issued for the Kyoto units, see Haites (n 17 above), 12.

[87] Betz and Stafford (n 46 above), 101; D Bodansky, 'Linking U.S. and International Climate Change Strategies' (2002), 4 <http://www.pewclimate.org/docUploads/us_international_strategies.pdf>, accessed 30 November 2008; Ellis and Tirpak (n 24 above), 20.

[88] Sterk and others, 'Ready to Link Up? Implications of Design Differences for Linking Emissions Trading Schemes' (2006) Jet-Set Working Paper I/06, 63–5 <http://www.wupperinst.org/Uploads/tx_wibeitrag/ready-to-link-up.pdf>, accessed 30 November 2008: under such a mechanism, units sold from a scheme covered by the Kyoto Protocol to participants in jurisdictions which are not parties to the Protocol would be stripped of their Kyoto status; while the domestic unit would flow into the foreign scheme, its Kyoto status would be retained in a 'gateway', allowing units to be sold back into the covered scheme to the extent that they can be reverted to Kyoto units with the stripped status in the Gateway; a similar solution could be implemented for unilateral links, although the stripped Kyoto units would then simply be retired, see Bodansky (n 87 above), 4.

As of late, attention has also been directed to the interface of emissions trading and international trade law. Mostly, however, concern in this area is focused on specific issues, such as the imposition of border adjustments on foreign manufacturers of imported products.[89] As an analysis of relevant treaties reveals, the free trade disciplines set out for trade in goods and services are unlikely to constrain a link between emissions trading markets;[90] while certain complementary activities, such as the provision of brokerage services, may fall within the scope of the trade liberalization regime, greenhouse gas units as such—and hence also their trade across a market link—are not currently governed by international trade law.[91]

5.2 Compliance with regional and national law

When linking trading schemes established at the national and regional level, it stands to reason that municipal law—that is, the domestic law of states and constituent entities, and of organizations of regional economic integration such as the European Community—will be of relevance on a number of levels. First and foremost, municipal law will be important in that its sets out the regulatory architecture of the trading scheme, with all the related administrative and institutional features. But municipal law can also have a bearing on emissions trading, and hence linking, through all the other rules and principles affecting trade in allowances and associated economic activities, be it by way of fiscal and accounting requirements, supervision of financial markets, or regulation of exchanges, banking, and insurance services. Also, interactions may occur with the wider body of environmental and energy legislation already in place in any jurisdiction. And finally, the institutional powers and responsibilities set out under municipal law may determine which entities have the ability to engage in transboundary linking arrangements. While implications will obviously differ for each trading scheme, the following case studies can serve as examples of the various legal issues that can potentially arise.

[89] For an overview, see M Mehling, N Meyer-Ohlendorf and R Czarnecki, 'International Trade Policy in a World of Different Carbon Prices' in European Parliament (ed), *Competitive Distortions and Leakage in a World of Different Carbon Prices* (European Parliament: Strasbourg, 2008), 23–34; H van Asselt, F Sindico and M Mehling, 'Global Climate Change and the Fragmentation of International Law' (2008) 30 Law & Policy 423, 433.

[90] A different assessment may apply under North America Free Trade Agreement (NAFTA), which could pose an obstacle to linking between NAFTA member states and third states under exclusion of other NAFTA members, see E Quinn, 'Market Convergence through the Back Door: Inadvertent Integration of the World's Carbon Markets under NAFTA' (2008) 2 Carbon and Climate Law Review 181.

[91] See J Werksman, 'Greenhouse Gas Emissions Trading and the WTO' (1999) 8 Review of European Community and International Environmental Law 1, 2; ZX Zhang, 'Greenhouse Gas Emissions Trading and the World Trading System' (1998) 32 Journal of World Trade 219, 225.

5.2.1 Case study: United States

In the United States, a range of procedural and substantive considerations may acquire relevance when establishing a link between a domestic emissions trading scheme and foreign schemes.[92] If a trading link is to be established unilaterally or by way of mutually reciprocal domestic legislation, the legislative procedure for domestic lawmaking will apply.[93] A distinct procedure applies if the link is to be implemented through a formal treaty, however. Under the federal Constitution, only the President 'shall have Power, by and with the advice and consent of the Senate, to make treaties, provided two-thirds of the Senators present concur'.[94] Where linking is to occur between a foreign scheme and a scheme adopted at the level of individual states, this may have implications for the admissibility of the linking arrangement; indeed, the Constitution expressly prohibits any State from entering into a 'treaty, alliance, or confederation' or from entering 'without the Consent of Congress ... into any Agreement or Compact ... with a foreign Power.'[95] In essence, this provision denies the constituent states international legal personality, limiting their ability to participate in diplomatic relations and altogether barring them from the conclusion of an international treaty.[96]

While the federate states may thus be precluded from entering into a formal treaty with other jurisdictions, they are empowered to adopt a binding 'compact' or 'agreement' with the consent of Congress. And even in the absence of Congressional endorsement, individual states may, under certain circumstances, enter into a compact or agreement with foreign powers. As the United States

[92] Linking may also become relevant between different domestic schemes, for instance between various schemes at the regional or state level and a federal trading scheme. Some of the legal considerations discussed below, for instance regarding international legal personality and the ability to conclude treaties, only apply to links with foreign schemes; yet the conceptual definitions outlined earlier in this chapter apply equally to domestic links.

[93] At the federal level, for instance, this will usually require a member of Congress to introduce a bill for consideration; following committee and subcommittee approval, the bill will be reported to the full legislature. If both legislative bodies—the House and the Senate—approve the bill in identical form, it becomes enrolled and sent to the President for signature into law. Congress may override a presidential veto through a two-thirds vote in Congress by a quorum of members in both the House and Senate; see DJ Oleszek, *Congressional Procedures and the Policy Process* (7th edn, CQ Press: Washington, DC, 2007), 17–22.

[94] US Constitution art II § 2, clause 2. With certain exceptions, international treaties do not become binding law until Congress has enacted implementing legislation, see N Purvis, 'Paving the Way for U.S. Climate Leadership – The Case for Executive Agreements and Climate Protection Authority' (2008) Resources for the Future Discussion Paper 4/2008, 9–10 <http://www.rff.org/News/Features/PublishingImages/DP-08-09.pdf>, accessed 15 June 2009.

[95] See US Constitution art I § 10.

[96] L Henkin, *Foreign Affairs and the United States Constitution* (2nd edn, Clarendon Press: Oxford, 1996), 149–150; the scope of this restriction is wide, covering all binding international arrangements 'regardless of title, designation, or form', see R E Dalton, 'United States' in D B Hollis, M R Blakeslee and L B Ederington (eds), *National Treaty Law and Practice* (Martinus Nijhoff: Leiden, 2005), 766.

Supreme Court has declared, a compact with a foreign power requires Congressional approval only if it tends 'to the increase of political power in the States which may encroach upon or interfere with the just supremacy of the United States.'[97] As the Restatement Third of the Foreign Relations Law of the United States comments, 'agreements involving local transborder issues, such as agreements to curb a source of pollution . . ., have been considered not to require Congressional consent.'[98] Accordingly, it appears possible, albeit not certain, that a linking agreement could be adopted without federal endorsement by way of a state compact or agreement. Ultimately, however, Congress can always supersede such state arrangements by legislation.

Should the preceding options prove unfeasible, the federate states can always resort to amending their internal legislation with a view to including rules on the mutual recognition of foreign units. Since neither party is legally bound to maintain its law, reciprocal legislation adopted concurrently by two or more jurisdictions does not constitute a treaty, nor a compact or agreement requiring Congressional consent. Affording a means of circumnavigating the constraints of international and constitutional law, such reciprocal recognition could be based on an informal understanding setting out the substantive provisions required to create an operational trading link. Any institutional responsibilities could be assigned to a private body established and funded by the respective participants, obviating the need for recourse to international law. Reciprocal arrangements have been concluded with foreign powers in the past, occasionally also to overcome procedural constraints.[99] Once the federal legislature adopts legislation on greenhouse gas emissions trading that is sufficiently dominant or pervasive to 'occupy the field',[100] however, continued operation of state or regional schemes might be pre-empted by the supremacy of federal law.[101]

[97] *Virginia v Tennessee* 148 US 503, 519, 13 S Ct 728, 734, 37 L Ed 537 (1893); Consent to an agreement is thus only required if the agreement tends to give the state elements of international sovereignty, interferes with the full and free exercise of federal authority, or deals locally with a matter on which there is or might be national policy, see Henkin (n 92 above), 155.

[98] American Law Institute, *Restatement of the Law, Third: The Foreign Relations Law of the United States* (American Law Institute: St. Paul, Minn., 1987) § 302, Comment f.

[99] Almost all states have, for instance, promulgated 'parallel uniform policy declarations' on child support enforcement, under which the state will give effect to child support order of a foreign jurisdiction if the foreign country has one and gives effect to support orders of the state, see GF DeHart, 'Comity, Conventions, and the Constitution: State and Federal Initiatives in International Support Enforcement' (1994) 28 Family Law Quarterly 89; in *Robertson v General Electric Co.*, 32 F2d 495 (4th Cir 1929), cert denied, 280 US 571 (1929), finally, Congress used reciprocal legislation as a substitute for a treaty with Germany to which the Senate failed to consent, with Germany, in turn, also adopting legislation.

[100] Henkin (n 96 above), 157–8.

[101] See E Haites and M Mehling, 'Linking Existing and Proposed GHG Emissions Trading Schemes in North America' (2009) 9 Climate Policy 373–88.

5.2.2 Case study: European Union

In Europe, the supranational *acquis* constituting European Community law spells out a number of substantive rules and principles that could affect a link between the European emissions trading scheme and foreign schemes. Among these, the directive establishing the European emissions trading scheme specifies certain conditions for a link to trading schemes in third states, and sets out a mandate to explore agreements with 'third countries listed in Annex B to the Kyoto Protocol which have ratified the Protocol to provide for the mutual recognition of allowances between the Community scheme and other greenhouse gas emissions trading schemes'.[102] Also, the directive empowers the Commission to 'draw up any necessary provisions relating to the mutual recognition of allowances' under such an agreement. A concrete mandate such as the one set out in the directive strongly determines the nature and scope of any link. On the one hand, it defines linking as a desirable objective and obviates the need to justify any related action. At the same time, the mandate also defines the conditions for its exercise, including material and formal restrictions on the scope and partners of the link.

Under the terms of the foregoing mandate, for instance, links with regional and sub-federal schemes as well as with national schemes in countries that have not ratified the Kyoto Protocol are currently precluded. While this mandate is phrased in positive terms, and thus does not categorically rule out the adoption of a linking agreement under different terms than those set out in the directive,[103] it is unlikely that the European Community would pursue efforts beyond the scope of an explicit mandate. In late 2008, a comprehensive revision of the trading scheme took account of this obstacle, however, by introducing greater flexibility in the creation of trading links for the period beyond 2012. Under the new mandate, eligibility of suitable linking partners will be expanded to 'compatible mandatory greenhouse gas emissions trading systems with absolute emissions caps established in any other country or in sub-federal or regional entities.'[104] Again, the crucial

[102] Directive 2003/87/EC of the European Parliament and of the Council of 13 October 2003 establishing a Scheme for Greenhouse Gas Emission Allowance Trading within the Community and Amending Council Directive 96/61 [2003] OJ L275/32, art 25(1).

[103] Moreover, the directive setting out the mandate is only derived legislation, without prejudice to the powers conferred on the European Community under its constitutive treaty; accordingly, the Council could theoretically move forward on the basis of its conferred powers and without observing the constraints imposed by the mandate.

[104] Additionally, '[n]on-binding arrangements may be made with third countries or with sub-federal or regional entities to provide for administrative and technical coordination in relation to allowances in the Community scheme or other mandatory greenhouse gas emissions trading systems with absolute emissions caps', see Directive 2009/29/EC of the European Parliament and of the Council of 23 April 2009 amending Directive 2003/87/EC so as to improve and extend the greenhouse gas emission allowance trading scheme of the Community, [2009] OJ L140/63, Art 25(1a) and (1b).

determination of what constitutes a 'compatible' scheme will be inherently political in nature, not legal.

Aside from the foregoing limitations, the mandate also sets out procedural requirements by expressly referring to a process for the adoption of international agreements between the European Community and third states or international organizations.[105] Essentially, this process grants the Commission a power to negotiate international agreements, subject to prior authorization by the Council. Once the negotiations result in an agreement on the text of the agreement, the Commission submits the text to the Council for approval. Ratification of the international agreement—a vital precondition of its entry into force—occurs through a Council decision. In the case of a linking agreement, the Council would have to decide by qualified majority, and with the participation of the European Parliament.[106] Due to shared competences between the European Community and its Member States, the link would probably be adopted as a mixed agreement to which both the Community and the Member States are parties.[107]

If the trading link between the European trading scheme and another scheme is established unilaterally or by way of mutually reciprocal domestic legislation, the applicable legislative procedures in the Community will be the same governing the adoption of the original scheme. In the case of the emissions trading directive, the legislative basis was Article 175(1) of the EC Treaty, which sets out a general mandate for action on environmental protection. It calls upon the European Commission to submit a proposal to the European Parliament and the Council, which votes by qualified majority and in co-decision with the European Parliament. While the mandate and procedures it calls for can be amended, such an amendment will typically be subject to the same procedural requirements governing the adoption of the mandate itself, and thus necessitate a protracted legislative or diplomatic effort that can be politically undesirable.

6. Conclusions

As they have evolved to date, carbon markets are proving to be complex entities, embedded in sophisticated networks of contingent interests, traditionally diverse approaches to governance, and distinct regulatory constraints.

[105] EC Treaty (Treaty of Rome, as amended), Art 300.

[106] For a detailed description of the negotiating process, see T Delreux, 'The European Union in International Environmental Negotiations: A Legal Perspective on the Internal Decision-making Process' (2006) 6 International Environmental Agreements: Politics, Law and Economics 231, 236; D Thieme, 'European Community External Relations in the Field of the Environment' (2001) 10 European Environmental Law Review 252, 255.

[107] Nearly all agreements on environmental protection have been concluded by way of mixity, including the Framework Convention on Climate Change and the Kyoto Protocol; see generally, R Leal-Arcas, 'The European Community and Mixed Agreements' (2001) 6 European Foreign Affairs Review 483, 494.

Reconciling the inevitable differences between two or more trading schemes in pursuit of greater integration is not primarily a task for lawyers or the law,[108] but an inherently political assignment. For lawyers, the challenge will be to translate the negotiated consensus into legally viable arrangements, observing applicable rules of domestic, regional and international law.

As the foregoing chapter has shown, a link can be created with a fairly straightforward decision on the acceptance of foreign units. Legal challenges—both procedural and substantive—tend to follow once the link is expected to satisfy qualified requirements in terms of its formality, its binding force, and institutional ramifications. Clearly, an international treaty will offer the greatest degree of certainty and predictability, yet it also faces the greatest range of formal and substantive constraints at the level of both international and municipal law. Again, overcoming any such obstacles will largely be a matter of political will and legislative priorities; as market integration deepens, participants may nonetheless wish to explore new options, such as a multilateral link with a centralized institutional architecture. By that point, finally, a link will also be measured against normative precepts beyond the rules and principles of positive law; driven by formal and informal cooperation between states, regions, provinces, local municipalities, and business,[109] the emerging global carbon market will also need to demonstrate its legitimacy as an effective means of addressing climate change, and offer sufficient transparency and accountability to stakeholders.

[108] Clearly, lawyers and the courts—particularly in common law jurisdictions—will assume an important role in operationalizing trading links at the interface of law and politics, notably where the legislator has overlooked potential challenges, left them for resolution under the law, or consciously decided not to pursue a political determination.

[109] Egenhofer and others (n 5 above), 8.

6

Private Actors in International and Domestic Emissions Trading Schemes

Jolene Lin

1. Introduction

The Kyoto Protocol (KP) is the first multilateral environmental agreement to give private actors a significant role to play in the implementation and enforcement of treaty obligations undertaken by States. To enable States to meet their treaty obligations, principally the legally binding greenhouse gases (GHG) emission reduction targets set out in Annex B of the KP, the Protocol includes the 'Flexible Mechanisms' or 'Kyoto Mechanisms'—Joint Implementation (JI), Clean Development Mechanism (CDM) and emissions trading.[1] The Kyoto Mechanisms seek to harness market forces and direct private investment flows towards climate change mitigation. As a result, an international carbon market has emerged. Further, several States have elected emissions trading (or have plans to do so) as a key component of their climate change policies. The European Union, for example, launched the European Union Emissions Trading Scheme (EU ETS) in January 2005.[2] The EU ETS is the largest multi-country, multi-sector GHG emissions trading scheme to date. It covers over 11,500 installations across the EU, which represents close to 50% of the EU's carbon dioxide emissions. Following Australia's ratification of the KP at the end of 2007, steps to implement the national Carbon Pollution Reduction Scheme are underway. The scheme, which the government hopes will commence in 2010, will cover all six GHGs (and not just carbon dioxide) and sectors including transport, waste, fugitive emissions from oil and gas production, and certain industrial processes.[3] Finally, in the United States, President Barack Obama has publicly stated that he plans to implement an economy-wide GHG cap-and-trade program and

[1] Joint Implementation, the Clean Development Mechanism, and Emissions Trading are established under Arts 6, 12, and 17 of the Kyoto Protocol respectively.

[2] European Commission's DG Environment website: <http://ec.europa.eu/environment/climat/emission/index_en.htm>, accessed 10 November 2008.

[3] Australian Government, Department of Climate Change website: <http://www.climatechange.gov.au/emissionstrading/index.html>, accessed 10 November 2008.

there are currently almost a dozen different proposals under consideration by Congress.[4]

The use of emissions trading at the international, regional and state levels to address climate change has created opportunities for private actors to participate in environmental protection (reputational effect, corporate social responsibility/ socially responsible investment), to meet their legal obligations under domestic environmental laws cost-effectively, and to the extent that carbon markets represent a new frontier in financial innovation, carbon trading presents a business opportunity for project developers and financial institutions. Accompanying these opportunities are risks that private actors face when they participate directly in the global carbon market or in local/ regional schemes. Identifying and understanding these risks is important for sound risk management. Astute risk management benefits not only private actors involved in the carbon markets. It is also vital for the systemic stability of the nascent market, which in turn is crucial for realizing the environmental benefits of emissions trading. This chapter therefore analyzes the risks and opportunities for private entities in the international climate regime and domestic emissions trading schemes respectively. This chapter does not cover the voluntary carbon market as this is discussed elsewhere in this book.[5]

The chapter begins with an overview of carbon markets and the ways in which private actors are involved. The focus is on private activity in the primary carbon market, and not the secondary carbon market, given that most of the risks faced by private actors in the secondary market are not very different from those faced by entities involved in trading securities and derivative financial products, except perhaps as regards the degree of uncertainty associated with the underlying carbon asset which differs from most commodities. Section 3 focuses on the risks involved in project-based transactions, which is the type of transactional activity that private actors are most likely to engage in. The CDM is the best example. In a market where the creation or production of the asset is not in the control of market players, but in the hands of a regulator, the risks of regulatory delay and uncertainty must be treated as significant commercial risks. As such, the problems of a weak institutional framework for the governance of the CDM are no longer academic concerns, but bear real practical and financial implications. While the carbon regulatory infrastructure requires prompt reform, private actors in the meantime have to manage the risks and rewards for various parties in their contractual arrangements. Section 4 discusses some of the risks that private actors face in the EU ETS. A comparison between the risks faced by private actors in the international carbon market and those faced in the EU ETS

[4] For a comparison of proposed bills, see <http://www.pewclimate.org/federal/analysis/congress/110/cap-trade-bills>, accessed 1 December 2008; the Obama-Biden Plan: Energy & Environment, <http://change.gov/agenda/energy_and_environment_agenda>, accessed 1 December 2008.

[5] See chapter 23 below.

will also be made. Section 5 looks briefly at the new issues and risks raised by Phase III of the EU ETS, and section 6 concludes the discussion.

2. Emissions Trading and the Carbon Market

2.1 What is a carbon transaction?

A carbon transaction takes place when one party pays another party for GHG emission reduction assets, which the buyer can then use to meet its objectives in relation to climate change mitigation. Such objectives can include compliance with domestic law (for example, in the case of a company that is regulated by the EU ETS) or international law (a State that has to comply with its KP emissions reduction target), or financial gain (for example, carbon funds buy emission reduction assets to transact in the secondary market). In other words, a carbon transaction involves the sale and purchase of GHG emission reduction assets, and the distinction between different types of carbon transactions lie in the *type of asset* that is being transacted between the parties.

An *allowance-based transaction* is one in which the buyer purchases emission allowances issued by the regulator of a cap-and-trade scheme, such as Assigned Amount Units (AAUs) under the KP or European Union Allowances (EUAs) under the EU ETS.[6] In a cap-and-trade scheme, the regulator determines an acceptable level of emissions, thus establishing a 'cap' for the scheme. The cap is usually a target that is lower than 'business as usual' emissions levels and the long-term objective is to reduce the cap in order to achieve environmentally more stringent standards. Once the cap is established, allowances that permit the emission of a stipulated amount of pollutant will be allocated amongst firms. The state may allocate the allowances to each sector or facility, which is known as 'grandfathering', or allowances may be allocated by auction.

The amount of emissions permitted by the aggregate quota is equivalent to the cap. A firm that wishes to emit more than its quota has to purchase additional allowances or it risks exceeding its quota, thereby incurring criminal and/or civil liability.

A *project-based transaction* involves a buyer purchasing emissions reduction assets from a project that can demonstrate that GHG emissions were reduced against the 'baseline scenario', a representation of GHG emissions under normal

[6] In accordance with Article 3(7) of the Kyoto Protocol, each Annex I Party is assigned AAUs equal to the percentages inscribed for it in Annex B during the Protocol's first commitment period (2008–2012). Articles 3(10) and 3(11) provide for the subtraction and addition of AAUs and Emission Reduction Units resulting from emissions trading and JI respectively. Article 3(13) allows for the banking of excess AAUs to future commitment periods. An AAU is equal to one metric ton of carbon dioxide equivalent (tCO_2e). The European Union Allowance is the allowance in use under the EU ETS as defined in Article 3(a) of Directive 2003/87/EC. It is also equal to one tCO_2e.

circumstances, for example, a project generating electricity from renewable sources instead of coal which would have produced more GHGs. This is otherwise known as the 'additionality requirement' in the CDM: CDM project activities must result in reducing or absorbing (sequestering) GHGs that are 'real and measurable and would not have occurred in the absence of the proposed project activity'.[7] The CDM and JI are the most prominent examples of project-based transactions. Each tCO_2e of GHG that is not produced as a result of the project translates to an Emission Reduction Unit (ERU) or Certified Emission Reduction (CER) in JI and the CDM respectively. To avoid confusion, project-based emissions reductions will henceforth be referred to as 'credits' in this chapter.

Cutting through the potential confusion caused by these acronyms, AAUs, EUAs, ERUs and CERs are all units or currencies of carbon trading. The acronyms help us identify the legal framework that creates the unit and regulates the activities surrounding its use. The risks associated with buying allowances or credits also differ. It is therefore important that we are clear about the type of asset that is being transacted.

2.2 Defining the carbon market

The primary market consists of the allowances market (AAUs and EUAs) and the market in credits (ERUs and CERs). A secondary market is emerging as the large potential of the carbon market has attracted financial actors who have, in turn, introduced financial instruments that have the potential to attract further capital towards climate change mitigation. These financial instruments include carbon delivery guarantees, derivatives, carbon-linked bond transactions, and monetization of future carbon receivables.[8] The secondary market does not receive detailed consideration in this chapter because the risks that private actors face in this market are not necessarily carbon-specific.[9] The nature of the activity and products in the secondary market render it more suited to regulation by securities law.[10] Figure 6.1 below illustrates the actors in the carbon market. The discussion in this chapter is primarily concerned with the sale and purchase of credits and allowances between suppliers and consumers.

[7] UNFCCC, Report of the Conference of the Parties on its seventh session, Marrakech, 29 October–10 November 2001). Addendum part two: Action taken by the Conference of the Parties. Vol II. FCCC/CP/2001/13/Add 2, at 20. For a full discussion of additionality see chapter 12 below.

[8] Capoor & Ambrosi *State and Trends of the Carbon Market 2008* (The World Bank: Washington DC, 2008), 64–5.

[9] But see also chapter 15 below.

[10] As for whether allowances are regulated by securities law, various EU jurisdictions have come to different conclusions in their implementation of the Emissions Trading Directive (Directive 2003/87/EC) depending on whether the jurisdiction classifies allowances as intangible patent-like assets (eg Finland, in which case the allowances are not regulated by securities law) or financial instruments (eg Sweden, in which case securities law applies): see discussion in Karoliina Anttonen, 'Breathing Life into the Carbon Market Legal Frameworks of Emissions Trading in Europe' (2007) 16 European Environmental Law Review 96.

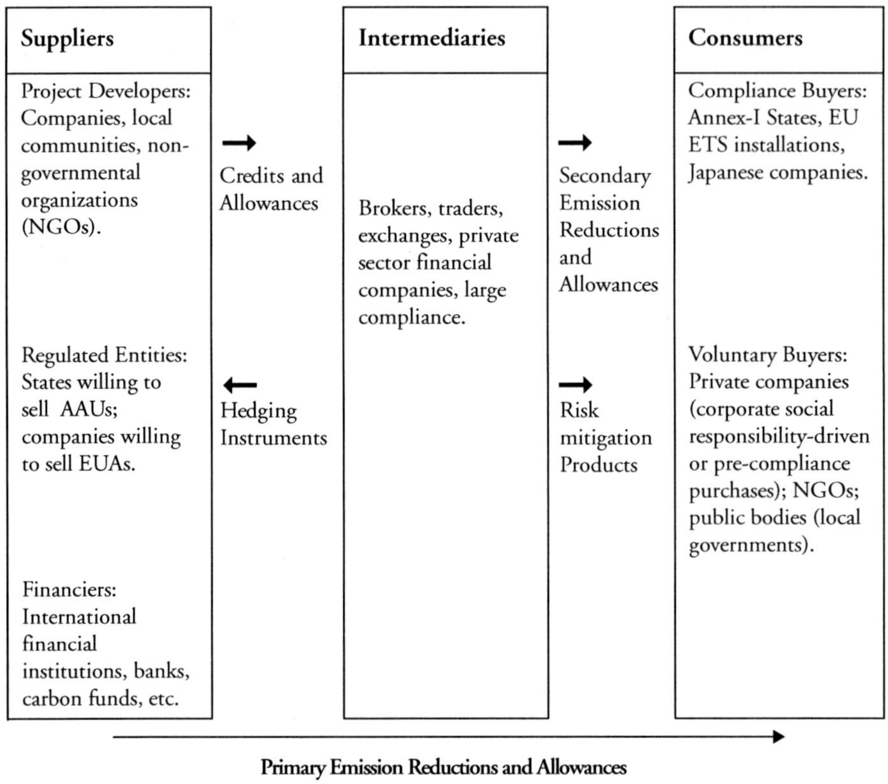

Figure 6.1. Actors in the carbon market

3. Risks and Opportunities in the Global Project-Based Carbon Regime

Allowances are homogenous assets created by law, and therefore are relatively risk-free for private entities that are buying emission reduction assets for compliance purposes. Private actors are not involved in creating allowances. It is in the realm of carbon projects and the emission reductions that result from these projects that private actors have a significant role to play.[11] As the CDM

[11] Article 6(3) of the Kyoto Protocol states, 'A Party included in Annex I may authorize legal entities to participate, under its responsibility, in actions leading to the generation, transfer or acquisition under this Article of emission reduction units.' and Article 12(9) of the Kyoto Protocol states 'Participation under the clean development mechanism, including in activities mentioned in paragraph 3(a) above and in the acquisition of certified emission reductions, may involve private and/or public entities, and is to be subject to whatever guidance may be provided by the executive board of the clean development mechanism'. These articles provide the Parties to the Kyoto Protocol the legal basis to authorize the participation of private entities in the generation and transfer of carbon rights.

accounts for most project-based market activity (87% of volumes and 91% of value transacted) and to avoid confusion, the discussion in this section will focus on the CDM only.[12] The CDM permits non-State actors to carry out projects that will lead to the issuance of CERs, and empirical data demonstrates that the private sector, not States, is driving the development of the CDM market.[13] This section identifies the regulatory and legal risks that private actors face in CDM projects.

The risks that are inherent in CDM activity stem primarily from the fact that the asset class is generated by an international regulatory process. Therefore, regulatory risks such as the timing of issuance of credits, approval delays, and extensive bureaucratic demands, are significant for private actors involved in the CDM. The CDM regulatory regime today is under-developed and requires significant reform in order to meet market demand. Until reform is underway, private entities will have to try to manage these regulatory risks through contractual arrangements.

Furthermore, while the CDM is a creation of an international treaty, the relationships between private parties and/or States in relation to underlying CDM projects and the sale and purchase of resulting CERs are governed by domestic law. Most jurisdictions have not passed laws to specifically regulate CDM activity. As a result, lawyers have resorted to prevailing contract law, tort law, property law, tax law and so on to allocate the risks and benefits of CDM activity amongst the parties. Some of these issues, such as taking security over carbon assets, will be discussed in detail in later chapters. This chapter will therefore merely touch on these matters briefly as part of the overview of the risks that private actors face in the global carbon market.

3.1 Legal risks

3.1.1 The right to review Executive Board decisions

Private actors participating in the CDM have been frustrated by the lack of transparency and predictability in the Executive Board's decision-making.[14] Faced with decisions that they consider to be incomplete or incorrect, they do

[12] Capoor and Ambrosi (n 8 above), 19.

[13] Private companies have been the most active buyers in the CDM and JI market, accounting for 79% of the aggregate volume transacted in 2007: Capoor and Ambrosi (n 8 above), 23.

[14] IETA, Position paper to COP12/CMP 2 (on file with author); 'Strengthening the CDM', Position Paper for COP 11/CMP 1, <http://www.ieta.org/ieta/www/pages/getfile.php?docID=1132>; IETA letter dated 6 October 2006 to the Chair of the Executive Board regarding the communication with the Board and its panels; World Bank, 'Reforming the Clean Development Mechanism', Background Paper for the Steering Committee Meeting, World Bank Carbon Finance Host Country Committee, <http://carbonfinance.org/Router.cfm?Page=DocLib&CatalogID=5668>, accessed 10 August 2007; Center for Clean Air Policy (Washington DC), Summary of an informal workshop on streamlining the CDM, CMP 1, <http://www.ccap.org/international/Summary%20of%20Day%201%20&%202%20Discussions.pdf>, accessed 10 February 2008.

not enjoy basic procedural rights, such as the right to be heard and to appeal against an Executive Board decision—which magnifies the difficulties caused by the unpredictability of the Board's decisions.[15] These decisions or its omissions, of course, have significant impact on project developers and their investments. Affected private entities also lack direct access to the Kyoto Protocol Conference of the Parties/Meeting of the Parties (COP/MOP) and can only make themselves heard through their governments.[16] Private entities that perceive themselves to be directly affected by the decisions of an administrative body therefore do not have access to a court or independent tribunal that reviews these decisions.[17]

It is arguable that future developments will include reforming the way the Executive Board operates, and setting up a mechanism that gives private parties an opportunity for independent review of decisions and timely resolution of disputes.[18] In the meanwhile, the review of administrative CDM decisions by national courts may be an option.[19] Whether such lawsuits would eventually be dismissed would depend on the substantive law of the State concerned. Private actors currently have limited recourse in the event of a wrongful decision by the Executive Board and this factor should be adequately reflected in the risk allocation amongst the parties involved in developing a CDM project.

[15] The need to provide for administrative due process was recognized by the UNFCCC negotiators and is reflected in the Marrakech Accords and in other decisions of the CMP. The Marrakech Accords provide for the independence of the Executive Board (Decision 3/CMP.1, Annex, para 8(c) and (f)), for open sessions and participation in CDM Executive Board meetings (Decision 3/CMP.1, Annex, para 16 and 17), and the rules of procedure for the Executive Board including participation in requests for review of DOE decisions initiated by the Executive Board (Decision 3/CMP.1, Annex, para 8(j), (k) and (o)). However, the Accords do not establish any formal right of review of any Executive Board decisions. Neither do they establish formal rights to hearings.

[16] The COP/MOP is the decision-making body of the KP. Article 13(4) of the KP states that the COP/MOP '... shall keep under regular review the implementation of this Protocol and shall make, within its mandate, the decisions necessary to promote its effective implementation'.

[17] Ernestine E Meijer, 'The International Institutions of the Clean Development Mechanism Brought Before National Courts: Limiting Jurisdictional Immunity to Achieve Access to Justice' (2007) 39(4) New York University Journal of International Law and Politics 925. See also chapter 13 below.

[18] It has been argued that in order to succeed, the CDM must provide a review mechanism that will give investors the requisite legal certainty: B Stewart, 'US Administrative Law: A Model for Global Administrative Law?'(2005) 68 Law & Contemporary Problems 63–109.

[19] Ernestine E Meijer, 'The International Institutions of the Clean Development Mechanism Brought Before National Courts: Limiting Jurisdictional Immunity to Achieve Access to Justice' (2007) 39(4) New York University Journal of International Law and Politics. Meijer argues that national courts are a suitable forum for addressing the lack of accountability at the international level. There is, however, the risk that a review of CDM procedures by national courts will jeopardize the coherence of the mechanism. The CDM is unlikely to survive as a global mechanism if it were subjected to litigation in various Parties and, consequently, differing judicial interpretations of the rights and obligations under the Kyoto Protocol. Streck & Lin have therefore argued that the best way to address the frustrated expectations of participants in the carbon market is to provide for an independent review mechanism: Charlotte Streck and Jolene Lin, 'Making Markets Work: A Review of CDM Performance and the Need for Reform' (2008) 19(2) European Journal of International Law 409–42.

3.1.2 Legal challenges by local stakeholders

The CDM is intended not only to help Annex I Parties meet their emission targets in a cost effective way, but also to promote sustainable development in developing countries.[20] In the past few years, CERs from large industrial HFC-23 decomposition projects have dominated the CDM, leading some commentators to criticize the CDM for failing to promote sustainable development.[21] The share of CERs from HFC-23 decomposition projects has decreased from their 2005 peak, and the number of clean energy projects (renewable energy, fuel switching and energy efficiency) has been increasing, thereby addressing concerns about the environmental benefits of the CDM.[22] Nonetheless, there are local communities and civil society in CDM host countries who remain concerned as to whether the CDM promotes sustainable (environmental, social and economic) development.[23] Project owners and DOEs should ensure that proper consultation and requisite environmental impact assessments are conducted so as to avoid potential disputes with local communities. Under the CDM Modalities and Procedures, the project developer must ensure that 'the public, including individuals, groups or communities affected, or likely to be affected, by the proposed clean development mechanism project activity' participate in designing the project.[24] The proposed project must be made available to the public for comments and a summary of received comments (and how these comments were taken into account) must be included in the Project Design Document (PDD). In the event that, following a preliminary assessment of the environmental impacts of the proposed project, the host country or the project participants consider these impacts to be significant, an environmental impact assessment (EIA) must be conducted.[25] The EIA results must be included in the PDD.[26]

Local stakeholders affected by a CDM project activity can seek judicial or administrative review of the executive decision to give host country approval to a CDM project that does not comply with local environmental and planning laws. Judicial review of the adequacy of the EIA consultation process is also a possibility.[27] Alternatively, stakeholders may use political channels to demand that their government (the host government) exercise its power of review to preempt

[20] See Article 12(2) of the Kyoto Protocol.

[21] See, for example, Michael Wara, 'Measuring the Clean Development Mechanism's Performance and Potential', Working Paper No 56, July 2006, Program on Energy and Sustainable Development at the Centre for Environmental Science and Policy, Stanford University.

[22] See analysis in Capoor & Ambrosi (note 8 above), 28–9 on project types.

[23] See case studies in PH May, E Boyd, F Veiga and M Chang, 'Local sustainable development effects of forest carbon projects in Brazil and Bolivia: A view from the field' (2004) International Institute for Environment and Development, London.

[24] Annex to Decision 17/CP.7/CMP.1, para 1(e). [25] Ibid, para 37(c). [26] Ibid.

[27] See, for example, *Greenpeace Ltd (on the application of) v Secretary of State for Trade and Industry* [2007] EWHC 311.

the issuance of CERs by the CDM Executive Board on the grounds that the DOE had been incompetent in conducting the consultation procedure prior to validating the PDD.[28] Such legal proceedings can give rise to considerable delay, which project owners can avoid by taking proper and adequate steps to procure the support of local stakeholders, or at least achieve compliance with relevant environmental and planning laws.

3.1.3 The legal nature of a CER

Purchasers of CERs will be concerned that they have legal title to these credits. In order to establish title, it will be necessary to determine if a CER constitutes property. Whether a CER is a form of property also bears implications, *inter alia*, for taxation issues arising from a sale and purchase of CERS (which very much depends on the domestic tax laws of the seller) and the treatment of the asset in insolvency proceedings. This section examines two English cases regarding the treatment of waste management licenses in insolvency proceedings to support the analysis that CERs are likely to be treated as property under common law.

The issue in *Environment Agency v Stout* was whether a waste management licence issued under the Environmental Protection Act 1990 could be disclaimed by an appointed liquidator in accordance with s 178 of the Insolvency Act 1986.[29] The Environment Agency argued that the licence was not 'property' within the terms of s 436 of the Insolvency Act and therefore did not come within the ambit of s 178. Justice Neuberger rejected the agency's argument and held that a waste management licence was 'property' within the terms of s 436 of the Insolvency Act.[30] The judge referred to the criteria laid down by Lord Wilberforce in *National Provincial Bank Limited v Ainsworth*:

> Before a right or an interest can be admitted into the category of property, or of a right affecting property, it must be definable, identifiable by third parties, capable in its nature of assumption by third parties, and have some degree of permanence or stability.

He was satisfied that the waste management licence in question satisfied this test.[31] Further, Justice Neuberger considered the facts that the legislation creating the waste management licenses provided for their transfer and that there is in fact a market for such licenses, provided support for the classification of the license as property.[32]

[28] See discussion below, 'Risks faced by Designated Operational Entities (DOEs)'.

[29] [1999] Env LR 407.

[30] Section 436 contains the following definition: 'Property includes money, goods, things in action, land and every description of property wherever situated and also obligations and every description of interest, whether present or future or vested or contingent, arising out of, or incidental to, property.'

[31] *National Provincial Bank Limited v Ainsworth* [1965] AC 1175 at 1247G–1248A.

[32] At p 415 of the judgment.

Similar facts arose in *In Re Celtic Extraction Ltd* and the Court of Appeal upheld the decision that a waste management licence came within the definition of 'property.'[33] The court considered the authorities and laid down some guidelines for determining if '... there is to be conferred on an exemption from some wider statutory prohibition the status of 'property.'[34] First, there must be a statutory framework conferring an element of discretion exercisable within that framework. Secondly, the exemption must be transferable. Thirdly, the exemption or license will have value. Based on this test, *allowances* such as EUAs are likely to be treated as property by the English courts. As for CERs, it can be argued that the Marrakech Accords and other applicable rules promulgated by the CDM Executive Board constitute the relevant statutory framework, thereby satisfying the first element of the three-pronged test. The other two requirements are easily met. Further, it should be noted that it is an established principle that the word 'property' is not a term of art but takes its meaning from its context.[35] In the likely event that the CERs were purchased for compliance purposes, and therefore serve the same function as allowances, this reinforces the argument that CERs should receive like treatment to allowances and therefore be classified as property. Finally, CERs are definable, identifiable by third parties, capable in their nature of assumption by third parties, and have some degree of permanence or stability (they cannot be destroyed except, arguably, through extinction by treaty), thereby satisfying Lord Wilberforce's test set out above. There is also clearly a market for CERs. Therefore, a CER is likely to be treated as property under common law.

How legal title to carbon credits can be established and transferred may be affected by the local property and land laws of the host CDM jurisdiction. While forestry projects are quite another kettle of fish, it is worth noting how certain jurisdictions have provided for the legal treatment of credits generated from forestry projects to see if lessons can be drawn.[36] Few countries have laws that define carbon rights. Australia is one of the few countries that have examined how to recognize the right to own sequestered carbon under their property law.[37] New

[33] [2000] 2 WLR 991.

[34] At p 489 of the judgment. The authorities considered by the Court of Appeal included *Attorney General of Hong Kong v Nai-Keung* [1987] 1 WLR 1339 (in which the Privy Council considered that textile export quotas were property within the definition in the Theft Ordinance of Hong Kong and therefore capable of being stolen) and *In re Rae* [1995] BCC 102 concerning an entitlement to be considered for the grant of a fishing licence. Warner J decided that the 'entitlement' was within the definition of 'property' as a present interest in property, namely the fishing vessels.

[35] See *Nokes v Doncaster Amalgamated Collieries Ltd* [1940] AC 1014, 1051; *Kirby v Thorn EMI plc* [1988] 1 WLR 445, 452.

[36] For extensive discussion, see Issue (2008) 2(3) Carbon and Climate Law Review which focuses on forestry carbon.

[37] Australian Greenhouse Office, 'Planning Forest Sink Projects—A Guide to Legal, Taxation and Contractual Issues' March 2005, <http://www.climatechange.gov.au/land/forestsinks.html>, accessed 15 June 2009.

South Wales, South Australia, Tasmania, Western Australia, and Victoria have introduced specific legislation recognizing the right to own carbon sequestered from vegetation on a specific area of freehold land. The name of this right varies across the states, but will hereby be referred to as a 'carbon sequestration right' for convenience. These states allow for the carbon sequestration right to be registered on the land title under the Torrens System. Once an interest is recorded on a land title, it attains 'indefeasibility of title' and is superior to unregistered interests.[38] Registered carbon sequestration rights 'run with the land' and are binding on future owners of the land for the period of registration. This provides buyers of carbon sequestration rights a degree of certainty. It also ensures that future land owners are aware of the existence of the carbon sequestration right, and covenants relating to maintenance of the vegetation on the land can be negotiated.

Where specific legislation on carbon sequestration rights does not exist, for example, in the Australian Capital Territory and the Northern Territory, the right to sequester carbon may be created as a personal right between the contracting parties. Such a right cannot be registered on the land title, and therefore does not bind future owners of the land. The project developer can mitigate its risks by requiring the landowner to provide prior notification of any intention to sell the land, and assignment of the agreement between the project developer and the present landowner to the future landowner.

3.2 Regulatory risks

3.2.1 Delay

It can take between one to two years for a project to proceed from the validation stage to registration, excluding the six months or so that it takes to procure the validation services of a Designed Operational Entity (DOE).[39] The length of time required for a CDM project to obtain registration by the CDM Executive Board (registration is a prerequisite for the issuance of CERs) has become a significant concern for project owners. Such delay can entail significant financial costs for project owners who have to maintain consultancy and legal services throughout the process. It also puts at risk project financing arrangements which are dependent on an expected delivery schedule. Furthermore, when registration of the CDM project is a condition precedent in an agreement for the sale and purchase of CERs, the failure to register by the stipulated date (usually the seller's obligation) can give rise to the right for either party to terminate the agreement without liability to the other party. Potential CDM project

[38] For discussion on the Torrens system and the concept of indefeasibility of title, see Samantha Hepburn, *Australian Principles of Property Law* (3rd edn, Routledge-Cavendish, 2007).

[39] Capoor and Ambrosi (n 8 above), 21; David Fogarty, 'Bureacracy rules delay Indonesia CDM projects' Reuters (23 October 2008), <http://uk.reuters.com/article/oilRpt/idUKSP40505720081023>, accessed 28 November 2008.

developers should be advised to take the length of time required for registration into consideration during negotiations and to err on the side of caution when agreeing on the date by which registration must be completed.

3.2.2 Host country risk

Before a project can proceed to the stage of validation by a DOE, the project participants must obtain the following approval by the host country Designated National Authority (DNA): (1) written approval of voluntary participation and (2) confirmation by the host country that the project will further its national sustainable development objectives.[40] Host country risk refers to the risk that the project fails to obtain, or faces significant delay in obtaining, approval by the DNA and/or that the CDM regulations in the host country hinder the development of the project. Project developers should be aware that some jurisdictions have promulgated domestic CDM laws that bear implications for the choice of project activity, structuring, and financing. These laws may add to the already significant host country risks that exist across the board in various CDM destinations. China is a case in point.[41] The *Measures for Operation and Management of Clean Development Mechanism Projects in China* forms the domestic regulatory framework for CDM projects in China.[42] It is clearly stipulated that the priority project activities are energy efficiency improvement, renewable energy, and methane recovery and utilization.[43] Proposals to carry out projects in the 'non-priority' categories may therefore be more strictly scrutinized by the DNA under the 'expected sustainable development effectiveness' requirement.[44] Article 24 creates incentives to undertake the priority CDM activities—the Chinese government levies a 65% tax on the CERs generated by hydroflurocarbon (HFC) and perflurocarbon (PFC) projects, and a 2% tax on the CERs generated by priority project activities and forestation projects.[45] Only 'Chinese funded or Chinese-holding enterprises within the territory of China' are eligible to carry out CDM projects with 'foreign partners'.[46] (Foreign project developers should require the inclusion of this eligibility requirement in the seller's representations and warranties clause in the contractual agreements and exercise caution in their due diligence). Article 17 refers to the 'Chinese funded or Chinese-holding enterprise' as the 'project owner' who has to '... protect state

[40] Decision 17/CP. 7/CMP.1, para 40(a).

[41] For detailed discussion of the CDM in China, see chapter 22 below.

[42] The Measures came into effect as of 12 October 2005, replacing the *Interim Measures for Operation and Management of Clean Development Mechanism Projects in China*. The Measures are available online at the official CDM in China website <http://cdm.ccchina.gov.cn/english/index.asp>, accessed 1 November 2008.

[43] Ibid, Art 4.

[44] Ibid, Art 15(8).

[45] Ibid, Art 24.

[46] This has posed difficulties for Hong Kong companies which are keen to improve the energy efficiency of their mainland operations and qualify for the CDM as the requirement of 'Chinese funded or Chinese-holding enterprises' has been interpreted to exclude Hong Kong companies; see Robin Kwong, 'China block on carbon trading for HK groups', *Financial Times*, 28 March 2008.

and business confidential information in accordance with relevant laws and regulations'. This should be read in light of the recently promulgated (on 1 May 2008) *Measures for Disclosure of Environmental Information (for Trial Implementation)* and the *Regulations of the People's Republic of China on Open Government Information*, as well as the *Official Secrets Law*.

3.2.3 Risks faced by Designated Operational Entities (DOEs)

So far, we have focused on the risks faced by buyers and sellers of CERs. The role of DOEs in CDM governance is also an interesting aspect of private sector involvement in the global climate change regime.[47] DOEs perform auditing and verification functions. As such, they serve as third-party independent guardians of the integrity (both procedural and environmental) of the CDM. Due to the large number of projects that require validation and verification, the current demand for DOE services far exceeds the number of specialists that are qualified to perform them. As such, there are significant growth opportunities in the DOE sector. The risks that DOEs face, such as their liabilities for erroneous validation, appear to be relatively manageable, at least on paper. However, a closer examination shows that DOEs face difficulties in performing their functions because of the lack of guidance on verification and validation standards and methodologies, the inherent uncertainty involved in predicting GHG reductions by projects, and current technical and knowledge constraints. These uncertainties have a negative impact on the relationship between the DOEs and the CDM Executive Board. The lack of trust between these regulators of the CDM increases the risks that DOEs face in operating in the CDM market. It also increases the regulatory risks that investors face when projects are subjected to a protracted verification process because the Executive Board hesitates to rely on the DOE reports.[48]

DOEs are accredited by the CDM Executive Board and designated by the COP/MOP on the basis of meeting a set of requirements including relevant expertise, financial stability, insurance coverage and resources.[49] The DOEs are critically involved in two stages of the CDM project cycle: (1) the validation stage and (2) verification. At the validation stage, the responsibility of a DOE is to validate a CDM project by independently evaluating the project design against the CDM requirements, including a substantive review of the baseline

[47] A list of all accredited DOEs can be found on the UNFCCC website: <http://cdm.unfccc.int/DOE/index.html>, accessed 31 March 2009.

[48] See discussion in MW Wara and DG Victor, 'A Realistic Policy on International Carbon Offsets', Program on Energy and Sustainable Development at Stanford University, Working paper #74, April 2008, at 19. At the same time, there are concerns that DOEs are susceptible to capture by their clients. The enforcement functions that the DOEs perform 'can [therefore] take place in a club-like atmosphere of repeat players with a stake in maintaining their reputation within an industry'; David Takacs, 'Carbon into Gold: Forest Carbon Offsets, Climate Change Adaptation, and International Law' (2009) 15 Hastings West-Northwest Journal of Environmental Law & Policy 39 at 75.

[49] Annex to Decision 3/CP. 7/CMP.1 para 3(c), 26 and 27, and Appendix A.

and monitoring methodology, and assuring that an adequate monitoring plan is in place to safeguard against the overstatement of emissions reductions.[50] At the verification stage, another DOE verifies the amount of emissions reductions. It then submits a report together with a request for issuance of CERs to the Executive Board.[51] Unless the Executive Board, the project participants, or one of the State Parties involved exercise their power of review, which is limited to fraud, malfeasance, or incompetence of the DOE, the CERs are issued.[52]

The failure of a DOE to meet the accreditation criteria can lead to immediate suspension or withdrawal of their accreditation.[53] The burden is on the DOEs to comply with applicable CDM Modalities and Procedures (Marrakech Accords) as well as other relevant decisions of the COP/MOP and the Executive Board.[54] DOEs also bear the risk of the erroneous issuance of excess CERs. In the event that a review by the Executive Board shows that excess CERs have been erroneously issued, the responsible DOE bears the costs of the review and has to transfer an amount of CERs equivalent to the number of excess CERs issued to the Board's cancellation account.[55]

Hart compared the validation estimates and verified results of 175 CDM projects that had issued CERs as of 1 May 2007 in order to assess the risk associated with validation/verification error.[56] On average, there tended to be overestimation of the amount of CERs that would be issued by a project at the validation stage. For the first 175 CDM projects that issued CERs, the validation procedure overestimated the number of CERs produced by approximately 27%. This large error rate increases the risks of investing in CDM projects. For DOEs, the error rate fuels contentions that the DOE verification system is dysfunctional and worsens the working relationship between the Executive Board and the DOEs. In interviews that Hart conducted with DOEs, project owners, and advisers, the leading explanation given for this validation/verification error was inadequate technology or methodology to measure emissions reductions. Several firms identified the use of inappropriate assumptions in the validation stage and conservative assumptions in the verification stage as

[50] The DOE's scope of work is set out in Section E of Decision 3/CP. 7/CMP.1. The criteria that a Project Design Document must meet are set out in Appendix B, Decision 3/CP.7/CMP.1.

[51] Information on the issuance of CERs is found in Section J of the Marrakech Accords.

[52] Thus, much of the power to create CERs rests with the DOEs, which are private sector entities contracted by the project participants; B Kingsbury, 'Global Environmental Governance as Administration: Implications for International Law' in D Bodansky et al, *The Oxford Handbook of International Environmental Law* (OUP: Oxford, 2007) at 75.

[53] Annex to Decision 3/CP.7/CMP.1, para 21.

[54] The Marrakech Accords were adopted by the 7th session of the UNFCCC COP held in Marrakech, Marrocco, in December 2001 and confirmed by the 1st session of the COP/MOP in Montreal in December 2005 FCCC/KP/2008/8/Add.1 Decision 3/CMP.1 (Modalities and procedures for a clean development mechanism as defined in article 12 of the Kyoto Protocol).

[55] Annex to Decision 3/CP.7/CMP.1, para 22.

[56] The discussion in this paragraph relies on C Hart, 'The Clean Development Mechanism: Considerations for Investors and Policymakers' (2007) 7 Sustainable Development Law & Policy 41 at 42.

potential factors influencing validation/verification error. Project sponsors tend to be optimistic at the validation stage, while DOEs conducting the verification may use conservative assumptions in accordance with best practices. Several respondents noted that the Executive Board has not provided adequate guidance for validation and verification procedures. There have been frequent revisions to the CDM methodologies, which have greatly increased the uncertainties.

The difficulties that the DOEs face in carrying out their functions increase the risk that they fall foul of the CDM rules governing their accreditation. Risk mitigation would ultimately centre on efforts to improve the working relationship between the Executive Board and the DOEs, such as improving communication channels between the Executive Board, DOEs, project owners, and advisors. It would also be in the interests of the Executive Board and DOEs to work together more closely to clarify the validation and verification procedures.

4. Risks and Opportunities in the EU Emission Trading System

In the CDM, projects carried out by private actors generate GHG emission reductions that are 'translated' into credits issued by a regulatory body. The EU ETS, in contrast, is a 'top-down' process. Through the National Allocation Plans (NAP), which require approval by the European Commission, Member States determine the number of allowances that should be issued to the installations within their jurisdiction that are covered by the EU ETS.[57] The Member States also determine how the allowances will be allocated amongst economic sectors and to the installations within these sectors. The role of the installations is to ensure that they have enough allowances to account for their actual emissions. Non-compliance is therefore the main risk faced by private actors in the EU ETS.

The EU ETS is a cap-and-trade program that targets carbon intensive industries and combustion activities across the European Union. Directive 2003/87/EC sets out the framework of the scheme, which Member States are required to transpose into national regulations.[58] In the UK, the EU ETS is implemented by the Greenhouse Gas Emissions Trading Scheme Regulations 2005 (the UK regulations).[59] The UK has been chosen as the focus of discussion in this section because some interesting issues arise from the interaction between

[57] In Phase III, Member States will no longer be required to draw up National Allocation Plans as allocation of allowances will be done at the EU level. For further detailed discussion of the EU ETS, see chapter 16 below.

[58] Article 5 of the Treaty Establishing the European Community articulates the principle of subsidiarity, whereby the Union does not take action (except in the areas which fall within its exclusive competence) unless it is more effective than action taken at the national, regional or local level. It is closely related to the principles of proportionality and necessity, which require that any action taken by the Union does not exceed what is necessary to achieve the Treaty's objectives. The administrative implementation of the EU ETS is therefore to be left largely to the Member States.

[59] SI 2005/925 as amended by SI 2005/2903, SI 2006/737, SI 2007/465 and SI 2007/3433; available online at <http://www.opsi.gov.uk/stat.htm>, accessed 3 December 2008.

the various policy instruments that the UK government has introduced to address climate change. The Climate Change Act 2008 will increase carbon constraints on private actors in the UK.[60]

4.1 Regulatory risks

4.1.1 Regulatory burdens

Private actors regulated under the EU ETS are at risk of being subjected to overlapping regulatory schemes and laws. Currently, the suite of instruments being used to address climate change includes the EU ETS, climate change agreements (CCAs), and the proposed Carbon Reduction Commitment (CRC). These instruments overlap with one another to some extent, as well as with other measures such as the Integrated Pollution Prevention and Control (IPPC) policy, the Large Combustion Plant Directive (LCPD), Combined Heat and Power (CHP) policies, the Renewables Obligation, and waste management policies. Feedback from stakeholders has prompted the regulators to undertake a review of the climate change instruments with a view towards streamlining regulation.[61]

4.1.2 Risks faced by verifiers

Verifiers perform almost identical functions to the DOEs in the CDM. In the UK, only verifiers that have been accredited by the UK Accreditation Service are allowed to provide verification services for installations under the EU ETS.[62] Apart from being liable to the client (the owners of the installation whose emissions are being verified) under general contract law and tort law for erroneous or negligent verification, a verifier also commits an offence under the UK Regulations if he:

> ... make[s] a statement which he knows to be false or misleading in a material particular, or recklessly to make a statement which is false or misleading in a material particular, where the statement is made ... as part of the verification of a report required under a monitoring and reporting condition.[63]

4.1.3 Expansion in scope

In November 2006, the European Commission launched a review of the EU ETS, as required under Article 30 of Directive 2003/87/EC, with the aim of proposing amendments to the scheme from 2013 onwards. The proposed amended

[60] The Climate Change Act 2008 became law on 26 November 2008; a copy of the legislation is available online at <http://www.opsi.gov.uk/acts/acts2008/pdf/ukpga_20080027_en.pdf>, accessed 3 December 2008.

[61] Consultation on the recommendations of the Climate Change Simplification Project, online: <http://www.defra.gov.uk/corporate/consult/cc-instruments/index.htm>, accessed on 3 December 2008.

[62] See DEFRA, 'EU Emissions Trading Scheme: Guidance on Annual Verification Version 3' (12 February 2007), Section 3.5 'UKAS Responsibilities'; online: <http://www.defra.gov.uk/environment/climatechange/trading/eu/pdf/annverifguide.pdf>, accessed 3 December 2008.

[63] Regulation 38(1)(f)(v) of the UK Regulations.

Directive was published on 23 January 2008 as part of the European Commission's broader Climate and Energy package.[64] The Commission has proposed including a number of additional industrial activities and gases from 2013 onwards, including carbon dioxide emissions from petrochemicals production, ammonia and other chemicals, carbon dioxide and perflurocarbon emissions from the production of aluminium (both primary and secondary), and methane from active coal mines. The inclusion of non-industrial sectors such as transport, agriculture and heat is a possibility. This continues the general trend of increasing the scope of the EU ETS (the scheme had already expanded its scope in Phase II).

4.2 Legal risks

4.2.1 The right to review decisions

An installation operator who has been aggrieved by a decision by the regulator may appeal the decision or seek judicial review.[65] *Cemex* was an application for permission to apply for judicial review, the claimant challenging the allocation of EUAs to its cement plant under the UK Phase II NAP (2008–2012).[66] *Arcelor Atlantique ET Lorraine and Others* was a constitutional challenge of the decree that was passed to implement Directive 2003/87/EC in France.[67] The applicants, companies in the steel-making industry, argued that the decree infringed the constitutional principles of legal certainty, the right to property and freedom of enterprise, and the principle of equality. However, neither of these decisions clarify nor establish points of law that may be of interest to private actors regulated by the EU ETS.

4.2.2 Legal nature of an EUA

The analysis of the legal nature of a CER also applies to EUAs under English law. Other legal risks that a private actor under the EU ETS might face include State aid and competition, which are addressed elsewhere in this book.[68]

4.3 Infrastructure risk

EU Allowances (EUAs) are to be issued to the installations in February each year. Delayed issuance of EUAs disrupts short-term and long-term carbon

[64] 'Proposal for a Directive amending Directive 2003/87/EC so as to improve and extend the greenhouse gas emission allowance trading system of the Community', online: <http://ec.europa.eu/environment/climat/emission/ets_post2012_en.htm>, accessed 3 December 2008.

[65] Appeals are addressed in Part 5 of the UK Regulations 2005.

[66] *Cemex UK Cement Limited v The Department for Environment, Food and Rural Affairs, Department of Trade and Industry, Lafarge Cement United Kingdom, Castle Cement Limited, Buxton Lime Industries Limited* [2006] EWHC 3207 (Admin).

[67] *Arcelor Atlantique ET Lorraine and Others before the French Conseil d'Etat (State Council) (Grand Chamber)* [2007] 2 CMLR 28.

[68] See chapters 16 and 17 below.

management plans and trading activity. Installations that have entered into agreements to sell or purchase EUAs will not be able to honour their obligations until such time as they do receive their allocation. In 2008 (first year of Phase II of the EU ETS), the UK government delayed issuing EUAs until the EU's Community Independent Transaction Log (CITL) was linked to the UNFCCC's International Transaction Log (ITL), which tracks the transfer of KP units.[69] The linking of the CITL, ITL, and Member States' registries was completed on 16 October 2008.[70] As this was a one-time occurrence, installations should be able to expect timely issuance of EUAs in the future. Nonetheless, the risk of a failure in the trading infrastructure therefore causing delay in the issuance or transfer of EUAs remains a significant one which ought to be addressed in the legal documentation.

An operator of an installation is legally required to surrender sufficient allowances by 30 April each year to account for its actual emissions from the previous year. Apart from EUAs, the KP units (CERs and ERUs) can also be used to meet compliance. Operators that wish to use KP units should hedge against the risk that the CERs and ERUs may not be issued on time or that the allowances delivered fall below the predicted output.

4.4 Market risks: allocation

The allocation of EUAs to installations covered by the EU ETS raises issues of fairness, business competitiveness, and regulatory certainty. The allocation process relies on the availability of reliable technical data and historic emissions data. Factors such as lower- than- usual emissions during the commissioning or trial period for newly installed equipment will affect the baseline against which EUAs are allocated. Installation operators should establish contact with their regulators (directly or through trade associations) to address these issues early on. While it might appear to be in an operator's interest to obtain as many allowances as possible, in the long run, it will be to the benefit of all parties (and the environment) if a stringent cap is adopted from the outset. Taking the UK ETS as an example, over-allocation led to regulatory intervention to withdraw allowances, which reduced certainty in the market.[71] Over-allocation was also a

[69] 'EUA issuance delay deals another blow to EU ETS', 7 March 2008, online: <http://www.carbon-financeonline.com/index.cfm?section=lead&action=view&id=11081>, accessed 3 December 2008. The concern was that, if the EUAs were issued before the CITL was linked to the ITL, a lengthy and complicated reconciliation process would have to be undertaken after the link is established. In the absence of the link, the ITL cannot record transactions within the EU ETS and yet EUAs are effectively AAUs for 2008–2012 (concurrently Phase II of the EU ETS as well as the KP's first commitment period).

[70] EUROPA, Community Transaction Log Latest Information, 28 October 2008 entry, online: <http://ec.europa.eu/environment/ets/>, accessed 4 December 2008.

[71] ENVIROS Consulting Limited, 'Appraisal of Years 1–4 of the UK Emissions Trading Scheme', December 2006, report commissioned by Department for Environment, Food and Rural Affairs (DEFRA) at 1 and 22.

problem in Phase I of the EU ETS, criticisms of which have focused considerable attention to addressing this issue. In Phase II, the UK will auction 7% of allowances as well as any surplus from the New Entrant Reserve (NER) and allowances from closures, up to the 10% limit set by Directive 2003/87/EC.[72]

5. The EU ETS Phase III

On 17 December 2008, the European Parliament adopted the Revised EU ETS Directive which is part of the EU climate package.[73] The Revised EU ETS Directive seeks to introduce changes that will address the weaknesses that had limited the efficacy of the scheme in Phases I and II. The proposed changes will mitigate some of the risks that private parties in the EU ETS currently face, while introducing others. This section will shed light on some of these issues.[74]

5.1 The '20 in 2020' target

The Revised EU ETS Directive provides that Phase III of the EU ETS will run from 2013 to 2020. The directive reflects the EU's independent commitment of a 20% reduction of GHG emissions by 2020, irrespective of any post-Kyoto international agreement.[75] In the event that an international agreement in which '... other developed countries would commit themselves to comparable emission reductions and economically more advanced developing countries contribute adequately according to their responsibilities and respective capabilities' is reached, the directive will be amended accordingly.[76] The relatively long time-frame for Phase III will provide certainty and greater flexibility for abatement efforts.

5.2 Auctioning and allocation free of charge

The Revised EU ETS Directive provides that auctioning will be the basic principle for the allocation of allowances in Phase III.[77] For example, allowances

[72] DEFRA, 'EU Emissions Trading Scheme: Approved Phase II National Allocation Plan 2008–2012', online: <http://www.defra.gov.uk/environment/climatechange/trading/eu/pdf/nap-phase2.pdf>, accessed 4 December 2008.

[73] 'Position of the European Parliament adopted at first reading on 17 December 2008 with a view to the adoption of Directive *2009/.../EC* of the European Parliament and of the Council amending Directive 2003/87/EC so as to improve and extend the greenhouse gas emission allowance trading system of the Community,' online: <http://ec.europa.eu/environment/climat/emission/ets_post2012_en.htm>, accessed 10 March 2009.

[74] For a more detailed discussion, see chapter 16 below.

[75] Paragraph 3 of the Recitals.

[76] Ibid.

[77] Auctioning is seen as the allocation method that best ensures efficiency, transparency, and avoids undesirable distributional effects. It also complies with the polluter-pays principle. See Paragraph 15 of the Recitals.

for the power sector will be fully auctioned in Phase III, although free allocations will be given for district heating and high-efficiency co-generation. This reflects the view that mitigation opportunities for the power sector come at lower costs compared to other sectors, that the power sector is able to pass on the costs of allowances to consumers, and that the sector is relatively insulated from competition outside the EU.[78] Other sectors will experience a gradual transition towards auctioning, starting with 80% allocation free of charge in 2013.[79] This percentage will be reduced by equal amounts annually, finally reaching 30% free allocation in 2020. The ultimate objective is to have no free allocation of allowances in 2027. However, as the discussion later in this book indicates, in light of the current global economic crisis, several EU Member States have successfully secured certain exclusions from the plan to phase out free allocations to address carbon leakage concerns and national circumstances.[80] Given the substantial scope of these exclusions and pending the European Commission's decision on the actual number of exempted sectors, auctioning may not emerge as the principal method of allocation in Phase III, as intended. Previously, the free allocation of allowances meant that operators did not have to concern themselves with acquiring allowances from the market. With the transition to auctioning, operators in the non-excluded sectors face an additional business cost and new exposure to (allowance) price volatility risks. Risk management strategies could involve spreading the risks through hedging instruments, borrowing and banking allowances, and investing in long-term abatement strategies, for example.

5.3 Expanded scope

The Revised EU ETS Directive extends the scope of the scheme to new sectors and gases. This includes GHG emissions from petrochemicals, ammonia and aluminium industry, and perflurocarbon (PFC) emissions from the aluminium sector. Aviation is dealt with in another directive.[81] The inclusion of aviation in the EU ETS affects both EU and non-EU aircraft operators.[82] This is a good example of how domestic/regional climate change regulations can significantly affect private parties who may not be within the jurisdictional control of the concerned State(s) but are forced to comply in order to carry on business operations in a carbon-constrained and globalized environment.

[78] Paragraph 19 of the Recitals.
[79] Paragraph 21 of the Recitals.
[80] See chapter 16 below.
[81] Directive 2008/101/EC of the European Parliament and of the Council of 19 November 2008 amending Directive 2003/87/EC so as to include aviation activities in the scheme for greenhouse gas emission allowance trading within the Community Text with EEA relevance (OJ L008, 13/01/2009 P 0003–0021).
[82] For details, see chapter 16 and also chapter 28 below.

5.4 Use of CERs and ERUs

In Phase III, EU operators will face restrictions on the extent to which they can use credits generated by the CDM and Joint Implementation (JI) to meet their EU ETS obligations. The Revised Directive provides that an operator may only use CERs and ERUs for compliance to the extent that (1) the level of CER/ERU use allowed by its Member State for Phase II (2008–2012) has not been exhausted or it has been granted an entitlement to use these credits under the new Article 11a(8); and (2) that the credits must stem from project types which were eligible for EU ETS compliance during Phase II. These restrictions are meant to ensure that the total use of credits does not exceed 50% of the EU-wide reductions of the existing sectors over the period 2008 to 2020 and 50% of the EU-wide reductions below the 2005 levels of new sectors and aviation over the period 2013 to 2020.[83] The Revised Directive also seeks to ensure that only credits from 'projects that provide real, verifiable, additional and permanent emissions reductions and have clear sustainable development benefits and no significant negative environmental or social impact' are used by operators for compliance.[84] The criteria that will be applied are not stated in the Revised Directive. The consequent reduced demand for CERs and ERUs will bear ramifications for the prices of credits, the viability of the CDM (or the demand for 'high quality' credits may drive the market towards greater environmental integrity), and compliance costs for EU ETS operators (depending on the price differentials between EUAs and the Kyoto credits).

6. Conclusion

The establishment of emissions trading schemes at the international, regional and State levels to address climate change is an experiment in global environmental governance that has created opportunities for the participation of private actors. The nature of emissions trading schemes is such that the active involvement of private actors is crucial for the development of the carbon market. Accompanying these opportunities are risks that private actors face when they participate directly in the global carbon market or in local/regional schemes. This chapter has sought to analyse some of these risks. Given the wide range of issues that could be addressed, this chapter could be no more than a broad-brush introduction to a range of issues, some of which are taken up in much greater detail in the rest of this book. What this chapter has aimed to achieve is to shed light on the role of the private sector in emissions trading schemes, specifically the KP and the EU ETS, and to highlight some of the issues that should inform private sector activity in the carbon markets.

[83] Article 11a(8). [84] Paragraph 29 of the Recitals.

PART III

THE KYOTO PROTOCOL MECHANISMS

7

International Emissions Trading and Green Investment Schemes

*Sander Simonetti and Rutger de Witt Wijnen**

1. Introduction

The Kyoto Protocol has introduced the first ever global scheme of emissions trading. Parties included in Annex I to the United Nations Framework Convention on Climate Change (UNFCCC) with greenhouse gas (GHG) emission limitation and reduction commitments inscribed in Annex B to the Kyoto Protocol (Annex B Parties) can achieve their commitments not only by taking domestic measures, but also by making use of the so-called 'flexible mechanisms' of the Kyoto Protocol: International Emissions Trading (IET), Joint Implementation (JI) and the Clean Development Mechanism (CDM). Although the mechanism of IET under Article 17 of the Kyoto Protocol is mostly referred to in the context of trading Assigned Amount Units (AAUs), it covers the international transfer of all Kyoto emission rights. The first part of this chapter sets out the legal framework of IET. The second part describes the principles and challenges of IET under Green Investment Schemes, where traded AAUs are 'greened' by investing the purchase price in environmental activities in the seller country.

2. International Emissions Trading

2.1 Emission rights

The use of the term 'emissions trading' in Article 17 of the Kyoto Protocol is somewhat misleading. Strictly speaking, Article 17 is not about trading but only about transferring. Trading is buying and selling, whereas transferring is delivering/accepting the sold/bought assets. Furthermore, the assets transferred

* The authors are members of the energy and carbon practice group of the international law firm De Brauw Blackstone Westbroek NV. They were on the drafting committee that developed the Manual and Model Agreement for the Sale and Purchase of Assigned Amount Units under a Green Investment Scheme for the European Bank for Reconstruction and Development.

under Article 17 are not emissions as such, but emission *rights*. There are four types of transferrable emission rights under the Kyoto Protocol: Assigned Amount Units (AAUs), Removal Units (RMUs), Emission Reduction Units (ERUs), and Certified Emission Reductions (CERs).[1] An emission right can be defined as the right to emit into the atmosphere a certain quantity of a specified substance during a defined period of time.[2]

2.1.1 Right to emit

Annex B Parties may only emit GHGs into the atmosphere if they have sufficient emission rights to set off against their emissions. At the beginning of the relevant commitment period, each Annex B Party is allocated an Assigned Amount, ie a limited number of AAUs.[3] The GHG emissions in such Party's territory during the commitment period may not exceed its Assigned Amount, unless additional emission rights have been acquired through the use of the flexible mechanisms. At the end of the commitment period, the Party must retire (surrender) emission rights (AAUs, RMUs, ERUs, and/or CERs) for the purpose of demonstrating its compliance with the Kyoto commitments. It is then assessed whether the number of emission rights retired matches the Party's aggregate GHG emissions during the commitment period.[4]

2.1.2 Specified substance

The Kyoto Protocol is about the reduction of GHG emissions. Six gases fall within the scope of the Protocol: carbon dioxide (CO_2), methane (CH_4), nitrous oxide (N_2O), hydrofluorocarbons (HFCs), perfluorocarbons (PFCs), and sulphur hexafluoride (SF_6).

2.1.3 Certain quantity

In order to make the six GHG gases of the Kyoto Protocol compatible, the concept of global warming potential has been introduced. Each gas has a certain rating based on its power to accelerate global warming. The rating index has been provided by the Intergovernmental Panel on Climate Change (IPCC) in its second assessment report of 1995.[5] On the basis of this index, each gas is

[1] Decision 2/CMP.1, para 6, provides that all of these types of rights may be used by Parties to meet their emission limitation and reduction commitments under the Kyoto Protocol.

[2] R de Witt Wijnen, 'Emissions Trading under Article 17 of the Kyoto Protocol', in D Freestone and C Streck (eds) *Legal Aspects of Implementing the Kyoto Protocol Mechanisms: Making Kyoto Work* (Oxford, 2005) 404 *ff.*

[3] The Assigned Amounts under the Kyoto Protocol are calculated with a view to reducing overall GHG emissions by at least 5% below 1990 levels by 2012.

[4] Para 13 and para 14 of the Annex to Decision 13/CMP.1 (Modalities for the accounting of assigned amounts under Art 7, para 4 of the Kyoto Protocol).

[5] For a complete overview of the Global Warming Potentials, see Table 1 in the UNFCCC guidelines on reporting and review, FCCC/CP/2002/8.

translated into CO_2 equivalents. Each Kyoto emission right (AAU, RMU, ERU, and CER) represents one metric ton of CO_2 equivalent.

2.1.4 Defined period of time

The first commitment period under the Kyoto Protocol runs from 1 January 2008 through 31 December 2012.[6] The Protocol provides that commitments for subsequent periods shall be established in amendments to Annex B.[7] However, to date no international agreement has been reached on GHG emission limitation and reduction commitments for the post-2012 period.

2.2 The Kyoto trading systems

The flexible mechanisms of the Kyoto Protocol are a unique combination of cap-and-trade and baseline-and-credit systems. Cap-and-trade refers to the system where Annex B Parties are allocated an emissions cap, the Assigned Amount. Parties that emit less than their cap may sell their surplus AAUs and Parties whose emissions exceed their cap must buy additional emission rights.

JI and CDM are baseline-and-credit systems. Under both systems, emission rights can be earned (ERUs and CERs, respectively) by participating in emission reduction projects abroad. Each emission right earned represents one metric ton of CO_2 equivalent reduced by the relevant project. At the start of the project, a baseline is established by calculating the amount of emissions that would occur in the absence of the project (the business-as-usual scenario). The difference between this baseline and the actual (lower) emissions as a result of the project is converted into tradable emission rights.

2.3 International Emissions Trading under Article 17

As mentioned above, IET under Article 17 of the Kyoto Protocol concerns the transferring and acquiring of emission rights. The material rules for IET have been laid down in the Modalities, Rules and Guidelines for Emissions Trading (the IET Modalities).[8] Pursuant to para 2 of the IET Modalities, not only AAUs, but also RMUs,[9] ERUs, and CERs can be transferred and acquired. The common denominator of these units is that each is an emission right issued

[6] Article 3, para 7, of the Kyoto Protocol. The actual deadline for the Parties' compliance with their emission limitation and reduction commitments under the Kyoto Protocol will be set at a later date. The annual GHG inventory for the last year of the commitment period (2012) is to be submitted in 2014. When this inventory is reviewed, an additional 100-day true-up period will start, during which Parties have the opportunity to undertake final transactions necessary to achieve compliance with their commitments. See paras 2.2 and 2.3 of the Kyoto Protocol Reference Manual on Accounting of Emissions and Assigned Amounts, published by the UNFCCC Secretariat in February 2007.

[7] Article 3, para 9, of the Kyoto Protocol. See also Decision 1/CMP.1.

[8] Annex to Decision 11/CMP.1.

[9] RMUs or Removal Units are emission rights which can be issued to reflect a removal of GHGs by sinks on the basis of to Article 3, para 3, of the Kyoto Protocol.

pursuant to the relevant rules of the Kyoto Protocol (or rules promulgated pursuant thereto) and each represents one metric ton of CO_2 equivalent. Para 2 of the IET Modalities also corrects an imperfection in Article 17, as it uses the phrase 'transfer and/or acquire' rather than 'trading'. Article 17 does not regulate the trading of Kyoto emission rights. Everybody, whether a State or not, whether a Party to the Kyoto Protocol or not, can trade Kyoto emission rights. But if such trade leads to an international transfer, such transfer can, as a general rule, only be effected by Parties to the Kyoto Protocol.[10]

2.3.1 Eligibility requirements

Just being a Party to the Kyoto Protocol is not sufficient to be able to internationally transfer or acquire Kyoto emission rights. First of all, only Annex B Parties may transfer or acquire Kyoto emission rights. This is a logical consequence of Article 17 which states that emissions trading may be participated in for the purpose of fulfilling emission reduction commitments under Article 3 (as only Annex B Parties have such commitments). Furthermore, Article 17 provides that any emissions trading must be supplemental to domestic actions for the purpose of meeting quantified emission limitation and reduction commitments under that Article. This requirement is not further elaborated, but its rationale is that Annex B Parties may not achieve their compliance with the Kyoto Protocol solely through IET (ie simply compensating all of their excess emissions by buying additional emission rights). The basis for compliance with the Kyoto obligations must be in domestic action.

In addition, para 2 of the IET Modalities provides that an Annex B Party is only eligible to transfer and/or acquire Kyoto emission rights if it is in compliance with the following eligibility requirements:

(a) it is a Party to the Kyoto Protocol (indeed a redundant requirement, but listed in para 2 nevertheless);
(b) its assigned amount has been calculated and recorded;
(c) it has in place a national system for the estimation of anthropogenic emissions by sources;
(d) it has in place a national registry;
(e) it has submitted annually the most recent required inventory;
(f) it submits the supplementary information on assigned amounts;

all in accordance with the relevant provisions pursuant to the Kyoto Protocol.

An Annex B Party shall be considered to meet the eligibility requirements after 16 months have elapsed since the submission to the UNFCCC Secretariat

[10] These are generally sovereign States. However, a regional economic integration organization can also become a Party to the Kyoto Protocol. The European Community has made use of this option. In this chapter, we focus on the role of States in IET.

of its report to facilitate the calculation of its Assigned Amount, unless the enforcement branch of the compliance committee finds that the Annex B Party does not meet the eligibility requirements.[11] The UNFCCC Secretariat maintains a publicly accessible list of Parties that meet the eligibility requirements.[12]

The enforcement branch of the compliance committee can suspend a Party's eligibility to participate in emissions trading, if it finds that the country does not meet one or more of the eligibility criteria.[13] If a Party's eligibility to participate in IET has been suspended, it may submit to the enforcement branch of the compliance committee a request to reinstate its eligibility, which will only be granted when an expert review team confirms that there is no longer a problem with respect to the eligibility of the Party concerned.[14]

2.3.2 Legal entities

Legal entities may participate in the flexible mechanisms of the Kyoto Protocol.[15] Para 5 of the IET Modalities provides that a Party that authorizes legal entities to transfer and/or acquire emission rights under Article 17 will remain responsible for the fulfilment of its obligations under the Kyoto Protocol and must ensure that such participation is consistent with the applicable rules. A legal entity may, after due authorization, only transfer and/or acquire Kyoto emission rights if the authorizing Party is eligible to do so. Each Party must maintain an up-to-date list of the legal entities that it has authorized to participate and make such list publicly available.[16]

The IET Modalities make no distinction as to the country of incorporation or domicile of the legal entity. Therefore, an Annex B Party may authorize a legal entity not incorporated or having its domicile in that Party's jurisdiction to participate in IET, even if the legal entity is incorporated or has its seat in the jurisdiction of a non-Annex B Party.

[11] See para 3 of the IET Modalities.

[12] Pursuant to para 3 of the IET Modalities, see <http://unfccc.int/kyoto_protocol/compliance/enforcement_branch/items/3785.php>. As of 1 January 2009, the only Annex B Parties that are not eligible are Australia (eligibility expected on 11 July 2009) and Croatia (eligibility expected on 27 December 2009).

[13] Decision 24/CMP.1, Section XV, para 4.

[14] Decision 24/CMP.1, Section X, para 2. This has happened in the case of Greece. In April 2008, the enforcement branch of the compliance committee found Greece to be in non-compliance with the national system requirement of para 2 sub (c) of the IET Modalities and declared Greece not eligible to participate in the flexible mechanisms of the Kyoto Protocol. Greece then submitted a revised plan and a request for reinstatement. On 27 October 2008, the enforcement branch decided that there no longer continued to be a question of implementation with respect to Greece's eligibility, and that Greece was now fully eligible to participate in the flexible mechanisms.

[15] For JI and CDM, this is provided in para 3 of Art 6 and para 9 of Art 12 of the Kyoto Protocol.

[16] In practice, however, we have found that such lists are often not available. Some countries seem to assume that legal entities that have opened an account in their national registry are thereby automatically authorized to participate in IET. However, such automatic authorization is often not adequately laid down in national regulations.

2.3.3 Transfer procedures

Transfers and acquisitions between national registries are made under the responsibility of the Parties concerned in accordance with the provisions of Decision 13/CMP.1 (Modalities for the Accounting of Assigned Amounts).[17] Pursuant to these Modalities, each Annex B Party must set up a national registry to ensure the accurate accounting of the issuance, holding, transfer, acquisition, cancellation and retirement of Kyoto emission rights. All transfers, including transfers initiated by legal entities which have been authorized to participate in IET, are made through the national registries of the relevant Annex B Parties (see Figure 7.1). A participating legal entity will therefore need to have an account in the national registry of the relevant authorizing Annex B Party. A transfer of Kyoto emission rights by a legal entity authorized by Party 1 to a legal entity authorized by Party 2 is, at the same time, a transfer of such emission rights by Party 1 to Party 2. The total number of Kyoto emission rights in the national account of Party 1 will be decreased by the number of Kyoto emission rights transferred by the legal entity and these will be added to the national account of Party 2.

As indicated in Figure 7.1, transfers of emission rights can also take place within a registry. Such transfers are not covered by Article 17, as these transfers are a purely domestic matter.

2.3.4 Commitment Period Reserve

Pursuant to para 6 of the IET Modalities, each Annex B Party must maintain in its national registry a Commitment Period Reserve, which reserve should not drop below 90% of that Party's Assigned Amount, or 100% of five times its

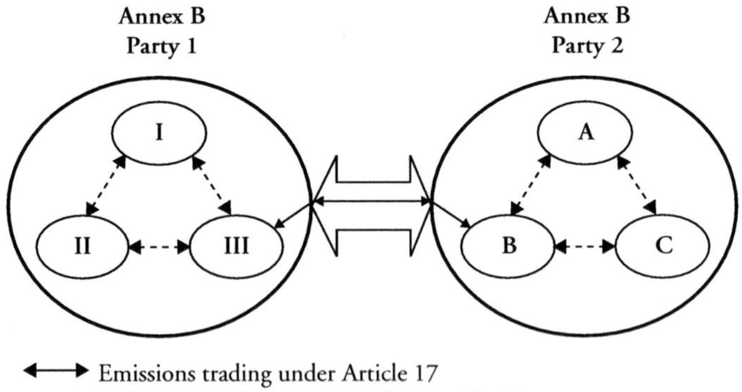

Figure 7.1. Transfer of Kyoto emission rights

[17] See para 5 of the IET Modalities.

most recently reviewed inventory, whichever is lower.[18] Therefore, a transfer under Article 17 cannot proceed if, as a result thereof, the aggregate number of Kyoto emission rights in the national registry of the transferring Party would fall below the Commitment Period Reserve.[19]

For legal entities participating in IET, the Commitment Period Reserve may be an obstacle frustrating the transfer for reasons outside the control of such legal entities. If a seller has sufficient emission rights in its account to honour its obligations to transfer a certain quantity to the account of a buyer in another jurisdiction, the transfer cannot be completed if it would at a national level decrease the number of emission rights in the national account of the authorizing Annex B Party below the level of the Commitment Period Reserve.

2.3.5 International Transaction Log

Transfers of emission rights between Parties are effected through and tracked by the International Transaction Log (ITL).[20] The transferring Party directs its national registry, in which the relevant emission rights are held, to transfer specified emission rights (AAUs, RMUs, ERUs, or CERs) to a specific account in the acquiring Party's registry. Subject to notification by the ITL that there are no discrepancies pertaining to the transfer, the transfer is completed when the relevant emission rights are removed from the transferring account and recorded in the acquiring account.[21]

The ITL performs an automated check to make sure, *inter alia*, that both Parties involved in the transaction fulfil the eligibility criteria and that the transferring Party complies with its Commitment Period Reserve obligation.[22] The outcome of the check is notified to the initiating registry (the registry of the transferring Party) and the acquiring registry (the registry of the acquiring Party). If the ITL detects a discrepancy, the initiating registry must terminate the transaction and notify both the ITL and the acquiring registry of the termination.[23] Should the initiating registry fail to terminate the transaction, the transferred emission rights are not valid for use towards compliance with the

[18] If a Party's Commitment Period Reserve is below the required amount (which can be updated during the relevant commitment period), the UNFCCC Secretariat will notify the Party, which must then bring its holdings of emission rights to the required level within 30 days of such notification (para 9 of the IET Modalities).

[19] For examples of Commitment Period Reserve calculations, see R de Witt Wijnen (note 2 above), 413.

[20] Established pursuant to para 38 of the Modalities for the Accounting of Assigned Amounts; see also Decision 12/CMP.1.

[21] Para 40 of the Modalities for the Accounting of Assigned Amounts.

[22] Para 42 of the Modalities for the Accounting of Assigned Amounts.

[23] Para 43(a) of the Modalities for the Accounting of Assigned Amounts. The ITL will also send the discrepancy record to the UNFCCC Secretariat for consideration as part of the review process for the relevant Party or Parties under Art 8 of the Kyoto Protocol.

Kyoto Protocol until the problem has been corrected and any questions regarding the transaction have been resolved.[24]

3. Green Investment Schemes

3.1 Background

As described above, Annex B Parties can utilize the flexible mechanism of IET to sell parts of their Assigned Amount or to buy additional emission rights. Traditionally, countries with economies in transition (mostly in Central and Eastern Europe), which suffered economic downturn in the 1990s, have been considered the primary suppliers of AAUs for the IET market. However, concerns that their surpluses resulted from the state of their economies rather than genuinely planned emission reduction policies undermined the perception of the environmental integrity of the mechanism, raising concerns that countries would simply sell their 'hot air' and dissipate the proceeds in their State budgets.[25] Public pressure and the need to ensure accountability of the use of taxpayer money have compelled seller and buyer countries to consider possibilities to enhance the environmental component of IET, introducing the idea of Green Investment Schemes (GIS), whereby the seller country is obliged to 'green' the AAUs sold by reinvesting the sale proceeds in GHG emission reduction projects or other projects beneficial to the environment. This would result in 'gourmet AAUs' with credible and documented greening to the satisfaction of the buyer.[26]

Some believe that AAU trading under GIS could play the role of a safety valve in the Kyoto market, balancing supply and demand, especially if there are disappointing delivery rates for JI and CDM projects.[27] In addition, GIS could provide a new and significant source of GHG mitigation financing in seller countries, offering the opportunity to address key climate change mitigation priorities that could not (sufficiently) be financed through other carbon market mechanisms and allowing governments to place emphasis on areas where early investment is crucial for the transition to a sustainable de-carbonized economy.[28] However, there may be little time left for utilizing this window of

[24] Para 43(b) of the Modalities for the Accounting of Assigned Amounts.

[25] M Khovanskaya (ed), *Developing a Green Investment Scheme in Romania* (Regional Environmental Center, 2006), 18.

[26] I Ramming, 'AAU Trading and the Impact on Kyoto and EU Emissions Trading. Before the Flood or Storm in a Tea-Cup?', in *Greenhouse Gas Market Report 2008* (IETA, 2008), 140 *ff.*

[27] See 'Point Carbon, The Kyoto balance: Saved by the AAUs', Carbon Market Analyst, 18 September 2008; and Ramming (n 26 above), 141. In comparison to JI, GIS has the added advantage that the double-counting rules of the European Linking Directive (2004/101/EC) do not apply, since AAUs as such cannot be used for compliance in the EU ETS, in contrast to emission rights generated by JI projects (ERUs).

[28] D Ürge-Vorsatz (ed), *Green Investment Schemes: Maximizing their benefits for climate and society*, Climate Strategies Study (Cambridge, 2008), 16–19 [hereafter Climate Strategies].

opportunity, as buyer countries will wish to use the acquired AAUs for compliance purposes during the first Kyoto compliance period, which ends on 31 December 2012.[29] After half a decade of moderate progress, the actions on GIS have recently accelerated, with several countries implementing GIS legal frameworks and announcing GIS deals. To facilitate the expansion of this relatively new segment of the international carbon finance market, the European Bank for Reconstruction and Development (EBRD) has developed a manual and model agreement for GIS transactions.[30]

The concept of GIS poses a number of interesting legal and political questions.[31]

3.2 Nature of the GIS Agreement

An AAU transaction under a GIS can be described as a 'hybrid' of two mechanisms[32]:

(1) IET: the process of sale and purchase of AAUs followed by the transfer of the AAUs sold to the buyer; and

(2) Domestic Greening: the process of implementing greening activities in the country of the seller with the revenues of the AAU sale.

Whilst the first component, IET, is regulated by Article 17 of the Kyoto Protocol, the IET Modalities and other COP/MOP Decisions, domestic greening activities are not covered by international regulation.[33] From a legal perspective, the greening component of the GIS deal is a self-imposed, binding commitment by seller countries to fulfil conditions set by buyers. As there is no international regulation with respect to the greening, this is at the discretion of the two governments involved and countries have great flexibility in drawing up their GIS plan. However, the absence of a robust international legal and institutional framework also makes it harder to ensure the environmental integrity of the GIS.

[29] See n 6 above.

[30] The Manual and Model Agreement for the Sale and Purchase of Assigned Amount Units under a Green Investment Scheme are publicly available at: <http://www.ebrd.com/country/sector/energyef/carbon/mccf/index.htm#gis>.

[31] This chapter only discusses GIS transactions between sovereign States. However, private entities may also have an interest in GIS if they are authorized by an Annex B Party to participate in IET and/or to use AAUs for compliance purposes under national or international emission trading schemes. Such private entities may face issues similar to the ones described in this chapter. In addition, GIS may also become relevant in the context of greening CERs (or ERUs) from certain – less popular – CDM (or JI) project types. For instance, certain parties consider HFC projects to be questionable from an environmental perspective. However, once registered under the CDM, such HFC projects can generate CERs for a substantial period of time. Therefore, in order to make such CERs more acceptable to the market, the idea of 'greening' the generated CERs (or rather the CER proceeds) has come up.

[32] Climate Strategies, 30.

[33] G Peszko et al, 'Latvia's Participation in International Emissions Trading', Options Study (The World Bank, 2007), para 3.2.

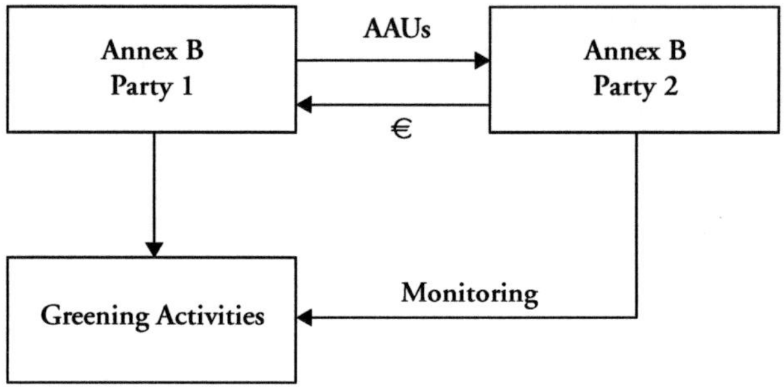

Figure 7.2. Basic structure of a GIS transaction

Therefore, the buyer will usually wish the greening activities to be monitored and checked against the agreed GIS plan (see Figure 7.2).

3.2.1 Treaty or commercial contract

A GIS transaction may be structured either as a treaty under public international law or as a commercial contract under private international law. Certain countries may have a preference for structuring the agreement as a commercial contract to avoid potentially lengthy ratification processes associated with adopting a treaty. It depends on applicable national (constitutional) law, however, whether a country can enter into a commercial agreement with respect to AAUs.[34] In certain countries, AAUs qualify as State assets that can only be disposed of after parliamentary vote, ie by way of treaty or national law.[35]

3.2.2 Applicable law and dispute resolution

If the parties choose (and are allowed) to enter into an international private law contract, they would usually include a choice of law clause, possibly selecting the law of a neutral third country, as well as a jurisdiction clause or other dispute resolution mechanism. Arbitration would typically be the preferred method for dispute settlement.[36] The most commonly used arbitration forums in the emissions trading market are probably the Permanent Court of Arbitration (PCA), the International Chamber of Commerce (ICC) and the United Nations Commission on International Trade Law (UNCITRAL). However, certain

[34] As the legal status of AAUs (which is also relevant for their tax and accounting treatment) differs from country to country, no univocal legal qualification can be given. In any case, AAUs and other emission rights as such do not qualify as financial instruments within the meaning of the EU Markets in Financial Instruments Directive (MiFID). However, certain derivative contracts with respect to emission rights may qualify as financial instruments; see L Bolhuis and S Simonetti, 'Financial Law in the Energy and Carbon Practice', *Nederlands Tijdschrift voor Energierecht* (Dutch Energy Law Journal), 2009/1, 4 *ff.*

[35] Cf Khovanskaya (n 25 above), 43.

[36] D Ratliff, 'Arbitration in Flexible-Mechanism Contracts', in D Freestone and C Streck (eds) *Legal Aspects of Implementing the Kyoto Protocol Mechanisms: Making Kyoto Work* (Oxford, 2005) 377 *ff.*

countries may be prevented by their national laws from entering into contracts including a choice of law (for a different law than their own) or an arbitration clause, as this would constitute a waiver of jurisdictional immunity.[37]

If the GIS deal is structured as a treaty, the Vienna Convention on the Law of Treaties of 1969[38] will apply. In case of a dispute under such GIS treaty, the question arises as to whether this qualifies as a dispute concerning the 'interpretation or application' of the UNFCCC or the Kyoto Protocol, in which case the dispute settlement procedures of the UNFCCC[39] would apply. If the dispute concerns the transfer of the AAUs, it would fall within the scope of the UNFCCC dispute settlement provisions, as AAU transfer is basically the application of Article 17 of the Kyoto Protocol (IET). However, if the dispute concerns other elements of the GIS agreement, such as payment, greening or monitoring, it would probably not be covered by the UNFCCC dispute settlement provisions.

3.3 Greening issues

Greening is the process that links the sold surplus AAUs, which have limited environmental merit, with activities that can deliver climate or other environmental benefits.[40] This is the core of the GIS agreement: the AAU proceeds are earmarked for greening activities in the seller country. Without the greening component, the agreement would be a 'simple' IET transaction. The GIS agreement may include a greening plan or schedule, in which parties elaborate the agreed greening activities, project selection methods, fund allocation structure, monitoring procedures and other greening matters. Some seller countries[41] have national laws and regulations on GIS (prescribing the types of projects on which GIS funds may be spent, the way these funds may be disbursed, etc). If the buyer is comfortable with these, the GIS agreement could refer to the relevant GIS law where applicable. If there is no GIS law in place, or if the parties wish to deviate from such law (if possible), the greening plan will have to be more detailed.

[37] Sovereign states generally enjoy immunity from suit; see J Klabbers, *An Introduction to International Institutional Law* (Cambridge, 2002), 146 *ff.*

[38] 1155, UNTS 331.

[39] Article 14 of the UNFCCC (to which Article 19 of the Kyoto Protocol refers) provides that Parties are committed to seeking to settle a dispute through negotiations or any other peaceful means of their own choice. If after twelve months following notification by one Party to another that a dispute exists between them, the Parties concerned have not been able to settle their dispute, it shall be submitted, at the request of any of the Parties to the dispute, to conciliation. Parties also have two options for binding settlement of their disputes, namely submission of the dispute to the International Court of Justice (ICJ) and/or arbitration. The UNFCCC refers to annexes on arbitration and conciliation to be adopted by the COP 'as soon as practicable'. However, these annexes have never been adopted.

[40] Climate Strategies, 46; see also K Tangen et al, *A Russian Green Investment Scheme: Securing Environmental Benefits From International Emissions* (Trading Climate Strategy, 2002); and W Blyth and R Baron, *Green Investment Schemes: Options and Issues, Organisation for Economic Co-operation and Development* (International Energy Agency: Paris, 2003).

[41] Such as Hungary, Latvia and Ukraine. See Climate Strategies, para 5.2.

3.3.1 Types of greening

The greening activities under a GIS may consist of 'hard' greening (emission reduction projects), 'soft' greening (other projects beneficial to the environment), or a combination thereof. Typical hard greening activities include investments in concrete emission reduction projects, such as renewable energy projects and retrofitting of old buildings.[42] Soft greening may include the setting up of an institutional and legal framework, environmental education and capacity building, research and technology development and adaptation to climate change.[43]

3.3.2 Additionality

Most buyers wish to be assured of additionality, ie that the GIS revenues do not replace budgetary spendings already present, planned or foreseen, or are used towards compliance with present or planned EU regulations or other international or national commitments in the pipeline. As this would indirectly mean that the revenues are not spent on GHG emission reductions, the additionality of investments spurred by AAU sales needs to be clear to the buyer, and is key to the environmental integrity of GIS.

Climate Strategies[44] identifies three additionality criteria: (i) environmental/climate additionality (new environmental benefits will arise); (ii) legal additionality (there is no obligation under law to materialize the project/investment); and (iii) financial additionality (there is no double financial support for the same emission reduction).[45] As most AAU selling countries are expected to be subject to more stringent emission reduction commitments after 2012 (in addition to increasingly stringent targets related to energy efficiency, renewable energy generation, biofuels, and other related environmental goals), it is likely that a substantial amount of reductions will take place in a business-as-usual scenario. This means that complying with buyers' additionality requirements will probably require actions in areas that are not easily tackled by other instruments.[46]

3.3.3 Greening ratio

Some buyers may be satisfied if the seller simply proves that all AAU proceeds are invested in greening activities. However, other buyers may also require the

[42] Climate Strategies, 46–7. [43] Peszko (n 33 above), para 3.6. [44] At 47.

[45] Certain financial additionality is regulated in EU Member States. EU Regulation 1083/2006 sets out legal additionality requirements, expressing that the same environmental achievement may not receive financial contribution under different legal rights, and that the State government may not use the same reduction under different legal titles for collecting money. A breach of this rule may lead to infringement procedures against the Member State in question under general EU legal rules. However, emission reductions above legally mandated levels might be triggered with additional funding, if there are no specific requirements for such in the given EU-funded support scheme.

[46] Climate Strategies, 35–6.

seller to guarantee a certain greening ratio, ie that each AAU sold is linked to a certain amount of GHG emission reductions generated by the projects in which the proceeds are invested.[47] If the parties agree that each transferred AAU must eventually be matched by a GHG reduction of one ton of CO_2 equivalent, the greening ratio is 1:1. This would only be possible if the greening plan consists of hard greening and the relevant projects can deliver measurable and quantifiable emission reductions.[48]

3.3.4 Project selection and fund allocation

The greening plan should also include the way in which GIS funds will be allocated to green projects and how project selection is to take place. Projects could be selected via public tenders, but GIS schemes may also follow a bottom-up disbursement approach, whereby projects have to be initiated and proposed by investors through an open application procedure.[49] If so desired, the buyer could have a role in the selection process. For instance, if project selection is effected via tenders, a representative from the buyer country may have a seat on the selection committee. Alternatively, the buyer may be granted the right to approve or veto the final award of the funds.

There are several methods to allocate the proceeds of the AAUs to the ultimate beneficiaries thereof, ie the parties implementing the relevant greening activities.[50] Fund allocation can be structured through grants, loans, credit guarantees, and equity.[51] Loans, credit guarantees and equity may be 'renewable' ways to distribute the GIS funds, as subsequent interest and redemption payments and other proceeds may be reinvested in greening. In the case of credit guarantees, the funds are used to provide (collateral for) guarantees for credits granted by other institutions, eg to the developer of green projects that applies for a bank loan. When the project developer repays the loan to the bank, the GIS funds may be reused for greening purposes. If the funds are used to take an equity share in green start-up projects, the stakeholder would be entitled to a corresponding share of the revenues. In addition, the relevant projects may later be sold in the market and yield a profit. These dividends and/or sale proceeds may also be reinvested in greening.

[47] Peszko (n 33 above), para 3.3.

[48] Climate Strategies, 49.

[49] Climate Strategies, 53.

[50] Beneficiaries could include private companies, non-profit organizations, central and local authorities, NGOs, physical persons, and government-owned or municipality-owned companies; see Climate Strategies, 54.

[51] Climate Strategies, 53; see also P Stoyanova, *Energy efficiency through Green Investment Schemes, The case of the Bulgarian building sector* (Department of Environmental Sciences and Policy, Central European University, Budapest, 2006). An alternative method (which will not be further discussed in this chapter) may be to allocate AAUs directly to green project developers and/or enterprises, in order for these entities to sell the AAUs themselves and subsequently use the proceeds; see Peszko (n 33 above), para 6.58.

3.3.5 Monitoring and verification

As indicated above, buyers may demand a certain greening ratio and require that the GHG emission reductions generated by the projects are verified. This would entail the definition of a business-as-usual scenario and the calculation of a baseline against which the additionality of the projects and the emission reductions can be measured.[52] Most parties will agree that appropriate monitoring of expenditure is an essential element of the GIS, the success of which depends on the trust of buyers in the scheme's environmental integrity and ability to meet the intended goals. The GIS will need to strike a delicate balance between the need for monitoring, verification, and review of projects, and the minimization of transaction costs.

The buyer may want to do the monitoring itself, or to have it done by the seller or an independent third party (which may be a private entity such as the Designated Operational Entity in CDM or the Accredited Independent Entity in track 2 JI). If the buyer is not directly involved in the monitoring, it may wish to be kept informed and periodically updated about the implementation of the greening activities. One of the options may be to use track 1 JI monitoring and verification procedures. However, it has been noted that as in track 1 the monitoring and verification are fully delegated to the JI host country, these procedures leave room for very 'discrete' monitoring and verification, which may not satisfy the GIS parties.[53]

3.3.6 Timeframe

A distinction must be made between the timeframe for the disbursement of the GIS funds on greening projects and the period within which these projects should 'deliver' their environmental benefits. If a greening ratio has been agreed, the latter could be called the 'crediting period' (similar to JI and CDM). In GIS, the duration of this crediting period, potentially applicable co-funding, and the expected greening ratio jointly determine the cost-effectiveness limit under which projects are financially viable.[54]

The transfer of AAUs under the GIS agreement will have to take place before the end of the Kyoto compliance period, because the buyer needs these emission rights for compliance purposes.[54a] However, since the greening activities under the GIS are not covered by Kyoto regulations, these are not bound to any commitment period and could be extended beyond 2012. This would entail that initially 'ungreened' AAUs are transferred to the buyer,

[52] Climate Strategies, 50.

[53] Climate Strategies, 50; see also J Väyrynen and F Lecocq, 'Track One JI and Greening of AAU's: How Could It Work?', in *Legal Aspects of Implementing the Kyoto Protocol Mechanisms: Making Kyoto Work*, 159 *ff*.

[54] Climate Strategies, 51.

[54a] Unless under a new international agreement (possibly a Copenhagen Protocol), AAUs from the first Kyoto compliance period may be carried forward ('banked') to the post-2012 period.

which are to be greened later. The downside of such approach for the buyer may be that if the greening activities do not deliver as envisaged, the buyer will have used 'hot air' for compliance purposes, which involves a reputational and credibility risk. However, Climate Strategies[55] notes that if all aspects of the GIS need to be completed by the end of the first commitment period, this would substantially increase the risk that the revenues cannot be spent in an optimal way.

3.3.7 State aid and competition

The earmarking of GIS funds for greening purposes and the allocation of these funds to certain greening projects or entities may raise questions of State aid and competition, especially within the European Union.[56]

The treaty establishing the European Community[57] prohibits any aid granted by or through a Member State which has the potential to distort competition and trade between Member States.[58] Generally, any measures provided from State resources (eg grants and subsidies) to an undertaking engaged in economic activity may constitute State aid, which may only be allowed under certain circumstances. The funding of greening activities under a GIS may fall within the scope of the exemptions of the Community Guidelines on State Aid for Environmental Protection.[59] It should then be ensured that the aid measures will result in a higher level of environmental protection than would occur without the aid, that the positive effects of the aid outweigh its negative effects in terms of distortions of competition, and that no cross-funding will take place. If these rules are not taken into account, the seller country may violate EU State aid rules and thus risk a possible infringement procedure. The aid has to be approved by the European Commission on a case-by-case basis.[60]

The buyer in a GIS transaction will usually require the seller to warrant that the greening activities under the GIS will not constitute a breach of national or international State aid, competition and public procurement rules.

[55] At 52.

[56] AAU trading and GIS may also raise certain tax issues, but these will not be discussed in this chapter, as the treatment of AAUs for accounting and tax purposes differs from country to country, depending on the legal status of AAUs under national laws and regulations; cf the Background Paper for the International Workshop on the Legal Nature of Emissions Reductions by the Foundation for International Environmental Law and Development (FIELD), at: <www.field.org.uk/library/archive>.

[57] Articles 87 and 88.

[58] Khovanskaya (n 25 above), 57.

[59] OJ 2001/C37/03.

[60] Following the formal investigation procedure as laid down in Art 6 of Regulation (EC) No 659/1999, the European Commission may decide to close the procedure with a decision in accordance with Art 7 of that Regulation. If the European Commission finds the aid to be incompatible with the principle of fair competition on the internal market and in violation of EU law, it has the authority to force the Member State to abolish the measure and to recover the aid from the beneficiary in order to restore the situation which existed before the aid was granted.

3.4 Other contractual issues

When the buyer and the seller have agreed on the greening activities to be implemented under the GIS, a settlement mechanism should be devised which tunes the parties' delivery and payment obligations to the greening plan.

3.4.1 Risks

Several risks must be taken into account when drafting the GIS contract, including authority and enforceability risks[61], payment risk, and commercial contingencies such as price and currency risks. The most important risks on the part of the buyer are the delivery risk (the risk that the AAUs will not be transferred) and the greening risk (the risk that the greening activities will not be implemented as envisaged).[62]

The delivery risk could materialize if, for instance, the ITL refuses to effect the AAU transfer for reasons of ineligibility or breach of Commitment Period Reserve (see para. 2.3.5 above). Eligibility of both parties to participate in IET will usually be a condition precedent for the GIS contract to enter into force. However, if one or both of the parties loses eligibility at a later date and is not eligible at the moment an AAU transfer under the GIS contract is due, the delivery cannot proceed. Similarly, if an AAU transfer would cause the aggregate amount of emission rights in the national registry of the seller to fall below its Commitment Period Reserve, the delivery cannot be effected. Of course, delivery could also be frustrated by the seller breaching its contractual delivery obligations[63] or due to technical failures at the national registry or the ITL. The latter may constitute force majeure. In any case, the buyer runs the risk of having insufficient emission rights to comply with its obligations under the Kyoto Protocol and/or having to buy alternative emission rights towards the compliance deadline (possibly at a much higher price).

The greening risk includes the possibility that the GIS funds are not properly disbursed to the agreed greening activities and the possibility that these activities do not generate the envisaged GHG emission reductions or other environmental benefits. In both cases the environmental credibility of the GIS is challenged and the buyer runs the risk of ending up with ungreened AAUs, which may damage its reputation, especially if the AAUs are used for Kyoto compliance purposes.

[61] These risks could be mitigated by including a condition precedent in the GIS contract to the effect that satisfactory legal opinions must be produced to confirm the authority of the persons signing the contract, as well as the enforceability of the agreement in the jurisdiction of the seller and the buyer.

[62] See also Väyrynen and Lecocq (n 53 above), 168–9.

[63] To hedge political and government performance risks, certain financial risk mitigation instruments are available (eg from the World Bank and the EBRD); see Khovanskaya (n 25 above), 44.

3.4.2 Settlement scenarios

To mitigate delivery and payment risks and to accommodate any other concerns of the parties, several payment and delivery settlement scenarios can be envisaged. These can be divided into two groups: (i) payment and delivery upon milestones; and (ii) upfront payment and delivery.[64]

The first type of settlement scenario requires the parties to specify certain greening milestones and a corresponding AAU delivery schedule in the greening plan.[65] These can be related to fund disbursement (ie certain greening projects must be funded by a certain date) or greening output (ie the projects must achieve certain GHG reductions or other environmental benefits by a certain date). In this type of scenario, the greening activities commence upon entry into the GIS agreement, but the AAUs and the purchase price are divided into certain tranches that are triggered by the greening milestones. Upon achievement of a greening milestone, the relevant tranche of AAUs is transferred to the buyer and the corresponding portion of the purchase price is paid to the seller (see Figure 7.3). If the buyer and the seller wish to mitigate payment and delivery risks, they could choose to structure payment and delivery through a custodian and escrow agent.[66]

The advantage of delivery and payment upon greening milestones is that the greening will have already taken place when the buyer receives the AAUs and the

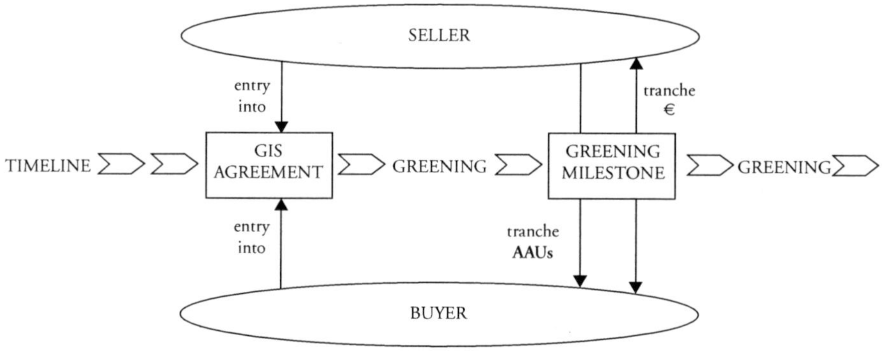

Figure 7.3. Payment and delivery upon greening milestones

[64] These different scenarios are also incorporated in the EBRD Manual and Model Agreement for the Sale and Purchase of Assigned Amount Units under a Green Investment Scheme (see n 30 above).

[65] Cf Peszko (n 33 above), para 5.17.

[66] In that case, the buyer transfers the purchase price to an escrow account with an escrow agent and the seller transfers the AAUs to a custody account with a custodian. Subsequently, the relevant tranches of AAUs and portions of the purchase price are released upon achievement of the greening milestones. This way, the AAUs will be outside the jurisdiction of the seller and any subsequent loss of eligibility or breach of the Commitment Period Reserve by the seller will no longer affect the delivery.

seller receives the purchase price.[67] Therefore, there is no risk of ungreened AAUs[68] and the contract would not have to include complex remedies for greening defaults (ie defaults in the implementation of the greening activities). The disadvantage is, of course, that the seller will have to prepay the greening activities, only to be reimbursed later, once the greening milestones have been achieved. However, this may be facilitated by using the underlying GIS contract as collateral to obtain external financing. Another disadvantage of this scenario is that all greening milestones will have to be achieved before the end of the current compliance period, as otherwise the seller can no longer use the AAUs to comply with its Kyoto obligations.

The second type of settlement scenario is based on upfront payment and delivery, whereby transfer of the AAUs to the buyer and the purchase price to the seller is effected prior to the greening. The seller will be responsible for implementing the greening activities in accordance with the greening plan and the monitoring thereof will take place as agreed between the parties (see para. 3.3.5 above). In case of a greening default, contractual remedies will apply. These will usually not provide that the—ungreened—AAUs must be returned to the seller (as the buyer still needs the AAUs for compliance purposes), but rather that the buyer is to be compensated for the lack of greening. This could be effected by obliging the seller to implement replacement greening activities, or to repay the purchase price to the buyer, which the buyer itself may then reinvest in greening, either in its own territory or elsewhere. These scenarios could also include a partial advance payment and subsequent payments upon submission of verified proof of achieved greening results.[69] The upfront payment and delivery scenarios could be combined with the concept of a greening agent, ie a third party which implements the greening plan on behalf of the seller, which could be either a national implementation agency in the jurisdiction of the seller or an international institution. Such greening agent would work under a separate contractual relationship with the seller, based on a cooperation agreement or other legal instrument.[70] If the greening agent is a trustworthy international institution, the buyer and seller may even agree that the seller will be deemed to have complied with its greening obligations under the GIS contract once the purchase price is paid to this international greening agent (either directly by the buyer or via the seller[71]). The agreement with the greening agent should then

[67] Climate Strategies notes (at 11) that certain seller countries may have 'relatively low financial discipline and major budgetary problems'.

[68] Cf Peszko (n 33 above), para 5.16.

[69] See Khovanskaya (n 25 above), 42. However, some parties may argue that this would effectively set a price for 'hot air' (the advance payment representing the price for ungreened AAUs and the subsequent payments merely being a 'greening bonus').

[70] Possibly, the buyer could also be a party to this agreement and thereby have a direct relation with (and remedies against) the greening agent.

[71] Payment by the buyer directly to the greening agent may not be feasible for certain seller countries whose national laws prescribe that the proceeds of an AAU sale must always flow through the national budget.

provide for remedies in case the greening agent does not implement the greening activities in accordance with the greening plan. Such remedies may include repayment of the relevant funds by the greening agent to the seller (or directly to the buyer), to be reinvested in alternative greening activities.

The advantage of the upfront payment and delivery scenarios is that the proceeds of the AAUs can be directly invested in greening (either by the seller or through a greening agent). In addition, these scenarios are more flexible in terms of extending the GIS timeframe, both in respect of fund disbursements[72] and crediting period, if applicable.[73] The disadvantage of such scenarios is that the remedies in case of greening defaults may be difficult to enforce. This may be especially cumbersome if the GIS timeframe has been extended beyond 2012, as the AAUs will then have already been surrendered for compliance under the assumption that their greening would take place at a later date. If such greening then fails to occur under the GIS contract, the buyer may have a hard time finding credible alternative greening solutions.

4. Summary

The first part of this chapter describes the mechanism of international emissions trading under Article 17 of the Kyoto Protocol. Various basic concepts are explained, including the definition of transferable emission rights, eligibility requirements, participation of legal entities, transfer procedures, the Commitment Period Reserve and the International Transaction Log. The second part of the chapter describes the principles and legal issues associated with emissions trading under green investment schemes, whereby the seller of emission rights commits to use the proceeds thereof for greening purposes.

[72] Climate Strategies states (at 118) that extending the timeframe for funds disbursements would be important for optimizing the climate effectiveness of the GIS.

[73] Climate Strategies notes (at 9) that if a strict greening ratio close to 1:1 is required and the crediting period does not account for emission reductions earned beyond the end of the first commitment period, investments would probably be limited to 'very low hanging fruit' (ie investments that would be made through JI or other policy mechanisms anyway).

8

A Mechanism with a Bright Future: Joint Implementation

Jelmer Hoogzaad and Charlotte Streck

1. Introduction

Although the term Joint Implementation itself is not mentioned anywhere in the Kyoto Protocol, Article 6 of the Protocol establishes the basis for what has become known as 'Joint Implementation'. Jointly with the Clean Development Mechanism (CDM), JI belongs to the project-based flexible mechanisms defined by the Kyoto Protocol.[1] JI involves the transfer of Emission Reduction Units (ERUs) between Annex I parties that have been generated by projects developed by Parties to the Protocol or authorized legal entities.[2]

The marginal abatement costs for greenhouse gases differ largely based on where the abatement is made. Therefore, from the point of view of economic efficiency, emissions should be reduced where such reductions are cheapest to achieve. Joint Implementation is based on the economic reasoning that international trade in carbon emission rights can reduce mitigation costs. Joint Implementation allows project sponsors to invest in projects that reduce greenhouse gas emissions *vis-à-vis* an established baseline scenario and to sell the difference between the baseline emissions and the project emissions with the consent of the government hosting the project. Through its approval, the government agrees to convert Assigned Amount Units (AAUs)[3] into ERUs in an amount equivalent to the verified emission reductions achieved by the project, and transfer them to the registry accounts of the investing country or private entity.

[1] The 'flexible mechanisms' are three trading mechanisms as defined in Art 6 (Joint Implementation), Art 12 (Clean Development Mechanism), and Art 17 (International Emission Trading). Through these mechanisms Parties to the Kyoto Protocol with an emission target can acquire Assigned Amount Units (AAUs) from other Annex I parties Certified Emission Reductions from authorized CDM projects or Emission Reduction Units (ERUs) from authorized JI projects.

[2] The Kyoto Protocol limits the availability of Joint Implementation to Annex B parties that are also members of Annex I of the United Nations Framework Convention on Climate Change (UNFCCC).

[3] The Kyoto Protocol has allocated 'assigned amounts,' which represent the quantity of greenhouse gas emissions a country is allowed to emit in a defined first compliance period (2008–2012).

Joint Implementation has been described as a wallflower, mimicking its much more popular sibling, the CDM. The CDM is viewed as the 'young debutante that swept the climate change world off its feet and is currently everyone's darling' while JI is 'hoping to pick up some of the other's charm, and not have to leave the ball alone'.[4] However this view does not do justice to JI. The mechanisms past performance is not as poor as the limited number of registered projects indicate and its potential for the future is greatly underestimated. The fact that JI projects only earn ERUs during the Kyoto Protocol's commitment period (2008–2012),[5] has delayed the implementation of JI projects. But it has also triggered some interesting transactions assigning value to the emission reductions (so-called 'early credits') generated before 2008 for both the acquiring and the selling party. The European Emission Trading Scheme (EU ETS) has had lasting effects on the scope of JI, as the double-counting provisions of the EU ETS have excluded a large number of projects based on energy generation from obtaining ERUs. Despite these limitations in time and scope, JI has created opportunities, in particular where parties are encouraging the development of projects through the adoption of JI guidelines under 'Track 1' of the mechanism. JI also describes a model of a mechanism that mobilizes emission reductions and assigns carbon credits within sectors, installations, or countries that operate under an international emission cap. As more countries are expected to accept such caps, the geographical scope of JI may expand in the same measure as the CDM loses area.

This chapter first describes the current state of play for JI, explaining the procedural requirements for a successful participation in JI, including providing insights in the so-called Track 1 JI, a modality of the JI mechanism that is often overlooked due to its national rather than global character. Next, the claim that that JI has to accept a life as wallflower compared to the more glamorous CDM is examined and argued to be untrue. Finally, the post-2012 future of JI is discussed, elaborating upon options for JI reform as well as proposals to introduce new elements to expand the mechanisms' sectoral coverage.

2. Background

From the definition of the mechanism in Article 6 of the Kyoto Protocol in 1997, it took another four years before the Conference of the Parties to the UNFCCC (COP) could reach an agreement on the implementation guidelines

[4] Dane Ratliff, 'Joint Implementation: Tracking Recent Developments' in David Freestone and Charlotte Streck (eds), *The Kyoto Protocol: Current Legal Status of Carbon Finance and the Flexible Mechanisms* (2007) 15 Environmental Liability 56–64, 56.

[5] This puts JI projects at a disadvantage compared with the CDM, which allowed for a prompt start before the entry into force of the Protocol and authorized the earning of credits starting in the year 2000.

and participation requirements, which define the rules under which JI projects can be designed and implemented. After years of complicated negotiations, which also saw the US pulling out from the process, it was the successful outcome of COP7 (which took place between 29 October and 9 November 2001 in Marrakech), which provided the international community with the rulebook needed to implement JI and CDM projects—the so-called 'Marrakech Accords'.[6] The Marrakech Accords contain a number of guidelines for the implementation of Article 6 and recommend them for adoption by the first COP of the UNFCCC serving as the Meeting of the Parties to the Kyoto Protocol (COP/MOP) after the entry into force of the Kyoto Protocol.[7] Yet, compared to the modalities governing the CDM, the JI text included in the Marrakech Accords is vague and incomplete. Since JI did not officially start until 2008, many of the rules guiding JI have been established relatively recently. The JI Supervisory Committee (JISC) established to monitor and register JI activities jointly with the different Parties that adopt JI regulations, helps to interpret the Accords.

JI is divided in two different tracks. Track 1 allows host countries, provided that they are parties to the Kyoto Protocol, to determine how to verify the Track 1 JI projects they are hosting. The host country government can define both the criteria that eligible JI projects should meet and also the rules for verification of the emission reductions created by these projects. 'Track 2' operates in a different way. In structure and operations it is similar to the CDM. Similar to the CDM's Executive Board, the JISC is the UNFCCC body that develops the procedures and guidelines[8] for this track. For Track 2 projects, the JISC defines the rules of the game and independent third parties check whether these rules have been applied correctly and the projects and subsequent emission reductions meet all relevant JI criteria.

The guidelines for JI Track 2 have been continually developed by the JISC since February 2006. Since this date the JI pipeline has grown to around 190 projects[9] at the time of writing, of which 170 have been submitted for 'determination',[10] while six[11] have reached 'final determination' by the JI Supervisory Committee (JISC). Table 8.1 shows that the great majority of JI projects are located in the Russian Federation and Ukraine. As the large amount of Russian projects each individually account for a relatively large emission

[6] See FCCC/10/CMP.1, <www.unfccc.int>, the Marrakech Accords where adopted and confirmed by the first Meeting of the Parties to the Kyoto Protocol in 2005.

[7] The decision on Implementation Guidelines for JI Projects in contained in the draft decision FCCC/CP/2001/16/CP.7, which was confirmed by the COP/MOP in 2005 as FCCC/10/CMP.1 7.

[8] FCCC/10/CMP.1.

[9] UNEP Risø JI pipeline, dated 1 January 2009 <http://cdmpipeline.org> accessed 18 March 2009 [hereinafter: UNEP Risø pipeline].

[10] Determination is the JI equivalent of validation under the CDM: independent third party confirmation that the project design meets all requirements and guidelines for JI.

[11] UNFCCC web-site: <http://ji.unfccc.int/ji_projects>, accessed 11 March 2009.

Table 8.1. Market share of host countries in the JI project pipeline.[12]

Country	ERUs up to 2012 (ktonne CO_2e)	Market share	Projects	Market share
Russia	196,235	61%	92	48%
Ukraine	57,366	18%	31	16%
Germany	17,474	5%	7	4%
Poland	16,459	5%	10	5%
Romania	10,005	3%	7	4%
Hungary	8,233	3%	10	5%
Other	15,454	5%	33	17%

reduction potential, the country has, in market share terms, 61% of ERUs (up to 2012).

In addition to the projects submitted to the JISC under the CDM-like Track 2 procedure, 30 projects have been directly approved by the authorities of the host country and are being developed under the streamlined Track 1 procedure. The Track 1 projects are located in Hungary, Germany, France, New Zealand, Romania, and Ukraine.

Figure 8.1 shows the project types and their market share in terms of expected ERUs up to 2012. Nearly half of the reductions stem from reducing emissions of methane from gas distribution systems, coal mines, landfills, and oil extraction. Reduced emissions of N_2O, HFCs, and PFCs constitute an additional quarter and about one eighth of the reductions in the pipeline are from energy efficiency projects. Only 7% of JI projects are renewables while in the CDM their share is 35%.

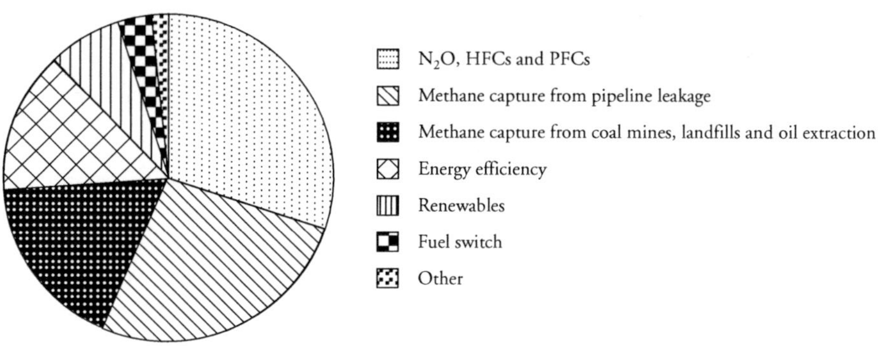

Figure 8.1. Share of different project types in the JI pipeline, in terms of ERUs up to 2012.[13]

[12] Calculations based on information from the UNEP Risø JI and CDM Pipelines dated 1 January 2009.

[13] Calculations based on information from the UNEP Risø JI Pipeline, dated 1 January 2009.

The disproportionate share of methane leakage reduction comes from the large emission reduction potential of projects in Russia's gas distribution system. Although these projects dominate the JI project pipeline, the lack of Russian host country approval prevents Russian JI projects from progressing to formal approval in the JI project cycle.

3. Pioneering Track 1 JI

Track 1 or 'fast track JI' is the verification of emission reductions according to guidelines and procedures defined by the host country government. In the definitions provided by the Conference of parties, JI Track 1 is similar to International Emissions Trading under Article 17 of the Kyoto Protocol, it is however project-based and requires emission reductions to be linked to a particular project activity. Since Track 1 does not rely on the JISC for final approval, the fees charged by the JISC also do not apply, potentially lowering a project's transaction costs.[14] Despite these advantages, Track 2 has been the preferred mechanism for project developers as long as few countries had adopted Track 1 guidelines for JI projects. This is expected to change as more countries adopt and build a track record in implementing Track 1 procedures.

Track 1 activity has so far been limited to a few countries: Hungary, Germany, France, New Zealand, Romania, and Ukraine. Out of the 32 Annex I countries that can benefit from Track 1,[15] this is a fairly meagre result. It also means a lost opportunity as Track 1 allows countries to develop domestic emission reduction opportunities and thereby complement their domestic policies. New Zealand, Germany, and France have successfully pioneered the opportunities and have benefitted from the degree of freedom and ability to reduce transaction costs that Track 1 offers.

The main incentive for the development of Track 1 procedures by host country governments came from the early movers in the JI market. Although the JI procedures from the JISC stem from 2006, the first JI projects were developed in 2000. At that time, project developers had little more than Article 6 of the Kyoto Protocol to guide them in the calculation of emission reductions, determination of project additionality, and development of monitoring plans. When the JISC launched its procedures and guidelines in 2006, some of these early project documents failed to comply with the new procedures. Some project

[14] Anna Korppoo and Olga Gassan-zade, *Joint Implementation: looking back and forward* (Climate Strategies, Cambridge, 2008) 18 <http://www.climatestrategies.org/reportfiles/ji_synthesis_report_final_201008.pdf>, accessed 18 March 2009.

[15] This value is based on the list of JI countries on the UNFCCC website: <http://ji.unfccc.int/ji_parties/index.html>, accessed 11 March 2009.

owners decided to revise their documents[16] while others pushed host governments to adopt Track 1 procedures.

A clear example of a country that successfully used Track 1 for the development of JI projects is Hungary. The country registered eight Track 1 projects on the JI section of the UNFCCC website, an achievement matched only by Germany. Out of the eight projects, only one complied with the JISC procedures. The JI documentation of the remaining projects are based on templates or guidance from the World Bank Prototype Carbon Fund, the Dutch ERUPT program, the Austrian JI/CDM Programme, or a tailor made format from the project developer. These JI documents adhere to any international standard but comply with the Hungarian Track 1 procedure.[17]

The Projects to Reduce Emissions (PRE) programme, launched by New Zealand in 2003 describes an interesting example of ways to use Track 1 to support domestic action towards Kyoto compliance.[18] This program was launched well before the JISC formulated its JI Track 2 guidance. Therefore the New Zealand government developed domestic Track 1 procedures for project determination and the verification of emission reductions. The basis of PRE was a tender assigning AAUs or ERUs to project developers that offered to reduce GHG emissions within New Zealand. An innovative aspect in the PRE approach was that it awarded the credits to 'projects offering the most reduction in emissions in exchange for the least number of emission units requested'.[19] This way, the New Zealand government used the tender to mobilize domestic emission reductions. Meanwhile, the difference between the credits issued and the reductions achieved supported the government in achieving its Kyoto target. The program was assessed in 2005 as part of a review of the New Zealand climate change policies. The review conclusions questioned the 'additionality' of the projects, and basically stated that it was unclear to what extent the projects under the programme would not have occurred without incentives from PRE. As it was unclear whether indeed the amount of reductions issued exceeded the level that would have been achieved without the incentives from the PRE programme, the review team recommend ending the PRE programme.[20]

[16] An example of such a project is 'Utilization of Coal Mine Methane at the Coal Mine named after A.F. Zasyadko'. This project was initially developed according to the Guidelines from the Dutch Emission Reduction Unit Purchase Tender (ERUPT). When the Track 2 procedures were operational the project documents were revised to meet the Track 2 requirements. At the time of writing the Zasyadko project was the only JI project with ERUs in final verification stage. See the section on JI projects at: <http://ji.unfccc.int/JI_projects/index.html>, accessed 11 March 2009.

[17] Information based on an analysis of project information at the UNFCCC website: <http://ji.unfccc.int/JI_projects/index.html>, accessed 11 March 2009

[18] See <http://www.mfe.govt.nz/issues/climate/policies-initiatives/projects/>, accessed 18 January 2009.

[19] <www.beehive.govt.nz/node/20700>, accessed 18 January 2009.

[20] Ministry for the Environment, Review of climate change policies, November 2005 <http://www.mfe.govt.nz/publications/climate/policy-review-05/policy-review-05.pdf> accessed 18 January 2009.

Since 2007 France has also used the Track 1 framework to develop domestic JI projects. The French government requested Caisse des Dépôts, a state-owned bank aimed at implementing projects with a public-interest,[21] to implement a domestic JI scheme to 'permit trading of trading of greenhouse gas emission reduction credits from outside the (...) EU ETS'.[22] The scheme targets projects that lack governmental financial assistance or projects where existing government assistance is insufficient. The objective of the scheme is to benefit from the lower abatement costs of domestic JI projects compared to the abatement costs that installations under the EU emission trading system face when reducing their emissions.[23] Caisse des Dépôts launched its first tender in October 2007, which initiated the development of around 20 domestic offset projects in a large variety of sectors. By issuing ERUs for 90% of the certified emission reductions achieved by the projects, these projects also contribute to France's efforts to maintain compliance with its Kyoto target.

The majority of the projects in the French Track 1 pipeline are bundles of small energy efficiency and fuel switch projects identified by energy service companies. The French finance ministry estimates that there is potential to reduce up to 15 million tonnes of CO_2 equivalents through domestic JI in the period 2008–2012. Caisse des Dépôts has guaranteed to purchase all ERUs from these projects based on non-public pre-determined pricing arrangements.[24]

Germany has also established a programme for JI projects that operate in the country. An interesting achievement of the program is that it adopted a framework for the Programme of Activities (PoA)[25] well before COP/MOP-4 requested the JISC to prepare PoA guidelines for JI.[26] Through this framework, Germany has been hosting the first programmatic JI projects.[27] One restriction of the

[21] <http://www.caissedesdepots.fr/spip.php?article57>, accessed 18 March 2009.

[22] Aude Bodiguel, Yves André and Benoît Leguet, 'Domestic Offset Projects: Implementing the Kyoto protocol', Paris, May 2008. <http://www.caissedesdepots.fr/IMG/pdf_08-05-14_Note_commune_AB-BL_ADEME-CDC_VF-EN.pdf> accessed 18 January 2009.

[23] Symposium Climate and domestic offset projects: Why such a policy and how to implement it?, Bercy, December 2005. <http://www.caissedesdepots.fr/spip.php?page=special&id_article=586>, accessed 18 March 2009.

[24] Aude Bodiguel, Yves André and Benoît Leguet, 'Domestic Offset Projects: Implementing the Kyoto protocol' (Paris, May 2008).

[25] *Handbuch für JI-Projekte mit Deutschland als Gastgeberland (‚JI-Inland')-Leitfaden für Antragsteller*, Version 1.1, November 2008. <http://www.dehst.de/cln_099/SharedDocs/Downloads/DE/JI__CDM/JI-CDM__JI__Manual__Inland__deutsch,templateId=raw,property=publicationFile.pdf/JI-CDM_JI_Manual_Inland_deutsch.pdf>, accessed 18 March 2009. Three programmatic Track 1 JI projects, all in Germany, have been determined at the date of writing (UNEP Risø Pipeline).

[26] Decision -/CMP.4 Guidance on the implementation of Article 6 of the Kyoto Protocol (Advance unedited version) reads in para 6: '*Requests* the Joint Implementation Supervisory Committee to develop, as soon as possible, definitions, forms, guidelines and procedures for projects under programmes of activities implemented under the verification procedure under the Committee, bearing in mind the work of the Executive Board of the clean development mechanism in this area.' <http://unfccc.int/files/meetings/cop_14/application/pdf/cmp_ji.pdf> accessed 18 March 2009.

[27] See <http://www.dehst.de>, accessed 18 January 2008.

German Domestic JI program is that Germany does not authorize projects promoted by an entity from an investor country that 'is unwilling to admit project activities within its national territory under comparable conditions.'[28] This criterion resulted in the cessation of investments in German projects under the Emission Reduction Unit Purchase Tender (ERUPT), a Dutch programme for the purchase of ERUs, as the Netherlands do not approve domestic JI projects.[29]

Despite these examples that demonstrate the possibilities and potential of Track 1, the Track 1 activity remains small and limited to countries with a low host country risk and comparatively strong governance.

4. JI and the CDM

The performance of JI is often compared with the CDM. By the time of writing this chapter, 1,300 CDM projects have achieved registration out of a total of 4,475 projects under development.[30] Also in terms of emission reductions in the pipeline, the potential from JI projects is dwarfed by the CDM. In February 2009, the JI project pipeline contained only 67 million tCO_2e/year while the CDM project pipeline represents around 600 million tCO_2e/year, outperforming JI by nearly ten times. The picture changes to the advantage of JI when looking at the reductions from these two trading mechanisms in comparison with the aggregated amount of GHG emitted in the relevant host countries. The Annex I countries had a total annual emissions of nearly 11 billion tCO_2e in 2006. They host around 190 JI projects whose emission reductions cover nearly 0.7%[31] of their annual emissions. Despite numbering

[28] German Emissions Trading Authority (DEHSt), German Manual for JI Host Country Approval-Guidance for Applicants, Version 1.1, Berlin 2008. This requirement is referred to as 'Reciprocity of JI rules with investor country'. The Manual states that 'Latest at the time of the first determination of the emission reduction (Verification), a Letter of Approval of an investor country has to be provided. (...), only investor countries are acceptable which allow JI projects on their own territory under comparable conditions.'

[29] *Staatsblad van het Koninkrijk der Nederlanden*, Jaargang 2006, 189, Artikel 16.46c, Wet Milieubeheer. <http://wetten.overheid.nl/BWBR0003245/Hoofdstuk16/Titel162/Afdeling1627/Artikel1646c/geldigheidsdatum_18-03-2009>, accessed 18 January 2008.

[30] Calculations based on the UNEP Risø CDM pipeline dated 1 January 2009.

[31] Calculations based on the UNEP Risø JI Pipeline dated 1 January 2009 and the GHG data from <www.unfccc.int>. The percentage is based on the annual emissions from Annex I parties eligible for JI (all Annex I parties excluding Turkey, Belarus and the USA) and comparing these with the annual emission reductions from all JI project in the pipeline.

over 4,500, the CDM projects perform only a little over three times better by covering about 2.3%[32] of the total emissions from non-Annex I countries.

The remaining disparity in the numbers between JI and CDM projects is partially attributable to the fact that the CDM enjoyed a 'prompt start' authorized by Article 12 of the Kyoto Protocol which allowed the registration of projects starting in 2001. This means that the CDM's Certified Emissions Reductions (CERs) obtained during the period between 2000 and 2008 can be used to assist in achieving compliance during the first commitment period. JI on the other hand can only generate ERUs during the first commitment period, namely between 2008 and 2012.

Furthermore, differences on the institutional level provide explanation for the difference in performance of JI and CDM. While the CDM is managed or administered on the international level by a governance structure that is largely independent from host country performance, administrative capacities and GHG accounting frameworks, JI depends on emission limitation targets, allocation of AAUs, registries, and the functioning of almost all other elements of the infrastructure of the Kyoto Protocol. Carbon credits generated by JI are simply AAUs in disguise, authorized for private trade, but generated by the host country which converts AAUs into ERUs based on the evidence of verified emission reductions. As a result, JI is environmentally more robust than the CDM, but institutionally more vulnerable.

JI is also highly dependent on the functioning of domestic institutions. In order to participate in JI, countries have to fulfil a number of accounting requirements and compliance has to be continuous throughout the implementation of a JI project. Countries have to elaborate project approval procedures and be able to account for ERUs as part of their AAU accounting. Participation in JI also depends on political will, which became very clear in the case of Russia. As of July 2009 the Russian Federation has not yet approved a single JI project. This is, among others, due to the late finalization of their JI rules in January 2008 and the fact that the Russian Designated Focal Point for JI was closed for project submissions for four months between May and September 2008 following the presidential elections. These delays have tamed expectations that Russian JI projects will ever deliver ERUs and if they do, probably not at the expected scale.[33] The continuous delays in the approval of Russian JI

[32] Calculations based on information from the UNEP Risø CDM Pipeline dated 1 January 2009, the GHG data from <www.unfccc.int>, the Kazakhstan 1998 National Communication and the estimate of the global GHG emissions in 2004 by the Netherlands Environmental Assessment Agency at <http://www.mnp.nl/en/dossiers/Climatechange/TrendGHGemissions1990–2004.html>, all sites visited on 18 March 2009. The percentage is based on the total annual reductions in the CDM pipeline divided by the annual emissions from CDM eligible host countries. The latter is estimated by taking the global emissions and deducting the emissions from Annex I Parties as well as the emissions from Parties not eligible under the Kyoto Protocol with a large economy, being Turkey and Kazakhstan. Global emissions are estimated at 45 Gtonne CO_2e/year of which emissions from CDM eligible host countries is 26 Gtonne CO_2e/year.

[33] PointCarbon 2008b.

projects has frustrated investors which—with few exceptions—have discontinued project identification in Russia.

As a consequence of the relatively small number of projects, the international administration of the Track 2 stream of JI is not overly burdened and remains comparatively efficient.[34] The fire of criticism that has been opened on CDM and its operational effectiveness does not apply to JI, or applies to a much lesser extent. NGOs pay little attention to the mechanism since environmental integrity is less of an issue because ERUs are technically not offsetting any emissions. The relationship between project participants, JISC and accredited independent entities responsible for determination of projects and emission reductions (the JI equivalent of CDM's validation and verification) is characterized by a cooperative spirit. Problems that are identified, such as delays in the accreditation of independent entities, are relatively quickly solved. Given these distinctive characteristic of JI, contrasted with the current criticisms of the CDM, it seems likely there will be renewed interest in JI.[35]

The early JI projects were sponsored by countries facing a stringent emission target under the Kyoto Protocol or by international organizations that had established funds sponsored by these countries for Kyoto compliance purposes. Before the establishment of the JI Guidelines by the JISC in 2004, the market was dominated by sovereign buyers with a compliance gap. This included the Netherlands, Austria, Denmark, Spain, Italy, and Japan and their respective purchase programs and funds. These funds were managed by multilateral banks such as the International Bank for Reconstruction and Development (IBRD) and the European Bank for Reconstruction and Development (EBRD).[36] Following the adoption of the EU ETS and the increased guidance from the JISC, private entities also entered the market.

5. JI—The Promise of Oblivion

Joint Implementation reflects the importance of flexibility in the compliance system of the Kyoto Protocol. Flexibility in achieving emission reductions was key in the Kyoto negotiations and JI represents this consideration. Much of the public debate and international negotiations are currently absorbed by the discussion on the future of the CDM. By leaving out examination of the JI mechanism, an opportunity is being lost. JI broadens the scope of ways to reduce

[34] This section borrows from C Figueres and C Streck, 'Great Expectations, Enhanced Financial Mechanisms for Post 2012 Mitigation', Background paper for the 2010 World Development Report, on file with the authors.

[35] The JISC modified the rules for the accreditation of independent entities at its 13th session in Poznán, Poland, to overcome a bottleneck in the accreditation which would have hampered project validation for years to come. The problems had been identified by project developers and independent entities and triggered swift action of the JISC.

[36] The Netherlands EBRD Carbon Fund was established on 27 October 2003, see <http://www.ebrd.com/new/pressrel/2003/130Oct27.htm>, accessed 18 March 2009.

emissions and reduces the costs of compliance for those countries that have assumed emission reduction targets. Consequently, it is both likely and important that a mechanism that rewards emission reductions in a capped environment continues to form part of a post-2012 legal framework. Over the longer term, more countries could choose to assume emission reduction commitments for the whole of their economies (by graduating to become Annex I countries) or by assuming some type of sectoral emission reduction targets. Although such targets are unlikely to be negotiated in 2009, they will almost certainly be on the agenda for future climate negotiations. They may begin as voluntary, no-lose targets, and eventually become binding. The relevance of JI or a JI-like mechanism in this context is to incentivize private investment in GHG abating projects ahead or in parallel to government programs.

All in all, JI has suffered from an unfortunate lack of attention, which may well change once its valuable role in a post-2012 climate regime is fully understood. There are a number of reasons why JI is an important mechanism for enabling deeper emission reductions. First, JI offers the possibility of mobilizing additional emission reductions. JI broadens the scope and reduces the costs of compliance. It also encourages a broad set of actors (in Annex I countries) to engage in mitigation of climate change even if no dedicated climate change policies are in place.

Secondly, JI mobilizes financing for projects and programs that have little or no access to international financing sources. JI can also provide an efficient financing mechanism for a range of policy priorities such as renewable energy deployment, efficiency targets, or forestry policy. A viable and comprehensive JI mechanism can help to channel international financial flows into clean technology thus rewarding countries that assume emission reduction targets.

A third reason is that a post-2012 JI would enable the continuity of the existing legal framework. A JI mechanism would allow the rolling over of CDM project activities from an uncapped into a capped environment. This would provide certainty to existing CDM projects for the duration of the approved crediting period even if they are hosted by a country that assumes national or sectoral emission caps. JI can complement domestic emission trading schemes by including offsets from sectors that do not fall under the emission trading scheme. Furthermore, the linking of JI with national emission trading schemes allows capped installations to access emission reductions outside of the scheme. JI can thus act as a price valve for emission trading schemes while not compromising the overall environmental integrity of trading schemes adopted by Parties which have accepted emission limitation obligations.

In order to live up to these expectations and fulfill these functions, the current performance of JI deserves a closer look. Even though JI's performance is often said to be far below that of the CDM, the call for reform of JI is much less strong. However, to be able to accommodate future challenges JI does need to be reformed. In particular the reforms need to meet the challenges posed to countries that might decide to graduate from a CDM to a JI environment by moving to Annex I or by assuming targets for certain sectors. Some suggestions for improvement are set out below.[37]

5.1 Effectiveness and efficiency

The efficiency of JI can further be increased by revising a number of rules, in particular the concept of project additionality. Additionality in the context of AAU-backed JI projects has less relevance than in the case of the CDM and it is thus possible to focus in the case of JI on environmental additionality. This means that the additionality of a project could be measured differently than in the investment or barrier additionality test of the CDM. Additionality could instead be established by the ability of a project to deliver emission reductions below a conservative emission baseline. The establishment of baselines could be streamlined by applying a number of commonly agreed international principles or host country national priorities.

The approval procedure under the JISC is a lengthy process. Table 8.2 shows that from the start of determination to final registration takes an average 430 days while under the CDM it takes 400 days.[38] Although both periods are clearly too long, it is particularly unsatisfying that JI determination takes so much time. There are a few possible explanations for this. One is that the JI Track 2 procedure is new and there are only six registered projects so far. Learning-by-doing can make the procedure more efficient as more projects in the JI cycle advance to final determination. Another explanation is that it is very time consuming to redraft the project documents for early projects into the format that is now needed to ensure they comply with Track 2 requirements. Finally, whereas retroactive crediting under the CDM is restricted to early projects, JI project developers can also credit the emission reductions achieved before the date at which the project achieved final determination. As a result, CDM project developers have an incentive to register their project as soon as possible in order to minimize the amount of reductions achieved before registration. JI project developers do not have that incentive.

The combination of both a stringent additionality concept and a time consuming registration procedure creates a large barrier for project developers. The CDM Executive Board formulated guidance for the demonstration and assessment of additionality. This guidance is also often used for JI projects. Among the options listed in this guidance a popular approach in both CDM and JI is demonstrating that 'the proposed CDM project activity is unlikely to be the

[37] This section borrows on C Figueres and C Streck (see n 34 above).

[38] UNEP Risø JI and CDM Pipelines dated 1 January 2009. These estimates do not take into account the time needed for the development of a Project Design Document or a new methodology.

Table 8.2. Average time needed for projects and methodologies in different parts of the JI and CDM procedures.[39]

Step in the project registration or methodology	Average time required (days)
From the start of public comments as part of JI project determination under Track 2 to a final determination by the JISC (positive decisions only)	436
From the start of public comments as part of the validation of a CDM project to submission for registration	281
From submission for registration of a CDM project to a final decision on project registration by the CDM Executive Board (positive decisions only)	128
From initial CDM methodology submission to the final decision by the EB (positive decisions in round 22 only)	305

most financially/economically attractive (...) or is unlikely to be financially/economically attractive.'[40] Although not explicitly required, this implies that revenues from the sale of ERUs or CERs are crucial in making the project feasible. The combination of the importance of revenues from the sale of compliance instruments and the time it takes to register projects and secure these assets could prove detrimental. If registration of a project takes on average more than 400 days and may even take as long as 700 days (see Table 8.2) in the case of a CDM project that requires approval of a new methodology, this means that CDM and JI are only attractive for projects that can wait more than one or two years before they get certainty on a future revenue stream that is crucial to the feasibility of the project. The majority of the registered CDM and JI projects were developed when there was still optimism that the time needed for the CDM cycle would shorten as experience with the CDM accumulated. Unfortunately that is not the case, making it increasingly difficult for CDM and JI consultants and buyers of carbon credits to convince project developers to invest time and effort in obtaining CDM or JI registration and to accept the uncertainties that come with it.

5.2 Pioneering new areas

JI baseline and monitoring methodologies are project specific, a characteristic in which JI differs from the CDM. The emission reduction calculations and project design of a CDM project has to be based on a methodology that has been

[39] Calculations based on the UNEP Risø JI and CDM Pipelines dated 1 January 2009.

[40] Executive Board of the Clean Development Mechanism 39th Meeting: Report Annex 10 'Methodological Tool "Tool for the demonstration and assessment of additionality"' (Version 05.2) (May 2008). Out of the six JI projects that achieved final determination under Track 1, five used an investment analysis according to the CDM guidance.

approved by the Executive Board. Under JI the situation is different. A JI project submitted for determination can be based on existing CDM methodologies or on a new approach developed by the project owner. Without a separate approval procedure for methodologies and projects, JI can be far more effective in exploring new technologies and sectors that bear emission reduction potential. So far project developers have taken a conservative approach and have largely relied on CDM methodologies. Parties that establish Track 1 procedures and the JISC could decide to encourage the development of simpler—or simply different—methodological approaches.

5.3 Encouraging programmatic JI

In 2008, COP/MOP-4 requested that the JISC should develop procedures for JI Programmes of Activities to be implemented under Track 2 JI.[41] The development of programmatic JI projects will allow Annex I countries to achieve emissions reductions in sectors not covered by domestic or regional emissions trading schemes a cost-effective way. When adopting a programmatic JI framework, the JISC has an opportunity to overcome the unnecessary restrictions faced by programmatic CDM projects. For example, one drawback to programmatic CDM is that it is restricted to projects to which only one methodology is applied.[42] In JI, procedures for methodology and for project approval are combined and project developers are free to define their own methodological approach or apply combinations of CDM methodologies. Another restriction on PoAs in the CDM is the fact that the starting date of project implementation cannot be before the date of project registration.[43] With the time needed for CDM validation and registration averaging around 400 days this requirement is a serious barrier for PoAs under the CDM. The JISC can overcome these shortcomings by allowing greater flexibility in the design and implementation of JI programmes.

5.4 Encouraging sectoral JI

Sectoral JI takes the idea of programmatic JI a step further. Sectoral approaches define additional emission reductions as reductions below a sectoral benchmark or baselines. Sectoral benchmarks can make the calculation and estimation of emission reductions more straightforward although it could require more work upfront when defining the baseline level.[44]

[41] Decision -/CMP.4, Guidance on the implementation of Art 6 of the Kyoto Protocol, para 6.

[42] Procedures for registration of a Programme of Activities as a single CDM project activity and issuance of Certified Emission Reductions for a Programme of Activities (Version 02), para 2.

[43] Glossary of CDM terms, Version 04.

[44] Wolfgang Sterk, 'From Clean Development Mechanism to Sectoral Crediting Approaches—Way Forward or Wrong Turn?' (May 2008) 1 JIKO Policy Paper 16–17. <http://www.wupperinst.org/uploads/tx_wibeitrag/CDM_sect_crediting.pdf>, accessed 19 March 2009.

In essence, allowing for sectoral approaches under JI can increase the participation of countries in mitigation efforts, targeting key emitting sectors without losing the rigour required to measure the environmental integrity of the mechanism.

5.5 Strengthen governance and legal framework

The JI process would further benefit from a clarification of the roles of the JISC relative to the accredited independent entities and the UNFCCC Secretariat, in particular where those roles differ from the corresponding allocation of responsibilities under the CDM. In order to protect the interests of private participants, a clear set of strengthened administrative rules should guide the interaction of the JISC and the Secretariat with independent entities and project participants. An appeal mechanism would further protect the rights of private participants *vis-à-vis* the international bodies.[45]

5.6 Reduce host country uncertainty

The dependence of JI on host countries adds a degree of political volatility to the system. JI approval procedures often delay project development and create uncertainty. Investors have little confidence in host country's ability to successfully support JI projects in the long term and actually transfer the ERUs to the buyer's registry accounts when verification has been finalized. This issue will be of less importance once JI host countries have proven willing and able to deliver. Reducing this uncertainty and improving political stability in the host countries may also alleviate part of the investors' reluctance to invest in projects whose feasibility relies heavily on JI revenues. Another aspect of the host country uncertainty is obtaining approval in some JI host countries that show high levels of corruption.[46]

6. Post-2012 Framework

The key challenge for the UNFCCC negotiations is developing a solid basis for a post-2012 framework, work that is on the agenda for COP/MOP-5 in December 2009. Since the call for reform of JI is less strong and its political priority is lower than that of the CDM, the negotiators have received little support from science and policy advisors on this issue.

[45] In relation to similar proposals for the CDM, see further chapter 26, below.
M v Unger, C Streck, 'An Appellate Body for the Clean Development Mechanism: A Due Process Requirement', CCLR 1 (2009), 29.

[46] Anna Korppoo, 'Report of the UNFCCC Technical workshop on Joint Implementation', Bonn 13-14 February 2007.

At the COP/MOP-4 in December 2008 in Poznan, some parties expressed their views on post-2012 JI. The main forum for these discussions is the 'Ad hoc working group on further commitments for Annex 1 Parties under the Kyoto Protocol' (AWG-KP), established by the COP/MOP in December 2005.[47] This working group is elaborating 'possible improvements' to the flexible mechanisms including JI.[48] The topics under discussion are the graduation of CDM projects to JI in countries that adopt an emissions cap, eligibility of nuclear facilities, and improving the definition of project eligibility and particularly the additionality concept.

It is anticipated that after 2012 there will be a gradual expansion of the number of Parties and sectors with emission targets, hence the development of modalities and procedures for the treatment of CDM projects in countries that become eligible to host JI projects. The AWG-KP proposed two alternative options. The first option is to keep such projects as CDM projects.[49] This raises questions on adjusting the AAU allocation for the reductions achieved by existing CDM projects or cancellation of AAUs upon CER issuance. The second option is to avoid AAU issues by graduating the CDM projects into JI.[50] However, this in turn raises questions on the crediting period and whether the CDM projects can also be transferred to Track 1.

The second item on the agenda is the eligibility of nuclear facilities under JI. The AWG-KP proposed to either maintain the position that nuclear power is not eligible under JI or to make nuclear power eligible under JI and allow Annex 1 parties to use for Kyoto compliance 'emission reduction units issued for such projects, on the basis of emission reductions achieved during the second commitment period'.[51] This relates to the discussion as to whether nuclear power can be used to meet Kyoto targets. At COP/MOP-4 the International Atomic Energy Agency (IAEA) distributed a booklet discussing various advantage and disadvantages of nuclear power as a means to mitigate climate change. The IAEA concludes 'that the need to mitigate climate change is one of the salient reasons for considering nuclear power in national energy portfolios' but also states that 'the nuclear sector needs to provide adequate responses to the concerns still surrounding its use'.[52] The discussions on this topic could open the door for a compromise where nuclear energy might become eligible under very strict conditions. A compromise could be to allow energy efficiency measures in nuclear power plants, aimed at improving the efficiency of the conventional power generation part of the plant. This would enable investors to

[47] See <http://unfccc.int/kyoto_protocol/items/4577.php>, accessed 19 March 2009.

[48] FCCC/KP/AWG/2008/INF.3 (2008) 1.

[49] FCCC/KP/AWG/2008/INF.3 (2008) 13 'Introduce modalities for treatment of clean development mechanism project activities upon graduation of host Parties'.

[50] FCCC/KP/AWG/2008/INF.3 (2008) 13.

[51] FCCC/KP/AWG/2008/INF.3 (2008) 13.

[52] IAEA 'Climate Change and Nuclear Power' (2008) 42 <http://iaea.org/OurWork/ST/NE/Pess/cop14.htm>, accessed 19 March 2009.

improve the efficiency of old nuclear power plants, thereby increasing their power output while maintaining or even lowering consumption of nuclear material and production of nuclear waste.

Another item discussed by the AWG-KP is ways to ensure environmental integrity and to assess additionality of JI projects. Options include the development of positive or negative lists of project types on the basis of technologies, host party criteria or the scale of the project. Questions raised include the applicability of these criteria to Track 1 projects, procedures for periodic review of the list and how to develop the first version of the list.[53] In contrast with the CDM, these discussions do not include the various options for sectoral approaches. This is surprising since JI could also benefit from expanding the regulatory framework with sectoral options. However, this will only be effective if they reduce the time and effort needed to achieve final determination and obtain at an earlier stage a degree of certainty on the projects' ability to generate valuable compliance instruments.

The final item on the agenda is to include co-benefits as criteria for the final determination of projects. One suggestion is to stimulate projects that can actually demonstrate co-benefits. These co-benefits could be determined by the independent entities. Co-benefits could include technology transfer or the delivery of environmental services as a side-effect of the project.[54]

In addition to the AWG-KP discussions, the 'Ad hoc working group on long-term cooperative action under the convention' (AWG-LCA) established by the COP -13 invited Parties to express their 'Ideas and proposals on paragraph 1 of the Bali Action Plan'. The Bali Action Plan is the roadmap to the development of the post-2012 framework.[55] In their response many Parties support markets as a means to mobilize capital, to keep the costs of abatement low and to enhance technology transfer. The European Commission goes even further by stating that expanding the carbon market 'must be at the heart of a post 2012 agreement'. The European Commission is in favour of extending the global carbon market to include developing countries, a position also supported by South Korea and Norway. [56]

At the same time, the European Commission, Japan and Australia regard improvements in JI, CDM, and IET indispensible. Parties have expressed the view that markets can only be a key vehicle for financing mitigation if 'existing and new market mechanisms meet a high standard of environmental integrity'. Others call for enhancing the geographical distribution of project activities and a reduction of the administrative costs. Australia suggests the 'automatic approval

[53] FCCC/KP/AWG/2008/INF.3 (2008) 14.

[54] FCCC/KP/AWG/2008/INF.3 (2008) 14.

[55] FCCC/CP/2007/6/Add.1 Report of the Conference of the Parties on its thirteenth session, held in Bali from 3 to 15 December 2007, (March 2008) Decision 1/CP.13, Bali Action Plan.

[56] FCCC/AWGLCA/2008/16/Rev.1, AWG-KP, Fourth session 'Ideas and proposals on the paragraph 1 of the Bali Action Plan', Poznan, January 2009, para 73–5.

for technical aspects of certain well-recognized technologies', but rejects a full waiver of the additionality test.[57] While under the AWG-KP sectoral approaches are only mentioned in the context of CDM and IET, in the AWG-LCA only Norway, Japan and Canada identify sector-based approaches as an option to explore.[58] However, concepts proposed under IET could also be possible under JI. An important difference between the two mechanisms is that under IET AAUs are being traded while the trading unit under JI is the ERU. By means of the linking directive[59] ERUs have been granted access to the EU ETS market whereas operators of EU ETS installations can not use AAUs for compliance. Although IET provides a lot of flexibility, the existing proposals suggest introducing a sectoral trading scheme approach. By introducing a sectoral approach in IET, the negotiators create guidance on the so called 'greening' of AAUs. Introducing sectoral approaches under Track 2 may provide a framework that triggers renewed interest in JI.

7. Conclusions

Overall, JI currently lacks the momentum and economic importance of its sister mechanism—the CDM. Many established CDM actors have little or no knowledge about JI and the way in which it functions. Discussions on JI take place in small circles and so far the mechanism has not been able to shrug off the image of a limited tool of interest only to those concerned with a few niche players in Central and Eastern Europe. Despite these perceptions, JI deserves more attention and care as a mechanism that may stimulate investments in GHG abatement projects in countries that have assumed binding or voluntary, enforceable or no-lose, economy-wide or partial, emission targets. Review and reform of JI should therefore be firmly placed on the post-2012 negotiation agenda and lessons learned from implementing JI should be developed and disseminated, including to developing country negotiators. JI is a very flexible tool for reducing emissions beyond what domestic policies can capture. Both economies in transition and also countries with a tight emissions cap under the Kyoto Protocol have implemented JI schemes to support their efforts to reduce the emission of GHGs. In some cases these initiatives have also supported the country's efforts at Kyoto compliance by issuing fewer ERUs than the amount of emissions mitigated by the project.

It is important that negotiators for the post-2012 climate regime acknowledge the opportunity presented by JI as a trading mechanism with the potential to

[57] FCCC/AWGLCA/2008/16/Rev.1, (2008) paragraph 74, 76, 77, 158.

[58] FCCC/AWGLCA/2008/16/Rev.1, (2008) paragraph 77.

[59] Directive 2004/101/EC of the European Parliament and of the Council of 27 October 2004 amending Directive 2003/87/EC establishing a scheme for greenhouse gas emission allowance trading within the Community, OJ L338, 13.11.2004.

provide long-term flexibility to carbon market participants. In particular, the future JI mechanism should be expanded to include countries that agree at the climate negotiations to adopt an emissions cap after 2012.

Moreover, the future regime should be designed to improve investors' confidence while maintaining the environmental integrity of the system. It should consider why Annex 1 Parties and project developers have so far not taken the initiative to explore all the opportunities that JI offers, and try to develop ways to overcome this reluctance. JI may benefit from the development of a framework that provides more guidance on how to explore new areas and lowers the barriers for new entrants. An example of this is the use of sectoral baselines in combination with limiting the additionality requirement to proving environmental additionality. Policy makers now face a difficult challenge. They must strike a balance between providing more guidance while maintaining the flexibility that host countries and project developers are still exploring.

9

Joint Implementation Transactions: An Overview

Anthony Hobley and Carly Roberts

1. Introduction

Joint Implementation (JI) is often seen as the poor relation to the Clean Development Mechanism (CDM). Why should this be the case? JI and CDM are built on the same concept—namely, in return for investment in a project situated in another country, an Annex I Party receives emissions allowances arising from the emissions reductions of that project, yet CDM is to date much more successful.

A principal difference between CDM and JI is that, whereas CDM is mainly regulated by an international authority, JI was intended to be regulated primarily by the Annex I Parties hosting the JI projects (Host Countries) with more limited international supervision. This difference of approach flows from the fact that JI projects take place within an Annex I Party cap, as opposed to CDM projects which are outside of such caps.

The underlying rationale for this difference in regulation is that for JI, the Host Country has to give up valuable Assigned Amount Units (AAUs) for each Emission Reduction Unit (ERU) created.[1] If these do not represent true tonne for tonne reductions then that Host Country would be giving away AAUs (potentially required for compliance) without corresponding greenhouse gases (GHG) emissions reductions. The Host Country should therefore be concerned to ensure that the ERU represents a true and additional emissions reduction. In theory it is self-policing. This is a key reason why it was decided that JI should have significantly more national and less international oversight than CDM.

There are clearly some critical differences between JI and CDM projects. Such differences are likely to be more pronounced between Track 1 JI which is exclusively governed by Host Countries and CDM which is principally governed by the CDM Executive Board. As a result of the involvement of the international oversight by the JI Supervisory Committee (JISC), Track 2 JI is

[1] Discussed above in chapter 8.

often described as closely resembling CDM. This is in our view misleading. There are some differences between the regulatory infrastructure governing Track 2 JI projects and that governing CDM projects which can be critical to commercial parties.

This chapter will build on the introduction to JI in chapter 8 by Hoogzaad and Streck, by re-examining the key differences between JI and CDM, framed more from the point of view of the prospective investor who is perhaps more used to CDM. It will seek to tackle some of the common myths and misconceptions which often arise due to such existing experience of the CDM.

Drawing upon this comparative analysis the chapter will draw out the key differences which give rise to the distinct risk profile of JI. This will then be used to offer practical advice on the key issues which need to be addressed both in due diligence and contractual drafting for JI projects, as distinct from CDM projects. Chapter 8 looks to the future in challenging the claim that JI is of little importance compared with the CDM. An examination in this chapter of an investor's risk profiling of JI as compared with CDM, and a look at JI contracting, perhaps show why, at present, JI investment has not scaled the heights of the CDM market. However, the authors of this chapter share Hoogzaad and Streck's optimism about the future of JI and intend the risks and contracting points detailed in this chapter to help provide a roadmap to JI, rather than a warning away.

2. Joint Implementation in Practice

Article 6 of the Kyoto Protocol provides for JI as a means for companies in Annex I Parties with a Kyoto target (the Investor Country) or the Investor Country itself to undertake GHG abatement or sequestration projects which reduce the emissions of GHGs in another Annex I Party (the Host Country). Such projects are then credited with ERUs which can be used by the investing company or Investor Country for compliance with their own emissions targets under the Protocol. The CDM mechanism operates on a similar basis, the difference being that the Host Country in respect of CDM is a non-Annex I Party (ie a country without a target under the Kyoto Protocol).

2.1 JI Host Countries

There are various Annex I Parties whose GHG emissions during the Kyoto Commitment Period will not exceed their emissions targets under the Kyoto Protocol.[2] These countries include the successor states of the former Soviet

[2] Figures from UNEP: <http://cdmpipeline.org>, accessed 10 April 2009.

Union and Eastern European states. Such countries may wish to trade any surplus AAUs with other Annex I Parties and JI projects are seen as an environmentally sound way of engaging in the trade of AAUs.

The current leading JI Host Countries are Russia, Ukraine, Bulgaria, Poland, and Lithuania. However, in the last six months there has been a sharp increase in JI project activity in Germany.[3] Various other countries, including Latvia, Romania, Czech Republic, Slovakia and New Zealand, are also acting as Host Countries for the purposes of JI. We have considered the emerging JI rules in a number of these countries.

There are various advantages in acting as a JI Host Country:

- GHG reducing projects established on the territory of the country receive additional income in respect of a variety of projects which helps ensure that the projects are financially viable;
- the projects result in a reduction in GHG emissions as measured against the baseline, thus helping reduce the GHG emissions trajectory in the Host Country; and
- the Host Country benefits from increased exposure to and exchange of knowledge with foreign investors.

2.2 Current JI pipeline

The international rules for JI provide two sets of JI procedures, referred to as 'Track 1' and 'Track 2'. Figure 9.1 provides a brief summary of the steps which need to be taken in respect of Track 1 and Track 2 JI projects before the Host Country will issue the resulting ERUs.

Track 2 has so far proved more popular than Track 1.[4] From the perspective of an Investor Country, the involvement of the JISC and Accredited Independent Entities (AIE) in Track 2 can create a perception of more certainty concerning compliance with specific JI procedures. This is seen to reduce delivery risks. From the perspective of the Host Country, Track 1 can create additional obligations, namely the requirement to fulfil all eligibility requirements continuously and the responsibility to validate the project and verify the emissions. Another advantage of Track 2, for both investor and seller, is that ERUs transferred under Track 2 are not subject to the requirements of the Commitment Period Reserve.[5]

[3] As at March 2009; see further chapter 8.

[4] Although a conversation with UNEP Risø indicates that precise statistics on the number of Track 1 JI projects is difficult to gather and verify so the actual number of Track 1 JI projects being developed is uncertain.

[5] The Commitment Period Reserve is the required portion of AAUs which the Annex 1 Party must hold in its registry, under Art 17 of the Kyoto Protocol. It amounts to the lower figure of: 90% of the country's AAU total or five times its most recent annual emissions inventory. Its purpose is to reduce the risk of overselling.

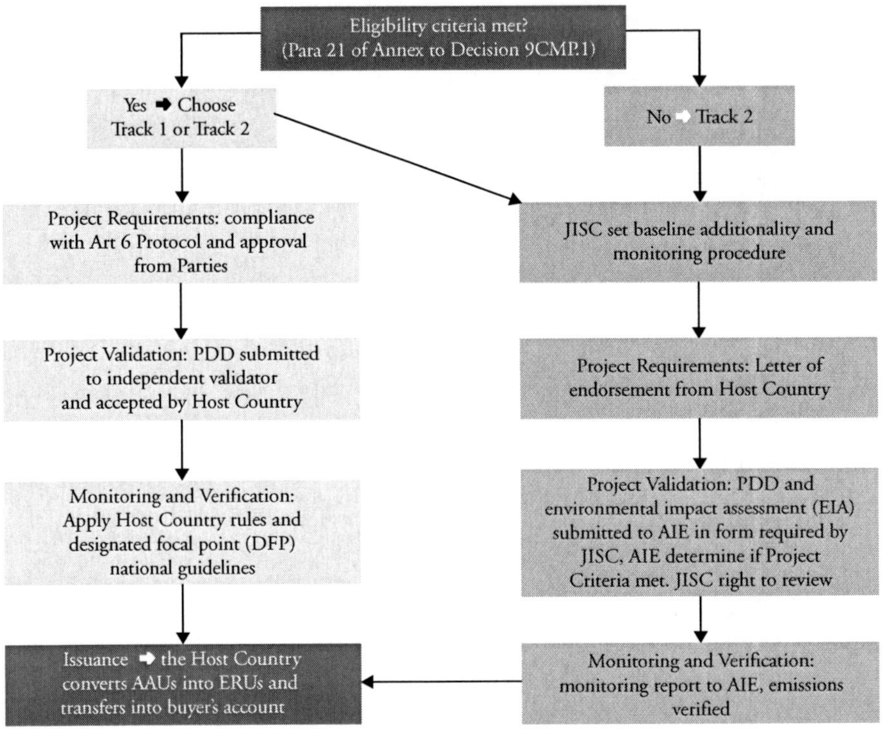

Figure 9.1. Summary of steps required for Track 1 and Track 2 JI projects

3. JI and CDM

Table 9.1 summarizes some of the key regulatory structural differences between JI and CDM.

The principal difference between CDM and JI is the fact that, whereas CDM is regulated principally by an international authority, the regulation of JI is undertaken mainly by the Host Countries and international supervision is limited. Although the JISC exercises a certain level of control over Track 2 JI, both tracks were intended by the Marrakech Accords to be much more governed by domestic regulation than the CDM is. The high involvement of the Host Countries in JI relative to the CDM has a direct impact on the risk profile faced by potential purchasers of ERUs. Faced with risks different to, and in many cases potentially more serious than, the regulatory risks connected with a CDM transaction, parties interested in purchasing ERUs from JI projects need carefully to consider their appetite for assuming these risks and mechanisms for dealing with such risks.

In comparison, the regulation of CDM transactions is more established and less subject to change. In view of this, it is perhaps no surprise that, among the

Table 9.1. Key regulatory structural differences between JI and CDM

Factor	CDM	Track 2 JI
Overall regulatory control	CDM Executive Board exercises overarching regulatory control over project approval process.	Host Country exercises high degree of control, even when subject to JISC requirements.
Project approval application	Relative consistency of rules from country to country, eg whether project is in Chile, Thailand, or Uzbekistan around 80% of the rules and processes are the same.	Rules increasingly following CDM but higher Host Country requirements and lack of consistency from country to country, eg JI rules vary materially between Russia and Ukraine.
Registration of Project	CDM Executive Board registers project.	AIE makes Determination, over which JISC has right of review (although through appraisal process JISC are currently reviewing all projects).
Number of Units	Theoretically no limit on CERs.	Host Country may impose caps, eg in the Russian JI rules.
Issue of Units	CERs issued by CDM Executive Board following verification.	ERUs issued by Host Country following verification and conversion of AAUs.
Delivery of Units	Buyer able to use focal point role to control forwarding of CERs.	Focal point concept now established under JI, but it does not provide buyer a right to direct delivery of ERUs—Host Country controls delivery.
Track Record	Post-2000 CERs can be used. CDM projects registered since 2004.	ERUs can only be used from 2008. JI projects registered since 2007.
Second commitment phase	Post-2012—Yes.	No provisions for post-2012.

flexible mechanisms of the Kyoto Protocol, CDM has until now dominated at the expense of JI.

4. JI Transaction Risk Assessment

There is no harmonized international framework in respect of JI projects, with rules varying between Host Countries. Although the potential for differences in procedure is greater in respect of Track 1 JI, there are material differences between the applicable Track 2 JI requirements of different Host Countries. Examples of Track 2 JI requirements specific to a particular Host Country include Russia's policy concerning sectoral caps on ERUs (that is, the limiting of numbers of ERUs within different sectors) and Ukraine's initial intention to require parties to use a template emissions reduction purchase agreement. Investors looking to manage sovereign and other risk by spreading carbon credit investment across different countries may prefer the consistency of alternatives such as CDM, where international approval and issuance procedures are harmonized and have been in place since 2002.

As noted above, JI was intended to be mainly regulated by the Host Countries, rather than by an international authority as is the case for CDM. This concept is a theme through most of the risks discussed below. It means that for JI, it is the Host Country which is responsible for and in control of the issuance and delivery of the ERUs.

4.1 Delivery risk

As ERUs are issued by the Host Country there is—compared to the CDM—a higher Host Country regulatory risk and therefore delivery risk in respect of JI, resulting from the issuance of ERUs which could be influenced by the Host Country's changing political and economic situation. There is also currently a widespread absence of provisions dealing with the transfer of ERUs from Host Country to Investor Country, so it is often unclear how this will work in practice. Currently only one JI project has reached the stage of issuing ERUs, so there is little practice or precedent for guidance.

As indicated by Figure 9.2, following verification of the emissions, the Host Country will convert the relevant number of AAUs in its national registry into ERUs.

The following mechanism for the issuance and transfer of ERUs is generally less clear and will vary from country to country. It is critical to buyers and investors in JI projects to have clarity on the precise regulation, rules and reselling mechanics established in each Host Country for the issuance and delivery of ERUs. Many such investors will have experience of the process for issuance and forwarding of CERs from CDM projects. They may have become used to the ability to control the forwarding through the focal point concept prescribed in the CDM Modalities of Communication.[6] This contrasts with JI, where a higher degree of regulatory uncertainty on this critical point must be accepted. This is important because it goes to the heart of delivery risk, which is fundamental to the risk evaluation in any primary CDM or JI transaction.

4.2 Host Country and political risk

JI is controlled by the Host Country government and is not as market-led as the CDM. It is therefore more dependent on the timetable of the Host Country in

[6] Under the CDM project, participants nominate a party that is authorized to communicate with the CDM Executive Board regarding the issuance and distribution of CERs.

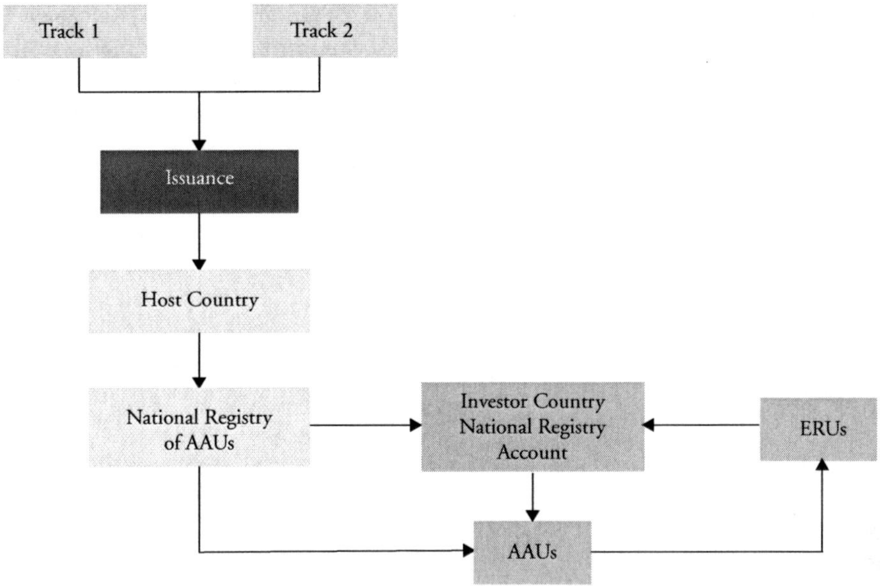

Figure 9.2. Conversion of ERUs to AAUs

developing JI procedures. Both Track 1 and Track 2 JI require the Host Country to prepare and implement relevant legislation and regulation. Before an Annex 1 Party can become eligible to host a JI project the national JI Rules must be notified to the UNFCCC Secretariat. In common with any other domestic legislation, the legislative process is directly affected by political factors. Such political considerations can materially affect the content and timing of such legislation.

JI cannot be implemented without the support of the lawmakers. The implementation of legislation and regulation in respect of JI can therefore be delayed as a result of the government focusing on other legislative areas or in connection with a change in government. Indeed, changes of government in both Russia and Ukraine, which together comprise the majority of the potential JI market, have previously resulted in delays in the implementation of JI legislation.

More specifically, the policies of the Host Country government can affect the timing and content of legislation. Investment by foreign companies can sometimes lead to opposition within a country, and the sale of ERUs can be seen as an example of the sale of State assets. Governments may therefore be influenced by such opposition and act carefully in order to ensure that they are not criticized for allowing the sale of assets on too generous terms.

The delay of some governments in implementing JI legislation and regulations has also been connected with a lack of conviction or enthusiasm of such governments in respect of the benefits of JI. For example, it has been suggested that the lengthy implementation of JI legislation in Russia (which is not yet complete and has yet to result in an approved JI project), is partly the result of the recent boom in oil and gas exports which probably renders the revenues from JI projects unattractive by comparison. The fact that the potential JI Host Countries can also sell surplus AAUs via the alternative flexible mechanism of international emissions trading in accordance with Article 17 of the Kyoto Protocol may also affect their political will to push through JI legislation. Similarly, many Eastern European countries which have joined the European Union have concentrated their efforts on the EU Emissions Trading Scheme at the expense of developing a framework for JI.

By way of contrast, many aspects of the CDM process, including project registration of the project and issuance of CERs, are controlled by a single international body, the CDM Executive Board. This gives parties to a CDM transaction (relative to JI) greater certainty that each CDM transaction should be handled in a similar or identical manner. Although each Host Country has its own rules in respect of its internal approval of the CDM project, there is relative consistency in approach between Host Countries.

Relatively speaking, this certainty in the CDM compares favourably with JI projects, where the project approval, registration of the project and issuance of ERUs are each controlled by the Host Country. Each Host Country is free to adopt its own laws and regulations governing these aspects, and this has resulted in material differences between Host Countries. For example, Russian legislation requires a participant in a JI project to invest in the project in accordance with Russian investment law—purchase of ERUs alone is insufficient to qualify as an investor. This has a direct impact on the structure and documentation for a Russian JI project. Furthermore, Russian JI rules provide the ability to impose caps on the number of ERUs which can be issued in respect of certain sectors. They also permit the Russian authorities to withdraw registration of an approved JI project on wide grounds.

In many Host Countries, the issue of the legal status of ERUs has not yet been adequately dealt with, leading to uncertainty as to ownership and tax treatment of ERUs. Contrast this with the CDM where this is less of an issue because the CERs are not created in the Host Country. Rather they are created in the CDM registry administered by the UNFCCC secretariat in Bonn under international law. Any interested purchaser of ERUs must therefore undertake careful due diligence of the applicable laws and regulations in the relevant Host Country in order to ensure that the project documentation correctly identifies and reflects the Host Country's requirements.

4.3 Involvement of different authorities

Another consequence of the authority of each Host Country to implement its own JI rules has been the involvement of various domestic authorities, rather than a single authority. By way of illustration, in order to be registered, a Russian JI project requires the involvement of a variety of different bodies. An applicant for a Russian JI project will submit its application to the Ministry of Economic Development in its capacity as 'designated focal point'. The designated focal point will in turn ask for the opinion of relevant (potentially as many as nine) federal executive authorities. The application will subsequently be considered by a Committee established by the designated focal point and comprised of representatives of the Federal Executive Authorities. Finally, the designated focal point will submit projects to the Russian government quarterly for approval. If approved, an 'authorized body' will subsequently control implementation of the JI project. The involvement of different bodies clearly increases regulatory risk, not to mention the risk of delays, especially where there is any uncertainty in respect of the exact division of responsibilities between the authorities. By way of contrast in the Ukraine, where fewer bodies are involved, the regulation of JI projects was originally supervised by the Ministry of Environmental Protection. However, responsibility has now moved to the National Environmental Investment Agency which was established in April 2007.

In addition to the number of authorities involved, it is important that the relevant authorities are adequately staffed and funded to ensure that they can apply the relevant regulations promptly and effectively. Delays in issuance of letters of approval in Host Countries have indicated that additional resources may be required, although the involvement of the National Environmental Investment Agency is anticipated to improve matters in the Ukraine.

4.4 Missing regulation

A related point is that, even in those countries such as Russia which have implemented and notified their JI Rules to the UNFCCC secretariat, the JI legislation is not yet complete. In particular, neither Ukrainian nor Russian legislation currently regulates the mechanics for the issuance and transfer of ERUs to a purchaser's account. Potential buyers must therefore be aware that they are becoming involved in a regulatory environment which will be subject to ongoing change and clarification.

4.5 Slow start/early finish

For the purposes of CDM, CERs produced from as early as 2000 under the prompt start provisions have been recognized and the domestic and international regulation of CDM transactions has been applied over a period of approximately six years, resulting in relative certainty about the application of the regulations. However, the fact that JI has only been available from 1 January 2008, combined with the relative slowness of the Host Countries to implement legislation, has meant that many Host Countries have only recently implemented sufficient legislation to govern JI projects. For example legislation for the approval of JI projects in Ukraine was only brought into force in August 2006 and Russia followed as recently as May 2007, and in both cases further regulation was necessary to implement the approval procedures. As a result, as with any new laws or regulations, there is likely to be a learning period during which all involved parties work out how to apply the JI rules, which is likely to result in the process taking longer than anticipated. Until such time as Host Country JI rules have been consistently applied over a period of time, there will remain some uncertainty about how the rules will be applied. An ERU purchaser will need to take this regulatory risk into account when negotiating the transaction documents.

5. Drafting Contracts for Sale and Purchase of JI Emission Reductions

In the authors' view, the JI market has not yet achieved a high enough volume of contracting to enable reference to a 'standard' form of JI emission reductions purchase agreement (ERPA) (called an investment agreement in some jurisdictions to conform to local laws), or to the 'market position' in negotiating such an agreement. JI ERPAs need to be much more tailored to each JI country than is the case for CDM ERPAs. Given the much smaller number of JI projects and uncertainties around Host Country JI regimes, primary JI ERPAs almost without exception contain 'non-firm' or 'non-guaranteed' delivery obligations.[7] That is, a JI project developer will not commit to pay liquidated damages upon failure to deliver a specified number of ERUs.

Given the limited examples of JI contracts, parties will sometimes wish to use a CDM ERPA as the basis for the purchase of ERUs from a JI project. A CDM ERPA should only be used with caution as the starting point for purchase of ERUs. Material changes are required to a CDM ERPA in order to make it suitable for a Track 1 or a Track 2 JI project.

Issues to which attention should be given in drafting a JI ERPA as distinct from a CDM ERPA include:

- different JI procedures in each Host Country;
- regulatory risks from uncertainties regarding incomplete Host Country JI regulations;
- fundamental differences between CDM and JI procedures and terminology;
- less protection (particularly ability to control issuance of credits) resulting from a party being nominated as focal point for all communications with the regulatory bodies;

[7] Like many primary CDM ERPAs.

- addressing the distinction between Track 1 and Track 2 JI;
- ability to switch between Tracks 1 and 2; and
- additional conditions relating to the implementation and notification to UNFCCC of the Host Country JI rules.

As discussed previously, unlike CDM, JI procedures can operate differently from Host Country to Host Country, even under Track 2. This usually requires bespoke changes to an ERPA on a jurisdiction by jurisdiction basis. In Russia for example, care should be taken to reflect the formal 'investor' role which is required to be taken by a primary buyer of ERUs from a Russian JI project. This investor role could potentially attract greater project liability than would be expected by an arms length buyer of CERs from a CDM project. It seems that investors in Russian JI projects are required to make some form of actual investment beyond just purchasing ERUs.

5.1 Use of templates

From a legal drafting perspective, CDM primary ERPAs are becoming sufficiently standardized that buyer entities can develop preferred templates for use in negotiation, rather than requiring bespoke drafting in the case of each transaction. Distinctions between templates seen in the primary CDM market tend to be limited to China/non-China CDM ERPAs and firm/non-firm CDM ERPAs, with additional discretionary clauses such as a clause for options over post-2012 CERs. However, in the JI market the differences between specific Host Country JI regimes coupled with the (to date) lesser amount of JI projects available, make it difficult for a primary buyer entity to develop a template JI ERPA for use in JI investments across different jurisdictions. A single JI ERPA that included drafting optionality to account for the differences between regimes as diverse as those of the Ukraine, Poland, and Russia would be time consuming to prepare and unwieldy in application. With few exceptions, buyers are not currently investing in sufficient volumes of JI projects to justify the development of such a template. Bespoke country-by-country JI ERPAs seem more common.

Another approach is to draft JI ERPAs very widely to allow for both differences between Host Country JI regimes, and uncertainties where Host Countries have not finalized the specifics of JI regimes. While using such a document may seem more time and cost efficient than drafting a bespoke ERPA for each JI transaction in a new jurisdiction, it is not a legally conservative approach to rely on a widely drafted contract. A wide term in a contract risks being void for uncertainty in the most extreme case and at the least could leave the buyer or seller with an unexpected liability. Although we have not yet seen JI examples, in CDM disputes unhappy parties to the ERPA have exploited such wide and uncertain provisions to renegotiate prices and other commercially agreed terms.

5.2 Drafting for regulatory risks

As described above, there is a perception that the greater reliance on national regulation of JI when compared to the UNFCCC regulation of CDM, creates higher regulatory risk for JI projects than for CDM. This perception can be exacerbated when one of the contracting parties is closely associated with the relevant national government. Less subjectively, the fact that various Host Country JI regimes are not yet fully approved or comprehensively set down in regulation, creates obvious risks for an investor. Following due diligence, the parties need to carefully consider how to deal with any such uncertainty in the ERPA, for example in respect of delivery of ERUs. The drafting will need to be wide enough to cover possible outcomes, whilst still protecting respective interests. A wide contract can carry the risks referred to above, so care should be taken in striking the appropriate balance.

Another issue to consider in drafting is that unlike CDM ERPAs, the JI ERPA may require conditions relating to the implementation and notification to UNFCCC of the Host Country JI rules.

5.3 Change of law, law governing the contract, and dispute resolution

Given the regulatory risk and uncertainty, attention should be given to the 'change of law' clause and dispute resolution clause in a JI ERPA, as well as to practical enforceability against the counterparty.

English law coupled with an appropriate type of arbitration such as through the International Chamber of Commerce (ICC), seems to have become the default choice of law in many CDM ERPAs. English law is often favoured by commercial counterparties entering into cross border contracts, due in part to the robust privity of contract that English law allows when compared to many civil and other systems of law. The use of arbitration rather than the English courts helps avoid issues of proving jurisdiction in a transaction that will frequently have tenuous or non-existent links to the UK. Locating an arbitration elsewhere than in London can sometimes help overcome concerns about a lack of neutrality if one of the counterparties is English. However, it will be interesting to see if English law remains the default option in the medium to longer term if the US engages with CDM or an equivalent.

Leaving aside the limited number of JI projects on which to base such an assertion, it does not seem to be as automatic when negotiating a JI ERPA that the governing law is English. Given the increased role of national legislation in implementing JI, it is understandable that parties sometimes argue for the relevant Host Country law as the governing law. However, any such 'home ground' advantages in interpreting the relevant local JI principles should be weighed against the current incompleteness of many JI regimes and the parties' confidence in the particular Host Country legal system. Parties negotiating a JI ERPA do often come back to English law and arbitration as the more neutral approach.

5.4 Drafting for JI project cycle

The project cycle and terminology for JI differs from CDM and this should obviously be accounted for if adapting a CDM ERPA for use as a JI ERPA. Some terminology and project cycle differences between JI and CDM include:

- the UNFCCC supervising body is the JI Supervisory Committee (JISC) rather than the CDM Executive Board;
- the Crediting Period for JI is between 1 January 2008 and 31 December 2012, rather than the seven year (twice renewable) or ten year Crediting Periods for CDM;
- the third party verifier of a JI project and its units is an Accredited Independent Entity (AIE), unlike CDM projects and its units which are verified by a Designated Operational Entity (DOE);
- JI projects have their Project Design Document (PDD) 'determined' and CDM projects have their PDD 'validated';
- 'determination' for a JI project is final after 45 days, whereas 'validation' for a CDM project is final after eight weeks;
- a JI project achieves 'final determination' rather than 'registration';
- the relevant unit is an ERU rather than a CER;
- units are issued by the Host Country rather than by the UNFCCC supervising body.

5.5 Focal point role in JI

The focal point role for CDM and JI identifies the party responsible for communicating with the relevant UNFCCC regulator. In the CDM market, this role has evolved commercially as an important right which is subject to negotiation between buyer and seller at the contracting stage, because the focal point controls distribution of CERs. The middle ground position in a CDM ERPA is often for both buyer and seller to act as joint focal point, so that neither party can cause distribution of CERs without the other party's signature.

By contrast in negotiating a JI ERPA, a party with a CDM background that is new to JI may place a high importance on securing the focal point role. However, it should be kept in mind both that the Host Country rather than the JISC controls issuance of ERUs and that the recently created role of focal point does not carry the right to direct the delivery of ERUs as is the case with CERs from

CDM projects. It is unlikely therefore to be considered of the same practical benefit to be the party responsible for acting as focal point with the JISC, as it would be for a CDM project.

5.6 Track 1 and Track 2

As at the end of 2008, Track 1 JI is not an option in many JI Host Countries so JI forward ERPAs tend to commence under a Track 2 framework. However, Track 2 JI procedures are expected to be more expensive than Track 1 procedures. If a project owner/seller is carrying the costs of JI procedures, then it will often wish to build in the ability to switch from Track 2 to Track 1 JI procedures should these become available in the relevant Host Country. The project owner/seller would typically prefer that this ability is unilateral. The investor/buyer's position is often that it does not wish the project owner/seller to have a unilateral right under the ERPA to switch to Track 1. This is often because the investor/buyer may have concerns about the unknown future form and/or expected rigour of Track 1 JI procedures in the relevant Host Country (or the perception of third-party buyers in relation thereto), compared to the Track 2 procedures, and the impact these may have on the desirability and pricing of the resultant ERUs.

These different negotiating positions are sometimes resolved by the investor/buyer having the right to veto a switch from Track 2 to Track 1, subject to the investor/buyer carrying any cost differences between Track 2 and Track 1 procedures if the project owner/seller requests a switch which the investor/buyer denies. Such a clause is often expressed quite generally, where it may be more advisable to specify exactly which costs the investor/buyer would be responsible for in such a scenario, or at least a framework for calculation where the details of the Host Country's proposed Track 1 procedures are not sufficiently clear.

If the investor/buyer is to be responsible for the costs of JI procedures then once again the counterparty would usually prefer that the investor/buyer did not have the unilateral right to decide whether to switch to Track 1 procedures, as the project owner/seller may not have certainty at the time of entering into the ERPA as to how this would affect its project responsibilities.

5.7 Secondary JI contracting

Secondary trading of Kyoto units is discussed more comprehensively in chapter 15 of this book. It is worth noting though that a secondary trading document should not treat ERUs in the same manner as CERs, particularly where pricing and calculation of liquidated damages is concerned, given the much lower secondary liquidity of ERUs.

A 'back-to-back' secondary JI ERPA between the primary investor/buyer of ERUs and a second buyer should be approached with care. (Incidentally, some would argue this should not be called a secondary document as it is not truly addressing secondary market trading. We use the term here for convenience to distinguish from the primary ERPA with the project owner.) Like a similar secondary CDM ERPA, we do not consider it advisable to simply 'back-to-back' the terms of the primary ERPA in the secondary ERPA. Particularly from the buyer's point of view, it is recommended that thought is given to how the second buyer can control the relationship between the primary buyer and seller. The primary buyer will often not wish to give the second buyer the same obligations concerning the project as the primary buyer has received in its contract with the primary seller. The second buyer can thus be left with limited options for recourse if the primary seller defaults in managing the project. One solution is to include some form of ability for the second buyer to require the primary buyer to enforce against the primary seller in the situation where the primary seller's default has caused a failure in the production of ERUs, and to pass on a share of any damages to the second buyer. This may not be accepted by many primary buyers. A softer alternative is a mechanism for the second buyer to share in any damages that a primary buyer happens to receive from enforcing a primary ERPA.

6. How to Approach JI Project Due Diligence

The above analysis of the risks connected with a JI transaction emphasizes that any potential purchaser of ERUs should undertake thorough due diligence in respect of the JI regulation of the relevant Host Country. Insufficient due diligence may result in the failure to identify regulatory risks associated with a transaction and the consequences of such risks. On the one hand, this could result in parties proceeding with a transaction and later incurring time and money dealing with a risk which was not reflected in the transaction documents. On the other hand, a risk which was identified but not properly examined could dissuade the parties from proceeding with a transaction, even though the risk could possibly have been successfully dealt with or mitigated.

Local legal counsel, acting together with the purchaser's main legal counsel (if separate), should therefore be instructed to carry out due diligence on the following issues. Please note that this is in addition to any technical, financial and legal due diligence in respect of the JI project itself.

- The existing Host Country JI regulations should be examined in detail in order to identify the obligations of the parties and the potential risks.
- Any relevant areas which are not currently governed by JI procedures should also be identified. For example, various Host Countries have yet to introduce procedures in connection with the issue and transfer of ERUs.

- Any legislation and regulation which is still under consultation should be identified. Legislation under consultation is prone to change and therefore, even where due diligence has previously been carried out, potential purchasers should update their due diligence. For example, there are a number of draft laws in the Ukraine which have been subject to material amendments during consultation and, once passed, will significantly alter the current JI structure.
- To the extent that JI procedures are in place, an examination should be undertaken as to how such procedures are being applied in practice. For example, whether the procedures are being applied consistently by the Host Country authorities and whether there are any delays in the intended timetables.
- The legal status of ERUs under Host Country law should be confirmed. It will affect whether ERUs are an asset with ownership rights which are capable of being transferred. The legal status of ERUs will also affect the tax and accounting treatment of such ERUs.

Risks identified during the course of due diligence will then need to be dealt with. Appropriate provisions should be included in the transaction documents in order to remove, mitigate or allocate liability for such risks. The risks may also be used in as a negotiation tool, for example to reduce the price or obtain more favourable terms for the purchaser.

7. Conclusion

As has been explained in both this chapter and chapter 8, JI has had a much slower start than its elder sibling CDM. In this chapter we have tried to examine some of the reasons for this, particularly from the practitioner's perspective. It is clear there are some particular aspects of the JI mechanism's regulatory design which, relative to the regulatory design of the CDM mechanism, create higher risks for investors and buyers. However, in our view JI is a valuable tool and a 'learning by doing' exercise for similar or new mechanisms in the post-2012 world. As we move towards COP 15 in Copenhagen there is increasingly talk of making CDM more 'JI like' in the context of existing countries who may take on caps or targets, whether sectoral or of other types. In this context it is important to examine and understand why JI has been slower to develop than CDM and apply these lessons to improve and enhance it for the post-2012 regime. Because the success of such an enhanced JI mechanism will depend upon investors and buyers it is critical to examine why these have not flocked to the JI mechanism in the same numbers and degree as they have to the CDM mechanism.

The regulation of JI and consequently the risks involved differ between Host Countries thus any such analysis needs to consider a number of such countries.

However, to date there are still only relatively small numbers of 'determined' JI projects and so such lessons need to be considered in a statistically limited context.

This also means that any buyer who wishes to purchase ERUs from more than one Host Country will need to examine the risks connected with each Host Country separately—the lessons learnt in respect of one jurisdiction will not necessarily be relevant in another jurisdiction. By comparison with CDM where the majority of the rules are the same for a CDM project from Chile to China, this seems an obvious but key point to take away.

Even a party who intends to purchase ERUs from a single Host Country should be aware that in most Host Countries, the domestic legislation and regulations governing JI projects are not yet complete and are subject to addition and amendment. Until such time as the relevant Host Country JI rules are completed, a potential purchaser will need to re-examine the potential risks in respect of each separate transaction and would also be advised to keep track of any developments during the course of a transaction which could affect the risk profile of such transaction.

Investors and buyers in JI projects are clearly more exposed to Host Country regulatory and political risk than they are with CDM, especially in relation to the issuance of ERUs. As noted, there is considerably less harmonization in the rules for JI than for CDM which means that investors are faced with spending more time and resources understanding the regulatory framework from country to country. All of this alone probably would not have delayed JI as much as has been found in practice. However, these issues are coupled with potential Host Countries themselves being slow to embrace and support the mechanism domestically, resulting in significant delays in the development of national rules and regulations. Even at the time of writing the Russian Government has still not finalized the rules for the issuance and delivery of ERUs.

Against the backdrop of these issues, in the last 12 months significant progress has still been made. Since the first JI project was sent for public comments on 27 October 2006 there are now 166 Track 2 and 26 Track 1 JI projects in the UNEP Pipeline with one project rejected. Of these, 159 are now at the determination stage and 31 are listed on the UNEP Risoe pipeline as 'registered'.[8] In addition, in recent years more established jurisdictions such as France, Germany, and New Zealand have begun to embrace the JI mechanism. In such countries where investors have greater confidence in the investment environment these issues may prove to be much less of a deterrent.

It is clear that from a practical perspective buyers and investors in JI projects not only need to spend more time understanding the domestic JI regulatory regime but need to tailor their JI ERPAs for the rules in each country to a much

[8] Figures from UNEP: <http://cdmpipeline.org>, accessed 10 April 2009.

higher degree than is the case with a typical CDM ERPA. JI does provide an excellent case study for an international offset mechanism within the context of caps or targets—whether sectoral or economy wide. However, when one talks of making CDM more 'JI-like' it is important that the regulatory design aspects of JI which have deterred investors and buyers are improved in the reform of the CDM.

10

The CDM Project Cycle and the Role of the UNFCCC Secretariat

*Maria Netto and Kai-Uwe Barani Schmidt**

1. Introduction

The Kyoto Protocol to the United Nations Framework Convention on Climate Change (UNFCCC) provides for three 'market-oriented mechanisms'—joint implementation (JI); emissions trading (ET); and the clean development mechanism (CDM). The main purpose of the three mechanisms is to maximize the cost-effectiveness of climate change mitigation by providing an opportunity to reduce greenhouse gas (GHG) emissions inexpensively abroad, rather than at home.

In particular, regarding the CDM, Parties to the Convention aimed at breaking new ground in promoting international investment in climate friendly technologies for the benefit of developing countries. The CDM aspires to help developing countries (non-Annex I Parties) achieve sustainable development and the ultimate objective of the Convention, while also aiding industrialized countries (so-called Annex I Parties) to meet their emissions targets under the Kyoto Protocol. Under the CDM, industrialized countries may implement projects in developing countries that will reduce GHG emissions such as carbon dioxide and methane.

The incentive to invest in CDM project activities comes from the opportunity of crediting emission reductions from projects in developing countries. This is carried out in the form of certified emission reductions (CERs) which may be used by industrialized countries towards meeting their emissions targets under the Kyoto Protocol. Certified Emission Reductions can also be traded on carbon markets. In other words, this means that part of an investment for an energy project in a developing country, which reduces emissions beyond what would have been the case without the CDM, can be generated by the sale of CERs. As governments establish similar emissions targets for their private sectors, companies will have an incentive to be engaged in the CDM.

* The views expressed herein are those of the authors and do not necessarily reflect views of the United Nations and the UNFCCC Secretariat.

The Kyoto Protocol entered into force in February 2005. This occurred when the requirement that 55 Parties to the Convention ratify (or approve, accept, or accede to) the Protocol, including Annex I Parties accounting for 55% of that group's carbon dioxide emissions in 1990,[1] was satisfied.

Prior to the Kyoto Protocol entering into force, Parties to the Convention, through the 'Marrakech Accords,'[2] agreed to facilitate *the prompt start of the CDM* by adopting its 'Modalities and Procedures'[3] and by establishing the CDM Executive Board. The CDM has therefore been operational since November 2001.

For the UNFCCC Secretariat, the coming into operation of the CDM meant a significant change from its role of supporting Parties' negotiations regarding the legal text for modalities to a more interactive role in supporting the implementation of CDM. The challenge also presented a need to work alongside and aid different actors, from the private and public sectors, all interacting under the CDM.

This new role implied, inter alia, the need to:

- maintain a neutral position while supporting different interactions amongst diverse actors;
- implement an electronic workflow management system involving actors scattered/spread around the world;
- maintain a reliable and accurate information system;
- respond to outside demands in an even more expedient manner;
- implement procedural steps with a short turnover time;
- adapt administrative procedures to a financial relationship with the private sector; and
- create a number of operational procedures.

This article will provide an overview of the CDM actors and the steps in the CDM project cycle with an analysis of the role of the Secretariat in supporting these stages.

2. CDM Actors

The CDM allows for a unique partnership where both intergovernmental processes and the private sector play an important role. While the intergovernmental process and the Executive Board of the CDM are given a regulatory role, the

[1] Article 25 of the Kyoto Protocol.

[2] See Report of the Conference of the Parties Meeting as the Parties meeting in its first session, Decision 3/COP/MOP.1, contained in document FCCC/KP/CMP/2005/8/Add.1.

[3] See Decision 3/CMP.1 'Modalities and procedures for a clean development mechanism, as defined in Article 12 of the Kyoto Protocol', document FCCC/KP/CMP/2005/8/Add.1.

operational tasks are the responsibility of project participants and designated operational entities (DOEs)—mainly from the private sector. As the following discussion will demonstrate, the structure of the CDM is decentralized and private sector oriented.

2.1 COP/MOP

The Kyoto Protocol stipulates that the CDM shall be subject to the authority and guidance of the Conference of the Parties serving as the meeting of the Parties to the Protocol (COP/MOP) and be supervised by an Executive Board of the CDM.[4] The COP/MOP is thus the overall governing body of the CDM. Under the prompt start phase of the CDM (before the Kyoto Protocol entered into force in 2005) the Conference of the Parties of the Climate Change Convention (COP) assumed the responsibilities of the COP/MOP.[5] However, when the Protocol entered into force the COP/MOP took over authority and responsibility.[6]

The COP/MOP reviews the annual reports of the Executive Board and provides guidance to the Board on its operations.[7] In addition, the COP/MOP assists in the funding arrangements of CDM project activities, as necessary, and reviews the regional and sub-regional distribution of DOEs and distribution of CDM project activities.

2.2 Executive Board

The Executive Board is comprised of ten members and ten alternate members. This amounts to one member and one alternate member from each of the regional groups of the United Nations, alongside two members and two alternate members from Annex I Parties,[8] two members and two alternate members from non-Annex I Parties, and also one member and one alternate member from the small island developing States.

Although the members are nominated by constituencies and elected by the COP/MOP, they must act in their personal capacity.[9] This is one of the innovations of the CDM; the members of the intergovernmental body (the Board)

[4] See Art 12(4) of the Kyoto Protocol.

[5] Paragraph 2 of decision 17/CP.7, document FCCC/CP/2001/13/Add.2.

[6] 3/CMP.1 'Modalities and procedures for a clean development mechanism, as defined in Article 12 of the Kyoto Protocol', document FCCC/KP/CMP/2005/8/Add.1.

[7] See the Reference/Documentation section in the UNFCCC CDM website for the list of decisions on guidance by the CMP to the Executive Board (<http://cdm.unfccc.int>, accessed 6 April 2009).

[8] The annex to decision 3 CMP.1 on modalities and procedures of a CDM stipulates that when referring to 'Parties' it refers to Parties that have ratified the Kyoto Protocol.

[9] Paragraph 8(c) Decision 3/CMP.1 Annex—'Modalities and procedures for a clean development mechanism, as defined in Article 12 of the Kyoto Protocol' FCCC/KP/CMP/2005/8/Add.1.

are not to represent individual interests of governments, nor have any pecuniary or financial interests on projects or companies involved in the implementation of the CDM.

The Board operates by the rules of procedure adopted by the COP/MOP.[10] These rules cover a number of issues including the Board's voting rules and quorum, terms of members and alternates and their re-election, cause for termination of membership, chairmanship, avoidance of conflict of interests by Executive Board members, number and form of Board meetings, confidentiality provisions, attendance by observers at Board meetings, transparency provisions, public availability of information, and use of outside expertise.

As mentioned above, the Executive Board was established to supervise the CDM and undertakes a mainly 'regulatory' function under the CDM. As will be examined in the description of the project cycle below, the Board is responsible for:

- making recommendations to the COP/MOP with regards to modalities and procedures for a CDM;
- developing guidance to project participants and DOEs on the implementation of the CDM;
- implementing the accreditation process of DOEs, including overseeing and developing procedures for accreditation, reviewing accreditation standards and making recommendations to the COP/MOP, accrediting companies as DOEs and making recommendation on the suspension/withdrawal of DOEs;
- approving baseline and monitoring methodologies necessary for assessing and quantifying emission reductions by proposed project activities;
- registering CDM project activities;
- issuing CERs and maintaining a CDM registry for tracking CERs;
- maintaining and making publicly available a repository of approved rules, procedures, methodologies and standards relating to the CDM.

2.3 Panels

The modalities and procedures for a CDM stipulate that the Board may 'establish committees, panels or working groups to assist it in the performance of its functions. The Executive Board shall draw on the expertise necessary to perform its functions, including from the UNFCCC roster of experts. In this context, it shall take fully into account the consideration of regional balance'.[11]

[10] Annex I to Decision 4/CMP.1, FCCC/KP/CMP/2005/8/Add.1.

[11] Paragraph 18 of Decision 3/CMP.1 Annex—'Modalities and procedures for a clean development mechanism, as defined in Article 12 of the Kyoto Protocol' FCCC/KP/CMP/2005/8/Add.1.

The panels operate following general guidelines and specific terms of reference and guidance developed by the Board.[12] They may be created for temporary and specific tasks, this was the case when the panel developed recommendations on simplified modalities and procedures for small-scale CDM project activities. Alternatively, they may fill a standing function, such as the panels on accreditation and methodologies.

Similar to the rules for Board members and alternates, panel members must act in their individual capacity. The panels' recommendations to the Board are publicly available, unless they address confidential or proprietary information. The terms of reference of the panels specify their size, terms, mandates, competence criteria for their experts and whether experts are entitled to receive fees for specific work. Panel experts are selected from public calls for candidates in accordance with the competence criteria set out in the terms of reference. Panels may, in addition, work with the assistance of outside experts that undertake specific tasks such as desk reviews of proposed methodologies or assessments of companies applying for accreditation.

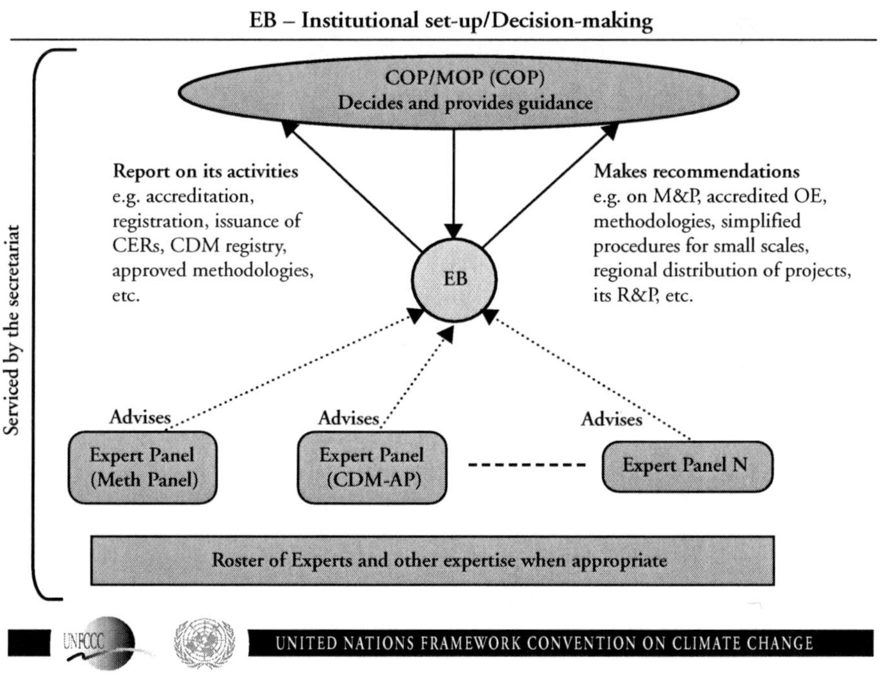

Figure 10.1. Institutional set-up and decision making of Executive Board
Source: UNFCCC Secretariat

[12] General guidelines for panels as well as the specific terms of references of different panels and guidance by the Board are available under the section on Panels and Reference on the UNFCCC CDM website: <http://cdm.unfccc.int>.

2.4 Designated Operational Entities

A Designated Operational Entity (DOE) under the CDM is either a domestic legal entity or an international organization, which is accredited and designated (on a provisional basis until confirmed by the COP/MOP) by the Executive Board to perform three key functions:[13]

- check for completeness and forward to the Board proposed new methodologies;
- validate and subsequently request registration of a proposed CDM project activity using an approved methodology;
- verify emission reductions of a registered CDM project activity, certify as appropriate and request the Board to issue CERs accordingly.

The role of the DOEs is another innovative tool of the CDM. The companies accredited as DOEs are expected to verify that the adequacy and adherence of project activities adhere to the CDM modalities. A DOE request to register a project activity or to issue CERs will not be implemented unless a request for review is triggered within a set time frame. These companies are part of the institutional infrastructure of the CDM and play a key role in the credibility of system.

The delegation of responsibility by the COP/MOP to DOEs implies that the accreditation process implemented by the Executive Board needs to be a thorough and detailed one, including the possibility of spot checking. It also implies that DOEs should be neutral and not participate in the design of project activities and methodologies.

2.5 Project participants

The Kyoto Protocol stipulates that participation under the CDM may involve private and/or public entities, and is subject to whatever guidance may be provided by the Executive Board.[14]

The modalities and procedures for the CDM refer to the term project participants without defining it. Taking into consideration the roles of project participants described in the modalities and procedures, the Board has further clarified that a project participant is either a Party involved or a private and/or public entity authorized by a Party to participate, under the Party's responsibility, in CDM project activities. Project participants take decisions on the allocation of CERs from the CDM project activity under consideration.[15] They also propose and implement a given CDM project activity.

[13] Entities applying for accreditation (applicant entities) may carry out some activities as DOEs, including submitting proposed new methodologies, for the purposes of witnessing whether they are qualified to undertake functions of DOEs.

[14] See Art 12(9) of the Kyoto Protocol.

[15] See the CDM-PDD Glossary of terms available under the reference/documents section in the UNFCCC CDM website: <http://cdm.unfccc.int>.

Project participants may also propose to the Board new baseline and monitoring methodologies to account for emission reductions. This is one of the important innovations of the CDM—methodologies for accounting for reductions are developed from a bottom-up approach.

2.6 Designated National Authorities

The CDM modalities and procedures state the need for a country to establish a Designated National Authority (DNA) in order to participate in CDM project activities.[16] The main tasks to be carried out by the DNA are defined in the CDM modalities and procedures. The DNA should,[17] among other tasks:[18]

- provide written approval to the project participants of voluntary participation of the Party to the CDM project activity;
- in the case of a host Party, include in the written approval a confirmation that the CDM project activity assists it in achieving sustainable development.

2.7 The public

Public transparency and scrutiny play an important role in ensuring the credibility of a CDM. All the information, which is the basis for decision-making, is made publicly available. In addition, during two stages of the CDM project cycle, the public are invited to give comments:

- when designing a project activity, project participants should call for comments by local stakeholders;[19] and
- before validating a project activity, a DOE must make the project design document of a proposed project activity publicly available and invite public comments on validation requirements from Parties, stakeholders, and UNFCCC accredited non-governmental organizations.[20]

[16] See paras 29 and 37(a) of Decision 3/CMP.1 Annex—'Modalities and procedures for a clean development mechanism, as defined in Article 12 of the Kyoto Protocol' FCCC/KP/CMP/2005/8/Add.1.

[17] See para 40(a) of Decision 3/CMP.1 Annex—'Modalities and procedures for a clean development mechanism, as defined in Article 12 of the Kyoto Protocol' FCCC/KP/CMP/2005/8/Add.1.

[18] Updated information on role/functions of DNAs is made available in the DNA section of the UNFCCC CDM website: <http://cdm.unfccc.int>.

[19] In this regard, the Executive Board has further clarified that 'an invitation for comments by local stakeholders shall be made in an open and transparent manner, in a way that facilities comments to be received from local stakeholders and allows for a reasonable time for comments to be submitted. In this regard, project participants shall describe a project activity in a manner which allows the local stakeholders to understand the project activity, taking into account confidentiality provisions of the CDM modalities and procedures.' See the section on reference/documentation of the UNFCCC CDM website: <http://cdm.unfccc.int>.

[20] Detailed 'procedures on public availability of the CDM project design document and for receiving comments as referred to in paragraphs 40(b) and (c) of the CDM modalities and procedures' have been developed by the Executive Board. In accordance with these procedures DOEs shall make PDDs available through a link to the UNFCCC CDM website. See the section on reference/documentation in the UNFCCC CDM website: <http://cdm.unfccc.int>.

In addition to these opportunities, the public, including Parties, NGOs, stakeholders, and institutions are able to make comments on the documents on which the Executive Board agreed to seek feedback from the public. Current calls for input and comment can be found through the main page of the UNFCCC CDM web site. At the moment, public input is sought in different areas of work of the Board such as:

- work plans for specific tasks and/or development specific technical recommendations/guidance;
- proposed new methodologies;
- applicant entities and accreditation procedures.

2.8 Secretariat

The UNFCCC Secretariat aids the COP/MOP and the Executive Board, performing functions assigned to it accordingly. In this role, the Secretariat should directly or indirectly support and link all actors described above (Figure 10.2 below). This is achieved by providing means for actors to communicate and interact in a cost effective manner, ensuring the rules adopted by the COP/MOP are known and that the Executive Board, panels, experts and entities are able to work professionally.

The Secretariat maintains a web-based information system providing up-to-date information on the implementation/operation of the CDM. This

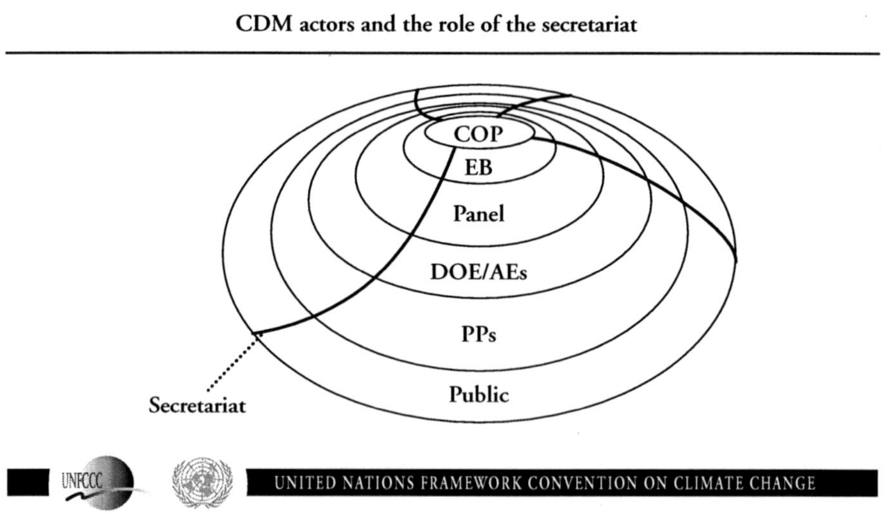

Figure 10.2. CDM actors and the role of the Secretariat
Source: UNFCCC Secretariat

includes Board/panel meeting reports, relevant decision/guidance of COP/MOP, guidance of the Board, procedures, calls for public input, repository of approved methodologies, modalities for different types of project activities (eg small scale, afforestation and reforestation), and registered project activities.

The Secretariat is expected to retain a neutral role in supporting the CDM actors and not make recommendations or judgements on specific CDM project activities.

3. CDM Project Cycle

3.1 Introduction to the cycle

The CDM project cycle is, in essence, very similar to the cycle of any other project. First, there is a design phase, followed by a review where the project design is confirmed to meet preset design features. Next, there is the implementation phase where the project is monitored, with a verification stating if performance/output of the project is certified to meet requirements and the final phase where a document is issued to confirm the existence of the output.

During the negotiations over the design of the CDM, the initial positions ranged from the view that most of the decisions should be taken by the supreme body of the intergovernmental process (COP/MOP) to a completely private sector based system. This led to the novel design of an intergovernmental mechanism, where the infrastructure includes private/public sector companies (designated operational entities (DOEs)) which take decisions unless objected to, within a short period of time, by a 10-member Board. Figure 10.3 shows the steps for the actors of the project cycle. It illustrates that the majority of steps are undertaken by the private sector and that areas where the Executive Board becomes involved are 'automatic' unless a review is triggered by three of the 10 Board members or by the governments which are project participants.

The project cycle is, in its major steps, the same for all types of CDM project activities. Differences exist in modalities for small-scale and afforestation and reforestation CDM project activities. These include different requirements for validation and monitoring plan and deadlines. This chapter describes the project cycle as described in the CDM modalities and procedures without highlighting the differences for small scale and afforestation and reforestation CDM project activities. The differences are described in the modalities of CDM project activities.[21]

[21] The 'Simplified modalities and procedures for small-scale CDM project activities' and 'Modalities and procedures afforestation and reforestation project activities under the CDM' are available in the reference/documents section of the UNFCCC CDM website: <http://cdm.unfccc.int>.

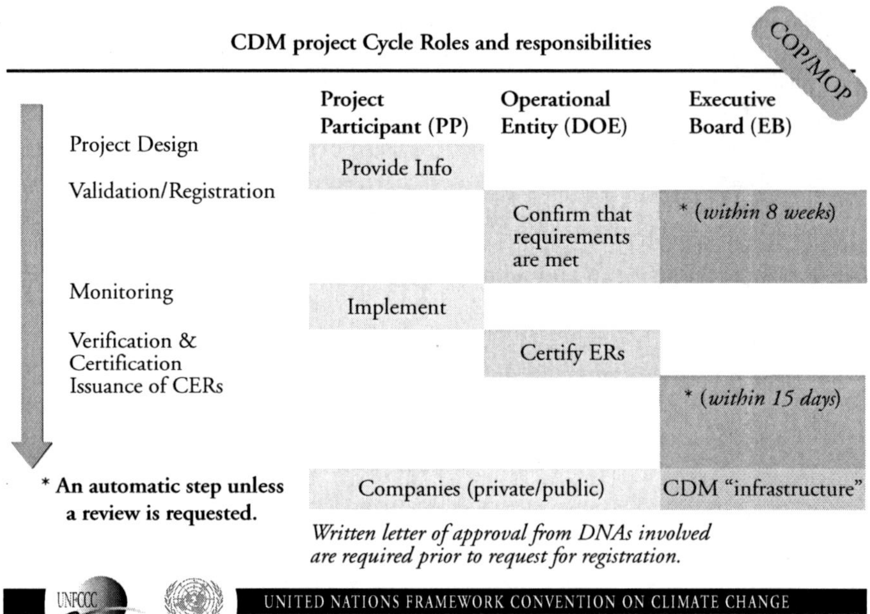

Figure 10.3. CDM project cycle roles and responsibilities
Source: UNFCCC Secretariat

There are two inputs to the project cycle which are key to its operation:

- the DOEs are required to be accredited by the Executive Board and designated by the COP/MOP;
- methodologies need to be approved by the Executive Board for determining the reductions in emissions that are additional to any that would have occurred in the absence of the certified project activity.

3.2 Accreditation of Operational Entities

The Executive Board has the mandate to accredit entities and recommend to the COP/MOP their designation as operational entities. The Executive Board has established an accreditation procedure[22] which describes the process of accreditation covering the application, assessment of whether applicant entities meet the accreditation standards and requirements, and recommendation of a decision to the Executive Board by a panel of accreditation experts.

The accreditation is valid for three years. The DOE can seek re-accreditation, during or after the expiration of the three year term. The DOE may be assessed

[22] See the section on procedures in the UNFCCC CDM website: <http://cdm.unfccc.int>.

on the basis of 'spot checks' which can be triggered in accordance with the accreditation procedure. The accreditation, provided by the Executive Board, is sector specific[23] without any geographical limitation. In accordance with a phased accreditation approach, a DOE may, for a period of time, be limited to perform validation/registration functions before it will be also accredited to perform the verification function. As this is a critical function, the Executive Board keeps this process under close review and adopts measures to ensure the high quality of service of the DOEs. [24]

The role of the Secretariat in this process is to provide logistical, managerial and substantive support to the work of the Board, the panel, assessment teams and applicant entities, as applicable. For example the Secretariat:

- maintains a web-based database and communication system permitting the Board, applicant entities, panel members and assessment teams, distributed globally, to communicate and manage their work flows;
- maintains a roster of accreditation assessment team experts;
- prepares background documentation to assist decision making by the Board and the panel related to the accreditation process;
- recommends options for and implements procedures related to the accreditation process;
- organizes physical meetings of the panel;
- coordinates and supports the work of assessment teams on-site;
- participates at assessments; and
- administers contracts and fee payments.

3.3 Approval of new methodologies

As referred above, one innovative feature of the CDM is that the modalities provide for a bottom-up approach. This allows for utilization of the large potential of private sector knowledge and ideas with regard to the development of methodologies for determining the reductions in emissions that are additional to any that would have occurred in the absence of the certified project activity (baseline methodologies) and monitoring of such reductions (monitoring methodologies). The Executive Board approves such methodologies based on a recommendation by a panel of methodology experts (also known as the Meth Panel).[25]

[23] See list of sectoral scope(s) for accreditation in the UNFCCC CDM website: <http://cdm.unfccc.int/DOE/scopes.html>, accessed 06 April 2009.

[24] For the latest version of the accreditation, see the UNFCCC CDM website: <http://cdm.unfccc.int>.

[25] The Board may establish additional panels for treating methodological issues for specific project types such as small-scale or afforestation and reforestation.

Project participants may put forward methodologies by submitting to a DOE a proposal supported by information on the planned project which intends to make use of the new methodology. After an assessment of the proposal, involving two external experts hired by the Meth Panel, comments by the general public and optional feedback by project participants on a draft recommendation of the Meth Panel, the Panel makes a final recommendation to the Executive Board for its consideration. If the Board approves the methodology, it is included in the repository of approved methodologies against which DOEs can validate requests for registration by project participants of projects as CDM project activities.[26]

In addition to this bottom-up approach, the Board may undertake efforts to provide consolidations of existing approved methodologies to streamline and ease their use or to develop methodologies and to provide guidance to methodology development in areas where methodological proposals are not successful or forthcoming.

The role of the Secretariat in this process is to provide logistical, managerial and substantive support to the work of the Board and the Meth Panel. For example the Secretariat:

- maintains a web-based database and communication system permitting the Board; DOEs, Panel members and experts distributed globally to communicate and manage their work flows;
- maintains a roster of experts to undertake desk reviews;
- undertakes quality checks and ensures consistency among methodologies;
- prepares background documentation to assist decision making by the Board and the Panel;
- recommends options for, and implements procedures related to, the process of approving new methodologies;
- organizes physical meetings of the Panel; and
- administers contracts and fee payments.

4. CDM Project Cycle Steps

This section describes the five steps of the CDM project cycle, and the role of the UNFCCC Secretariat in each.[27]

[26] See the section on procedures in the UNFCCC CDM website (<http://cdm.unfccc.int>) for detail on 'Procedures for submission and consideration of a proposed new methodology'.

[27] A guide to how to do a CDM project, including specification for afforestation and reforestation as well as small-scale project activities is available in the section 'Project Activities' of the UNFCCC CDM website: <http://cdm.unfccc.int>.

4.1 Design of a CDM project activity

The actors in the project design phase are the project participants. Based on the modalities and procedures for the CDM and the guidance by the Executive Board, the project participants have to design the project activity to ensure that it will be validated and registered as a CDM project activity. One of the most difficult design tasks for a project participant is to secure that the proposed project activity uses approved methodologies for baselines and monitoring, and applies it correctly to the proposed activity. Project participants must submit information on their proposed CDM project activity using the CDM-PDD. Project participants are guided by the modalities and procedures for a CDM[28] as well as by the Executive Board.[29]

If during the design phase project participants decide that a new methodology is required they may propose it, as described above.

The role of the UNFCCC Secretariat is to ensure that information on approved methodologies and guidance for design of project activities is easily available in a cost effective manner. Although the Secretariat is to provide information on the generic level of design it will not provide input on particular design or eligibility questions.

4.2 Validation/registration

The DOE is the key actor in this phase. Validation is the process of independent evaluation of a project activity by a DOE against the requirements of the CDM on the basis of a CDM-Project Design document (CDM-PDD).[30] The sections 'Validation and Registration' as well as 'Verification' of the modalities and procedures for a CDM describe the key functions of DOEs.

After being contracted by project participants, a DOE is to check that validation requirements[31] have been met after reviewing the CDM-PDD and any supporting documentation prepared and submitted by the project participants (including ensuring that the baseline and monitoring methodologies used are approved by the Board).

[28] In particular paras 43 to 52 (regarding baselines), 53 to 60 (regarding monitoring) and Appendix B of Decision 3/CMP.1 Annex—'Modalities and procedures for a clean development mechanism, as defined in Article 12 of the Kyoto Protocol' FCCC/KP/CMP/2005/8/Add.1

[29] For more detail see the section on reference/documentation in the UNFCCC CDM website: <http://cdm.unfccc.int>.

[30] See the latest version of the CDM-PDD in the reference/documents section of the UNFCCC CDM website: <http://cdm.unfccc.int>.

[31] Validation requirements are stipulated in para 37 of the annex to Decision 3/CMP.1 Annex—'Modalities and procedures for a clean development mechanism, as defined in Article 12 of the Kyoto Protocol' FCCC/KP/CMP/2005/8/Add.1. See also subsequent guidance provided by the EB reference/document as well as Designated Operational Entities' section of the UNFCCC CDM website: <http://cdm.unfccc.int>.

The DOE must make the CDM-PDD publicly available.[32] Subsequently, it should receive public comments by Parties, stakeholders and UNFCCC accredited non-governmental organizations and make these publicly available also.

After the deadline for receipt of comments, the DOE determines whether the project activity should be validated. The DOE informs project participants of its determination on the validation of the project activity. If the DOE determines the proposed project activity to be valid, it is submitted to the Board a request for registration[33] in the form of a validation report. This shall include the CDM-PDD, the written approval of the host Party, and an explanation of how it has taken due account of comments received. Before submitting the request for registration, the DOE has to have received the written approval(s) from DNAs of Parties involved.[34] The validation report is made publicly available.

Registration Table 10.1. is the formal acceptance by the Executive Board of a validated project as a CDM project activity. 'The registration by the Executive Board shall be deemed final eight weeks after the date of receipt by the Executive Board of the request for registration, unless a Party involved in the project activity or at least three members of the Executive Board request a review of the proposed CDM project activity.' [35] The request for review 'shall be related to issues associated with the validation requirements.' In response to requests by project participants and the COP/MOP, the Board has established a roster of experts to assist it in this task (Registration and Issuance Team, RIT) and to enhance quality and consistency by one expert reviewing each request submitted. The Secretariat prepares a summary note for each request to facilitate the decision making of members and the Board. Registration as a CDM project activity is the prerequisite for the verification, certification, and issuance of CERs related to that project activity.

The role of the UNFCCC Secretariat is to ensure that information on, *inter alia*, public availability of CDM-PDDs, validation reports, requests for review and registrations is provided through the UNFCCC CDM website in an accurate and cost effective manner. The Secretariat serves as a communication

[32] The Board has elaborated 'Procedures on public availability of the CDM-PDD and for receiving comments as referred to in paras 40(b) and (c) of the CDM modalities and procedures (version 04)'. The current version is available in the reference/documentation section of the UNFCCC CDM website: <http://cdm.unfccc.int>.

[33] Detailed 'procedures for registration of a proposed CDM project activity' and level of fees for project registration have been elaborated by the Executive Board. See 'How to do a project activity' and 'CDM registry' in the UNFCCC CDM website: <http://cdm.unfccc.int>.

[34] The modalities and procedures for a CDM does not specify whether the written approval by DNAs shall be obtained before or after making a PDD publicly available (please see paras 40(a) and (f) 37(a) and 28 to 30 of Decision 3/CMP.1 Annex—'Modalities and procedures for a clean development mechanism, as defined in Article 12 of the Kyoto Protocol' FCCC/KP/CMP/2005/8/Add.1.

[35] Detailed 'Procedures for review as referred in paragraph 41 of the CDM modalities and procedures' have been elaborated by the Executive Board. See the section on reference/documentation in the UNFCCC CDM website: <http://cdm.unfccc.int>.

channel between DOEs and the Executive Board if guidance is required by DOEs in order to perform their functions. For example the Secretariat:

- maintains a web-based database and communication system permitting the Board, DNAs, DOEs and project participants, distributed globally, to communicate and manage their work flows;
- prepares background documentation to assist decision making by the Board;
- recommends options for and implements procedures related to the process of validation and registration;
- administers registration fee payment.

4.3 Monitoring

Monitoring is implemented by the project participants in accordance with the monitoring plan contained in the registered CDM-PDD. The modalities and procedures for a CDM provide guidance for the establishment of a monitoring plan which is based on an approved methodology.[36] The preparation by project participants of a monitoring report in accordance with the monitoring plan (contained in the CDM-PDD) is a pre-condition for verification and certification by a DOE and the issuance of CERs by the Executive Board.

The UNFCCC Secretariat has no specific role in this step.

4.4 Verification/certification of the CDM project activity

Verification is the periodic independent review and ex-post determination by the DOE of the monitored reductions in anthropogenic emissions by sources of greenhouse gases that have occurred as a result of a registered CDM project activity during the verification period. Certification is the written assurance by the designated operational entity that, during a specified time period, a project activity achieved the reductions in anthropogenic emissions by sources of GHGs as verified.

Similar to the validation/registration phase, the DOE is the key actor in this phase. As a first step, the DOE makes the monitoring report received from the project participants publicly available and proceeds with the verification steps. The DOEs have to establish a verification report which is made publicly available[37] and based on this report, the DOE must 'certify in writing that, during the specified time period, the project activity achieved the verified amount of reductions in anthropogenic emissions by sources of greenhouse gases that

[36] See paras 53 and 54 of Decision 3/CMP.1 Annex—'Modalities and procedures for a clean development mechanism, as defined in Article 12 of the Kyoto Protocol' FCCC/KP/CMP/2005/8/Add.1.

[37] See para 62 of Decision 3/CMP.1 Annex—'Modalities and procedures for a clean development mechanism, as defined in Article 12 of the Kyoto Protocol' FCCC/KP/CMP/2005/8/Add.1.

would not have occurred in the absence of the CDM project activity. It [the DOE] shall inform the project participants, Parties involved and the Executive Board of its certification decision in writing immediately upon completion of the certification process and make the certification report publicly available'. [38] In response to requests by project particpants and COP/MOP, the Board has established a roster of experts that assist it in this task (Registration and Issuance team) to enhance quality and consistency by one expert reviewing each request submitted. The Secretariat prepares a summary note for each request to facilitate the decision making of members and the Board.

The role of the UNFCCC Secretariat is to ensure that information on, *inter alia*, public availability of monitoring plans, validation reports and certification reports is provided through the UNFCCC CDM website in an accurate and cost effective manner. The Secretariat serves as a communication channel between DOEs and the Executive Board if guidance is required by DOEs in order to perform functions. For example, the Secretariat:

- maintains a web-based database and communication system permitting the DOEs and project participants, distributed globally, to communicate and manage their work flows;
- prepares background documentation to assist decision making by the Board;
- recommends options for and implements procedures related to the process of verification and certification.

4.5 Issuance

The Certification report submitted to the Board by a DOE constitutes the request for issuance of CERs. It is considered final after 15 days, unless a Party involved in the project activity or at least three members of the Board request a review of the proposed issuance of CERs. Similar to the 'automatic' step at registration, the scope of such a request has been limited. The review is limited to issues of fraud, malfeasance or incompetence of the DOE.

Once the issuance is considered final and upon instruction by the Executive Board, the CDM registry administrator issues the CERs. This occurs within the CDM registry which is maintained to ensure the accurate accounting of the issuance, holding, transfer and acquisition of CERs by non-Annex I Parties. At this stage a share of the CERs issued is withheld to cover administrative costs and to assist developing country Parties that are particularly vulnerable to the adverse effects of climate change to meet the costs of adaptation.[39]

[38] See para 63 of Decision 3/CMP.1 Annex—'Modalities and procedures for a clean development mechanism, as defined in Article 12 of the Kyoto Protocol' FCCC/KP/CMP/2005/8/Add.1.

[39] For more detail on share of proceeds and registration fee, see 'How to do a project activity' under the Project Activities Section in the UNFCCC CDM website: <http://cdm.unfccc.int>.

The role of the UNFCCC Secretariat is to ensure that information on, *inter alia*, requests for issuance, issuance and related project information is provided through the UNFCCC CDM web-based information system in an accurate and cost effective manner. The Secretariat provides operational support with regard to the implementation in a timely manner of the review procedure. The Secretariat maintains and manages the CDM registry by executing the request for issuance by the Board, the establishment and services related to accounts of project participants and linking to the international transaction log.

5. Conclusions

In summary, the CDM project activity cycle provides a number of innovations for a multilateral mechanism. It is implemented to assist countries to meet legally binding commitments by linking a mutual interest to help countries to achieve sustainable development.

In designing the mechanism, important issues for private sector involvement are addressed, such as:

- the 'bottom up' approach in defining criteria and rules (project participants can make proposals for methodologies);
- the involvement of the private sector (project participants, designated operational entities);
- decentralized decision making (eg unless a request for review is triggered, the decision to request registration of a project activities by one of the many DOEs is implemented by the Secretariat);
- basing decision-making on expert panels established through a public call for experts and bearing in mind the need for incorporating knowledge from all regions of the world.

As the Secretariat to the CDM Executive Board, the UNFCCC Secretariat is requested to support a day-to-day process of implementation and decision making in addition to its role of helping with the negotiation processes. The Secretariat has also assumed a new role for the UN— in supporting a globally operating accreditation body. This new role is providing additional challenges with regard to the:

- treatment of public as well as confidential information of a competitive nature (eg the accreditation process);
- need to avoid conflict of interest situations;
- need to ensure fair and equitable treatment of all actors and not to create competitive advantages by, for example, providing information/advice to one private sector actor and not to another;
- need to respond to shorter deadlines/turn over times.

In addition, operational experience shows that the Secretariat needs to assume more and more substantive expert support functions for the decision making in all three functions, bringing a new role to the Secretariat in this process. The CDM is a very dynamic mechanism and the role of the Secretariat is to allow it to optimize its processes. The UNFCCC CDM website provides an accurate and authoritative source of the latest state improvements to processes and roles. All in all, the CDM is a challenging experience for all actors involved and may also provide a useful experience for other multilateral agreements.

11

Trying to Catch up with the Executive Board: Regulatory Decision-Making and its Impact on CDM Performance

Matthias Krey and Heike Santen

1. 2008—Clean Development Mechanism in Crisis?

The Clean Development Mechanism (CDM) is a great success. Up to late December 2008, the CDM Executive Board (EB) had issued emissions credits for 233 million tonnes of CO_2 as Certified Emission Reductions (CERs), coming from 430 CDM projects.[1] Over 1250 registered CDM projects forecast generation of 1.4 billion CERs by end of 2012—an amount that the most ardent optimists would have regarded as a fairy tale dream in 2001 when the CDM was started.

At the same time, 2008 was undoubtedly the year when the CDM has faced the strongest criticism from stakeholders. 2008 started with stock exchange listed CDM project developers blaming the EB for an allegedly inefficient CDM registration and CER issuance process. This was due to a dramatic fall in the share prices of these companies because the volumes of CER delivered from their project portfolios were much lower than had been forecasted at listing. The most turbulent year ever in the CDM's short history ended with the suspension of the largest auditor for CDM projects by the EB. Det Norske Veritas (DNV) was found to be deficient in several aspects and thus lost its accreditation as a Designated Operational Entity (DOE). This means that a substantial number of projects were stuck until the ban on DNV was finally lifted. In between those events, project developers, DOEs, and regulatory decision-makers blamed each other for the uncertainty and unpredictability surrounding CDM project registration and CER delivery.

It is the aim of this chapter to highlight some of the main reasons for uncertainty regarding CDM project performance, to assess the complaints brought forward, and to summarize briefly the proposals for reform that have

[1] UNFCCC (<http://cdm.unfccc.int>), December 2008.

been made up till late 2008. We also discuss whether COP/MOP has lived up to the expectations of the private sector stakeholders in CDM projects and what still remains to be done. In the first section we describe the overarching principles of CDM rule-making and how it impacts on CDM project development. The following section focuses on the situation in which project developers as well as CER buyers find themselves when they need to predict if, when, and how many CERs a project will deliver in the future. We do so based on empirical and anecdotal evidence and the experiences of project developers, buyers, and CDM consultants. In the third section we describe typical reasons for the uncertainty that surrounds the CDM potential of a proposed greenhouse gas (GHG) mitigation project at an early stage of project development, eg at Project Identification Note Stage (PIN). The fourth section exemplifies the challenges of determining the CDM potential of a project that has been submitted for validation with a corresponding Project Design Document (PDD), and in section five we look at reasons for the uncertainty surrounding the issuance of CERs after there has been CDM registration of a project. Section six provides an overview on proposals for reform and the decisions made by the COP/MOP in December 2008 designed to improve CDM performance.

2. Understanding CDM Rule-making and its Implications

It is important to understand that the CDM is based on a complex and highly dynamic body of legal text. The main CDM principles were outlined in the 1997 Kyoto Protocol and further detailed in the 2001 Marrakech Accords.[2] The general design of the CDM as defined in both agreements, the general nature of CDM projects, the basic principles of emission reduction accounting, and the eligibility rules, can be seen as a pretty stable and fixed pillar of the CDM regulatory framework with a relatively long shelf-life.[3] However, these generic principles needed to be specified in more detail. This task was allotted to institutions at different levels within the UNFCCC system—the Conference of the Parties acting as the Meeting of the Parties to the Kyoto Protocol (COP/MOP), the EB, EB panels, and the UNFCCC Secretariat. The Marrakech Accords are decisions of the COP/MOP and have the same legal and regulatory standing as any COP/MOP decision before or since.

Among the institutions deciding on CDM regulatory issues, the COP/MOP represents the highest level. However, the COP/MOP only meets once a year and has a full agenda covering all aspects of the Kyoto Protocol implementation.

[2] UNFCCC (Kyoto Protocol To The United Nations Framework Convention On Climate Change, 1997); and UNFCCC (The Marrakech Accords and The Marrakech Declaration, 2001).

[3] Although changes, especially of the Marrakech Accords, are feasible in theory, substantial changes are not likely to happen. Nevertheless, some amendments of the Marrakech Accords have been made over time.

Therefore, it is not able to take many decisions on CDM matters and generally 'rubber-stamps' technical items submitted to it by the EB, such as the EB rules of procedure, small-scale project rules, and rules for review of registration and CER issuance.[4] Although the COP/MOP is formally responsible for decisions, there is a clear tendency that those decisions are *de facto* taken by the EB. Occasionally, COP/MOP decisions also take the form of lists of tasks for the EB to perform.

The EB takes discretionary decisions in several formats. The largest number of decisions is reflected in the EB meeting reports and their annexes. In addition to that, the EB occasionally publishes documents entitled 'guidance.' Sometimes controversial issues are referred to the COP/MOP by the EB. However, the bulk of technical work is realized by (currently six) panels and working groups established by the EB. These panels essentially fulfill the same functions for the EB that the EB fulfills for the COP/MOP.

Despite this sophisticated and hierarchical work-flow within the CDM institutions, it should be noted that the CDM itself is characterized by high levels of transparency and openness towards stakeholder inputs. In fact, the concrete rules for CDM projects are largely based on stakeholder inputs and consequently reflect the outcomes and standpoints of a broader technical and political discussion. Stakeholders are explicitly invited to contribute to the development of the CDM rules in several ways. This bottom-up process manifests itself, for example, in the fact that project developers need themselves to propose new baseline and monitoring methodologies for CDM projects[5] and also that the project proponents can themselves propose revisions of approved methodologies. In addition to that, all proposed new methodologies and CDM projects are published and may be commented on by stakeholders. There are a number of practical aspects of the CDM rules and issues (eg Programme of Activities, additionality testing, inclusion of carbon capture and storage in CDM, etc) that are the subject of controversial discussion and criticism by stakeholders; here the EB has specifically invited public inputs.

In practice, however, this bottom-up approach has considerable drawbacks. All contributions have to be channelled, processed, assessed, and converted into regulatory documents by the CDM institutions. The ever growing CDM project pipeline and the tremendous increase of requests for revisions of approved methodologies and proposals for new methodologies, means a higher workload for the actors within the institutional framework. Although the CDM EB has

[4] The corresponding proposals of the EB were confirmed at COP11, Decision 4/CMP.1. A unique exception is the definition of eligibility of land for afforestation and reforestation which has been decided at EB 22 (CDM Executive Board, 22. Meeting Report, Annex 16, 2005) and EB 26 (CDM Executive Board, 26. Meeting Report, Annex 18, 2006) and where COP/MOP 2 did not support these decisions (1/CMP.2, para 25).

[5] While this bottom-up concept originally only referred to large-scale methodologies, the initial top-down development of small-scale methodologies was also modified into a bottom-up process.

repeatedly increased the staff within the UNFCCC Secretariat that check consistency of rules and their application, nevertheless delays in processing submissions—be they registration requests, issuance requests, or revisions of monitoring plans—have been growing.

Besides the challenges of the workload generated, this stakeholder-based approach also frequently ends up in 'learning-by-doing'. Up until now, the EB has not had a clear doctrine of precedent and it is not necessarily bound by its previous decisions, as we will show in the following sections. On the one hand, this allows the EB to correct misguided developments and incorporate the latest experiences and scientific evidences into the rules. On the other hand, this inevitably leads to reduced predictability and certainty in its regulatory decision-making and is increasingly regarded as a major hurdle for efficient CDM operations.

In recent years the EB has been responsive to hurdles and drawbacks in the CDM regulatory framework and has repeatedly adjusted the structure and the terms of reference for implementing the CDM regulatory framework. For instance, the Registration and Issuance Team (RIT) was set up in early 2006 as a second layer of independent assessment of CDM projects after the first controversies about an overly lax registration policy had surfaced, together with criticism of the perceived lack of quality of the DOEs and, especially, of their validation reports.

The following three sections illustrate the very practical impacts of the above described regulatory decision-making process on today's project participants in terms of predicting, if, when and how many CERs a project will deliver in the future.[6]

3. Uncertainties at Early Stage

At a very early stage of project development, project participants usually demonstrate the CER generation potential they are willing to sell forward to the CER buyers by using a PIN (Project Idea Note) or on draft Project Design Document (PDD). PIN/PDD quality varies and good quality PINs/PDDs outline the design of the project and the applicable approved methodology in a sound manner. Even though there is no applicable and approved methodology available (such as for N_2O reduction from new capacity of nitric acid plants or CO_2 emission reductions from carbon capture and storage (CCS)), such projects

[6] In the following the term project participants is used for illustration of the impact of regulatory uncertainty. Besides this group there are many other actors in the CDM market that equally depend on regulatory certainty on the supply side of the CDM market (eg financial institutions, CDM project aggregators, brokers etc). In this respect project participant should be read as a synonym for any entity in the CDM market that depends on predictable and reliable regulatory decision-making.

are nevertheless still being planned because of the huge emission reduction potential from such activities.

The early movers in the market that initiated the first projects had to face a situation with only a handful of methodologies approved and a limited understanding on how the regulator would think and act. Has this situation changed to any great extent? What level of guidance and certainty does the EB provide to CER buyers who want to assess the long-term CER generation potential of projects that are at a planning/feasibility stage and that may be submitted for registration within the next one to two years?

An initial CER volume estimate requires an understanding of the framework for GHG accounting. This framework is specified by the 81 approved large-scale and 42 small-scale baseline and monitoring methodologies.[7] As we show in the following discussion, project participants are well advised to consider those methodologies as dynamic. We have assessed the duration of validity of the different versions of a set of 10 prominent approved methodologies. The selection of methodologies followed a cross-sectoral approach and focuses on methodologies with a high revision frequency. The set includes the following project types: industrial gases (AM0001); alternative waste treatment (AM0025); landfill gas (ACM0001); renewable power generation for the grid (ACM0002); fuel switch in cement manufacture (ACM0003); manure management (ACM0010); small-scale energy efficiency (AMS-II.C); and two afforestation and reforestation methodologies (AR-AM0007/8). The results are displayed in Figure 11.1.

The EB at its 28th meeting in December 2006 decided that a methodology should only be changed every six months as it recognized the challenge that frequently changing versions of methodologies poses to project developers.[8] However, a little less than a year later, in October 2007 the EB at its 35th meeting abolished this decision.[9] This rare case of a 'U-turn' in decision making showed that the EB was not able or willing to stop methodology revisions; and since then revisions have become even more frequent. The assessment of the validity of versions of the reviewed methodologies in Table 11.1 shows that the majority of the versions have remained valid for less than half a year.

Two methodologies are in their ninth version already. More than 40% of the versions were valid for less than 90 days and 20% valid for less than two months! It needs to be pointed out that the majority of these revisions of methodologies did not actually result in a major impact on CER volumes or in a limitation of their applicability to underlying project types. Most of the revisions only entail slight corrections or amendments in methodology wording that have a very

[7] UNFCCC approved large and small scale methodologies (as at March 2009) can be found at <http://cdm.unfccc.int/methodologies/PAmethodologies/approved.html> and <http://cdm.unfccc.int/methodologies/SSCmethodologies/approved.html>, accessed 31 March 2009.

[8] CDM Executive Board 28. Meeting Report, Annex 16, 2006.

[9] CDM Executive Board 35. Meeting Report, Annex 13, 2007.

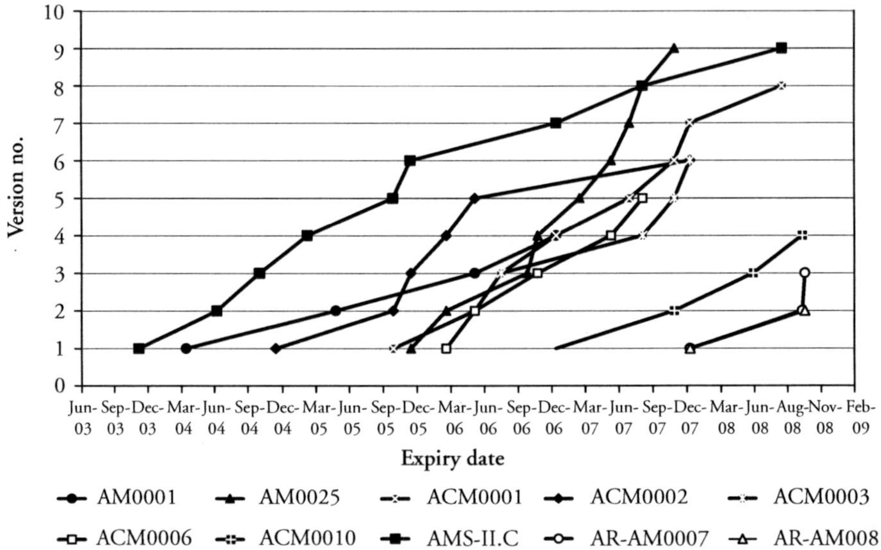

Figure 11.1. Development of versions of 10 prominent approved CDM methodologies over time

limited impact on most projects at an early planning stage. Having said this, there have been examples of major implications from methodology revisions. For example, the EB decided in July 2007 to phase-out the methodology ACM0004 for waste heat recovery projects—a very prominent CDM project type until that date.[10] The EB merged ACM0004 with some other proposed methodologies, to create a new consolidated methodology with much broader applicability conditions (ACM0012). However, there have been strong complaints about the appropriateness of that methodology as it reportedly cannot be applied because parts of the methodology cannot be consistently applied to waste heat and pressure projects.[11] Another example is the revision of the methodology ACM0002 for grid-connected renewable power generation. As the revision also had major implications for projects undergoing validation, the issue is covered in detail in the next section. At the same time, incoherencies among different methodologies exist and have an impact on CER volume estimates.

A very illuminating example is the situation in which project participants in 'energy efficient lamp distribution' projects found themselves before 16 August

[10] ACM0004 has not been active since 5 July 2007 and was replaced by ACM0012 <http://cdm.unfccc.int/methodologies/DB/3NL6ELY805NDZHF4YZRPK94MAPALUB/view.html>, accessed 31 March 2009.

[11] International Emission Trading Association (IETA), 'State of the CDM', 2008.

Table 11.1. Duration of validity of 10 prominent approved CDM methodologies

Validity (days)	No of versions	%
<=30	2	3.9%
31–60	9	17.3%
61–90	10	19.2%
91–120	1	1.9%
121–150	4	7.7%
151–180	3	5.8%
>180	23	44.2%
Total	52	

2008. Before that date project participants could choose to develop a project and disseminate the lamps to households under the large-scale methodology AM0046[12] or the small-scale methodology AMS-II.C. The former was approved in February 2007 and allowed emission reductions for avoided technical transmission losses to be credited. This part of the CO_2 emission reduction calculation in India, for instance, can contribute up to 20% of total CER volumes. In February 2007, version 8 of AMS-II.C was valid, and this did not allow project participants in 'small-scale lamp distribution' projects to consider technical transmission losses when calculating reductions. More puzzling was the fact that until six months before the approval of AM0046, the previous version of AMS-II.C had allowed projects to take such losses into account. Finally, in August 2008, one-and-a-half years after the approval of AM0046, the EB removed the inconsistency and since then both methodologies have permitted technical transmission losses. During the period of inconsistency it was not clear whether the EB would decide in favour, or against, the possibility of increasing eligible emission reductions from such projects.

Another and more recent example of regulatory inconsistency and the resulting uncertainty for project participants is the different methodological approach of accounting for project emissions for the same activities in two different methodologies. It is common scientific knowledge that the application of chemical fertilizer is associated with considerable GHG emissions.[13] The emissions occur during fertilizer production (CO_2 from energy consumption and N_2O emissions from nitric acid production) as well as after the application of fertilizer as N_2O emissions from soils. At its 42nd meeting in September 2008, the EB stated that GHG emissions from fertilizer application 'are insignificant' and thus decided to remove project emissions due to fertilizer application from all approved afforestation and reforestation methodologies.[14] The

[12] *Lighting retrofit for residential use.*

[13] Intergovernmental Panel on Climate Change (IPCC). Guidelines for National Greenhouse Gas Inventories, Vol. 3—Industrial Processes and Product Use, chapter 3—Chemical Industry Emissions, 2006.

[14] CDM Executive Board 42. Meeting Report, paragraph 35, 2008.

appropriateness of the decision will not be disputed here. What needs to be mentioned, though, is that in parallel and since December 2006, project participants undertaking cultivation of oilseed plants and subsequent use of the oil in transportation are forced to deduct emissions due to the application of fertilizer from their emission reductions. As long as this inconsistency persists, changes to the methodologies concerned can be expected. The result of the current situation is considerable uncertainty about the levels of emission reductions that can be expected from afforestation and reforestation as well as from oilseed-based biofuel transportation projects.

For many years there has also be considerable uncertainty around the eligibility and the expected CER volumes from several other high-potential CDM project types that have no approved methodology. In the case of carbon capture and storage, a project type that has been considered by several multinational companies as well as several oil and gas exporting countries for many years, the EB did not want to take a decision and referred the decision to COP/MOP, which has also been unable to decide.[15] The topic has now become one element of the negotiation package for the Copenhagen COP in December 2009. Similarly, projects designed to reduce the use of non-renewable biomass were stalled for more than two years, before COP 13 in Bali decided to accept this project type.[16] Another recurring hot regulatory topic is the eligibility of N_2O emission reduction projects in new nitric acid and adipic acid production facilities. For both industries, the approved methodologies are only applicable if the plant had operated commercially before the end of 2005 (nitric acid) or the end of 2004 (adipic acid). The first attempt to broaden the applicability to production capacity implemented after those dates was made in September 2007 for nitric acid. Since then several requests for revisions of the relevant methodologies have been unsuccessful. Only in September 2008, did the EB request the Methodology Panel to draft guidance and clarifications on the expansion of industrial gas recovery projects to new facilities.[17]

The above examples illustrate two important points. There are a significant number of PINs or draft PDDs circulating in the market for which regulatory decision-making has not been able to provide certainty with regard to the carbon reduction value of the underlying project ideas for a number of years. In several cases, the high revision rates of the applicable methodologies and the inconsistency of EB decisions across approved methodologies has added considerably to the uncertainty surrounding the possibility of achieving emission reductions for projects due to start in the next year or two. This means that for many project ideas, the information provided in a PIN or draft PDD by a project

[15] CDM Executive Board 22. Meeting Report, paragraph 23, 2005.

[16] UNFCCC COP/CMP 3, Decisions and Resolutions, Decisions 2/CMP.3—Further Guidance Relating To The CDM, 2008, <http://unfccc.int/documentation/decisions/items/3597.php?such=j&volltext=/CMP.3#beg>, accessed 31 March 2009.

[17] CDM Executive Board 42. Meeting Report, paragraph 27, 2008.

participant (that does not perhaps understand much about methodologies and their application) is not always fully sufficient to estimate accurately the carbon potential of a CDM project idea. Fewer inconsistencies in the application of approved methodologies would constitute a considerable improvement of the situation in which the current actors in the CDM market find themselves.

4. Uncertainties after Submission for Validation

Once the PDD is developed and the project has been submitted for validation, the key questions for project participants are when the project will be registered and how many CERs the project will generate after registration. Registration probability and the timing of registration depends on a number of factors of which the very obvious ones are: the quality of the project, the quality of the documentation (PDD and associated documents) and the quality of the CDM consultant preparing the documentation. Those factors are clearly in the hands of the project participants and shall not be discussed here. In the event that projects submitted for validation receive a request for review of registration and are therefore delayed or are rejected, eg on grounds of incomplete substantiation of additionality or early consideration, it is clearly not the regulatory framework that should get the blame.

However, projects in the validation stage can be impacted in relation to the timing of registration in a number of ways. The average period between submission for validation and final registration currently lasts about one year. This does not reveal why projects are stalled and whose responsibility it is to move things faster. Nevertheless, there is currently an understandable debate about the appropriateness and the efficiency of certain procedural steps in which the UNFCCC Secretariat and the DOEs are involved, that make it almost impossible to predict when the registration of projects will be achieved.

Before a project can be submitted for registration, the DOE has to issue a final validation report that confirms that the project is in full compliance with the CDM rules and procedures. Over the second half of 2008, there has been growing frustration among project participants that DOEs have not been able to process validations with the speed that the developers expect.[18] The reasons reported by project participants for the delay of validation on the side of the DOEs have been incompetence, cumbersome internal procedures, long response time, undercapacity, and rule-setting by DOEs (beyond those specified by the EB).

Once the validation report is complete, the project can be submitted for registration. All project participants focus heavy criticism on what is called the 'completeness-check', whereby the UNFCCC Secretariat checks whether all the

[18] Hoogzard, J et al, 'A call to reform' (October 2008) 5 Carbon Finance 16.

necessary documents have been submitted and the registration fee has been paid. Only after the completeness check is completed is the validated PDD formally uploaded to the UNFCCC website and the official eight-week registration period starts (four weeks for small-scale projects). The 'completeness-check' has not been defined in the CDM procedures and is the only step in the CDM process that does not have a formal deadline. In 2008, the UNFCCC Secretariat kept project participants waiting for several months before projects were officially submitted for registration, reaching up to a year in extreme cases, and there were unpredictable differences between project types. The average delay attributed to the 'completeness check' since the inception of the CDM is around 40 days. Part of the reason why this check takes such a long time is that there is a backlog of unusually high numbers of requests for registration submitted to the UNFCCC Secretariat by DOEs during the months June, July, and August 2008. What is most interesting, is that the high number of submission rates was caused by the decisions of the EB to revise two methodologies namely ACM0004 (March 2008 effective) and ACM0002 (August 2008 effective).

The implications of the revisions of those methodologies, the way that the implications are linked to other aspects of the CDM process, and the fact that the overall situation has resulted in huge uncertainty in the market in terms of timing of registrations as well as uncertainty in terms of CER volumes is illustrated by the following.

The EB decided that once a methodology has been revised, the previous version of the methodology may only be applied for projects undergoing validation for eight months after the submission of the relevant project for validation. [19] This means that the UNFCCC Secretariat must receive the validated PDD and the validation report for its 'completeness-check' within eight months from the submission for validation of a project. The most serious impact of a methodology revision yet seen in CDM was the case of ACM0002 (Renewable power generation for the grid). The EB at its November 2007 meeting decided that version 6 of the methodology should be replaced by a new version.[20] The deadline for application of version 6 for project undergoing validation was 13 August 2008.[21] Version 7 of ACM0002 referred to a newly established 'tool' that had much stronger requirements for data availability to calculate the emission factor of the grid. Specifically, the 'tool' requires annual plant-specific fuel consumption and power generation data from the fossil fuel fired power plants connected to the relevant grid.[22] If only power generation data is

[19] CDM Executive Board 35. Annex 13.

[20] CDM Executive Board 35. Meeting Report, paragraph 24, 2007

[21] UNFCCC, ACM0002: Consolidated methodology for grid-connected electricity generation from renewable sources <http://cdm.unfccc.int/methodologies/DB/5YDYJJEA3C8SLA0FDNQ-BOLBKVGAGO7/view.html>, accessed 31 March 2009.

[22] CDM Executive Board 35. Meeting Report, Annex 12, 'Tool to calculate the emission factor for an electricity system,' 2007.

available, then very conservative default efficiencies of the plants have to be applied. Project participants recognized that in the top two CDM destinations China and India, the currently published grid emission factors from the DNAs would not be valid after 13 August and that there was considerable uncertainty if and when the DNAs would recalculate the new grid emission factor based on the 'tool'. In China especially it was evident that the 'tool' included much higher default efficiencies than the Chinese DNA had previously used in its calculations. Chinese project participants therefore rightly sensed that their projects undergoing validation would see a considerable reduction of CER volumes for every MWh generated if these projects were not submitted for their 'completeness-check' before 13 August 2008. Therefore, in August 2008, the UNFCCC secretariat received 250 submissions of requests for registration, an increase of 400% compared to the average submission rates in previous months of 2008. It is very likely that this strong increase was exclusively due to project developers of renewable power projects—with the great majority being Chinese projects—submitting before the deadline. Two thirds of those projects had not been uploaded for the official request for registration by December 2008. In the 30 days period prior to 13 August 2008, 45 ACM0002 projects were submitted for registration compared to an average in the same period in 2008 of only 16. Between 13 August and 1 November 2008 only two ACM0002 projects had been registered, an average of less than one project in a 30-day period.

Currently, the grid emission factors of China and India are not yet clear. The EB meeting in October 2008 took a decision that has potentially far-reaching consequences for renewable energy projects in those countries.[23] It stated that validators should first check the appropriateness of the DNA calculations. Essentially this means that the DNAs of China and India will soon undergo validation. At the time of writing[24] it is not clear what the outcome will be. In November 2008, the Indian DNA had already recalculated and published the new grid EF factor values, which did not change substantially because in India both power generation and fuel consumption data is available.[25] But the Chinese DNA has not yet recalculated the emission factor. The key problem with the Chinese data is that they do not include power-plant-specific efficiencies. In this case, default power plant efficiencies have to be used which are much more conservative than the factors used to date. For subcritical coal plants built before 2000, the efficiency in the 'tool' is set at 37% and for plants built after 2000 at 39%. Supercritical plants have to use 45%. Previously, the DNA value for newly built coal plant efficiencies was 35.82%.[26] If the Chinese coal plants have to base

[23] CDM Executive Board 43. Meeting Report, para 64, 2008. [24] March 2009.

[25] Central Electricity Authority (CEA), Baseline Carbon Dioxide Emissions from Power Sector, Version 4, Sept 2008, <http://www.cea.nic.in/planning/c and e/Government of India website.htm>, accessed 31 March 2009.

[26] National Development and Reform Commission (NDRC) China <http://cdm.ccchina.gov.cn/english>, accessed 31 March 2009.

calculations on 45% efficiency instead of 35.82%, their emissions factor would decrease by 200g CO2/kWh. The impact on the grid emission factor in the different provincial grids would be a reduction in emission factor of 0.1 to 0.2 tCO2/MWh, which means a reduction in CER volumes for those projects of around 10–20%.

Project participants are currently very concerned about request for reviews almost becoming a *de facto* standard and that a considerable proportion of projects submitted for registration do not automatically achieve registration status within the official eight-week period (four weeks for small-scale projects). The International Emissions Trading Association reports that between June 2007 and August 2008, 40% of all projects requesting registration have received a request for review.[27] However, it must be noted here that request for reviews can have various well justified reasons (eg missing additionality, methodology not applicable, etc) and there have been and there will continue to be good reasons for project review.[28] Reviews that project participants are most concerned about are, for example, those where the EB requests more information as to how the DOE validated certain aspects of the project as well as those circumstances where guidance that was not available at the date of submission for validation is used to prompt requests. The potential of such retroactive application of guidance obviously increases the longer a project is delayed during validation, the completeness-check, and the potential request for reviews. The effect of a request for review can be hazardous as the project can be rejected by the EB or the review process delays registration for a minimum of a few weeks or a maximum of months.

The above discussions illustrate that the uncertainty regarding the timing of registration and the CER volumes to be expected from projects that are submitted for validation can be considerable. The reasons for this uncertainty can be various and cannot be exclusively attributed to the EB, for example, the duration of 'completeness-check' which does not have a formal deadline, the revision of methodologies and requests for reviews. The example of ACM0002 illustrates that revisions can have a substantial effect on the CER volumes to be expected, even if projects have already been submitted for validation.

5. Advanced Stage/Post Registration

Although project participants could assume that the greatest hurdles to CER generation are cleared when registration is finally achieved, there are still significant uncertainties ahead after registration and many CDM projects face significant problems in CER issuance. Some of these problems relate to regulation whereas others are caused by problems in project implementation.

[27] International Emission Trading Association (IETA), State of the CDM, 2008.
[28] See chapter 12 on 'Interpreting the Additionality of CDM Projects' below.

With respect to regulation, issuance requests are being reviewed by the EB more often than in the early days of the CDM. In November 2008 for instance, the EB decided about 55 issuance requests of which two were rejected, one was withdrawn after having received a request for review, and 22 were requested to be reviewed.[29] In fact, the share of issuance requests that are requested to be reviewed has strongly increased throughout 2008. While in the first half of 2008, around 20 to 30% of all issuance requests received a request for review, this share grew to 40% in September to November 2008. This is a clear indication of the increased scrutiny and stringency of the EB with regard to the issuance of CERs. Interestingly, only a few of these review requests have had considerable consequences for the number of issued CERs. The issuance rate, comparing total requested CERs with the issued CERs, was always close to 100%. However, 13 out of 427 issuance requests in 2008 were rejected or had to be withdrawn by the project participants and some CERs were only issued with a substantial reduction of the requested amount.

Looking at the reasons for review mentioned in the requests for review, the main reasons were quality problems with the submitted monitoring and validations reports, for example they lacked an explanation of the special operating conditions, lacked proof of diligent verification of the facts claimed in the monitoring reports, and inconsistencies between the different reports and their annexes. This clearly indicates a general lack of competence in the CDM market, but reveals also that the existing procedures and rules do not always provide comprehensive guidelines for project participants and verifiers.

The main reasons for lower than expected issuances of CERs are performance problems with the technology used in the CDM project. The average CDM project delivers only 96% of the CERs predicted in the PDD.[30] However, the performance in terms of expected versus achieved CER generation varies greatly for different project types. While CDM projects involving the reduction of industrial GHG emissions have on average performed well (>100%), many landfill gas projects have faced real problems with CER delivery. On average, landfill gas projects have only generated 37% of the expected CERs. It has to be noted however that the CER performance does not only vary greatly if different project types are considered, but that also the performance of projects within one project category might be significantly different. This is a clear indication of the fact that many factors influence performance and the regulatory framework and EB decisions are only some of these.

In 2008, two requests for CER issuance from landfill gas flaring/energy utilization projects were rejected and another two requests were withdrawn as the EB doubted that the realized discrete measurements of the methane fraction of

[29] CDM Executive Board, 44. Meeting Report, 2008.

[30] UNEP Riso Centre, CDM/JI Pipeline Analysis and Database, <http://cdmpipeline.org>, accessed 31 March 2009. In the following all performance figures are based on this source.

the landfill gas were in line with the methodology. This meant that a quarter of all issuance requests for landfill gas projects in 2008 were not successful due to problems of correctly measuring the methane fraction. Up to version 8, ACM0001 (methodology for landfill gas) allowed the periodical measurement of the methane fraction, if a statistically valid number of samples were taken and if measurement was done at a 95% confidence level. However, no further guidance was provided on how to determine if sampling was statistically valid. Furthermore, the statistical test can only be performed ex-post and might result in a negative outcome if the variation of the methane content was higher than previously assumed. Based on a request for clarification that mentioned this issue (AM_CLAR 0095),[31] ACM0001 was revised and now requires a minimum of four quarterly measurements per year if these periodical measurements are the preferred option (as per version 9.1). Then, the lower bound of the 95% confidence interval obtained from the periodical measurements should be used to estimate baseline methane emissions. The projects for which the issuance requests had been put under EB scrutiny used a discrete measurement frequency of at least once a day up to hourly during working time, which the EB considered to be insufficient to suitably reflect the variations in the methane content.

There are also other methodologies where the required monitoring is complex, challenging, and not sufficiently explicit, leaving much room for interpretation and reviews. This refers for example, to AM0025 for alternative treatment of solid wastes, AM0046 for the distribution of efficient light bulbs etc. Although no projects under AM0025 have so far requested the issuance of CERs, it can be assumed that the adopted monitoring procedures may also lead to an increased number of reviews, as the methodology requires numerous parameters to be monitored and also vaguely demands measurements to be performed at 95% confidence level.

In the light of the increasing number of review requests during issuance, some stakeholders suspect that RIT members are dedicated to find at least one error in each monitoring/validation report. In fact, some extremely formal correction requests have been observed. One issuance request (project 0498, issuance of 14 November 2008) received a request for review as the grid emission factor used in the monitoring report was 883g CO2/kWh, while the corresponding value mentioned in the registered PDD was 0.88318kg CO2/kWh.[32] The subsequent correction of the report lead to a reduction of *one* requested CER!

Even if all the issues raised during a request for review can be sufficiently clarified and no substantial reduction of issued CERs takes place, a request for

[31] UNFCCC, Requests for clarification on application of approved methodologies by DOEs, AM_CLA_0095, 15 April 2008. <http://cdm.unfccc.int/methodologies/PAmethodologies/Clarifications/index.html?p=3>, accessed 31 March 2009.

[32] UNFCCC, 2008, Project: 0498 SIDPL Methane extraction and Power generation project—Issuance Request, Request for Review 1 and 2, <http://cdm.unfccc.int/Projects/DB/DNV-CUK1152256546.37/IPROCESS/DNV-CUK1196855079.62/view>, accessed 31 March 2009.

review can still pose significant problems to a project participant, as any review request will result in delayed issuance. While the issuance process took about one month in late 2008, issuances that had received a request for review needed on average three months to be processed. In some exceptional cases, the final decision on the issuance request and/or the issuance only happened more than a year after the submission of the issuance request.

Generic performance problems can be due to over-optimistic parameter choices in the PDD or to genuine technological underperformance. Regarding the former, many methodologies require the ex-ante estimation of achievable emission reductions using methodologies and default factors taken from the IPCC Guidelines for National Greenhouse Gas Inventories. While this procedure, on the one hand, facilitates the elaboration of PDDs, it is on the other hand frequently inappropriate as the specific local circumstances might not always be suitably reflected by the guidelines meant for the assessment of nationwide conditions. Then, the ex-ante estimation might result in a considerable misjudgement and in later under- (or over-) performance. This methodology-inherent uncertainty, as to what extent the ex-ante estimation procedure as per the methodology suitably reflects the specific project conditions, can be found in many waste-related methodologies (sectoral scope 13). This is one reason for the dramatic underperformance of many landfill gas project activities. Furthermore, the performance might be strongly influenced by the timelines of the regulatory processes. Increased lead times during registration (see section 3 above) and delays in issuance (see above), can result in CER losses—at least if the project has already been commissioned at the originally envisaged start of crediting period—or substantial delay in delivery.

From the above, it can be concluded that many of the problems that have occurred in the post-registration phase can be attributed to negligence and to a lack of understanding of the CDM-related requirements of the project owners, to technical and organizational difficulties and to other issues not directly related to CDM rules. However, it has to be noted that appropriate and clear guidelines, procedures and rules for monitoring and verification are a key element for the successful generation of CERs. In the past, obviously, the existing set of rules and procedures did not always provide enough certainty to project participants and verifiers. Recent decisions might alleviate this uncertainty. The Validation and Verification Manual, finally adopted after long internal processing at the 44th meeting of the EB in November 2008,[33] will surely provide clearer guidance for project participants and verifiers on verification requirements and processes, and might help to reduce the number of requests for reviews and reviews undertaken during issuance.

[33] CDM Executive Board 44. Meeting Report, Annex 3, Validation and Verification Manual (version 01), 2008.

In conclusion, there are two key issues that strongly impact CER issuance success: first of all, a realistic ex-ante estimate of the achievable emission reduction and secondly, appropriate operation, monitoring and reporting during the operation of the project. In both steps, project participants can mitigate their specific performance risk by sound engineering and diligent project implementation. However, the CER performance will still to a certain extent continue to be influenced by EB actions and decision-making.

6. Proposals for CDM Reform and Related COP/MOP Decisions

The turbulent year 2008 justifies a short summary of the proposals for regulatory reform of the CDM that have been made by prominent CDM actors, such as the International Emissions Trading Association (IETA)[34] as well as the UNFCCC Secretariat.[35] Besides, prominent CDM stakeholder representatives have made individual suggestions to the 'CDM media.'[36] Two major categories of proposals were made regarding the governance structure of the CDM as well as concrete procedural recommendations.

In terms of governance structure, suggestions included the following:

- clear timelines for every process step in the registration and issuance procedure in order to increase predictability;
- redefinition of the role of the EB as CDM policy maker with clear guidelines; especially with regard to new approaches for additionality determination and baseline setting (eg benchmarks) as well as oversight of sampling for project reviews;
- redefinition and expansion of the role of the UNFCCC Secretariat to act as the technical support unit for the EB; including project reviews;
- development of an appeal mechanism operated independently from the EB where project participants could challenge EB and Secretariat decisions; and
- expansion of the Methodology Panel to a full-time body with the mandate to develop its own methodologies.

Procedural recommendations were made on issues such as transparency of the reasons for and outcomes of project reviews, and the avoidance of the retroactive

[34] International Emission Trading Association (IETA), 'State of the CDM', 2008.

[35] Compare UNFCCC, 'Synthesis of views from Parties on ways and means to enhance equitable regional and subregional distribution of projects under the clean development mechanism [. . .]', FCCC/KP/CMP/2008/INF.4, 2008; and UNFCCC, 'Elaboration of possible improvements to emissions trading and the project-based mechanisms under the Kyoto Protocol', FCCC/KP/AWG/2008/INF.3, 2008.

[36] M Lehman, et al, 'In defence of the DOEs' (2008) 5 Carbon Finance 17.

application of guidance, as well as closer interaction between the EB and the DOEs to ensure that DOEs understand and implement the intention of the EB decisions.

7. Conclusions

The layout of the CDM rule-making offers so far unprecedented and far-reaching stakeholder participation and transparency, but the associated, sometimes bureaucratic and lengthy procedures definitely meet their limits when facing the huge project inflow the CDM saw in 2008. Uncertainty is an immanent feature of the current structures and procedures of rule-making. At the same time we have shown that it is not always regulatory decisions which are to blame for sub-optimal CDM performance. In many instances, project participants and DOEs alike can enhance their contribution to CDM performance by putting in place appropriate internal project management procedures. With regard to the proposals for CDM reform in terms of governance structure and concrete procedural recommendations, the negotiators at COP/MOP14 in December 2008 were only able to decide on small procedural reforms. The COP/MOP mandated the EB to set timelines for decisions on methodology revisions and deviations from methodologies, as well as on approval of monitoring plan revisions. Negotiators agreed that the EB should not apply guidance retroactively, especially so as not to make projects suffer from the lengthy 'completeness-checks.' The check itself should be speeded up, the COP/MOP decided. COP/MOP also requested the EB to find ways of interacting with project proponents directly, and not only through DOEs, as is mostly the case today. In the view of many project participants, projects would move much quicker through the registration and issuance process, if the interaction between the EB and the Secretariat/Panels on the one hand and project participants on the other were improved. The initial draft text regarding the introduction of an appeal institute in the CDM and an increased role for the UNFCCC Secretariat in decisions on specific projects was removed from the CO/MOP decision.

In conclusion, COP/MOP14 did not provide the grand structural reform that many market participants had hoped to see. Instead, first small but positive decisions were made that will help increase the speed of project registration and CER issuance. However, the decisions do not provide the much required consistency, transparency, and accountability of the mechanism and hence more crisis talk can be expected for 2009.

12

Interpreting the Additionality of CDM Projects: Changes in Additionality Definitions and Regulatory Practices over Time

*Axel Michaelowa**

1. The CDM as an Offset Mechanism, its Stakeholders, and the Challenge of Additionality

The Clean Development Mechanism (CDM) is the only policy instrument under the Kyoto Protocol that mobilizes greenhouse gas emissions reductions in developing countries. The basic idea of the CDM is based on the principle of offsetting emissions—one tonne of CO_2 eq reduction from a project in developing countries generates an emissions credit (Certified Emissions Reduction, CER) of 1 t CO_2 eq that can be used by industrialized countries for compliance with their Kyoto targets. To ensure the environmental integrity of the overall Kyoto Protocol system, it is important that each CER represents a real emissions reduction as it is not covered by an equivalent reduction from the host country's emission budget. This is due to the fact that developing countries do not have emission commitments under the Kyoto Protocol. Fictitious CERs generate an inflationary effect on the emissions budget of the entire Kyoto system. Therefore, a complicated regulatory system has been set up since 2001 to check CDM project proposals and guarantee the environmental integrity of the CDM. At the core of this system, the CDM Executive Board (EB) decides about the technical rules for projects. Over time, the EB has created a number of supporting panels that provide technical expertise and prepare EB decisions (see Figure 12.1). Independent 'Designated Operational Entities' (DOEs) audit project proposals (Project Design Documents, PDDs) to validate their conformity with the CDM rules. Each project is then checked by the EB before it is formally registered as CDM project.

* My thanks to the UK Department of Environment, Food and Rural Affairs (DEFRA) for financing two studies on CDM methodologies and CDM reform in 2007 and 2008, elements of which have been used in this article.

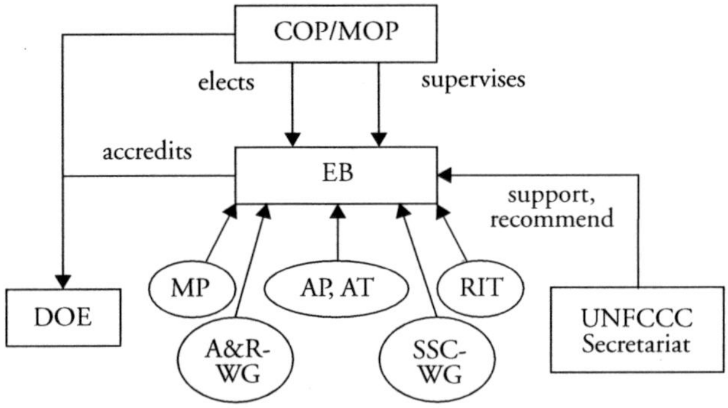

Key: **AP**— Accreditation Panel; **A&R WG**— Afforestation and Reforestation Working Group; **AT**— Assessment Team; **MP**—Methodology Panel; **RIT**— Registration and Issuance Team; **SSC-WG**— Small-scale Working Group

Figure 12.1. CDM institutions

The concept of the 'additionality' of the emission reduction achieved by a CDM project lies at the heart of the CDM rules. The essential idea underlying the concept of additionality is that the emissions reductions of a CDM project would not have happened under 'business-as-usual'. However, as in the tale of the blind man and the elephant where the blind man touches different parts of the large animal and thus thinks he encounters several different animals, there are widely differing views about additionality. Some stakeholders argue that the concept of additionality does not make any sense and that any project reducing emissions compared to a baseline should get CERs. For an economist observer this is a bit like picking up a €20 bill lying on the sidewalk—and then claiming an extra payment from a bank for bringing this bill back into circulation. The other extreme is the demand that no profitable project should be credited. Again, for the external observer, this position also does not make sense as profitable projects may not materialize due to availability of more profitable alternatives, unavailability of capital, or other barriers. So the CDM should accept profitable projects as long as the project developer can show that these projects would not happen without the incentive from CER sales.

The CDM involves an array of stakeholders—project developers and consultants, CER buyers, national and international regulators and environmental NGOs. The interests of most of these stakeholders are aligned to maximize the amount of CERs, as sellers, buyers and national regulators all profit from generation of high CER volumes[1]. Of course, the CER price falls if the supply of CERs increases. Many actors do not take this effect into account, thinking that

[1] See A Michaelowa, 'Joint Implementation—the baseline issue' (1998) 8 Global Environmental Change, 81–92.

they will be able to sell before the price decreases. The only exception is the environmental NGOs who want to protect the integrity of the Kyoto Protocol. However, many of the NGOs lack the technical expertise to identify cases where CERs are not justified at all or are overestimated.

The interests of project developers and CER buyers are strongly aligned, as both sides want to maximize CER volumes. These actors have always argued that it is impossible to gauge reasons why project developers invest in a project. Therefore, any CDM project whose emissions are below the baseline should be entitled to earn CERs.[2] This transfers the determination of additionality into baseline-setting and obfuscates the problem, eg by setting a benchmark (see further discussion of benchmarks as additionality determination tool below).

The guiding question of this chapter is how the concept of additionality has evolved over time both in its theoretical definition and its actual implementation regarding registration of CDM projects. It starts with a discussion of the inability of the negotiators of the Kyoto Protocol and the Marrakech Accords to agree on an operational definition of additionality. The unexpected effort of the CDM Executive Board to introduce principles of additionality testing and the development of the consolidated additionality tool are discussed in section 3. Recent developments that allow the use of benchmarks and default discount factors to take into account non-additional projects, instead of using the additionality tool, are assessed in section 4. Section 5 describes how projects have been reviewed and rejected over time, with the EB developing 'fashions' of rejecting certain project types without prior warning. The fate of cement blending and waste heat recovery projects is assessed in detail. Given this situation, several proposals for reform of additionality testing have been made, which are dealt with in section 6. Section 7 discusses what would happen if additionality testing was scrapped. Section 8 summarizes the lessons of five years of additionality determination of CDM projects.

2. The Initial Interpretation of Additionality

According to the Kyoto Protocol, the emission reductions resulting from CDM projects have to be 'real, measurable and long-term',[3] and they have to be 'additional to any that would occur in the absence of the certified project activity'.[4] Despite a long debate by researchers and interest group representatives about the concept of additionality,[5] negotiators in the UNFCCC process were unable to agree on an interpretation of the additionality criterion. They therefore repeated the

[2] See H Rentz, 'Joint Implementation and the question of additionality—a proposal for a pragmatic approach to identify possible Joint Implementation projects', in (1998) Energy Policy 4, 275–9 or IETA, 'IETA Position on additionality' (Geneva, 2006).

[3] Kyoto Protocol, Art 12, 5(b)

[4] Ibid, Art 12, 5(c)

[5] For a summary of the arguments, see S Greiner and A Michaelowa, 'Defining Investment Additionality for CDM projects—practical approaches', in (2003) Energy Policy 31, 1007–15.

wording of the Kyoto Protocol in the Marrakech Accords of 2001 which state that 'a CDM project activity is additional if anthropogenic emissions of GHGs by sources are reduced below those that would have occurred in the absence of the registered CDM project activity'.[6] The main reason for the inability to decide were substantial differences in views, lack of understanding of the issue especially on the side of developing country negotiators, and the feeling that the issue was too technical to be decided at the level of the Conference of the Parties. Many negotiators did not see the critical role of additionality in safeguarding the environmental integrity of the Kyoto Protocol, as they struggled to understand the theoretical background of the Kyoto mechanisms.

3. The EB Cuts the Gordian Knot

As it had become clear that the COP/MOP would never be able to agree on a technical definition of additionality the EB was left with the task—and had the courage—to define detailed rules for additionality. This process is by no means finished. While the COP/MOP has repeatedly stressed that key additionality rules defined by the EB are not mandatory, it has not cancelled any additionality rules specified by the EB.

For small scale projects,[7] additionality testing was defined by the 7th meeting of the EB (EB 7) in January 2003.[8] This was due to the fact that the CDM rules and baseline methodologies were to be streamlined by the EB for small-scale projects, whereas for large-scale projects, they were to be proposed by the project developers and only then decided by the EB. At that time, no large-scale baseline methodology had yet been proposed. Overall, the aim was to facilitate life for small-scale project developers, and thus the proposed rules were understood to be on the lenient side. The EB specified the aim of the additionality test for small-scale projects was to show that the project would not have occurred anyway and it enumerated four types of barriers used to prove additionality—investment barriers, technological barriers, barriers due to prevailing practice, and other barriers. Investment barriers are defined as the availability of a financially more viable alternative that would have led to higher emissions. The technological barrier requires the existence of a less technologically advanced alternative that involves lower risks due to the performance uncertainty or low market share of the new technology adopted for the project activity and so would lead to higher emissions. The prevailing practice barrier says that prevailing practice or existing regulatory or policy requirements would lead to the implementation of a technology with higher emissions. Other barriers include

[6] UNFCCC (2001): The Marrakech Accords, FCCC/CP/2001/13/Add.2, Bonn.

[7] The Marrakech Accords had decided that renewable energy projects below 15 MW capacity, energy efficiency projects saving less than 15 GWh per year and all other projects with project emissions of less than 15,000 t CO_2 eq should benefit from simplified rules.

[8] EB 7, attachment A to Annex 6, later confirmed by decision 4/CMP.1.

institutional barriers or limited information, managerial resources, organizational capacity, financial resources, or capacity to absorb new technologies. Project participants have to demonstrate to a validator that the project would otherwise not be implemented due to the existence of one or more barriers listed. A short explanation of each barrier is provided.

When subsequently the first large-scale methodology submissions were imminent, the EB decided that a baseline methodology has to demonstrate that a project activity is additional and is therefore not the baseline scenario.[9] It added later that this could be done through a flow-chart, a series of questions, a qualitative/quantitative assessment of different potential options describing why the non-project option is more likely, a qualitative/quantitative assessment of one or more barriers facing the proposed project, and proof that the project type is not common practice in the proposed area of implementation, and not required by recent/pending legislation or regulations.[10] As these options were seen as too broad and could not be operationalized properly, in late 2004 a consolidated additionality tool was agreed as a general framework for additionality testing, but with a voluntary character.[11] It contained a mandatory series of steps to avoid the risk that project developers simply choose the test that suits them best (see Figure 12.2):

0. (Only applicable to early start projects[12]): Evidence that the incentive from the CDM was seriously considered in the decision to proceed with the project.
1. Identification of alternatives to the project.
2. Investment analysis, with the aim of determining whether or not the project is the most economically or financially attractive alternative.

Instead of doing the investment analysis, project developers could carry out step 3:

3. Barrier analysis, with the aim of checking for the existence of prohibitive barriers

The following steps had to be done by all users:

4. Common practice analysis.
5. Impact of CDM registration. Subsequently, the EB clarified that this step could be applied qualitatively.[13]

The consolidated tool was revised in November 2005. The revision determined that step 0 would be fulfilled even if the project could only state that the objective was to mitigate climate change. In the second revision of the tool in February 2007,[14] steps 0 and 5 were removed. Step 0 was thought to be

[9] EB 8, Annex 1. [10] EB 8, Annex 1. [11] EB 16, Annex 1.
[12] An early start project was a project that had started after 1 January 2000 but before the date of registration of the first CDM project (8 November 2004). [13] EB 17, para 17.
[14] EB 29, Annex 5.

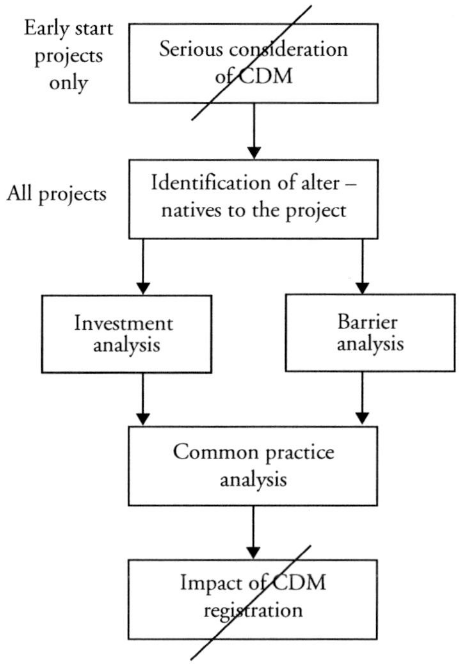

Figure 12.2. Steps in the consolidated additionality tool and changes over time

superfluous as the EB thought projects would now routinely be submitted to the CDM before they were actually built and implemented (for a subsequent reversal of this decision, see the discussion on 'serious consideration' below). Removal of step 5 was explained by the feeling that this step would amount to a duplication of steps 2 and 3. Given the problems that the EB had encountered with interpretation of the barrier test, the barriers that could be assessed under step 4 were now defined more clearly (see Box 12.1).

It was also clearly stated that if the CDM does not remove the identified barriers that prevent the project, the project is not additional. Moreover, it was

Box 12.1. Examples of barriers that prevent project implementation

- Similar activities have only been implemented with grants or other non-commercial finance terms.
- No private capital is available from domestic or international capital markets.
- Process/technology failure risk in the local circumstances is significantly greater than for other technologies.

added that DOEs should carefully assess and verify the reliability and credibility of all data, rationales, assumptions, justifications, and documentation provided by project participants and document their conclusions transparently in the validation report. As described in section 5, in practice this has not happened until very recently.

A small revision was undertaken in December 2007, focusing on the eligibility of benchmarks in the investment test. In the fourth revision in May 2008, it was made clear that an internal company benchmark is only acceptable if the project developer is the only company that would be able to implement the project (eg a retrofit of an existing plant). Such a benchmark has to be consistent with the investment behaviour of the company in the previous three years. Internal company documentation is not acceptable as proof of the existence of barriers. These changes were made as a reaction the haphazard and non-transparent benchmark choices by several project developers. Common practice analysis is now only to include projects that are 'operational'.

The EB has approved many methodologies that require the use of the consolidated additionality tool, as the use of the tool was proposed by the methodology developers. Methodology developers did not want to run the risk of suggesting new approaches for additionality determination. So the application of the additionality tool become *de facto* mandatory. However, the mandatory character of the tool has repeatedly been rejected by the COP/MOP.[15] In the same spirit, the COP/MOP has consistently repeated the need for new proposals to demonstrate additionality.[16] But as nobody has yet come up with such proposals, the additionality tool still retains its precedence.

In October 2006, EB 27 approved a further 'Combined tool to identify the baseline scenario and demonstrate additionality'.[17] This tool is only applicable if the project developer is able to implement directly all potential alternative scenarios,[18] which is often not the case—for example a wind power plant developer does normally not have enough funds to set up a coal power plant. This tool has not therefore been applied frequently. It has four steps (see Figure 12.3) and eliminates the possibility of choosing between the investment and the barrier test:

1. Identification of alternative scenarios.
2. Barrier analysis, with the aim to eliminate alternative scenarios which are prevented by the identified barriers.
3. Investment analysis for the remaining alternatives.
4. Common practice analysis.

[15] Decision 12/CP.10, paras 9, 28; decision 7/CMP.1, with the latter stating that this even applies to methodologies that require use of the tool.
[16] Decision 7/CMP.1, paras 25–7, decision 1/CMP.2, para 16(c).
[17] EB 27, Annex 9.
[18] EB 28, Annex 14.

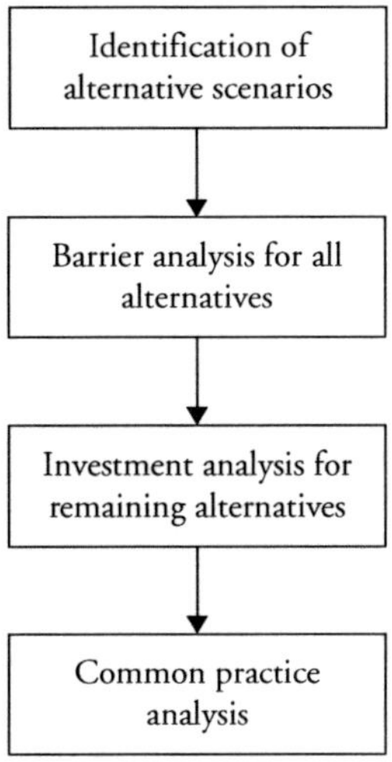

Figure 12.3. Steps in the combined baseline scenario identification and additionality tool

Originally, the applicability conditions were more stringent as options had to be 'under the control' of the project developers, ie the developer would be the only actor able to implement them.

Overall, it can be said that the consolidated additionality tool has become the *de facto* standard for additionality testing under the CDM and has increased in stringency over time. This is surprising given that many interest groups would benefit from a more lenient application of the additionality rules and that the COP has repeatedly called for alternative additionality tests.

As no project developer has been willing to submit new additionality tests, the EB has started to embark on simplified additionality determination methods. This was not done by specifying new additionality tests but through new baseline methodologies for specific project types.

4. Benchmarks and Free Rider Default Additionality Parameters

For energy efficiency projects, two revolutionary forms of additionality testing have been allowed in late 2008 in the context of two newly approved baseline methodologies.

The small-scale methodology—II.J—for projects distributing energy efficient lighting does not require an additionality test and applies a discount of 5% to cater for lamps that would have been bought by users even without the CDM project. The Methodology Panel had proposed a discount of 15%.

The methodology AM 0070 for energy-efficient refrigerators defines a benchmark. Any refrigerator with an efficiency that beats the benchmark is automatically deemed to be additional. The benchmark is defined as the lower level of the best 20% of refrigerators sold in the host country or the average efficiency of the manufacturer's refrigerators produced during the last three years before the start of the project. To take into account the autonomous energy efficiency improvement during the lifetime of the project, the benchmark has to be reduced by 3.5% for each year of the crediting period or can be monitored *ex post*. The idea that the benchmark could serve as additionality test was introduced by the Methodology Panel; the methodology developer had proposed to apply the consolidated additionality tool. The possible role of benchmarks in additionality determination is discussed further in section 7 below.

5. Case Law—Lessons From Review and Rejection of Projects

When one looks at the additionality tests, the non-specialist will have difficulty in recognizing the changes made over time. However, registration practice of the EB with regard to additionality determination has changed substantially, sometimes even dramatically in the last three years. When the first CDM projects came in for registration from late 2003, the EB lacked the manpower to do a thorough assessment of their additionality. Moreover, the EB trusted the DOEs to weed out non-additional projects in the validation process. The validators did not however assess the project developers' arguments in a critical manner. Therefore, between 2005 and 2007 large numbers of non-additional projects were validated and registered.[19] However, as it became clear that many non-additional projects were being validated and automatically registered, the EB introduced a new layer of checks. In early 2006, after the first controversies about an overly lax registration policy and the poor quality of the validation reports surfaced, the EB set up the Registration and Issuance Team (RIT) for assessment of Project Design Documents (PDDs) and validation reports. The RIT consists of a number of experts who are randomly allocated to projects submitted for registration. From a number of 20, its membership was successively expanded to 34. The introduction of the RIT immediately led to the identification of many problematic projects and the first rejections of projects due to lack of additionality. Moreover, the UNFCCC Secretariat commissioned two reports on the additionality of projects. These reports were never published but according to well-informed sources they confirmed the RIT assessments. A

[19] See A Michaelowa and P Purohit, 'Additionality determination of Indian CDM projects. Can Indian CDM project developers outwit the CDM Executive Board?' Discussion Paper CDM-1, (Climate Strategies, London, 2007) as well as L Schneider, 'Is the CDM fulfilling its environmental and sustainable development objectives? An evaluation of the CDM and options for improvement' (Oeko Institut: Berlin, 2007).

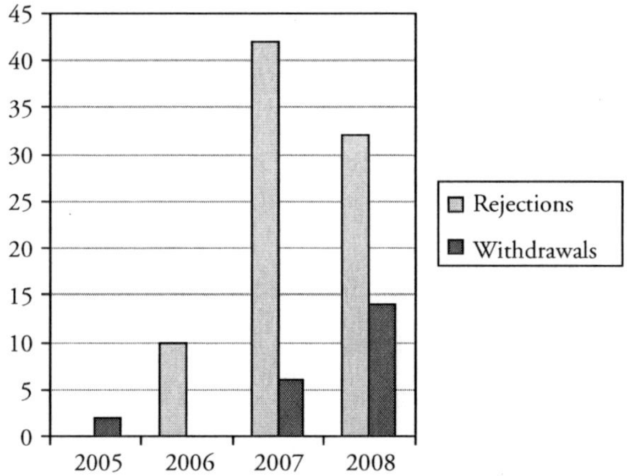

Cut-off date of analysis: 1 December 2008

Figure 12.4. Rejections and reviews over time (as a percentage of submissions for registration

further increase in the stringency of additionality testing took place in early 2007 when the UNFCCC Secretariat had hired enough staff to allow another layer of scrutiny after the RIT. This shows some mistrust of the RIT assessments. In fact, the RIT members had never undergone any training, nor has the EB given any guidance to RIT members. The consequence was a rapid rise in requests for reviews, reviews, and rejections (see Figure 12.4). In mid-2007, the media started to report on low credibility of emissions credits from CDM projects[20] which put pressure on the CDM regulators at all levels. Therefore, the cumbersome three-layer process of scrutiny has so far been retained.

Lack of additionality accounted for 57% of rejections by December 2008. Challenges abound in all steps of the additionality test, as discussed below and shown in Figure 12.5. While PDD quality has on average improved considerably since 2005, the interpretation of the additionality test has also been strengthened.

5.1 Problems with the investment test

Regarding the calculation of financial parameters for the investment analysis, often not all investment costs and revenue parameters, discount rates, and time horizon are specified in the PDD that are required to derive the internal rate of return (IRR) transparently for all relevant alternatives. Therefore, in May 2008 a

[20] N Davies, 'Abuse and incompetence in fight against global warming', *The Guardian*, 2 July 2007.

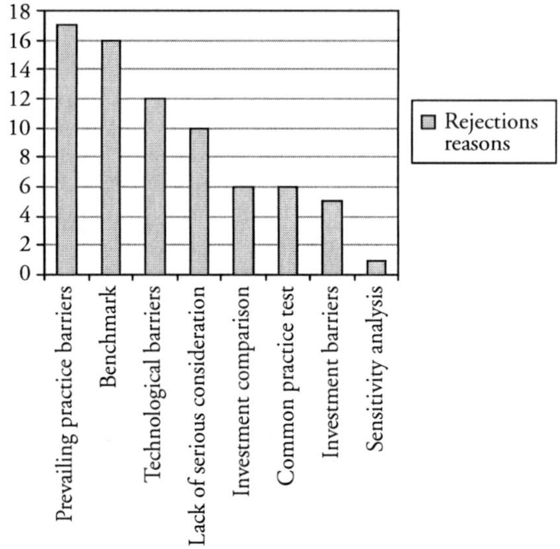

Cut-off date of analysis: 1 December 2008

Figure 12.5. Specific reasons for rejections of projects due to lacking additionality

detailed interpretive guidance on the implementation of the investment test was added to the additionality tool.[21] It specifies that all data should be of the time when the decision to invest was taken. The investment test should be applied for a period of at least 10 but less than 20 years. The length of the period has to be justified. The residual value of the equipment at the end of the period has to be taken into account. Risk premiums applied for returns on equity cannot be derived from general stock market returns. Sensitivity analysis is only to be done for 'heavy' parameters that generate more than 20% of either costs or revenues. The validator has to assess the reasonable character of the variation; it should generally be at least 10%. This guidance should enable developers to prevent simple mistakes that in the past led to rejection of several projects.

An example of the situation prevailing in the early days of the CDM is given by two wind power projects (UNFCCC no 0221 and 0224) which were the first projects to be rejected in 2006. While the PDD argued that the projects were commercially unattractive, the project developer Bajaj Auto described the bundle of projects as follows in its public annual report for 2001/2002: 'The project is extremely beneficial on a standalone basis and has a payback period of three years with an internal rate of return in excess of 28%. In addition to hedging Bajaj Auto's power costs, this investment also provides sales tax incentives and

[21] EB 35, Annex 35.

Box 12.2. Registration of wind power projects despite high IRR

A 125 MW wind project in the Indian state of Karnataka (UNFCCC no. 0315) applied the benchmark test and argued that its IRR was 7.3%. The IRR calculation did not take into account that wind energy investments attract accelerated depreciation of 80% in the first year and get a 10-year income tax holiday IRR in PDD. IRR with these tax benefits and realistic investment costs was calculated by an independent observer to be 22%, which was communicated to the EB during a review. Nevertheless, the project was registered.

an income tax shield.'[22] CDM or carbon credits were not mentioned in the report. However, a large number of wind projects with similar characteristics that had avoided bragging in publicly available documents, were registered (see Box 12.2).

Waste heat recovery projects in heavy industry were generally registered until mid-2008 even though some projects exhibit investment characteristics that put their additionality in doubt. One particularly striking and large example is discussed in Box 12.3; similar problems have come up with several projects in China.

5.2 Challenges in specifying benchmarks

An issue that has gained prominence in late 2008 is the choice of a benchmark for the investment analysis. Benchmarks have been contested right from the beginning of investment testing. Initially, internal company hurdle rates were used without any underlying documentation. A key benchmark problem—the question for which industry sector a benchmark is appropriate—surfaced in the case of a biomass power plant in Malaysia (UNFCCC no 1014) in 2007. The project proponent had argued that the minimum IRR used in the palm oil

Box 12.3. Waste heat recovery projects registered despite artificial transfer pricing

The JSW Vijayanagar Steel plant in India uses waste gas for electricity production (projects UNFCCC no 0325 and 350). JSW Steel operates the steel plant, JSW Energy the power plants. Due to regulatory reasons in the Indian power sector, JSW Steel charges JSW Energy a transfer price for the waste gas equal to the coal price that would have been paid for coal delivered to the power plants, despite being able to provide the gas at no cost apart an investment in a gas storage tank which pays off after just 100 GWh of electricity produced from waste gas. The projects were registered without a review in early 2007.

[22] Bajaj Auto, 'Annual report 2001–2002 (Pune, 2002).

industry should be a valid benchmark for an investment in the electricity supply industry, based on the underlying assumption that 'project activities under similar conditions developed by the same company should be allowed to use the same benchmark'. However, according to the EB, the proponent failed to properly demonstrate why and the project was rejected. Since EB 42, some large Chinese waste heat recovery projects were rejected because they used the benchmarks of the industry sectors of which the project developers were part (eg steel industry, coke industry), which are higher than the power industry benchmark. Traditionally, waste heat recovery projects in the Chinese iron and steel industry have used a benchmark of 13% IRR, and those of the coke industry 12%. The EB's interpretation is now that if the power is predominantly exported to the grid the power sector benchmark of 8% has to be used.

The EB has also explicitly addressed the level of certain benchmarks. In the case of renewable power projects in India, the EB rejected the commonly used 16% benchmark as inappropriate.[23] Since then the inflow of such projects has been reduced substantially. The 2008 guidance for investment testing specifies that benchmarking is only allowed if the baseline scenario does not require an investment; it is generally allowed in projects that could use grid electricity. The benchmark has to be appropriate to the type of IRR chosen, ie lending rates for project IRR and required returns on equity for equity IRR. The latter rule is leading to a lot of difficulty in a situation where the share of equity and loans is not finally known.

5.3 Problems with the barrier test

The barrier test was envisaged to allow projects that are highly attractive on paper, but face prohibitive non-monetary barriers to get CDM registration. Initially project developers often listed a host of barriers in very general terms and did not provide an explanation why the barriers listed were prohibitive. Generic arguments about technological and prevailing practice barriers have often not been found adequate by the EB. For example, the argument that boilers able to fire bagasse face a technological barrier in Brazil or India has not been accepted due to the high number of companies able to install such boilers. Due to a substantial number of rejections of projects using the barrier test from 2007 onwards, project developers have refrained from using the barrier test. In particular, projects using methodology ACM 0005 on blending of cement with fly ash or slag suffered a sudden jump in their rejection rate. Their number has subsequently diminished dramatically (see Figure 12.6).

A typical rejection case is shown in Box 12.4.

Several baseline methodologies now prohibit use of the barrier test and the EB is contemplating guidance that would considerably strengthen the barrier test for

[23] EB 40, para 40.

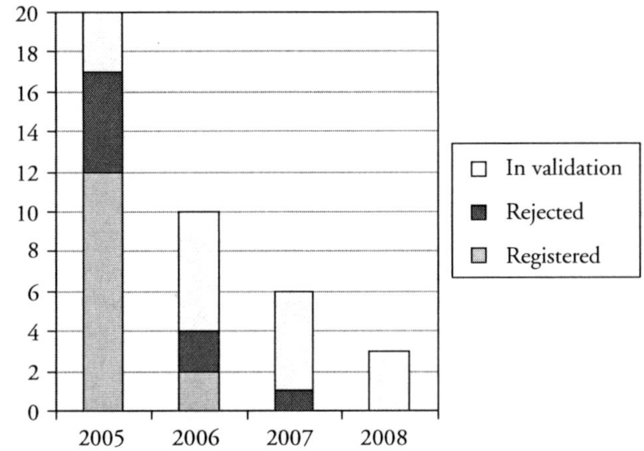

Figure 12.6. Cement blending blues

Note: The figure shows the number of projects submitted each year for validation comments.

highly profitable projects. In the version proposed by the Methodology Panel, project developers would have to show either that the project is a first-of-its-kind in the relevant region or country, that at least one barrier cannot be directly alleviated or otherwise affected by the high financial revenue of the project but will be alleviated by the CDM, or that bank loans, other debt or equity financing could only be obtained after the benefits of the CDM have been taken into account.

Box 12.4. Reasons for rejection of cement blending project

A project blending with blast furnace slag in Brazil (UNFCCC no. 0754) listed the following barriers:

- Development of logistics for additives supply is costly and difficult.
- Use of slag increases the production costs of cement.

However, the project developer argued in an external report that use of additives enhanced profitability. When a request for review was made, the company argued that long distance transport of slag increased its costs. However, the data provided could not corroborate this argument.

5.4 Lack of 'serious consideration'

CDM project developers have always had to report the starting date of the project activity in the PDD, which is defined in the glossary of CDM terms as 'implementation or construction or real action of a project activity'. The project start date was important in the early phase of the CDM when projects had to prove that they did not start before 1 January 2000.[24]

Moreover, projects submitted for registration before a certain cut-off date, which was moved by the COP and the EB several times,[25] were allowed to get CERs from their start date, even if they would be registered much later. Step 0 of the consolidated additionality tool had specifically been made for such 'early start' projects, which had to prove 'that the incentive from the CDM was seriously considered in the decision to proceed with the project activity. This evidence shall be based on documentation (preferably official, legal and/or other corporate) that was available at, or prior to, the start of the project activity'. However, due to capacity constraints, the EB did not check the evidence in detail and no 'early start' project was rejected due to lack of 'serious consideration'. In September 2005, the EB stated that a project can be registered as a CDM project after it has started, but would get CERs only from the date of registration.[26] This led to many submissions of projects that had started many years earlier.

Obviously, the additionality of a project is doubtful if its developer has been able to implement the project and suddenly discovers the CDM years later. Thus lack of proof that the CDM was seriously considered when initiating a project has come up as an argument for rejections since early 2008. EB 38 in March 2008 stated that 'several project activities, which were already in operation, could not be registered due to a lack of credible evidence on the role of CDM at the starting date of the project activity' and 'reaffirmed, as indicated in section B5 of the "Guidelines for completing the project design document", that project activities which have a starting date before the date of validation shall provide evidence that the incentive from the CDM was seriously considered in the decision to proceed with the project activity'.[27] In August 2008, a new definition of 'start date' was given as the date on which the project developer has committed to expenditures related to the implementation or related to the construction of the project.[28] If a project claims to have been stopped and only been resuscitated by the CDM, 'cessation of project implementation must be demonstrated by means of credible evidence such as cancellation of contracts or revocation of government permits'[29]. Such resuscitations have frequently been invoked in the case of Chinese hydropower plants.

[24] In fact, two large projects (a gas flaring reduction project in Equatorial Guinea, UNFCCC no 0972, and a gas power plant in Argentina, no 0443) were rejected due to having already started in the late 1990s.

[25] Initially, the Marrakech Accords had set the cut-off date at end of December 2005. It was moved to March 2007 by a decision of COP 12.

[26] EB 21, para 63.

[27] EB 38, para 59.

[28] EB 41, para 67.

[29] For such projects, investment analysis should reflect the economic decision making context at point of the decision to recommence the project. Therefore capital costs incurred prior that date should be determined as recoverable value of the assets (EB 41, Annex 45, para 7).

5.5 Problems with the common practice test

With respect to the common practice test, developers sometimes use a very narrow definition for assessment of similar projects. Somewhat surprisingly, problems with the common practice test have been invoked only rarely. EB practice shows that projects that are deemed important by an EB member due to their large contribution in the CDM pipeline of their country are not rejected even if exhibiting very doubtful additionality characteristics. This has been the case with hydro and gas power plants in China as well as waste heat recovery in the Indian steel sector. However, in May 2008 validator TÜV Süd publicly rejected the validation of a Chinese hydro project after it became clear during a request for review that the project developer had manipulated the project start date as well as the investment calculations.

5.6 The Validation and Verification Manual

Although the EB has released further guidance on the implementation of the additionality tool, according to project developer lobby groups, there is still uncertainty and a lack of clarity among the project developers.[30] This is of even higher importance as the number of projects with a more complex demonstration of additionality, like demand-side energy-efficiency or transport projects, is likely to increase in the near future. The EB decided to address this situation by adoption of a Validation and Verification Manual (VVM),[31] which contains a set of requirements designed to make validation of additionality more credible. The validator has to use local knowledge as well as sectoral and financial expertise for the additionality check.[32] It has to show how it checked the authenticity of documentation.[33] Serious consideration is covered in a whole section 'Prior consideration of the clean development mechanism'.[34] For assessment of the investment test, expertise in relevant accounting practices is required.[35] Barrier validation has to be done using independent sources.[36] In the case of a high number of similar projects, 'essential distinctions' to the CDM project have to be proven to fulfil the common practice test.[37] If a region different from the host country is chosen for common practice testing, the explanation for this choice has to be assessed.

6. Current Proposals for Additionality Reform

Ever since the agreement of the Kyoto Protocol, project developers, researchers, NGOs, and CDM regulators have fought over the interpretation of

[30] IETA 2007, 'State of the CDM, Position on CDM for COP/MOP3' (Geneva, 2007).
[31] EB 44, Annex 3. [32] VVM, para 93. [33] VVM, para 95.
[34] VVM, paras 96–102. [35] VVM, para 109(a). [36] VVM, para 115(a).
[37] VVM, para 118(c).

additionality. Despite the repeated statements of the Conference of the Parties that the additionality tool is only a voluntary instrument and that other approaches for additionality determination are also possible, no one has submitted a new approach and the tool is now universally applied. With the Validation and Verification Manual, a further streamlining of additionality testing is achieved. With growing experience, further shortcomings of current additionality testing will be discovered. They can be addressed by incremental reforms. Some stakeholders would prefer a radical overhaul of additionality testing. Both paths are discussed below.

6.1 Incremental reform of the additionality tool—the WWF proposal

A very detailed proposal submitted by WWF in 2007 builds on the consolidated additionality tool.[38] It demands that the tool should be made mandatory. All parameters used in the investment test should be checked by a local expert for this type of projects contracted and paid directly by the CDM EB. The discount rate used for conversion of future costs and revenues in the investment test should be the average of the one-year benchmark interbank rate and low-risk government bond rate with a lifetime similar to the project of the most liquid government bond market denominated in the currency which is used for the majority of the project investment. The investment barrier should require written proof from the three largest commercial banks in the host country and one international commercial bank that they are not willing to provide a loan or other financing to the project despite its high IRR. Moreover, the financing agent of the project should provide a letter with a detailed explanation why it became interested in funding the project under CDM. The investment test should be done on the basis of the investment comparison test, for which the project developer has to use alternatives that are real investments done in the last five years. If it has not done any investments, publicly available data from comparable investments have to be used. Only if such data are not available can the benchmark test be used. The benchmark should be a financial product (eg government bond) with a lifetime comparable to that of the project. If such a product does not exist, publicly available rates of return of commercial investments with a similar risk structure and lifetime as the project should be used. Generally, equity IRR calculated for the entire lifetime of the project is the most appropriate parameter to be used in the investment test, as long as IRR is not limited by host country regulation. In the investment test, minimum performance parameters for specific technologies should be defined. Transfer prices

[38] A Michaelowa, 'Submission to the UNFCCC regarding best practice examples on the demonstration of additionality to assist the development of project design documents, in particular for small-scale project activities', (2007), available at <http://cdm.unfccc.int/public_inputs/dev_pdds/cfi/dyibgvnxf91f7o2wxvd1wbyumlhady>, accessed 15 June 2009.

between companies belonging to the same group should be disregarded or the investment test done at the group level.

Also, the barrier test can be improved considerably. The technology failure risk barrier should require a statement by an internationally accredited technical certification body about the process /technology failure risk in the circumstances of the host country in terms of downtime during the crediting period. On the basis of this downtime and the resulting loss of output, the investment test should have to be applied. The small-scale projects' 'other barrier' test should require confirmation by a local expert for this type of project who is contracted and paid directly by the CDM EB. The small-scale project technological barrier test should require proof that the project will be less financially attractive than a less technologically advanced alternative. Moreover, the barrier regarding non-availability of skilled labour should require proof that no education/training institution in the host country provides the needed skill and that no expatriate workers with these qualifications could reasonably be hired in that host country due to security reasons. The lack of infrastructure barrier should require proof showing that the infrastructure is not there and that the project would not pass the investment test if the project developer had to provide the infrastructure himself.

The common practice test should require that except for publicly supported projects, no project of this type is currently operational in the host country. It should only be applied to projects that have undergone and passed the barrier test. The broadest possible definition of each technology should be used for the common practice test.

While some projects could face problems in providing the required documentation and transaction costs would rise due to the need to hire local experts, the WWF proposal would avoid the problematic principle of self-documentation by project developers, which has been one of the major reasons of media criticism of the CDM.

6.2 Benchmarking[39]

Different stakeholders from industry have proposed benchmarking as a tool for the demonstration of additionality.[40] Most of the proposals imply a double purpose of the benchmark. As illustrated in Figure 12.7, projects beating the benchmark would automatically be considered additional, while the benchmark would also serve as the baseline against which emission reductions achieved by

[39] This section benefited from input by Nicolas Müller and Martina Jung (Ecofys).

[40] See eg the Cement Sustainabiliy Initiative described in B Vanderborght, 'CO_2 cement industry: Getting it right', presented at IEA Forum, 01 October 2007, Paris, available at: <http://www.iea.org/textbase/work/2007/tracking/vanderborght_cement.pdf> and World Business Council for Sustainable Development (WBCSD), The Greenhouse Gas Protocol Initiative, Calculation tools (2008), available at <http://www.ghgprotocol.org>, accessed 10 January 2009.

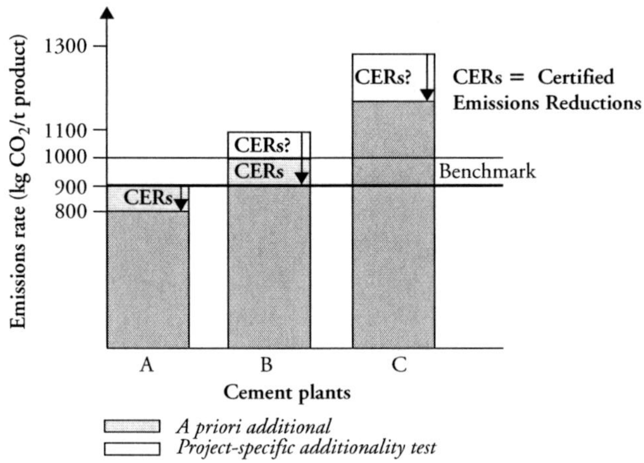

	kg/ton	kg/ton	kg/ton
Benchmark level	1000	1000	1000
Actual emissions	900	1100	1300
Emissions reduction	100	200	200
Emissions after projects	800	900	1100
CERs a-priori additional	100	100	–
CERs project-specific additional test	–	≤100	≤200

Figure 12.7. Common additionality and baseline benchmark in the cement industry
Source: Bedoya, M (2005): The Cement Sustainability Initiative—Benchmark approach to CDMs, presentation held at COP11/MOP 1, Montreal.

the project would be calculated.[41] This approach has been retained in the case of the methodology AM 0070 for energy-efficient refrigerators (see above).
The focus of such an approach is on the selection of the baseline scenario. It assumes that a project is automatically additional if an approved baseline methodology has been followed and the emissions in the project case are lower than those in the baseline scenario.[42] The concept of a single benchmark for baseline and additionality tests can be considered as problematic, though, as it is not able to avoid the generation of 'fake' CERs due to projects which would have happened anyway.

In a more stringent version of the same concept, the additionality benchmark would be lower than the one determining the baseline (the black line in Figure 12.7). Those projects beating the additionality benchmark would be considered automatically additional and would use the higher baseline benchmark for the

[41] See eg D Broekhoff, 'Expanding Global Emissions Trading: Prospects for Standardized Carbon Offset Crediting' (IETA: Geneva, 2007).
[42] IETA, 'IETA Position on additionality' (Geneva, 2006).

calculation of emission reduction (project A in Figure 12.7). Further criteria, such as the use of particular technologies or methods, could be applied. If the project, however, does not reach the additionality benchmark but achieves emission reductions below the baseline benchmark, it could be considered that the project could demonstrate its additionality on a project-specific basis (project B in Figure 12.7). Then, CERs could be generated if the project proves to be additional. Projects not reducing emission intensities below the baseline benchmark could not earn any credits.

However, further questions arise as to the application of the project-specific additionality test in combination with benchmarking. When benchmarking multiple and possibly interrelated measures—eg waste heat recovery and using fly ash as an additive at a cement plant—their emission-reduction activities and additionality are 'bundled' at a facility level. Although the coverage of multiple measures is an advantage of the benchmarking approach, it is likely to check the additionality of a single component of such bundled efforts. In summary, it can be concluded that, even when separating the baseline and additionality benchmark, it is not guaranteed that projects reducing below the additionality benchmark would not have happened anyway.

The stringency level of the additionality benchmark is one of the essential elements with regard to using benchmarks for additionality testing. The level of the additionality benchmark could be set conservatively in order to minimize the risk of non-additional projects.[43] This, however, relies on the assumption that additionality is closely correlated with low carbon intensity, which is not necessarily the case. In spite of this, it can be assumed that, in general, a more stringent benchmark allows fewer non-additional projects to generate CERs. However, at the same time, setting the additionality benchmark at a more stringent level increases the number of additional projects which are no longer eligible. If such an additionality benchmark is to be considered at all, the stringency level would have to be justified by a detailed analysis drawing on a similar argumentation to that used in the current additionality tool. In setting the stringency level, one should try to balance on the one hand the additional amount of non-additional projects with, on the other hand, those opportunities lost because otherwise additional projects become non-additional. Furthermore, it might be necessary to set country or region-specific benchmarks in order to take into account specific circumstances in countries and regions. For a number of projects, the generation of CERs is doubtful, as there is a range of uncertainties regarding the demonstration of additionality of a project. A rather lax additionality benchmark would, of course, ensure that more of the respective projects would be eligible as CDM projects. The volume of CERs would in this case,

[43] See PROBASE, 'Procedures for Accounting and Baselines for JI and CDM Projects, Final Report' (Groningen, 2003): available at <www.jiqweb.org/probase/prob_fr.pdf>, accessed 15 June 2009, and M Lazarus, S Kartha and S Bernow, 'Key Issues in Benchmark Baselines for the CDM: Aggregation, Stringency, Cohorts, and Updating' (Tellus Institute, Stockholm Environment Institute: Boston, 2000).

however, exceed the real emission reductions achieved. One possible option could, for example, be to combine a lax benchmark with discounting of project emission reductions. The expected weaker signal for CERs might be sufficient for such projects, provided that there is a lower uncertainty regarding the project registration. The global volume of emission reductions achieved would still be equal or lower than the volume of CERs generated.

There is also an ongoing debate as to how to treat new plants when compared to existing plants. In some sectors new plants might be built with best available technology under business as usual. Thus, applying a benchmarking for new plants based on existing plants bears the risk of crediting emission reductions happening under business as usual. To avoid this, the level of performance required for the newly built plants to be additional as well as their crediting should be extremely stringent. The average level of the region is unlikely to be appropriate for testing additionality, as the average is the medium techno-economic optimum of the region without CDM. An adjustment to this level or to intensity barely lower than the average is unlikely to be additional, even for retrofitted plants. If there is an autonomous development of a certain technology leading to new plants being built with the best available technology, this might be a reasonable test for additionality. This is, however, not very likely. Such a stringent level, however, bears the risk of excluding projects which might have been additional—based on a project-specific additionality determination. This is because multiple barriers exist which prevent plants (new or retrofits) to actually achieve such levels. For retrofits, this level is likely to be too stringent. As such, the stringency level would defeat the purpose of attracting developers of emission-reduction projects. An intermediate solution would be the 'currently used best technology' criterion. Only new plants better than the former champion would be additional. For new plants, this stringency level might be reasonable in certain cases, depending on the sector and country/region-specific circumstances. For sectors in which greenfield plants are normally built at the best achieved level (proven technology of the former generation), this level might not, however, guarantee that projects are really additional. Furthermore, as mentioned above, there are continuous efficiency improvements under business as usual which might lead to a situation that the additionality benchmark based on new plants in the past has already become common practice. With such a benchmark, there is a risk that the whole generation of future new plants to be built in developing countries will automatically be additional and generate CERs. Such an approach might be justified but only if regional circumstances can be identified which lead to a performance of new plants below the best achieved level. Depending on the percentile chosen, a top percentile approach or hybrid approach can be a suitable stringency level for any new plants to be built. For example, a top 10 percentile approach often captures the most recent plants and can therefore be an appropriate level for additionality if new plants are not built at best achieved levels. A top percentile approach is certainly appropriate for retrofitted plants if the level is determined with a lot of care. Ideally, such a percentile should be neither too generous, nor too stringent for the determination of the additionality. However, if no regional differences are taken into account, an additionality benchmark for retrofitting projects might lead to very different and unfair results for different regions or countries.

From the discussion of the different stringency levels, it becomes apparent that an accurate determination of additionality through benchmarks needs some level of disaggregation (consideration of regional circumstances, different benchmarks for new and retrofitted plants, etc). Considering that a higher level of disaggregation increases transaction costs, a disaggregated additionality benchmark which might be considered sufficiently accurate might only in some sectors have significant advantages as compared to a project-specific additionality test. The advantages of the benchmarking approach include decreased uncertainty with regard to project registration and project development costs. In turn, benchmarking can be cost-effective if there is sufficient potential for the generation of CERs in the respective sector. Benchmarking offers moreover the possibility of correcting for perverse incentives, as benchmarks can be adjusted over time according to the development of commercially available technologies.

Another issue with regard to using benchmarks for testing additionality of CDM projects is that benchmarks are backward-looking. As a consequence, a benchmark used under the CDM will have to be updated on a regular basis in order to appropriately reflect the situation in the present. The main parameter to be determined here will be the frequency of the update of the additionality benchmark. Updating requirements will have to take into account the need to balance accuracy, technical and administrative costs of updating, and concerns about investor uncertainty.

7. Do Away with Additionality in The Long Term?

The key decisive issue for the long-term success of the CDM is the resolution of the additionality issue. The EB is now rejecting projects due to lack of additionality, but these projects are generally small and other—mostly large—projects with similar argumentation continue to pass. Despite the weak performance of validators on this issue, it was only in late 2008 that the first validator was suspended. Until mid-2008, project developers were becoming increasingly bold, submitting projects that had started long ago. Especially noteworthy are Chinese hydropower plants that report unrealistically low plant load factors and therefore manage to show low profitability. A proposal is currently pending with the EB to exempt all renewable energy projects from the common practice test. If all renewable plants in developing countries were submitted and registered under the CDM, there would be an explosion of CER supply (see Table 12.1).

Table 12.1. Renewable energy increase and related CER potential

Technology	Currently in CDM pipeline (GW)	Increase (GW) pa	Annual CER potential (million)	CER potential 2013–2020 (million)
Large hydro (China)	13.1	8	35	1260
Hydro (other countries)	9.2	4	14	496
Wind (China)	8.4	3	6.6	508
Small hydro (China)	2.4	2	9	324
Wind (India)	3.4	1.5	2.8	102
Wind (other countries)	2.6	0.5	0.8	29
Bagasse	3.4	0.5	2.0	72
Other biomass	3.6	0.5	2.6	92
Sum	*44.1*	*20*	*73*	*2883*

Source: Own calculations based on UNEP Risø Centre (2008) and recent expansion rates of renewable energy.

8. Additionality—Can the EB Stem the Tide of Non-additional Projects?

Had the EB followed the vague additionality guidance of the Marrakech Accords, the CDM would have been swamped with non-additional projects right from the outset. However, the EB took bold decisions very early to interpret additionality in the context of small-scale project rules, developed the consolidated additionality tool in 2004 and upheld it against repeated decisions of the COP that stressed the voluntary character of the tool. However, in the first three years of project registration, many projects were registered that had doubtful additionality attributes. In a series of three regulatory changes, the scrutiny of PDDs was sharpened—first the RIT was introduced in March 2006, then a second layer of Secretariat check of PDDs started in earnest in early 2007, and in late 2008, the VVM was adopted. The additionality test was significantly strengthened. As a consequence, rejection rates have increased to 10%, the majority of which are a result of missing additionality. A disturbing feature of the regulatory system are sudden 'fashions', where from one day to the next certain project types always get rejected or a certain regulatory parameter suddenly takes centre stage. Figure 12.8 shows these trends.

A key reason for the emergence of such 'fashions' is the 'arms race' between the project developers and the regulators. Regulators learn to understand and detect arguments which have low credibility as more projects of a particular type with generally critical additionality enter the registration pipeline. Project developers try to apply the rules to the letter and frequently find loopholes.

With the acceptance of benchmarks and default discount factors that replace project-specific additionality testing for certain project types, a new chapter in

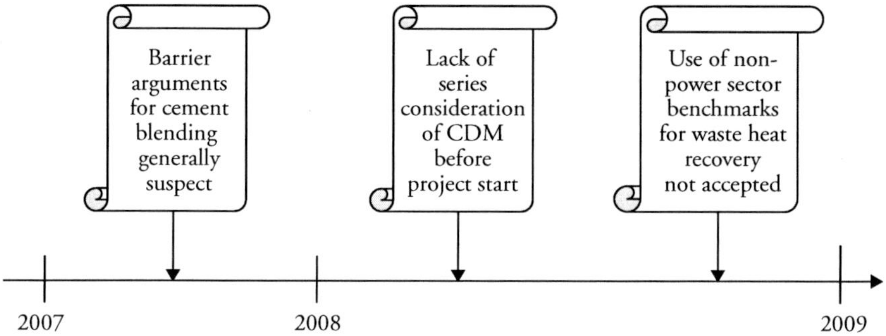

Figure 12.8. Additionality 'fashions' of the EB

Source: Bedoya, M (2005): The Cement Sustainability Initiative—Benchmark approach to CDMs, presentation held at COP11/MOP 1, Montreal.

additionality determination has been opened. It is still too early to determine whether these new additionality tests can remain credible over time. It is clear that there will not be a magic solution for the additionality conundrum. The only decisive solution would be participation of all countries in a system with mandatory emissions commitments. In such a system, additionality would no longer matter for environmental credibility.

13

Responsibility for the Environmental Integrity of the CDM: Judicial Review of Executive Board Decisions

*Christina Voigt**

1. Introduction

This chapter discusses the responsibility of the Executive Board (EB), a UN body set up under the Kyoto Protocol (KP) to govern the environmental integrity of the Clean Development Mechanism (CDM). As there is increasing evidence that a high level of environmental integrity has not yet been achieved, the issue dealt with in this chapter is the possibility for judicial review of EB decisions, both within the international climate regime by a review panel (internal review) and separate from it (external review).

This issue is the flipside of another question already dealt with in legal literature: do private participants in the CDM have a remedy against an administrative decision of a CDM institution if that decision has affected private commercial interests?[1] Yet, while complaints from private legal entities suffering economic losses have sparked a lively debate as to the immunity of CDM institutions,[2] and have led to involvement of the United Nations Office of Legal

* The author would like to thank Jolene Lin, Francesco Sindico, Professor Hans Christian Bugge, and the editors of this book for helpful comments. All mistakes remain the author's sole responsibility.

[1] E Meijer, 'The International Institutions of the Clean Development Mechanism Brought before National Courts: Limiting Jurisdictional Immunity to Achieve Access to Justice' (2007) 39 NYUJ Int'l L and Pol 873–928; C Streck and J Lin, 'Making Markets Work: A Review of CDM Performance and the Need for Reform' (2008) European Journal of International Law 409–42; and J Werksman, 'The "Legitimate Expectations" of Investors and the CDM: Balancing Public Goods and Private Rights under the Climate Change Regime' (2008) 1 Carbon and Climate Law Review 95–104.

[2] See UNFCCC 'Privileges and Immunities for Individuals Serving on Constituted Bodies under the Kyoto Protocol: Implementation of Decision 9/CMP.2' (13 November 2007) UN Doc FCCC/KP/CMP/2007/2; UNFCCC 'Privileges and Immunities for Individuals Serving on Constituted Bodies Established under the Kyoto Protocol' (4 October 2006) UN Doc FCCC/SBI/2006/21.

Affairs,[3] not much attention has been directed to vested legal interests in the environmental integrity of the CDM.

The involvement of private market investors in the CDM can have the effect of diverting the mechanism from delivering strong environmental results. The environmental integrity of the mechanism not only depends on strong rules for, and clear guidance to, market participants, but it also depends on the possibility of review of the decisions of UN-CDM institutions for compliance with these rules.

2. The Importance of the CDM

As a market mechanism of the UN climate change regime the CDM has ostensibly been a success story. By encouraging climate projects in developing countries that have no quantified greenhouse gas (GHG) emission reductions targets, the CDM has facilitated a global partnership between countless actors to finance emission reductions in such countries. The mechanism has become a vital element of the international carbon market. It has in a very short period mobilized and channelled significant investment flows into the developing world. Most importantly, it has 'put a price on carbon' in developing countries. So far, it stands out as the only instrument to bridge the North-South gap which otherwise is a distinct feature of the current regime.

At the time of writing over 1400 projects have been registered with the Executive Board and more than 200 million Certified Emission Reductions (CERs) have been issued. It is expected that these projects will deliver 1,460,000,000 CERs by 2012.[4] Its central role in the carbon market ensures that the CDM remains an important post-2012 climate agreement option.[5]

3. Environmental Integrity

3.1 The significance of environmental integrity

In its current state, the CDM is an offset mechanism. It allows industrial (Annex-I) countries, to use emission reduction units (Certified Emission Reductions or

[3] UN Doc FCCC/KP/CMP/2005/6, and Office of Legal Affairs, United Nations, 'Memorandum: Transmitting the Response on Privileges and Immunities for Individuals Serving on Constitute Bodies Established under the Kyoto Protocol to the UNFCCC', UN Doc FCCC/SBI/2006/6, Annex.

[4] 'CDM: CDM Home' <http://cdm.unfccc.int/index.html>, accessed 23 February 2009.

[5] See B Mueller and P Ghosh, 'Implementing the Bali Action Plan: What Role for the CDM?' (Comment) (October 2008) Oxford Institute for Energy Studies: <http://www.oxfordenergy.org/pdfs/comment_1008-2.pdf>, accessed 2 February 2009; L Schneider, 'A Clean Development Mechanism with Atmospheric Benefits for a Post-2012 Climate Regime' (Discussion Paper) (25 September 2008) Öko-Institut <http://www.oeko.de/oekodoc/779/2008-227-en.pdf>, accessed 2 February 2009. Also Decision -/CMP.4 Further guidance relating to the clean development mechanism. See also chapter 26 below.

CERs) generated by projects implemented in developing countries to comply with their binding emission reduction targets under the Kyoto Protocol. In addition, European installations can use CERs to comply with their CO_2 emission caps under the EU emission trading scheme (EU ETS). Successful CDM projects in developing (non-Annex I) countries raise reduction targets in Annex-I countries with otherwise set emission caps. This effect makes it essential that, to maintain the integrity of the CDM and the entire carbon market, each CER corresponds to the reduction of one tonne CO_2 equivalent (CO_2 eq) which is real, measurable, long-term and additional. These requirements have, so far, exclusively determined the 'environmental integrity' of the CDM.

The market created by the CDM engages participants who each seek ways to maximize the number of CERs generated by a CDM project at lowest costs. It has been noted that rather than competing like typical market players, participants in the CDM have a common interest in demonstrating a greater level of emissions reductions from a project and exchanging a greater amount of CERs.[6] It is because of this situation and the need for a corrective system of checks and balances, that the governance of the CDM, including stringent regulation and independent validation and verification, becomes crucial.

Environmental integrity is also crucial to the legitimacy of the CDM as a tool in the global climate regime. The rules for the CDM are set out in Article 12 of the Kyoto Protocol and have been further elaborated in the CDM Modalities and Procedures (MP).[7] The importance of environmental integrity has already been made clear by the first Conference of the Parties serving as Meeting of the Parties to the Kyoto Protocol (CMP.1) when adopting the Marrakech Accords. In decision 2/CMP.1, the Parties emphasized that 'environmental integrity is to be achieved through sound modalities, rules and guidelines for the mechanisms, sound and strong principles and rules governing land use, land use change and forestry activities, and a strong compliance regime.'[8] CMP.3 demanded that the EB, the main administrative body in the CDM, 'strive to take concrete actions to improve, and where possible simplify, the operational aspects of the clean development mechanism, such as the review processes, ensuring that its environmental integrity is not affected'.[9] Similarly, CMP.4 encouraged the EB 'to

[6] C O'Brian and KN Ortega, 'Three Years after Implementation, the CDM Moves Markets, and Matures in New Directions' (Latham and Watkins, Client Alert 658, 2 January 2008, 5). <http://www.lw.com/resources.aspx?page=clientalertdetail&publication=2070>, accessed 2 February 2008.

[7] UNFCCC Decision 3/CMP.1, Annex I: Modalities and Procedures for a Clean Development Mechanism (30 March 2006) UN Doc FCCC/KP/CMP/2005/8/Add.1.

[8] 'Report of the Conference of the Parties Serving as the Meeting of the Parties to the Kyoto Protocol on its First Session' (Montreal 28 November—10 December 2005) UN Doc FCCC/KP/CMP/2005/8/Add.1, 4.

[9] UNFCCC Decision 1/CMP.3 'Report of the Conference of the Parties Serving as the Meeting of the Parties to the Kyoto Protocol on its Third Session, held in Bali from 3 to 15 December 2007, Part Two: Action Taken by the Conference of the Parties Serving as the Meeting of the Parties to the Kyoto Protocol at its Third Session' (14 March 2008) UN Doc FCCC/KP/CMP/2007/9/Add.1.

approve, more methodologies with broad applicability conditions, in order to increase the availability of different technologies and measures, ensuring ease of use without jeopardizing the environmental integrity of the clean development mechanism'.[10]

Ensuring environmental integrity is imperative to the success of the CDM. Given the CDM's important role in creating a global carbon market, the environmental integrity of the CDM is also crucial for the overall environmental performance of this market.

3.2 Definition of environmental integrity

Environmental integrity of the CDM is generally understood as requiring emission reductions generated by the CDM to be real, measurable, long-term, and additional to any that would have occurred in the absence of the project. Particularly, the requirement of 'additionality' stands central in discussions on environmental integrity. In simple words, an emission reduction project is 'additional' when the project goes beyond business as usual und thus deserves to be rewarded with tradable credits.[11] If a CDM project is not additional but nevertheless leads to the issuance of CERs, those credits result in an increase in global GHG emissions since they enable Annex I countries or companies to emit more GHGs than otherwise permitted under to the Kyoto Protocol or their domestic cap, whereas the emission reductions from the project in the non-Annex I country would have occurred anyway.

Demonstration of additionality is not straightforward. The controversial issue is that the issue as to whether emissions reductions would also have occurred in the absence of the project is counterfactual and needs to be measured against an assumed baseline. A baseline for a CDM project activity is a hypothetical reference case (the most plausible alternative scenario to the implementation of the CDM project activity), representing the volume of greenhouse gases that would have been emitted if the project were not implemented.[12] This baseline is used to determine the volume of greenhouse gas emission reductions achieved by a project activity.

In order to demonstrate additionality the so-called 'additionality tool' established by the CDM Executive Board or the 'combined tool for baseline selection

[10] Decision -/CMP.4 (2008) 'Further guidance relating to the clean development mechanism', para 33.

[11] Article 12.5 of the Kyoto Protocol defines a CDM project as additional if 'anthropogenic GHG emissions are reduced below those that would have occurred in the absence of the registered project activity.' Kyoto Protocol to the United Nations Framework Convention on Climate Change (adopted 11 December 1997, entered into force 16 February 2005) 37 ILM 22 art 12.5.

[12] A 'baseline scenario' for a CDM project activity is defined in 3/CMP.1, Annex, para 44 as: 'the scenario that reasonably represents the anthropogenic emissions by sources of greenhouse gases that would occur in the absence of the proposed project activity.' (3/CMP.1, Annex (n 7), para 44). See further chapters 11 and 12 above.

and demonstration of additionality' (combined tool) can be used.[13] These tools involve an analysis of various barriers facing the implementation of projects to show that if not for the sale of CERs, a project activity would not be implemented. It must also be shown that implementation of the project was not required under any mandatory law or policy that is widely enforced.

Yet, while much attention around environmental integrity of the CDM has been directed to the additionality criterion, *other* (direct or indirect) environmental effects have so far not been included into the definition of environmental integrity. The current design of the CDM credits projects which lead to GHGs reductions. Direct or indirect impacts of the project activity on the environment are neither credited nor discredited. This is particularly controversial with regard to projects that potentially can lead to decline in water quality or quantity of rivers,[14] and/or have an impact on land or river ecosystems, or affect biological diversity.[15]

On the EU level, the European Parliament recently suggested the adoption of quality criteria for the use of CERs in the EU ETS. From 2013 onwards, only high quality CERs from countries which have ratified a future international agreement on climate change shall be accepted in the Community scheme.[16] Such high quality CERs are 'credits from projects used by operators [that] represent real, verifiable, additional and permanent emission reductions and have clear sustainable development benefits and *no significant negative environmental or social impacts*'.[17] In other words, from 1 January 2013 quality measures may be applied to restrict the use of credits from specific project types. EU institutions suppose that such a quality control mechanism is needed to assure the environmental integrity of future project types.[18]

[13] UNFCCC CDM Executive Board 39th Meeting, 'Annex 10: Methodological Tool 'Tool for Demonstration and Assessment of Additionality' (Version 05)' <http://cdm.unfccc.int/eb/039/eb39_repan10.pdf>, accessed 6 February 2009).

[14] This is particularly relevant in the context of large hydro-power projects.

[15] See for impacts in the context of reforestation and afforestation projects: Y Smith, 'Afforestation and reforestation in the clean development mechanism of the Kyoto Protocol: implications for forests and forest people' (2002) 2 Int. J. Global Env Issues 3/4, 322–43; and J Mackensen, 'Environmental and socio-economic impacts of afforestation and reforestation activities (AR) under the Clean Development Mechanism (CDM), UNEP, <http://www.adb.org/Documents/Events/2003/reg_Workshop_Forests_Climate_Change/E&S-Impacts-paper-Mackensen.pdf>, accessed 23 February 2009.

[16] 'Position of the European Parliament adopted at first reading on 17 December 2008 with a view to the adoption of Directive 2009/.../EC of the European Parliament and of the Council amending Directive 2003/87/EC so as to improve and extend the greenhouse gas emission allowance trading system of the Community', P6-TC1-COD(2008)0013.

[17] Paragraph 29 (my emphasis). Ibid. See also paras 31 and 32. Ibid. 'Position of the European Parliament adopted at first reading on 17 December 2008 with a view to the adoption of Decision No .../2009/EC of the European Parliament and of the Council on the effort of Member States to reduce their greenhouse gas emissions to meet the Community's greenhouse gas emission reduction commitments up to 2020', P6_TC1-COD(2008)0014, para 11.

[18] 'Questions and Answers on the revised EU Emissions Trading System', MEMO/08/796, Brussels, 17 December 2008, <http://europa.eu/rapid/pressReleasesAction.do?reference=MEMO/08/796>, accessed 24 February 2009.

No such qualitative criteria have yet been adopted at the UN level. If environmental integrity is to be a central element of the CDM, it is necessary to define it in a wider sense rather than exclusively in the context of additionality, including environmental benefits or, at least, the absence of negative environmental impacts. As a starting point, environmental integrity of the CDM ought to be defined so as to require emission reductions that are real, measurable, long term, and additional, *and* which do not lead to environmental harm directly or indirectly caused by the project activity. It needs to be noted that, while the current definition of environmental integrity is based on the text of Article 12 KP and the implementing COP decisions, the proposed expansion of the definition goes beyond the existing understanding. Yet, nothing prevents a legal interpretation of 'environmental integrity' so as to include the absence of negative environmental impacts caused directly or indirectly by a CDM project.[19]

Avoiding detrimental environmental side-effects of a CDM project is not only a requirement of environmental integrity but can also be linked to the CDM's purpose of contributing to the sustainable development of developing countries.[20] In contrast to environmental integrity of CDM projects, which is to be achieved through sound modalities, rules and guidelines, and ensured by UN CDM bodies, no such legal security exists for sustainable development benefits. Grounded in concerns relating to national sovereignty, it remains the prerogative of CDM host countries to determine whether a project contributes to sustainable development. No common indicators or international standards have been accepted for measuring sustainable development. According to an evaluation by the Dutch Government, practice has shown that host countries have not been very critical in approving CDM projects with little or no added 'sustainable' value. They often define the CDM's contribution to sustainable development in line with existing national development strategies, which more often than not are based on economic growth rather than sustainability.[21] Although some countries, such as China, India, and Brazil, have set up local standards for sustainable CDM projects, they either fail to include verifiable indicators to measure the outcome, or do not have the means for monitoring or enforcing the sustainability benefits.[22] Two recent reports also highlighted

[19] In case an expanded definition of environmental integrity of the CDM, ie requiring that projects lead to real, verifiable, additional and permanent emission reductions and cause no negative environmental impacts, is seen as to exceed the interpretative boundaries of the existing legal framework, a COP/MOP decision, possibly even an amendment to the KP, would be necessary.

[20] The Gold Standard Foundation established voluntary sustainable development quality criteria, including social, economic and environmental criteria, for CDM projects. See <http://www.cdmgoldstandard.org>, accessed 6 February 2009. These include, inter alia, environmental criteria, such as effects on water quality and quantity; air; emission of other pollutants; soil condition (quality and quantity); and impacts on biodiversity (species and habitat conservation).

[21] The Netherlands' Ministry of Foreign Affairs (IOB Evaluation) (April 2008) 'Clean and Sustainable? An Evaluation of the Contribution of the Clean Development Mechanism to Sustainable Development in Host Countries'.

[22] P Castro and A Michaelowa (2008) 'Empirical Analysis of Performance of CDM Projects', Climate Strategies Report, 8; and IOB Evaluation) (n 21 above).

the fact that local stakeholders, and partly also host countries Designated National Authorities (DNAs), who are in charge of writing a letter of approval for each project application, are not sufficiently informed about the additional, economic, and social benefits that could accrue from sustainable CDM projects.[23]

The UN CDM institutions, ie the EB and accredited Designated Operational Entities (DOEs), do not assess or review projects for sustainable development benefits. The monitoring of these benefits is not structurally integrated into the CDM, nor are procedures envisaged to enforce the benefits.[24]

3.3 The problem of environmental 'disintegrity'

Despite the encouraging growth of the CDM, the low degree to which environmental integrity—even in the narrower sense of additionality—has been ensured in practice begs for caution. A recent analysis of the environmental performance of the CDM revealed serious deficiencies in the way in which additionality has been assessed in the period from 2005–2007.[25] It estimated that by the end of 2007, additionality was unlikely or questionable for about 40% of the registered CDM projects and about 20% of all issued CERs.[26] This is consistent with a number of studies, which have shown that the CDM offsetting market is far from perfect in ensuring environmental integrity.[27]

In other cases, projects can potentially lead to the relocation or displacement of people and affect the livelihood of local communities, lead to the destruction of estuary and forest ecosystems, and/or have significantly negative impacts on

[23] Castro and Michaelowa (n 22 above) and IOB Evaluation (n 21 above).

[24] IOB Evaluation (n 21 above). See also, C Voigt, 'Is the Clean Development Mechanism Sustainable? Some Critical Aspects' (2008) 8 Sustainable Development Law and Policy 2, 15–21; C Voigt, 'The Deadlock of the Clean Development Mechanism: Caught Between Sustainability, Environmental Integrity, and Economic Efficiency' in B Richardson and others (eds), *Climate Law and Developing Countries: Legal and Policy—Challenges for the World Community* (Edward Elgar, forthcoming in 2009).

[25] L Schneider, 'Is the CDM Fulfilling its Environmental and Sustainable Development Objectives? An Evaluation of the CDM and Options for Improvement' (Report Prepared for WWF) (5 November 2007) Öko-Institut <www.oekoinstitut.de/oekodoc/622/2007-162-en.pdf> accessed 2 February 2008. Mass media has also covered the issue. See N Davies, 'Truth about Kyoto: Huge Profits, Little Carbon Saved' *The Guardian* (London, 2 June 2007) UK News and Analysis 6; N Davies, 'The Inconvenient Truth about the Carbon Offset Industry' *The Guardian* (London, 16 June 2007) UK News and Analysis 14; M Gregory, 'The Great Carbon Bazaar' *BBC News* (4 June 2008) <http://news.bbc.co.uk/2/hi/business/7436263.stm>, accessed 3 May 2009.

[26] Schneider (n 25 above), 44.

[27] See, eg MW Wara and DG Victor, 'A Realistic Policy on International Carbon Offsets' (2008) Stanford University Program on Energy and Sustainable Development Working Paper No 74/2008 <http://pesd.stanford.edu/publications/a_realistic_policy_on_international_carbon_offsets>, accessed 3 February 2009; A Michaelowa and P Purohit, 'Additionality Determination of Indian CDM Projects: Can Indian CDM Project Developers Outwit the CDM Executive Board?' (2007) <www.noe21.org/docs/Michaelowa-teripress-2007>, accessed 3 February 2009; and MW Wara, 'Is the Global Carbon Market Working?' (2007) 445 Nature 595.

biodiversity.[28] This applies, in particular, to wind farm projects, hydro-power projects, and afforestation and reforestation projects, although this list is not exhaustive.

This concerning state of environmental 'disintegrity' must be remedied if the mechanism is to achieve long-term climate benefits. The flaws can (partly) be linked to the colliding interests informing the governance of the mechanism. The protection of the CDM's contribution to sustainable development and its environmental integrity is counterweighted by demands for procedural efficiency and economic feasibility on the other.[29] Another reason for concern is the absence of stringent environmental performance safeguards formulated at the international level. Without such safeguards or standards, there is the danger that Annex I countries can use offsets that do not meet basic CDM requirements to meet their obligations and that CDM projects cause detrimental environmental side effects.

4. Responsibility of the Executive Board for Environmental Integrity

The EB supervises and administers the CDM under the authority and guidance of the COP/MOP, to which it is accountable.[30] The EB has a broad range of competence when supervising the implementation of the CDM according to Article 12.4 of the Kyoto Protocol. It approves or rejects new methodologies for baselines and monitoring plans, and is responsible for the accreditation and suspension of the DOEs, which support the EB in governing the CDM. DOEs are private, domestic entities, or international organizations that are responsible for different stages in the CDM project cycle. They are the 'extended arm' of the EB and perform either validation of proposed projects (ie assessment whether a potential project meets all the eligibility requirements of the CDM) or verification and certification of emission reductions (ie assessment of whether and how much the project has achieved in GHG reductions). Often, however, DOEs are paid by the project developers and their role has been subject to substantial criticism.[31]

Upon requests by the DOEs, the EB decides whether to register a proposed project as a CDM project or to request its review.[32] Registration by the EB is the

[28] International Rivers, 'Comments on Changuinola 1(Chan 75) Large Hydro Project (Panama) 11 August 2008' <http://www.internationalrivers.org/node/3215>, accessed 6 February 2009. Other projects which were criticized for negative social and environmental impacts were Plantar (Brazil), El Canada Dam (Guatemala), and Durban landfill (South Africa). (Information from CDM watch; site no longer available).

[29] See C Streck and TB Chagas, 'The Future of the CDM in a Post-Kyoto World' (2007) 1 Carbon and Climate Law Review, 53; C Streck, 'The Governance of the CDM: the Case for Strength and Stability' (2007) 2 Environmental Liability 15, 91?100.

[30] 3/CMP.1, Annex (n 7 above) paragraph 5. [31] See L Scheider (n 25 above) 19–24.

[32] Registration is generally considered granted unless at least three members of the EB or a State Party involved in the project activity requests the review of a project (Decision -/CMP.1' Modalities and procedures for a clean development mechanism as defined in Article 12 of the Kyoto Protocol, 27(a)).

formal UN-acceptance of a DOE validated project as a CDM project activity. The EB also has the mandate to issue CERs. Issuance requires verification and certification by a DOE other than the one which validated the project, that the emission reductions as set out in the project design document have taken place in the verified period. The DOE then requests the EB to issue a certain number of CERs. The issuance is considered final unless three or more members of the Board or a State Party involved in the project request a review. The EB is further responsible for ensuring access to information and the transparency of the CDM process, for clarifying and interpreting the decisions of the COP/MOP, and for preparing annual reports, technical notes, and decision papers for review and adoption by the COP/MOP.

Therefore, the Board—with the support from DOEs—is *the* institutional entity responsible for the environmental integrity of the CDM. Additionality is a principal condition for the environmental integrity of a project under the CDM, therefore only projects that are additional are eligible and can (in principle) be registered. As we have seen previously, not all projects passing the threshold of EB registration meet this quality. Furthermore, responsibility for environmental integrity could be said also to include the responsibility to ensure that CDM projects have no negative environmental impact on, for example, ecosystems or biological diversity. Yet not much attention has been directed towards this end.

There are various reasons why CDM projects with a poor environmental score, either with respect to missing additionality or wider environmental impacts or both, are being put through the CDM registration process. Additionality problems can be linked to systematic flaws of the CDM and reform proposals to this extent are abundant.[33] Reasons for poor environmental performance in a wider sense are related to the governance of the CDM, in particular—but not only—the operations of the EB. Here, questions of a more general type arise with respect to the scope of environmental integrity which is to be governed by the EB, the Board's competences to avoid and address (potential) environmental impacts caused by a CDM project, and the EB's possible responsibility and liability for environmental harm.

Governance concerns in a narrower sense include possible conflicts of interests and bias of EB members, limited expertise in dealing with highly technical and complex issues (eg approving new baseline and monitoring methodologies), the cumbersome approval of individual projects, time constraints, information

[33] See for the suggestion of sectoral approaches, eg M Lehmann and L Larmuseau, 'The Flexible Mechanism CDM and JI in the Post 2012 Regime' (DNV Position Paper) (June 2008); K Holm Olsen and J Fenham, 'A Reformed CDM?Including New Mechanisms for Sustainable Development' (UNEP Risø Centre Perspective Series) (November 2008) <www.cd4cdm.org/Publications/Perspectives/ReformedCDM.pdf>, accessed 3 February 2009; W Sterk, 'From Clean Development Mechanism to Sectoral Crediting Approaches? Way Forward or Wrong Turn?' (2008) Wuppertal Institute JIKO Policy Paper No.1/2008. Others suggest standardized baselines and regional approaches. See IETA, 'State of the CDM 2008: Facilitating a Smooth Transition Into a Mature Environmental Financing Mechanism' (Report) (October 2008) <www.ieta.org/ieta/www/pages/getfile.php?docID=3111>, accessed 3 February 2009.

deficits, predictability of decisions, and interpretations of the EB,[34] and influences of political and national interests of EB members.[35] Again, there are various reform proposals for strengthening the governance of the CDM, which have been discussed elsewhere.[36]

So far, the critique of the EB's decision-making has sparked some improvements. The UNFCCC Secretariat has been given a strengthened advisory role towards the EB.[37] Since April 2007, all requests for registration are initially assessed by the Secretariat which issues a summary note of the appraisal and forwards it to the EB.[38] This note makes it easier for the EB to decide whether to request review of the project. As a consequence, the rate of requests for review has increased from about 16% in 2006 to about 23% in 2007.[39] Importantly, the share of projects that actually were reviewed increased from about 6% in 2006 to about 18% in 2007.[40] In addition, also the percentage of projects which were rejected as a result of the review process increased significantly. IETA notes that request for review has now become an almost standard step in the project registration process.[41]

Despite these changes, some of the abovementioned problems continue to prevail. Possible action to address these shortcomings includes a procedure to review the Board's decisions.

5. Possibilities for Judicial Review of Executive Board Decisions

Within the CDM there exists an asymmetric interest in successful projects which preferably lead to many and cheap CERs. While cheap CERs normally come from very clearly additional projects;[42] these projects score badly in terms of

[34] For an overview of due process concerns and suggestions for reform see Streck, 'The Governance of the CDM' (n 29 above) 91–100.

[35] See F Flues, A Michaelowa and K Michaelowa, 'UN Approval of Greenhouse Gas Emissions Reduction Projects in Developing Countries: The Political Economy of the CDM Executive Board' (2008) CIS Working Paper No. 35/2008 <http://www.cis.ethz.ch/publications/publications/WP35_Flues_UN.Approval.CDM>, accessed 3 February 2009. The authors argue that EB decisions can be expected to be highly politicized. Approving projects and methodologies is linked to national interests of EB members. EB membership of the country concerned raises the chances of a project being approved.

[36] See, eg Streck, 'The Governance of the CDM' (n 29 above), 98–100; IETA (n 33 above), 10–20.

[37] On the role of the secretariat see A Merrill, 'What is the Secret with the Secretariat? Decision Making by the CDM Registry and Issuance Unit' (2007) IETA Greenhouse Gas Market 2007 114–17 <www.ieta.org/ieta/www/pages/getfile.php?docID=2735>, accessed 3 February 2009.

[38] UNFCCC CDM Executive Board 29th Meeting, 'Annex 14: Terms of Reference and Procedure for a Registration and Issuance Team (RIT) (version 5.1)' <http://cdm.unfccc.int/EB/029/eb29_repan14.pdf>, accessed 6 February 2009, para 25.

[39] L de Jonge, 'Clean development mechanism—results and recent developments' (Presentation given at a seminar on the CDM, Ministry of Finance, Oslo, Norway, 22 May 2008) (on file with author).

[40] T Kopperud and P Molin, 'CDM—current status and challenges' (Presentation at Det Norske Veritas Climate Change Services, SN Power CDM Meeting, 26 November 2008) (on file with author).

[41] IETA (n 33 above), 17.

[42] In the past, these included in particular HFC-23 destruction projects.

providing sustainable development benefits and/or avoiding negative environmental impacts. Project developers, host governments, Annex I governments, and most CER buyers all have an inherent interest in either profit maximization or reduction of compliance costs and are therefore not guarantors of the environmental integrity of projects in the wider sense.[43] In this situation, it is the *sole* responsibility of the EB—with support by DOEs—to ensure environmental integrity in terms of additionality *and* absence of negative environmental impacts.

At the same time, it is crucial for the CDM to attract a critical mass of investors and project developers in order to make a significant contribution to the overall reduction of greenhouse gases. This requires economic feasibility coupled with efficient, predictable and transparent processes.[44] At the heart of this lies a tension between the demands of the market and the need to ensure the environmental integrity of the projects required by the nature of the CDM as an offset mechanism.[45]

The EB has to strike a balance between meeting the demands of economic efficiency and safeguarding the environmental quality of the mechanism. In making these decisions, the EB has a certain margin of discretion; but it is important to keep in mind that non-additional projects can water down the integrity of the entire global carbon market and may eventually lead to increased global emissions.

5.1 Why review Executive Board decisions?

The reason for focusing on the EB is its function as the superior executive organ of the CDM.[46] It is the 'constituted body' responsible for the functioning of the CDM and has the final decision-making authority to register projects and to issue CERs. Yet, there is no judiciary to check the legality of EB decision-making. It is therefore timely to discuss the necessity and possibility of judicial review of EB decisions in more detail.

Reviewing the decisions of the EB is not a new idea.[47] In several cases, project participants, whose project registration was rejected by the EB, threatened to sue the EB in order to recover alleged financial losses.[48] In reacting to this situation, the CMP requested the Executive Secretary to take a number of actions—especially in response to concerns raised by non-State legal entities involved in

[43] See A Michaelowa, 'Ökologisch integer?—Inwieweit liefern CDM-Vorhaben tatsächlich zusätzlichen Klimaschutz?' (Expertenanhörung 'Faule Klimaschutzzertifikate?', Berlin 4 September 2007) (on file with author).

[44] IETA (n 33 above), 9.

[45] See Werksman (n 1 above), 95–104.

[46] While the EB will be fully accountable to the COP/MOP, it acts as a regulatory authority when supervising the implementation of the CDM. Streck & Lin (n 1 above) 417.

[47] See, eg Streck, 'The Governance of the CDM' (n 29 above); Streck and Lin (n 1 above) 2008.

[48] In 2007, the UNFCCC Secretariat noted 12 threats of legal proceedings against the EB (FCCC/KP/CMP/2007/2). See, eg M Wilder and I Millar, 'Is Suing the Executive Board the Last Resort?' CDM & JI Monitor (28 May 2008).

the CDM—to minimize the risks of disputes, complaints, and claims against individuals serving on constituted bodies, including the EB, established under the Kyoto Protocol.[49] Already in 2006, the Executive Secretary of the UNFCCC Secretariat recommended establishing 'dispute settlement arrangements for addressing any disputes, complaints, or claims against constituted bodies or individuals serving on constituted bodies or Expert review teams under the Kyoto Protocol'.[50] Although no formal legal proceedings have been initiated, the issue led to a debate among scholars as to the necessity and feasibility of bringing claims against the EB before national courts.[51]

Often it is the *exercise* of the rules on environmental integrity, especially on additionality, which gives rise to concern. Decisions to register projects under the CDM which turn out not to be additional can partly be linked to systematic flaws of the CDM. However, they are partially caused by mal-governance of the CDM, eg in situations where the EB, faced with high complexity, limited expertise, conflicting interests and/or time constraints, acts outside its delegated authority, including its authority to ensure the environmental integrity of the CDM, or without legal foundation, or where decisions taken are based on factually incorrect technical or scientific conclusions.

Where a State Party's interest is affected, the Party can seek redress under the non-compliance mechanism of the Kyoto Protocol.[52] This possibility, however, does not extend to non-State participants, even if they are directly involved in the CDM; nor does it extend to other entities directly or indirectly affected or likely to be affected by a CDM project. These entities could include employees of the project and their families, local and indigenous communities, or the concerned public in general.[53] Neither does the non-compliance mechanism allow for cases brought against CDM institutions.

These entities need to be granted access to review a decision particularly in cases where the determinations of the EB infringe sufficient interest, or even the rights of private or public legal entities, or where the Board failed to take all measures as required by the legal framework in order to avoid such injury.

The CDM creates a much closer contact between international institutions and private entities, diminishing the role of states.[54] In this situation it is important to uphold the 'Rule of Law' by creating conditions for due process,

[49] UNFCCC 'Privileges and Immunities for Individuals Serving on Constituted Bodies Under the Kyoto Protocol' (2 March 2007) Decision 9/CMP.2.

[50] UNFCCC Subsidiary Body for Implementation, 'Privileges and Immunities for Individuals Serving on Constituted Bodies Established Under the Kyoto Protocol: Note by the Secretariat' (4 October 2006) FCCC/SBI/2006/21.

[51] See Meijer (n 1 above); Werksman (n 1 above).

[52] UNFCCC Conference of the Parties Serving as the Meeting of the Parties to the Kyoto Protocol, 'Annex: Procedures and Mechanisms relating to Compliance under the Kyoto protocol' (30 March 2006) Decision 27/CMP.1 Annex.

[53] The public concerned in this context should include the public affected or likely to be affected by, or having an interest in, the decision-making of the EB with regard to ensuring environmental integrity in the narrower and in the wider sense, including non-governmental organizations promoting environmental protection.

[54] Meijer (n 1 above), 891.

stakeholder participation, and access to justice—especially by guaranteeing fair and transparent review of EB decisions, which potentially can affect a wider range of interests.

5.2 Scope of review

The EB has broad competences when supervising the implementation of the CDM according to Article 12.4 of the Kyoto Protocol. Within its mandate, the EB acts both as a legislative and as an administrative institution. By carrying out legislative or quasi-legislative functions, the EB does not directly impact participants in the CDM. Some of the administrative decisions of the EB, however, directly affect non-State participants in the CDM. By deciding on the registration of projects, approving new methodologies, issuing a certain amount of CERs and accrediting or suspending a DOE, the Board decisions have a direct bearing on the interests of non-State entities, such as private banks or companies or entrepreneurs.

While the decisions of the Board are not *de jure* legally binding, they have been accepted as *de facto* binding by participants in the CDM, including both State and non-State participants, such as private project developers and investors. In this context, the EB has been compared to a domestic market regulator who has the responsibility for granting licences to carry out a certain activity within its regulatory purview.[55]

In addition, EB actions can also have an indirect effect on third persons, ie non-addressees of Board decisions. So is it possible, for example, that an EB decision to register a project with little or no environmental integrity, can affect *other* participants in the global carbon market with a vested interest in stringent environmental performance of CDM projects. Such participants include, for example, investment banks with green investment portfolios, which are bound to invest in projects with high environmental credibility or governments as compliance buyers that committed themselves to purchasing CERs exclusively from projects with high environmental quality. At a minimum, non-additional projects and projects causing negative environmental impacts could tarnish the reputation, goodwill and trustworthiness of these institutions. At worst, for these participants credits from potentially environmentally harmful projects could be valueless for the purpose of compliance (eg under the new EU ETS rules) and could amount to financial losses.

Moreover, EB decisions do not address local communities' interests (unless they are project participants). Yet, it is perceivable that high impact projects with questionable environmental integrity, such as large hydroelectric projects, afforestation and reforestation (AR) projects, or adipic acid factories which cut back on their nitrous oxide emissions, could affect local communities by

[55] Streck and Lin (n 1 above), 417.

potentially harming air or water quality, causing water shortages, or even forcing the relocation of people. Although there are requirements for stakeholder participation and environmental impact analysis, these are only of procedural character. A CDM project may well go ahead despite significant stakeholder concerns.[56]

It is the sole responsibility of the EB—with support of the DOEs during the validation process—to ensure environmental integrity of CDM projects when applying the general rules to specific projects and entities. It is important that the EB exercises its functions in a consistent manner in compliance with the regulatory framework and in an unbiased and impartial manner. The inclusion of a check on the broader environmental compliance is possible only if the responsibilities of the EB also include ensuring environmental integrity of CDM projects in a wider sense. Depending on the interpretation of 'environmental integrity', such broader environmental responsibility can be understood as already being part of the existing rules.[57] However, where the interpretation of the existing legal framework leads to the conclusion that it falls outside the scope, it would have to be included by a COP/MOP decision or by an amendment to the KP in order to establish the EB's competence in this respect.

In order to ensure the legality and legitimacy of EB decision-making, EB decisions need to be exposed to legal scrutiny. In situations involving participants or stakeholders, including NGOs, having a sufficient interest or, possibly, whose rights have been impaired by an EB decision, these entities should have access to a review procedure before an independent and impartial body, to challenge the substantive and procedural legality of that decision.

A sufficient interest of stakeholders can, for example, be said to exist where the environmental impacts of a CDM project affect their livelihoods, or their health or wellbeing, eg access to fresh water. Furthermore, non-governmental organizations promoting environmental protection can be deemed to have such an interest in situations where CDM projects cause significant environmental

[56] In order to be validated by a DOE, the project activity must meet the validation requirements set out in 3/CMP.1, Annex, para 37. These include, inter alia, proof that comments by local stakeholders have been invited, a summary of the comments, and a report on how due account was taken of any comments received (3/CMP.1, Annex, para 37(b)). 'Stakeholders' means the public, including individuals, groups or communities affected, or likely to be affected, by the proposed project activity (3/CMP.1, Annex, para 1(e)). Furthermore, the Project Design Document needs to describe the environmental impacts of the project. For a proposed afforestation and reforestation project activity, this includes documentation of the environmental impacts, including impacts on biodiversity and natural ecosystems, and impacts outside the project boundary. This analysis should include, where applicable, information on, inter alia, hydrology, soils, risk of fires, pests, and diseases. However, only if a negative impact is considered significant by the project participants or the host Party, a statement that project participants have undertaken an environmental impact assessment (EIA), in accordance with the procedures required by the host Party, is necessary. It is worth noting, that an EIA is contingent upon existing national law, exempting project developers from such assessment where the law of the CDM host does not contain EIA obligations.

[57] See above, n 19.

impacts, even if no direct harm to humans is caused. As a consequence, it is proposed that a review mechanism is established.

Such review could cover the assessment of the substantial and procedural legality of EB decisions, including, for example, that (a) certain determinations of the EB are *ultra vires* the Board's delegated authority, or that some of the decisions and/or interpretations of COP/MOP decisions had been taken without legal foundation; (b) decisions taken are based on factually incorrect technical or scientific conclusions; (c) EB Members have a conflict of interest concerning decisions taken; (d) an alleged breach of confidentiality had occurred; (e) the conduct of a member of the EB is not in conformity with the operational policies, procedures and practices, which has resulted in violations of procedural rights of private and/or legal entities; or (f) the decisions, recommendations or other actions of the EB are biased or made improperly.[58] Finally, review could cover claims that the EB decisions are inconsistent, which affect predictability and legal security.

In this way, a review procedure could deal with the determination of the Board's accountability for environmental harm directly or indirectly caused by the project, or for harm to local people and stakeholders, and the issues of adequate and effective remedies and injunctive relief.

The challenge is that the Board's delegated authority rests on broad rules, providing it with a wide margin of discretion. In this context, judicial review could replace broad terms with those of a narrower content. The concern is that such 'judge-made law' might affect the carefully negotiated balance between law and politics.[59] However, this is not an exclusive situation in the context of review of EB decisions; also in other areas of international law legal review can lead to a certain degree of judicial activism without putting the entire regime at risk.[60]

5.3 The forum for review

Review of EB decisions is a crucial element in increasing legitimacy and credibility of the CDM. While no dispute settlement arrangement yet exists within the framework of the CDM, three different avenues for dispute settlement or review procedures will now briefly be examined: external review, internal review, and national review.

5.3.1 External review

The first possibility is to bring a claim against the EB before an already existing judicial or quasi-judicial institution outside the UNFCCC/Kyoto System. A

58 FCCC/SBI/2006/21 (n 2) para 14.

59 J Klabbers, 'On Rationalism in Politics' (2005) 74 Nordic J Intl L 405, 427.

60 For a discussion in the context of WTO dispute settlement see L Bartels, 'The Separation of Powers in the WTO: How to Avoid Judicial Activism' (2004) 54 ICLQ 861; S Essermann and R Howse, 'The WTO on Trial' (2003) 82 Foreign Affairs 130.

note prepared by the UNFCCC Secretariat concluded that none of the dispute settlement arrangements established by the United Nations would be suitable for this purpose.[61] The arrangements which the Secretariat considered included arbitration administered by an established arbitral tribunal such as the International Chamber of Commerce (ICC), investigation by a Claims Commission, and the United Nations Administrative Tribunal. These arrangements were considered either unsuitable or as requiring substantial revision to make them fit for this purpose.

Arbitration is, however, a common method of dispute settlement and should be given further consideration. While the ICC normally applies its own arbitration rules rather than any provisions adopted by the COP/MOP, arbitration administered by the Permanent Court of Arbitration (PCA) might provide a more suitable setting. The PCA has adopted optional rules tailored to address environmental disputes, which differ from UNCITRAL rules in order to reflect the particular characteristics of disputes having a natural resources, conservation, or environmental protection component.[62]

However, arbitration needs to be pre-arranged. The usual basis for arbitration is a contractual or treaty-based relationship where the contracting parties include an arbitration clause and agree on the terms of arbitration. In the context of the CDM, no such contractual relationship exists between private participants and the EB, let alone between stakeholders and the Board. It has therefore been suggested that private entities and CDM institutions be required to sign a general arbitration agreement before CDM participation commences.[63] Unfortunately, this still leaves stakeholders without access to arbitration.

5.3.2 Internal review

Second, there is the possibility of creating a new internal review mechanism by a decision of the COP/MOP. Such an arrangement was seen as the preferred alternative by the UNFCCC Secretariat.[64] A review mechanism could function similarly to the WTO dispute settlement system. This would require establishing non-standing panels out of a roster of experts in the various fields of the

[61] FCCC/SBI/2006/21 (n 2 above).

[62] Permanent Court of Arbitration, 'Optional Rule for Arbitration of Disputes Relating to Natural Resources' (adopted 19 June 2001 and 16 June 2002). Different from UNCITRAL rules, the PCA environmental rules can be used in 'agreements, contracts, conventions, treaties, the constituent of an international organization or agency or reference upon consent by the parties by a court' (Article 3.c), or in a submission agreement. These rules apply to 'all concerned citizens', are supposed to have the ability to bring together highly diverse legal systems and cultures, and can accommodate multiple parties whether they are states, IGOs, NGOs or private entities ('any claimant or respondent' (Article 15.2) instead of 'either party' (UNCITRAL). See D Ratliff, 'The PCA Optional Rules for Arbitration of Disputes Relating to Natural Resources and/or the Environment' (2001) 14 Leiden J Intl L 887.

[63] Meijer (n 1 above) 897. Meijer suggests that '[t]his general clause would cover all the situations (contractual and non-contractual) in which the two parties interact and where conflicts might arise.' Ibid.

[64] Ibid.

CDM. In order to meet the requirements of fairness, equity and impartiality, current or former members of the EB should not qualify as panel members. The review procedure would be initiated by the receipt of a complaint and a request to set up a panel.

Important in this regard is the question of who would have the right to bring a complaint. Standing should be granted to those natural or legal persons that can claim to be or likely to be affected by a CDM project. This includes the addressees of EB decisions—eg project developers, host countries, DOEs. In addition, other persons (natural and legal) with a sufficient interest in the environmental integrity of the CDM should be allowed to initiate a review procedure. Participants in the global carbon market (for example other Annex I governments, other host governments or CERs buyers for 'green investment funds') could potentially be said to have such sufficient interest in the environmental integrity in the narrower sense of additionality. The environmental effectiveness of the entire carbon market and the price of tradable allowances can become substantially watered down by the inflow of below-standard CERs.

CERs which do not correspond to additional emission reductions can increase overall GHG emissions. When linked to the global carbon market, this effect can impinge on climate change mitigation strategies elsewhere, even making them redundant. Project participants can thus be assumed to have a sufficient interest in having environmental integrity of the CDM ensured by the EB. Otherwise, they see the stringency of their climate mitigation efforts jeopardized by non-additional CDM projects.

Similarly, other CDM host countries with CDM projects of high environmental quality might have concerns about competitive advantages and discrimination and therefore have an interest in seeing equally stringent CDM projects being implemented elsewhere.

Finally, there are those with an interest in wider environmental integrity, ie those interested in the prevention of environmental damage either directly or indirectly linked to the CDM project. This group would most likely comprise stakeholders[65] and NGOs who either represent stakeholder interests or have an interest in environmental protection. Third party rights should apply to these persons and enable them to bring a complaint against decisions which affect them.

Moreover, standing could also be extended to all State Parties to the Kyoto Protocol, not just those who participate in the particular CDM project and/or claim injury or loss. Instead, a general legal interest in the protection of the atmosphere needs to be acknowledged so as to allow a State Party to exercise rights of legal protection on behalf of the international community as a whole. This would include a right of standing in the review process, similar to an *actio popularis*, where all parties can be said to have an interest in GHG emissions

[65] See for definition of 'stakeholders', n 56 above.

remaining within the overall target set out under the Kyoto Protocol.[66] Increased GHG emissions by the influx of non-additional credits could affect the carefully negotiated balance of rights and obligations under the Kyoto Protocol. Broadening the competence to invoke a claim of non-integrity of an EB decision would give recognition to climate protection as a global community interest, rather than the narrow commercial interest of CDM project participants.

The establishment of a review procedure is intended to grant access to justice to the public concerned with matters of the CDM's environmental integrity. It is understood that this includes a wider group of persons than those that could maintain that their rights have been impaired. Whether such a broad circle of potential claimants is manageable or even desirable is subject to debate. However, given the significance of environmental integrity to the CDM, given also the potential for adverse effects of some project activities, and the many, diverse, and complex ways in which a CDM project could affect persons and communities, it seems important to establish inclusive access to a review procedure.

Such a development would be recognizant of the changing nature of international society, international law and the role of non-state actors. Opening a potential review mechanism to claims for these groups would strengthen the CDM's legitimacy by providing all interested participants and stakeholders with a much needed forum for dispute resolution.[67]

In order to ensure an independent review process, it has been suggested that a panel should be given the power of investigation, while ensuring an open and transparent process.[68] The panel function is then to investigate the substantive and procedural legality of EB decisions or, in other words, whether the EB acted in accordance with its obligation to ensure and govern the environmental integrity of the CDM.

Finally, the panel should have the power to issue binding decisions. Further appeal to an Appellate Body could be envisaged. While the possibility to appeal the decision of the panel might increase the legitimacy of the mechanism, it might also increase its costs. It is expected that by introducing a one-step review mechanism due process and legitimacy concerns are already adequately met.

Another, less stringent, alternative would be to set up an internal inspection panel to investigate the adherence of the EB with its own policies and procedures. A similar investigative procedure has been established by the World Bank when creating its Inspection Panel.[69] An internal CDM inspection panel could react to complaints by CDM participants or stakeholders who are or may be affected by a CDM project. Different from the dispute settlement mechanism

[66] P Sands, *Principles of International Environmental Law* (2nd edn, Cambridge University Press: Cambridge, 2003), 187.

[67] See also PR Kalas and A Herwig, 'Dispute Resolution Under the Kyoto Protocol' (2000) 27 Ecology L Q 53, 132.

[68] For more detail on the features of a review mechanism see Streck and Lin (n 1 above), 439.

[69] For more information see World Bank, 'The Inspection Panel' <http://www.theinspectionpanel.org>, accessed 6 February 2009.

described above, an inspection panel would not be in the position to render a binding decision. Rather, based on the information gathered during the course of the investigation, the panel could issue a report, including proposals to the EB for remedial action.

Finally, there remain the alternatives of setting up of a standing expert committee or an ombudsman to hear complaints from affected participants or stakeholders. The function of these bodies would be of a more mediatory nature, attempting to find a reconciliatory solution without adjudication.

5.3.3 Review in national courts

Third, remedy against EB decisions can potentially be sought in national courts. Private participants in the CDM which claim their rights violated by, for example, non-registration of their projects, non-approval of a new baseline methodology or the issuance of less than demanded or no CERs, may wish to challenge these decisions of the EB. Depending on the domestic legal system in question, it may be possible that a private legal entity which claims its rights violated seeks recourse to domestic courts.[70] Threats to bring claims against the EB and its members occur regularly, although no lawsuit has yet been filed before a national court. So far, these claims concerned financial losses as well as damage to reputation due to negative publicity.[71]

The issue of bringing a lawsuit against the EB before a national court raises at least three important questions: first, whether the EB comprises of an institutional quality which makes it possible to be a party to a legal proceeding; second, which rights could possibly have been violated by EB decisions; and, third, the issue of immunity of EB members.

The legal personality of the Executive Board

Legal personality is a prerequisite for being a party to legal proceedings before domestic courts (*ratione personae*). Though the legal qualification of the EB is not straightforward, some writers seem to suppose the EB to have the legal personality of an international organization (IO).[72]

In a traditional sense, international organizations have broadly been defined as 'forms of cooperation founded on an international agreement creating at least one organ with a will of its own, established under international law'.[73] While it is arguably possible to classify the COP/MOP as an international organization,[74]

[70] See Meijer (n 1 above), 899. [71] FCCC/KP/CMP/2007/2, 13 November 2007.

[72] See, for example, Meijer (n 1 above). Meijer implicitly recognizes IO quality of the EB when discussing the possibility to initiate legal proceedings against the Board in domestic courts.

[73] H Schermers and NM Blokker, *International Institutional Law: Unity Within Diversity* (Martinus Nijhoff Publishers: Leiden, 1995) 23.

[74] See R Churchill and G Ulfstein, 'Autonomous Institutional Arrangements in Multilateral Environmental Agreements: A Little Noticed Phenomenon in International Law' (2000) 94 AJIL 623.

it is questionable whether the EB fits within the ambit of this definition. On the one hand, the self-governed, treaty-based nature, could possibly speak for an IO quality of the EB, or, at least, for the applicability of the law of international organizations. The law of international organizations is dynamic and various institutional arrangements included in MEAs have been considered IOs despite their special characteristics.[75] On the other hand, however, the ad hoc nature of the EB and the apparently conscious decision of the Kyoto Member States not to create formally an international organization speak against the classification of the EB as an IO in the traditional sense. It seems therefore convincing to suppose that the EB does not have a legal personality.

However, the doubt surrounding the qualification of the EB as an IO could cause a court to deny this quality to the EB (while still recognizing it as a legal person), thereby relieving the court of any potential immunity considerations and upholding its jurisdiction.

Rights violations

The second issue concerns the scope *ratione materiae* of remedies against the EB. In other words, what accountability does the EB have to non-State entities involved in a CDM that could give those entities the right to seek redress? Although decisions of the EB are directly addressed to private participants, this does not necessarily mean that they correspond to or infringe those entities' rights or legitimate expectations. On the other hand, the EB does not issue decisions that are addressed to local stakeholders, but this does not prevent these individuals and others from claiming accountability of the EB for environmental integrity of CDM projects.[76]

The legal connection between the EB as a possible bearer of an obligation owed to private entities and these entities as holders of respective rights, however, is an intricate issue. Scholars have long sought to establish the link between infringements of rights and the actions of international organizations. Werksman refers to similar attempts in the field of alleged human rights violations by operations of the World Bank and of the International Monetary Fund or by the activities of multinational companies.[77] He points to the long-standing effort to establish accountability of these institutions for human rights violations, which has not yet resulted in a clear accountability regime. In comparison, he notes, 'government officials, international civil servants and academics appear to have embraced the notion of private rights and public duties under the climate regime with disturbing ease'.[78] In fact, very little has been written about the source of accountability of the EB. Most scholars simply presuppose the existence of a duty–right relationship between the EB and private entities and focus instead on the question of immunities.[79]

[75] Ibid, 658–9.

[76] See generally K Wellens, *Remedies Against International Organizations* (OUP: Oxford, 2002) 34.

[77] Werksman (n 1 above), 103. [78] Ibid.

[79] Meijer (n 1 above) 886, 894, 900, 906.

Private commercial interests of CDM investors are not protected under the current CDM legal framework, to the degree that profit expectations of CDM participants give rise to legitimate expectations that, when frustrated, amount to expropriation with the result that compensation can be claimed.

The accountability of the EB toward local communities and stakeholders for (possible) damages to their land, or to their health by air or water contamination, or for infringement of various human rights (for example their right to life and health, physical integrity and security, their right to an adequate standard of living, right to property, or right to a safe and healthy environment) is yet another issue which warrants intellectual recourse. International financial institutions and development banks have consistently rejected the notion that they have a direct obligation owed to individuals affected by their operations. But this question remains contentious.

Yet an argument in favour of an obligation of the EB toward local communities and affected stakeholders to avoid environmental damages could be derived from the Board's duty to ensure the environmental integrity of each CDM project. If environmental integrity is defined in a broader sense (as mentioned above) as to include the absence of environmental damages and side-effects, then it is arguable that an actionable third party right applies to affected individuals, which would enable them to seek redress for their damages.

Immunities

Most of the discussions triggered by threats from private legal entities on the accountability of the EB have focused on the question of immunity of the EB and its members. EB members act in their personal capacity and do not have the immunities granted to UN staff, nor do they retain any sovereign immunity as government officials of their home state. Neither the Kyoto Protocol nor the UNFCCC confer immunities on the members of bodies established under those agreements. The only exception is immunity granted to the members of the EB under German law according to the headquarter agreement between Germany and the UNFCCC Secretariat for decisions taken in Germany (and other countries that have entered into conference agreements with the UNFCCC Secretariat to hold COP meetings). Outside these countries EB members generally act without immunity, that is, it is possible for a member to be liable under domestic laws for its actions and decisions. The COP/MOP, aware of this, has, for a couple of years, reviewed this issue in great detail, although without any final decision.[80] The safest way to address this situation is to establish an effective regime of privileges and immunities through an amendment of the KP.[81] However, the

[80] See FCCC/KP/CMP/2007/2 (n 2 above); FCCC/SBI/2006/21 (n 2 above).

[81] The issue was again on the agenda for the Subsidiary Body for Implementation at its twenty-ninth session in Poznan, December 2008. See UNFCCC, 'Privileges and Immunities for Individuals Serving on Constituted Bodies Established Under the Kyoto Protocol: Implementation of Decision 9/CMP.2, Note by the Secretariat' (29 September 2008) UN Doc FCCC/KP/CMP/2008/10.

strong focus of the debate on immunities has overshadowed the two other questions just discussed: the legal personality of the EB and the question of existence of any actionable rights of private entities. Moreover, as Werksman notes, by focusing on the issue of immunity, the COP/MOP might implicitly have acknowledged that the EB has a duty towards private actors, which then are entitled to respective actionable compensatory rights.[82] Other scholars have attempted to redefine the doctrine of immunity, especially by suggesting a more nuanced approach which does not grant immunity to an international organization when engaging in administrative decision-making.[83]

Both doctrinal efforts and decision-making by the COP/MOP could remedy the situation of non-immunity of EB members, but they do not address the question of accountability of the EB in the first place. In order to provide a workable solution, this is the first step that must be taken.

Seeking remedies before a national court warrants a note of caution. Suing the EB in a national court does not mean that this is the best or even a suitable forum for addressing the accountability of the EB. The danger is that different courts might arrive at different conclusions and give diverse interpretations of the same rights and obligations under the Kyoto Protocol.[84] Consequently, the already very complex legal framework of the CDM would be put at risk of incoherence and further fragmentation.

6. Conclusions

There is discontent in the CDM. On the one hand, project developers and CDM investors feel that the system governing the CDM and the decisions of the EB are too strict, time consuming, inconsistent and expensive. On the other hand, environmental groups, NGOs and local stakeholders consider the system too lax, providing too many loopholes, and thereby reducing environmental effectiveness and causing undesirable environmental side-effects. In this situation, it is the EB that is the last-instance decision-making body; it has the final say. Because of the accumulation of power in the EB and the direct effect that its decisions have on participants in the CDM, a safety check is surely necessary. The introduction of an internal judicial procedure to review EB decisions would not only provide a check on the legality of the EB decision-making; it could also hold the Board accountable for environmental integrity of the CDM and thereby increase the legitimacy and credibility of the CDM. Environmental integrity in this context covers both real, measurable, long-term, and additional GHG reductions, as well the absence of 'collateral' environmental damages.

[82] Werksman (n 1 above) 103. [83] See Meijer (n 1 above) 907f.

[84] See Streck and Lin (n 1 above) 428. Cf Meijer (n 1 above) 913–14. Meijer argues that litigation in national courts 'would allow for the building of a consistent set of CDM-related case law'. Ibid.

An internal review panel would also strengthen the rule of law in the CDM by introducing due process requirements and by giving access to justice to affected persons. In this regard it is important to establish broad competences for bringing complaints. At the same time, the availability of a review process for EB decisions would effectively pre-empt the possibility of success in seeking redress in national courts. The risks of potentially widespread 'nationalization' of CDM decisions by review by national tribunals would put the CDM governance at risk of fragmentation, lack of coherence and unpredictability.

The inclusion of a review mechanism would not be intended to establish yet another bureaucratic level in the CDM project process. Rather, it should run parallel to it and provide the possibility of safety checks, where necessary. Such a review arrangement needs to be a neutral mechanism; meaning that the scope of review includes claims of environmental 'disintegrity' based on individual or collective interests of stakeholders as well as claims of infringements of private (commercial or other) rights. The prospect of objective and neutral review of EB decisions for their substantial and procedural legality is intended to increase the legitimacy of the CDM. It should therefore be feasible to receive the necessary political support and accept for the establishment of such mechanism.

The establishment of a review mechanism would not only provide a much-needed forum for dispute resolution to all interested CDM participants and those who are affected by a CDM project. The mere possibility of holding the EB accountable for its decisions would also trigger a move towards greater accountability. It would lead to better governance, institutional changes and improvements in the regulatory framework. Greater accountability increases credibility, which is important if the CDM is to play a significant role in a new global climate agreement.

14

Carbon Contracting

Martijn Wilder and Louisa Fitz-Gerald

1. Introduction

There exist several types of carbon credits under the international, regional and national emissions trading schemes, and as to be expected, there are many different approaches to contacting for the sale and purchase of these assets. Agreements for the purchase and sale of carbon credits will vary in structure and complexity, influenced by a range of factors including:

- The type of credit being traded – Is it regulatory-based or contractually-based? Is it governed by international rules or a domestic or regional scheme? Is it a project-based credit or an allowance unit?
- The nature of the trading market – Is it a primary or secondary trade? How well are the market rules developed? Are the buyers and sellers sophisticated traders?

These differences have so far precluded the development of a single standardized contract for carbon commodity transactions, although some freely available templates, such as CERSPA[1] and the International Emissions Trading Association (IETA)[2] Emissions Reduction Purchase Agreement (ERPA) have been developed and market standard contracts based on the International Swaps and Derivatives Association (ISDA) models, such as for the trade of European Allowances, are commonplace.

Despite these differences, the maturing of the carbon market has led to a crystallization of various contractual norms for the purchase and sale of carbon credits, for example around responsibility for costs and dispute resolution. Moreover, all emission reduction purchase agreements must deal with key commercial issues, such as the nature of the commodity, the transfer of legal title to the commodity, and the consequences of events of default, particularly failure to deliver.

[1] <http://www.cerspa.org>. CERSPA, however, is not intended to be a one-size-fits-all contract, but rather a reference or tool to advance negotiations, given the need to tailor contractual terms to each specific transaction. For example, buyers requiring carbon assets for compliance obligations under domestic emissions trading schemes may be willing to pay a higher price for guaranteed deliver of credits.

[2] <http://www.ieta.org>, accessed 10 March 2009.

This chapter provides an overview of some of the options for contracting for carbon by focusing on the following:

- a history of carbon contracting approaches;
- the key issues involved in carbon contracting;
- some recent contracting issues; and
- some approaches to the structuring of financing arrangements for carbon projects.

2. History of Approaches to Carbon Contracting

2.1 The Pre-Kyoto period

In the period prior to the entry into force of the Kyoto Protocol, when emissions trading rules for both the international and various national emissions trading schemes were still undergoing development, carbon contracts emerged to allow parties to start trading what were thought of generically as 'carbon rights' or later 'emission reductions.' Parties doing projects that reduced emissions were sought out and early market participants such as the World Bank through its Prototype Carbon Fund,[3] the Netherlands government[4] and companies like Ecosecurities and CO2.com used largely bespoke bi-lateral agreements negotiated between individual parties (although over time these parties developed their own model ERPAs)

The key issue parties faced was defining what was being purchased and drafting documents that captured the concept of legal rights to a physical activity of reducing or sequestering greenhouse gas emissions from an activity in the hope that they would become more clearly defined tradeable commodities in the future. The concept of ERs and VERs remains and even terms like 'Verified Emission Reduction Agreement' or 'VERPA' emerged. In this regard, VERPAs are simply a contract under which 'emission reductions' are purchased. A sample clause would be, for example:

Emission Reduction means all existing and future legal and beneficial rights arising from one greenhouse reduction, including any right, interest, credit, entitlement, benefit, allowance, certificate or registrable right arising from or in connection with that greenhouse gas reduction.

Over time the expectation that such rights would be independently verified by a third party led to the term Verified Emission Reductions or VERs. The early

[3] See David Freestone, 'The World Bank's Prototype Carbon Fund: Mobilising new Resources for Sustainable Development' in (Sabine Schemmer-Schulte and Ko-Yung Tung (eds.) *Liber Amicorum for Ibrahim S.I. Shihata* (Kluwer Law Int: The Hague, 2001), 265–341

[4] Daniel van der Weerd, 'CERUPT and ERUPT Contracts' in David Freestone and Charlotte Streck (eds), *Legal Aspects of Implementing the Kyoto Protocol: making Kyoto work* (Oxford University Press: Oxford, 2005), 313–328.

contracts used by the World Bank for its Prototype Carbon Fund and companies like Ecosecurities took such an approach. The intention was that as the market developed, such rights would then become the basis for some form of future compliance right, although there was no guarantee that this would be the case. Buyers simply took this risk and if the Kyoto Protocol did not enter into force, sellers would still be paid for, and be required to deliver, ERs, even if such ERs never acquired any real compliance value for the buyers.

In the absence of any real frameworks these early ERPAs had to include significant detail around the development of the project itself, the process for generating emission reductions, monitoring abatement, verifying the abatement and design an appropriate process (in the absence of any issuance of credits under the international regime) for the transfer and delivery of such VERs. They also had to be flexible enough in their drafting to cater to such requirements as may be imposed in the future for such ERs to become CERs or ERUs. Given the pre-Kyoto nature of these contracts, they also invariably included contractual terms for converting the ERs into Kyoto credits —CERs and ERUs— once this was possible to do. This usually involved obligations on the seller to do all the things reasonably necessary to achieve such 'conversion' although the additional costs of doing so would be paid for by the Buyer. Interestingly we have seen companies who contracted to sell Australian-based forestry credits under the state trading regime in NSW now looking to transition such credits to the proposed national scheme.

The relatively small size of this early market and predominance of a few key carbon funds such as the World Bank's Prototype Carbon Fund and the Dutch Government's CERUPT and ERUPT funds resulted in a relatively standardised contracting approach. Similar definitions and concepts were used and a similar approach was adopted to risk allocation between seller and buyer.

These early ERPAs have formed the basis of today's more modern ERPAs. However, VERPAs are still used for the sale and purchase of emission reductions used in the voluntary market and for emission reductions from CDM projects before they are registered and able to sell the pre-registration emission reductions as ERs or VERs rather than CERs. In this circumstance we are seeing numerous guidelines and standards emerge for the voluntary market which again makes contracting easier as the VER being purchased simply needs to meet a particular standard. Buyers of these types of credits might be in non-Kyoto countries and so do not need fully-compliant Kyoto credits for compliance purposes, or may seek to use them for 'carbon neutrality' purposes or in retail schemes where consumers are the ultimate buyers.

2.2 The entry into force of the Kyoto Protocol

With the finalisation of the Kyoto Protocol in 1997, the notion of more defined rights such as Certified Emission Reductions (CERs) from Clean Development Mechanism (CDM) projects and Emission Reduction Units (ERUs) from Joint

Implementation (JI) projects emerged. However, even then the Protocol still needed to enter into force and the specific rules and procedures for creating CERs and ERUs did not emerge until many years later with the Marrakech Accords in 2001. Parties continued to buy and sell ERs and future CERs and ERUs under bi-lateral bespoke ERPAs and in nearly all cases this involved primary contracts between the project developer/owner and a buyer.

Typically, ERPAs in today's project-based credit market include a range of Kyoto-specific terms to assign responsibility between the buyer and seller for meeting international rules for registration, verification and certification of the project, monitoring of the emission reductions, and issuance of the credits by the CDM Executive Board, and apportioning risk between the buyer and seller in relation to failure to meet these rules for any reason. They incorporate the rules and procedures to do with project participants and nominating the Focal Point and they are able to rely on an established and operational Independent Transaction Log and National Accounts. With such increased certainty, ERPAs although all fairly similar remain largely one-on-one bilateral private contracts negotiated between private parties.

While these early contracts all focused on ERs, VERs, ERUs and CERs, the establishment of the European Union Emissions Trading Scheme introduced into the market permits that were issued to regulated companies to allow them to emit greenhouse gas emissions (European Union Allowances or EUAs). Unlike project based credits, EUAs are compliance instruments issued by governments and so once issued can be freely traded. In such trades there is no need for lengthy contractual provisions dealing with project based issues. As such these units are much more suited to very standard documents with negotiations focusing on commercial issues around price and volume.

With the growth of EU demands for CERs and increasing market participants, more generic ERPAs emerged in the market. IETA and the World Bank promoted standard ERPAs and others like CERSPA were released. These were increasingly modified and adjusted as the international rules developed and the uncertainty around CERs and ERUs dissolved with the entry into force of Kyoto in 2005. In addition, once the CDM Executive Board began to issue CERs the ability to buy issued CERs on a spot basis and for market players to trade them meant that contracting clauses concerning the specific aspects of verification and monitoring could be removed; for once a CER had been issued it had already been through this process.

Furthermore, with the growth of secondary trading in CERs, ERUs and European Emissions Trading Units (EUAs), secondary ERPAs emerged (SERPAs). Such SERPAs for the forward purchases of non-issued CERs still reflected project based ERPAs that were back-to-back with the underlying ERPA, (unless the first Buyer from the project was prepared to take all the delivery risk and simply sell CERs on a forward guaranteed basis). However, where SERPAs were dealing with EUAs or issued CERs then the contracts rapidly became far more simplified and standardized with a number of market standard contracting approaches emerging (see below). When contracting for an EUA or an issued CER held in existing National Registry accounts, issues concerning the CER creation process and matters dealing with project participants become largely redundant.

Finally, in the current financial situation there are also an increasing number of parties wishing to get out of current contracts owing to the falling CER price and this has placed great tension on ERPAs and the relationship between parties.

2.3 Post-Kyoto—more sophisticated transactions

Now that the carbon markets are reaching a level of maturity and gaining in liquidity, more complex and sophisticated sale and purchasing approaches are emerging, such as the use of trading platforms, the increasing involvement of banks and brokers as intermediaries, tenders, sale of portfolios of CERs through information memorandums, e-obligation) to purchase CERs. Also parties are simply trading options—ie the right, but not the obligation, to buy CERs from another party in the future.

In addition, with increased competition for CERs from projects, buyers have also at times offered advance payments to projects and agreed to fund the costs of the CDM development aspects of projects (ie, drafting the Project Design Document, hiring of a validator) and recovering this from the future payments. While the later has become common place in ERPAs, advanced payments are less frequent and will usually require some high degree of comfort and some form of guarantee. The Asian Development Bank has under its Asia Pacific Carbon Fund provided significant upfront payments for CERs—but it usually also finances the projects in question.

While fully developed derivate and secondary markets are still a few years away, the increasing prevalence of these structures—which are not in any way new, but just the application of existing commodity trading practices to a new commodity—has recently seen a move towards even more standardized contracts and transactions. These new markets and players are demanding clear and explicit definitions and terms to minimize risk exposure and increase market liquidity. Not surprisingly in the case of the secondary market, significant trading is now done under the International Swaps and Derivatives Association's (ISDA) Master Agreement and specially-drafted schedules and confirmations to cater for the unique aspects of EUAs or CERs.

IETA, ISDA and the European Federation of Energy Traders (EFET) have each released tailored agreements for the trading of EUAs under the EU ETS that fit under their existing standardized commodity transaction agreements. The IETA, EFET and ISDA documents were each drafted for different purposes, and as such each have different focuses. The IETA contract is a specialized EUA trading document and does not envisage the trading of any other commodities. The EFET Appendix relevant to trading EUAs is in fact an appendix amending a larger electricity trading agreement and therefore contemplates that parties will generally be electricity retailers or traders. Finally, the ISDA Confirmation of OTC Physically Settled EU Emissions Allowance Transaction is designed to allow trading of EUAs under an ISDA Master Agreement which can function as either a single product or a multi-product master agreement encompassing a variety of cash and physically-settled transactions involving various underlying assets. Generally the ISDA is likely to be the most familiar agreement for major financial institutions and corporations accustomed to trading in other commodities and products.

The structure of the ISDA-style documentation allows buyers and sellers—who expect to enter into several transactions with each—to enter into a single master agreement which will govern the terms of their transactions, but subject to any special conditions as set out in the confirmation for each transaction. This structure provides certainty to the parties in terms of the general terms, but enough flexibility to include amending conditions in a confirmation to the extent this is deemed appropriate or necessary, for example, if there are subsequent changes in the rules, or concepts require specific definitions in line with regulatory rules for the creation and delivery of the carbon credit.

Interestingly, the market is now also seeing the adoption of these ISDA-style agreements for the secondary sale of project-based credits (such as CERs), This is despite allowances—which do not rely on complex baseline and credit rules and processes for verification and certification—lending themselves far more easily to ISDA-style commodity agreements. However, the structure of these agreements, where schedules and confirmations are used routinely to add to or vary terms of the master agreement, are now being used to incorporate the peculiarities of project-based rules and how the parties will share the associated risks between them.

ISDA themselves have recognized the growing demand for ISDA documentation to be prepared for multiple markets and allowances. While an EUA annex is well established, the ISDA Emissions Working Group meeting held on 13 January 2009 in London, decided that with respect to non-EU Registry documentation, the consensus was that this now represents the members highest documentation priority in this field. There was some discussion as to whether there should be jurisdictional annexes or one generic amendment. Most members seemed to prefer the generic approach if it is possible. Switzerland, Japan, Australia and New Zealand were identified as priority jurisdictions for the development of jurisdictional annexes, if this approach is pursued.

It is interesting to note that in anticipation of the commencement of the Australian emissions trading scheme, market participants have begun the forward sale of future Australian permits and CERs. In nearly all cases CERs are being sold under ISDA with the parties being electricity companies and financial institutions well experienced in ISDA. Furthermore, it is worth noting that under the emissions trading regime established in the State of New South Wales (NSW) in Australia, which has been operating for many years, the project-based credits, NSW Greenhouse Abatement Certificates, both bespoke contracts and ISDA documents have been used on a regular basis.

3. Key Issues in Carbon Contracting

The purchase and sale of carbon credits, whether it takes place under international, regional or national regulatory frameworks, involves a complex mix of contracts, tax, finance and securities law principles, as well as the integration of international and national law, and private and public legal regimes.

For example, while Kyoto units such as certified emission reductions (CERs) are created under international law, the legal status of such credits in the place where they are created, held, bought or sold is almost entirely dependent on the domestic law of the relevant jurisdiction, including property, tax and accounting law. Contracts for the purchase and sale of CERs allow sufficient flexibility to accommodate the constantly evolving international rules, but also provide the certainty required by domestic laws to bind the parties in contract.

In transacting to sell carbon credits, the key issues lie in negotiating and determining:

- the nature of the commodity which will be transacted (i.e. contractually defined emission reductions, CERs, ERUs, VERs, EUAs);
- the commercial terms, including price, on which it will be sold; and
- the risks and warranties that each party is prepared to accept and provide in the transaction, including the remedies for any breach of obligation or warranty.

Table 14.1 sets out issues which must be defined and managed in any type of contract to sell the rights or benefits associated with a reduction in greenhouse gas emissions, regardless of the type of project structure or contract which is ultimately adopted.[5]

4. Contracting Approaches

As already noted, most emission reduction purchase agreements, especially in the case of primary deals, are separately negotiated, taking into account the specific circumstances of the individual project, the domestic laws of the jurisdiction in which the project is hosted and the commercial requirements of each

[5] Adapted from M Wilder and M Willis, 'Carbon Contracts, Structuring Transactions: Practical Experiences', in David Freestone and Charlotte Streck (eds), *Legal Aspects of Implementing the Kyoto Protocol* (Oxford University Press: Oxford, 2005), 295–312.

Table 14.1 Issues to be defined and managed

Issue	Comment
Nature of the credit	Where a contract is for the sale of CERs or ERUs under the Kyoto Protocol, the credits can be simply defined by reference to the relevant international rules. However, where a contract is transacting credits in the absence of any statutory or legal regime (eg to supply the growing demand for voluntary carbon offsets), the precise definition will need to be defined in the ERPA. In addition, a prudent purchaser would seek to draft the contract to include sufficient flexibility to require the seller to provide all necessary assistance for the purchaser to create whatever statutory right may flow from the contractual right in the future that suits their purposes. The critical components of any contractual definition of emission reductions (other than in contracts for Kyoto credits) must include the following: — a reference to the physical reduction or sequestration in levels of greenhouse gas emissions; — a clear definition of the greenhouse gases to be included (generally being those under the Kyoto Protocol); — a clear unit of measurement usually being per tonne of carbon dioxide equivalent; — reference to the particular and clearly defined project creating the reductions; and — the basis against which such reductions are occurring (ie against a pre-determined baseline).
Quantity of credits	The contract will need to clearly specify the total number of rights to be delivered. This may be, for example '100% of all CERs arising from the Project from the time of Registration until 31 December 2012' or 'The first 1 million CERs arising from the Project'. If the seller will may 'guarantee' the delivery of a minimum volume of credits, in order to minimise its risk and exposure for failing to produce sufficient numbers of credits, the seller may want to ensure that it retains some form of 'buffer' and therefore perhaps guarantees no more than 70% of its total carbon stock on a forward basis, with an option to the buyer to purchase the remainder on a spot basis. This will provide some protection in the event that the modelling of the delivery volumes proves to be incorrect or for various reasons the supply of CERs is less than what was anticipated (as has often been the case).
Legal title	The seller will need to demonstrate and warrant that they have clear legal title to the right and the ability to transfer the rights to the buyer without any third party consents or further government approvals. Whilst most host country governments have accepted the general proposition that a private sector entity that expends its own resources to develop an emission reduction project should be entitled to the benefits arising from such emission reductions (i.e. the credits), some host countries have sought to regulate the profit sharing of revenues from credit sales, and the nature of entities which are entitled to develop CDM projects. China, for example, specifies that developers of CDM projects in China must be Chinese owned-or-controlled and that the project developer must share a proportion of the CER revenues with the Chinese government.

Issue	Comment
Price and Payment	Once a price has been agreed, the method of payment will be critical. In some cases, the buyer may provide advance payment for the delivery of a portion of the anticipated credits, and this approach is becoming more popular as the CDM market becomes more competitive. The parties may choose from a number of options in terms of structuring payment, including: — a fixed price per credit; — pegging the credit price to the prevailing market price (e.g. the price of Allowances in the EU ETS); or — imposing a floor or ceiling price on a market-indexed credit price.
Transfer of legal title	Central to the contract will be the point in time to which legal title to the rights would be transferred. Although the logical point for transfer of legal title to CERs would occur upon receipt in the buyer's account, a seller may wish to retain beneficial title until it receives payment.
Failure to deliver	In the event that the seller fails to deliver the requisite volume of credits, then the contract will need to provide mechanisms to address this. Common mechanisms in carbon trading contracts for delivery shortfall include: — no consequences, provided that the shortfall was not the result of the seller's 'wilful default' (e.g. the sale of the credits to a third party); — an obligation to use best endeavours to make good the shortfall during the remainder of the term of the contract; — an obligation to procure replacement credits; — an obligation to pay a 'cash make good' for any short-term shortfall (e.g. to pay the buyer the difference between the agreed credit price and the prevailing market price, so that that buyer can purchase replacement credits for the short term, whilst the contract remains on foot); or — the right for the buyer to terminate the agreement and recover from the seller 'buyer's replacement cost' (equal to the difference between the credit price and the prevailing market price for the entire undelivered contract volume). Where the contract contains remedies along the lines of the first two bullet points, the seller does not guarantee the delivery of any minimum volume of credits. However, where the contract contains any of the final three remedies, the seller essentially guarantees to the buyer the delivery of a minimum volume and agrees, if there is a shortfall in delivery, to incur the cost of providing replacement credits or paying the buyer the market price difference.

of the parties. Furthermore, many sophisticated buyers tend to prefer to use their own form of contract which has been adapted to their own risk profiles and which may contain certain provisions that are seen as more attractive than other buyers rather than rely on generic templates.

Therefore, although a number of precedent carbon credit trading contracts have been developed, these are rarely used verbatim in a primary transaction, although they do provide a useful indicator of a market standard where a particular clause of a tailored contract is subject to negotiation.

In recent years, approaches to carbon contracting have been influenced by the increased competitiveness amongst buyers to identify attractive projects. To transact carbon credits, potential buyers and sellers have employed a variety of approaches. These include:

- implementing a competitive tender process for the sale of carbon credits from one or more potential projects on the basis of a set of standard terms and conditions;
- inviting buyers to provide their proposed ERPA terms by way of a bid for the purchase of carbon credits from specific projects;
- a sale of options to purchase agreed volumes of carbon credits at a later date at an agreed or market price, usually with a strike price to exercise the option itself; or
- transacting carbon credits through the procedures established in an online auction process (such as that managed by the Asia Carbon Exchange);
- issuing a request for particular types of carbon credits (e.g. CERs or ERUs) from the market;
- inviting the submission through a buyer's web site of potential projects through a brief project description, and individually negotiating the purchase of emission reductions from such projects;
- brokering the transaction of carbon credits through a trusted broker with access to local markets, in return for a brokerage commission or percentage of credits; or
- a combination of one or more of the above.

Certain of the above options are more suitable for long-term forward trades of CERs, which have dominated the market to date, with many buyers seeking to pass through the risks they assume under a forward ERPA by way of a 'back-to-back' agreement with a secondary purchaser.

In terms of 'secondary' or 'spot' trades of emissions credits (where those credits have already been issued and are capable of being electronically transacted immediately), it is likely that participants will tend to utilize standard agreements to trade these credits in a similar way to other types of commodities and derivatives.

5. Interplay Between International and Domestic Laws: Impacts on Contracting

As already noted, carbon contracting involves negotiating an often complex interplay between international rules and domestic laws, both of which impact upon the rights of project participants. Property rights and the ownership of carbon credits provide useful examples of this.

The international rules for the CDM establish a process whereby the entities responsible for implementing a project activity may, with the approval of a Party to the Kyoto Protocol, be listed as 'project participants' in that project activity, either at the time that the project is registered or afterwards. The Executive Board may only issue CERs to the registry accounts of entities who are participants in the relevant project.

However, the international rules have no mechanism for verifying that the entities listed as project participants are the entities who can actually claim legal title to the underlying emission reductions. Unless legal title to ERs, VERs, CERs and Allowances is determined under domestic law it is purely a private matter and the Executive Board deliberately has no mandate to verify the legal entitlement of project participants to CERs. The process envisaged by the international rules seems to assume that any entity that may wish to challenge the claim of project participants to emission reductions and the resultant CERs would raise a challenge during the stakeholder consultation process prior to registration of the project. If such a challenge is not raised and resolved and again it would need to be dealt with as a private matter between the parties, then there is a *prima facie* assumption that the project owner has the right to receive full legal title to CERs from the project.

There are some exceptions to this, such as in China and South Africa where CERs are the property of the State and only assigned over once a Host Country Letter of Approval (LOA) is issued. However, while in such cases the holder of the LOA can be said to have the right to take ownership of and trade the CERs, in most countries the LOA carries no weight as to legal title but simply confirms participation in a project. Furthermore, often in the due diligence undertaken by the buyer it is revealed that the seller does not always own, or have the ability to sell, the CERs, possibly because they belong to another corporate entity of the seller or they have some form of collateral over them. In this regard due diligence on the seller is very standard.

In some circumstances third parties may challenge the legal title of project participants to carbon credits. In addition, if a party that has entered into a forward transaction to receive a stream of credits from a project becomes insolvent, the administrator and creditors will need to determine who is entitled under domestic law to receive the credits to which that entity would be entitled. For this reason, any contracts purporting to transact emission reductions or credits should specifically deal with the issue of legal title, even where the relevant legal framework under which those credits are created (eg the Kyoto Protocol) operates on a working assumption in respect of legal title to such credits, as discussed above.

The effect of this lack of clarity is to render it essential that carbon contracts clearly define the carbon credit and assign ownership in a manner that minimizes the risk of challenges to legal title. Such a definition would include, for example:

- a clearly identifiable and definable right, which exists within a clear legal or policy framework;

- clear ownership over the right, to the exclusion of all other potential claimants;
- a clear point at which legal title transfers from seller to buyer;
- an unconditional warranty from the seller that it holds unencumbered title to the underlying emission reductions and the resultant carbon credits;
- irrevocability of transfer of title, or revocability only subject to certain pre-established criteria; and
- transferability of the carbon commodity itself (including an agreed process by which such commodity is deemed to be transferred).

Although such provisions cannot prevent conflicting claims to legal title arising, they assist the buyer to establish to third parties that it is a bona fide purchaser for value, and would also enable the buyer to claim damages from the seller if the buyer's title is challenged.

6. Managing Risk through Carbon Contracts

All projects entail risks which must be managed through contractual arrangements. The constantly evolving regulations and political developments surrounding the Kyoto Protocol increase the risk profile for undertaking emission reduction projects in comparison to other types of projects (such as general construction or infrastructure projects). In addition, emission reduction projects will also face risks that are common to any project in developing or industrialised country. Such risks include:

- the risk that an event beyond the parties' control (eg a force majeure event) may impact the ability of the project to operate and produce credits;
- the risk that the market for the credits may move in such a direction as to create an incentive for one of the parties to breach the long term carbon commodity supply contract;
- the risk that the market for the other products created by the project (ie electricity) may rise or fall so that the project is no longer financially profitable;
- the risk of cost overrun from the project, such as due to delayed construction;
- the risk that the parties may not be able to establish legal title to the emission reductions on which the CERs/ERUs are based;
- the risk that greenhouse gases may not be properly monitored and accounted for so that the project will not generate as many CERs/ERUs as expected; and
- the risk that the CDM or JI project may face community or NGO opposition due to issues such as the development of a baseline or the nature of the project participants.

These types of risks should be mitigated through efficient management of the project and contractual clauses between the different parties to the project, including the project developer, the construction contractor, the project operator and the carbon commodity purchaser.

As with any general commercial transaction, the greater the risk involved in purchasing credits, the lower the price that a purchaser will be willing to pay and the stricter the conditions it will endeavour to place on the seller. Risk allocation will be dealt with through contracts between the various participants, financiers and purchasers of credits through measures such as:

- conditions precedent to the entry into force of a contract (for example, a condition precedent in a CDM ERPA to the effect that the project must achieve registration before the delivery and purchase obligations become binding);
- guarantees from parent companies or governments involved as to the financial stability of the project developer;
- force majeure clauses covering risk contingencies such as natural disasters or change in law or policy making the project unfeasible or leading to the seller failing to deliver the contracted amount of credits;
- laying off particular risks to other parties such as contractors or the verifier for the project;
- adequate warranties and indemnities in contracts;
- procuring insurance for various project and credit risks; and
- providing rights of termination in certain events.

The types of clauses appropriate will depend on the nature of the project, the type of credits being transacted and the level of involvement of the carbon commodity purchaser in the underlying project (ie whether it will be taking equity in the project or merely purchasing credits).

7. Recent Contracting Issues

7.1 The Impact of the new Modalities of Communication

The Executive Board has recently issued a draft standard 'Modalities of Communication' (MOC), the document which parties to an ERPA sign to nominate project participants and nominate a focal point. The new MOC is a departure from the past practice and it remains unclear how it sits alongside existing contractual arrangements. Nonetheless, it is clear that when new project participants are added or removed a new MOC must be signed and lodged with the Executive Board. Existing contracts will then need to be modified.

It should also be noticed that the MOC now introduces three new categories and creates a concept of shared focal point. This shared focal point does not really work as it allows the other party at any time to go and withdraw from the EERPA and to add and remove the other project participants and focal point role without the Buyer's consent which would breach the international rules and UK law.

7.2 Voluntary credits and the concern for quality

As noted earlier in this chapter, verified emissions reductions (VERs) emerged as an instrument for defining greenhouse gas emissions abatement or sequestration rights in contracts written before the rules of many international (in particular the CDM) and domestic regulated schemes had been settled.

However, the growth in the consumer-based voluntary offset market, increasing criticism of the environmental benefits of non-verified credits and the reputational risk a company faces if credits they have purchased are exposed as not being backed up by genuine emissions reductions, have led to an increasing tendency amongst voluntary credit buyers to favour quality over price and purchase VERs even when they have no intention of later converting them for use in a regulated scheme. VERs continue to be in high demand in markets such as the US where the details of anticipated mandatory emission trading schemes have not been settled as they offer the potential to be converted for use in those future schemes.

An additional element required for the contractual definition of a VER is the process for third party verification and the standard against which the credit is being verified. In response to the increasing concern about the quality of voluntary credits there has been a proliferation in the number of voluntary carbon standards against which verification can occur. Prominent standards include the Voluntary Carbon Standard and the Gold Standard, both administered by NGOs, and the Greenhouse Friendly standard administered by the Australian Government.[6]

Another key criticism levelled at VERs is that there are no central registries for VERs at the national or global levels. This gives rise to the potential for double counting or sellers offering the same credits for sale multiple times. By contrast, CERs generated under the CDM, for example, are tracked on an international registry. For companies purchasing VERs it is important that they exercise extreme care in their due diligence process for assessing sellers. A number of VER registries have emerged which offer a mechanism for assisting buyers to track the history of a credit, but which cannot guarantee that the credits have not been sold off-registry.

8. Conclusion

Rights to reductions in greenhouse emissions (and that form the basis of a CER), and the demand for them, are the result of regulatory structures adopted

[6] For an extensive discussion of the voluntary sector see chapter 23 below.

by law makers in an attempt to harness market forces to drive lowest cost abatement of greenhouse gas emissions. Notwithstanding the statutory or contractual basis of credits, they are ultimately commodities or assets and many existing contractual structures and risk mitigation techniques can be drawn upon in order for parties to agree and articulate commercial terms for their sale and purchase through 'contracting for carbon.'

However, as the existence and value of credits depend on constantly evolving legal frameworks, special consideration of the risks peculiar to carbon commodity transactions should be at the forefront of parties' minds when negotiating and setting contractual terms.

Contractual clauses, such as remedies for delivery failure and force majeure, are crucial in managing the types of contingencies that can arise in this market. Emission reduction project developers and purchasers of credits must stay abreast of legal and regulatory developments and reassess their existing contracts and contracting strategies in light of such developments, to ensure that they adequately manage and price the risks arising from this regime.

15

The Secondary Market for Emissions Trading: Balancing Market Design and Market Based Transaction Norms

Andrew Hedges

1. Introduction

At the heart of the Kyoto Protocol is the belief by signatories that, by and large, market forces can be harnessed to achieve environmental goals; that a well designed cap-and-trade scheme can identify lower cost abatement opportunities faster and more cost effectively than a command and control system. However, it is without doubt that 2008 was a watershed moment, not only for the global economy but also in the recognition of the capacity for poorly regulated markets to create systemic risks. The consequent enhanced focus on regulation by governments and competent authorities makes it an apt time to consider how appropriately designed emissions trading schemes can create the conditions for successful environmental outcomes. The EU Emission Trading Scheme (EU ETS) is a leading example how to strike a balance between the necessity for a sufficiently certain market design to underpin its growth and the critically important need to allow private sector mechanisms to emerge and evolve. Section 1 of this chapter will therefore consider the design elements of the EU ETS that provide market participants with the certainty necessary to underpin their participation, with a particular emphasis on the reforms to be implemented from 2013. Section 2 will then consider the market mechanisms and norms that have been developed or applied to provide those participants with the confidence that key risks are addressed. Section 3 considers some of the particular issues that arise in the context of the interplay between secondary markets such as the EU ETS and the primary markets where emission reduction credits under the flexible mechanisms of the Kyoto Protocol are generated. Examples of new types of transactions will be used to demonstrate the innovation that such interaction drives.

2. Design Elements for the Creation of a Successful Secondary Market

The manner in which the EU ETS has addressed the key elements of a robust multi-State emissions trading scheme can be considered in light of the following thematic statements relating to such a scheme, namely that it should be:

- comprehensive in scope;
- capable of creating scarcity;
- predictable in application;
- robust in regulation and enforcement; and
- technically robust.

2.1 Comprehensive in scope

The likely depth of the market can be increased by ensuring that there is comprehensive application of an emissions trading scheme across a sufficiently wide number of both: (i) countries or States; and (ii) key sectors and installations responsible for the relevant greenhouse gas or gasses being targeted. Ensuring a deep market with multiple participants (in particular, beyond those solely with compliance obligations) achieves the following key outcomes: (i) the risk of market manipulation by a few participants is reduced; (ii) the potential for leakage through the shift of affected industries to unregulated markets in the same region is also reduced; and (iii) the likelihood that the price signal generated by trading is a reliable indicator for investment decisions is enhanced.

The development and implementation of the EU ETS remains an exceptional example of European cooperation. Within a period of three years[1] a scheme was implemented that covered some 12,000 installations.[2] The recent passage of the amendments to the EU ETS Directive that will take effect from 2013[3] (the Amending Directive) represent, despite the presence of a number of negotiated

[1] The terms of what became 'Directive 2003/87/EC of the European Parliament and of the Council of 13 October 2003 establishing a scheme for greenhouse gas emission allowance trading within the Community' (EU ETS Directive) were proposed in October 2001 with final approval by the Council of Ministers in July 2003. The EU ETS Directive was formally published in October 2003.

[2] Including emitting sectors such as: power stations and other combustion plants; oil refineries; coke ovens; iron and steel plants; and factories making glass, cement, lime, bricks, ceramics, pulp, paper, and board.

[3] The text of the Amending Directive was approved by the European Parliament on 17 December 2008 and was adopted by the Council of Ministers on 6 April 2009. In this chapter, references to the Amending Directive are to the agreed European Parliament text ('Position of the European Parliament adopted at final reading 17 December 2008 with a view to the adoption of a Directive amending Directive 2003/87/EC').

compromises, a continuation of the commitment to subsume national interests in pursuit of a common goal. In particular, the European community has taken the steps necessary to further widen and deepen the comprehensive application of the EU ETS by:

2.1.1 The Inclusion of New Sectors

Being the carbon dioxide emissions from petrochemicals, ammonia and aluminium; and

2.1.2 The Inclusion of New Gases[4]

Being nitrous oxide emissions from the production of nitric, adipic and glyoxal and glyoxylic acid production[5] and perfluorocarbons from the aluminium sector.[6]

The above is in addition to the inclusion, under previous amendments to the EU ETS Directive[7], of carbon dioxide emissions from all flights which arrive at or depart from an aerodrome situated in a member state from 1 January 2012. The recent Amending Directive also flagged the possibility of further extending the Scheme to the shipping sector by 2013 if no alternative mechanism has been developed internationally and approved by the European Union by 31 December 2011.[8]

The EU ETS is implemented by way of Phases. The first Phase of the Scheme ran from 2005 to 2007. The second and current Phase began in 2008 and ends in 2012. Phase III will run from 2013 to 2020. Analysis of the successes and failures of the first Phase of the Scheme often refers to it as a 'learning by doing' phase. Critically, though, learning can only occur where there is a functioning market. The fact that Phase I was able to identify teething issues was made possible by the quick emergence of a functioning market. This occurred, in part, due to the Scheme being comprehensive enough to generate the critical mass necessary to drive the development of the market infrastructure discussed in Section 2.

[4] The EU ETS currently only covers carbon dioxide emissions, with the exception of the Netherlands, which has opted in nitrous oxide emissions.

[5] Annex 1 of the EU ETS Directive (as amended by the Amending Directive) under the sub-heading Chemical Industry.

[6] Annex 1 of the EU ETS Directive (as amended by Amending Directive) under the sub-heading Production and processing of metals. Note that this covers primary production and secondary production where combustion installations with a rated thermal input exceeding 20 MW are operated.

[7] See Directive 2008/101/EC amending Directive 2003/87/EC so as to include aviation activities in the scheme for greenhouse gas emission allowance trading within the Community.

[8] See Recital (3) of the Amending Directive: the criteria are there is no international agreement including international maritime emissions through the IMO approved by the member states and/ or such agreement through the UNFCCC has not been approved by the Community by 31 December 2011—in which case the Commission will make a proposal with the aim of its entry into force by 2013.

2.2 Capable of creating scarcity

The objective of an emissions trading scheme such as the EU ETS is not only to reduce absolute emissions but also to prompt investment in emission reductions within installations and the development of new low carbon installations. That objective is undermined where there is insufficient scarcity of allowances to underpin a strong price signal for such investments over a sufficient time period to justify them. The EU ETS is based upon the allocation of emission allowances in respect of each Phase of the Scheme which are less than a baseline of historically verified emissions, thus creating the required level of scarcity. Furthermore, the Scheme is intended to demonstrate to subject installations that such scarcity will extend and deepen over time. For instance, the recent Amending Directive confirmed that the third Phase of the Scheme will extend through to 2020 and seek to achieve at least a 20% overall reduction of greenhouse gases compared to 1990.[9]

However, the impact of the financial crisis in 2008 and attendant recessions within a number of key Western economies, including Europe, have resurrected the question of whether an unfettered market-based scheme can generate a strong long-term carbon price signal. In essence, the decline of economic activity in Europe during the current Phase has resulted in significant falls in the price for allowances within the Scheme (as market participants align the allocation of allowances for the Phase with the lower emissions profile such economic downturn implies). This has occurred despite the steps taken by the European Commission prior to the commencement of this Phase to address the drivers of the price collapse in Phase I.[10] Those steps included: (i) issuing detailed guidance to member states outlining the rigour with which it would assess proposed member state caps; and (ii) forcing further reductions on member states that it considered failed to comply with that guidance.

The structure of the EU ETS and the decisions made by the European community in respect of Phase III have the capacity to demonstrate that an appropriately designed scheme can generate a long-term price signal despite short-term volatility. Key elements that should achieve this in the medium term are as follows.

2.2.1 Structure of Phase III cap

From 2013 the total number of allowances will decrease annually in a linear manner such that the allowances allocated will reduce to 21% below actual EU ETS emissions in 2005. This is arguably sufficient to drive meaningful scarcity in Phase III despite the current economic conditions.

[9] Consider also that the decrease of allowances under the EU-wide cap for Phase III of 1.74% per year will continue to apply beyond the end of Phase III and will determine the cap for Phase IV (2021 to 2028) and beyond.

[10] In April 2006 the price of allowances within the Scheme collapsed after market participants gained access to actual emissions data that demonstrated the allocation of allowances for the phase were in excess to the actual emissions occurring within that Phase.

2.2.2 Banking

Unlike the first Phase of the EU ETS, allowances from the current Phase can be used or 'banked' in the next Phase. This means that the prices for allowances in Phase II should not be confined to the current economic conditions but also reflect the supply and demand balance for Phase III.

2.2.3 Auctioning

In Phase I and II of the Scheme, the overwhelming proportion of allowances have been distributed to compliance entities for free. This has distortionary effects on the economic efficiency of the Scheme. Phase III will utilize the auctioning of Allowance allocations to a far greater extent (particularly in the power generation sector). Analysts have argued that this will probably stabilize the volatility of price movements in that Phase.[11]

2.3 Predictable in application

The challenge faced by member states and regulators such as the European Commission by the price falls mentioned above is to remain committed to the principle that, having established the parameters for the market, the operation of that market should thereafter identify long-term price signals without a floor or cap. That is, to refrain from either seeking to artificially prop up the price for allowances within the EU ETS in order to drive the creation of a low carbon economy or intervene to limit the strength of such price signals. To do so would undermine a fundamental tenant of the design of the EU ETS, namely that there should be no ex-post adjustments to allowances following the allocation decision for a Phase. In a well designed and regulated market, the predictable application of the market parameters without further adjustment to meet political goals is key to investment certainty. This view has also underpinned the European Commission's long held the view that price caps and floors undermine the capacity of an emissions trading market to generate an economically efficient price for carbon. [12]

[11] Market analysis by Deutsche Bank.

[12] For instance, the collapse in the price of carbon in the current Phase has not led the Commission to condone calls for price floors. It remains of the view that 'a stable and predictable regulatory framework is vital for market stability' and that any issues revealed by such price falls will be addressed through existing means such as: the determination of the cap on emissions in the EU ETS Directive well in advance of the start of Phase III, the linear reduction factor for the cap on emissions which continues to apply also beyond 2020 and the extension of the trading period from five to eight years. Refer: Europa Press Releases RAPID, 'Questions and Answers on the revised EU Emissions Trading System' (17 December 2008) <http://europa.eu/rapid/pressReleasesAction.do?reference=MEMO/08/796&format=HTML&aged=0&language=EN&guiLanguage=en>, accessed 4 April 2009. The Commission expressed similar views in its review of the functioning of Phase I of the EU ETS (despite the price collapse that occurred). See European Commission, 'Building a Global Carbon Market—Report pursuant to Article 30 of Directive 2003/87/EC' (Communication from the Commission to the Council, the European Parliament, the European Economic and Social Committee and the Committee of the Regions) (November 2006) 6 <http://ec.europa.eu/environment/climat/emission/pdf/com2006_676final_en.pdf> accessed 5 April 2009.

Negotiations on the shape of Phase III were focused on apprehension of extreme prices and the impact of a carbon price on the international competitiveness of particular industrial sectors rather than the consequences of a low carbon price. Whilst, overall, the design of Phase III remains substantially true to the economic efficiency pursued by the European Commission in its original reform proposals, there are elements which cause some concern. Those include the following.

2.3.1 Mitigation of excessive prices

The Amending Directive enables the injection of additional allowances into the market where, for more than six consecutive months, the allowance price is three times greater than the average price of allowances during the two preceding years. The means of doing so will either be through bringing forward planned auctions or the auctioning of up to 25% of the then remaining new entrants reserve.[13] Whilst described as a measure to mitigate price fluctuations, these measures remain a potential market intervention tool that will need to be taken account of by market participants.

2.3.2 Proposals to adjust the functioning of the carbon market

The Amending Directive also provides the European Commission with the capacity to make proposals to adjust the EU ETS where it has evidence that the carbon market is not functioning properly.[14] In light of the consistent approach of the European Commission to push for the predictable and economically efficient operation of the market, this is a power unlikely to be exercised lightly. However, it remains a positive right to potentially intervene to adjust previously agreed parameters.

Another critical example of where a consistent and predictable approach underpins the operation of an economically efficient emissions trading scheme is in the allocation of allowances. An ongoing issue has been the manner in which allowances are allocated to entities with compliances obligations. Aspects of that issue include the fact that allowances have been in almost all cases issued to compliance entities for free and that the proportion of such allowances some sectors have received is excessive (whether by design or methodological deficiencies). Free allocation generates the conditions for windfall profits in sectors where the marginal cost of allowances can be passed through to consumers (although the actual allowances were distributed for free).[15] A common theme in European Commission assessments of member states' allocation decisions in both

[13] Amending Directive, para 29 (inserting a new Art 29a into the EU ETS Directive).

[14] Amending Directive, para 28 (replacing Art 29 of the EU ETS Directive).

[15] K Neuhoff, K Keats & M Sato, 'Allocation, incentives and distortions: the impact of EU ETS emissions allowance allocations in the electricity sector' (2006) 6 Climate Policy 73–91. See also D Ellerman and P Joskow, 'The European Union's Emissions Trading System in Perspective' (Pew Centre on Global Climate Change, May 2008) 27–9.

the first and second Phase was that those decisions favoured certain sectors over others and thereby could amount to unwarranted state aid.[16]

Significant progress has been made to address these issues in Phase III. In particular, the introduction of auctioning of a significant portion of allowances from 2013 to key emission sectors such as power generation is a positive resolution of one of the more acrimonious debates in the reform of the EU ETS.

The scope of auctioning in Phase III will be implemented in line with the following principles.

2.3.3 Power sector

100% of allowances for the power generation sector in a member state will be auctioned unless it is located in a country meeting certain criteria,[17] in which case there will be transitional free allocation to allow for modernization of the electricity sector.[18] Under this transitional process auctioning will rise from 30% in 2013 to 100% by 2020.

2.3.4 Industrial sectors

Subject to the exclusions below, 20% of allowances for the industrial sector will be auctioned in 2013, rising to 70% in 2020.[19]

2.3.5 Industrial sectors exposed to leakage

There will be continued free allocation of 100% of allowances to those sectors that meet certain criteria[20] designed to identify those sectors that are exposed to

[16] Although in the decision of *Federal Republic of Germany v Commission of the European Communities* (7 November 2007, T-374/04) it was commented that the Commission had reached similar conclusions regarding anti-competitive impacts without sufficient evidence. See the further discussion at n 33.

[17] Amending Directive, para 11 (inserting Art 10c into the EU ETS Directive). The criteria are: (i) electricity network is not directly or indirectly connected to the EU main grid in 2007; (ii) electricity network was, in 2007, so connected but only through a single line with a capacity less than 400MW; or (iii) where in 2006 the GDP per capita at market prices did not exceed 50% of the average GDP per capita of the EU. Analysis by Deutsche Bank has indicated that the countries meeting at least one of these criteria are: Bulgaria, Cyprus, the Czech Republic, Estonia, Hungary, Lithuania, Malta, Poland, and Romania.

[18] Amending Directive, para 11 (inserting Art 10c into the EU ETS Directive). Member states meeting the criteria are to provide the European Commission with a national investment plan consisting of information on the upgrading of the infrastructure and clean technologies, as well as on the diversification of their energy mix and sources of supply. It must be equal to the market value of the free allocation, as far as possible.

[19] Amending Directive, para 11 (inserting Art 10c into the EU ETS Directive—see para 2 thereof).

[20] Amending Directive, para 11 (inserting Art 10a into the EU ETS Directive—see para 12 thereof). The relevant sector or sub-sector will need to meet both of the following criteria: (i) combine direct and indirect additional costs induced by the EU ETS Directive must raise production costs by at least 5% of Gross Value Added; and (ii) the Non-EU trade intensity (defined as the ratio between the total value of exports to non-EU countries plus the value of imports from the non-EU countries and the total market size for the Community) of the sector or sub-sector is above 10%.

the risk of carbon leakage due to the cost increases that would be imposed by auctioning.

As will be discussed further in Section 2, the design and implementation of auctioning will be of significant interest to market participants. Vigorous debate regarding such design issues will continue prior to the European Commission's target of 30 June 2010 to adopt a Regulation that will provide the rules and conditions for ensuring efficient, coordinated auctions without disturbing the market within Europe.[21] The use of a Regulation as the vehicle for such harmonization will enhance the legal effectiveness of the Commission's position.

For those sectors that will continue to enjoy free allocation during Phase III, the Amending Directive has also sought to enable the European Commission to tackle the divergence in approach taken to the methodology for such allocation. By 31 December 2010 the European Commission will develop common rules to harmonize the process. The Amending Directive has flagged that these should utilize ex-ante benchmarks[22] as well as the most efficient technologies.[23] Considering the difficulties other attempts at comprehensive benchmarking rules have faced, it is likely that the process to generate such rules will be difficult.[24] Furthermore, there are precedents for certain member states to adopt strained interpretations of rules under the EU ETS when faced with pressure from industrial sectors. As the final decision on harmonized rules for free allocation will leave particular sectors unhappy, there is a good chance that similar issues will arise.[25]

[21] For a comprehensive discussion of the issues that will be canvassed in detail as the Regulation develops see, Yvonne Hofmann, 'Auctioning of CO2 Emission Allowances in the EU ETS' (Ecofys report for European Commission) (October 2006) <http://ec.europa.eu/environment/climat/emission/pdf/etsreview/ets_co2_emission_auctioning.pdf>, accessed 7 April 2009. See also C Hepburn, M Grubb, K Neuhoff, F Mathes & M Tse, 'Auctioning of EU ETS Phase II Allowances: How and why?' (2006) 6 Climate Policy 137–160 <http://www.econ.cam.ac.uk/eprg/pubs/tsec/hepburn.pdf>, accessed 7 April 2009.

[22] Ex-ante benchmarks are determined prior to the start of the relevant Phase. Such a benchmarking process does not provide for ex-post adjustments based on actual production.

[23] Benchmarks focused on the most efficient technologies (as opposed to those skewed or 'grandfathered' to account for the existing performance of a sector as whole) reward operators that have taken early action to improve efficiency whilst creating incentives for less efficient operators to either improve performance or acquire emissions allowances from the market.

[24] See the discussion of common challenges to benchmarks in B Buchner, C Carraro & D Ellerman, 'The Allocation of European Union Allowances: Lessons, unifying themes and general principles' (September 2006) Centre for Energy and Environmental Policy Research 12–14 <http://www.feem.it/NR/rdonlyres/8B7B8607-FA89-4984-96BD-C4F7042348D5/2103/11606.pdf>, accessed 7 April 2009. See also D Ellerman and P Joskow, 'The European Union's Emissions Trading System in Perspective' (May 2008) Pew Centre on Global Climate Change 36 (n 24 above) <http://www.pewclimate.org/docUploads/EU-ETS-In-Perspective-Report.pdf>, accessed 7 April 2009.

[25] A good example is the ongoing issues associated with the definition of combustion installation. This has only been resolved through further definition in the Amending Directive. For a wider discussion of the factors likely to influence the Commission's approach see D Phylipsen, A Gardiner, Tana Angelini & Monique Voogt, 'Harmonisation of Allocation Methodologies' (October 2006) European Commission <http://ec.europa.eu/environment/climat/emission/pdf/etsreview/harmonisation.pdf>, accessed 7 April 2009.

The Amending Directive does, however, contain some guidance which may assist narrow debate, including: (i) calculating benchmarks by reference to products rather than inputs;[26] and (ii) using (as the starting point for the benchmark analysis for a sector or sub-sector) the average performance of the 10% most efficient installations in a sector or sub-sector in the Community in the years 2007 to 2008.[27]

A further area where consistency of regulation can impact on the emergence of a functioning market is in relation to the nature of the commodity itself. Section 2 will briefly consider issues such as the financial services treatment of allowances. However, it is important to recognize that where a scheme such as the EU ETS creates the risk of differential regulatory treatment of the types of allowances traded within the scheme there is a risk of market fragmentation. The treatment of international offsets such as certified emission reductions arising from hydro-electric projects having a capacity greater than 20MW is a good example of how this can arise. Unlike offsets arising from nuclear or forestry projects under the flexible mechanisms of the Kyoto Protocol (which were expressly stated as being not acceptable for compliance purposes in the EU ETS) a political compromise was reached in respect of large hydro projects. That was to provide that where a member state was exercising its role under the flexible mechanisms of the Kyoto Protocol to approve the participation of an entity in respect of such a project, it would assess whether the criteria and guidelines of contained in the World Commission on Dams 2000 Final Report would be respected.[28] Although this requirement is not focused on the point of compliance it has caused a residual uncertainty regarding the risk that one or more member states could introduce a control at the point of compliance. Considering the legal difficulties that a member state would face in doing so during the current Phase, it is arguable that the market concern regarding this issue is over-stated. However, the existence of this uncertainty regarding the treatment of international offsets issued in respect of large hydro projects has caused a ripple effect throughout the emissions trading market in Europe. This has resulted in a degree of fragmentation in the market, indicated by the lower price for such international offsets and the increasing prevalence of buyers seeking to exclude sellers from supplying such offsets under market transactions.

2.4 Robust regulation and enforcement

Market confidence in an emissions trading system is based in part on the extent to which the market forms the view that there is both a capable regulator with the power to ensure the overarching design is adhered to and that its decisions

[26] By focusing on products the aim is to ensure harmonized free allocation applies equally to all producers of the same product, regardless of local circumstances or industrial inputs.

[27] Amending Directive, Art 10a, paras 1 and 2.

[28] Article 11b, 6 of the EU ETS Directive.

are not subject to political interference or likely to be consistently overturned by judicial challenges.

The European Commission's decisions in relation to the National Allocation Plans proposed for Phase II of the Scheme demonstrated the importance of a strong regulator but also the difficulties posed by attempting impose such regulation so under a decentralized system. When the Commission's guidance to member states failed to result in proposals that included significant reductions, the Commission reacted strongly by stating that it would assess the proposed plans in a manner guaranteeing sufficient scarcity.[29] In practice, this meant significant cuts to member states' proposed caps to ensure an overall cut of approximately 6% below verified emissions for 2005. To do so, however, required the Commission to take a robust view of the scope of its mandate under the EU ETS Directive.[30] This has been the focus for those member states which had their proposed caps significantly reduced. A number have applied to the Court of First Instance to have the Commission's decision in respect of their proposed cap annulled.[31] Importantly though, no expedited procedures were granted.[32] In most cases the relevant member states therefore adjusted their caps to align with the Commission's decision. The majority of cases have not been withdrawn, thus creating a residual risk regarding the caps imposed by the Commission.[33]

If the design of the EU ETS succeeds in creating real scarcity and the attendant high price for carbon, there is likely to be a significant rise in judicial

[29] J Sepidus, 'Scarcity and Allocation of Allowances in the EU Emissions Trading Scheme—A legal analysis' (September 2007) NCCR Working Paper, 16 <http://papers.ssrn.com/sol3/papers.cfm?abstract_id=1088726> access 7 April 2009. See also R Betz, K Rogge & J Schleich, 'EU Emissions Trading: An early analysis of national allocation plans for 2008–2012' (2006) 6 Climate Policy 361–94.

[30] Particularly those aspects of Article III of the EU ETS Directive relating to a cap taking the potential to reduce emissions and that it must comply with the rules on competition and State aid.

[31] Current cases pending before the Court of First Instance are Bulgaria, Czech Republic, Hungary, Romania, Poland, Latvia, and Estonia. Slovakia withdrew its case after reaching a resolution with the Commission. By Spring 2009 the Polish and Estonian cases have been heard but no judgment given. Many of the pleadings submitted allege not only procedural failings but that it had exceeded its powers of evaluation under the EU ETS Directive. See Sepidus (n 29 above) 19.

[32] Poland applied to suspend operation of the decision, but the application was dismissed by the European Court of First Instance on 9 November 2007 (Case T-183/07). Undiscouraged, Poland launched legal proceedings on 25 May 2008 to challenge the Commission's decision to cut its proposed EUA allocation by 26.7%.

[33] The decisions in Case T-374/04 *Federal Republic of Germany v Commission of the European Communities* (7 November 2007) indicates that if these cases proceed there is the potential for the Commission's approach to be subjected to adverse judicial comment. In particular, although the Court will not seek to take the place of the Commission where it has had to make complex economic or ecological assessments, it will assess closely whether it is vitiated by manifest error or misuse of powers, that it did not exceed the bounds of discretion and that procedural guarantees have been observed. See the above-mentioned case where it was found that the Commission had failed to substantiate factually or legally its argument that increases to the new entrant's reserve were anti-competitive or that the Annex III and Art 9 of the EU ETS Directive required the allocation of allowances under a National Allocation Plan to be absolute.

challenges to the decisions of regulators such as the European Commission and equivalent regulators at the member state level. The reforms agreed for Phase III have, in certain respects, pro-actively sought to reduce the potential for this situation to arise by further centralizing the role of the European Commission. However, there remain a number of likely points for ongoing issues. Those include:

- differences between the European Commission and member states on the implementation of allocation auctions for Phase III;
- differences between member state regulators and compliance entities regarding the application of benchmarking rules; and
- differences at both the European and member state level regarding the implementation of the rules relating to industries exposed to carbon leakage.

Another critical layer to ensure the functioning of an emissions market concerns the financial penalties that apply if a compliance entity fails to meet its obligation to submit allowances matching its verified annual emissions. Where such financial penalties are not sufficiently high, they will act as an indirect price cap. This should not occur within the EU ETS as there is high penalty for non-compliance (currently €100 per tonne of additional CO2 emitted that is not covered by a surrendered allowance)[34] and an on-going obligation on the defaulting entity to surrender in the subsequent year allowances sufficient to cover the excess tonnes. However, such penalties must also be consistently imposed on those entities breaching the scheme rules in order to ensure compliance entities are under a real incentive to participate. The approaches of certain member states in this respect have undermined market engagement by compliance entities within those member states. For instance, enforcement in Italy against non-compliant installations has not been as vigorous as the designers of the EU ETS envisaged.

Robust monitoring and verification of emissions from covered installations is also essential in order to ensure the legitimacy and effective functioning of an emissions trading scheme. This is an area where a consistent regulatory approach within the EU ETS has been lacking. As the scarcity of allowances deepens and the consequent cost of compliance cost increases, this will become a more critical issue. The capacity to improve the differential regulatory approach taken across the EU ETS at the member state level has been agreed under the Amending Directive by providing that a Regulation will be adopted by 31 December 2011 for the monitoring and reporting of emissions based upon the existing principles in Annex IV to the EU ETS Directive.[35] That date is also the deadline for the adoption of a Regulation relating to the verification of emission reports and the

[34] The Amending Directive for Phase III includes a mechanism to increase this penalty in line with CPI.

[35] Amending Directive, para (16) amending Art 14 of the EU ETS Directive.

accreditation and supervision of verifiers, including accreditation conditions, mutual recognition, and the evaluation of accreditation bodies.[36]

2.5 Technically capable

Emissions trading markets differ from other commodity markets in that the 'commodity' is created by regulation and has no physical representation other than an electronic record in a registry system. The importance of the ability of such technical systems to operate in a secure and reliable manner is often underestimated. As will be discussed in Section 2, physical systems relating to the transfer and holding of allowances are a key concern of market participants. This concern has driven the development of industry-wide contractual positions to manage issues associated with the technical systems underpinning the market.

The underestimation of the importance of such systems to market participants can also be identified in the context of significant changes to existing systems. The painful development and implementation process for the connection of the Independent Transaction Log (ITL) to the European system of National Registries linked via the Community Independent Transaction Log (CITL) is a case in point. The failure to meet agreed timelines and provide reliable information regarding that process had a number of repercussions for the market for allowances in Europe, including:

- contractual uncertainty as counterparties to physical transactions for certain types of allowances developed bespoke provisions to address the possibility that the physical means of delivery would not be available at the time for delivery;
- industry costs associated with the development of a common position to address the issue associated with the delay in the linking of the ITL to the CITL and the deployment of that position across all existing trading arrangements;[37] and
- delays in the physical issuance of allowances in 2008 as some member states elected to delay such issuance until the issues with the linking of the ITL to the CITL were resolved.

Whilst reforms introduced by the Amending Directive for Phase III (such as the shift to a single European registry) are aimed at enhancing the technical reliability of the Scheme, any such significant transition bears with it a similar risk of delay and market disturbance. This will be accentuated by the need for that

[36] Amending Directive, para (17) amending Art 15 of the EU ETS Directive.

[37] The difficulty of achieving such consensus should not be under-estimated. It is usually only where a significant market disruptive event is imminent that industry groups begin to operate cohesively, thus leaving limited time to agree a common position and then implement that position across the existing trading documentation between parties.

technical transition to occur within the context of a series of other reforms occurring in a similar timeline. For instance the introduction of auctioning will likely result in different approaches to auction platforms, the introduction and operation of which may impact on the allocation of allowances in 2013. Finally, once operational, the registry will need to be highly reliable in light of the fact that, as the central registry, any failure of it will impact immediately on the entire market.

3. Market-driven Components of Secondary Market Trading

An emissions trading scheme having the key characteristics outlined in Section 1 ensures that a large portion of significant emission sectors recognize that the requirement to offset their emissions is unavoidable and thus must be managed. The components of the secondary market discussed in this Section evolved in order to enable such installations not only to acquire allowances but to manage the price and delivery risks associated with their engagement with the market for allowances.

The nature and development of the market components operating in the EU ETS closely follow those developed in other commodity markets. This was enhanced by the nature of the 'learning by doing' aspect of first Phase of the EU ETS. For many installations the quantity of allowances allocated and the fact that such allocation was for free meant that, if it elected, an installation could rely upon its allocation of allowances to offset its annual emissions. Many elected to do so. Power generation companies (particularly those in deregulated markets such as the United Kingdom) did not, because of both their experience in trading in similarly structured commodity markets and, as significant emitters, their greater need for managing their compliance position.[38] The market has therefore been dominated by power companies managing their compliance position and financial institutions servicing their needs (as well as trading for their own profit generation). Those participants had the knowledge and risk management tools developed through participation in well-functioning commodity markets to quickly apply them to the new market for allowances. As a consequence the existing market components from such commodity markets have become the norm for the emissions trading market in Europe.

The deployment of existing market components and norms used in commodity trading assumes that an allowance is akin to a physical commodity and should be treated similarly for regulatory purposes. The Scheme itself did not

[38] On price interaction in the electricity sector see J Reinaud, 'CO2 Allowance Electricity Price Interaction: Impact on Industry's Electricity Purchasing Strategies in Europe' (IEA Information Paper (February 2007) <http://www.bvsde.paho.org/bvsacd/cd63/price-interaction.pdf>, accessed 7 April 2009.

legislate the legal nature of an allowance. By doing so the legal norms of each member state were respected. However, this has meant that market participants that wish to engage in the trade of such allowances must assess the regulatory treatment of doing so on a jurisdiction by jurisdiction basis. The evolution of the treatment of emissions trading from a financial service perspective is illustrative.

In most member states, the deployment of existing market components common to commodity trading was assisted by the similar regulatory treatment (or lack thereof) given to emissions trading as to commodity trading under domestic financial services regimes. For instance, the domestic regulator in the United Kingdom, the Financial Services Authority, made the critical decision prior to the commencement of the Scheme that the intangible nature of an allowance did not justify a different approach to their regulatory treatment of other tangible commodities. The Financial Services Authority therefore regulates the trade of allowances in the same manner as they do other commodities such as oil or natural gas. In essence, its regulation of such markets arises not in relation to the underlying commodity but in relation to derivative instruments in respect of the underlying allowance and the purpose for which such instruments are being utilized. The practical consequence is that market participants that confined their participation in the market to certain types of transactions, such as spot transactions or forward transactions that are entered into for commercial purposes, could remain outside the ambit of financial services regulation. Broadly speaking, those entities that go beyond those areas will be obliged to be authorized (meaning that they will have to comply with the applicable rules of the Financial Services Authority) or ensure that they access the market via an authorized entity. For key sectors such as power companies operating in the United Kingdom, the overall impact of this was that the extension of their trading operations to include allowances was unlikely to require them to adjust their approach to financial services regulation. This was particularly important for companies which had structured their trading operations in a similar manner for other markets such as power and gas. However, for participants that intended to be actively engaged in multiple jurisdictions, it was necessary to assess the treatment of the underlying commodity and derivatives in respect of it under the financial service regulation of other member states.[39]

This situation has been significantly impacted by the implementation of the Markets in Financial Instruments Directive or MiFID.[40] The types of transactions relating to allowances that are subject to such regulation has, along with

[39] For instance, the implementation of the Scheme in Sweden included the characterization that the allowance itself was a financial instrument.

[40] Directive 2004/39/EC of the European Parliament and the Council of 21 April 2004 on markets in financial instruments and amending Council Directives 85/611/EEC and 93/6/EEC and Directive 2000/12/EC of the European Parliament and the Council and repealing Council Directive 93/22/EEC.

the trade of commodities more generally, been made subject to a more consistent pan-European approach. In essence, the MiFID regime applies to derivative contracts which relate to a limited set of underlying classes (including commodities and emissions allowances[41]) and which meet certain other criteria, such as whether the derivative contract in question is traded on a regulated market or multilateral trading facility, whether it is financially settled, or whether it has the characteristics of other derivative financial instruments.[42] For financial institutions engaged in derivatives trading relating to allowances this has the advantage of improving their ability to operate on a cross-border basis without being subject to regulatory requirements additional to the MiFID requirements and those of their home jurisdiction.[43]

3.1 Understanding the market for emissions trading in Europe

Understanding the market for emissions trading in Europe can be best approached through an analysis of two types of trades and two market segments. The types of trades are spot transactions and forward transactions, and the segments of the market are the Over-The-Counter (OTC) market and the exchange market.

3.1.1 Spot transactions

Spot contracts are for a specified quantity of allowances where delivery will occur immediately or within short period of time after the transaction is agreed. Its importance as a transaction type derives in part to the approach of financial services regulation discussed above. Spot contracts for physical delivery (that is, a transaction which is settled by means of the actual delivery of the allowances) are generally not treated as a regulated activity.

[41] Although the treatment is broadly similar, MiFID did not characterize an emissions allowance as a commodity, but as an 'exotic' derivative, a separate type of underlying class that includes freight rates and climatic variable in addition to emissions allowances.

[42] J Herbst, S Lovegrove, & M Woodward, 'Commodity Derivatives' in M Elderfield (ed) *A Practitioner's Guide to MiFID* (City and Financial Publishing, 2007) 313–18; C Bates, 'Scope, Authorisation and Passporting' in M Elderfield (ed) *A Practitioner's Guide to MiFID* (City and Financial Publishing, 2007) 52–5.

[43] Although it should be noted that the domestic application of MiFID has, in certain cases, been somewhat convoluted. For instance, the United Kingdom implemented MiFID by adding the MiFID definitions of commodity derivatives to the existing definitions of regulated derivatives covered by the existing UK regulatory regime. As a consequence, where an activity is being carried out in the UK, the financial services regulation assessment will need to consider whether the activity is regulated under the UK regulatory regime. If it is, it remains regulated post-MiFID. If it is not regulated under the UK regulatory regime, the second step is to determine whether it would be regulated under the MiFID definitions.

3.1.2 Forward transactions

Forward transactions are for a specified quantity or quantities of allowances where delivery and payment are at point in time in the future sufficiently distant that it is no longer a spot transaction. The term 'forward' is used to refer to such transaction occurring in the OTC market (discussed below). Within the OTC market the customary forward delivery date for each calendar year is 1 December. Transactions where settlement occurs in the future on an exchange are referred to as 'futures' (discussed below). The settlement date for such futures contracts is not fixed but generally falls around mid-December.

3.1.3 OTC Market

This segment of market refers to transactions which are negotiated and entered into between two parties (either directly or via the services of a broker). Whilst there is no obligation to do so, most market participants use common contractual templates. These are designed as contractual frameworks by which authorized representatives of the parties can thereafter agree multiple transactions for allowances. Depending on the type of framework or 'master' agreement used, such transactions can cover physical forward delivery or a range of financial derivatives associated with the underlying commodity.

3.1.4 Exchange market

This segment involves a number of regulated electronic trading platforms with standardized contracts determined by the exchange. They enable the real-time anonymous matching of buyers and sellers under such contracts. Performance and credit risk to such transactions is removed by the exchange ensuring, via a clearing house, that the non-performance by one party to a contract will not affect the other party. Different exchanges offer different types of contracts but include both spot and futures contracts.

These components and their interaction are considered further below. However, an overarching theme common to both the design elements of a market discussed in Section 1 and the market components in this Section is that of certainty. In the former case it is the certainty that the market is capable of defining a commodity that has a value driven by economic and physical factors rather than political ones. In this Section it is the certainty that a variety of transactions can be created between market participants with the confidence that the obligations associated with such transactions will be satisfied or, where disrupted, such disruption is dealt with on a similar basis across the market. This is most clearly seen in the treatment of credit issues and the management of physical delivery failures.

3.2 OTC Market

Any market participant wanting to engage in the regular trade of emissions allowances in the OTC market will need to establish trading relationships with a number of counterparties. To do so will likely require the market participant to elect to use one or more of the three main industry agreements used for documenting OTC transactions. Those are: the Emission Trading Master Agreement (ETMA) published by the International Emissions Trading Association (IETA); either one of the framework agreements published by the European Federation of Energy Traders regarding power of gas, but in each case including an Allowance Appendix; or an ISDA Master Agreement including an Emissions Annex to be included in the Schedule to it, published by the International Swaps and Derivatives Association.

Each of the industry framework agreements can be understood as involving three conceptual layers. The first is the overarching contractual terms that will govern the relationship between the parties (for instance, credit support, representations and warranties, termination events, and dispute resolution). The second layer concerns the provisions designed to govern the physical delivery and receipt of allowances (for instance, opening and maintenance of registry accounts, force majeure, grace periods for delivery failure, compensation for failed deliveries and payment timing). The final layer is the specific terms agreed in relation to each individual transaction entered into under the framework agreement (for instance, the number of allowances, the delivery date, and the price payable). The intent of the first two layers is to resolve all of the key issues between the parties when first entering into the framework agreement, thus leaving the authorized representatives of each party to agree the terms for each transaction without the need for further negotiation of legal documentation.

Although broadly similar, it has taken some years for the industry groups that maintain these documents to harmonize their approach to key aspects relating to the delivery and settlement of emission transactions. Whilst these differences persisted they created risks for market participants that entered into transactions under different framework agreements. To avoid such issues arising again, the industry response to the impact of issues such as the delay of the connection of the ITL to the CITL[44] involved a more integrated development process across the key industry groups. That process extended over 2007 and 2008 and eventually resulted in a common position being adopted by all of the key industry groups.[45] It is

[44] There were also potential issues relating to the transition of national registries developed for the EU ETS to satisfy certain criteria necessary for them to used for the cross-border transfer of allowances under the Kyoto Protocol.

[45] In essence, that position was to introduce a force majeure type event known as a Suspension Event that would have the affect of delaying the physical settlement of the transaction until the relevant issues were resolved. If there is no such resolution then, broadly speaking, the transaction will be terminated at the end of second Phase without any compensation being payable. If the issue is resolved and the transaction proceeds, a cost of carry is applied to increase the unit price for the affected allowances (calculated by reference to the length of the delay).

noteworthy that emerging emissions trading markets such as that due to commence in Australia appear likely to run a similar risk of a long harmonization process between competing templates. Early movers that have documented forward trades prior to the commencement of the Australian scheme are now using their approach as a proposed industry standard.[46] At the same time other industry bodies are seeking to develop template OTC documentation.[47] From a European perspective, particularly for those entities with international trading desks, it would be beneficial if new markets such as Australia adopted OTC documentation broadly similar to those existing in the EU ETS (such as the IETA, ETMA or ISDA Master Agreement). To the extent that alternative forms evolve, a strong effort should be made by industry bodies to ensure that key provisions relating to delivery and settlement are harmonized from the outset.

Whilst the establishment of an OTC trading relationship governed by an industry template will thereafter enable transactions to be quickly entered into, the process to do so can take a significant period of time. Most active market participants have a set of preferred provisions relating to one or more of the OTC framework agreements. Such positions will include preferred positions in respect of those elements of a framework agreement where an election can be made (for example, in relation to payment timing or methods for calculating compensation in the case of a force majeure type event resulting in termination).[48] A preferred position may also encompass additional clauses or required amendments to the framework agreement (such as replacing subjective tests for credit related default triggers with objective triggers such as a credit rating downgrade). For market participants with multiple trading relationships, ensuring the terms of the framework agreements are on the same terms is an important risk management tool.[49] This driver creates difficulties for parties to compromise on their preferred terms.[50] The tension between strongly held

[46] Author's discussions with industry participants.

[47] The Australian Financial Markets Authority has initiated a documentation development process that will likely develop ISDA-based documentation.

[48] By way of example, from a pool of 15 fully engaged OTC European market participants: only one elects for there to be no payment in the case of such termination; another four prefer a method using market quotations and the rest prefer to elect to use a method using a broad concept of loss. (Author's private survey of preferred positions of market participants for IETA ETMA.)

[49] For instance, by enabling a party to have the confidence that the only circumstances where a default could be triggered in respect of their financial wherewithal is following a credit rating downgrade.

[50] A tendency enhanced by the desire of market participants to establish a preferred position and then have that implemented by documentation teams. Those documentation teams often have limited authority to grant concessions, thus creating delays as negotiators unable to shift their position continue to negotiate.

preferred positions can be most clearly seen in the divergent positions often taken by financial institutions and compliance buyers. Areas of particular tension include: (i) whether a failure to deliver should attract a potential liability for the Excess Emissions Penalty[51] that the compliance buyer could be exposed to due to the delivery failure;[52] (ii) the length of time any suspension event may continue before a transaction is terminated; and (iii) the tendency of large financial institutions to require margining over other types of credit support, such as parent company guarantees.

3.3 Exchange Trading

Electronic exchanges for the trade of emission allowances serve an important function in the EU ETS. They provide market liquidity through:

- an electronic platform that can match up buy and sell orders;
- guaranteed anonymity for participants;
- reduced default risk through clearing mechanisms;
- rapid execution of transactions; and
- transparency of the intra-day and closing average price of transactions.

Exchange trading emerged quickly after the commencement of the Scheme due to the ability of existing exchanges operating in European energy markets to apply the same electronic order matching systems to the new commodity. Different exchanges offered different products, broadly distinguishable by those offering spot contracts and those offering futures contracts. The European Climate Exchange (ECX)[53] has emerged as the dominant exchange for the trading of futures contracts. The dominant exchange for spot contracts is Bluenext.[54] The ECX is an apt example of the key components of an emissions exchange and its regulation.

[51] The Excess Emissions Penalty is the fine imposed on an installation having compliance obligations under the EU ETS that does not submit allowances matching its annual verified emissions. It is set at €100 for the current Phase. Importantly, payment of the Excess Emissions Penalty does not remove the obligation to submit allowances meeting the relevant shortfall.

[52] This has been a recurrent tension since the commencement of the Scheme and has driven the evolution of relevant provisions in industry documentation: for instance, there now exists a mechanism whereby a period of time before a compliance deadline can be agreed by the parties as an EEP Risk Period. Any failure to deliver during this period may result in a liability for the EEP on the delivering party at fault.

[53] The European Climate Exchange is a member of the Climate Exchange Group plc group of companies (which also includes the Chicago Climate Exchange and Chicago Climate Futures Exchange). ECX carbon contracts are listed for trading on the ICE Futures Europe (the former International Petroleum Exchange. All contracts are cleared by ICE Clear Europe.

[54] BlueNext was incorporated on 21 December 2007 by a partnership between NYSE Euronext and Caisse des Dépôt.

Market participants may trade on the ECX by two means: direct trading and order-route trading.

3.3.1 Direct trading

This is the route taken by large-scale traders and involves becoming a either clearing or non-clearing member. In each case the member will need to execute participation agreements binding it to the exchange rules. Non-clearing members will also need to enter into a clearing agreement with the clearing house for the exchange. Trading fees are applied on a per transaction basis (but at a lower rate than for market participants using the route described below).

3.3.2 Order-route trading

This involves placing orders (either electronically or via phone) via a member of the exchange. The customer's contractual relationship is with the member who provides such access. Trades arising from this access are registered in the name of the member, thus enabling the customer to remain anonymous.

The role of the clearing house is critical to the structure of an exchange. The clearing house acts is the counterparty to each transaction. That is, the matching of a seller and a buyer via the exchange creates a sale obligation between the seller and the clearing house and a purchase obligation between the buyer and the clearing house. Buyers and sellers remain anonymous and have no exposure other than to the clearing house. The cost for obtaining the benefit of the removal of credit and default risk is the obligation on the part of participants is not only the trading fees but also the obligations to comply with a margining system. This involves the obligation to pay an initial margin each time a new position is opened and comply with ongoing variation margining. The variation margin involves the daily calculation of the difference (either positive or negative) between the participant's position on the exchange and the relevant market price for that trading day. Cash amounting to the participant's exposure must be transferred the next day to the clearing house of its position will be automatically closed. Members that provide participants with access to the ECX, pass through such margining obligations to those participants, often requiring a sum to be held on deposit to meet margin calls. The costs associated with exchange trading along with the above margining system have supported the dominance of the OTC market over the exchange market in Europe. However, as international financial turmoil accelerated in 2008 it brought a marked increase in exchange trading as market participants sought to reduce their exposure to the performance risk of their counterparties.[55]

[55] Volumes on the ECX in the latter half of 2008 almost tripled from the same period in 2007. See B Farey, Bloomberg News, 12 January 2009. This is also enhanced by the ability of parties to an OTC transaction to transfer its settlement to an exchange such as the ECX.

3.4 Impact of future developments

The design, timing, and harmonization of approach across Europe to the auctioning of allowances in Phase III have the potential to have a significant impact on the secondary market. In assessing how this will play out, the auction regime developed and implemented by the UK government in respect of its Phase II auctions provides an example of an attempt to meet similar objectives to that given to the European Commission. A key element was the approval of 'Primary Participants' that were required to meet certain criteria, including their capacity to offer access to the auction for no cost to entities with compliance obligations (Indirect Participants).[56] In effect, this limited the likely Primary Participants to large financial institutions. The access each Primary Participant was obliged to offer Indirect Participants was at no cost but brought with it a number of obligations. Those included: identity checks; performance obligations; document retention for two years; the establishment of internal information barriers; and, importantly, managing credit risk in respect of Indirect Participants.[57]

Whilst the auction conducted by the UK government under this scheme managed to attract four Primary Participants, the lack of a cost recovery mechanism for the obligations imposed on Primary Participants limited the attractiveness of the role.[58] There was also no positive incentive to locate and approach Indirect Participants (in fact, the lack of cost recovery could be seen as a disincentive for doing so). The outcome was that a mechanism designed to grant wide-spread access to an auction did not fully achieve its goals, as Indirect Participants found there was a limited range of Primary Participants available.

This raises an important tension that will increase during the development of the Regulation in respect of Phase III auctions. Whilst certain parameters will be necessary to ensure open access to the auction regime, it is clear that if these are overly restrictive or burdensome they are likely to reduce competition between qualified service providers such as the Primary Participants mentioned above. It is arguable that the design process for this Regulation will test whether the underlying principles discussed in this chapter remain correct, namely that regulators such as the European Commission will establish the market parameters but otherwise allow the market to operate to identify and fill the niches created by such parameters.

[56] The Community Emissions Trading Scheme (Allocation of Allowances for Payment) Scheme 2008, a Treasury Scheme made under the Finance Act 2007, c11.

[57] The UK Government's objective in limiting such cost recovery was actually to allow a full range of buyers to participate at low cost, particularly small compliance buyers. See 'Consultation on Proposed UK Auction Design for Use during Phase II' (Defra, December 2007).

[58] Authors' private research. See also Defra, 'Summary of Feedback of first UK auction in Phase II of the EU Emissions Trading System' (January 2009) <http://www.defra.gov.uk/ENVIRONMENT/climatechange/trading/eu/pdf/feedback-summary-160109.pdf>, accessed 7 April 2009. The UK auction rules were subsequently changed to include a small service fee for Primary Participants.

The above is one example of the impact of the reforms agreed for Phase III to impact on the secondary market in Europe. The potential impact of other reforms such as the move to a central European registry were also discussed in Section 1. A debate that arose in the development of the Amending Directive was whether international offsets from certain types of projects should be excluded from use for compliance purposes in Phase III. This was not entirely resolved as the Amending Directive allows a European regulatory procedure to determine which, if any, should be so excluded. For reasons similar to those discussed in Section 1 regarding the treatment of CERs from large hydroelectric projects, the residual risk that CERs from certain types of projects could be barred from use in the EU ETS will, until resolved, cause on-going disturbance in the consistency of treatment of various types of allowances as a fungible commodity. The longer this continues the more likely it is that market participants will look beyond the offset being acquired to the source project. Finally, the use of CERs in Phase III will also involve, in practice, the exchange of such CERs for European issued allowances. The method by which this will occur and the allocation of risk regarding delays associated with such an exchange will also be a matter for debate within industry groups.

4. Transaction Evolution Through Primary and Secondary Market Interaction

As the global carbon markets have rapidly expanded in recent years so has the terminology of the market. The broad terms 'primary market' and 'secondary market' are examples. They capture the interplay between the markets where the generation of emission reduction credits such as CERs and ERUs are physically located and secondary markets such as those discussed in the preceding Sections. Between them lies a fertile area of interaction that drives innovation in the design of transactions.

Each of the elements discussed in Sections 1 and 2, whether government or market initiated, has the objective of creating the conditions by which each trade of a volume of allowances within the EU ETS is a firm transaction with little likelihood of failure. Where physical events occur there are agreed mechanisms to provide certainty as to their contractual treatment. Similarly, there are contractual compensation mechanisms to address a failure to deliver. This certainty is not replicated in the primary market. This creates an inherent tension in the transition from the primary to secondary market. We will not look in detail at the primary market and the various risks that may impact on the development of a successful emissions reduction project.[59] However, the existence of those risks combined with the development of market practice in key countries such as China[60] has ensured that a significant number of primary market transactions are on a

[59] See further chapters 9, 11 and 12. [60] See chapter 22.

non-firm basis. That is, whilst a project developer will be obliged to diligently implement and operate the emission reduction project, the remedies for non-delivery are likely to be limited. Furthermore, the requirement to provide credit support by either the seller or the buyer has not been prevalent (although this has been changing in recent years).

4.1 Relevant market conditions

The above interplay has created the following market conditions, which are relevant to the emerging transaction types discussed further below. These are generalizations for which many exceptions can be found. However, they remain broadly accurate.

(1) A range of small to medium carbon originators emerged that have been successful at locating projects. These originators often have the benefit of a number of primary market contracts in their favour for the delivery of CERs. Almost all of those contracts will be on a non-firm basis.

(2) Primary market originators can sell forward on a non-firm basis into the European secondary market but such sales will be at a discounted level due to lack of certainty regarding the delivery of such CERs. For volumes to be sold on a firm basis, the seller will need to be able to meet the requirements of the secondary market discussed in Section 2. Critically, this includes the ability to provide sufficient credit support to cover the exposure of the buyer to the risk of non-delivery, the cost of which for some originators may undermine the value of the transaction.

(3) Market participants that had expanded their focus in recent years from the secondary market to identifying opportunities to purchase directly in the primary markets have discovered that successfully doing so is difficult, particularly for compliance buyers with limited experience of operating in the key primary markets.

4.2 Emergence of innovative transactions

These conditions have caused the emergence of innovative transactions. A particular trend has been increasing occurrence of primary market originators and secondary market participants mutually recognizing that each has an expertise the other party requires. Aggregation structures have therefore been developed between primary market originators and secondary market participants in order to provide: (i) primary market originators with better access to secondary market pricing for carbon credits without having to establish a significant trading capacity in such markets; and (ii) secondary market participants with cost effective access to primary market volumes.

Relatively simple structures involve a secondary market participant agreeing that, subject to certain conditions being met regarding the nature of the project identified and the contract under which the primary market volumes will be acquired, it will acquire the volumes from that project on a back to back basis. The price payable under that transaction provides the originator with a base return and the potential for a share of any upside arising from the onward sale of such credits. Common issues that arise in these transaction structures include:

- adequately defining the conditions under which the secondary market participant will be bound to acquire the volumes located by the primary originator;
- the scope of exclusivity arrangements between the parties; and
- the mechanism by which any shared upside is to be calculated and distributed (for instance, whether on the basis of realized returns or by reference to market pricing at or around the time of delivery).

More complex legal issues arise where the commercial structure creates the necessity for the secondary market participant to expose itself to the risk of loss where the originator does not perform. Examples of where this could arise include where a prepayment structure is included that further rewards the originator for the notional volumes brought to the transaction. An alternative example is where the generation of secondary forward sales from the aggregated volumes will involve the forward sale of a portion of the estimated volumes on a firm basis.

In both cases the secondary market participant has exposed itself to two key risks. The first relates to the performance of the projects themselves. By prepaying for a notional volume or entering into forward transactions the secondary market participant has been required to form a view on the likely delivery profile of the volumes acquired by the originator—thus exposing itself to potential losses if those volumes do not arise. Secondly, where the originator fails to perform its obligations (in an extreme case, by becoming insolvent) the secondary market participant is exposed to the risk that it will lose the benefit of the contractual chain by which it would have received delivery of the volumes it requires. This is because the volumes due to the secondary market participant will be transferred from the primary market developer to the primary market originator and then to the secondary market participant. The provision by the originator of customary security arrangements such as a letter of credit is generally not palatable in these circumstances, because the underlying intent of such aggregation structures is to minimize just those type of collateral requirements that an originator would need to meet under full engagement with the secondary market. Alternative transaction structures have therefore been developed to provide with the secondary market participant with adequate comfort that the risk of non-performance is adequately managed.

The most common technique used to date has involved novating all or part of the portfolio of primary market contracts that the originator has the benefit of to a special purpose vehicle (that is, a separate company established for the purpose of holding such contracts). A key benefit associated with this approach is the enhanced ability to put in place an adequate security package over that new vehicle and its assets (being the novated primary contracts). Additional protection can be obtained by ensuring that the new vehicle is owned and managed by an entity separate from the primary market originator. Common issues arise though in respect of such transaction structures. The primary one is the practical delay issues associated with novating a number of primary market contracts to the holding vehicle. This is due to the need for each such novation to be agreed with a primary market seller (which may have no incentive to provide its consent). Further structuring considerations include: determining how the new holding vehicle will fund its ongoing financial obligations under the primary market contracts; the role of the primary market originator in continuing to manage the contracts held by the vehicle (including in particular under what circumstances its management role can be terminated); and the level of control required by the secondary market participant over the holding vehicle.

The above issues have been satisfactorily resolved in a number of transactions involving significant aggregated volumes of CERs that were either sold to a single secondary market participant or privately marketed to a number of such participants. The commercial and legal issues with documenting such transactions tend to require large volumes in order to justify the investment of resources by the parties involved. It could therefore be argued that the impact of the collapse in the secondary market prices in Europe in 2009 could have a limiting effect on further significant aggregation structures. This would be incorrect though. For instance, as the margin between the primary and secondary market reduces, aggregation transactions are likely to arise where the role of financial institutions in developing and brokering such transactions is reduced in favour of transactions that bring a number of sophisticated primary market project owners and secondary market compliance buyers together. This is the benefit of a functioning international market based around well-designed cap-and-trade schemes. As the economic incentives shift, new solutions are found.

PART IV

CARBON TRADING OUTSIDE KYOTO: REGIONAL SCHEMES

16

The European Union Emissions Trading Scheme

*Markus Pohlmann**

1. Introduction

1.1 Overview

On 29 April 1998, the European Community (EC), one of the pillars of the European Union (EU),[1] and the initial 15 EU member states signed the Kyoto Protocol.[2] Under the Kyoto Protocol,[3] the EC and the EU member states are committed to jointly reduce their greenhouse gas (GHG) emissions[4] during the period 2008–2012 by a total of 8% below 1990 GHG emissions levels; this is often referred to as the 'EU bubble'. In order to better reflect differing GHG emission reduction potentials among EU member states, the 15 EU member states entered into the Burden Sharing Agreement in June 1998. Under the Burden Sharing Agreement, they translated the 8% emission reduction target of

* Markus Pohlmann is a legal counsel at the World Bank. The views expressed in this chapter are the views of the author and do not necessarily represent the views of the World Bank.

[1] The EU was formally established in Maastricht by the Treaty on European Union dated 7 February 1992 (TEU); it is based on three pillars, namely (1) the European Communities (consisting of the European Community (EC, which was formerly known as 'European Economic Community') and the European Atomic Energy Community (Euratom)), (2) a common foreign and security policy, and (3) cooperation in the field of justice and home affairs. The EU has a single institutional framework for all three pillars. The main legislative authority is vested in the European Parliament, the Council of the EU (Council), and the European Commission.

[2] The EU itself was not endowed with legal personality in the TEU; therefore, the treaty-making power in the area of environmental policy on the EU level rests with the EC. But since the EC is not endowed with exclusive treaty-making competence in the area of environmental policy, the EU member states themselves were also able to negotiate, sign, and ratify the UNFCCC and the Kyoto Protocol (under the so-called 'mixed' competence). The EC and the EU member states ratified the Kyoto Protocol on 31 May 2002. However, following a long struggle for survival, the Kyoto Protocol did not enter into force until 16 February 2005.

[3] Kyoto Protocol to the United Nations Framework Convention on Climate Change (adopted 11 December 1997, entered into force on 16 February 2005) (1998) 37 ILM 22 (Kyoto Protocol).

[4] The Kyoto Protocol covers six major GHGs: carbon dioxide (CO_2), methane (CH_4), nitrous oxide (N_2O), hydrofluorocarbons (HFCs), perfluorocarbons (PFCs), and sulphur hexafluoride (SF_6).

Table 16.1 27 EU member states and their targets under the Burden Sharing Agreement (initial 15 EU member states (EU-15)) and the Kyoto Protocol (new 12 EU member states (EU-12))

EU-15	Burden Sharing Agreement target (%)	EU-12	Kyoto Protocol target (%)
Austria	−13	Bulgaria	−8
Belgium	−7.5	Cyprus	No target
Denmark	−21	Czech Republic	−8
Finland	0	Estonia	−8
France	0	Hungary	−6
Germany	−21	Latvia	−8
Greece	+25	Lithuania	−8
Ireland	+13	Malta	No target
Italy	−6.5	Poland	−6
Luxembourg	−28	Romania	−8
Netherlands	−6	Slovakia	−8
Portugal	+27	Slovenia	−8
Spain	+15		
Sweden	+4		
United Kingdom	−12.5		

the EC under the Kyoto Protocol into differentiated emission reduction or limitation targets of each EU member state.[5] Since 2004, 12 additional countries have joined the EU, all of which, except for Malta and Cyprus, have ratified the Kyoto Protocol and have accepted individual GHG emission reduction targets of 6% or 8% thereunder (see Table 16.1).[6]

Soon after signing the Kyoto Protocol, the EU started exploring the potential of an emissions trading scheme to help its member states comply with its GHG emission reduction targets under the Burden Sharing Agreement and the Kyoto Protocol. Despite the Kyoto Protocol not being in force and strong opposition from powerful EU member states, such as Germany and France, the European Union Emissions Trading Scheme (EU ETS) was finally adopted in October 2003 with the objective of reducing EU GHG emissions in a cost-effective and economically efficient manner.[7]

[5] The Burden Sharing Agreement was rendered legally binding among EU member states by Council Decision (2002/358/CE); Council Decision (EC) 2002/358 of 25 April 2002 concerning the approval, on behalf of the European Community, of the Kyoto Protocol to the United Nations Framework Convention on Climate Change and the joint fulfilment of commitments thereunder [2002] OJ L130.

[6] These new EU member states are not covered by the Burden Sharing Agreement but, except Malta and Cyprus, have individual emission reduction targets under the Kyoto Protocol.

[7] Council Directive (EC) 2003/87 establishing a scheme for greenhouse gas emission allowance trading within the Community and amending Council Directive 96/61/EC (EU ETS Directive) [2003] OJ L275/32 [hereafter: EU ETS Directive 2003/87].

The EU ETS was set up as a cap-and-trade regime initially only covering CO_2 emissions. It is a downstream private entity-based emission trading regime, ie the point of regulation is the installation releasing GHGs into the atmosphere.[8] Under the EU ETS, each covered installation located in the EU is issued an overall capped amount of tradable EU allowances (EUAs) for a certain trading period; each installation is required to annually measure, monitor, and report its CO_2 emissions and annually surrender an amount of EUAs equal to the amount of its CO_2 emissions during the preceding year. The EU ETS started operations on 1 January 2005 with the first trading period occurring between 2005 and 2007 (Phase I); this is often referred to as the 'pilot' or 'trial' phase. Phase I was followed by a second trading period that takes place between 2008 and 2012 (Phase II). As of today, the EU ETS covers all 27 EU member states in addition to Norway, Iceland, and Liechtenstein, which were all linked to the EU ETS as non-EU member states.[9]

After only a few years of operation and despite a lack of prior experience in market-based environmental policy instruments at the EU level,[10] the EU ETS has grown from what was initially perceived as a 'policy experiment' into the flagship of EU climate policy. In many ways, it serves as a role model and testing ground for the development of other national, regional, and international GHG emissions trading schemes worldwide. Today, the EU ETS is the world's largest multi-national, multi-sector, private entity cap-and-trade GHG emissions trading regime and it is the world's largest market for GHG emission credits. In October 2008, it had a monthly record trade volume of 355 million EUAs, thus, bringing the total number of EUAs traded in 2008 to around 2.2 billion.[11] Although the establishment of the EU ETS was motivated by the Kyoto Protocol, the EU ETS is independent of the Kyoto Protocol and is intended to continue to exist and operate even if the international community fails to agree on a post-2012 international agreement on climate change.

[8] In contrast, an 'upstream' emission trading regime would require upstream sources, such as entities involved in fossil fuel production, preparation, delivery and sales, to hold allowances for emissions embedded in the fuels they sell.

[9] Decision of the EEA Joint Committee (EEA) 146/2007 of 26 October 2007 amending Annex XX (Environment) to the EEA Agreement [2007] OJ L 100/92. The linkage of the EU ETS with Norway, Iceland and Liechtenstein (all of which are members of the European Economic Area (EEA) but not the EU) is based on an agreement reached between these countries and the European Commission on 26 October 2007 and was implemented through the incorporation of the EU ETS Directive into the EEA agreement by way of EEA Joint Committee Decision No 146/2007.

[10] At the time the EU ETS Directive was drafted and negotiated, EU institutions and stakeholders mainly built upon the experiences made by the USA with its company-based cap-and-trade emissions trading regimes under the US Clean Air Act of 1990 (national regime covering SO_2), and under the Californian Regional Clean Air Incentives Market of 1994 (regional regime covering SO_2 and NO_X). In addition, the UK contributed the experience made under its national voluntary company-based cap-and-trade emissions trading scheme (covering all six Kyoto GHG) which started in April 2002.

[11] As a comparison, during the same period in 2007, about 1 billion EUAs were traded within the EU ETS (*Point Carbon News*) 'Carbon Market Europe' (Oslo, 7 November 2008) vol 7).

This chapter will provide the reader with some historic and legal background information on the EU ETS, will put the EU ETS in the context of overall EU climate policy, and will explain and discuss the cornerstones of the EU ETS and its linkages to the Kyoto Protocol and other emission trading schemes in light of the current amendments to the EU ETS structure for the post-2012 era.

1.2 Background and legislative history

During the negotiations of the Kyoto Protocol, the EU and its member states opposed the introduction of any market-based flexible instruments into the international climate regime. Most EU member states were used to domestic command-and-control measures, such as energy taxes or voluntary agreements and the idea of having a free market regulate access to a global public good appeared suspicious. Nevertheless, in order to ensure that the Kyoto Protocol would enter into force, the EU conceded, and agreed for two project-based baseline-and-credit mechanisms (the Clean Development Mechanism (CDM) and Joint Implementation (JI)) and an international cap-and-trade emissions trading mechanism (IET) to be included in the final text of the Kyoto Protocol. With respect to IET rules and procedures, the Kyoto Protocol was very vague. It only stated that principles, modalities, rules, and guidelines for emissions trading still need to be defined and that those countries with a Kyoto target 'may' participate in emissions trading for compliance purposes under the Kyoto Protocol.

In June 1998, the European Commission started discussing emissions trading as part of an EU climate strategy and it stressed the EU's potential to 'play an important role in meeting commitments at less cost, thereby safeguarding the competitiveness of EU industry'.[12] The EU institutions' shifted towards market-based climate policy instruments for the following reasons: first, the EU wanted to prepare itself for a potential IET that was expected to become operational by 2008; second, key personnel in the Environment Directorate-General of the European Commission were replaced by economists who favoured economic policy instruments; third, the EU became increasingly aware of the need of market-based mechanisms to achieve its Kyoto target without compromising EU industry's competitiveness; and fourth, the earlier proposal of a directive on an EU-wide carbon/energy tax became more and more likely to fail.[13]

In March 2000, the European Commission took the initiative and presented a Green Paper on GHG emissions trading in the EU making a strong case for

[12] Commission (EC),'Climate change—Towards an EU Post-Kyoto Strategy' (Communication to the Council and Parliament) COM (98) 353 final, 3 June 1998.

[13] J Birger Skjærseth and J Wettestad, *EU Emissions Trading: Initiation, Decision-making and Implementation* (Ashgate Publishing: Aldershot, 2008) 74, 99.

the introduction of an EU-wide emissions trading regime.[14] Shortly thereafter, in June 2000, the EU launched its first European Climate Change Program (ECCP) as a multi-stakeholder consultative process to develop an EU-wide strategy to achieve its Kyoto target. Along with a wide range of additional EU climate policy measures,[15] the EU ETS was on the top of the agenda.[16]

In October 2001, the European Commission proposed a draft directive to the European Parliament and the Council establishing an EU-wide CO_2 emissions trading scheme.[17] Following US rejection of the Kyoto Protocol in March 2001, the EU faced increased pressure not only to demonstrate international climate policy leadership by saving the Kyoto Protocol, but also to set up a legal framework that would allow the EU ETS to begin in 2005. After two reading sessions in the European Parliament, in which around 80 amendments were initially requested (23 of which found their way into the final text of the directive), the European Parliament and the Council officially adopted Directive 2003/87/EC establishing the EU ETS on 13 October 2003 (EU ETS Directive).[18] Since the EU ETS was adopted by way of a directive,[19] each EU member state had to transpose the directive into national law by 31 December 2003.[20]

On 23 July 2003, the European Commission proposed an amendment to the EU ETS Directive incorporating a link of the EU ETS to project-based CDM/JI credits under the Kyoto Protocol, namely certified emission reductions (CERs) and emission reduction units (ERUs). Following intensive discussions on

[14] Commission (EC), 'Green Paper on Greenhouse Gas Emissions Trading within the European Union' COM (2000) 87 final, 8 March 2000.

[15] Eg promotion of renewable energy (Directive 2001/77/EC), biofuels (Directive 2003/30/EC), biomass and cogeneration of heat and electricity (Directive 2004/8/EC), energy efficiency in buildings (Directive 2002/91/EC), industrial and agricultural installations (Directive 1996/61/EC) and household appliances, reduction of CO_2 emissions from new passenger cars, strategies to shift transportation from road to rail and water, minimum taxation of mineral oils, coal, natural gas and electricity (Directive 2003/96/EC), afforestation and reforestation measures, reduction of methane emissions from landfills (Directive 1999/31/EC), reduction of N2O in soils (Directive 1991/676/EEC), product eco-design requirements (Directive 2005/32) and various research, development and climate change awareness campaigns.

[16] On 24 October 2005, the EU launched a second ECCP by setting up various working groups to review the first ECCP, explore further EU climate policy measures and develop recommendations to the European Commission in the areas of the EU ETS, climate change impact and adaptation, carbon capture and geological storage, aviation and CO_2 emission reductions from light-duty vehicles.

[17] Commission (EC), 'Proposal for a Directive of the European Parliament and of the Council Establishing a Scheme for Greenhouse Gas Emission Allowance Trading within the Community and Amending Council Directive 96/61/EC' COM (2001) 581 final, 23 October 2001. The proposal was based on Art 175(1) and the co-decision procedure in Art 251 of the Treaty Establishing the European Community dated 25 March 1957 (EC Treaty), providing for qualified majority voting in the Council and the European Parliament's right to veto the proposal.

[18] Council Directive (EC) 2003/87.

[19] A directive under EU law is a legislative act, which is not self-executing but instead identifies a certain legally binding set of objectives while giving member states some leeway on how to transpose those objectives into national law.

[20] EU ETS Directive 2003/87, Art 31(1).

quantitative and qualitative restrictions of the use of CERs and ERUs under the EU ETS, European Parliament and Council adopted Directive 2004/101/EC on 27 October 2004 (Linking Directive).[21]

On 13 November 2006, the European Commission issued a report which set out the terms of reference for a review of the EU ETS (ETS Review).[22] On the basis of that report, the European Commission convened four meetings in 2007 within the framework of the ECCP to discuss potential amendments to the EU ETS. At the same time, in March 2007, the heads of EU member states agreed to achieve the following EU climate goals by 2020:

- 20% GHG emission reduction below 1990 emission levels; the GHG emission reduction target will be increased to 30% if other developed countries commit themselves to comparable emission reductions and economically more advanced developing countries commit themselves to adequate contributions in a post-2012 international agreement on climate change;
- 20% share of renewable energy in EU energy consumption;
- 20% increase in energy efficiency; and
- 10% share for biofuels in petrol and diesel.

On the background of these events, on 23 January 2008, the European Commission adopted a 'climate action and renewable energy package' consisting of various legislative proposals on how to achieve the proclaimed EU climate goals. As part of this package, the European Commission proposed to amend the EU ETS Directive as of the beginning of the trading period starting 1 January 2013 (Phase III) so as to improve and extend the EU ETS post-2012 (Amendment Proposal)[23] and to have the aviation sector covered as additional sector under the EU ETS (Aviation Proposal).[24] Both amendments aim to help covered sectors achieve a 21% GHG emissions reduction by 2020 compared to verified

[21] Council Directive (EC) 2004/101 establishing a scheme for greenhouse gas emission allowance trading within the Community, in respect of the Kyoto Protocol's project mechanisms [2004] OJ L338/18 [hereafter: Linking Directive].

[22] Commission (EC), 'Building a Global Carbon Market–Report pursuant to Article 30 of Directive 2003/87/EC' (Communication) COM (2006) 676 final, 13 November 2006.

[23] Commission (EC),'Proposal for a Directive of the European Parliament and of the Council amending Directive 2003/87/EC so as to improve and extend the greenhouse gas emission allowance trading system of the Community' COM (2008)16 final, 23 January 2008 [hereafter: Amendment Proposal].

[24] Commission (EC),'Proposal for a Directive of the European Parliament and of the Council amending Directive 2003/87/EC so as to include aviation activities in the scheme for greenhouse gas emission allowance trading within the Community' COM (2006) 818 final, 20 December 2006 [hereafter: Aviation Directive].

2005 emission levels.[25] On 17 December 2008, the European Parliament approved some additional amendments to the initial Amendment Proposal endorsing a final political compromise reached among the heads of EU member states on 12 December 2008. The Aviation Proposal was adopted as Directive 101/2008/EC on 19 November 2008 (Aviation Directive), and the final version of the amendment to the EU ETS Directive was adopted on 6 April 2009 and published in the EU official Journal on 5 June 2009 (EU ETS Amendment).[26]

2. Cornerstones of the EU ETS

2.1 Centralization v Decentralization

The EU ETS is a mandatory private entity-based cap-and-trade system.[27] Although the European Commission initially preferred in its Green Paper a more centralized approach, the EU ETS evolved into a heavily decentralized system leaving most of the crucial EU ETS policy aspects (including cap-setting, EUA allocation, method of allocation, operation of registries, monitoring, reporting and verification of annual emissions, enforcement, and limits on the use of CDM/JI credits) to the responsibility of each EU member state. The EU ETS Directive ensures a certain level of harmonization among EU member states by setting out certain EU ETS rules, procedures, guidelines and criteria and by allowing the European Commission to review and reject EU member states' cap-setting and allocation decisions as reflected in the respective national allocation plans. However, the current EU ETS can be described as a system of 30 largely independent, but inter-linked national emissions trading schemes, which have agreed to make their allowances tradable within the EU ETS while adhering to certain common EU rules, procedures, guidelines and criteria.[28]

[25] Commission (EC),'Proposal for a Decision of the European Parliament and of the Council on the effort of member states to reduce their greenhouse gas emissions to meet the Community's greenhouse gas emission reduction commitments up to 2020' COM (2008) 17 final, 23 January 2008. According to a Decision proposal by the European Commission dated 23 January 2008, EU member states must cut GHG emissions from sectors not covered by the EU ETS, such as road transport, housing, agriculture, forestry and waste, by an average total of 10% from 2005 emission levels by 2020 setting up differentiated emission reduction targets among EU member states between +20% and -20% (so-called 'Effort Sharing' Decision).

[26] Directive 2009/29/EC of the European Parliament and of the Council of 23 April 2009 amending Directive 2003/87/EC so as to improve and extend the greenhouse gas emission allowance trading scheme of the Community <http://eur-lex.europa.eu/LexUriServ/LexUriServ.do?uri=OJ:L:2009:140:0063:0087:EN:PDF>. [hereafter: EU ETS Amendment]

[27] As opposed to a baseline-and-credit system this approach was widely considered more manageable and environmentally predictable and in line with the IET structure envisioned under the Kyoto Protocol.

[28] See also A D Ellerman and P L Joskow, 'The European Union's Emissions Trading System in perspective' (Report) (May 2008) The Pew Center on Global Climate Change Report <http://www.pewclimate.org/docUploads/EU-ETS-In-Perspective-Report.pdf>, accessed 18 May 2009; C Streck, 'Implementing the Kyoto Protocol: The European Emissions Trading Scheme' (2005) 1 Intl E LC N 4; J Birger Skjærseth and J Wettestad (n 13 above), 155.

The establishment of a mainly decentralized EU ETS structure was driven by the EU principles of subsidiarity[29] and proportionality.[30] Both principles shall ensure that EU member states remain sovereign decision-makers that have ceded only limited authority to EU institutions. In accordance with these principles and subject to future review and reassessment, EU legislators agreed in 2003 that the objectives of the EU ETS can be sufficiently achieved at EU member state level and that a more centralized and harmonized regime would not be necessary to establish 'a high level of protection and improvement of the quality of the environment' as one of the EC Treaty's objectives.[31] Following a review and reassessment of the decentralized approach in the context of the ETS Review, it seems that EU legislators have changed their minds; they now move towards a more centralized and harmonized approach with its EU ETS Amendment and the Aviation Directive.

2.2 Coverage

2.2.1 Status quo

As of 2005, the EU ETS covered only CO_2 emissions from large emitters in the power and heat generation industry[32] and in selected energy-intensive industrial sectors, such as combustion plants, oil refineries, coke ovens, iron and steel plants and cement, glass, lime, ceramics, and pulp and paper factories.[33] However, as of 2008, EU member states are allowed to have additional activities, installations, and GHGs—not listed in Annex I to the EU ETS Directive—covered by the EU ETS, subject to prior approval by the European Commission (Opt-in Rule).[34] In addition, EU member states may allow operators of covered installations to form a pool of installations from the same activity and to jointly

[29] The subsidiarity principle requires that in areas which do not fall within the EC's exclusive competence (such as environmental policy) the EC 'shall take action only if and in so far as the objectives of the proposed action cannot be sufficiently achieved by the member states and can therefore, by reason of the scale or effects of the proposed action, be better achieved by the EC' (EC Treaty (Treaty of Rome, as amended), Art 5(2)). It ensures that the EC does not take action (except in the areas which fall within its exclusive competence) unless it is more effective than action taken at national, regional, or local level.

[30] The proportionality principle states that 'no action by the EC shall go beyond what is necessary to achieve the objectives of this [EC] Treaty' (EC Treaty, Art 5(3)).

[31] EC Treaty, Art 2.

[32] Eg combustion installations are only covered in case their rated thermal input exceeds 20 MW.

[33] See Annex I to the EU ETS Directive.

[34] EU ETS Directive 2003/8, Art 24(1); eg France and the Netherlands applied to the European Commission for an opt-in of N_2O emissions from the production of fertilizers, and the UK and Norway, the latter being linked to the EU ETS as a non-EU member state as of 2008, have applied for the inclusion of carbon capture and storage activities as of 2008.

surrender EUAs for its combined verified emissions in a preceding year.[35] The possibility of EU member states to apply to the European Commission for a temporary exclusion of certain covered installations from the EU ETS (Opt-out Rule) expired on 31 December 2007.[36] As of today, around 11,000 installations in 27 EU member states account for about 45% of EU-wide total CO_2 emissions or about 30% of its total Kyoto GHG emissions. Further, Norway, Iceland and Liechtenstein's GHG emissions are covered under the EU ETS because they are member countries of the European Economic Area.

2.2.2 Aviation and other sectors and GHG

Aviation

The European Commission became increasingly concerned that the emission trends in the aviation sector could undermine the effectiveness of EU climate policy. Between 1990 and 2005, aviation sector's (domestic and international) CO_2 emissions increased by nearly 90% and air traffic forecasts predict that by 2020 it will double its CO_2 emissions from present emission levels. As a result, the European Commission proposed in its Aviation Proposal dated 20 December 2006,[37] and the European Parliament and the Council informally agreed on 27 June 2008, to have the civil aviation sector covered as additional sector under the EU ETS as of 2012. On 19 November 2008, the Aviation Proposal was finally adopted as the Aviation Directive.

According to the Aviation Directive, as of 1 January 2012 almost all (EU and non-EU) aircraft operators[38] must participate in the EU ETS, which will cover as aviation activities all flights arriving at or departing from any EU airport irrespective of their ultimate destination.[39] The Aviation Directive is limited to

[35] Such operators have to nominate a trustee who would be responsible for surrendering EUAs equal to the total verified emissions from all installations in the pool and be subject to any penalties under the EU ETS for failure to do so. In the event that the trustee fails to pay any penalty, each operator of an installation in the pool will be held liable for any excess emissions from its own installation.

[36] EU ETS Directive, Art 27.

[37] See n 24 above.

[38] An 'aircraft operator' is generally defined as a person who operates an aircraft at the time it performs an aviation activity listed in Annex I of the EU ETS Directive or, where that person is not known or is not identified by the owner of the aircraft, the owner of the aircraft. Only certain flights listed in Annex I of the EU ETS Directive are excluded from EU ETS coverage, such as certain flights for military, governmental, search, rescue, fire-fighting, humanitarian, emergency medical service, and training and scientific research purposes as well as flights of 'commercial air transport operators' (ie operators that, for remuneration, provide scheduled or non-scheduled air transport services to the public for the carriage of passengers, freight or mail) that carry out less than 243 flights in each of three consecutive four-month periods or whose total CO_2 emissions are less than 10,000 tons of CO_2 per year.

[39] According to Aviation Directive, Art 3(b), the Commission must develop guidelines on the detailed interpretation of the aviation activities listed in Annex I to the EU ETS Directive by 2 August 2009. Furthermore, flights departing from non-EU countries that land in the EU may become exempt from EU ETS coverage, in the opinion of the European Commission, if the non-EU country has adopted equivalent measures to reduce the climate change impact of such flights.

CO_2 emissions only. Instead of having each EU member state allocate a CO_2 emissions cap for its respective aviation sector at national level, the Aviation Directive includes an EU-wide CO_2 emissions cap-setting mechanism for the aviation sector at EU level by reference to a baseline of historic average CO_2 emissions from aviation in the EU from 2004 to 2006. The EU-wide CO_2 emissions cap for the aviation sector will be 97% in the year 2012, and 95% in Phase III, of the average 2004–2006 aviation emissions level.[40] The final allocation and issuance of EUAs to each (EU or non-EU) aircraft operator will follow certain decisions made at EU level (eg regarding cap-setting, overall amounts of EUAs to be auctioned or allocated for free, and benchmark-setting for free EUA allocation), but will be implemented by the respective EU member state endowed with administrative power over an aircraft operator (the so-called Administering member state).[41] The Aviation Directive determines that, in the year 2012 and from 1 January 2013, 85% of the amount of EUAs allocated to aircraft operators will be issued free of charge and 15% will be auctioned off.[42] In order to allow new aircraft operators to enter the EU ETS as of 1 January 2013, 3% of the total quantity of EUAs to be allocated for Phase III and any subsequent trading period shall be set aside in a special reserve for new entrants or fast-growing aircraft operators. Covered aircraft operators must surrender an amount of EUAs every year equal to the CO_2 emissions of their flights during the previous year, and they will be subject to the same monitoring, reporting and verification requirements and non-compliance penalties applicable to all other operators under the EU ETS Directive (see below). In the event of non-compliance of an aircraft operator and the failure of other enforcement measures, the Administering member state may (as a last resort and pending prior approval by the European Commission) revoke or suspend the aircraft operator's authorization to operate. In addition, aircraft operators shall be allowed to use CDM/JI credits for EU ETS compliance purposes.[43].

[40] The initial Aviation Proposal was less strict and, for trading periods up until 2022, required CO_2 emissions of the aviation sector to be capped at 100% of its average 2004–2006 emissions.

[41] An EU member state is endowed with administrative powers over an aircraft operator if that EU member state has granted the operating license for such aircraft operator or that EU member state has the greatest estimated attributed aviation emissions from flights performed by that aircraft operator in the 'base year', ie the first calendar year of operation (for aircraft operators starting operation in the EU after 1 January 2006) or the calendar year starting on 1 January 2006 (for all other aircraft operators).

[42] The amount to be auctioned off is envisioned to be increased over time. While each EU member state shall decide on the use of any auction revenues, the Aviation Directive recommends that the auction revenues 'should' be used by EU member states to fund climate change mitigation activities within or outside the EU and requires EU member states to report to the European Commission on any such expenditures.

[43] According to the Aviation Directive, each aircraft operator may use CDM/JI credits during the year 2012 up to an amount of 15% of the number of EUAs it is required to surrender for the year 2012. For Phase III and any subsequent trading period, the percentage limit for CDM/JI credit imports for the aviation sector shall be reviewed and published by the European Commission at least six months prior to the start of each trading period.

By 1 December 2014, the European Commission must review the functioning of the Aviation Directive in relation to aviation activities and propose, if necessary, amendments to the European Parliament and the Council. In the meantime, however, various legal challenges to the Aviation Directive have already been threatened and it remains to be seen what impact such challenges will have on the scheduled review process.[44]

Other sectors and GHG emissions

According to the EU ETS Amendment, in 2013 EU ETS coverage will be extended to several additional sectors, activities, and GHGs to enhance the scheme's environmental effectiveness, increase its market liquidity, and to introduce new and additional abatement opportunities. As of 2013, the EU ETS will also cover additional sectors such as the chemical and aluminum sectors. In addition, there are several additional GHG emissions, such as CO_2 emissions from petrochemicals, ammonia and aluminum, nitrous oxide (N_2O) emissions from acid production, and perfluorocarbon (PFC) emissions from the aluminum sector. The EU ETS' coverage is estimated to increase by up to 100 $MtCO_2$ or up to 4.6% of EUAs issued in Phase II, as a result of the inclusion of these additional sectors and GHG emissions. Furthermore, EU ETS coverage will extend to CCS installations that capture, transport and geologically store GHG emissions to the extent permitted under a corresponding directive on CCS (as proposed by the European Commission on 23 January 2008[45] and pending final adoption by EU legislators) allowing covered installations to reduce their GHG emissions by way of CCS. Finally, in order to increase the cost-effectiveness for small combustion installations, EU member states shall be allowed to exclude from EU ETS coverage installations with emissions of less than 25,000 tCO_2e/year and combustion installations with a rated thermal input of less than 35 MW[46] in each of the three preceding years, provided that,

[44] The Aviation Directive has been widely criticized by the aviation industry. In particular, the 'extra territorial' inclusion of non-EU aircraft operators has been strongly opposed by international airline groups and foreign governments, especially the US government; as a result, they have threatened legal and other retaliatory action. The European Commission argues that the 'extra territorial' approach is necessary not to distort competition between EU and non-EU aircraft operators and it feels confident that it has not violated any international aviation or trade law. Under WTO law, the Annex on Air Transport Services to the General Agreement on Trade in Services of 1995 ('GATS') expressly exempts 'any traffic rights (...) or any services directly related to the exercise of traffic rights' from GATS coverage. However, even if the non-discrimination principles of 'most favoured nation' (Article II GATS) and 'national treatment' (Article XVII GATS) apply, the claimant will have to make a case that the Aviation Directive does, in fact, discriminate against its aviation sector by applying different standards to EU and non-EU aircraft operators.

[45] Commission (EC), 'Proposal for a Directive of the European Parliament and of the Council on the geological storage of carbon dioxide and amending Council Directives 85/337/EEC, 96/61/EC, Directives 2000/60/EC, 2001/80/EC, 2004/35/EC, 2006/12/EC and Regulation (EC) No 1013/2006' COM (2008) 18 final, 23 January 2008.

[46] Combustion installations with a rated thermal input of less than 20 MW will continue to be excluded.

such installations are subject to the equivalent of other GHG emission reduction measures.[47]

Most notably, however, other crucial sectors, such as, road transportation, housing, waste, shipping, agriculture and forestry are currently not intended to be included in the EU ETS as of 2013, since the European Commission continues to question the suitability of EU ETS for such sectors or their ability to monitor and report emissions accurately.

2.3 Permits and EUAs

2.3.1 Permits

As of 1 January 2005, each installation[48] in any EU member state covered by the EU ETS is prohibited from undertaking certain activities listed in Annex I to the EU ETS Directive[49] and emitting GHG specified therein, unless its respective operator[50] holds a GHG permit that allows the installation to undertake such activities and to emit listed GHG.

The GHG permit is an authorization issued by a competent authority in an EU member state that allows an operator to emit certain GHG from all or part of one or more specific installations, provided the competent authority is satisfied that: (1) in case of multiple installations, the installations are on the same site and operated by the same operator; and (2) the operator is capable of monitoring and reporting GHG emissions accurately.[51] Therefore, each GHG permit is a public authorization or license that is issued by way of an administrative act, and is linked to one or more specific installations; hence the GHG permit cannot be traded. Each GHG permit contains, among others, monitoring and reporting requirements, as well as an obligation on the operator to

[47] EU ETS Amendment, Art 27; currently, the largest 7% of covered installations account for about 60% of total EU ETS emissions while the smallest 14% of covered installations account for only 0.14%. According to the European Commission, this new exclusion option may end up excluding up to 4,200 small installations accounting for about 0.7% of total EU ETS emissions from EU ETS coverage.

[48] According to EU ETS Directive, Art 3(e), the term 'installation' means a stationary technical unit where one or more activities listed in Annex I to the EU ETS Directive are carried out and any other directly associated activities which have a technical connection with the activities carried out on that site and which could have an effect on emissions and pollution.

[49] Such covered activities include energy activities of energy-intensive combustion installations, mineral oil refineries and coke ovens, production and processing activities of certain ferrous metals such as metal ore, pig iron and steel, mineral industry activities of cement clinker and lime production and other furnaces, as well as glass and ceramic products manufacture and other activities of pulp, paper, and board production.

[50] According to EU ETS Directive, Art 3(f), the term 'operator' means any person who operates or controls an installation or, where thus provided for in national legislation, to whom decisive economic power over the technical functioning of the installation has been delegated.

[51] EU ETS Directive, Art 6(1).

surrender a number of EUAs equal to the total verified emissions of the installation(s) during the preceding calendar year by 30 April of any given year.[52]

2.3.2 EUAs

In addition to the permit requirement, as of 1 January 2005, each operator of an installation in any EU member state covered by the EU ETS is also prohibited from emitting excess GHG without an equivalent number of EUAs on an annual basis. Each EUA entitles its holder to emit one ton of CO_2 equivalent and is valid for emissions during the entire trading period in which it is issued.[53] As opposed to a permit granting the operator of each covered installation a site-specific, non-transferable license to undertake certain activities and emit GHG, an EUA can be held by any person[54] (public/private entities, individuals, NGOs, etc.) that has a holding account in an EU member state's national emission registry. The EUA entitles the respective holder to either emit an equivalent amount of GHG (ie for compliance purposes), or transfer the EUA to another person's national emission registry account within the EU, or in a third country, provided that the non-EU country recognizes EUAs. Instead of using the EUA as described above, a permit holder may also cancel the EUA, which effectively removes it from the market and, therefore reduces the overall market cap of tradable EUAs.[55] The operator of each installation is obligated to surrender, by April 30 of any given year, a number of EUAs equal to the total verified emissions from that installation during the preceding calendar year. After those EUAs are surrendered, they are cancelled.

2.3.3 Banking and borrowing

The EU ETS Directive allows for banking and borrowing of EUAs within each trading period. Due to the fact that EUAs are valid for emissions during the entire trading period for which they are issued, the operator of an installation has the flexibility to use its EUAs during each trading period when needed. If a covered installation has emitted fewer tons of covered GHG in a preceding year than its operator holds, the operator may use such excess EUAs for compliance purposes in the current or any of the following years of the same trading period (this is called 'banking'). If an installation has emitted more tons of covered GHG in a preceding year than its operator holds, the operator may use the EUAs that are issued to it for the current year (this is called 'borrowing'). Borrowing within a trading period is made possible because EUAs for the current year are issued by EU member states on 28 February of each year,[56] whereas

[52] EU ETS Directive, Art 6(2).
[53] EU ETS Directive, Art 3(a).
[54] EU ETS Directive, Art 19(2).
[55] EU ETS Directive, Art 12(1).
[56] EU ETS Directive, Art 11(4).

operators of each installation are required to surrender EUAs for its emissions during the preceding year on April 30 of each year.[57]

With respect to banking and borrowing between different trading periods, the EU ETS Directive is more restrictive. Due to the fact that EUAs are only valid for emissions during the trading period for which they are issued, EUAs that are issued for a subsequent trading period cannot be borrowed and used for compliance purposes in the current trading period. Although EU member states were allowed to provide for banking of EUAs between Phase I and Phase II on a national basis,[58] most EU member states (except for France and Poland) decided not to allow banking between these trading periods so as to mitigate the risk of having a potential EUA over-supply in the 'trial' Phase I period that could have spilt over into the Phase II period, hence inflating the national emission cap for a period that coincides with the first commitment period under the Kyoto Protocol. Banking of EUAs between Phase II and any subsequent trading period is made mandatory.[59]

2.3.4 Legal nature of EUAs

The legal nature of EUAs has been subject to extensive discussions among legal experts.[60] Although an early draft proposal by the European Commission of the EU ETS Directive defined an 'allowance' as being an 'administrative authorization',[61] such harmonized legal characterization on the EU level was later rejected by the Legal Service of the European Commission due to a perceived conflict with the EU principle of subsidiarity.[62] The EUAs' legal status is, therefore, to be decided by each EU member state in the context of its respective national laws. As a consequence, holders of EUAs will have to turn to domestic legislation and judicial precedents to find out the level of legal authorization, ownership, obligations, liabilities, and/or protection provided for under domestic law for any

[57] EU ETS Directive, Art 12(3). [58] EU ETS Directive, Art 13(2)[2].

[59] EU ETS Directive, Art 13(3)[2].

[60] Eg M Wemaere and C Streck, 'Legal Ownership and Nature of Kyoto Units and EU Allowances', in David Freestone and Charlotte Streck (eds), *Legal Aspects of Implementing the Kyoto Protocol Mechanisms: Making Kyoto Work* (OUP: Oxford, 2005); M J Mace, 'The Legal Nature of Emission Reductions and EU Allowances: Issues Addressed in an International Workshop' (2005) 2 Journal of European Environmental & Planning Law 123–34; J Button, 'Carbon Commodity or Currency? The Case for an International Carbon Market Based on the Currency Model' (2008) 32 Harvard Environmental Law Journal 571–96.

[61] The reason for the initially proposed classification of EUAs as 'administrative authorizations' may be that, during the initial drafting phase, the European Commission looked very closely to existing US emissions trading schemes such as the emissions cap-and-trade systems established under the Clean Air Act (CAA) of 1990 and the Regional Clean Air Incentives Market (RECLAIM) of 1994. Title IV, Sections 402(3) and 403(f) of the CAA as well as Rule 2007(b)(1) of the RECLAIM regulation expressly state that an allowance/credit is a 'limited authorization' to emit SO_2/NO_x and shall not constitute a 'property right' (or, in case of RECLAIM, a 'security or other form of property'). In both cases, it is further stressed that nothing shall limit the authority of the respective regulator to condition, limit, suspend or terminate allowances/credits or the authorization to emit.

[62] See n 29 above.

kind of EUA-related activity in any given legal context (such as contract, property, security, insolvency, state aid, accounting, or tax law).

However, taking the determination of the legal status of an EUA aside, the final text of the EU ETS Directive, as well as Regulation[63] 2216/2004 for a standardized and secured system of registries (as of 1 January 2012, repealed and replaced by Regulation 994/2008 adopted on 8 October 2008[64] (together the 'Registry Regulations')) prescribe certain legal characteristics of EUAs that apply EU-wide. The legal characteristics of an EUA may be viewed as 'property rights'[65] and include the following:

- EUAs can be held by any person with an account in a national emission registry;
- an EUA allows its holder to emit one ton of CO_2 equivalent into the atmosphere;
- EUAs can be transferred to another person's national emission registry account either within the EU or in a third country that recognizes EUAs, provided that the transfer is not blocked or terminated or registry procedures are not temporarily suspended in accordance with the Registry Regulations;
- EUAs are only valid and cannot be revoked during the trading period for which they are issued;
- EUAs can be banked and borrowed within trading periods and banked (but not borrowed) between Phase II and any subsequent trading period;
- EUAs can be surrendered and cancelled in accordance with the Registry Regulations.

[63] Commission Regulation (EC) 2216/2004 of 21 December 2004 for a standardised and secured system of registries pursuant to Directive 2003/87/EC of the European Parliament and of the Council and Decision No 280/2004/EC of the European Parliament and of the Council [2004] OJ L 386/1. Adopted by the Council in conjunction with the European Parliament or by the European Commission alone, a regulation is a general measure that is directly—ie without having to be transposed into national law—binding in all its parts. Unlike directives which are addressed to EU member states and have to be transposed into national law, and decisions which are for specified recipients, regulations are addressed to everyone and apply directly and immediately without having to be transposed into national law.

[64] Commission Regulation (EC) 994/2008 of 8 October 2008 for a standardised and secured system of registries pursuant to Directive 2003/87/EC of the European Parliament and of the Council and Decision No 280/2004/EC of the European Parliament and of the Council [2008] OJ L 271/3. This revision of Regulation 2216/2004 is supposed to, among others, enable the completion of EU ETS transactions without the involvement of the International Transaction Log and shall assure the independence of the EU ETS as of 2012. It will also facilitate the inclusion of aviation activities from 2012 and the ability of the EU ETS to link to other emissions trading systems.

[65] The term 'property rights' is not identical with private ownership but is an economic concept to define a certain (more or less limited) bundle of rights attached to the use of a public good. The 'property rights' approach, therefore, defines a certain proprietary position of the respective user of a public good without implying any specific legal nature of such proprietary position.

In addition, the Registry Regulations allow EU member states to permit a holder of an EUA to use it as security. According to the Registry Regulations,[66] EU member states may allow account holders in its registry to nominate so-called 'additional authorized representatives'. In this case, the account holder shall not be allowed to submit any EUA transaction request to the registry administrator before the additional authorized representative has given its consent to the envisioned transaction. This mechanism endows the additional authorized representative with a veto power and, therefore, with a security interest in the EUAs held in the respective holding account.[67]

As of today, EU member states are still hesitant to expressly state the legal status of EUAs in their national laws. As a publicly issued but privately tradable asset with the objective of helping EU member states manage a public good and comply with its international emission reduction obligations, an EUA has a unique dual (public-private) character that does not seem to easily fit into existing national legal terms. On one hand, as a publicly issued asset with an underlying objective of public concern, national legislators seem to be reluctant to endow a holder of an EUA with express private legal ownership and strong constitutional protection against any kind of future interference or limitation by public authorities. They seem to be afraid that any such endowment may deprive public authorities of the flexibility to create, cancel, withdraw or suspend EUAs that have been improperly allocated and, therefore, to effectively exercise its public task to manage access to and use of air as a public good.[68] On the other hand, as a privately tradable asset, national legislators seem to be reluctant to characterize an EUA as a public authorization, permit, or license to emit GHG, since such legal terms are reserved under existing command-and-control environmental policies for granting a person a (non-tradable) license to use a public good under certain conditions. Due to its uniqueness in every given national legal context, EUAs are likely to have the status under any domestic legal regime of an entitlement 'sui generis'.[69] For a more comprehensive analysis of the legal nature of EUAs as determined at EU member state level, see chapter 2.

[66] Council Regulation (EC) 2216/2004 of 21 December 2004 for a standardised and secured system of registries pursuant to Directive 2003/87/EC of the European Parliament and of the Council and Decision No 280/2004/EC of the European Parliament and of the Council [2004] OJ L386/1, Art 23(2); Council Regulation (EC) 994/2008 of 8 October 2008 for a standardised and secured system of registries pursuant to Directive 2003/87/EC of the European Parliament and of the Council and Decision No 280/2004/EC of the European Parliament and of the Council [2008] OJ L271/3, Art 19(2).

[67] For example in UK, Austria, Sweden, and Finland, account holders are expressly allowed to register 'additional authorized representatives' for certain holding accounts in the national emission registry.

[68] MJ Mace (n 60 above), 124.

[69] In law, 'sui generis' is a legal term used to identify a classification that exists independently of other categorizations because of its uniqueness or due to the specific creation of an entitlement or obligation.

2.4 Cap-setting and national allocation plans

2.4.1 Status quo

Prior to each EU ETS trading period, each EU member state must develop a national allocation plan (NAP) stating, among others, the total quantity of EUAs to be allocated for a given trading period and the portion of EUAs retained in a reserve for new entrants, the amount of EUAs to be allocated to each operator of a covered installation, the method of EUA allocation (free and/or auctioned), and the percentage of CDM/JI credits allowed for compliance purposes.[70] The NAPs must be based on objective and transparent rules and criteria, including those listed in Annex III to the EU ETS Directive[71], and it must be published and the European Commission must be notified no later than by 31 March 2004 (for Phase I; so-called NAP I) or 18 months prior to the beginning of Phase II (so-called NAP II) and any subsequent trading period. The European Commission will review the NAPs and it may reject a NAP, or any aspect thereof, within three months following notification on the basis that it is incompatible with the criteria listed in Annex III to the EU ETS Directive or the required method of allocation. However, the European Commission must give reasons for any rejection. An EU member state may only allocate and issue EUAs to operators of covered installations if the European Commission has accepted subsequent amendments to the NAP.

The current delegation of the emission cap setting procedure to the national level means that the overall EU ETS emission caps for Phase I and Phase II are essentially the sum of 30 national emission caps as agreed between the EU member state and the European Commission in the respective NAPs. In order to assist EU member states in ensuring the accuracy of emission data monitored and reported by the operators of covered installations, the European Commission adopted in its Decision[72] 2004/256 on 29 January 2004, as succeeded, as of 1 January 2008, by Decision 2007/589 adopted on 18 July 2007, legally binding guidelines for monitoring and reporting of emissions within the EU ETS.[73]

[70] EU ETS Directive, Art 9.

[71] The European Commission issued specific guidance on how these rules and criteria are to be applied by EU member states for Phase I (Commission (EC), 'Communication from the Commission on guidance to assist member states in the implementation of the criteria listed in Annex III to Directive 2003/87/EC establishing a scheme for greenhouse gas emission allowance trading within the Community and amending Council Directive 96/61/EC, and on the circumstances under which force majeure is demonstrated' (Communication) COM (2003) 830 final, 7 January 2004) and Phase II (Communication COM (2005) 703 final dated 22 December 2005).

[72] A decision under EU law is a legally binding legislative act on a particular matter requiring one or more specified EU member states or EU citizen to take or refrain from taking a particular action.

[73] Commission Decision (EC) of 18 July 2007 establishing guidelines for the monitoring and reporting of greenhouse gas emissions pursuant to Directive 2003/87/EC of the European Parliament and of the Council [2007] OJ L229/1. Because experience with monitoring and reporting in Phase I and Phase II has shown some degree of divergence in how EU member states apply the guidelines, the European Commission is currently planning to replace the current guidelines by 31 December 2011 with a new regulation.

In March 2004, due to large uncertainties regarding the accuracy of available national emission data and the interpretation of what installations and activities are, in fact, covered, the cap setting process for Phase I can be described as a 'best guess exercise' giving the benefit of doubt to each EU member state's allocation estimate. The outcome of this exercise is well known. In 2006, the announcement of various EU member states to have over-allocated EUAs, ie to have allocated more EUAs than tons of CO_2 emissions verified in 2005, sent shockwaves through the market. The over-allocation together with the wide-spread prohibition of banking between Phases I and II resulted in a collapse of the price for EUAs casting serious doubt on the environmental integrity of the EU ETS.

During the NAP II review process, which also coincided with the first commitment period under the Kyoto Protocol, the European Commission was much more apt to strictly apply the cap-setting and allocation criteria in Annex III of the EU ETS Directive. Hence, as of October 2007, the European Commission has rejected proposed emission caps of 23 EU member states (accepting only the ones provided by Denmark, France, Slovenia, and the UK) cutting such caps by up to 55% (see Table 16.2 below).[74] Despite the efforts of the European Commission to minimize the risk of over-allocation for Phase II, the European industries' current severe economic downturn is expected to result in a drop of current EU emission levels, thus, increasing the risk of another market over-supply of EUAs by the end of Phase II. However, although prices for EUAs have already dropped from about EUR 30 in July 2008 to below EUR 12 in January 2009, market analysts have not forecasted another EUA price collapse similar to that of Phase I[75] since this time, operators are allowed to bank any EUA over-supply in Phase II to any subsequent trading period.

2.4.2 EU ETS Amendment

According to the EU ETS Amendment, several changes will become effective as of 2013. First, Phase III will last eight years from 2013 to 2020, instead of three years under (Phase I) and five years under (Phase II), in order to increase predictability and to provide operators of covered installations with greater flexibility in their emission abatement efforts. Second, the notion of NAPs will be abolished and replaced by a centralized single EU-wide cap-setting approach at EU-level; this EU-wide cap will be calculated for the year 2013 on the basis of average annual EUA allocations during Phase II and will be published by the European Commission on 30 June 2010.[76] The single EU-wide

[74] As a result, countries such as Poland, Bulgaria, Czech Republic, Estonia, Hungary, Latvia, Lithuania, and Romania have challenged the European Commission's NAP decision at the European Court of Justice (ECJ). ECJ decisions are not expected to be made before the end of 2009.

[75] *Point Carbon News*, 'Analysts turn bears for 2009' (Oslo, 18 December 2008)

[76] With respect to covered installations, the quantity of EUAs to be issued from 1 January 2013 will be adjusted to reflect the average annual quantity of EUAs issued to such installations during Phase II minus 1.74%.

Table 16.2. Overview of 27 EU member state NAPs and intervention level of the European Commission

EU member state	Emission cap agreed between EU member state and European Commission for Phase I[1]	2005 verified emissions	Emission cap proposed by EU member state in NAP II for Phase II	Emission cap allowed by European Commission for Phase II and percentage of allowed cap v proposed cap	CDM/JI credit limit allowed by European Commission for Phase II[4]
Austria	33	33.4	32.8	30.7 (93.6%)	10%
Belgium	62.1	55.58	63.3	58.5 (92.4%)	8.4%
Bulgaria	42.3	40.6[2]	67.6	42.3 (62.6%)	12.55%
Cyprus	5.7	5.1	7.12	5.48 (77%)	10%
Czech Republic	97.6	82.5	101.9	86.8 (85.2%)	10%
Denmark	33.5	26.5	24.5	24.5 (100%)	17.01%
Estonia	19	12.62	24.38	12.72 (52.2%)	0%
Finland	45.5	33.1	39.6	37.6 (94.8%)	10%
France	156.5	131.3	132.8	132.8 (100%)	13.5%
Germany	499	474	482	453.1 (94%)	20%[3]
Greece	74.4	71.3	75.5	69.1 (91.5%)	9%
Hungary	31.3	26	30.7	26.9 (87.6%)	10%
Ireland	22.3	22.4	22.6	22.3 (98.6%)	10%
Italy	223.1	225.5	209	195.8 (93.7%)	14.99%
Latvia	4.6	2.9	7.7	3.43 (44.5%)	10%
Lithuania	12.3	6.6	16.6	8.8 (53%)	20%
Luxembourg	3.4	2.6	3.95	2.5 (63%)	10%
Malta	2.9	1.98	2.96	2.1 (71%)	Not announced
Netherlands	95.3	80.35	90.4	85.8 (94.9%)	10%
Poland	239.1	203.1	284.6	208.5 (73.3%)	10%
Portugal	38.9	36.4	35.9	34.8 (96.9%)	10%
Romania	74.8	70.8[2]	95.7	75.9 (79.3%)	10%
Slovakia	30.5	25.2	41.3	30.9 (74.8%)	7%
Slovenia	8.8	8.7	8.3	8.3 (100%)	15.76%
Spain	174.4	182.9	152.7	152.3 (99.7%)	20%
Sweden	22.9	19.3	25.2	22.8 (90.5%)	10%
United Kingdom	245.3	242.4	246.2	246.2 (100%)	8%
TOTAL	2,298.5	2,122.16	2,325.34	2,080.93 (Ø 89.5%)	Ø 11.4% (without Malta)

Source: RAPID – EU press release (IP/07/1614) dated 26 October 2007

(1) All listed quantities are in units of million metric tons of CO_2

(2) Due to recent EU accession, this figure is not independently verified

(3) The 22% figure in the German NAP relates to the EUAs allocated free of charge rather than the total cap

(4) JI/CDM limit is expressed as a percentage of the member state's cap and indicates the maximum extent to which companies may annually surrender CDM/JI credits instead of EUAs.

cap will decrease annually by a linear factor of 1.74%[77] to ensure that the 21% emission reduction target (as compared to verified 2005 emission levels) for Phase III is achieved by 2020. In order to be able to allow new installations to enter the EU ETS post-2012 and to issue EUAs to such new entrants, 5% of the EU-wide cap for Phase III will be set aside at EU level as a so-called 'new entrants reserve'.[78]

The shift to a centralized cap-setting approach was regarded as imperative after the experiences made during NAP I and NAP II. In particular, the NAP process was perceived as being very time-consuming and cost-ineffective. Further, various national circumstances and interests have led to considerable differences among EU member states regarding the interpretation of covered activities, installation types, the amounts of EUAs allocated to like facilities (eg in the power sector), the applied methods of allocation, and the level of restrictions placed on the use of CDM/JI credits thereby creating distortions of competition within the EU market.[79]

2.5 Method of allocation (grandfathering v auctioning)

2.5.1 Status quo

During Phases I and II, it is the responsibility of each EU member state to determine the method of allocation, ie how many EUAs are to be allocated and issued to domestic operators of covered installations by way of free allocation on the basis of historic emission benchmarks (grandfathering) or auctioning. Although the EU ETS Directive allows for auctioning of EUAs of up to 5% in Phase I and up to 10% in Phase II, only three EU member states reported auctioning in Phase I and only ten in Phase II with an average auctioning rate of only 3.7%.[80]

[77] In case of the conclusion of an international post-2012 climate change regime, any agreed additional emission reduction targets at EU level will lead to a corresponding increase of the linear factor.

[78] Up to 300 million EUAs of the new entrant reserve shall be set aside until 31 December 2015 for the construction and operation of up to 12 commercial CCS prototype projects and innovative renewable energy projects. In the event that the new entrant reserve has not been fully allocated to new entrants by the end of Phase III, the remaining EUAs shall be auctioned at EU member state level.

[79] The European Commission, therefore, seems to have concluded that the objectives of the EU ETS cannot be sufficiently achieved by the NAP process at national levels (subsidiarity principle) and that action at EU level was necessary to achieve the environmental objectives of the EC Treaty (proportionality principle).

[80] European Environment Agency, Application of the Emissions Trading Directive by EU member states—reporting year 2007, EEA Technical Report, No 3, 2008, pp 57 and 61

2.5.2 EU ETS Amendment

According to the Final Amendment Proposal, auctioning will become the principal method of allocation in Phase III. As a general rule, the power sector (due to its ability to pass down costs to energy consumers) and CCS installations will be subject to full auctioning as of 2013. Other sectors will be subject to 20% auctioning in 2013 gradually increasing each year to 70% auctioning in 2020 'with a view to reaching' 100% auctioning by 2027.[81] Furthermore, EU member states 'should' use at least 50% of their auctioning revenues for measures, such as, GHG emission reduction, renewable energy, energy efficiency, CCS, forestry, public transportation, or adaptation.[82] With this policy shift, the European Commission responded to increasing criticism of the grandfathering option, in particular from the environmental community.

Critics of the grandfathering option have long argued that the free allocation of EUAs contradicts the 'polluter pays' principle as stipulated in EC Treaty Article 174(2), therefore, it does not provide proper incentives for domestic industries to kick-start investments in energy efficiency or renewable energy, but instead, creates the opportunity of so-called 'windfall profits', ie additional profits that operators in certain sectors, such as the power sector, make by passing on the market value of EUAs to energy consumers while obtaining EUAs for free. On the other hand, proponents of the grandfathering option argue that the increase of EU ETS compliance costs caused by auctioning could put energy intensive industries at a competitive disadvantage as compared with non-EU industries, resulting in the relocation of such industries outside the EU (so-called 'carbon leakage') rather than in GHG emission abatement measures.

However, in light of the current severe economic downturn of EU industries and to the disappointment of the environmental community, several EU member states pushed successfully for certain exclusions in the final version of the EU ETS Amendment from the afore-mentioned road map for the gradual phase-out of the grandfathering approach in order to address, in particular, special national circumstances and the risk of carbon leakage in certain sectors and sub-sectors.

For example, EU member states that lack interconnectivity of its national electricity network to the EU electricity grid or heavily rely on a single fossil fuel

[81] EU ETS Amendment, Arts 10 and 10(a); according to EU ETS Amendment, Art 10(2), 88% of the total quantity of EUAs to be auctioned will be distributed among EU member states in shares equivalent to the respective EU member state's share of overall EU-wide emissions verified in the year 2005 or the average emissions in the period 2005–2007, whichever is higher; 10% will be distributed to poorer EU member states for solidarity and growth purposes; and 2% will be given to EU member states whose GHG emissions in 2005 were at least 20% below their respective Kyoto base year emissions, ie eastern EU member states.

[82] EU ETS Amendment, Art 10(3); from a legal perspective, the use of the word 'should' (instead of 'shall') seems to imply that this provision has the status of a (non-binding) recommendation only. However, EU member states 'shall' regularly report to the European Commission on the use of such auction revenues.

in electricity production and have a GDP/capita of 50% or less of the average EU GDP/capita, ie Poland and some other eastern EU member states, are granted the option to issue to their power sectors, in 2013, up to 70% of EUAs allocated on the basis of their average annual emissions from 2005–2007 free of charge, and to gradually decrease to 0% free allocation by 2020.[83] In addition, district heating, as well as high efficiency cogeneration installations, will receive its EUAs 100% for free in respect of the production of heating or cooling. Finally, other covered sectors or sub-sectors deemed to be exposed to a significant risk of carbon leakage will be eligible in Phase III to receive up to 100% of allocated EUAs free of charge until a post-2012 international agreement on climate change is concluded, when this exception shall be reviewed. The European Commission is to determine the sectors or sub-sectors exposed to a significant risk of carbon leakage by no later than 31 December 2009, and after assessing the respective sub-sector's ability to pass down its additional EUA and electricity costs into product prices without a significant loss of market share to less carbon efficient non-EU installations and following certain criteria.[84] It must adopt EU-wide harmonized implementation measures for the free allocation of EUAs to such sectors and sub-sectors by 31 December 2010.[85]

In light of the substantial scope of exceptions agreed in the EU ETS Amendment and pending the European Commission's decision on the actual number of exempt sectors or sub-sectors, it remains to be seen if the EU ETS manages to reach its objective of auctioning having become the principal method of allocation in Phase III.

2.6 Compliance and enforcement

EU member states must ensure that, by 30 April of each year, the operator of each covered installation surrenders a number of EUAs equal to the total verified emissions from that installation during the preceding calendar year.[86] In the event that an operator of a covered installation does not surrender a sufficient amount of EUAs by 30 April of any given year to cover its verified emissions during the preceding year, each EU member state must ensure that the operator is held liable for the payment of an excess emissions penalty in the amount of

[83] In return for this exception, the EU member states concerned must submit to the European Commission a national investment plan with the objective of retrofitting and upgrading its electricity infrastructure and clean technologies and of diversifying its respective energy mix.

[84] In accordance with the EU ETS Amendment, Art 10(a)(15), a sector or sub-sector shall be deemed to be exposed to a significant risk of carbon leakage if: (1) the sum of additional costs incurred by the EU ETS Directive results in a substantial increase of production costs of at least 5% and the non-EU trade intensity (ie the ration between the total value of exports to and imports from non-EU countries and the total market size for the EU) is above 10%; (2) the sum of additional costs incurred by the EU ETS Directive results in a particularly high increase of production costs of at least 30%; or (3) the non-EU trade intensity is above 30%.

[85] EU ETS Amendment, Art 10a(1) and (13).

[86] EU ETS Directive, Art 12(3).

EUR 40 (Phase I), or EUR 100 (Phase II and subsequent trading periods) for each ton of CO_2 equivalent emitted but not matched with EUAs. In addition, payment of the excess emissions penalty does not release the operator from the obligation to surrender an amount of EUAs equal to those excess emissions when surrendering EUAs in the following calendar year.[87] In order to ensure that the excess emissions penalty remains sufficiently high over time to ensure compliance, the EU ETS Amendment increases the penalty relating to EUAs issued from 1 January 2013 onwards in accordance with the European index of consumer prices.

2.7 EUA transfers and registries

Each EU member state must maintain its own national registry in the form of a standardized electronic database to record the creation, transfer and surrender of EUAs, and this registry must be operated and maintained by a national registry administrator. In addition to operators of covered installations, other companies, and natural persons may also open an account in a national emission registry. A high degree of uniformity of national emission registries is ensured through the Registry Regulations. According to the data contained in the NAPs or, as of 1 January 2013, determined at EU level, the total quantity of EUAs will first be issued into the EU member state's holding account assigning a unique identification code to each EUA. Each operator of a covered installation will then be forwarded the amount of EUAs allocated to it by having equivalent EUA amounts transferred from the EU member state's account to the respective operator's holding account. Following such issuance, any subsequent trade of EUAs takes place between operator and personal holding accounts.

Transfers of EUAs among operators located in different EU member states are not only recorded in the respective national emission registries but are also reported to a central EU emission registry (the Community Independent Transaction Log (CITL)), which is operated and maintained by a central administrator. In the event that the central or registry administrator becomes aware of any EUA transaction process that does not conform to the requirements, specified in the EU ETS Directive or the Registry Regulations, it may block, terminate, or suspend any concerned EUA transaction.

3. Linkages of EU ETS to Kyoto Protocol and Other Emissions Trading Schemes

3.1 EU ETS and Kyoto Protocol

Although the EU ETS is structured to be able to operate independently from the Kyoto Protocol, the EU considers the EU ETS to be its main policy instrument

[87] EU ETS Directive, Art 16.

to help EU member states ensure compliance with its obligations under the Kyoto Protocol. Therefore, the EU ETS rules have been structured in line with the Kyoto Protocol rules and include various linkages to ensure a sound co-existence of both regimes.

3.1.1 EUAs and AAUs

As of 1 January 2008, the issuance of EUAs into the EU member state's holding account shall take place by converting an equal quantity of assigned amount units (AAUs) held by the EU member state, in accordance with Article 3 of the Kyoto Protocol, into EUAs and adding the EUA element to the unique identification code of each such AAU.

In other words, as of 1 January 2008, each issued EUA is linked to an AAU and any subsequent transfer of such EUA within or between registries is shadowed by a transfer of an AAU.[88] However, in order to secure the independence of the EU ETS from any post-2012 international agreement on climate change, as of 1 January 2012 a slightly different procedure will apply.[89] Instead of converting AAUs into EUAs, each issuance and subsequent transfer of EUAs shall be shadowed by a corresponding transfer of AAUs from an EU member state's holding account to a separate ETS AAU deposit account. In any case, this linkage between EUAs and AAUs in the Registry Regulations will ensure that any EU ETS transaction is properly reflected at the international level and each EU member state can, while operating EU ETS emission registries, also keep track of its status of compliance with its respective Kyoto target.

3.1.2 ITL and CITL

The registry system under the Kyoto Protocol includes an independent transaction log (ITL) to automatically monitor all transactions related to Kyoto emission credits and to ensure that transfers and other activities are consistent with Kyoto Protocol rules. The CITL is a supplementary transaction log and automatically monitors all EUA transactions under the EU ETS to ensure consistency with EU ETS rules. Each national emission registry under the EU ETS must inform the ITL or CITL of any proposed EUA, AAU, CER, or ERU transaction for final approval. If any inconsistency is detected or an invalid action is proposed by a national registry, the CITL or ITL will reject and terminate the proposed transaction. ITL, CITL, and the EU member states' emission registries were successfully connected on 16 October 2008, thus enabling EU ETS account holders to also import CDM/JI credits into their national registry holding accounts.

[88] See Directive 2216/2004. [89] See Directive 994/2008.

3.1.3 The Linking Directive

Status quo

The parties to the Kyoto Protocol are allowed to invest in CDM and JI projects in return for compliance CDM/JI credits, namely CERs (for CDM projects) and ERUs (for JI projects).[90] To ensure the environmental integrity of the Kyoto Protocol, the use of CERs and ERUs may only be supplemental to domestic emission reduction efforts (supplementarity principle).[91] In accordance with the Linking Directive, EU member states may allow operators of covered installations to acquire CERs (from 2005) and ERUs (from 2008), in accordance with Kyoto Protocol rules, and use them for compliance purposes under the EU ETS.[92] The objective of the Linking Directive is to increase the diversity of low-cost compliance options within the EU ETS, hence, leading to a reduction in the overall costs of compliance with the Kyoto Protocol while improving the liquidity of the market in EUAs. However, to ensure compliance with the supplementarity principle, each EU member state must limit the use of CERs and ERUs by specifying in its NAP a certain percentage limit by which CDM/JI credits can be used for EU ETS compliance purposes. As a general rule, operators of each installation should be allowed to use CERs and ERUs to supplement the EUA allocation for each installation by up to 10%. In assessing proposed limits in excess of 10%, the European Commission takes into account the effort that an EU member state has to make in order to reach its Kyoto target.[93] For Phase II, the European Commission has approved an average limit of the use of CDM/JI credits of about 11.4% (see Table 16.2 above) which translates into approx. 1.4 billion tons of CO_2 equivalent. CERs and ERUs can be held in operator and personal holding accounts and their import shall take place through the issue and immediate surrender of one EUA by the EU member state in exchange for one CER or ERU.[94]

In addition to the quantitative limit, the Linking Directive also establishes some qualitative limits on the underlying CDM/JI projects. In particular, the use of CDM/JI credits generated from nuclear facilities and from land use, or land use change and forestry (LULUCF) activities shall not be allowed for

[90] The Kyoto Protocol, Arts 6 and 12.

[91] Although the Kyoto Protocol rules do not define the exact limit of the use of CERs and ERUs for compliance purposes, they state that domestic action by countries with a Kyoto target shall constitute a 'significant element' of their emission reduction efforts. The 'significant element' criterion is generally understood to imply that at least 50% of GHG emission reduction efforts shall take place domestically.

[92] In contrast to CDM/JI credits, AAUs are not allowed to be imported and traded under EU ETS rules.

[93] See Commission (EC), 'Communication from the Commission to the Council and to the European Parliament on the assessment of national allocation plans for the allocation of greenhouse gas emission allowances in the second period of the EU Emissions Trading Scheme' (Communication) COM (2006) 725 final, 29 November 2006.

[94] Linking Directive, Art 11(a)(1).

EU ETS compliance purposes. With respect to hydro projects with a generating capacity exceeding 20 MW, EU member states must ensure, in its approval procedures, that relevant international criteria and guidelines, including such of the World Commission on Dams, will be respected by the project's activity.[95]

EU ETS Amendment Proposal

Although the European Commission acknowledged the cost-saving potentials of the Linking Directive, it was concerned that allowing too many CDM/JI credits into the EU ETS would undermine domestic mitigation efforts (including energy efficiency and renewable energy projects) or even allow domestic GHG emissions to increase. As a result, the European Commission proposed in its initial Amendment Proposal to adopt a rather limited approach towards the use of CDM/JI credits post-2012, including different sets of rules depending on whether a satisfactory post-2012 international agreement on climate change is concluded or not.[96]

Due to the current severe economic downturn of EU industries, several EU member states have pushed successfully for a more generous approach towards the use of CDM/JI credits post 2012 in the EU ETS Amendment negotiation process. Therefore, the EU ETS Amendment now allows for a maximum EU-wide use of credits of up to 50% of the EU-wide reduction efforts of existing sectors over the period 2008–2020, and up to 50% of EU-wide reduction efforts below 2005 emission levels of new sectors and aviation over the period 2013–2020. In this context, all existing operators may use credits during the 2008–2020 period up to a minimum of 11% of their respective allocations for Phase II. This minimum amount may be increased for existing operators with the lowest sum of free allocation and allowed use of CDM/JI credits during

[95] Linking Directive, Arts 11(a)(3) and 11(b)(6).

[96] According to the initial Amendment Proposal, until a satisfactory post-2012 international agreement on climate change is concluded by the EU, the use of CDM/JI credits during Phase III was supposed to be limited to the amounts of credits previously authorized by EU member states in their respective NAP II that have not been used up by the end of Phase II. Within this quantitative limitation, operators of covered installations were supposed to be allowed to use in Phase III: (1) CERs and ERUs issued before 2013 from CDM/JI projects accepted by all EU member states during Phase II; (2) CERs issued after 2012 from project types accepted by all EU member states during Phase II and which have been 'established' before 2013; and (3) CERs issued after 2012 from new projects accepted by all EU member states during Phase II that start after 2012 in least developed countries. Instead of expressly prohibiting certain project types, the requirement of allowing CDM/JI credits only from projects accepted by all EU member states placed an indirect qualitative restriction on the post-2012 use of CDM/JI credits and allowed single EU member states to unilaterally ban EU-wide the use of CDM/JI credits from certain project types in Phase III. Upon conclusion of a satisfactory post-2012 international agreement on climate change by the EU, however, operators of covered installations were supposed to be allowed to use an additional number of CDM/JI credits equivalent to half of the additional reduction effort agreed by the EU, provided that the host country of the underlying CDM/JI project has ratified such international agreement.

Phase II. New entrants, new sectors and aircraft operators in Phase III may use credits of up to a minimum of 4.5% (new entrants and new sectors) or 1.5% (aircraft operators) of its respective verified emissions during Phase III. To the extent of any quantitative limit on the use of credits, each operator may use in Phase III by way of exchange into EUAs: (1) CDM/JI credits issued in respect of emission reductions generated in Phase II but not yet surrendered; (2) CDM/JI credits issued in respect of emission reductions generated from 2013 onwards from projects that were registered before 2013; and (3) CERs issued in respect of emission reductions generated from 2013 onwards from projects started from 2013 onwards in so-called 'least developed countries'.

In the event of the conclusion of a post-2012 international agreement on climate change and in line with the initial Amendment Proposal, it is envisioned to allow for an additional use of credits (including additional types of project credits or mechanisms in the post-2012 international agreement) of up to half of the additional reduction effort agreed by the EU, provided that the host country of the underlying project has ratified such international agreement.[97]

From a quality perspective, only credits from projects that were eligible during Phase II shall be acceptable for use in Phase III. However, to ensure the environmental and economic integrity of post-2012 project types, from 1 January 2013, measures may be applied to restrict the use of credits from certain project types.[98]

3.1.4 Double counting

The European Commission was concerned that, in accordance with Kyoto Protocol rules, installations covered by the EU ETS could undertake or invest in CDM/JI projects, thus reducing or limiting emissions under the EU ETS while at the same time also generating CDM/JI credits. In order to avoid double counting under the EU ETS and the Kyoto Protocol rules, the Linking Directive states that EU member states hosting project activities shall ensure that no CDM/JI credits are issued for reductions or limitations of GHG emissions from installations covered by the EU ETS, unless an equal number of EUAs are cancelled by the operator of that installation (in case the project directly affects GHG emissions of the installation) or from the national registry of the EU

[97] In the event that a post-2012 international agreement on climate change has not been concluded by 31 December 2009, the EU may enter into bilateral or multilateral agreements with third countries enabling operators to use credits from 'high quality' projects in those third countries for EU ETS compliance purposes.

[98] Recitals (32) of the EU ETS Amendment, for example, suggests that, in the event of a conclusion of a post-2012 international agreement on climate change, only 'high quality CDM credits' shall be accepted under EU ETS rules.

host country (in case the project indirectly affects GHG emissions of the installation).[99]

3.2 EU ETS and other emissions trading schemes

In order to promote internationally emissions trading as an environmental policy instrument and to increase EU ETS market liquidity, the EU ETS Directive allows for the linking of the EU ETS with other GHG emission trading schemes worldwide to provide for the mutual recognition of its respective allowances.[100] Currently, various other countries and regions outside the EU have already established or are in the process of establishing GHG emission trading schemes.[101] Most of these countries and regions (including the European Commission and several EU member states) are members or observers of the International Carbon Action Partnership (ICAP) that was launched on 29 October 2007 with the objective to provide a forum for countries and regions that have or are in the process of establishing mandatory cap-and-trade emission trading schemes to share experiences and knowledge and to discuss issues related to the design, compatibility and potential linkage of regional carbon markets.

In particular with respect to existing and emerging federal and/or sub-federal US emission trading schemes, as of 1 January 2013 the EU ETS Amendment allows for the execution of agreements between the EU and other countries or sub-federal or regional entities to provide for the recognition of allowances between the EU ETS and other mandatory GHG emission trading schemes with absolute emission caps. Most notably, the ratification by that other country of any post-2012 international agreement on climate change shall no longer be required.

[99] Linking Directive, Art 11(b)(2)–(4); as specified in further detail in Commission Decision (EC) 780/2006 on avoiding double counting of greenhouse gas emission reductions under the Community emissions trading scheme for project activities under the Kyoto Protocol pursuant to Directive 2003/87/EC of the European Parliament and of the Council [2006] OJ L316/12.

[100] According to EU ETS Directive, Art 25, such linkage is conditional upon: (1) the execution of a corresponding agreement between the EU and a third country listed in Annex B to the Kyoto Protocol; and (2) such third country having ratified the Kyoto Protocol.

[101] For example, national and regional cap-and-trade CO_2 emissions trading schemes are existing or emerging in the following countries: Australia (expected to start in July 2011), New Zealand (started with the forestry sector in 2008 and will include other sectors thereafter), Japan (Japan Voluntary Emissions Trading Scheme; started in 2006), Canada (expected to start in 2010) and 10 north-eastern US states (known as the Regional Greenhouse Gas Initiative (RGGI); expected to start in 2009). In addition, in 2007, numerous bills designed to limit GHG emissions were submitted to the US Congress calling for the adoption of a federal cap-and-trade GHG emissions reduction system in the United States of America. During his campaign, President-elect Barack Obama voiced his support for the enactment of a multi-sector federal cap-and-trade emissions trading program. But in light of the current economic downturn, some believe that passage of corresponding legislation will have to wait until at least 2010. For all existing and proposed emission trading regimes compare: J Reinaud and C Philibert, 'Emission Trading: Trends and Prospects' (9 December 2007) (OECD Paper No.) COM/ENV/EPOC/IEA/SLT (2007) <http://www.iea.org/textbase/papers/2007/ET_Trends&Prospects.pdf>, accessed 18 May 2009.

4. Lessons Learned

After four years of EU ETS operation, several lessons have been learned by policy-makers on how to design an emissions trading scheme that promotes both cost-efficiency and environmental effectiveness. These lessons may also be valuable for other policy-makers currently in the process of reviewing or setting up similar emission trading regimes on a federal, sub-federal or regional level:

- A high level of harmonized rules and procedures seems necessary to minimize internal market distortions; in particular, harmonized and detailed emission monitoring, reporting and verification guidelines appear crucial to accumulate accurate and reliable data for cap-setting and allowance allocation purposes and to minimize the risk of over-allocation.
- The centralized setting of an overall emissions cap and a transparent and common process on how to reduce the overall cap over time appears to be crucial in order to achieve overall emission reduction targets, to create scarcity in the market, and to provide covered sectors, operators and installations with sufficient planning reliability.
- A centralized process of allocating allowances in accordance with strict common rules and criteria seems necessary to ensure equal treatment of covered sectors, operators and installations and to minimize the risk of national/state favoritism.
- In principle, auctioning appears to be the preferred method of allocation to make the polluter pay, to incentivize early domestic investments in energy efficiency or renewable energy and to minimize 'windfall profits'; however, certain limited exceptions seem necessary to address 'carbon leakage'.
- In principle, maximizing an emissions trading scheme's coverage of sectors and activities is beneficial to increase market liquidity and reduce price volatility; however, coverage of small emitters may not be cost-effective and emissions trading may not be the appropriate climate policy response for certain sectors and activities (eg private households, transportation, etc).
- Strong compliance and enforcement mechanisms (including penalty sanctions) seem mandatory to ensure a scheme's environmental integrity.
- Linking emission trading schemes is essential to not only further increase market liquidity and reduce price volatility but also to minimize the risk of market distortions between comparable installations located within and outside the emission trading scheme's territory.
- In principle, the import of CDM/JI credits appears to be important to ensure cost-effectiveness; however, an unlimited import of CDM/JI credits may inflate a scheme's overall emissions cap, reduce the price for allowances

and prevent covered sectors, operators and installations from investing in domestic climate mitigation measures.

- Emission trading is only part of a climate policy solution and should be supplemented by additional climate policy instruments.

5. Conclusions

The EU ETS will remain being the flagship of EU climate policy post-2012 in order to help EU member states reach the ambitious 21% EU emission reduction target by 2020. As of 1 January 2013, various additional EU ETS features will facilitate this goal while mitigating the risks of internal market distortions. Such new features include an extension of EU ETS coverage to new sectors, activities, and GHG and the establishment of a more centralized and harmonized EU ETS structure. By shifting key EU ETS policy decisions, such as cap setting, EUA allocation, auctioning, and import of project-based credits from the national to the EU level, the European Commission seems to have concluded that the objectives of the EU ETS cannot be sufficiently achieved in these policy areas at EU member state level (principle of subsidiarity) and that a more centralized and harmonized structure is necessary to achieve the overall environmental objectives of the EC Treaty (principle of proportionality). Interestingly, however, the European Commission has not applied the same rationale with respect to the determination of the legal nature of EUAs at EU level.

Despite the inclusion of more generous limits to the use of the grandfathering approach and project-based credit imports post-2012 in the EU ETS Amendment, the 'climate action and renewable energy package' of the EU is sending a strong signal to the rest of the world that the EU is committed to combatting climate change post-2012 and hopefully this will encourage other countries and regions not only to adopt similar emission trading schemes but also to negotiate an ambitious future international climate pact.

17

Emissions Trading before the European Court of Justice: Market Making in Luxembourg

Navraj Singh Ghaleigh[*]

1. Introduction

Anyone with even a nodding familiarity of global carbon markets recognizes the current primacy of the EU Emissions Trading Scheme[1] within them. With a trading volume of at least €28bn in 2007, the EU ETS represents approximately 70% of global traded volumes in carbon products[2] and 62% of physical volumes.[3] It is variously described as 'the main driving force of the global carbon market', 'the main driver for emissions reductions, both at home and in developing countries',[4] and 'the engine, perhaps even the laboratory, of the global carbon market'.[5] The EU ETS's trading volumes dwarf those of its rivals—the voluntary Chicago Climate Exchange, the New South Wales ETS, the New Zealand ETS, and the fledgling Japanese scheme—none of which has a volume equal to even 1% of the EU ETS.[6] The Scheme's position of primacy will remain unchallenged unless and until a federal US scheme[7] is established.

Within its own territory the Scheme is a significant policy instrument, with 40% of the EU's total GHG emissions within its regulatory ambit, representing approximately 11,000 of the EU's largest emitting installations. For the period

* I am grateful to Thomas Horsley and Lynne Ross for their excellent research assistance and to Mario Mendez, David Freestone, Charlotte Streck and Rachel Cohen for their acute comments and corrections. Opinions and errors remain the author's alone.

1 Directive 2003/87/EC of the European Parliament and of the Council of 13 October 2003 establishing a scheme for greenhouse gas emission allowance trading within the Community and amending Council Directive 96/61/EC, OJ 2003 L275/32. [Hereinafter 'EU ETS', 'the Directive' or 'the Scheme'.]

2 Point Carbon, K Roine et al, 'Carbon 2008—Post-2012 Is Now' (Oslo, 2008). The World Bank's estimate is €37bn, representing trades of more than 2bn EUAs. See World Bank, 'State and Trends of the Carbon Market 2008' (Washington DC, 2008).

3 World Bank, ibid.

4 Point Carbon, ibid.

5 World Bank, State and Trends of the Carbon Market 2008 (Washington DC, 2008).

6 Ibid, at 7.

7 WL Andreen, 'Federal Climate Change Legislation and Preemption' (2008) 3 Environmental & Energy Law & Policy Journal 261.

2008–2012 alone it is estimated to generate emissions reductions of 3.3% (139 $MtCO_2$ pa) from the base year of 1990 in the EU-15.[8] Whilst the European Climate Change Programme[9] extends to issues of fuel efficiency and quality, vehicular emissions, biofuels, renewables, and carbon capture and storage, it is no exaggeration to describe the EU ETS as the keystone in the architecture of the European response to global climate change.

For these reasons an understanding of the topography of EU ETS litigation is significant in and of itself. The legal challenges that the EU ETS, and decisions taken under it, face before the highest courts of the EU inform the manner in which the Scheme is interpreted and operates. Moreover, with two thirds of the 11,000 regulated entities involved in or planning to abate as a consequence of the scheme, its susceptibility to litigation is intimately linked to relevant investment decisions and the cost of capital.[10] Outside the EU, despite the failure of the Lieberman-Warner Bill in 2008, the aftermath of the US General Election has seen the prospect of a US federal emissions trading scheme edge closer in the form of the Waxman-Markey ('American Clean Energy and Security') Bill, creating the potential for comparative analysis and raising future issues of interoperability. At the level of international policymaking, the legal robustness of the Scheme is a key determinant for future iterations of carbon markets and cognate mechanisms. From the perspective of the EU, commitment to the scheme is not in question—the EU's post-2012 strategy firmly sees an emissions trading scheme at its heart, as do many member states and their sub-units. The precise nature of these architectural arrangements continues to evolve.[11]

The concern of this chapter is the discrete matter of challenges to the EU ETS brought before the European Court of Justice (ECJ) and the Court of First Instance (CFI)—the 'Community Courts'.[12] This chapter commences with a brief account of the prehistory of the EU ETS (Section 2), followed by its legal form and operation (Section 3) before addressing the Courts case law concerning Directive 2003/87/EC (Section 4). It should be noted that the focus herein on the Community Courts' case law is not to suggest a priority in importance over litigation that takes place purely at the domestic level. Rather, space precludes consideration of disputes before the Courts of all 27 member states and so national law will be discussed only tangentially.

[8] European Environment Agency, 'Greenhouse Gas Emission Trends and Projects in Europe 2008: Tracking Progress Towards Kyoto Targets', EEA Report 5/2008, 7.

[9] European Commission, ECCP, <http://ec.europa.eu/environment/climat/eccp.htm>, accessed 1 February 2009.

[10] Point Carbon (n 2 above), K Miles, 'International Investment Law and Climate Change: Issues in the Transition to a Low Carbon World' Society of International Economic Law, Online Proceedings Working Paper No. 27/08, available at <http://www.ssrn.com/link/SIEL-Inaugural-Conference.html>, accessed 1 December 2008.

[11] In particular, see Section 3 and conclusion below for details of Phase III EU ETS—COM (2008) 0016.

[12] For an introduction to the judicial branch of the EU's, its legal basis, role and jurisdiction, see Section 4 below.

2. Prehistory of the EU ETS

Given the EU's historical hostility to market mechanisms as solutions to pollution as recently as at the Kyoto Protocol negotiations,[13] the Scheme's global centrality is a turnaround of some moment. Since that time however, the EU has sought to position itself as a global leader in this policy area with market mechanisms as its primary tool.[14] The EU's warm embrace of market solutions to environmental problems is emblematic of its changing policy toolkit over the past decade.[15] For present purposes it suffices to note that prior to and continuing into the 1990s, the EU adopted a policy approach of 'regulatory environmentalism', premised on the assumption that reliance on free-market solutions would misallocate natural resources and produce inadequate incentives to prevent environmental degradation.[16] There also existed a secondary and emerging strain in EU policy that, as early as 1993 in the form of the Community's Fifth Environmental Action Programme, acknowledged the limitations of command-and-control regulation and the utility of market mechanisms to 'internal[ize] external environmental costs'.[17] This approach cohered somewhat better with the well-detailed preference of the US for environmental markets, which were deployed with success in the SOx/NOx contexts.[18] Also familiar is the influence that these American domestic policy successes had in the negotiations at Kyoto, the architecture of the Kyoto Protocol, and in particular the flexibility mechanisms contained in its Articles 6, 12, and 17.[19]

It is notable that having 'lost' the battle of ideas over the optimal means by which to tackle climate change, the EU subsequently embraced the new settlement with gusto.[20] The Kyoto Protocol committed the EU to an 8% GHG emission reduction by the end of 2012. Reductions were to be re-assigned to member states pursuant to its own 'Burden Sharing Agreement',[21] facilitated by

[13] See C Damro, I Hardie, and D MacKenzie, 'The EU and Climate Change Policy: Law, Politics and Prominence at Different Levels' (2008) 4 Journal of Contemporary European Research 185–9.

[14] See S Oberthür and C R Kelly, 'EU Leadership in International Climate Policy: Achievements and Challenges' (2008) 43(3) The International Spectator 35–50. At the time of writing, the EU's latest published approach to the Copenhagen Convention in December 2009 is contained in COM (2009) 39/3.

[15] See n 13 above.

[16] J Golub (ed), *New Instruments for Environmental Policy in the EU* (Routledge: London, 1998), 8.

[17] 'A European Community programme of policy and action in relation to the environment and sustainable development', Official Journal C138, 17/05/1993, p 5.

[18] See C Streck and MW Gehring, 'Emissions Trading: Lessons From SOx and NOx Emissions Allowance and Credit Systems Legal Nature, Title, Transfer, and Taxation of Emission Allowances and Credits' (2005) 4 Environmental Law Reporter 10219.

[19] NS Ghaleigh, 'The Environment and Anti-Americanism', in B O'Connor (ed) *Anti-Americanism: History, Causes, Themes* (Greenwood Press: Oxford, 2007) Vol 1, 139.

[20] The EC, as well as its member states, is a party to the Kyoto Protocol—Decision 2002/358, OJ 2002 L 130/1.

[21] Ibid, recital 10. For details of the member states' reductions under the Burden Sharing Agreement, see Table 16.1 in Chapter 16 herein and attendant text.

Kyoto Protocol Article 4(1).[22] Foremost amongst the jointly implemented responses of the EU is the Emission Trading Directive.[23] The Directive followed Commission consultations, studies and finally a 'Green Paper on Greenhouse Gas Emissions Trading within the European Union'[24] which acknowledged the EU's Kyoto obligations as well as the necessity that that process did not represent the outer limit of the EU's relevant ambitions. Accordingly the proposal was for a scheme whose industry sector coverage was substantially repeated in the final Directive,[25] though with a threshold of 50MW rated thermal input. In terms of quantum of allowances and distribution, the Green Paper saw a role for the Commission only for purposes of determining member states' internal allocations where the risk of national discrimination arose. The setting of total quantities of allowances was for the member states themselves. The question of free allocation (by grandfathering or benchmarking) or auctioning was left open whilst the need to avoid discrimination against new entrants was given consideration.

The response to the Green Paper from EU institutions, industry and NGOs[26] was unusually uniform and positive, facilitating—indeed requiring—swift legislative action. Introduced in October 2001 under the co-decision procedure,[27] the legislative proposal progressed quickly such that by December 2002 the Council had agreed a common position and textual agreement was found between the Council and the Parliament in the second reading.[28] Such (relative) speed is accounted for by the widespread national support for the measure and perhaps the EU's desire to send a signal to the Marrakech UNFCCC Conference of the Parties, asserting its leadership role.

In terms of the substance of the process, it is notable that the questions of industry sector coverage and thresholds attracted minimal contention.[29] Further, the task of setting total quantities of allowances to be allocated and the task of distributing those allowances, which were envisaged as separate in the Green Paper were combined in what were to become National Allocation Plans ('NAPs'). This created the risk that the 'level of ambition under the scheme risks being diluted as

[22] 'Any Parties included in Annex I that have reached an agreement [may] fulfill their commitments under Article 3 *jointly* ...' Italics added.

[23] Ibid. For an account of the Scheme's details, on which this section borrows, see J Robinson et al, *Climate Change Law: Emissions Trading in the EU and the UK* (Cameron May: London, 2007) Part I.

[24] COM (2000) 87 Final.

[25] See Section 3 'Legal Form' below.

[26] See in particular Council Conclusions (Environment) 22 June 2000, European Parliament Resolution on the Commission Green Paper on GHG Emissions Trading A5-0271/2000 OJ C197, 12.7.2001, p 400.

[27] The name given to the legislative process whereby measures are proposed by the Commission and jointly agreed by the Council and the European Parliament, pursuant to EC Treaty, Art 175(1).

[28] This is an unusual event for a measure with such broad and deep implications. Where such a consensus is not reached—the norm—a conciliation committee of both institutions is convened to negotiate a compromise.

[29] J Robinson et al, *Climate Change Law: Emissions Trading in the EU and the UK* (Cameron May: London, 2007), 63.

a result of industry lobbying over their allocations ... rather than ensuring that negotiation over the distribution [between and within member states] is a zero sum game with no impact on the environmental integrity of the scheme.'[30]

3. Legal Form

The EU ETS is in its basic structure a conventional cap-and-trade scheme.[31] At its heart a fixed number of allowances are issued which are divided into a quantity of pollutant which is emitted over commitment periods or phases. The level of resultant emissions is thus set to be equal to the established cap on emissions. Allowances are allocated to operators who are then obligated to monitor and report their emissions, and to surrender at the end each period an equal number of allowances to the units of pollution emitted. Penalties are attached to non-compliance. The scheme provides for the buying and selling of allowances between parties, whether regulated entities with obligations under the Scheme or third parties.

To this generic schema, the EU ETS's specific approaches to coverage and allowance should be noted. The Directive's coverage of activities, detailed in its Annex I, excludes aviation,[32] shipping and most contentiously the aluminium and chemical sectors, while including energy, ferrous metals, minerals, and pulp and paper. The Commission's Explanatory Memorandum to its original proposal justified the chemical exemption on the basis of its limited contribution to the EU's total CO_2 emissions (approx 1% of the total) and that the large number of installations (approx 34,000) would create significant administrative complexity to the Scheme.[33] The Memorandum remains silent on the exclusion of the aluminium sector. These choices have generated much subsequent controversy, not least before Community Courts.

Allowances, as we shall see, have been a source of at least equal controversy. Defined by Article 3(a) as the right to emit one tonne CO_2e during a specified period,[34] allowances are allocated and issued to installations by way of a

[30] Ibid, 65.

[31] The economic rationale for such mechanisms is the claim that they minimize the marginal abatement cost of arriving at a particular level of pollution—the locus classicus for which is J H Dales, *Pollution, Property and Prices: an essay in policy-making and economics* (Toronto UP: Toronto, 1970), drawing, inter alia, on the Nobel prize winning insights of Ronald Coase's 'The Problem of Social Cost' (1960) 3 Journal of Law and Economics 1–26.

[32] Amendments to scope of the Directive to include aviation have recently been adopted—see Directive 2008/101/EC of the European Parliament and of the Council of 19 November 2008 amending Directive 2003/87/EC so as to include aviation activities in the scheme for greenhouse gas emission allowance trading within the Community (OJ L8, 13.1.2009, 3–21).

[33] COM (2001) 581 Final, OJ C75E, 26.3.2002, 33 at Part 11.

[34] 'Allowance' means an allowance to emit one tonne of carbon dioxide equivalent during a specified period, which shall be valid only for the purposes of meeting the requirements of this Directive and shall be transferable in accordance with the provisions of this Directive'.

two-stage process. Stage one requires member states to develop National Allocation Plans 'stating the total quantity of allowances that it intends to allocate for that period and how it proposes to allocate them ... based on objective and transparent criteria, including those listed in Annex III'.[35] Such NAPs are subject to Commission approval, only after which may member states definitively determine the total quantity of allowances and the allocation of the same amongst installations. [36] As with the question of sectoral scope, the details of such matters have greatly exercised the ECJ and are discussed below.

The EU ETS has been implemented in phases—2005 to 2007 and 2008 to 2012—which coordinate with the Kyoto Protocol compliance period, with Phase III to run from 2013 to 2020. Phase I is commonly described as a learning-by-doing phase, allowing member states to get acquainted with a novel system, to make progress towards their Kyoto Protocol commitments and towards meeting their particular CO_2 goals pursuant to the Burden Sharing Agreement.[37] It has been decided that the Scheme will be extended to other greenhouse gases and installations in its Phase III.[38]

As is well known, the 'trial period' of Phase I was characterized by a price collapse in late April 2006 after the publication of the verified emissions data by member state after member state revealed that emissions were significantly below their allocations to installations. Early 2006 pre-announcement over the counter prices were slightly over €30/ton, by mid-May had fallen to approx €15/ton and then to near zero from early 2007 until the end of Phase I. In a sense it is inaccurate to characterize this as a market failure—it might be argued, the market reacted precisely as it ought to have by adjusting when information that changes expectations was made available. Once aggregate emissions, and the resulting demand for allowances were known, the fact of over-allocation had its predictable price consequences.[39] Thereafter, Phase II forward contracts dominated the markets' attention, with December 2008 EU Allowances (EUAs) ranging between €12–25 per tonne, remaining within the €20-24 band for the majority of 2007. Upon the commencement of Phase II, such prices remained durable (at around €20–25 for most of 2007), revealing the price of emitting GHG in the EU but also sending a strong signal to CDM and JI project

[35] Directive, Art 9(1). [36] Ibid, Art 9(3).

[37] Council Decision 2002/358/EC concerning the approval, on behalf of the European Community, of the Kyoto Protocol to the UN Framework Convention on Climate Change, and the joint fulfillment of commitments thereunder, OJ L130, 15.05.2002, p 1.

[38] Proposal for a Directive of the European Parliament and of the Council amending Directive 2003/87/EC so as to improve and extend the greenhouse gas emission allowance trading system of the Community, COM (2008) 16 Final.

[39] In the view of the Commission, the 'swiftly corrected market price of allowances demonstrat [es] convincingly that the carbon market is working.' Ibid at p 2. There is however also a strong argument that over-allocation was accompanied by over-abatement: AD Ellerman & BK Buchner, 'Over-Allocation or Abatement' A Preliminary Analysis of the EU ETS Based on the 2005–06 Emissions Data' (2008) 41 Environmental Resource Economics 267–87, at 270.

developers that emission reductions generated through projects which generate carbon credits would find a robust market in the EU ETS.[40]

A consequence of the Phase I price collapse was its impact on the design of Phase II. The Commission's approach to the Phase II caps has been described as 'unquestionably tough',[41] being much tighter than in Phase I in an overt attempt to create demand for emission reductions, whether generated within the EU or in non-Annex I countries. The Phase II cap for EU 27 is 2,098 Mt/yr, cutting member states' suggested allocations in NAPs by 245 Mt/yr (10.4%). The largest absolute cuts were in Poland (76Mt), Germany (29Mt), Bulgaria (25Mt) and the largest relative cuts in Baltic states (average 37%).[42] These figures represent a cut of 130MtCO_2 (6.0%) below 2005 verified emissions and 160MtCO2 (7.1%) below 2007 verified emissions.

Whilst the cuts in MSs allowances are deep, the pain has been considerably eased by Phase II's 'credit limits' (the maximum CDM/JI volumes that can be purchased for compliance purposes) which vary according to member states from 10% in most cases, up to 22% for Germany.[43] Coupled with tightness of allocations, this creates the possibility for sizable offset/credit imports.[44] Nonetheless it remains the case that the EU ETS is projected to reduce EU-15 emissions by 139MtCO_2 pa during 2008–12 (a 3.3% reduction from the 1990 baseline). The ongoing impact of such measures is demonstrated by the fact that whilst the 2006 emissions of only four of the EU-15 were lower than their Kyoto target (France, Germany, Sweden, and the UK); that figure is expected to rise to 12 out of 15 by 2010 (Denmark, Italy, and Spain being the miscreants).[45] Such is the strength of these projections that the operation of Phase II in much of 2008 saw relatively strong prices between €19–29/ton. Since the onset of the global recession and as at February 2009, that price had halved.

EU ETS Phase III has recently taken concrete form.[46] Unlike its predecessors, Phase III will run for an eight-year period commencing 1 January 2013 in

[40] See World Bank (n 2 above) at 7. [41] Ibid.

[42] See Point Carbon (n 2 above) at 28, Table 1.

[43] Facilitated by the 'Linking Directive' EC 2004/101, OJ 2004 L228/18.

[44] Although outwith the scope of this paper, large-scale credit imports create at least two concerns. Firstly, reliance on emissions reductions made in CDM/JI projects whose ability to achieve *actual* emissions reductions continues to be questioned—see M Wara & DG Victor, 'A Realistic Policy on International Carbon Offsets', Stanford Program on Energy and Sustainable Development Working Paper No 74, April 2008—raises questions of effectiveness and thereby market and public confidence. Perhaps more importantly, the EU's policy of engagement at Copenhagen with developing economies is to offer unilateral emissions reductions of 20%, or 30% if matched by such parties. If EU MSs are able to count credit imports as 'domestic', the threshold is raised to a near impossible level for developing economies, who face the prospect of having to reduce their own emissions by 30% *plus* the quantity of emission reductions imported into the EU.

[45] European Environment Agency, 'GHG Emission Trends and Projections in Europe', 2008.

[46] Directive 2009/29/EC of the European Parliament and of the Council of 23 April 2009 amending Directive 2003/87/EC so as to improve and extend the greenhouse gas emission allowance trading scheme of the Community <http://eur-lex.europa.eu/LexUriServ/LexUriServ.do?uri=OJ:L:2009:140:0063:0087:EN:PDF.>

contrast with the five-year durations of Phases I and II. The emissions cap will be set by the Commission and features a steady trajectory towards 2020, aiming to reduce emissions by 21% overall,[47] based on annual reductions of 1.74%, but the figures are subject to modification by the Commission during the detailed implementation phase, in order to meet the overall target of 20% by 2020 against a 1990 baseline.[48] The cap is then divided among member states according to emission levels under the ETS and subject to a redistribution mechanism. Emission allowances will increasingly be allocated by auction, in contrast to Phase II where free allocation has been the norm.[49] It is estimated in aggregate 50% of allowances will be auctioned from 2013.[50]

Phase III exhibits a higher degree of harmonization, partly in response to criticism of Phases I and II. This is evident in the EU-wide cap being determined by the Commission and harmonized rules for transitional free allocation. Although these measures benefit EU ETS participants by creating a more level playing field, these are achieved by the Commission exercising a higher degree of control in implementing the scheme.

4. The Community Courts' Case Law

In its relatively short life a notable feature of the EU ETS Directive is the sheer number of legal challenges to which it, and Commission Decisions taken under it, has been submitted. By way of (admittedly unscientific) comparison the EU's Regulation (EC) No 2037/2000 on Substances that Deplete the Ozone Layer has generated 25 actions before the Community Courts, and the Air Quality Framework Directive (96/62/EC) gave rise to 17. By comparison the EU ETS Directive has (as at the end of 2008) resulted in 40 proceedings before the Courts.[51] That number includes procedural actions as well as full judgments and a good number which remain pending. The analysis that comprises the remainder of this chapter organizes the case law into the following categories: challenges to the validity of the Directive;

[47] Commission 'Question and Answers on the revised EU Emissions Trading Scheme' MEMO/08/796 Q5. Note the 21% reduction reflects the proportionately higher percentage contribution by EU ETS sectors to the EU's 20% emissions reduction target than other sectors and is expressed against a baseline of 2005, MEMO/08/796.

[48] MEMO/08/796 Q9.

[49] The limited practice of auctioning is evident from the analysis of past and present auction activity in the EU ETS reported by DG Environment at <http://ec.europa.eu/environment/climat/emission/auctioning_en.htm>, accessed 18 March 09 and see also MEMO/08/796 at Q5 which states less than 4% of allowances have been auctioned in Phase II.

[50] MEMO/08/796 at Q13.

[51] These figures are arrived at by searching the Court's official case database <http://curia.europa.eu/jurisp/cgi-bin/form.pl?lang=en> for the precise citation of the legal instrument. As at 31 December 2008, '2003/87/EC' returned 84 hits, which when 'cleaned' to avoid double counting of multiple procedural steps in a single action, arrives at 40 actual actions.

Article 226 infringement proceedings; Article 230 challenges to Commission decisions on the NAPs in Phase I and Phase II; and a category of miscellaneous cases.

As to the Community Courts themselves, it should be recalled that the uniform application of Community law requires a Community court system. The European Court of Justice and its inferior court, the Court of First Instance, consist of judges from member states. The former is the highest judicial authority in the matter of Community law and is, pursuant to Article 220 EC, tasked to 'ensure that in the interpretation of [the] Treaty the law is observed.' This entails *inter alia* monitoring the application of Community law both by Community institutions when implementing the Treaties and by member states and individuals in relation to their obligations under Community law.[52]

In the environmental context, one area issue that has consistently arisen is that of access to the courts for individuals who feel that their 'environmental' rights have been infringed and whether adequate 'judicial protection' is afforded. This is a particularly pressing issues for individuals' (including enterprises) actions against decisions of the Commission and whether they are admissible in Article 230 proceedings—a 'judicial review' procedure which imposes an admissibility test of 'direct and individual concern'. In the environmental context, where the *public* interest, rather than *specific* or *private* interests, commonly dominates the test can be difficult to satisfy.[53]

4.1 Article 234—preliminary references challenging the validity of the Directive

A critical set of questions arose in the litigation initiated in the French *Conseil d'État* [54] which then sought a reference for a preliminary ruling under Article 234 EC in *Société Arcelor Atlantique.*[55] The applicant is the world's largest volume producer of steel and as such its primary argument is the predictable one that the Directive is discriminatory in its scope of application in that it excludes sectors in direct competition with steel producers, such as installations producing aluminium and plastics. It also raised secondary objections before the *Conseil d'État,* arguing on proportionality grounds that since there were limited technical opportunities to reduce emissions beyond those achieved in steel sector

[52] P Craig and G de Burca, *EU Law: Text, Cases and Materials* (4th edn OUP: Oxford, 2008), 66–76 and EC Treaty, Art 220, 226–45 and Treaty on European Union, Art 46.

[53] See J H Jans & HHB Vedder, *European Environmental Law* (3rd edn, Europa Law Publishing: Groningen, 2008), chapter 5.3.

[54] Decision No. 287110 of 8 February 2007 *Société ARCELOR Atlantique et Lorraine* pp 1519–1540. For a discussion of the national law implications of this case - which have attracted much comment in France—see
J-M Sauve, 'Judging the Administration in France: Changes Ahead?' (2008) Public Law 531–45.

[55] Case C-127/07 *Société Arcelor Atlantique et Lorraine and Others v Commission* [2008] ECR 00, OJ 2009 C44/8.

since 1990, that its Article 43 EC freedom of establishment rights[56] was infringed by the Directive, likewise its fundamental property rights, and that the Directive imposed obligations with unforeseeable financial implications and as such were inconsistent with legal certainty.

In its ruling the ECJ adopted a narrow interpretive approach, adhering closely to the specific reference question asked of it by the national court, namely whether the Directive was compatible with the principle of equal treatment.[57] This question was answered in the affirmative with reference to the exclusion of the plastics and aluminium sectors with respect to the included steel sector.

In determining the question of equality of treatment the Court first considered the comparability of the situations. This is to be assessed in light of the subject matter and purpose of the Directive and the Court noted that 'the possible existence of competition between those sectors cannot constitute a decisive criterion'.[58] Given the Directive's subject matter and purpose, differing sources of CO_2 emissions are comparable. Sectors of the economy that emit such pollution can therefore contribute to the functioning of the Community trading scheme and so in the instant circumstances (steel versus aluminium and plastics) are in a comparable situation and, in principle therefore must be treated equally. Having established comparability, differential treatment on the part of included sectors is established by virtue of the disadvantages associated with compliance with the Scheme.

At the heart of the judgment lies the debate over justification of this differential treatment. Mindful that comparable situations may be differently treated by virtue of objective and reasonable criteria (proportionate and compatible with the aim of the legislation), the burden is on the Community institutions to (1) demonstrate the existence of a justification and (2) furnish the Court with information enabling it to verify the justification.

To this end the Community institutions submitting observations pointed to the novelty and complexity of the Trading Scheme and the provision for subsequent legislative review in support of its decision to limit the initial scheme to CO_2 and sectors making the most significant contribution to the overall emissions of that pollutant. The Court conducted its review on the basis of these considerations and the evidence set out by the Community institutions in support of its decision in which context, the ECJ reiterated its established general principles of judicial review that: (1) the EC legislator enjoys broad discretion in

[56] A subset of free movement of capital, one of the 'four fundamental freedoms' of the European Community—along with free movement of goods, persons and services—freedom of establishment comprises the entitlement of economic operators (whether persons or enterprises) to carry on an economic activity in a stable and continuous way in one or more member states.

[57] The Court's approach is certainly defensible in that they addressed the question that was asked of them by the national court, and only that question. However, the Court in such rulings has certainly been known to take a broader approach. There seems to be no consistency between where the Court does not deal with questions put by the litigants in domestic disputes (because the national court did not ask the specific question) and where it does. (Craig and de Burca, n 52 above).

[58] Ibid, para 36.

respect of political, economic and social choices; (2) where the latter choices involve complex assessments and evaluations, the EC legislator may favour a 'step-by-step' approach to regulation; (3) the EC legislator must take into account all the facts and technical/scientific data available at the time in reaching its decisions.[59] As applied to the Directive, the ECJ thus found that the preferred 'step-by step' approach was within the limits of the legislator's discretion but that the legislator's preferred incremental approach to Emission Trading does not release it from the demands of the principle of equal treatment (ie in deciding the order of inclusion in the Scheme etc).

The ECJ then examined the contested sectors individually, examining whether their initial exclusion was in fact justified in accordance with the above principles. In respect of chemicals the ECJ accepted that exclusion could be objectively justified and in particular that the relatively diffuse nature of the sector would impose a significant administrative burden were it to be included, and that the advantages to the scheme of including this sector from the outset may not have outweighed the difficulties. Exclusion from the Scheme was thus justified on the basis of 'administrative feasibility'. The ECJ also accepted that exclusion could be objectively justified for the aluminium sector and that the legitimacy of 'administrative feasibility' again meant that the legislator was not required to base its decisions solely on a threshold for emissions. However, the Court's reasoning here differs from that underpinning its appraisal of the chemical sector. The justification for the exclusion of the aluminium sector is not considered in isolation. It is not simply a review of a cost-benefit analysis as in the former case. The Court refers to the respective total emissions of both the aluminium and the steel sector:

> The difference in the levels of direct emissions between the two sectors concerned is so substantial that the different treatment of those sectors may . . . be regarded as justified[60]

The distinction in the Court's approach is of course very subtle. In effect, the ECJ's appraisal of the exclusion of aluminium sector also comes down to a cost-benefit analysis. However, the question is why is the ECJ 'comparing' the two sectors in its reasoning if competition considerations are excluded?[61] It is unclear why this is not irrelevant for the assessment of whether or not the exclusion of that particular sector could be considered objective justified within the bounds of the legislator's discretion. It appears that the Court cannot shake off 'competition concerns' absolutely.

4.2 Article 230 challenges to Commission Decisions on the National Allocation Plans

In terms of volume, it will be no surprise that challenges to Commission decisions on NAPs form that largest part of the Courts' EU ETS docket.

[59] See paras 57–59, ibid, for authorities cited. [60] Ibid, para 72. [61] Ibid, para 36.

Such annulment actions are addressed by phase in the below analysis, and chronologically within phase.

4.2.1 Phase I Challenges

The first of these actions issued from the UK in late 2005, with the United Kingdom raising an action for annulment of the Commission's refusal in a letter to consider amendments to the UK's original NAP to increase its allowance. [62] The UK maintained that its original NAP was merely 'provisional' while the Commission adopted a Decision in respect of the original NAP and rejected the UK's amendments. The core question was whether the Commission was entitled to reject as inadmissible amendments to allocations submitted by a member state after publishing a Decision on the NAP?

The Court of First Instance held that Commission may not restrict a member state's right to propose amendments to its NAP prior to that member state issuing a definite decision on its allowances and allocations under Article 11(1). (By implication this applies to Phase II and beyond under Article 11(2).) Any amendments to a NAP made by a member state must be notified and accepted by the Commission before they may form the basis for the member state's final decision under Article 11(1). The contested Decision was thereby annulled. The CFI's decision is supplemented by a series of important comments on the respective roles and powers of the Commission and the member states under the Directive, in particular that the Commission's review under Article 9(3) is strictly limited to an assessment of NAP against the criteria articulated in Art 10 and Annex III of the Directive—there is 'no other ground for rejection'.[63] The Commission is not entitled to restrict amendments to a NAP to those necessary to overcome incompatibilities it identified in its Decision on that NAP—Article 9(3) does not set out any such limits to the permissible amendments member states may make to their NAP. Particular attention is drawn to the public consultation process in Article 11, which would be rendered meaningless if amendments were limited to correcting incompatibilities identified by the Commission in its Decision. Member states may propose amendments after expiry of the Commission's three-month review period and any such amendments are still subject to the Commission's (limited) review and, in any case, must be made before the Article 11(1) decision is taken. The ruling appears to 'clips the wings' of the Commission which perhaps sought to carve out a more dynamic role for itself in the NAP process. The CFI rejected its arguments on economic stability as 'unsubstantiated' and 'at the very least, exaggerated.' The Commission is not

[62] Case T-143/05 *United Kingdom v Commission* (joined with T-178/05), Case T-178/05 *United Kingdom v Commission.*[2005] ECR II-4807.

[63] But note comments of CFI in the subsequent ruling in Case T-374/04 *Germany v Commission* [2007] ECR II-4431, below, in which the Court stated that member states may base their NAPs on criteria other than Annex III where such are transparent and objective. It is then for the Commission to demonstrate that such infringe the Directive and/or EC law.

entitled to constrain member states in the exercise of their rights and as such is limited to reviewing only NAPs and amendments against Article 10 and Annex III.

The CFI's ruling in *EnBW Energie Baden Wurttemberg v Commission*[64] has had an even greater direct impact on Courts' jurisprudence in this area. The case arose from the disgruntled German power station operator's challenge to one of the specific 'ex post' transfer rules in the German NAP under which an operator decommissioning an old power plant and replacing it with a (cleaner) new one may continue to enjoy the (larger) allowance it had in respect of the older plant for four years. The applicant argued that the transfer rule constituted illegal State Aid and that its principal rival, RWE, would acquire, free of charge, excessive allowances under the transfer rule owing to its replacement of conventional installations. These allowances could be sold on the market, conferring an unjustified competitive advantage on that operator. With several nuclear installations to decommission, the applicant operator would not benefit in a like manner. Its additional allowance was thereby capped.

The Commission's response—as might have been anticipated—was to raise a plea of inadmissibility,[65] with Germany intervening to support this plea. This is the first case to rule on the admissibility of Article 230 EC actions by individuals against Commission Decisions on NAPs pursuant to Article 9(3) of the Directive. The CFI found the application inadmissible for want of locus standi. According to established case law, Article 230 EC requires that annulment of Decision would itself confer an advantage on the applicant, ie would result in contested NAP/transfer rules no longer enjoying authorization, but the CFI found that the Commission does not enjoy a general power of authorization under Article 9(3) of the Directive—the NAP is presumed lawful, subject to Commission's review of Article 10/Annex III. The CFI stated that the purpose of Commission review procedure is to provide legal certainty for the member states, resolve disputes quickly and to ensure that, during the relevant trading period, the NAP does not risk being challenged by the Commission under Article 226 EC. It is the decision of the member state itself, pursuant to Article 11(1) of the Directive, that affects the legal status of the individuals concerned. With no need for general authorization by Commission and its contested Decision not affecting the rights of the applicant, the applicant does not have locus standi, leading to a finding of inadmissibility.

This concept of 'direct concern' for the purposes of Article 230(4) EC as articulated in *EnBW Energie Baden Wurttemberg v Commission* is very much the orthodox position of the Court, though this is not to say that it is uncontroversial. The ECJ had recently affirming this very restrictive interpretation of 'direct concern' in *Union de Pequenos Agricultores v Council.*[66] In so doing, the

[64] Ibid.
[65] For the established case law on this question, see Jans and Vedder (n 53 above) at 209–14.
[66] Case C-50/00 [2002] ECR I-6677.

ECJ rejected Advocate General Jacobs' attempt to broaden standing and in the subsequent case of *Jego-Quere*[67] the ECJ also, implicitly, overruled the CFI's more cautious attempt to loosen the test. This remains a controversial area in general, with serious concerns over the adequacy of the Community system of legal protection.[68]

EnBW Energie Baden Wurttemberg v Commission casts a long shadow over the EU ETS case law. In *Drax Power and Others v Commission* [69] the operators of a UK power station sought an annulment of Commission's revised (post T-178/05) Decision on the UK NAP, alleging the Commission's Decision contradicted the CFI's ruling in Case T-178/05, that the Commission wrongly concluded that 30 September 2004 was cut-off point for amendments to NAP by member states (that is, three months before the deadline in Article 11(1) for final decision by member states based on NAP). The Commission again raised a plea of inadmissibility and the CFI again reiterated its conclusions in *EnBW Energie Baden Wurttemberg*—that it is the decision of the member state under Article 11 (1) of the Directive that 'directly effects' the legal situation of the applicants. Inadmissibility again followed. Similarly in *US Steel Košice v Commission*,[70] where the monopoly steel producer contested the Commission Decision on the Slovakian NAP. The primary objection was that Commission abused its powers of review under Article 9(3) of the DIR. Following the submission of the NAP, the Commission and the Slovakian Government entered into bi-lateral negotiations, with Commission pushing for a reduction in the total allowance submitted. The Slovakian representative accepted a reduction (in a letter), which brought its total within its Kyoto Protocol commitments. The Commission raised a plea of inadmissibility for want of locus standi, reasoning as above in *EnBW Energie Baden Wurttemberg*. The CFI stated that there was nothing in the Directive in principle to prevent bilateral negotiations between the Commission and member states after the NAP was submitted and prior to Commission Decision. Also, even if it did affect the legal status of applicant directly, CFI held the Commission's Decision was not of direct and individual concern to the applicant. Again, the decision granting rights to individual installations is made by the member state under Article 11 of the Directive. The only remedy for the applicant, where appropriate, is before national courts under national law.

A rather different set of issues arose in *Germany v Commission*.[71] The focus here is specifically on the provision in the German NAP for 'ex post allocation adjustments' which may be made in certain defined cases to correct intervening over-allocations (ie where an operator winds-up an installation). The Commission Decision declared these provisions of the NAP incompatible inter alia with

[67] Case C-263/02 [2004] ECR I-3425 on appeal from the CFI.
[68] See generally Craig and de Burca (n 52 above), 509–28.
[69] Case T-130/06 [2007] ECR II-67.
[70] Case T-489/04 [2007] ECR II-127.
[71] Case T-374/04 [2007] ECR II-4431.

point 10 of Annex III of the Directive.[72] It was argued that allocations to individual installations must be determined in advance and may not be subsequently reallocated after a member state makes its final decision under Article 11 (1) of the Directive. The Commission also declared that the provisions on ex post adjustment for new entrants were discriminatory, favouring this group, contrary to criterion 5 of Annex III.[73]

The CFI started from the position that, as a Directive, 2003/87 EC leaves to the member state the choice of form/method to achieve the binding result it prescribes (see Article 249). Contested 'ex post adjustments' are not covered by the Directive and so in principle, it falls within member state competence to provide such, subject to the Commission demonstrating that they infringe the Directive/EC law. The CFI examined literal, historical, contextual and teleological interpretations of the criterion, finding that Article 11(1) does not expressly exclude such adjustments and that Article 9(3) also permits member states to base NAP decisions on criteria other than Annex III where these are transparent and objective. Commission guidelines on Annex III also do not expressly exclude possibility of ex post adjustments. The Court's central focus was on teleological interpretation—whether the Directive's objectives and sub-objectives preclude the possibility of 'ex post adjustment' mechanism. The CFI found that, in principle, ex post adjustments linked to relatively small reductions in production output may frustrate the Directive's objective, as such deter installations from reducing production, but that the Commission had not submitted evidence to this effect. With respect to the Commission's finding of discrimination contrary to criterion 5 of Annex III for the 'new entrants' rule, the CFI repeated the principle of non-discrimination (that comparable situations must not be treated differently . . .) and found that Commission had made only general statements and not shown that installations concerned were comparable or different. Finally, CFI confirmed the fundamental nature of the duty to support Decisions with adequate reasoning in Article 253 EC, which it views as finding specific expression in Article 9(3) of Directive 2003/87 EC. Accordingly the CFI found an infringement and was generally highly critical of the lack of evidence advanced by the Commission in support of its objections. The Commission Decision was annulled.

Other cases relating to Phase I were, at the time of writing, still pending. Both *Romania v Commission*[74] and *Bulgaria v Commission*[75] are applications for annulment of Commission Decision on NAPs for the year 2007. Both applicants argue that the Commission infringed Articles 9(1), (3) and 11(2) by

[72] The Directive, Criterion 10 of Annex III: 'the [NAP] shall contain a list of the installations covered by this Directive with the quantities of allowances intended to be allocated to each.'

[73] The Directive, Criterion 5 of Annex III: 'The [NAP] shall not discriminate between companies or sectors in such a way as to unduly favour certain undertakings or activities in accordance with the requirements of the Treaty, in particular Articles 87 and 88 thereof.'

[74] Case T-484/07 OJ 2008 C51/57.

[75] Case T-500/07 OJ 2008 C64/51.

determining unilaterally, and in accordance with its own method, the total allocation for Romania and Bulgaria respectively. This is the key objection which finds repeated expression in pending references (see the applications which follow below).[76] A single Phase I case was removed from the Register—*Associazone Italiana Tecnico Economica del Cemento and Others v Commission*[77] in which the applicants contested Commission Decision on the Italian NAP for the first trading period, claiming it sanctioned State Aid contrary to Article 87 EC (also point 5 of Annex III).

4.2.2 Phase II Challenges

The spectre of admissibility and challenges to NAPs feature even more strongly in Phase II than I. Given the Court's long standing case law on such matters, the reasons for such actions are not transparent, although suggestions are made in the conclusion below.

The formula that 'Commission Decision do not affect the legal situation of the applicant for the purposes of Art 230 EC' is repeated with some regularity in the Phase II case law, starting with *Fels-Werke GmbH and Others v Commission*[78] in which the applicant sought an annulment of part of Commission Decision on German NAP declaring its provisions on 'allocation guarantees' incompatible with Annex III. The applicant argued that the Commission misapplied Annex III and that the guarantees do not constitute State Aid. The Commission's plea of inadmissibility was granted by the CFI on the 'usual terms', and was upheld in the applicants appeal to the ECJ in *Saint-Gorbain Glass Deutschland GmbH and Others (FelsWerke . . .) v Commission.*[79]

In *US Steel Košice v Commission* [80] the monopoly steel producer sought an annulment of Commission Decision on Slovakian NAP arguing, as above, that the Commission exceeded its authority under Article 9(3), conducting its own assessment to determine the total emission allowance for that member state. The applicant claimed the Commission's misuse of its powers was motivated by a desire to achieve a scarcity of resources to increase the market price of tradable allowances.[81] Specific to the position of the Slovak Republic, the applicant contested the Commission's interpretation of the conditions for the provision of State Aid to the applicant pursuant to Title 4 point 2(a) of Annex to Accession Treaty. This provision sanctions State Aid to the applicant until the end of fiscal year 2009, up to a maximum of \$500m, and is conditional on production

[76] The Bulgarian cases also raises the following objections pertaining to the infringed duty of loyal cooperation (EC Treaty, Art 10), the Decision being delivered after expiry of the three-month deadline, and that the Decision is insufficiently reasoned.

[77] Case T-371/05 OJ 2005 C296/36.

[78] Case T-28/07 [2007] ECR 11–98.

[79] Case C-503/07 [2008] ECR 11–48.

[80] Case T-27/07 [2007] ECR 11–128.

[81] A similar argument is made by P Street (2007) 9 Environmental Law Review 268–9, noting the political pressure exerted by the Commission, particularly in respect of the Commission's Phase II NAP review, in his critical assessment of the EU ETS.

limits. The Commission argued that the production limits continued to apply irrespective of whether or not the maximum aid amount is reached before the end of the fiscal year 2009. As well as granting the Commission's plea of inadmissibility on the usual terms, the CFI also rejected the argument that the Decision constituted a reviewable Decision on the application of the Treaty provisions on State Aid. The Commission's review under Article 9(3) was held to be only a provisional assessment and not a Decision for the purposes of Article 87 EC.[82] Near identical claims were made, and rejected on near identical grounds in the case of *CEMEX UK Cement v Commission*.[83] Further, a set of actions seeking the annulment of the Commission Decision on the Polish NAP all raise similar arguments and while they have not been joined formally, they were heard on the same date and dismissed as inadmissible, on the usual terms, by Order of the CFI on 23 September 2008. [84]

The application by multiple Polish energy companies (and supported by more of them) in *BOT Elektro Belchatow and Others v Commission*[85] was dismissed as inadmissible. Following earlier rulings post *EnBW Energie Baden Wurttemberg*, the Court reiterated that it is the decision of the member state pursuant to Article 11(2) of the Directive that affects directly the legal situation of the applicant. The fact that the latter—national—authority must comply with the Commission Decision on the NAP does not alter this fact.[86] This is a rather formalistic mode of reasoning.

The Court's remarks at paragragh 49 are entertaining in this context:

> The fear expressed by the applicants that the size of the reduction, as approved by the contested Decision, in the total quantity of allowances would necessarily lead to a proportional reduction in their individual allowances refers to an entirely hypothetical event.

[82] The appeal to the ECJ against Case T-27/07, *US Steel Košice v Commission* Case C-6/08P [2008] ECR 1–96, was dismissed on the basis that it is only when the member state adopts a definite decision under Art 11 that an operator is legally granted a specific quantity of rights. Therefore the Commission's Decision does not affect the legal situation of the applicant for the purposes of Art 230 EC.

[83] Case T-13/07 [2007] ECR 11–146. The sought annulment by the cement producer to the Commission's Decision on the UK's phase II NAP alleged that Commission did not object to (ie approved) an under-allocation by the UK to the applicant in respect of one its plants, thereby discriminating in favour of cement manufacturers in competition with the applicant and constituting illegal State Aid. As a consequence, Case T-313/07 *CEMEX UK Cement v Commission* was removed from the Court's register.

[84] Case T-193/07, OJ 2008 C301/36 *Gorażdże Cement v Commission*. The argument was that Commission infringed Arts 9(3) and 11(2) and inter alia the principle of cooperation by applying its own method to determine Poland's total allocation and imposing this on Poland. In effect, this restricted Poland's total allowance to a level markedly lower than that notified, which had been consistent with that member state's Kyoto obligations. The other dismissed cases were Case T-195/07, OJ 2008 C301/36 *Lafarge Cement v Commission*, Case T-196/07, OJ 2008 C301/37 *Dyckerhoff Polska v Commission*, Case T-197/07, OJ 2008 C301/37 *Grupy Ożarow v Commission*, Case T-198/07, OJ 2008 C301/38 *Cementownia 'Warta' v Commission*, Case T-199/07, OJ 2008 C301/38 *Cementownia 'Odra' v Commission*, Case T-203/07, OJ 2008 C301/39 *CEMEX Polska v Commission*.

[85] Case T-208/07, OJ 2008 C327/25.

[86] Ibid, paras 33–5.

The Court further examines the second criterion of Article 230(4)—the requirement that the measure leaves the addressee no discretion in its implementation. Here the CFI held that member states do retain a margin of appreciation in that they can submit amendments to the Commission and remain free to impose stricter limits than those approved by the Commission (citing Articles 175–6 EC as authority for the latter)!

The following Phase II annulment actions remained pending as of January 2009: *Czech Rep v Commission,*[87] *Hungary v Commission,*[88] *Estonia v Commission,*[89] *Latvia v Commission,*[90] *Lithuania v Commission,*[91] *Romania v Commission,*[92] *Buzzi Unichem v Commission,*[93] and *Poland v Commission.*[94]

[87] Case T-194/07, OJ 2007 C199/38. Annulment sought of Commission Decision on its NAP on the basis that the Commission had exceeded its authority and infringed Arts 9(3) and 11(2) by applying its own method for fixing the maximum quantity of allowances for that Member State.

[88] T-221/07, OJ 2007 C199/41. Annulment sought of Commission Decision on its NAP on the basis that Arts 9(3) and 11(2) of the Directive do not empower Commission to determine unilaterally the total quantity of emissions each Member State may allocate.

[89] T-263/07, OJ 2007 C223/12. Annulment sought of Commission Decision on its NAP on the basis the Commission had infringed Arts 9(3) and 11(2) by making manifest errors of assessment, relied on false assumptions and did not verify its data.

[90] T-369/07, OJ 2007 C269/66. Annulment sought of Commission Decision on its NAP on the basis the Commission had failed to publish its Decision within the three-month time limit, infringing Art 9(3), that Commission's expansive interpretation of Art 9(3) has encroached on its right to determine unilaterally its domestic energy policy, contrary to Art 175(2)(c) EC, that the Decision infringes criterion 1 of Annex III (not taking into account Latvia's Kyoto obligations) and finally, infringement of non-discrimination on nationality (from the perspective of Member States) in that the Commission's methods disadvantage Member States with low emissions.

[91] T-368/07, OJ 2007 C283/35. Annulment sought of Commission Decision on its NAP on the basis that Commission had been delivered after the three-month period, infringing Art 9(3), the Commission had exceeded its review power by overlooking NAP and the Kyoto Protocol and unilaterally determining total allowance in line with its own methods.

[92] Annulment sought of Commission Decision on its NAP on the basis the Commission infringed Arts. 9(1), (3) and 11(2) by determining, in accordance with its own method, the total allocation for Romania.

[93] T-241/07, OJ 2009 C6/30. Annulment sought of Commission Decision on the Italian NAP contesting the Commission's objection to the provision for ex-post allocation adjustments in the NAP, according to which installations may retain allowances in respect of plant closures due to product rationalization, ie transferring rights to remaining installations.

[94] T-183/07, [2007] ECR 11–152. Annulment sought of Commission Decision on its NAP on the basis that the contested Decision reduces the total CO2 limit proposed by Poland in its NAP by 26.7%, infringing the procedural demands of Art 9(3) as its Decision was taken after the expiry of the three-month period, that Commission substituted its own facts for those submitted in the NAP obtained by an inconsistent application of its own preferred economic model, that the Commission also failed to take into account the binding Kyoto Protocol in reaching its Decision, contrary to criteria 1, 2, and 12 of Annex III, that the Commission exceeded its powers in limiting the transfer of rights from Phase I to Phase II, that the Commission's failure to consult with it and not take its energy concerns into account affects Poland's energy security. The application for interim measures was rejected by the CFI for want of urgency (T-183/07 R, [2007] ECR 11–152) and the applicant subsequently sought suspension of measures in respect of Commission's Decision, rejecting its NAP. Judgment on merits is still pending before the ECJ.

4.3 Miscellaneous cases

For the sake of completeness[95] two further categories of cases require brief consideration, although neither has implications for the operation of the EU ETS as such.

The first category comprises *Abraham and Others v Region Wallonne and Others*[96] and *Evropaiki Dynamiki v Commission*.[97] The former concerns Directive 85/337 EEC on environmental impact assessments for infrastructure projects and cites *United Kingdom v Commission*[98] in the context of public participation affecting project decisions; the latter's applicant sought an annulment of the Commission's decision to award the contract to another tenderer following the rejection of its tender for support services for the system of registries established under Directive.

The second set of marginal cases concern infringement proceedings, both successfully brought by the Commission under Article 226 EC, seeking declaration that member states have failed to fulfill their obligations under Directives. The Court made such declarations against Finland and Italy owing to the non/incomplete transposition of the Directive within the deadline prescribed in Article 31.[99]

5. Conclusions

There has been a veritable 'explosion of climate change litigation in subnational, national and supranational tribunals under a wide range of substantive legal approaches'[100] in recent years. For lawyers and legal scholars this has generated temporal challenges as past and present emissions result in future impacts, as well as geographical and jurisdictional complexities. It has manifested itself in a great variety of claims including attempts to protect the skiing industry of New England,[101] the Inuit culture,[102] the Californian coastal natural heritage,[103] and most famously perhaps whether CO_2 is indeed an air pollutant for the purposes of the US Clean Air Act and how this determination ought to be made.[104] The implications of this litigation for national and international law

[95] See the methodology outlined in n 51 above.

[96] Case C-2/07, [2008] ECR 1–1197.

[97] Case T-406/06, OJ 2009 C6/24. Dismissed by the CFI.

[98] Case T-178/05, see n 62 above.

[99] Case C-107/05, [2006] ECR 1–10 *Commission v Finland* and Case C-122/05, [2006] ECR 1-65 *Commission v Italy*.

[100] HM Osofsky, 'Climate Change Litigation as Pluralist Legal Dialogue?' (2007) 43 Stanford Journal of International Law 181 at 184.

[101] *Connecticut v American Electric Power Co*, 406 F Supp 2d 265 (SDNY 2005).

[102] An Inuit Petition to the Inter-American Commission on Human Rights for Dangerous Impacts of Climate Change (2004), available at <http://www.ciel.org/Publications/COP10_Handout_EJCIEL.pdf.>

[103] *California v General Motors Corporation*, 2007 WL 2726871 (ND Cal Sept 17 2007).

[104] *Massachusetts v. EPA*, 127 S Ct at 1462.

is gaining increasing systematic attention,[105] and its spillover affects international negotiations.[106]

Such litigation stands in marked contrast to the rather technical foregoing analysis of the Community Court's EU ETS docket. This is explained in part by the fact that much climate change litigation is 'impacts' oriented and as such claimants are casting around for established legal tools (international human rights law, torts, administrative judicial review, takings and expropriation etc) that might be deemed to be applicable for the novel nature of claim they are making. The EU ETS litigation is not concerned with the impacts of climate change (declining snow packs, costs of adaptation to sea level rises etc) but rather the finessing of a new market mechanism from the perspective of key market actors within the established confines of EU law. Accordingly, traditional doctrines of non-discrimination, equal treatment, *locus standi* and legal certainty come to the fore. What is less traditional is the sheer volume of litigation that the Directive has attracted in its short life. This might be explained as a classical 'interest group theory of politics'[107] scenario, whereby a small number of well resourced actors affected by a regulatory change can be anticipated to organize themselves to respond to and challenge that change. Such an approach may explain the *Arcelor* litigation but it does not explain the slew of actions post-*EnBW Energie*.[108] Having been established that the Court's conventional, restrictive, approach to standing in judicial review actions applied equally to challenges to Commission Decisions on NAPs, it would be clear that any subsequent challenges on this basis would similarly fail. However, the challenges continued despite their near-zero prospects of success and the 'interest group theory of politics' provides no clues as to why. When looking at the member state source of challenges and the deepest of Phase II cuts, a certain degree of overlap, emerges with Poland, Germany, Bulgaria, and the Baltic states all featuring in both categories. Accepting the possibility that market actors in the energy, minerals and ferrous metals industries have been 'encouraged' by their national governments to challenge decisions, there is perhaps an element of both playing to national audiences (against 'Brussels' and 'Europe') and seeking to pressurize the Commission's decision making processes—the latter strategy at least seems to have been singularly unsuccessful.

[105] See inter alia Osofky (n 100 above); WCG Burns, 'Potential Causes of Action for Climate Change Damages in International Fora: The Law of the Sea Convention' (2006) 1(2) Journal of Sustainable Development Law & Policy, 27–51.

[106] DB Hunter, 'The Implications of Climate Change Litigation for International Environmental Law Making' American University Washington College of Law Research Paper No 2008-14.

[107] See M Olson, *The Logic of Collective Action* (Harvard University Press: Cambridge MA, 1965). An alternative approach might be to view the challenges as instances of 'voice' on the part of actors who have no possibility of 'exit' from the EU (given their sunk costs) and regardless of their 'loyalty'—see AO Hirschman, *Exit, Voice, and Loyalty: Responses to Decline in Firms, Organizations, and States* (Harvard University Press: Cambridge MA, 1970).

[108] N64 above.

The changes to the Scheme brought about by Phase III amendments play interestingly into these questions. It is arguable that there is a significant relationship between the case law findings and the challenges of EU enlargement. A disproportionate volume of the above challenges issue from the 'new' EU member states (the 'EU-10') which in many cases have struggled to adapt their ageing industrial stock to the requirements of the EU's ambitious energy and climate change policies. To this end, Phase III's scheme for the auctioning of allowances makes provision for 'redistribution mechanisms' by which 10% of allowances are to be redistributed to low per capita income member states as a solidarity mechanism.[109] It is not fanciful to infer that such a concession is a response to the EU-10 grievances expressed in part by the EU ETS case law.

In addition to those powers noted above, Phase III allocates even broader powers to the Commission, notably in the context of the new auctioning provisions. Auctioning will be introduced in stages, beginning with the electricity generation sector which will be subject to 100% auctioning from 2013. Optional and temporary derogations will exist for certain member states,[110] allowing for auctioning to build up from a minimum of 30% to 100% during Phase III and conditional upon modernizing generation plant and infrastructure[111] which also bleeds into the 'enlargement question'. For other sectors auctioning is to begin at 20% in 2013, reaching 70% in 2020 and 100% by 2027.[112] The Commission will develop and adopt harmonized rules for transitional free allocation by comitology by 31 December 2010.[113] Exceptions for sectors at risk of carbon leakage are to be assessed and determined by the Commission[114] with these sectors to receive 100% free allowances. The Commission will review these provisions (and level of free allowances) following any international agreement reached at COP15 to determine their impact eg of sectoral agreements.[115] In all these areas the Phase III arrangements have granted to the Commission a raft of new decision making powers in critical areas of policy—environmental, economic and industrial.

At the time of writing—less than three months after the agreement of Phase III—the consensus contained therein is already unravelling. Whilst Phase III was adopted as part of wider energy and climate package which emphasized issues of energy security and the EU's wider strategic external relationships, the deepening global recession has generated stronger budgetary pressure on member states

[109] N47 above, MEMO/08/796 Q7. A further 2% of allowances are redistributed to those member states which had achieved early progress against Kyoto Protocol reduction targets and proceeds of auctioning 300m allowances are to be set aside to subsidize the development of carbon capture and storage demonstration plant, or other innovative renewable energy technologies.

[110] The derogation is determined upon criteria including the interconnectivity of electricity grids, share of fossil fuels on generation mix and GDP/capita, from MEMO/08/796 Q7.

[111] MEMO/08/796 Q15.

[112] MEMO/08/796 Q7, Q15.

[113] MEMO/08/796 Q14. These will rely on benchmarks for process technologies, favouring carbon efficient technologies.

[114] MEMO/08/796 Q15.

[115] MEMO/08/796 Q15.

and weakened the political compact which existed in December 2008.[116] How this manifests itself is inevitably a matter of conjecture but candidates might include member states' (1) pressure on the financing elements of the package (eg subsidies for renewable energy, CCS and investment in large scale strategic infrastructure projects), (2) demands for wider definitions of sectors at risk of competitiveness issues and carbon leakage, and (3) call for limits on the EU's level of ambition for the UNFCCC negotiations.

In the face of such pressures the restrictive interpretation of the ECJ, in particular in respect of the admissibility issue, takes on a different light. What might have appeared to be a highly formalist Court, adhering to sometimes obscure procedural rules might be better characterized as one committed to holding member states to their political compacts. For the purposes of creating a stable market, such an approach generates a substantial certainty dividend. Attempts to dilute or resile from these commitments in Community Courts—whether raised by member states or enterprises in their territories—will not be encouraged by the Courts' 'bullet-proofing' of the Commission and its decisions. The extension of the Commission's powers by Phase III thus creates a position of some certainty in a market (like many others, perhaps) where there has been precious little.

[116] See conclusions from the Environment Ministers' and EcoFin meetings (respectively <http://www.consilium.europa.eu/uedocs/cms_data/docs/pressdata/en/envir/106429.pdf> and <http://www.consilium.europa.eu/uedocs/cms_data/docs/pressdata/en/ecofin/106572.pdf>, accessed 18 May 2009) in February/March 2009 and the Spring Council, 19–20 March 2009.

PART V

CARBON TRADING OUTSIDE KYOTO: NATIONAL AND SUB-NATIONAL SCHEMES

18

Emissions Trading in the US: Legal Issues

K Russell LaMotte, David M (Max) Williamson and Lauren A Hopkins

1. Introduction

Amid great expectations, the United States ratified the Framework Convention on Climate Change in 1992 but ultimately it declined to ratify the Kyoto Protocol. What began as reluctance by the Clinton administration to submit the Protocol for ratification—stemming in part from a US Senate resolution warning that presenting the Protocol for ratification would be futile[1]—developed into outright rejection of the Protocol by the subsequent Bush administration. Convinced that legally binding greenhouse gas (GHG) reduction targets would have 'wrecked' the US economy,[2] former-President Bush instead encouraged a focus on voluntary GHG initiatives. The perception among key state governments and non-governmental organizations that the federal government was shirking its responsibilities, however, led to the initiation of a number of sub-national regulatory initiatives. These initiatives have served to pressure the regulated community and the federal government to consider action domestically and to engage in climate negotiations internationally.

The interplay of regulatory activity between the states and the federal government has greatly influenced the development of environmental law in the United States. The political and constitutional challenges of a federal system have been one of the animating features of US environmental law. Climate change is no exception; indeed, because of its scope and breadth, the regulatory response to climate change at the state and federal level will undoubtedly raise unprecedented tensions in the state-federal relationship. At the time of writing (in mid-2009), those tensions are already beginning to emerge. Commitments under several sub-national trading initiatives are gearing up just as Congress and the new administration under President Obama are preparing for national and possibly even international action on climate change. Accordingly, many

[1] Byrd-Hagel Resolution, S Res 98, 105th Cong (25 July 1997).

[2] Associated Press, 'Bush: Kyoto treaty would have hurt economy' (30 June 2005) <http://www.msnbc.msn.com/id/8422343/>, accessed 2 February 2009.

essential features of US climate policy are in flux, and the precise nature of the interaction between national and sub-national GHG regulation remains uncertain. Recognizing that the situation is fluid, this chapter offers a guide to the key legal issues presented by national and sub-national GHG regulatory initiatives in the United States.

Section 2 of this chapter provides an overview of sub-national trading schemes at the regional and state level, where mandatory regulations creating carbon trading regimes are, as of this writing, most advanced in the United States. Section 3 surveys a number of legal issues that will play an essential role in the design and implementation of any cap and trade scheme in the United States, whether at the national or sub-national level. Section 4 addresses a series of constitutional challenges specific to state or regional programs that may limit the scope and perhaps the viability of these non-federal programs. Section 5 then highlights a number of prospective legal issues that are likely to arise in designing a new federal climate change statute, such as the relationship between an emissions trading market and existing US environmental laws and trade policy. This part also identifies the legal issues that may arise in the event that there is no new climate change-specific federal statute adopted, in which case the US Environmental Protection Agency (US EPA) may instead seek to generate a market for emissions trading through regulations adopted under the existing Clean Air Act.

2. Sub-National Greenhouse Gas Trading Initiatives

Although a number of state and regional GHG initiatives have developed independently, the various programs exhibit certain common features, most notably a combination of targeted 'command and control' restrictions on carbon emissions and a 'cap and trade' approach to limiting emissions from the electric utility sector and other traditional heavy-emitting 'smokestack' industries. Building on these similarities, preliminary discussions regarding the eventual merger of the principal regional programs have begun. While there continue to be differences among approaches, there is growing consensus that merging these systems is desirable provided certain design elements can be worked out, such as fair and effective allocation and/or auction of allowances, agreement on which sectors will be covered, and leakage of emissions outside the boundaries of the sub-national program. Whether such details can be resolved before the effort is potentially superseded by a federal program remains to be seen.

2.1 The Northeast Regional Greenhouse Gas Initiative

The most advanced of these state and regional programs, the Regional Greenhouse Gas Initiative (RGGI), became effective on January 1, 2009 after holding a series of emissions allowance auctions in late 2008. RGGI is a cooperative effort among ten US mid-Atlantic and Northeast states to reduce carbon dioxide (CO_2) emissions from large fossil fuel-fired electric utilities in two stages.[3] The program caps and stabilizes utility sector emissions at 188 million tons of CO_2 per year (with sub-allocations to individual states) from 2009 through 2014 and then requires a 2.5% annual reduction over four years for an overall 10% reduction by 2018, which is estimated to be approximately a 10% reduction below the region's 1990 emissions level. RGGI established a Memorandum of Understanding (MOU) among the participating states and adopted a model rule detailing the CO_2 emissions trading program and requirements for offset projects. Although each state must individually adopt the model rule, the program allows flexibility in the implementation of certain design features such as allowance allocation, source exemptions, and permitting.

While RGGI has been applauded as a first mover, it has also garnered criticism for its relatively modest reduction goals and over-allocation of emissions credits, which was compounded by a flurry of early fuel switching (generally from oil to natural gas) by electric utilities and a period of mild winter weather which further reduced fuel combustion demands. At the same time, by precipitating early action within regulated industry in the face of anticipated mandatory GHG cuts, RGGI has arguably accomplished exactly what it was designed to do: impose an early market signal to influence behavior within targeted emitting sectors. Notably, the RGGI states anticipate (and indeed, have pushed for) future federal climate legislation, and the MOU provides for transitioning into a federal program provided it is comparable to the goals of RGGI.

RGGI builds on a progenitor in Massachusetts that controlled emissions from the six largest electric generators in that state.[4] Although the Massachusetts regime was soon superseded when the state joined RGGI, it is worth noting that the program triggered a dispute with the nuclear energy industry, which challenged its exclusion from certain incentives given to low-carbon power sources. Although the suit was eventually dropped, it serves as a harbinger of the broader debate about whether low-carbon but environmentally controversial energy sources, such as nuclear energy and large hydropower, will be given favorable treatment under US climate programs.[5]

[3] For further information and the RGGI Model Rule, see <http://www.rggi.org>, accessed 10 February 2009. The 10 member states of RGGI are Connecticut, Delaware, Maine, Maryland, Massachusetts, New Hampshire, New Jersey, New York, Rhode Island, and Vermont.

[4] For further analysis on the Massachusetts Greenhouse Gas Initiative ('MGGI') see Beveridge & Diamond, PC, 'Massachusetts Adopts Greenhouse Gas Trading Program' (24 October 2006) <http://www.bdlaw.com/news-news-archive-83.html>, accessed 2 February 2009.

[5] See *Entergy Nuclear Generation Co v Massachusetts Dept of Envt Protection*, No 04-4643-A (Mass Super, filed 21 June 2007).

2.2 California Assembly Bill 32

The momentum for sub-national action was significantly accelerated by California's passage in 2006 of the Global Warming Solutions Act (better known by its Assembly Bill designation 'AB 32')[6] with the vigorous support of Governor Arnold Schwarzenegger. AB 32 caps state-wide GHG emissions at 1990 levels by year 2020. The law mandates the monitoring and annual reporting of GHG emissions by all sources 'of significance,' and delegates broad authority to the California Air Resources Board (CARB) to develop and implement GHG emission reduction measures in accordance with an aggressive series of deadlines, including final regulations to set GHG 'emission limits and emission reduction measures' by 1 January 2011. The AB 32 emissions regulations will become operative as of 1 January 2012, just as commitments under the Kyoto Protocol are winding down.

Significantly, AB 32 does *not* mandate that CARB use a market-based system for complying with the carbon emission limits.[7] AB 32 instead delegates that fundamental decision to CARB, which was directed to prepare a Scoping Plan detailing the strategies California will use to reduce GHGs, and then to issue programmatic regulations. Among the potential strategies that may be considered are market-based compliance mechanisms such as a cap-and-trade program.

Although CARB has not yet finalized the regulations implementing AB 32, it approved the Scoping Plan in December of 2008. The plan adopts a mix of command and control regulations combined with a market-based cap-and-trade regime. The cap covers approximately 85% of California's emissions and extends to the electricity, transportation fuels, natural gas, and industrial emitting sectors.[8] The plan includes a trading component, but initial trading will involve only electricity generators and large industrial facilities (with annual emissions greater than 25,000 metric tons of CO_2e). The plan incorporates auctioning of allowances and limited use of offsets (both within and outside of California) and will be designed to inter-operate with the Western Climate Initiative.

2.3 The Western Climate Initiative

Inspired by California's lead, a coalition of US states and Canadian provinces have formed an alliance known as the Western Climate Initiative (WCI). WCI's goals are to establish mandatory reporting of GHG emissions, impose regional limits on emissions, expand the market for carbon trading, and contain the

6 Cal Health & Safety Code §§ 38500–99 (2006).

7 Although the California law expressly provides that CARB may employ market mechanisms, the definition of 'market-based compliance mechanism' allows for a system other than a carbon dioxide emission cap-and-trade system. See ibid §§ 38505(k), 38562(c), and 38570.

8 California Air Resources Board, 'Assembly Bill 32 Scoping Plan' <http://www.arb.ca.gov/cc/scopingplan/document/scopingplandocument.htm> accessed 2 February 2009.

vexing problem of 'leakage,' whereby GHG emissions tend to increase in areas outside the regulated emissions cap.[9] WCI participants are now recommending a regional cap-and-trade program that would cover approximately 90% of GHG emissions in the region.[10] The program is scheduled to operate in phases, beginning in 2012 with the electricity and industrial sectors, and expanding in 2015 to include transportation, residential, and other commercial or industrial emissions not covered in the first phase. Allowances will be auctioned and the program will enable participants to purchase a limited number of offset credits. WCI partners also envision that the cap and trade program will integrate or work in conjunction with other US or Canadian sub-national trading systems. The WCI has proved an effective forum for the dissemination of ideas throughout a large part of North America and has increased pressure on the US federal government to respond to climate challenges with a comprehensive national solution.

2.4 The Midwestern Greenhouse Gas Reduction Accord

Similarly, nine Midwestern states and two Canadian provinces have formed a coalition known as the Midwestern Greenhouse Gas Reduction Accord (MGGRA) to draft a plan for GHG reductions in the largely agricultural and industrial US heartland.[11] Although MGGRA participants have held a series of meetings and prepared policy positions, the states have not yet committed to specific actions or reduction targets. According to the draft of design recommendations, released in December 2008, the MRRGA Advisory Group is considering reductions in the range of 15 to 25% below 2005 levels by 2020, and 60 to 80% below 2005 levels by 2050.[12] The Advisory Group is also exploring the details of a cap-and-trade system for the region that will maximize the region's strengths, including its capacity for biofuels and renewable energy production and potential carbon capture and sequestration.

[9] See ——, 'Western Climate Initiative' <http://www.westernclimateinitiative.org> accessed 2 February 2009. The participating states and provinces are Arizona, British Columbia, California, Manitoba, Montana, New Mexico, Ontario, Oregon, Quebec, Utah, and Washington. WCI observers include: the US states of Alaska, Colorado, Idaho, Kansas, Nevada, Wyoming; the Saskatchewan province of Canada; and the Baja California, Chihuahua, Coahuila, Nuevo Leon, Sonora, and Tamaulipas states of Mexico.

[10] Western Climate Initiative, 'Design Recommendations for the WCI Regional Cap-and-Trade Program' (23 September 2008) <http://www.westernclimateinitiative.org/the-wci-cap-and-trade-program/design-recommendations> accessed 15 June 2009 [hereafter: WCI Design Recommendations].

[11] See ——, 'Midwestern Greenhouse Gas Reduction Accord' <http://www.midwesternaccord.org/>, accessed 2 February 2009. Members of the Midwestern Accord are Iowa, Illinois, Kansas, Manitoba, Michigan, Minnesota, Wisconsin. Observers include Indiana, Ohio, Ontario, and South Dakota.

[12] ——, 'Midwestern Greenhouse Gas Reduction Accord: Preliminary Recommendations of the Advisory Group' (December 2008) <http://www.midwesternaccord.org/Meeting%20material%20pages/GHG-meeting-8/Accord_Draft_Recs_Dec08.pdf>, accessed 2 February 2009.

2.5 The Chicago Climate Exchange

While not a governmental program, the Chicago Climate Exchange (CCX) has proved a useful early model for carbon trading among companies in the United States.[13] CCX is essentially a demonstration project sponsored by major US GHG emitters and carbon offset project developers to field test a cap-and-trade system using a commodities trading market model. CCX covers multiple industry sectors through an agreement to reduce the six Kyoto GHGs by 6% over the period 2003 to 2010. Participation in CCX is voluntary, but once a party has joined, reduction commitments are contractually binding.[14] The CCX trades its own proprietary allowances, which are allocated to emitters according to an emissions baseline, allows trading of domestic offset credits, and in theory links to the European Union Emissions Trading Scheme (EU ETS) to allow trading of ETS allowances and by indirect extension, trading of CDM credits.[15]

3. Legal Issues in US Carbon Trading

The trading of GHG emissions rights has various antecedents in the United States, such as the leaded gasoline trading program, the federal Acid Rain Program, allotments of ozone-depleting substances pursuant to Montreal Protocol obligations, vehicle manufacturer certification trading, California's RECLAIM trading scheme for nitrogen and sulfur emissions, and, more recently, a federal renewable fuels standard,[16] and various state-level alternative energy portfolio standards. There are a number of legal issues common to trading programs that must be considered, some of which may need to be modified or clarified as applied to carbon trading. These issues relate broadly to carbon trading programs whether at the state, regional or federal level; however, much of the experience and commentary on these issues is drawn from the US federal context.

[13] See <http://www.chicagoclimateexchange.com>, accessed 2 February 2009.

[14] Chicago Climate Exchange, 'Chicago Accord' (2004), <http://www.chicagoclimatex.com/about/pdf/ChicagoAccord_050623.pdf>, accessed 2 February 2009.

[15] In practice, however, there has been little inter-continental trading as CCX credits have traded in the range of US $2 to $6, far below the pricing of EU ETS EUAs which have been valued as high as €30.

[16] See Energy Policy Act of 2005, Pub L No 109-58 (8 August 2005); Energy Independence and Security Act of 2007 (EISA), HR 6; Pub L No 110-140 (19 December 2007); see also Beveridge & Diamond, PC, 'New RFS Program Requirements for Renewable Fuel Producer and Importers' (Beveridge and Diamond Client Alert 18 July 2007) <http://www.bdlaw.com/news-202.html>, accessed 2 February 2009.

3.1 Property rights

Surprisingly, the basic notion of a 'right' to emit GHGs (sometimes disparagingly referred to as a 'right to pollute') and the nature of the emissions being traded remains subject to some uncertainty. Domestic law in each sovereign jurisdiction determines the legal status of the allowances traded and the corresponding property rights granted to the allowance holders. In the US tradition, property rights are 'bundles' of smaller sub-rights, such as the right to use, transfer, or preserve property, and which enjoy conditional protection from government interference.[17] The concept of trading carbon emissions as 'rights' has set up a tension wherein, on the one hand, a strong conception of property rights could trigger government compensation requirements under the US Constitution's 'taking clause'[18] if rights are altered (for instance, if allowances are promised to emitters, but the overall cap on emissions is subsequently lowered), yet, on the other hand, a weak conception of property rights could fail to vest participants with the certainty needed to invest in, transfer and trade in emissions allowances. Thoughtful definition of emissions rights will be necessary to ensure administrative flexibility while minimizing risk and uncertainty in emissions trading.

The US experience through the Clean Air Act's (CAA) pioneering Acid Rain Program established the contours of property rights in emissions trading.[19] Allowances under the Acid Rain Program do not constitute legal property rights; instead, they are termed 'limited authorization[s]' to emit certain levels of sulfur and nitrogen.[20] The CAA further states that nothing under the act or US law shall be construed to limit the government's authority to 'terminate or limit' the authorizations.[21] At the same time, and perhaps paradoxically, acid rain allowances may be 'received, held, and temporarily or permanently transferred.' The

[17] Eg, Markus W Gehring and Charlotte Streck, 'Emissions Trading: Lessons From SOx and NOx Emissions Allowance and Credit Systems Legal Nature, Title, Transfer, and Taxation of Emission Allowances and Credits' (2005) 35 ELR 10219, 10221; *Moore v Regents of the Univ of Cal*, 793 P 2d 479 (Cal 1990) (Mosk J, dissenting) ('[T]he concept of property is often said to refer to a 'bundle of rights' that may be exercised with respect to that object—principally the rights to possess the property, to use the property, to exclude others from the property, and to dispose of the property by sale or gift.'); Travis Allan & Kathy Baylis, 'Who Owns Carbon? Property Rights Issues in a Market for Green House Gasses' (2005) <http://works.bepress.com/cgi/viewcontent.cgi?article=1017&context=kathy_baylis>, accessed 2 February 2009.

[18] US Const amend V ('... nor shall private property be taken for public use, without just compensation').

[19] The Acid Rain Program was established under Title IV of the Clean Air Act Amendments of 1990. The program is administered by the US EPA and applies principally to emissions of sulfur dioxide and nitrogen oxides from large (>25MW) fossil-fuel electric generation units in the continental United States. Allowances are distributed to facilities according to a 1985–1987 actual emissions baseline. Facilities must hold sufficient allowances in a compliance account to equal annual emissions during a given calendar year. Failure to maintain the required allowances results in a $2,000/ton penalty, debiting of allowances to cover the deficit from the next allocation, and possible criminal and civil penalties. Clean Air Act Amendments of 1990, Pub L No 101-549, 104 Stat 2399 (1990) (codified at 42 USC §§ 7651–7651o).

[20] Clean Air Act, 42 USC § 7651b(f). [21] Ibid.

existing and proposed GHG cap and trade systems in the United States have incorporated similar language.[22]

The policy justification for denying allowances the legal status of property rights is two-fold. First, there is a concern that granting full property rights to emitters would restrict the government's ability to adjust emissions targets, for example, if the existing allocation program fails to achieve anticipated reductions in atmospheric concentrations of GHGs or if new scientific data shows that more vigorous cuts are needed. As noted, under the Fifth Amendment to the US Constitution, the government may not take private property for public use 'without just compensation'.[23] US courts have interpreted this clause to include regulatory takings, requiring the government in some situations to compensate property owners where regulations have significantly impaired the value of property.[24] A second justification, with particular currency in the environmental community, is the strong philosophical objection to granting participants a 'right to pollute'.[25] Artfully drafted definitions of emissions rights help protect the government against regulatory takings claims, while avoiding any grant of an express 'right to pollute,' and at the same time ensuring maximum transferability of allowances in the trading system.

Despite the constrained statutory definition of emissions allowances under the US Clean Air Act, experience under the Acid Rain trading program resulted in *de facto* property rights protections under the US system of common law interpretation of statutory mandates. In recognition of the value of Acid Rain Program allowances in business transactions and financial accounting, and despite the absence of a defined property right, US courts have ruled that rights to receive these allowances can be litigated in federal courts,[26] and that a

[22] For example, RGGI defines a CO2 allowance as a 'limited authorization' by the regulatory agency to emit up to 1 ton of CO2. The definition further states that no provisions should be construed to 'limit the authority of the regulatory agency or a participating state to terminate or limit such authorization' and adds that the limited authorization 'does not constitute a property right.' RGGI Model Rule, XX-1.5(c)(6)–(7) (revised 31 December 2008). Proposals for federal cap and trade legislation have followed this template closely. The Lieberman-Warner bill provided that: 'An emission allowance shall not be a property right ... Nothing in this Act or any other provision of law limits the authority of the United States to terminate or limit an emission allowance.' Lieberman-Warner Climate Security Act of 2007, S 2191 § 1201(c), 110th Cong (2007). Similarly, the American Clean Energy and Security Act of 2009 (also known as the 'Waxman-Markey bill') provides that an emissions allowance 'does not constitute a property right' and that nothing shall 'limit or alter' the authority of the US government to terminate or limit the allowances or credits. American Clean Energy and Security Act of 2009, HR 2454, 111th Cong § 721(c) (2009) [hereafter: 'Waxman-Markey Bill'].

[23] US Const amend V.

[24] Eg, *Penn Coal Co v Mahon*, 260 US 393 (1922) ('[W]hile property may be regulated to a certain extent, if regulation goes too far it will be recognized as a taking.').

[25] See, eg, Gehring & Streck (n 17 above), 10221.

[26] *Ormet Corp v Ohio Power Co*, 98 F 3d 799 (4th Cir 1996). In *Ormet*, the US Court of Appeals for the Fourth Circuit determined that disputes over ownership of emissions allowances under the Acid Rain Program presented a substantial federal question that could be heard by federal courts, notwithstanding the fact that the Clean Air Act did not confer property rights in allowances and did not create a private right of action for claimants of allowances.

state-law triggered diminution in value of these allowances constitutes 'injury in fact' sufficient to demonstrate standing.[27]

GHG emissions allowances under a federal climate bill are likely to benefit from these *de facto* property right protections even if cap and trade legislation ultimately incorporates the limited definition of allowances under the CAA Acid Rain Program. However, given some level of inherent uncertainty, participants in GHG trading markets should be aware of the limited nature of the 'rights' to allowances and should take care to structure their business transactions to clarify that the counter-parties to any transaction must recognize the right to emit GHGs as a thing of value which can be contracted for, regardless of whether it is technically a property right.

3.2 Contracting and transactions

(1) Even in the absence of statutorily defined property rights, parties to allowance trades are likely to be able to protect their rights and enforce trades through properly structured contracts. Under US law, private parties can typically create rights that are enforceable between themselves, even if the rights are not otherwise legally recognized, as long as there is sufficient 'consideration,' or in other words, both sides bring something of value to the transaction. US courts place great emphasis on the principle of freedom of contract, and generally prefer to honor the intent of the contracting parties as conveyed through the contract terms. Therefore, the parties to a carbon transaction should strive to ensure that the final contract fully reflects the parties' intent.

(2) The legal elements of such transactions are addressed elsewhere in this book and are beyond the scope of this chapter. However, in developing a carbon trading contract and structuring the trade in the United States, parties will need to familiarize themselves with the regulatory procedures that apply to the transaction as well as the potential business implications and tax rules (discussed below) governing the trade. All of the emerging carbon trading regimes in the US have proposed to use a system of project validation, emissions verification, and electronic-based registry to manage 'ownership' and trading of emissions allowances or credits, similar to the experience of the Acid Rain Program and Europe's ETS. However, as of this point the precise requirements for US programs, other than RGGI, can only be guessed at and companies will have to make decisions based on projections of most likely regulatory structure.

[27] *Clean Air Markets Group v Pataki*, 194 F Supp 2d 147 (NDNY 2002).

3.3 Taxation

Tax treatment of carbon trading markets in the United States remains somewhat uncertain, although precedents established by the Internal Revenue Service (IRS) in the US EPA's Acid Rain Program and other pollution trading scenarios may inform the IRS's approach to the use and trade of carbon market instruments. In the United States, taxes may be imposed at the federal, state, and sometimes regional or local level. Accordingly, the full tax implications of an individual market action will vary based on the jurisdiction. In addition, it is possible that sales or excise taxes (as opposed to income tax) could be imposed on carbon market transactions, or that allowances could be taxed as capital property. A full discussion of the tax consequences of market activities is beyond the scope of this chapter, and entities should consult experienced tax specialists; however, the principal tenets of basic federal income and capital gains tax law, as they have applied to allowance trading under the Acid Rain Program, are surveyed below.

Typically under the federal tax system, tax is assessed on net 'ordinary' income, while profits from the sale of property or assets are treated as 'capital gains' (or losses) based on the difference between the sale price and the seller's tax basis.[28] Basis is simply the cost of, or the money invested in, the property or asset, usually including transaction costs such as broker fees.[29] The tax treatment in a carbon cap and trade market will depend on whether allowances are freely distributed or auctioned, and, if the allowances are freely distributed, whether the IRS decides that it may tax the receipt of these allowances as income.[30]

In the Acid Rain Program, the US government distributed allowances without cost to regulated entities, and the IRS subsequently determined that the receipt of these allowances was *not* taxable as income.[31] Instead, the IRS treated the allowances as intangible capital assets, the value of which could not be depreciated (like, for example, machinery) but could be deducted in the year that the allowance was used by the regulated entity for compliance purposes. The basis of the allowance holder in this scenario is considered to be zero.[32] Under these rules, the sale of allowances by a regulated entity results in capital gain or loss, whereas transactions by allowance traders or investors (as opposed to regulated entities) are typically declared as ordinary income. With a zero tax basis, and thus nothing to offset gains, the full amount realized by the sale of these allowances on the trading market is taxed as a capital gain.[33]

[28] Ethan Yale, 'Taxing Cap-and-Trade Environmental Regulation' (2008) 37 JLS 535, 539.

[29] 26 USC § 1012; Gehring and Streck (n 17 above), 10230.

[30] Yale (n 28 above), 539.

[31] Tax rules applicable to the federal Acid Rain Program are described in Rev Rul 92-16, 1992-12 IRB 5 (23 March 1992) and Rev Proc 92-91, 1992-46 IRB 32 (16 November 1992); see also Announcement 92-50, 1992-13 IRB 32 (30 March 1992).

[32] Larry B Parker & Donald W Kiefer, 'Implementing SO2 Allowance Trading: Implications of Transaction Costs and Taxes' (12 March 1993) CRS Report 93-313 5. <http://ncseonline.org/nle/crsreports/air/air-11.cfm> accessed 10 March 2009; Gehring & Streck (n 17 above), 10225.

[33] Yale (n 28 above), 540.

Although these rules have been tested in the Acid Rain Program, it is unclear whether the IRS or state taxing authorities will extend this approach to GHG trading. Currently no IRS guidance exists regarding the tax treatment of GHG transactions under emerging sub-national trading programs. Given the differences between these emerging programs and the Acid Rain Program—particularly the auctioning, rather than free distribution, of allowances—participants should exercise caution in relying on the Acid Rain tax rules for GHG allowance trading. The fact that participants in many emerging trading programs will be required to purchase their allowances through an initial auction, for example, should result in the purchaser acquiring some tax basis in the allowances. As a result, gains from the subsequent sale of these allowances on the market would presumably be taxed at the applicable capital gains rate (usually between 10 and 15%) on the excess of the sale proceeds over the seller's basis.[34] If these allowances were used for compliance purposes (ie, retired in the EPA registry), rather than sold on the market, it is likely that the basis of the allowances (whether the company purchased the allowances from the initial auction or in a secondary market trade) would be recognized as an operating expense and allowed a tax deduction for that year.[35] Despite the differences in the emerging 'ecosystem' commodity markets, the IRS did apply the Acid Rain Program tax rules to trading of wetland mitigation credits, thus giving some indication of the likelihood of applying the rules to other ecosystem credits, including carbon.[36]

International trading of GHG allowances imposes another layer of tax planning. The IRS recently published its first guidance on emissions trading in over seventeen years.[37] The guidance addresses US tax treatment of gains from the sale of CO_2 allowances on the European market by 'controlled foreign corporations,' or foreign corporations that are majority-owned by US corporations or individuals.[38] The IRS characterized the allowances as 'property that does not give rise to income' under Subpart F of the IRC and concluded that the taxpayer was eligible for a deferral of US taxation. Although the guidance stems from a 'Private Letter Ruling' and therefore may not be used as binding precedent, it is an indication that the IRS is beginning to consider the US tax consequences of GHG trading.

While it is unclear at this juncture what tax rules will be applied to carbon allowance trading, it is possible that some components of the tax rules

[34] Ibid. If the seller does not realize a profit on the sale of allowances (i.e., incurs a capital loss), losses can usually be set off against current year capital gains, carried forward five years, or carried back three years. However, losses cannot generally be used to offset regular income, and accordingly, there is a risk that market participants would lose the value of the capital loss if they have no gains to offset. Gehring & Streck (n 17 above), 10225.

[35] See Parker & Kiefer (n 32 above), 6.

[36] IRS Priv Ltr Rul 9612009 (18 December 1995).

[37] Daniel R McKeithen and others, 'United States: First IRS Guidance Under Subpart F on US Tax Treatment of Gain From the Sale of Surplus CO_2 Emissions' (25 June 2008) <http://www.mondaq.com/article.asp?articleid=63262> accessed 2 February 2009.

[38] IRS Priv Ltr Rul 200825009 (7 March 2008).

formulated for the Acid Rain Program will persist. At the same time, tax treatment of GHG trading at the state and local level, despite RGGI's entry into force in early 2009, also remains unclear. Entities participating in any of the emerging trading regimes should plan for a variety of tax implications, and should consider the option of approaching the IRS as well as state tax authorities for administrative guidance or private letter rulings.[39]

3.4 Inter-system linkage

The emergence of independent trading systems raises the possibility of joining these systems through the concept of 'linkage.' Formal linkage occurs when one trading system enables its regulated entities to use allowances (direct linkage) or offset credits (indirect linkage) from another system in order to meet emissions reduction obligations.[40] These linkages can be one-way (one system recognizes another system's credits) or reciprocal for complete integration. Linkage theoretically carries the benefits of lowering overall abatement costs and increasing liquidity by broadening the market for allowances.[41] Linking systems may also mitigate leakage, or the migration of industrial sources of emissions to areas not subject to emissions caps.[42]

Despite these benefits, design differences among systems pose problems for effective linkage, and the technical, administrative and political hurdles associated with linkage can be prodigious, particularly for direct linkages involving mutual recognition of emission allowances for compliance purposes. For example, differences in design features may raise environmental integrity and competitiveness concerns if there is not general parity in the stringency of emissions reductions requirements, enforcement rules, standards for recognition of offsets (ie, emissions reductions outside of the cap), and allowance distribution or allocation systems.[43] There are considerable economic implications associated with linkage as well, including the possibility that costs could increase for facilities in areas with the lower relative abatement costs.

Whether and how to link the various emerging sub-national trading schemes with one another, and how to link those programs or a federal cap-and-trade

[39] See also Ernst & Young, Thought Center, Webcasts and Podcasts, 'Emission allowances and renewable energy certificates, Accounting and tax issues' <http://webcast.ey.com/thoughtcenter/default.aspx?prog=%7B8c1a84e5-68a1-4330-b04f-f1aacceebb43%7D>, accessed 6 February 2009.

[40] Judson Jaffe & Robert N Stavins, 'Linkage of Tradable Permit Systems in International Climate Policy Architecture' Discussion Paper 2008–07, Cambridge, Mass.: Harvard Project on International Climate Agreements (September 2008) <http://belfercenter.ksg.harvard.edu/files/StavinsWeb5.pdf>, accessed 2 February 2009.

[41] Eg, ibid, at 10.

[42] Eg, Brettny Hardy, 'The Missing Link: Greenhouse Gas Emissions Trading Between the European Union and the United States?' (2007) <http://hdl.handle.net/10161/289>, accessed 2 February 2009, at 17–18.

[43] For a list of design features and an assessment on the effect of different features on technical feasibility and environmental integrity in the context of EU ETS linkage with European national systems, see Sven Bode, 'Emissions Trading Systems in Europe: Linking the EU Emissions Trading System with National Programs' in Bernd Hansjürgens (ed) *Emissions Trading for Climate Policy: US and European Perspectives* (Cambridge University Press: Cambridge, 2005).

system with foreign emissions trading or emissions 'credit' systems (such as the CDM market), is thus a key issue in the US emerging compliance markets. It is also predominantly a policy question rather than a legal question. As a non-party to the Kyoto Protocol, for example, the United States faces comparatively fewer treaty law restrictions on its ability to recognize credits from other regimes, unlike Kyoto Protocol parties which cannot use US credits for compliance purposes. Linkage arrangements can in theory be worked out after a trading regime is established. Because many of these market dislocations are best addressed through harmonization, however, linkages ideally should be considered during the establishment of new emissions trading mechanisms so they can be factored into the overall architecture of the new market.

Fortunately, US policymakers and legislators have been conscious of these considerations. Linkage issues have arisen at all levels in the emerging emissions trading programs in the United States: between two or more sub-national programs (eg, between RGGI and WCI); between sub-national programs and international regimes (eg, between RGGI and the EU ETS); between sub-national programs and a future US federal program; and between a US federal and international regimes. Although the resolution of many of these issues remains in flux, several key developments are in place already.

3.4.1 Linkages between sub-national regimes

With respect to linkages *between sub-national regimes*, the emergence of The Climate Registry (TCR) as a common reporting platform appears to offer a solution to one of the primary technical issues that must be resolved in linkage, namely the need to identify a common trading and recordkeeping platform to avoid double-counting.[44] Although TCR was initially a purely voluntary program designed to ensure consistent and transparent GHG reporting, it has taken on a quasi-official status as states have increasingly endorsed it as the registry they intend to use in their GHG reduction programs. In addition, as part of its effort to publish a mandatory GHG reporting rule, the US EPA is working with TCR to harmonize reporting and recordkeeping practices among jurisdictions.[45]

3.4.2 Linkages between sub-national and international regimes

At least one sub-national program currently provides for the *unilateral recognition of overseas credits* for purposes of compliance. The RGGI program allows covered entities to 'link' with external programs outside of the United States to import

[44] For more on The Climate Registry see <http://www.theclimateregistry.org/>, accessed 2 February 2009.

[45] In the Preamble of the US EPA's proposed mandatory GHG reporting rule, the agency noted that the proposed rules monitoring and GHG calculation methodologies for many source categories are the same as, or similar to, the methodologies contained in state and regional reporting programs such as TCR. Mandatory Reporting of Greenhouse Gases; Proposed Rule, 74 Fed Reg 16,461 (10 April 2009). The agency also noted that it is considering the option of delegating authority for data collection under the rule to state agencies and requested comments on the role of states in implementing the rule. Ibid, 16,595.

external credits for recognition under the RGGI offset compliance requirements. If the average price of carbon exceeds \$10 in the RGGI trading market, covered entities will be able to use CO_2 offset allowances for the permanent retirement of GHG allowances or credits that have been issued pursuant to 'any governmental mandatory carbon constraining program outside the United States that places a specific tonnage limit on greenhouse gas emissions', CERs, or other emissions reduction credits issued under 'protocols adopted through the UNFCCC process'.[46] The current Western Climate Initiative design recommendations similarly contemplate unilateral indirect linkages through the recognition of offset credits from North American markets and the CDM. The recommendations provide, however, that WCI participants may impose additional qualifying criteria and specify that offset credits will not be accepted for GHG reductions in developed countries from sources that are covered by the WCI cap and trade program.[47] These one-way recognition provisions, which do not require the consent of the regime exporting the carbon credit, would appear to sidestep questions about the compatibility of sub-national linkage mechanisms that might otherwise arise under the US Constitution's Compact Clause (discussed further below).

By contrast, even *one-way linkages* may raise novel constitutional questions to the extent that they involve an agreement (eg, as an eligibility criterion) with a foreign state. California's Scoping Plan under AB 32 contemplates the adoption of agreements with developing countries 'to work jointly to develop minimum performance standards or sectoral benchmarks, backed by appropriate monitoring and accounting frameworks. Such agreements would encourage early action in developing countries toward binding commitments, and could also reduce concerns about competitiveness and risks associated with carbon leakage'.[48] As discussed further below, these one-way linkages are arguably more constitutionally suspect to the extent that they depend on active entanglements between sub-national authorities and foreign governments.

Similar issues arise with respect to sub-national programs, like the WCI, that contemplate the future development of *mutual recognition linkages* for allowances traded under other programs, based on agreements to be negotiated.[49] For example, California gave significant attention to the potential establishment of such two-way linkages between its cap-and-trade system and the EU ETS.[50] Indeed, in late 2008 Governor Schwarzenegger hosted a summit with various

[46] Regional Greenhouse Gas Initiative Model Rule, XX-10.3(b)(1) (revised 31 December 2008), <http://rggi.org/docs/Model%20Rule%20Revised%2012.31.08.pdf>, accessed 2 February 2009.

[47] WCI Design Recommendations (n 10 above) para 9.7.

[48] California Air Resources Board for the State of California, 'Climate Change Proposed Scoping Plan' (October 2008), 38 <http://www.arb.ca.gov/cc/scopingplan/document/psp.pdf>, accessed 2 February 2009.

[49] WCI Design Recommendations (n 10 above) para 13.5.

[50] Market Advisory Committee to the California Air Resources Board, 'Recommendations for Designing a Greenhouse Gas Cap-and-Trade System for California' (30 June 2007) <http://www.climatechange.ca.gov/publications/market_advisory_committee/2007-06-29_MAC_FINAL_REPORT.PDF>, accessed 2 February 2009.

countries, including China, India, Mexico, Brazil, and Indonesia, to explore these linking relationships. Although such mutual recognition schemes maximize the policy benefits of linking trading schemes, they are also potentially more vulnerable to constitutional challenge when they involve sub-national entities engaging in bilateral or multilateral binding legal agreements, particularly with foreign countries.

3.4.3 Linkages between US federal and sub-national regimes

With respect to *linkages between federal and sub-national regimes*, it is likely that most of these issues will be resolved through the absorption of sub-national schemes by the federal program.[51] Nevertheless, one key issue that must be resolved in designing the federal program is whether it will reward (ie, give credit for and thereby effectively link to) early action taken by covered entities under a state program.

3.4.4 Linkages between US federal and international regimes

More generally, *legislative proposals at the federal level* have addressed a wide range of linkage models with other trading systems as well as the CDM and offset credit markets. The various cap and trade bills introduced in the last Congress differ widely on how they treat various linkage opportunities. For example, some permit use of foreign allowances only and exclude the use of CERs; others would give recognition only to foreign offset credits including CERs; yet others give recognition only to foreign offsets that allow forest credits but exclude CERs, etc. These linkage issues, which will have a significant impact on the pricing of the global carbon market given the potential size of US demand, are thus merely one more open issue in the large mix of policy and political questions that lawmakers are addressing.

There is one common element across nearly all of these bills, however: non-US originating credits and allowances, to the extent they are recognized at all, would generally be permitted for compliance purposes only up to a quantitative limit (eg, 15% of a covered entity's compliance obligation).[52] Another common issue is that (unlike the California regime), these laws are generally not being drafted in a way that is calculated to increase the possibility of closely aligned

[51] For example, the Waxman-Markey Bill would allow entities to exchange allowances issued under the RGGI, WCI, or California trading programs for federal allowances. Waxman-Markey Bill (n 22 above), § 790(a).

[52] The Waxman-Markey bill for example, proposes to allow covered entities to offset up to two billion tons of emissions per year. This amount would be divided equally between domestic and international offset credits unless the domestic offset limit is not reached, in which case the US EPA may adjust the ceiling for international offset credits. Waxman-Markey Bill (n 22 above), § 722(d). In addition, international offsets could not be used for compliance under the bill unless the United States is a party to a bilateral or multilateral agreement with any country where an offset activity would take place. Ibid, § 743(b)(2)(A).

regimes with significant two-way linkages. Those opportunities, if they are feasible at all, will likely require complex negotiations and efforts to graft together the US system with its international analogues.

4. Potential Challenges to Sub-National Regulation

In response to the recent flurry of activity at the US regional, state, and local levels on climate change, questions have surfaced regarding whether these initiatives are compatible with the US Constitution and its system of federalism. Though the states have broad authority to regulate public health and welfare, the Constitution places limitations on their authority to regulate in certain areas where federal authority is pre-eminent. Emerging carbon trading regimes at the regional and state level are already being challenged by affected interests. For example, early in the RGGI development process, Edison Electric Institute (a trade association for the utility sector) raised many of the constitutional issues discussed below in comments it submitted regarding the RGGI MOU.[53] More recently, a power company sued to overturn the State of New York's administrative regulations implementing RGGI, raising several constitutional issues in its complaint.[54] Whether such initiatives will ultimately survive constitutional challenges depends on many unknowns, though perhaps most significantly it will depend on the scope and content of future federal legislation on climate change.

The issues implicated by sub-national regulation of climate change can be divided into four separate constitutional limitations on state action: limitations imposed by federal pre-emption of state law through the Supremacy Clause of the US Constitution; limitations imposed by the 'foreign affairs' power of the Constitution; limitations imposed by the Compact Clause; and limitations imposed by the 'dormant commerce clause' power.

4.1 Pre-emption

The doctrine of pre-emption flows from the Supremacy Clause, which states that federal laws are 'the supreme law of the land.'[55] Federal laws will pre-empt, and thus invalidate, state laws when: (1) federal laws explicitly state that they pre-empt any state law addressing the subject of the federal legislation;[56]

[53] Edison Electric Institute, 'Edison Electric Institute Comments on the Regional Greenhouse Gas Initiative Memorandum of Understanding' (20 March 2006) <http://www.rggi.org/docs/rggi-eeimou_comments032006final.pdf>, accessed 2 February 2009, at 17–26.

[54] BNA, 'Cogeneration Plant Sues New York to Overturn State's RGGI Regulations' (30 January 2009); *Indeck Corinth LP v Paterson*, filed 30 Jan 2009 (NY Sup Ct).

[55] US Const art VI, cl 2.

[56] The CAA, for example, contains an express pre-emption provision for state motor vehicle standards. 42 USC § 7543(a) ('No State or any political subdivision thereof shall adopt or attempt to enforce any standard relating to the control of emissions from new motor vehicles…').

(2) Congress 'intends federal law to occupy the field;' or (3) to the extent that the state law conflicts with a federal statute such that it is 'impossible for a private party to comply with both state and federal law' or the state law 'stands as an obstacle' to the accomplishment of Congress' goals.[57]

In the context of climate change, the viability of each of these pre-emption challenges will depend primarily on the scope and content of future federal GHG legislation. Challenges based on existing federal legislation are beginning to emerge. Pre-emption challenges have been mounted, albeit unsuccessfully, against certain state regulatory activity directed at GHGs, such as California's mobile source standards,[58] on the basis of existing federal legislation under the CAA. In addition, the recent challenge to New York's implementation of RGGI includes a claim that RGGI is pre-empted by the federal Public Utility Regulatory Policies Act (PURPA).[59] The lack of challenges to other sub-national initiatives in part is a reflection of ripeness—these programs have only begun to take effect. As commitment periods for these programs start up, state participants may expect to see additional judicial challenges. While it remains to be seen how courts will rule on these lawsuits, the absence of specific federal legislation on climate change could make it difficult for a court to conclude that Congress has either expressly or impliedly pre-empted such programs under the CAA or any other statute.[60]

This issue is therefore more likely to be resolved directly in new federal legislation on climate change. Although some of the federal bills have been silent on pre-emption, others would explicitly pre-empt all sub-national trading initiatives.[61] There will be significant pressure from industry to include such express

[57] Eg, *Crosby v NFTC*, 530 US 363, 372–3 (2000).

[58] *Cent Valley Chrysler-Jeep, Inc v Witherspoon*, 456 F Supp 2d 1160 (ED Cal 2006).

[59] 16 USC §§ 2601–2645. PURPA essentially ensures that certain power producers which sell electricity to the wholesale market receive the full avoided cost (generally the cost the other producers would have incurred had they generated or purchased the power from another source). Because RGGI's regulations may impose price terms that conflict with the avoided cost requirement under PURPA, the plaintiff in this lawsuit argues that RGGI conflicts with, and is therefore pre-empted by, federal law.

[60] President Obama released a memorandum on May 20, 2009 directing federal agencies to interpret pre-emption provisions narrowly and to pre-empt state law 'only with full consideration of the legitimate prerogatives of the States and with a sufficient legal basis for preemption.' Although this directive is not binding on Congress, it may affect the interpretation of federal legislation to the extent that Congress fails to expressly pre-empt sub-national trading initiatives. Memorandum for the Heads of Executive Departments and Agencies: Preemption, 74 Fed Reg 24,693 (20 May 2009).

[61] See, eg, Dingell-Boucher Discussion Draft § 733(b) ('[N]o State, local, or regional authority may adopt or enforce a program that caps the amount of greenhouse gases that may be emitted or sold, and that uses tradable emission allowances for the purpose of meeting that cap'). The Waxman-Markey bill contains a provision that would temporarily preempt sub-national cap and trade programs (between 2012 and 2017), and would allow entities to exchange allowances issued under the RGGI, WCI, or California trading programs for federal allowances. Waxman-Markey Bill (n 22 above), § 790(a). States would be permitted to continue to enact and enforce programs that do not involve trading as such, including programs (such as California's) that involve an overall cap on the amount of GHG emissions.

pre-emption mechanisms. It remains unclear, however, whether such express pre-emption language will prevail. A more likely result may involve a model similar to the existing Clean Air Act, wherein Congress set a national standard for automobile emissions but allowed California to undertake experimental deviation from the standard.[62] The remaining states have the option of selecting the national or California approach, and industry thus has only two different standards with which it must comply rather than a quiltwork of up to 50 different targets.[63] Another option would be to reward regional trading participants by allocating additional federal emissions allowances to those early moving states, and to ensure that covered entities that have participated in state-level reduction measures are given credit for such activity under a subsequent federal scheme.[64] Yet another possibility would be to allow federal and regional/state trading systems to link with a federal trading program by recognizing the regional/state allowances in the federal market.[65] While trade-offs exist for each option, express pre-emption of sub-national trading systems is likely to be politically challenging, leaving the likelihood that a threat of pre-emption challenge will hang over sub-national carbon schemes for some time until resolved by the courts.

4.2 Foreign affairs pre-emption

The US Constitution assigns foreign affairs powers solely to the federal government. 'The Constitution's foreign affairs provisions have been long understood to stand for the principle that power over foreign affairs is vested exclusively in the federal government.'[66] For example, the President has the power to make treaties and recognize foreign governments;[67] Congress has the power to regulate commerce with foreign nations.[68] Where Congress and the President have acted through a treaty, the treaty and related legislation may preempt state laws under the Supremacy Clause. Even where Congress and the President have not acted, courts have occasionally invoked the dormant aspect of these powers to limit state activity that interferes with federal foreign relations.[69]

One focal point of foreign affairs challenges to sub-national cap and trade systems will be whether these systems interfere with the ability of the federal government to speak with 'one voice' on climate change in foreign affairs. For

[62] One author has termed this 'modified federalism': Alice Kaswan, 'The Domestic Response to Global Climate Change: What Role for Federal, State, and Litigation Initiatives?' (2007) 42 USF L Rev 39, 80 (citing Ann Carlson, 'Federalism, Preemption, and Greenhouse Gas Emissions' (2003) 37 UC Davis L Rev 281, 313–18).

[63] Ibid.

[64] Jonas Monast, 'Integrating State, Regional, and Federal Greenhouse Gas Markets: Options and Tradeoffs, (2008) 18 Duke Envtl L & Pol'y F 329, 337–40, 345–46.

[65] Ibid, 340–42.

[66] *NFTC v Natsios*, 181 F 3d 38, 49 (1st Cir 1999), aff'd 530 US 363 (2000).

[67] US Const art II.

[68] US Const art I, § 8.

[69] The two most important cases in this area are *Zschernig v Miller*, 389 US 429 (1968), and *American Insurance Assn v Garamendi*, 539 US 396 (2003).

example, the Supreme Court has invalidated state laws in cases where the Court determined that the laws present a risk of disruption or embarrassment in foreign affairs, due in part to the intrusive evaluation by state courts (or regulatory authorities) of foreign legal systems.[70] Another key issue is whether the state measures in question involve areas where the states have traditionally exercised regulatory control over international activity (eg, in the field of taxation). If a state law falls within an area of traditional state competence, the law is less likely to be pre-empted even though it affects foreign affairs, unless there is a clear and substantial conflict with federal policy.[71] On the other hand, if a state law exceeds the scope of traditional state competence and has even a slight impact on federal foreign affairs interests, the law is likely to be pre-empted.[72] Yet another factor is whether the federal authorities have raised concerns about the risks of conflict with federal foreign affairs activity.[73]

In evaluating the vulnerability of the state climate laws, therefore, much will depend on the precise content of the laws and the ways in which they touch on foreign affairs. Under one theory, because climate change is a quintessentially global issue and the Executive Branch is engaged in multilateral negotiations to address climate change issues, virtually any state action in this field is an intrusion on the foreign affairs power of the national government. If federal policy is framed in this way, state and sub-national systems that reduce overall US GHG emissions could potentially weaken the President's ability to secure the participation of key developing countries in exchange for US commitments.[74] On the other hand, state and regional cap and trade systems could actually bolster the US negotiating position by increasing the 'commitment credibility' of the federal government.[75] As discussed above, moreover, linkage provisions in various sub-national programs raise foreign affairs pre-emption issues to the extent that they may require direct cooperation and engagement between states and the foreign authorities to which they are linking. The contours of this analysis will likely change with the emerging federal climate policy of the new administration. The degree to which state climate change laws might entail subjective evaluations of the adequacy or compatibility of foreign GHG control regimes or offset schemes will also be relevant.

[70] Eg, *Zschernig* (n 69 above).

[71] *Garamendi* (n 69 above), 420, n 11; 'Foreign Affairs Preemption and State Regulation of Greenhouse Gas Emissions' (2006) 119 Harv L Rev 1877, 1880.

[72] *Garamendi* (n 69 above), 420, n 11.

[73] See *Barclays Bank PLC v Franchise Tax Bd*, 512 US 298 (1994).

[74] EPA cited this as one of its reasons for initially declining to regulate the GHG emissions of motor vehicles. 'Foreign Affairs Preemption and State Regulation of Greenhouse Gas Emissions' (n 71 above), 1883, n 35 (citing Control of Emissions from New Highway Vehicles and Engines, 68 Fed Reg 52,922 (8 September 2003)).

[75] Richard B Stewart, Symposium, 'States and Cities as Actors in Global Climate Regulation: Unitary vs. Plural Architectures' (2008) 50 Ariz L Rev 681, 705.

4.3 Compact Clause

The Compact Clause of the US Constitution prohibits states from entering into compacts with other states or foreign nations without Congressional consent.[76] Although the Constitution does not define a 'compact,' five indicia of a compact have been distilled from US judicial law: (1) reciprocity, or the mutual exchange of privileges or benefits among states; (2) co-operation among state officials or legislators in developing the agreement; (3) establishment of a regional body to administer or enforce the agreement; (4) conditioning of state statutes on action by the other states; and (5) the ability of a state to modify or repeal its statute unilaterally.[77]

Even if an agreement among states, such as the RGGI MOU, rises to the level of a constitutionally suspect 'compact,' courts have held that not all compacts require congressional consent. Instead, the key concern is whether the compact would encroach upon or interfere with the 'just supremacy' of the federal government.[78] Courts have provided guidance on what will not constitute an interference with the just supremacy of the federal government. For example, the mere presence of federal interest in an issue covered by an interstate compact does not encroach upon federal supremacy.[79] Despite the seemingly broad language in the Constitution prohibiting interstate compacts without Congressional consent, judicial treatment of these compacts has substantially curtailed this limitation on state power. To date, no US court has invalidated an interstate agreement under the compact clause.

If an interstate agreement falls within the scope of the Compact Clause and is deemed to encroach upon the just supremacy of the federal government, the parties to the agreement will be required to seek Congressional consent. Historically, where states have sought permission before entering into interstate compacts, Congressional consent amounted to little more than a rubber stamp on state proposals.[80] As states have begun to cooperate on issues of greater complexity, Congress has increasingly used its authority under the Compact Clause to place conditions on granting consent, such as requiring the addition or inclusion of certain provisions in the compact.[81] This power could amount to a federal tailoring

[76] US Const art I, § 10, cl 3.

[77] 'The Compact Clause and the Regional Greenhouse Gas Initiative' (2007) 120 Harv L Rev 1958, 1973–76. One of the key cases on the Compact Clause, as well as the source of these indicia, is *Ne Bancorp, Inc v Bd of Governors of the Fed Reserve Sys*, 472 US 159 (1985).

[78] *Virginia v Tennessee*, 148 US 503, 520 (1893); see also *US Steel Corp v Multistate Tax Comm'n*, 434 US 452 (1978).

[79] *US Steel Corp* (n 78 above) 479 n 33; see also 'The Compact Clause and the Regional Greenhouse Gas Initiative' (n 77 above) 1963.

[80] Michael S Smith, 'Murky Precedent Meets Hazy Air: The Compact Clause and the Regional Greenhouse Gas Initiative' (2007) 34 BC Envtl Aff L Rev 387, 392.

[81] Claire Carothers, 'United We Stand: The Interstate Compact as a Tool for Effecting Climate Change' (2006) 41 Ga L Rev 229, 244 n 118 (listing various conditions of congressional consent).

of interstate compacts. A decision not to seek Congressional consent, therefore, is ultimately a choice between exposing an interstate compact to legal challenges versus potentially relinquishing to Congress substantial influence over the compact.

The sub-national schemes described above thus raise the question of whether they require Congressional consent. States participating in RGGI and the WCI, for example, have all signed an MOU but have not yet sought Congressional approval. Whether these MOUs amount to compacts requiring Congressional consent is unclear.[82] They would appear to reflect some of the indicia of an interstate compact, such as RGGI's presence of an overarching administrative body.[83] In a recent legal challenge to New York's implementation of RGGI, the plaintiff asserts that RGGI's MOU violates the Compact Clause by enlarging the participating states' political influence over the regulation of GHGs without Congress' authorization and encroaching on federal supremacy through interference with federal government's interest in regulating GHGs as pollutants.[84] Although Compact Clause challenges are rare and have so far been unsuccessful, cap and trade agreements among states may well invoke renewed judicial application of this constitutional doctrine.

4.4 Commerce Clause

The Commerce Clause of the US Constitution grants Congress the power to regulate commerce among the states.[85] The so-called 'dormant commerce clause' has been understood to provide 'protection from state legislation inimical to the national commerce [even] where the Congress has not acted'.[86] Under this doctrine, absent a compelling justification, a state may not adopt measures that facially discriminate against interstate or foreign commerce, even where the state seeks to advance a legitimate state policy objective.[87] The dormant commerce clause doctrine serves to restrain states both from adopting protectionist policies (ie, policies that disproportionately benefit in-state interests) and from adopting measures that result in excessive interference in foreign affairs.

The first question is whether a statute involves facial discrimination. '[A] statute that facially discriminates against interstate or foreign commerce will, in most cases, be found unconstitutional'.[88] The threshold is higher for statutes that

[82] But see Smith (n 80 above), 407 (concluding that RGGI is unlikely to require Congressional consent based on the Supreme Court's reasoning in *US Steel Corp*).

[83] Carothers (n 81 above), 250.

[84] *Indeck Corinth LP v Paterson*, filed 30 Jan 30 2009 (NY Sup Ct).

[85] US Const art I, § 8, cl 3.

[86] *Barclays* (n 73 above), 310 (quoting *S Pac Co v Arizona*, 325 US 761, 769 (1945)).

[87] Eg, *Kraft General Foods, Inc v Iowa Dep't of Revenue & Finance*, 505 US 71 (1992).

[88] *NFTC* (n 66 above), 67 (citing *Oregon Waste Sys, Inc v Dep't of Envtl Quality*, 511 US 93, 99 (1994)); see also *City of Philadelphia v New Jersey*, 437 US 617, 624 (1978) (noting that judicial review of facially discriminatory laws amounts to 'a virtually *per se* rule of invalidity' unless the state demonstrates that the law serves a legitimate local purpose and that no less discriminatory alternatives are available).

discriminate against foreign commerce than for those discriminating against interstate commerce alone, because of the potential effects on the 'efficient execution of the Nation's foreign policy'.[89] Notably, the presumption against state laws that discriminate against foreign commerce does not require a finding that such laws provide an advantage to in-state interests.

If a law facially discriminates, it will survive judicial review only if it demonstrably advances a legitimate local purpose that cannot be adequately served by reasonable nondiscriminatory alternatives.[90] On the other hand, courts will generally uphold a nondiscriminatory state law with 'incidental' effects on interstate commerce, unless the state law imposes a burden that is 'clearly excessive' compared to the state benefits it generates.[91] This balancing of interests for nondiscriminatory laws provides states an opportunity to justify laws that restrict interstate commerce provided there are countervailing benefits within the state.

It remains unclear how the major sub-national GHG initiatives such as RGGI, WCI, and MGGRA will fare under these standards. Although they do not appear to involve facial discrimination, they may nevertheless provoke dormant commerce clause challenges if the burden they place on interstate commerce outweighs the state benefits generated. Commonly cited state benefits of GHG regulation include protection against sea-level rise and the threat of frequent and higher intensity tropical storms, as well as mitigation of localized climate change impacts such as heat waves, flooding, or drought.[92] Despite these potential benefits relating to the protection of the health and welfare of state residents or resources, where state policy concerns are strongest, it may be difficult to demonstrate that the state initiatives will have a direct impact on local conditions.[93]

The burden of sub-national cap and trade programs on interstate commerce, on the other hand, will probably be characterized as a restriction on the movement of emissions-intensive products or services across state lines. For example, in response to concerns that RGGI's cap on electricity generators within the region will result in leakage, regulators are seeking ways to limit the importation

[89] *S Cent Timber Dev, Inc v Wunnicke*, 467 US 82, 100 n 7 (1984); see also Laurence Tribe, *American Constitutional Law* (3rd edn, University Textbook Series, 2000) ('If state action touching foreign commerce is to be allowed, it must be shown not to affect national concerns to any appreciable degree...').

[90] See *Japan Line Ltd v County of Los Angeles*, 441 US 434, 448–49 (1979) (stating that '[f]oreign commerce is preeminently a matter of national concern' and noting that '[t]he need for federal uniformity is no less paramount in ascertaining the negative implications of Congress' power to 'regulate Commerce with foreign Nations' under the *Commerce Clause*').

[91] Eg *Pike v Bruce Church, Inc*, 397 US 137, 142 (1970).

[92] See, eg, Daniel A Farber, Symposium, 'Climate Change, Federalism, and the Constitution' (2008) 50 Ariz L Rev 879, 897.

[93] Interestingly, the US Supreme Court in *Massachusetts v EPA* found that state had standing to challenge EPA's denial of its rulemaking petition because 'the rise in sea levels associated with global warming has already harmed and will continue to harm Massachusetts' 549 US 497, 526 (2007).

of electricity by out-of-state generators not covered by RGGI's emissions cap.[94] Attempts to regulate these generators, even if drafted such that the statute is facially non-discriminatory, may attract dormant commerce clause challenges if they restrict the sale of electricity by generators outside of the RGGI participating states.[95] In defense of potential dormant commerce clause challenges, the states will ultimately need to establish that sub-national cap-and-trade systems are not protectionist and that the local interest in mitigating climate change impacts outweighs any burden on interstate commerce.

5. Prospects for Federal Cap-and-trade Legislation

Following the United States' rejection of the Kyoto Protocol, US public opinion has slowly shifted in favor of action on global warming. This movement reflects a number of factors including educational campaigns by public interest groups as well as former Vice-President Al Gore's film 'An Inconvenient Truth', along with increased public alarm at certain highly visible manifestations of a changing climate, such as Hurricane Katrina and other intensified weather events. During the 2007–2008 Congress, no fewer than 10 comprehensive climate-focused bills were introduced, including the Lieberman-Warner bill which proceeded further in the legislative process than any previous comprehensive global warming measure before being defeated in the US Senate.[96] The current Congress has likewise seen significant legislative activity on climate change, including most notably as of this writing the passage of the Waxman-Markey bill, HR 2454, by the US House of Representatives on June 26, 2009.

Most of the legislative proposals have adopted a market-trading approach, although there are significant differences between the bills with regard to sectoral coverage, distribution of allowances, trajectory of emissions reductions, governmental oversight, and use of emissions offsets from unregulated sectors and international linkage. Despite the current global financial turmoil, there continues to be strong support from Congressional leaders and the new Obama Administration for action on climate change. At this point, prospects for a cap-and-trade system at the federal level appear good, although the process will take some time. The pace of legislative activity will likely be influenced by the approaching session of the parties to the UNFCCC to be held in Copenhagen Conference in December 2009. However, the realities of the US political system

[94] Heddy Bolster, 'The Commerce Clause Meets Environmental Protection: The Compensatory Tax Doctrine as a Defense of Potential Regional Carbon Dioxide Regulation' (2006) 47 BC L Rev 737, 738–9.

[95] Ibid.

[96] See Pew Center on Global Climate Change, 'Economy-wide Cap-and-Trade Proposals in the 110th Congress As of December 1, 2008' <http://www.pewclimate.org/docUploads/Chart-and-Graph-120108.pdf> accessed 13 February 2009. Bills in the US Congress expire at the end of each two-year congress, and thus climate legislation has had to be re-introduced in the 111th Congress, which began 6 January 2009.

and continuing resistance and skepticism among lawmakers from states with heavy industry and coal mining interests make it unlikely that federal legislation will be finalized until well into 2010.

Beyond overarching policy design issues of a cap-and-trade system (such as allowance allocation, sector coverage, use of offsets, etc), there are a number of legal issues that will either pose a challenge to federal climate legislation and/or will heavily influence the implementation of any cap and trade program.

5.1 International trade and competitiveness

One of the fiercest points of resistance to a federal climate program is the concern that increased costs of emissions controls will further undermine the economic competitiveness of US industry and lead to a transfer of wealth to non-Annex I developing countries (China, India, and others), thus exacerbating trade imbalances. At the urging of US labor unions and manufacturing interests, several proposed cap and trade bills have proposed a quasi-tariff system to impose a GHG 'equivalency' requirement on certain imported materials (such as cement, paper, and metals).[97] Importers could avoid the duty by showing that the country of origin is listed as having a comparable GHG reduction program, or by purchasing credits from a special reserve fund.[98] Significant attention has been focused on whether and how these tariff measures can be structured so as to avoid non-compliance with World Trade Organization obligations.[99] Regardless of how trade law issues are resolved, these proposals offer the prospect of establishing yet another demand driver for emissions allowances or offset credits that would be used to fulfill such equivalency requirements. However, trade issues will be one of the most challenging areas in which to find accord, given the high stakes and array of politically powerful stakeholders.

5.2 Potential Clean Air Act regulation

On April 2, 2007, the United States Supreme Court ended one chapter of the contentious climate debate by ruling in *Massachusetts v EPA*[100] that the US EPA

[97] See, eg, the Lieberman-Warner Climate Security Act of 2008, S.2191 § 6006 (as reported).

[98] For example, the Waxman-Markey bill contains provisions that would allow the President and US EPA to implement an 'International Reserve Allowance Program' requiring foreign entities in certain countries to submit allowances in conjunction with the importation of goods and products into the United States. Waxman-Markey Bill (n 22 above), § 766.

[99] Eg, Joost Pauwelyn, 'US Federal Climate Policy and Competitiveness Concerns: The Limits and Options of International Trade Law' (2007) <http://www.env.duke.edu/institute/internationaltradelaw.pdf>, accessed 2 February 2009 12; Katrin Jordan-Korte & Stormy Mildner, 'Climate Protection and Border Tax Adjustment: Economic Rationale and Political Pitfalls of Current US Cap-and-Trade Proposals' (June 2008) Forum for Atlantic Climate and Energy Talks (FACET) Analysis No. 1 <http://www.aicgs.org/documents/facet/jordan.faceta01.pdf>, accessed 2 February 2009.

[100] 549 US 497 (2007).

has authority to regulate carbon dioxide as a 'pollutant' under the federal CAA. The decision placed the US EPA under court order to assess whether GHG emissions 'endanger' public health and welfare, and if so, to promulgate regulations under the CAA to control industrial and vehicle emissions of CO_2 (and presumably other GHGs).

The US EPA under the recent Bush Administration delayed action on the required 'endangerment' finding and instead embarked on a lengthy public comment process (known as an 'Advanced Notice of Proposed Rulemaking') to solicit views of diverse stakeholders as to how GHGs could, and should, be regulated under the existing CAA absent Congressional amendment of the statute.[101] At the same time, in response to a series of challenges by non-governmental organizations objecting to the permitting of coal-fired electric plants, the US EPA Administrator ruled in December 2008 that carbon dioxide is not yet regulated under the CAA, and therefore new industrial plants do not have to install emissions control equipment to limit, capture, or destroy GHGs from smokestacks.[102]

In a policy shift set in motion by the new Obama Administration, the US EPA released in April 2009 its proposed finding that the combination of six key GHGs in the atmosphere 'endanger' public health and welfare. The US EPA further proposed to find that motor vehicle emissions of a subset of the GHGs contribute to the endangerment, triggering EPA authority to regulate these sources.[103] Although the proposed finding laid the initial groundwork for CAA regulation of GHG emissions from motor vehicles (and potentially additional GHG sources), US EPA Administrator Jackson has stated that the agency and the Obama Administration favor dedicated and comprehensive federal climate legislation.[104] Nonetheless, it remains a distinct possibility that, in the absence of Congressional action, the US EPA will regulate at least some sources of GHGs under the existing CAA.

While CAA regulation could take various forms, it is likely to involve limitations on GHG emissions and control technology to capture carbon dioxide or

[101] Regulating Greenhouse Gas Emissions Under the Clean Air Act, Advance Notice of Proposed Rulemaking, 73 Fed Reg 44,353 (30 July 2008) <http://www.epa.gov/fedrgstr/EPA-AIR/2008/July/Day-30/>, accessed 9 March 2009 [hereafter: ANPR]. The 'ANPR,' as it is known, is a veritable treatise on the options for regulating GHGs and runs some 500 pages.

[102] 'EPA's Interpretation of Regulations that Determine Pollutants Covered By Federal Prevention of Significant Deterioration (PSD) Permit Program', Memorandum from S Johnson (EPA Admin.) to Regional Administrators (18 December 2008) <http://www.epa.gov/nsr/documents/psd_interpretive_memo_12.18.08.pdf>, accessed 9 March 2009.

[103] Proposed Endangerment and Cause or Contribute Findings for Greenhouse Gases Under Section 202(a) of the Clean Air Act, 74 Fed Reg 18,886 (24 April 2009).

[104] Eg, 'EPA Finds Greenhouse Gases Pose Threat to Public Health, Welfare / Proposed Finding Comes in Response to 2007 Supreme Court Ruling' (17 May 2009) <http://yosemite.epa.gov/opa/admpress.nsf/d0cf6618525a9efb85257359003fb69d/0ef7df675805295d8525759b0056 6924!OpenDocument>, accessed 2 June 2009.

destroy other GHGs such as methane, nitrous oxide, or fluorinated gases. In addition to motor vehicles, the sources targeted in the proposed endangerment finding, future rulemaking actions are particularly likely to target other major sources of air pollution such as power plants. Industry has also raised a concern that an endangerment finding and regulation of GHGs under the motor vehicle-related provisions of the CAA could also require the US EPA to regulate individual sources as small as steam heating boilers at institutional facilities such as hospitals, schools, and large commercial buildings. Much of the regulation could be at the state level, potentially involving co-operative arrangements among states to jointly meet CAA requirements.

A critical remaining question is whether the US EPA has authority under the existing CAA to use market-oriented mechanisms to establish a full or partial cap and trade program. The US EPA has taken the position that it may have some authority to do so;[105] however, the agency's tentative views must be measured against recent federal court decisions that vacated a sulfur and nitrogen oxide trading program covering the eastern half of the United States and other judicial precedent.[106] Moreover, any climate regulation at the executive agency level is likely to be hampered by lawsuits, whereas under the US legal system, climate legislation passed by Congress would be comparatively immune to court challenges.

5.3 Interaction with NEPA

Activities in the United States, including carbon trading, could also be significantly affected by the National Environmental Policy Act (NEPA), a law that requires federal executive agencies to study the environmental impacts of major governmental decisions.[107] NEPA may influence government decisionmaking in at least two ways. First, unless Congress were to expressly grant exceptions in federal cap and trade legislation, NEPA's core requirement of an environmental study (called an Environmental Impact Statement or EIS) for significant federal actions could theoretically require the US EPA or other executive agencies, such as the Department of Energy or Department of Agriculture, to study the environmental consequences of a wide range of implementation actions. Government actions that could be subject to NEPA review include acceptance of traded emissions allowances into a regulated entity's compliance account,

[105] See, eg, ANPR (n 101 above), 44,439 (trading of emissions credits authorized under vehicle and engine requirements), 44,490 (emissions trading possible under New Source Performance Standards).

[106] See eg, *North Carolina v EPA*, No 05-1244, slip op (DC Cir, 11 July 2008).

[107] National Environmental Policy Act of 1969, Pub L No 91-190, 42 USC §§ 4321–4347 (1970). NEPA is considered one of the seminal charter statutes of the modern environmental movement in the United States and is administered by the prominent White House Council on Environmental Quality. See White House Council on Environmental Quality, 'About the Council' <http://www.whitehouse.gov/ceq/>, accessed 2 February 2009.

certification of carbon offset projects, and distribution of grants and incentives funded by sale of trading allowances. Although application of NEPA to carbon trading is an academic concern at this juncture, it is likely that stakeholders dissatisfied with a future cap and trade program or objecting to a particular agency decision will use NEPA as a bargaining chip in the political and legal process. Indeed, NEPA has been a focal point of environmental litigation over the last 40 years and is a powerful weapon when wielded by opponents to a particular project or governmental program in light of the significant time delay (an EIS can typically take two to five years to prepare and approve) and drain on resources (sometimes running into the millions of dollars) necessitated by NEPA review. Congress would be well advised to consider the NEPA process, and provide for waivers where appropriate, in order to forestall litigation and blocking maneuvers by disaffected opponents of carbon trading.[108]

Second, the federal NEPA statute and similar state-level environmental review laws have precipitated a recent flurry of lawsuits seeking to force consideration of GHG mitigation responses, particularly with regard to the permitting of fossil fuel fired electric utility plants and transportation-related emissions. For example, in *Center for Biological Diversity v National Highway Traffic Safety Administration*, a federal court rejected a federal agency's decision relating to vehicle fuel efficiency because the agency had not sufficiently studied the levels of GHGs associated with various alternative efficiency standards.[109] Another recent decision indicates that quantitative analysis of GHG emissions may be required for federal permit approvals. In *Mid States Coalition for Progress v Surface Transportation Board*, a court held that a NEPA study of the construction of railroad lines linking power plants to low-sulfur coal reserves in Wyoming's Powder River Basin failed to consider the impacts of GHG emissions increases associated with increasing the ready supply of coal to utilities.[110] After remanding the case, the court later upheld a revised study that quantitatively modelled CO_2 emissions at the national and regional scale.[111] Notwithstanding these court decisions, it remains unclear whether the cumulative impacts of GHG emissions

[108] See, eg, 'Brown Announces Groundbreaking Greenhouse Gas Reduction Plan' *California Chronicle* (San Francisco, 12 September 2007) <http://www.californiachronicle.com/articles/view/37524>, accessed 2 February 2009; see also Beveridge & Diamond, PC, 'Federal Court Allows Retrospective Environmental Mitigation Penalties' (2 December 2008) <http://www.bdlaw.com/news-422.html>, accessed 2 February 2009.

[109] 538 F 3d 1172 (9th Cir 2008); see also *Border Power Plant Working Group v US Dep't of Energy*, 260 F Supp 2d 997, 1029, 1033 (SD Cal 2003) (requiring a study of GHG emissions indirectly caused by connecting the US electricity grid to natural gas-fired power plants in Mexico); *Friends of the Earth v Mosbacher*, 488 F Supp 2d 889 (ND Cal 2007) (considering whether NEPA study required for Overseas Private Investment Corporation and Export-Import Bank loan guarantees and political risk insurance provided to oil-and-gas and power plant project participants).

[110] 345 F 3d 520, 550, 556 (8th Cir 2003).

[111] *Mayo Found v Surface Transp Bd*, 472 F 3d 545, 556 (8th Cir 2006) (ruling that STB adequately considered foreseeable adverse effects of increased GHG emissions for purposes of NEPA).

and climate effects associated with many projects will be significant enough to warrant an in-depth study of the impacts through the NEPA process. The judicial decisions thus far contain scant guidance as to when increases in GHGs are 'significant' or how decision makers can determine what global impacts will result from a single isolated project, such as a new highway or power plant. Given this uncertainty, it is likely that the White House Council on Environmental Quality will initiate a review of the NEPA guidelines to address their application to global warming effects.

Ironically, these procedural roadblocks may actually stimulate the carbon markets, as emitters have begun voluntarily purchasing large blocks of carbon offsets or have agreed to GHG emissions projects in order to appease opponents and obtain regulatory approvals. Moreover, although the federal NEPA law is purely a procedural requirement (ie, the federal agency must 'stop, look, and listen' before making a decision), some state-level review statutes, such as the California Environmental Quality Act[112] and New York's State Environmental Quality Review Act[113] can require mitigation of environmental impacts, thus potentially creating a need for offsetting emissions reduction projects.

Although federal agencies are not yet clearly under a mandate to study GHG emissions under NEPA, these agencies are increasingly including an assessment of climate implications in NEPA study documents. Some, but not all, of these studies have included a quantitative analysis of any increase in direct and indirect GHG emissions and a general discussion of the scale of such increases *vis-à-vis* effects on atmospheric concentrations and climate models. However, agencies continue to struggle with the further step of linking climate modelling to actual effects on natural resources, such as groundwater or forests. Notably for carbon markets, many of these studies have included consideration of emissions reductions and carbon offsets as part of the alternatives and mitigation analysis. A few examples of climate studies on recent major federal projects are summarized below.

5.3.1 CAFE standards

Following the court decision in the *Center for Biological Diversity* litigation, the National Highway Traffic Safety Administration prepared a final EIS to support

[112] Cal Pub Res Code §§ 21000–21177. GHG estimation requirements under state statutes such as the California Environmental Quality Act are already challenging project developers to account for the cumulative atmospheric impacts of individual projects. For example, a California judge ordered Wal-Mart to revise its EIS because the document 'simply ignore[d] the ... scientific and factual analysis regarding attainment of California GHG emission targets in its discussion of the cumulative impact of the project.' *Ctr for Biological Diversity v Town of Yucca*, No. CIVBS 800607 (Cal Super Ct 2009) <http://www.globalclimatelaw.com/uploads/file/Wal%20Mart%20Ruling.pdf>, accessed 2 June 2009.

[113] NY Envtl Conserv Law §§ 3-0301(1)(b), 3-0301(2)(m), 8-0113; NY Comp Codes R & Regs, tit 6, §§ 617.1–617.20 (2009).

its fuel efficiency (CAFE) standards for passenger cars and light trucks for model years 2011–2015.[114] With perhaps the most comprehensive analysis of project-related GHGs in an EIS to date, the EIS quantified the effect of the project-related GHG emissions on several aspects of the global climate but concluded that it was impossible to quantify how the project's impact on these aspects of the climate would in turn impact environmental resources.

5.3.2 Columbia River Crossing Project

A draft EIS prepared for a large transportation project including a mixture of bridge construction, highway improvements, public transit, and bicycle and pedestrian lanes linking Oregon and Washington states devoted eight pages to climate change, including the major sources of GHGs in each state, state initiatives to reduce GHG emissions, and quantification of emissions from five project alternatives, as well as a discussion of potential mitigation of CO_2 emissions.[115]

5.3.3 Imperial-Mexicali Transmission Lines

In considering the permitting of two transmission lines that would bring electricity from Mexican natural-gas-fired power plants into the California grid, the US Department of Energy prepared an EIS calculating the CO_2 emissions from the proposed action as a percentage of United States CO_2 emissions, and then as a percentage of global CO_2 emissions. As the emissions are less than 1% of annual United States CO_2 emissions, the agency concluded that '[t]he expected impacts to global climate change would be negligible'.[116]

5.3.4 Orlando Gasification Project

When considering whether to provide grant funding to a proposed coal-to-liquids demonstration plant in Florida, the US Department of Energy quantified the annual CO_2 emissions from the proposed project compared to CO_2 emitted globally from 2000 to 2005 and compared CO_2 emissions of liquid

[114] National Highway Transportation Safety Administration, Final Environmental Impact Statement, Corporate Average Fuel Economy Standards, Passenger Cars and Light Trucks, Model Years 2011–2015 (October 2008) <http://www.nhtsa.dot.gov/portal/site/nhtsa/template.MAXIMIZE/menuitem.43ac99aefa80569eea57529cdba046a0/?javax.portlet.tpst=4670b93a0b088a006bc1d6b760008a0c_ws_MX&javax.portlet.prp_4670b93a0b088a006bc1d6b760008a0c_viewID=detail_view&itemID=9629cf2be9b0d110VgnVCM1000002fd17898RCRD&overrideViewName=Article>, accessed 13 February 2009.

[115] Federal Highway Administration and Federal Transit Administration, 'Interstate-5 Columbia River Crossing Draft Environmental Impact Statement' (April 2008) 3-430 to 3-437 <http://drafteis.columbiarivercrossing.org/Default.aspx?SectionID=2>, accessed 9 March 2009.

[116] US Department of Energy, 'Final Environmental Impact Statement for the Imperial-Mexicali 230-kV Transmission Lines' (2004), 4-58 to 4-59 <http://www.gc.energy.gov/NEPA/finalEIS-0365.htm>, accessed 2 February 2009.

hydrocarbon fuels to conventional petroleum fuels. Central to this discussion is an analysis of carbon sequestration, and how it could reduce the CO_2 emissions from production of liquid transportation fuels from coal.[117]

5.3.5 Strategic Petroleum Reserve Expansion Project

In order to implement newly-passed legislation requiring expansion of Strategic Petroleum Reserve by adding an additional site and enlarging storage capacity at existing sites, the US Department of Energy analysed GHG emissions for the various project alternatives, including emissions from construction of the project and the operation of the project, concluding that 'the incremental emissions and climate change impacts of the proposed SPR site development are considered very small'.[118]

5.4 Carbon Reporting

For years, nongovernmental organizations have pushed for increased transparency and reporting of GHG emissions from industrial and commercial firms in order to foster more realistic consideration of GHG reductions through initiatives such as the Carbon Disclosure Project[119] and CERES' Global Disclosure Project.[120] This trend has increased in momentum in two significant ways. First, in 2007, the New York State Attorney General initiated an investigation of several electric utilities seeking to force enhanced disclosure of financial risks associated with climate change in those companies' financial statements, as required by the Securities and Exchange Commission (SEC). In recent settlements, some of the utilities have agreed to disclosures covering 'current carbon emissions; projected increases in carbon emissions from planned coal-fired power plants; company strategies for reducing, offsetting, limiting, or otherwise managing its global warming pollution emissions and expected global warming emissions reductions from these actions; and corporate governance actions related to climate change, including if environmental performance is incorporated into officer compensation.[121]

[117] US Department of Energy, 'Final Environmental Impact Statement for the Orlando Gasification Project' (2007) 4-60 to 4-61 <http://www.gc.energy.gov/NEPA/final-EIS-0383.htm>, accessed 2 February 2009.

[118] US Department of Energy, 'Site Selection for the Expansion of the Strategic Petroleum Reserve Final Environmental Impact Statement' (2006) <http://www.fossil.energy.gov/programs/reserves/publications/Pubs-SPR/2006_SPR_EIS.html>, accessed 2 February 2009 3-66 to 3-67.

[119] 'Carbon Disclosure Project' <http://www.cdproject.net/>, accessed 2 February 2009.

[120] Ceres, 'Engagement & Disclosure' <http://www.ceres.org/Page.aspx?pid=556>, accessed 2 February 2009.

[121] New York State Attorney General, 'Attorney General Cuomo, Joined by Vice President Gore, Announces Agreement with Major Energy Company, Dynegy Inc.' (Press Release New York 23 October 2008) <http://www.oag.state.ny.us/media_center/2008/oct/oct23a_08.html>, accessed 2 February 2009.

Significantly, the US EPA is currently writing regulations to require 'carbon footprint' reporting of GHG emissions from US industrial and commercial facilities under a Congressional mandate included in a 2007 appropriations bill.[122] This provision, promoted by influential US Senators Barbara Boxer and Diane Feinstein, requires the agency to issue regulations for 'mandatory reporting of greenhouse gas emissions above appropriate thresholds in all sectors of the economy of the United States'. The new law leaves the agency with broad discretion to determine which industries and sources will be required to submit reports, what levels of emissions will trigger reporting requirements, and what exemptions will be provided. The US EPA released a draft of its proposed mandatory reporting rule for GHGs in April 2009.[123] The rule requires suppliers of fossil fuels or industrial GHGs, manufacturers of vehicles and engines, and facilities that emit 25,000 or more metric tons of CO_2e per year to submit annual reports to the US EPA. The agency is also working with TCR (the quasi-official registry endorsed by some of the emerging sub-national compacts) to develop a data-sharing protocol that would theoretically allow companies to report emissions to TCR while having the reporting recognized by the US EPA and other registries.

The US EPA's forthcoming mandatory reporting rule portends a watershed point in the climate discussion in the US and will likely bring home the reality of GHG emissions regulation to many companies and sectors to a heretofore unprecedented degree. It will also provide the federal government with accurate data and reporting metrics that will support extensive regulation of GHG emissions across most economic sectors. The 'awakening' occasioned by mandatory reporting, even in absence of mandated reductions, could spark increased pre-compliance purchases of carbon offsets and demand for emissions reductions technology.

6. Conclusions

US climate policy is approaching a dramatic period of transition. The state and regional initiatives, at one time few and isolated, are growing in number and actively seeking ways to join forces on climate change and to pressure the federal government. With a powerful new Congress, President Obama's support, and continued pressure from the international community as well as from aggressive state and regional climate change initiatives, a national GHG cap and trade system could be on a fast-track to development. In addition, strong US

[122] Consolidated Appropriations Act of 2008, HR 2764, Pub L No 110-161 (26 December 2007). See also US Environmental Protection Agency, 'Greenhouse Gas Reporting Rule' <http://www.epa.gov/climatechange/emissions/ghgrulemaking.html>, accessed 2 February 2009.

[123] Mandatory Reporting of Greenhouse Gases (n 45 above), 74 Fed Reg 16,448.

leadership in the effort to reach a post-2012 global agreement is now a distinct possibility, although it is clear that the Obama Administration is focused first and primarily on setting the domestic GHG regulatory framework. Despite this palpable momentum, the prospect of mid-term Congressional elections in 2010 may weaken support for a federal climate bill in key parts of the United States that rely heavily on coal for power and manufacturing for an economic base. Struggling US and global economies are also likely to prompt industry challenges to costly GHG regulation at all levels of government and in the courts. Though the United States may well become a more active participant in the global effort to curb climate change, parochial concerns and the complex nature of the US federal-state system of government introduce myriad challenges. Much remains to be addressed before comprehensive action on climate change can become a reality.

19

Offsets in the Emerging US Cap-and-Trade Programmes

Kyle W Danish

1. Introduction to Offsets

As the United States moves forward from voluntary efforts to the establishment of mandatory cap-and-trade programmes for greenhouse gas (GHG) emissions, offsets have become a central issue of policy. Offsets can help minimize the total costs associated with GHG regulation; offsets also can provide other benefits, such as economic development and reduced pollution. However, incorporation of offsets into a cap-and-trade programme requires careful attention to policy design. US policy-makers are likely to be working through offset design issues and options over the next several years—from the enactment of legislation through the development of project-specific rules and procedures.

After providing a brief background on offsets, including a discussion of their value within a cap-and-trade programme, the use of offsets in various policy contexts, and different types of offset projects, this chapter discusses the most important design issues for an offset project programme: additionality of GHG emission reductions made, quantitative limits on offsets, addressing the risk of reversal of sequestered emissions, providing credit for early action offset projects, incorporating international offset projects, and projects aiming to reduce emissions from deforestation and forest degradation. Each section highlights how leading proposals for US federal cap-and-trade legislation have addressed these issues. The final section discusses the use of offsets in state and regional cap-and-trade programmes.

1.1 The role of offsets in a cap-and-trade programme

To understand the role of offsets, it is necessary first to understand the likely scope of an emissions cap. While a future US 'economy-wide' emissions cap could be expected to cover a large proportion of GHG emissions, there are a variety of sources and sinks of emissions that are not suitable for a cap or other direct regulation. Some sources, such as farms, might be too small for allowance-based regulation. In other cases, difficulties of measurement and monitoring could preclude across-the-board regulation. This obstacle would apply to methane emissions from agricultural practices, and carbon dioxide emissions associated with certain forms of land use. For these reasons, most proposals for a US cap would cover just under 90% of US greenhouse gas emissions.[1]

Even though across-the-board regulation is not feasible for these uncovered entities, there are many opportunities for low-cost emissions abatement at individual emission sources and sinks in these sectors. Furthermore, a substantial proportion of these opportunities are available at lower cost than options within the capped sectors.

It is possible to extend the cap-and-trade programme to reach these low-cost abatement opportunities through an offset project programme. An offset project reduces or sequesters emissions not otherwise addressed through an emissions cap or other regulatory programme.[2]

In a governmental cap-and-trade programme, an offset project is subject to verification and certification by a regulatory agency, which can delegate some elements of this process to accredited private verifiers. Then, a capped entity may use the emission reductions from the project as credits to cover its emissions in the cap-and-trade programme.

In this way, an offset project programme increases the pool of emissions abatement opportunities available to the cap-and-trade programme, improving the efficiency of GHG reduction efforts, and thereby reducing the cost. With offsets, a capped entity can meet its obligations by making reductions at its own facilities, by purchasing allowances from another capped entity, or by purchasing offset credits. The cost containment benefits of extending the cap-and-trade programme to incorporate offsets are measurable and significant. For example, in its analysis of the Lieberman-Warner Climate Security Act of 2008, which authorized capped entities to meet up to 30% of their compliance obligations through use of offsets, the US Environmental Protection Agency determined that, if the bill were modified to disallow any use of offsets, the cost of the programme would increase by 93%. By contrast, if the bill removed the 30% limit and allowed unfettered use of offsets, compliance costs would decrease by 71%.[3]

[1] The Dingell-Boucher discussion draft covered 88% of US greenhouse gas emissions. Pew Center on Global Climate Change, 'Summary of the Dingell-Boucher Discussion Draft' (2008) 2 <http://www.pewclimate.org/docUploads/Dingell-BoucherSummary.pdf>, accessed 31 March 2009. The Lieberman-Warner Climate Security Act of 2007 covered 87% of US greenhouse gas emissions. Energy Information Administration, 'Energy Market and Economic Impacts of S. 2191, the Lieberman-Warner Climate Security Act of 2007' (April 2008) <http://www.eia.doe.gov/oiaf/servicerpt/s2191/pdf/sroiaf(2008)01.pdf>, accessed 31 March 2009.

[2] Congressional Research Service, 'The Role of Offsets in a Greenhouse Gas Emissions Cap-and-Trade Programme: Potential Benefits and Concerns' (4 April 2008) 1 ('An offset is a measurable reduction, avoidance, or sequestration of GHG Emissions from a source *not covered* by an emission reduction programme') (emphasis in the original). In this chapter, the term 'emission reduction' also means 'emission avoidance, destruction, or sequestration' unless otherwise specified.

[3] Environmental Protection Agency, 'EPA Analysis of The Climate Stewardship and Innovation Act of 2007' (2007) <http://www.epa.gov/climatechange/downloads/s280fullbrief.pdf>, accessed 20 March 2009.

The critical test for an offset project is that it must generate emission reductions that are in excess of business-as-usual activities. This concept is often referred to as 'additionality'. Because use of an offset credit permits a capped entity to increase its emissions, it is imperative that the offset project generate reductions that were not going to occur anyway. As discussed in greater detail in Section 2 of this chapter, verifying a project's additionality is one of the most challenging aspects of offsets programme design.

1.2 History of offsets programmes

Offsets policies and programmes are not new. Since the mid-1970s, the US Environmental Protection Agency has authorized the use of offsets in programmes to address conventional air pollutants such as ground-level ozone. However, according to some analyses, these early offset programmes failed to meet their potential either for generating reductions or cost savings due to cumbersome rules and procedures.[4]

Offset projects have been central to the design and operation of the Kyoto Protocol and European Union Emissions Trading System (EU ETS) regimes. The Protocol establishes two offset project systems, the Clean Development Mechanism (CDM) and Joint Implementation (JI). Each allows countries to use credits generated by verified offset projects for compliance with the Protocol's emission reduction obligations. Under both the CDM and JI systems, national or international governmental institutions certify offset projects and issue credits, subject to established rules and procedures. These mechanisms, particularly the CDM, have yielded large markets for offset projects; the buyers mostly comprise European Union companies that are authorized to use CDM credits for complying with their emission targets under the EU ETS.[5]

There is also a voluntary carbon market, which consists of offset projects developed outside the auspices of the Kyoto Protocol or the EU ETS. Companies in the United States, anticipating future regulation, are among the largest buyers in the voluntary carbon market.[6]

In addition, as discussed in greater detail in Section 9 of this chapter, the various state and regional programmes under development in the United States

[4] A Denny Ellerman, Paul L Joskow, David Harrison, Jr, 'Emissions Trading in the US: Experience, Lessons, and Considerations of Greenhouse Gases' (Pew Center on Global Climate Change, May 2003) 8–9 <http://www.pewclimate.org/docUploads/emissions_trading.pdf>, accessed 19 March 2009.

[5] Karan Capoor and Philippe Ambrosi, 'State and Trends of the Carbon Market 2008' (The World Bank, May 2008), 29 <http://wbcarbonfinance.org/docs/State_Trends_FINAL.pdf>, accessed 19 March 2009.

[6] Katherine Hamilton, Milo Sjardin, Thomas Marcello, Gordon Xu, Ecosystem Marketplace and New Carbon Finance, 'State of the Voluntary Carbon Markets 2008' (8 May 2008) <http://ecosystemmarketplace.com/documents/cms_documents/2008_StateofVoluntaryCarbonMarket2.pdf>, accessed 19 March 2009.

each incorporate, or anticipate incorporating, offset project programmes. Finally, offset project programmes are a central element of the mainstream proposals for federal cap-and-trade legislation.

1.3 Scope of offset projects

Offset projects can be implemented in a number of domestic, uncapped sectors. In the agriculture sector, it is possible to implement projects that capture methane emissions from manure lagoons, sequester carbon by switching from conventional to no-till cultivation, or reduce nitrous oxide emissions associated with fertilizer use. In the forestry sector, offset projects can involve afforestation, reforestation, and forest management. Methane capture projects are available among landfills, coal mines and natural gas systems.

So long as developing countries are not subject to economy-wide emission limits, they offer a broad range of opportunities for offset projects, including not only the categories described above, but also renewable energy, energy efficiency, and industrial gas-related projects.

2. Additionality and Standards-Based Approaches

An important element of offset programme design is the approach used to determine a project's additionality and its emissions baseline. The conventional approach is a project-specific assessment, which relies on project-by-project determinations of whether a particular project or its emission reductions would have occurred even without the incentive provided by the sale of the carbon credits.

However, there is growing interest in an alternative to project-by-project assessments: a standardized approach. Under such an approach, a project would be deemed additional if it improved upon a particular performance requirement or technology benchmark. Such standardized approaches already are being used in the Regional Greenhouse Gas Initiative; the Climate Action Reserve programme, which was established by the California Climate Action Registry; and the Environmental Protection Agency's Climate Leaders Programme.

The two approaches have different advantages and disadvantages.[7] Project-specific assessments are more specifically tailored to individual circumstances. They therefore have a greater probability of accurately determining that a project is actually additional and precisely quantifying its emissions abatement benefits.

[7] Offset Quality Initiative, 'Ensuring Offset Quality: Integrating High-Quality Greenhouse Gas Offsets Into North American Cap-and-Trade Policy' (July 2008) 5–8 <http://www.offsetqualityinitiative.org/pdfs/OQI_Ensuring_Offset_Quality_7_08.pdf>, accessed 19 March 2009. See also chapter 12 above.

However, the greater attention to project-specific detail also entails drawbacks. Such an approach has more significant risk of inconsistency and subjectivity across projects. Project-specific assessments also involve greater administrative and transactional costs, thereby potentially limiting the flow of projects into the market.

Standardized approaches offer a set of objective criteria that apply to all projects of a particular type. A standard could take the form of a performance standard, such as a benchmark emission rate or factor for a particular activity. In such a case, a project would be deemed additional if it reduced emissions below the benchmark. For a land use project, the standard could be the prevailing type of land use in a particular area. A shift to a type of land use associated with lower emissions or increased sequestration would qualify for credits. A standard also could take the form of a technology benchmark, in which the use of a specific technology in a particular setting would be deemed additional—such as methane digesters at dairy farms. Standardized approaches offer certain comparative advantages, including greater objectivity; a streamlined process for developing and approving projects; and greater certainty for both investors and regulators.

Standardized approaches also have certain potential disadvantages. For example, the time saved in approving projects could be offset at least in part by the more time-consuming process of developing the standards in the first instance. In addition, by generalizing the additionality test and quantification process, a standardized approach implies acceptance of some amount of imprecision; the standard could under-credit certain projects and over-credit others.

3. Annual Quantity Limits on Offsets

Each of the major proposals for US cap-and-trade legislation that has incorporated offsets also has included some form of limit on the quantity of offsets available for use by capped entities in a particular year.

A quantitative limit has a number of potential rationales. To a certain extent, it reflects a lack of confidence in the ability of a regulatory system to distinguish 'good' offsets from 'bad' offsets. A quantitative limit minimizes, but in a very imprecise way, the exposure of the programme to non-additional offsets mistakenly approved by the regulators. However, any kind of broad-based numerical limit on offsets will throw out the good with the bad; the limit could exclude any number of offsets that reflect real emission reductions. Furthermore, to the extent that high quality offset projects are more costly to develop, then a quantitative limit could end up precluding a large number of such projects.

Another potential rationale for a quantitative limit on offsets is to promote technology development and deployment by sustaining the long-term allowance price in the capped system. Under this view, high allowance prices, and the resulting increase in energy prices, are needed to provide incentives for investment in low-carbon technology development. Some supporters of a quantitative limit on the use of offsets are concerned that, absent such a quantity limit, offsets could provide too much cost containment, depressing the allowance price and diminishing the incentive effect for technology. However, there are other mechanisms for sustaining the allowance price, including establishing a 'floor' price for allowances. The Boxer-Lieberman-Warner bill, for example, included such a floor price mechanism.[8] In addition, there are other, potentially more direct means of promoting development and deployment of technologies, including through the use of technology-specific subsidies or standards.

In some cases, quantity limits also appear motivated by the desire to stimulate emission reductions within the capped segment of the emission trading system. For example, as discussed in Section 9 below, the Western Climate Initiative has proposed an approach in which capped entities can use offsets to meet no more than 49% of the difference between business-as-usual emission levels and the emissions cap for that year. Thus, in each year, capped facilities would need to meet over half of their obligation through on-system reductions.

A quantity limit on offsets can take the form of a use limit or an issuance limit. A use limit focuses on the quantity of offsets that a capped firm or facility can use for compliance purposes. For example, the Dingell-Boucher discussion draft legislation has a limit on use. It provides that a capped entity may satisfy up to five percent of its annual compliance obligations in 2013 through 2017 with domestic or international offset credits. The limit increases over time. After 2025, the limit is 20% for domestic offsets and 15% for international offsets, for a cumulative total of 35%.[9]

The Boxer-Lieberman-Warner bill, by contrast, incorporates an issuance-based limit. Specifically, the bill would establish a limit on the amount of offset credits that the Environmental Protection Agency could issue in any particular year. Thus, under the Boxer-Lieberman-Warner bill, the Environmental Protection Agency could issue an amount of domestic offset credits not greater than 15% of the overall emissions cap.[10] The bill would establish corresponding issuance limits on international offsets.[11]

In one sense, a percentage limit on overall issuance could provide for a more substantial and efficient role for offsets than the equivalent percentage limit on use by individual entities. If one assumes that, under an entity-based limit on use, some entities would use less than their full entitlement, and that other entities otherwise would prefer to use more, then an issuance-based limit would

[8] Lieberman-Warner Climate Security Act of 2008, S Amdt 4825, 110th Cong § 524 (2008).

[9] Dingell-Boucher Discussion Draft § 712(c) (7 October 2008).

[10] Lieberman-Warner Climate Security Act of 2008, S Amdt 4825, 110th Cong § 302(b) (2008).

[11] Ibid § 321.

be more efficient because it could provide the most cost-effective distribution of offset use among all entities.

However, an issuance-based limit has some significant drawbacks. Under an issuance-based limit, the agency would have only a finite pool of offset credits to distribute in any particular year. The agency would have to determine rules for distributing these credits in the event that there were more qualifying projects than credits in any particular year. It is unclear what type of criteria it would use. Furthermore, it would be difficult for project developers to know where they stand in the queue. Therefore, a project developer could obtain approval for a project in a particular year only to find that, in some subsequent year, it submitted its paperwork too late to earn credits; thus, it would be denied credits even though its project had performed successfully. This kind of year-to-year, difficult-to-manage, uncertainty about crediting is likely to deter investment in offset projects.

By contrast, under an approach that limited the amount that each entity could use for compliance, a project developer at least would be assured of receiving offset credits every year in which its project performed. If the developer could not find a buyer for the credits because capped entities had all reached their use limits, the developer still could bank its credits for sale in a subsequent year. Accordingly, there is a strong argument that an entity-based use limit is more efficient than an otherwise comparable issuance-based limit.

4. Addressing the Risk of Reversals in Projects Involving Sequestration

Offset projects involving sequestration store carbon in the soil, in trees, or underground. Such projects are subject to a continued risk that some or all of the stored carbon will be released and emitted into the atmosphere, thereby negating the benefits of the project.

Such a 'reversal' could be unintentional, such as from a forest fire or an insect infestation. A reversal also could be intentional. For example, a landowner could abandon an offset project, switching from forest management to clear-cutting or from no-till agriculture to more conventional tillage techniques.[12]

The risk of reversal presents a particular design challenge for an offsets programme because, under the basic design of such a programme, the regulatory agency issues credits to an offset project on the assumption that the carbon sequestered in any particular year will never reach the atmosphere.

[12] For an outline of different types of intentional and unintentional reversals in agriculture and forestry projects, see Brian C Murray and Lydia P Olander, Nicholas Institute for Environmental Policy Solutions, Duke University, 'Addressing Impermanence Risk and Liability in Agriculture, Land Use Change, and Forest Carbon Projects' (October 2008) 3, Table 1 <http://www.nicholas.duke.edu/institute/offsetseries3.pdf>, accessed 20 March 2009.

There are a number of approaches for addressing reversal risk within an offsets programme. One approach is to issue a credit for sequestration projects that only has a temporary use for compliance, and for which renewal is conditioned on the continued effectiveness of the project. This approach was adopted by the Kyoto Protocol system in the context of the CDM. However, the CDM's experiment with temporary credits has not been successful. Buyers generally have shunned the temporary credits, and as a result there have been relatively few afforestation and reforestation projects registered under the CDM.[13]

To date, proposals for US federal legislation have adopted a liability approach similar to the CDM's temporary crediting scheme. Under this liability approach, the programme issues a project a credit for each ton sequestered in a particular crediting period. However, if there is a reversal then a particular actor is liable for repaying the programme with an amount of credits corresponding to the reversal.

One option under such a liability regime is to impose the repayment requirement on the ultimate holders of credits from a sequestration project that experiences a reversal.[14] In effect, this would mean that credits from sequestration project would have a default risk similar to a bond.[15] This kind of 'buyer liability' regime, however, has a number of flaws. Ideally, liability should be assigned to a party with the best ability to assess and manage the risk of reversals. In most cases, buyers will have far less information about and control over a project than the project developer. Indeed, in a robust carbon market, credits from an offset project could be bought and sold several times, implying that the ultimate holder of a credit might know very little about the underlying project that generated the credit. For these reasons, under a buyer liability regime, there is a strong likelihood that buyers would apply a very conservative discount to credits from sequestration-related projects, resulting in less investment than would be optimal. As discussed above, this appears to have been the experience with 'lCERs' issued under the CDM.

[13] UNEP Risø CDM pipeline, dated 1 March 2009, http://cdmpipeline.org/ accessed 26 March 2009. The temporary nature of CDM credits from forestry projects in addition to the exclusion of forestry credits from the EU ETS has seriously hampered the development of forestry projects under the CDM. See: Bernhard Schlamadinger, Benoit Bosquet, Charlotte Streck, Ian Noble, Michael Dutschke, Neil Bird, 'Can the EU emission trading scheme support CDM forestry?' (2005) 5 Climate Policy 199-2008; Jenny E Henman, Steven P Hamburg, Angel A Salazar Vega, 'Feasibility and Barriers to Entry for Small-Scale CDM Forest Carbon Projects: A Case Study from the Northeastern Peruvian Amazon,' (2008) 3 Carbon and Climate Law Review 254.

[14] The Boxer-Lieberman-Warner bill has both buyer liability, and also seller liability, which is discussed below. See Lieberman-Warner Climate Security Act of 2008, S Amdt 4825, 110th Cong § 306 (2008).

[15] Brian C Murray and Lydia P Olander, 'Addressing Impermanence Risk and Liability in Agriculture, Land Use Change, and Forest Carbon Projects' Nicholas Institute for Environmental Policy Solutions, Duke University, October 2008, 6 <http://www.nicholas.duke.edu/institute/offsetseries3.pdf>, accessed 20 March 2009.

A 'seller liability' approach, by contrast, shifts the risk to the landowner or project developer, either one of whom is likely to have better information about, and a greater ability to manage, the project than a buyer. The seller might also be better positioned to obtain insurance than buyers.

A seller liability regime, however, has its own problematic aspects. For example, it is possible that a seller could not afford to insure itself against the repayment requirement for a truly catastrophic carbon loss. Thus, bearing the liability for such risks, or even purchasing insurance against such liability, might be costly enough to deter otherwise sound projects.

In addition, a seller liability regime presents the risk that the seller simply is not present or available to comply with the repayment requirement. A reversal could occur decades after the commencement of the project—at a point at which the original selling entity is no longer in business. Land tenure could shift, and a new landowner could take over. A project developer could go out of business.

An alternative to approaches that assign liability to the buyer or seller is a regime that socializes the risk across all projects. One example of this type of socialized risk regime is a buffer reserve.[16] Under this approach, all offset projects involving sequestration would contribute a risk-weighted amount of credits to a common reserve. Then, if any project experiences a reversal, the agency would cancel a corresponding amount of credits in the reserve.

The buffer reserve approach would require the regulatory agency to determine for each project the percentage probability that it could experience a reversal. Then, before issuing credits to the project in any year, the agency would set aside a quantity of credits equal to the product obtained by multiplying the percentage probability of reversal for that project and the quantity of credits that otherwise would be issued for the project that year. The agency would take the credits that have been set aside and convey them to a buffer reserve. Credits in the buffer reserve could not be used to satisfy compliance requirements for regulated entities.

In the event of a reversal, the agency would cancel a corresponding amount of credits in the buffer reserve, thereby making the regulatory programme whole. There would be no specific action required of the project developer. Such a system would need to be coupled with strict reporting and auditing requirements.

A potential drawback of a buffer reserve is moral hazard. With a pooled insurance system in place, project developers might be less vigilant about preventing reversals. One option for minimizing moral hazard is to provide that the

[16] For examples of buffer reserves, please see California Climate Action Reserve, 'Revised Forest Project Protocol (Draft)' (December 2008) § 7.2 at 27 <http://www.climateregistry.org/resources/docs/protocols/project/forest/forest-revisions/draft-forest-project-protocol-december-2008.pdf>, accessed 19 March 2009; Voluntary Carbon Standard, 'Tool for AFOLU Non-Permanence Risk Analysis and Buffer Determination' (November 2008) <http://www.v-c-s.org/docs/Tool%20for%20AFOLU%20Non-Permanence%20Risk%20Analysis%20and%20Buffer%20Determination.pdf>, accessed 19 March 2009.

buffer reserve is available only for unintentional reversals, and that project developers or sellers are liable for intentional reversals.

Other, complementary mechanisms might be better suited for preventing intentional reversals. Such mechanisms could include a requirement that the landowner adopt legal restrictions that constrain how the land is used by the owner and successor owners. Examples include easements and deed restrictions.[17]

Regardless of the liability or insurance regime adopted, another question that arises with sequestration projects is the duration of time for which a project should be held accountable for reversals. The average residency life of a carbon dioxide molecule in the atmosphere is approximately one hundred years. Therefore, viewed in a simple way, a reversal negates some or all of the benefits of a sequestration project if it occurs any time during the one hundred year period after the project was commenced. However, as a matter of public policy, it may not be reasonable to impose risk management requirements of infinite duration. In addition, even the temporary sequestration of carbon for a finite but lengthy period of time could have significant value—for example, by allowing for an orderly, more efficient transformation to a zero-emitting energy system. These factors suggest that there should be some sort of rule of reason in establishing a period of time for reporting, liability, insurance, and land-use restriction requirements.

5. Crediting Early Action

Most of the proposals for federal legislation in the United States have included mechanisms to credit actions taken in capped sectors before enactment. However, the proposals have been less clear about the treatment of early offset projects. In particular, it is unclear to what extent early projects can come into the eventual federal offset programme and receive a prospective crediting period. It is also unclear whether they can earn retroactive credits, ie, credits for reductions achieved before the date that the federal offset standards come into effect.

Although the size of the US voluntary offsets market is uncertain because of limited data, the US Government Accountability Office has estimated that the supply of offsets in the market reached 10.2 million tons in 2007.[18] This number likely will grow significantly as the effective date for a federal cap-and-trade programme grows nearer. Importantly, starting in around 2002,

[17] Offset Quality Initiative, 'Ensuring Offset Quality: Integrating High-Quality Greenhouse Gas Offsets Into North American Cap-and-Trade Policy' (July 2008) 19 <http://www.offsetqualityinitiative.org/pdfs/OQI_Ensuring_Offset_Quality_7_08.pdf>, accessed 19 March 2009.

[18] United States Government Accountability Office, 'Carbon Offsets: The US Voluntary Market is Growing, but Quality Assurance Poses Challenges for Market Participants' (August 2008) 7 <http://www.gao.gov/new.items/d081048.pdf>, accessed 19 March 2009.

developers of offset projects began taking steps to verify and register their projects with a number of offset registries or exchanges, including the Climate Action Reserve of the California Climate Action Registry, the Chicago Climate Exchange, the Voluntary Carbon Standard, and the American Carbon Registry.[19] These programmes have mimicked the structure of a regulatory offsets programme, including the use of project standards, third-party verifiers, and registries to establish ownership.

In addition, as discussed further below, the United States has seen the emergence of a number of state and regional cap-and-trade programmes. Each of these programmes has incorporated offsets in one way or another, and therefore has provided another source of early projects for the eventual federal system.

Developing federal policy approaches for treatment of these early offset projects entails balancing two potentially competing objectives. On the one hand, there are cost containment benefits to ensuring that there is an ample supply of offsets available for regulated entities when they face their compliance obligations. It could take a number of years after the date of enactment before an offset programme is finalized in federal regulations. If project developers are forced to wait until the programme is in place to develop projects, there will be a risk that the system will be severely under-supplied at the first compliance deadline.

On the other hand, early projects present risks of non-additionality. The General Accountability Office has concluded that there is a considerable variance in additionality among voluntary carbon projects, even those that have been verified under one of the existing registry programmes. To be sure, many projects have been implemented in contemplation of carbon credits. Others, however, would have been undertaken anyway for other reasons.

The issues of prospective crediting and retroactive crediting involve somewhat different analyses of these questions. For prospective crediting, the analysis is less complicated because it is understood that a project will need to meet the ultimate federal offset programme standards in order to earn credits going forward. Nevertheless, there are some modest supplemental requirements that can be imposed to provide greater assurances of additionality. For example, Duke University's Nicholas Institute observes that a 'pre-compliance' market of projects in anticipation of regulation began in 2002.[20] Accordingly, that year could

[19] Lydia Olander and Brian Murray, 'Treatment of Early Agricultural and Forestry Offsets in a Federal Cap-and-Trade' Duke University Nicholas Institute for Environmental Policy Solutions, October 2008, 3–4 <http://www.nicholas.duke.edu/institute/offsetseries2.pdf>, accessed 19 March 2009; Climate Action Reserve of the California Climate Action Registry <http://www.climateregistry.org/reserve.html>, accessed 19 March 2009; Chicago Climate Exchange <http://www.chicagoclimatex.com>, accessed 19 March 2009; Voluntary Carbon Standard <http://www.v-c-s.org/index.html>, accessed 19 March 2009; American Carbon Registry <http://www.americancarbonregistry.org>, accessed 19 March 2009.

[20] Lydia Olander and Brian Murray, 'Treatment of Early Agricultural and Forestry Offsets in a Federal Cap-and-Trade' (Duke University Nicholas Institute for Environmental Policy Solutions, October 2008), 3 <http://www.nicholas.duke.edu/institute/offsetseries2.pdf>, accessed 19 March 2009.

be a start date for projects to be considered for entry into the federal offset programme. In addition, the programme could include a threshold requirement that early projects have been verified under one of the pre-existing registry programmes.

Retroactive crediting entails a more complicated analysis. To be sure, the programme could allow retroactive crediting only for reductions that meet the federal offset programme standards that eventually come into effect. The CDM has followed this approach, allowing registered projects commenced as early as 2000 to earn retroactive credits. While this approach would provide the strongest assurances of environmental integrity, it is likely to chill investment in early projects. Without foreknowledge of what the ultimate standards will require, many project developers and investors would wait until the standards were in place, resulting in a lack of supply of offsets during early compliance periods.

A second approach would be to provide retroactive credits to all projects that have been verified under pre-existing programmes irrespective of their compliance with ultimate federal offset standards—but provide the credits in the form of allowances set aside from the emissions cap, rather than offset credits. By awarding allowances from the cap instead of offset credits, this approach would remove any jeopardy to the emissions cap from non-additional projects. However, for the same reason, this approach would not provide cost containment benefits. To help contain compliance costs for regulated entities, offset projects have to be awarded credits additional to the allowances issued for the emissions cap. Otherwise, the programme merely is a different form of auctioning except that the payment for credits goes to entities in uncapped sectors.[21]

Other hybrid approaches are possible. For example, retroactive offset credits could be available only for projects that meet a subset of the pre-existing registry programmes that are deemed to be of relatively higher integrity. Such a subset could consist, for example, only of state or regional government programmes. Another approach would be to provide retroactive offset credits to any verified projects on a discounted basis, but provide full crediting to projects that eventually pass muster under the ultimate federal standards.

6. International Offsets

As discussed above, most of the studies that have examined the cost containment impact of offsets have concluded that the greatest cost-savings come from allowing international offsets into the US programme.

[21] Lydia Olander and Brian Murray, 'Treatment of Early Agricultural and Forestry Offsets in a Federal Cap-and-Trade' Duke University Nicholas Institute for Environmental Policy Solutions, October 2008, 8 <http://www.nicholas.duke.edu/institute/offsetseries2.pdf>, accessed at 19 March 2009.

International offsets, however, have become a particularly thorny issue in the design of the US programme. The concept has encountered some of the conventional scepticism in the US polity about any mechanisms that send money abroad, and the specific concerns about the appearance of providing aid to countries that are trade competitors, such as China and India. In Congressional debates, it is sometimes difficult to get a hearing for the point that broadening the US programme to bring in lower cost offsets from other countries would ease the regulatory burden on US industry from a climate change programme and thereby save US jobs.

In addition, Congress is debating the establishment of a US climate change programme at a time when there is a broad-based call for reform of the Clean Development Mechanism. A variety of criticisms have been lodged against the CDM.[22] Critics say that the CDM credits too many non-additional projects, and also that its project-by-project review process is cumbersome and costly. A growing chorus of experts and policy-makers believe that the CDM needs to be reconfigured to encourage advanced developing countries (such as China and India) to adopt emission limits at least for major emitting sectors.[23]

Bills introduced thus far in the US Congress have reflected different approaches on international offsets. The Boxer-Lieberman-Warner bill, for example, would direct the Environmental Protection Agency to promulgate regulations establishing a system for issuing offset credits for projects outside the United States.[24] Notably, the bill does not expressly limit the scope of acceptable offsets to those from the CDM, nor does it provide that reductions from CDM projects would be automatically acceptable. Furthermore, the bill would require international offsets to meet standards at least as stringent as those for domestic offsets. As discussed above, the bill would establish a limit on the quantity of credits that could be issued for international offsets. In addition, in a nod to concerns about trade and competitiveness, it would prohibit the Environmental Protection Agency from issuing credits 'for a project that competes directly with a United States facility.'[25]

The Dingell-Boucher draft legislation has a similar overall approach to the Boxer-Lieberman-Warner bill, including a limit on the quantity international offsets that could be used for compliance.[26] The Dingell-Boucher draft, however, does not include a restriction on offsets from facilities competing with US

[22] For examples, see Michael W Wara and David G Victor, 'A Realistic Policy on International Carbon Offsets' (April 2008) Stanford Programme on Energy and Sustainable Development Paper No. 74 <http://iis-db.stanford.edu/pubs/22157/WP74_final_final.pdf>, accessed 19 March 2009; Jeffrey Ball, 'U.N. Effort to Curtail Emissions in Turmoil' *Wall Street Journal* (12 April 2008) A1.

[23] United States Government Accountability Office, 'International Climate Change Programmes: Lessons Learned from the European Union's Emissions Trading Scheme and the Kyoto Protocol's Clean Development Mechanism' (November 2008) 38 <http://www.gao.gov/new.items/d09151.pdf>, accessed 19 March 2009.

[24] Lieberman-Warner Climate Security Act of 2008, S Amdt 4825, 110th Cong § 321 (2008).

[25] Ibid. § 321(c)(2)(B).

[26] Dingell-Boucher Discussion Draft § 762 (7 October 2008).

facilities. On the other hand, unlike the Boxer-Lieberman-Warner bill, the Dingell-Boucher draft would prohibit the issuance of offset credits for any international projects involving the destruction of hydrofluorocarbons.[27]

International offsets also featured prominently in a detailed set of recommendations for a US climate change programme issued in January 2009 by a coalition of environmental groups and large companies. The US Climate Action Partnership's 'A Blueprint for Legislative Action' recommends that any federal climate programme incorporate 'a transparent process for evaluating and approving international offsets.'[28] However, as a means of encouraging developing countries to limit their emissions, the Blueprint also recommends a mechanism that offers 'preferential access' to the US market for countries that are 'moving swiftly to reduce emissions broadly across the bulk of their national economies'.[29] In addition, the Blueprint further states that there should be a schedule to assure that over time offsets 'result in incremental reductions beyond a nationally appropriate country or sector-specific emission reduction commitment that covers a suitable share of a country's emissions.'[30]

7. Activities that Reduce Emissions from Deforestation and Degradation

In the United States, as in international discussions, there is significant focus on developing policies to encourage activities that reduce emissions from deforestation and forest degradation (REDD) in developing countries. Deforestation accounts for over 20% of global emissions of carbon dioxide annually. The costs of mitigating deforestation are quite low. In addition, there are potentially substantial co-benefits from REDD, including protection of biodiversity and poverty alleviation in indigenous forest communities.[31]

However, there are also unique concerns about REDD-based offset projects.[32] As with other forestry projects, reversals are a factor. In addition, a project that prevents deforestation in a particular place is vulnerable to the risk that it simply shifts the threat to another location, a phenomenon known as 'leakage'. Furthermore, there are concerns about whether a carbon market for REDD offset projects will be sufficiently protective of the rights of indigenous

[27] Ibid. § 762(b)(1).

[28] Climate Action Partnership, 'A Blueprint for Legislative Action' (January 2009), 9 <http://www.us-cap.org/pdf/USCAP_Blueprint.pdf> accessed 19 March 2009.

[29] Ibid, at 4.

[30] Ibid, at 9.

[31] See further chapter 27 below.

[32] For an overview of these issues, see Erin C Myers, 'Polices to Reduce Emissions from Deforestation and Degradation (REDD) in Developing Countries: An Examination of the Issues Facing the Incorporation of REDD into Market-based Climate Policies' Resources for the Future, December 2007, RFF Discussion Paper No. 07-50 at 11 <http://www.us-cap.org/pdf/USCAP_Blueprint.pdf>, accessed 19 March 2009.

peoples, and whether the marketplace will ensure an equitable share of the benefits for such communities.

Citing these concerns, some stakeholders have recommended avoiding exclusive reliance on offsets as a source of funding for deforestation avoidance, and instead relying on funding from the auctioning of allowances set aside from the emissions cap or other government aid. They reason that because an allowance set-aside, unlike offset credits, cannot be used to increase emissions from capped entities, such funding would not jeopardize the emissions cap if REDD activities turn out to be non-additional or associated with leakage. Such non-market-based funding also may be better suited for supporting forest protection capacity-building in developing countries.[33]

In addition, citing leakage concerns, some stakeholders prefer that a US programme address REDD on the basis of improvement across an entire national deforestation baseline rather than on the basis of discrete projects.

However, neither a government fund nor an allowance set-aside is likely to generate the same amount of financial resources as a carbon market. In addition, rainforest countries do not have forest baselines at this point, and imposing such a requirement might mean waiting several years for real action. Furthermore, advances in remote-sensing and other technologies allow for the implementation of strategies to identify and minimize leakage.

US federal legislative proposals have differed in the way they balance these various concerns. A bill introduced by Rep Ed Markey (D-MA), for example, would not provide offset credits for REDD activities at all, but instead would rely solely on funding generated through the auctioning of a set-aside of a portion of allowances to support REDD.[34]

The Dingell-Boucher draft legislation, by contrast, would provide offset credit for REDD activities. The legislation would require the Environmental Protection Agency to promulgate regulations for REDD offset projects. These regulations would have to include a requirement that the offset project adhere to 'widely accepted, environmentally sustainable forestry practices'; 'promote native species and conservation or restoration of native forests'; support the 'internationally recognized rights of indigenous and other forest-dependent people'; and recognize the 'right of free prior informed consent' of affected communities, 'if consistent with the applicable laws of the country involved'.[35]

Furthermore, while the Dingell-Boucher draft would credit REDD activities on a sub-national project basis, it also incorporates a mechanism to gradually shift countries with substantial deforestation issues (eg Brazil and Indonesia)

[33] Doug Boucher, 'Out of the Woods: A Realistic Role for Tropical Forests in Curbing Global Warming' Union of Concerned Scientists, December 2008, 5. <http://www.ucsusa.org/assets/documents/global_warming/UCS-REDD-Boucher-report.pdf>, accessed 19 March 2009.

[34] Investing in Climate Action and Protection Act, HR 6186, 110th Cong § 722 (2008).

[35] Dingell-Boucher Discussion Draft § 764(b)(2) (7 October 2008).

away from sub-national project-based crediting to an approach that only credits reductions in deforestation measured against a national baseline.[36]

8. Offsets in State and Regional Programmes

Moving into the vacuum created by inactivity at the federal level during the tenure of the Bush Administration, a number of US states have taken steps to develop their own GHG regulatory programmes. Offsets play a role in each of these programmes. As a result, the state programmes have at least three implications for the federal programme. First, they could be a source of 'early action' credits. Second, they are modeling different approaches to offset policy that could be adopted at the federal level. Third, the perceived performance of the state offset policies could affect political support for offsets in Congress.

The discussion below focuses on the three programmes that were in the most advanced stages of development by early 2009: (i) the Regional Greenhouse Gas Initiative (RGGI), which held its first allowance auctions in September of 2008;[37] (ii) California's 'AB 32'[38] programme, for which the California Air Resources Board issued its final Scoping Plan in December 2008;[39] and (iii) the Western Climate Initiative (WCI), for which the WCI working group issued its final Design Recommendations in September 2008.[40] On issues such as offsets, California has made a particular point of coordinating its programme with the WCI, of which it also is a participating member.

This section does not discuss a fourth regional programme, the Midwest Regional GHG Reduction Accord, because it was in a very preliminary stage of design at the end of 2008.[41]

What follow is a review of the offset policies of the RGGI, the AB 32 programme, and the WCI.

[36] Ibid, 764(f)(2).

[37] For further information and the RGGI Model Rule, see www.rggi.org, accessed 19 March 2009.

[38] Cal. Health & Safety Code § 38505(g).

[39] California Air Resources Board, 'Climate Change Proposed Scoping Plan' (October 2008), <http://www.arb.ca.gov/cc/scopingplan/document/scopingplandocument.htm>, accessed 19 March 2009.

[40] Christine Cordner and Lisa Weinzimer, 'Western Climate Initiative Releases Design Recommendations for Cap-Trade Programme' (2008) Electric Utility Week at 4; Western Climate Initiative, 'Design Recommendations for the WCI Regional Cap-and-Trade Programme' (23 September 2008) <http://www.westernclimateinitiative.org/the-wci-cap-and-trade-program/design-recommendations>, accessed 15 June 2009.

[41] Oregon, Washington, Massachusetts, New Jersey, and Florida also have been developing state-specific programmes. The Pew Center on Global Climate Change maintains a webpage that provides up-to-date information on the various state initiatives. See <http://www.pewclimate.org/states-regions>, accessed 19 March 2009.

8.1 The Regional Greenhouse Gas Initiative (RGGI)

Established in 2005, the RGGI was the first regional effort in the United States aimed at regulating GHG emissions. Ten Northeastern states are full participants in the RGGI: Connecticut, Delaware, Maine, Maryland, Massachusetts, New Hampshire, New Jersey, New York, Rhode Island, and Vermont. These states have agreed to a Memorandum of Understanding, which sets a regional emissions cap measured in short tons of CO_2 emissions, applicable only to emissions from the electric power sector.[42] The Memorandum of Understanding further apportions the regional cap into individual state caps, which are referred to as 'budgets'. The Memorandum of Understanding also provides the outline for a cap-and-trade programme.

Pursuant to the Memorandum of Understanding, each state has codified, through its own state laws and procedures, the RGGI Model Rule.[43] The Model Rule elaborates a detailed common framework of rules addressing allowance tracking, compliance, offset projects, and other issues.

The Model Rule has detailed provisions on offsets, including with respect to the amount of offsets an entity can use for compliance; the types of projects that are eligible; the standards for crediting; and the crediting period. The Model Rule provides that a covered source initially may use offsets to meet up to 3.3% of its compliance obligation in a compliance period, ie to account for 3.3% of its total emissions during the period.[44] However, the RGGI allows this percentage limit to increase to 5% in the event that the rolling average price of emission allowances (calculated over a 12-month period) reaches $7/t$CO_2$ (a 'stage one trigger event').[45] If the rolling average price reaches $10/t$CO_2$ (a 'stage two trigger event'), the use limit expands to 10%.[46]

This graduating limit approach is intended to help contain the costs associated with the RGGI programme. However, it is not clear how effective this approach would be. Most types of offset projects involve significant lead time, so project developers might not be able to bring significant new amounts of offsets into a RGGI market quickly in response to a trigger event.

The RGGI also incorporates a phased approach to the geographic sourcing of offset projects. Offset projects at sites within the RGGI member states are not subject to requirements separate from the basic requirements set forth in the Model Rule. However, offset projects in other states are eligible for consideration only if there is a memorandum of understanding in place with the host state

[42] Regional Greenhouse Gas Initiative, 'Memorandum of Understanding' (12 December 2005) <http://rggi.org/docs/mou_12_20_05.pdf>, accessed 19 March 2009.

[43] Regional Greenhouse Gas Initiative, Model Rule (5 January 2007, final with corrections) <http://rggi.org/docs/model_rule_corrected_1_5_07.pdf>, accessed 19 March 2009.

[44] Ibid XX-6.5(a)(3).

[45] Ibid; and Ibid XX-1.2(bd) (defining 'stage one trigger event'). All prices are adjusted for inflation.

[46] Ibid.

that has oversight responsibility for the project.[47] As of the end of 2008, the terms of such a memorandum of understanding had not been agreed by the RGGI member states.

The RGGI also addresses the possible use of credits from other countries. The Model Rule provides that, if there is a 'stage two trigger event', covered sources may use not only RGGI offsets up to the 10% limit, but also European Union Allowances from the European Union Emissions Trading System or credits from the Kyoto Protocol regime.[48]

For domestic offset projects, the RGGI has identified a finite list of eligible types. The five types are: landfill methane capture and destruction; reduction in emissions of sulfur hexafluoride (SF_6); sequestration of carbon through afforestation; end-use energy efficiency projects resulting in the reduction of CO_2 emissions from natural gas, propane, and heating oil; and methane reduction from farming operations.[49]

The RGGI end-user energy efficiency category likely would not be considered an eligible offset project category under an economy-wide cap-and-trade programme, which would regulate emissions from use of natural gas and petroleum. Thus, this type of project probably would not be considered an offset under the proposed California Air Resources Board and WCI programmes, nor under the major federal bills that were introduced in 2008.

A particular innovation of the RGGI has been its emphasis on using performance standards or benchmarks for offset projects, instead of project-by-project determinations. Thus, the RGGI Model Rule incorporates detailed methodologies and assumptions to be used for each type of offset project.[50] For example, a landfill methane capture and destruction project is eligible only if the landfill is not subject to the Clean Air Act New Source Performance Standards for municipal solid waste landfills.[51] Similarly, the Model Rule includes a schedule of benchmark SF_6 emission rates for different regions of the country, and provides that an SF_6 project will receive offset credits only to the extent that it exceeds the relevant benchmark.[52]

The RGGI Model Rule provides that the crediting period for an offset project is ten years, which may be renewed once through a separate application.[53] For afforestation projects, the crediting period is 20 years, which may be renewed twice.[54]

8.2 CARB Scoping Plan and WCI Design Recommendations

In 2006, California enacted legislation setting a limit on statewide GHG emissions. A.B. 32, the 'Global Warming Solutions Act,' required the state to

[47] Ibid XX-10.3(a)(2)(ii). [48] Ibid XX-6.5(a)(3); and XX-10.3(b).
[49] Ibid XX-10.3(a)(1). [50] Ibid. XX-10.5. [51] Ibid. XX-10.5(a)(1).
[52] Ibid. XX-10.5(b). [53] Ibid at XX-10.3(e)(1). [54] Ibid at XX-10.3(a)(2).

reduce GHG emissions to 1990 levels by 2020.[55] The legislation vested broad authority in the California Air Resources Board (CARB), working in cooperation with other state agencies, to develop and enforce regulations to meet the cap. The CARB regulations must be in effect starting in 2012. The legislation explicitly authorizes the use of a market-based approach.[56]

The legislation directed the CARB to issue a 'Scoping Plan' for the regulatory programme by no later than January 1, 2009, and to promulgate regulations by no later than January 1, 2011.[57] The Board issued its final Scoping Plan in October 2008.[58] In contrast to the RGGI, the California programme is 'economy-wide' and covers six GHGs.

The Western Climate Initiative (WCI) is a collaboration of seven US governors, representing the states of Arizona, California, Montana, New Mexico, Oregon, Utah, and Washington. The WCI also includes four Canadian premiers, representing the provinces of British Columbia, Manitoba, and—belying its 'Western' moniker—Ontario and Quebec. Collectively, the governors and premiers are referred to as the 'WCI Partners'.

In September 2008, the WCI Partners issued a document outlining design recommendations for a WCI regional cap-and-trade programme.[59] At this point, neither California nor the WCI has reached the same level of detail as the RGGI in the development of offset rules; both programmes contemplate further rule-making. The CARB Scoping Plan and the WCI Design Recommendations outline a concept for a limit on offset use, list certain types of offset projects for further consideration, and otherwise identify principles to guide further policy-making. Both documents reiterate the basic criteria for valid offsets: reductions or sequestration should be additional, real, verifiable, permanent, and enforceable.[60]

In the area of offsets, there has been a particular emphasis on integrating policy-making efforts between California and the WCI; however, CARB has made clear that it retains discretion to develop more stringent requirements than those adopted under the auspices of the WCI.[61]

CARB and the WCI have emphasized that the ability of regulated entities to use offsets and allowances from other systems should be subject to an overall limit. Both the Scoping Plan and the WCI Design Recommendations make the point that allowing unlimited offsets could reduce the environmental and health

[55] California Health & Safety Code § 38500 et seq (2007). [56] Ibid at § 38570(a).

[57] Ibid at § 38561(a).

[58] California Air Resources Board, 'Climate Change Proposed Scoping Plan' (October 2008) (see n 39 above).

[59] Western Climate Initiative, 'Design Recommendations for the WCI Regional Cap-and-Trade Programme' (23 September 2008) (see n 40 above).

[60] California Air Resources Board, 'Climate Change Proposed Scoping Plan' (October 2008) 37 (see n 39 above); Western Climate Initiative, 'Design Recommendations for the WCI Regional Cap-and-Trade Programme' (23 September 2008) (see n 40 above).

[61] California Air Resources Board, 'Climate Change Proposed Scoping Plan' (see n 39 above).

benefits associated with in-system emission reductions, and could delay the transition to lower-carbon energy production.[62]

Citing these rationales, the CARB Scoping Plan and the WCI Design Recommendations propose that regulated entities should be able to use offsets and allowances from other recognized trading systems to cover no more than 49% of their total emission reductions from 2012–2020. As described in the two documents, this limit would be measured against the reductions that otherwise would be required each year to bring emissions from the level of the 2012 cap down to the level of the annual emissions cap for a particular year.[63] The approach is intended to ensure that a majority of reductions come from entities covered by the emissions cap.

The CARB/WCI limit is measured in a different way than the RGGI limit. The RGGI percentage limit is the percentage of a covered source's total emissions that can be covered by offsets or allowances from other systems. The CARB/WCI limit is the percentage of the entire programme's total expected reductions that can be covered by offsets or allowances from other systems.

Many questions remain about how the 49% limit will work. For example, it is not clear whether the limit will apply to each year's cap, for a three-year compliance period, or across the duration of the 2012–2020 period. In addition, it is uncertain how the offsets limit will apply to an individual entity or facility. Furthermore, the WCI Design Recommendations do not clarify whether the 49% limit is an aggregate limit for the entire WCI programme or a maximum limit for each participating jurisdiction.

The WCI Design Recommendations also provide a list of project types 'as a priority for investigation and potential participation in the offset programme'; the Design Recommendations are careful to note, however, that the types are not at this point *per se* eligible.[64] The types are: agriculture (soil sequestration and manure management); forestry (afforestation/reforestation, forest management, forest preservation/conservation, forest products); and waste management (landfill gas and wastewater management).[65]

With respect to geographical parameters, the WCI document recommends giving special emphasis to offset projects located in the United States, Canada, and Mexico, provided that such projects are subject to requirements comparable to offset projects within WCI jurisdictions.[66] The Design Recommendations make clear that the WCI Partner Jurisdictions will not accept offset credits from

62 Ibid.; see also Western Climate Initiative, 'Design Recommendations for the WCI Regional Cap-and-Trade Programme' (23 September 2008) (see n 40 above).

63 California Air Resources Board, 'Climate Change Proposed Scoping Plan' (October 2008) 37 (see n 39 above); Western Climate Initiative, 'Design Recommendations for the WCI Regional Cap-and-Trade Programme' (23 September 2008) 38 (see illustration at Figure A) (see n 40 above).

64 Western Climate Initiative, 'Design Recommendations for the WCI Regional Cap-and-Trade Programme' (23 September 2008) 39 (see n 40 above).

65 Ibid.

66 Ibid.

projects in developed countries if those projects reduce emissions that, in the WCI, would be covered by the cap-and-trade programme.[67] On the other hand, WCI Partner Jurisdictions may accept offset credits from developing countries, including through the CDM; however, the WCI might impose additional quality criteria for such projects.[68] Finally, the Design Recommendations 'encourage' the development of offset projects within the WCI Partner jurisdictions in order to ensure that the WCI captures the collateral environmental, social, and health benefits of some types of projects.[69]

The CARB Scoping Plan discusses advantages of opening the California programme to international offsets. The Scoping Plan observes that, in addition to helping to lower compliance costs for California entities, allowing offsets from other countries could reduce emissions associated with the production of commodities that are imported into the state, such as cement.[70]

The Scoping Plan further explains that one concept under evaluation is the idea of limiting the sourcing of international offsets to countries that achieve certain national or sectoral emission benchmarks.[71]

9. Conclusion

Offsets will be a key element of the emerging US federal programme for mandatory regulation of greenhouse gas emissions. However, to achieve the full potential of an offsets programme, it will be necessary to work through a number of complicated—and, in some cases, controversial—policy design issues.

[67] Ibid 40. [68] Ibid. [69] Ibid.

[70] California Air Resources Board, 'Climate Change Proposed Scoping Plan' (October 2008) 37–8 (see n 39 above). [71] Ibid. 38.

20

Carbon Markets and Policy in Australia: Recent Developments

Martijn Wilder and Louisa Fitz-Gerald

1. Introduction

The election of the Australian Labor Party (ALP) to Government on 24 November 2007 resulted in a significant shift in Australian climate change policy. On its second day in office, the Government of Prime Minister Kevin Rudd ratified the Kyoto Protocol[1] and reaffirmed its intention to introduce an emissions trading scheme in Australia (the Carbon Pollution Reduction Scheme—CPRS) to commence in 2010.[2] It also set a longer term target to reduce emissions by 60% by 2050[3] and has announced plans to increase the share of renewable energy in our national generation mix to 20% by 2020.

This bundle of measures represents a complete turnaround of the position adopted by the previous Government, and establishes Australia as a global leader alongside the European Union in terms of climate change policy development. It also signals a clear commitment on the part of the Government to ensure that levels of anthropogenic emissions of greenhouse gas emissions are in accordance with those safe levels recommended by the Intergovernmental Panel on Climate Change, a commitment that has been borne out by policy details released throughout the year. Minister for Climate Change and Water, Penny Wong, commented recently that 'absolutely clearly, ... the time for playing political games with this issue is over'[4] suggesting that the current Government is aware

[1] Kyoto Protocol to the United Nations Framework Convention on Climate Change (1997) opened for signature 16 March 1998, 37 ILM 22 (entered into force 16 February 2005) (Kyoto Protocol).

[2] Kevin Rudd, Peter Garrett and Wayne Swan, 'Labor's Five Tests for an Effective Emissions Trading Scheme' (2007) Official Website of the Australian Labor Party <http://www.alp.org.au/media/0507/msCCloo301.php>, accessed 14 September 2008.

[3] Peter Garrett, 'Labor's Greenhouse Reduction Target – 60% by 2050 Backed By the Science' (2007) Official Website of the Australian Labor Party <http://www.alp.org.au/media/0507/msCC020.php>, accessed 14 September 2008.

[4] Penny Wong, 'Transcript of Press Conference, Commonwealth Parliamentary Offices Adelaide: Garnaut Review Draft Report' (Press Release, 4 July 2008) <http://www.environment.gov.au/minister/wong/2008/pubs/tr20080704a.pdf>, accessed 14 September 2008.

of the dire environmental and economic consequences that may stem from a failure to take appropriate action.

This chapter will examine the recent developments in climate policy in Australia, with a specific focus on the recently introduced legislation for emissions trading in Australia. The article will also consider the newly implemented mandatory greenhouse and energy reporting regulations, which provide detail on the mandatory corporate greenhouse emissions and energy reporting scheme, and the design proposals for an expanded national renewable energy target. The chapter will conclude by considering some recent developments in the voluntary carbon market in Australia, including the investigation by the Australian Competition and Consumer Commission (ACCC) into 'green' marketing.

2. A National Emissions Trading Scheme for Australia

Under the Kyoto Protocol, Australia is now required to keep annual emissions to within 108 per cent of 1990 levels between 2008 and 2012.[5] While Australia is on track to meet this target based on the Government's most recent reports,[6] the central policy measure to manage Australian emissions is the introduction of an emissions trading regime.

Designing an emissions trading scheme is a detailed and complex exercise that involves consideration of a wide range of economic, political, social, and environmental factors. While prior to the federal election in November 2007, both the former Liberal Government under John Howard and the new ALP Government under Kevin Rudd (as well as the States and Territories through the National Emissions Trading Taskforce—NETT) had committed to implementing an emissions trading scheme, the development of the detailed design of such a system had not, at the time of the election, been undertaken. While in opposition, the ALP also commissioned a study from Professor Ross Garnaut of the Australian National University, to look at the costs of climate change (similar to the Stern Review in the United Kingdom). The Garnaut Review[7] was one of the key inputs into the development of the scheme and like Stern, Garnaut pushed for very strong action and targets on climate change.

On 16 July 2008 the Federal Government released a Carbon Pollution Reduction Scheme Green Paper,[8] followed by a White Paper[9] in December

[5] Annex B, Kyoto Protocol.

[6] Department of Climate Change, 'Tracking to the Kyoto Target: Australia's Greenhouse Emissions Trends 1990 to 2008–2012 to 2020' (2008) <http://www.climatechange.gov.au/projections/pubs/tracking2007.pdf>, accessed 14 September 2008.

[7] <http://www.garnautreview.org.au>, accessed 14 September 2008.

[8] Australian Government Department of Climate Change, *Carbon Pollution Reduction Scheme Green Paper* (2008) <http://www.climatechange.gov.au/greenpaper/report/pubs/greenpaper.pdf>, accessed 14 September 2008.

[9] Australian Government Department of Climate Change, Carbon Pollution Reduction Scheme White Paper (2008) <http://www.climatechange.gov.au/whitepaper>, accessed 9 April 2009.

2008, setting out the Government's final design of the Carbon Pollution Reduction Scheme, including Australia's medium term target range for national emissions. Finally, on 10 March 2009 the Australian Government issued the exposure draft of the Carbon Pollution Reduction Scheme Bill 2009 (the CPRS Bill)—the law intended to introduce emissions trading across Australia.[10]

The CPRS Bill sets out the legal provisions designed to give effect to the policy positions in the Government's White Paper, albeit with some significant policy shifts announced in May 2009 (particularly the price cap on permits, set at AUD10, and a one-year delay in the commencement of the scheme). The Bill is part of a package of draft legislation that includes a bill setting out the consequential amendments to existing Acts (including the National Greenhouse and Energy Reporting Act 2007) and a bill establishing the scheme's governing body—the Australian Climate Change Regulatory Authority (Authority).

The key policy design features, together with some comments, are set out in the Annex to this chapter. However, in summary the CPRS is a very ambitious scheme that goes much further than the European Emissions Trading scheme. It is an economy-wide emissions trading regime, under which some 70% of permits will be auctioned. The CPRS Bill sets out the Government's new commitment to reduce Australia's emissions by 25% by 2020 if Australia is a party to a comprehensive international agreement that is capable of stabilizing atmospheric concentrations of greenhouse gases at around 450ppm CO_2-e or lower. It may secure some 5% of theses reductions through the purchase of REDD credits. A range of measures including some level of free permits will be issued to industries exposed to emissions intensive trade, and some power stations and auction revenues will be used to compensate households for increased electricity and fuel costs. The scheme will have the potential to link internationally in the future and already allows the unlimited purchase of CERs and ERUs but in the initial years will not link and will be subject to a price cap which increases over time. Finally, forestry projects are eligible to create units for trading in the scheme.

The scheme has been criticized by environmental groups as not ambitious enough and providing too much compensation. Industry response has varied but on the whole has been focused on maximizing compensation and the number of free permits they will receive. However, the Government's last minute changes to the scheme have won wide support and in reality the compensation that is provided to industry, including price caps, will be phased out over time. The real test will be to what extent the scheme actually delivers emission reductions and leads to changes in behaviour.

The package of CPRS bills has been passed by the Australian House of Representatives and now needs to pass the Senate. Whether or not the recent policy changes are sufficient to ensure passage of the legislation through the

[10] <http://www.climatechange.gov.au/emissionstrading/legislation/index.html>, accessed 9 April 2009.

Senate remains to be seen. While the changes clearly address some of the major concerns raised by the Coalition and the Greens, both parties are still expressing opposition to the CPRS.

Furthermore, the one-year delay from 1 July 2010 to 1 July 2011 and the $10 fixed price in 2011–12 have had a dampening effect on Australian carbon markets, with most businesses—many of which were, up until the recent policy changes, actively trading—now adopting a wait-and-see approach.

The Government has expressed its desire to achieve passage of the legislation in the winter sitting. If the Bill is not passed by the Senate and is then re-introduced after a period of three months, the Government will have a trigger allowing it to request a double dissolution election. There is currently no clear indication that it would use such a trigger.

In the event that the Government is not successful in having the CPRS legislation passed, but still commits to a reduction in national greenhouse gas emissions, it will need to find other measures to achieve this. The 'alternative' approach may well involve some form of carbon tax, regulation of greenhouse gas emitting activities, and most likely an increase in State-based regulation including a possible revisiting of an emissions trading scheme by the States.

3. Mandatory Reporting

The CPRS is underpinned by a National Greenhouse and Energy Reporting Scheme (NGERS). NGERS is critically important for the CPRS as the information reported under NGERS will inform emissions liabilities, caps and the number of permits that will be allocated to each eligible firm. Previous experience under Phase 1 of the EU ETS demonstrates the importance of using robust data to ensure correct allocation of allowances.[11]

The framework for mandatory corporate reporting of greenhouse gas emissions and energy production and consumption was set out in the National Greenhouse and Energy Reporting Act 2007 (Commonwealth) (the Act), passed in 2007. The National Greenhouse and Energy Reporting Regulations 2008 (Commonwealth) (the Regulations) released in 2008 set out particular definitions, criteria and rules that corporations should apply when reporting greenhouse and energy data under the Act.

The Act imposes mandatory annual reporting obligations on corporations in control of corporate groups that emit greenhouse gases, or produce or consume energy, in volumes that exceed particular thresholds. The reporting thresholds for corporate groups become progressively lower during the first three reporting years, as shown in Table 20.1.

An annual facility-level threshold of greenhouse gas emissions in excess of 25kt CO_2e (ie kilo tonnes of CO_2 or equivalent in other GHGs) or energy production

[11] See discussion of EU ETS above at chapter 16.

Table 20.1. Greenhouse gas emissions and energy production by year

Year commencing	Greenhouse gas emissions (in kt CO_2e)	Energy production or consumption (in TJ)
1 July 2008	125	500
1 July 2009	87.5	350
1 July 2010	50	200

or consumption in excess of 100TJ has applied since 1 July 2008, when reporting obligations under the Act commenced.

A 'facility' is defined in Section 9 of the Act as 'an activity, or a series of activities (including ancillary activities), that involve the production of greenhouse gas emissions, the production of energy or the consumption of energy and that . . . form a single undertaking or enterprise and meet the requirements of the regulations'.

The following are defined as a single 'facility':

- several activities that are conducted at a single site and together produce one or more products or services;
- transport sector facilities, even where they cross state and territory boundaries;
- any of the following activities, if undertaken by a single corporation: electricity transmission or distribution, gas or water supply, sewerage or drainage services, or telecommunications.

If a corporate group exceeds any of the above corporate group thresholds, then the corporation in control of the corporate group must register and report on behalf of the entire corporate group. If only a facility under the operational control of a member of the corporate group exceeds the facility threshold, the controlling corporation only has to report the greenhouse and energy data for that facility. Corporations that fail to comply with the Act face civil and, potentially, criminal penalties under the Act.

However, corporations are not required to report data for small facilities that, among other things, emitted less than 3ktCO_2e of greenhouse gases, and did not produce or consume more than 12TJ of energy, in a reporting year.

4. Other Government Policies

Emissions trading will remain at the centre of Australia's climate change policy over the next few years. However, the Government has a number of other policy measures to supplement emissions trading[12] including:

[12] Penny Wong, 'Climate Change – A Responsibility Agenda' (2008) Australian Government Department of the Environment, Water, Heritage and the Arts <http://www.environment.gov.au/minister/wong/2008/pubs/tr20080206.pdf>, accessed 14 September 2008.

- The increase of the existing Mandatory Renewable Energy Target (MRET) from 2% to 20% by 2020.
- The inclusion of a 'greenhouse trigger' of the Environment Protection and Biodiversity Act 1999 (Commonwealth), under which new projects of commonwealth significance with greenhouse emissions above a certain threshold will require environmental impact assessment approval.
- Introduction of the Tax Laws Amendment (2008 Measures No 1) Bill 2008 (Commonwealth) into the House of Representatives (2008 Bill) on 13 February 2008 under which taxpayers, subject to certain conditions, who carry on a business can claim an upfront tax deduction for expenditure on trees (that happen in addition to create carbon sinks).

5. Mandatory Renewable Energy Target

During the election campaign, the ALP pledged to significantly expand the national renewable energy target to 20% of the total energy mix by 2020. In December 2008, in fulfilment of that promise, the Australian Government and the Council of Australian Governments (COAG) released for public consultation draft legislation for an expanded national Mandatory Renewable Energy Target (MRET) (the draft Renewable Energy Amendment Bill 2008 (the MRET Bill)).

MRET was first introduced by the previous Liberal Government under Prime Minister John Howard and took the form of a tradeable renewable energy certificate (REC) scheme. It was designed to stimulate investment in renewable energy in Australia as well as having the additional benefit of reducing greenhouse gas emissions. The scheme was initially highly successful, but over time it became apparent that the relatively low target (9500 GWh by 2010) would quickly become saturated and the market would be oversupplied with RECs.

While in Opposition, the ALP argued that the existing target set by the Liberal Government was inadequate and that if elected it would increase the target to 20% of the total energy mix by 2020. Indeed, as the target was approached, many renewable energy companies withdrew from Australia. However, many of these companies have now returned and many others are arriving with the announcement of the revised target by the Rudd ALP Government.

The changes set out in the Bill are proposed to take effect from 1 July 2009, and the scheme would terminate on 1 January 2031.

The new target will require 45,000 gigawatt hours (GWh) of additional renewable energy (on top of the approximately 15,000 GWh of existing renewable energy). Annual targets ramp up slowly each year from 2010, and more quickly from 2015 to 2020.The targets then plateau at 45,000 GWh until 2024, before ramping down to 23,000 GWh by 2030. The phase-out of the scheme by 2031 reflects the assumption that, by the latter stages of MRET, carbon prices under the CPRS will be high enough to act as incentives for renewable energy.

In addition to an increased target, some other important changes are proposed—particularly in relation to assistance (in the form of free certificates or cash payments) for energy-intensive trade-exposed industries, and preferential treatment of small renewable energy plants (allowing them to create more than one certificate for each megawatt hour of renewable energy produced). However, the basic structure and operation of MRET is not proposed to change, and all currently eligible renewable energy plants will remain able to create RECs.

As is the case with the existing scheme, if a liable entity does not surrender sufficient certificates to meet its obligations by the deadline each year, it will have to pay a penalty for each missing REC. This penalty has not yet been determined, but the commentary to the Bill notes that it will be a fixed, non-indexed amount, to be set at a level just above projected peak REC prices.

In the time since it became apparent that MRET would become saturated and the recent announcement of the increased target, a number of states introduced state-based feed-in tariffs which have further driven investment in renewable energies. The issue of state-based feed-in tariffs is not addressed in the draft Renewable Energy Amendment Bill or the accompanying commentary, so those schemes may continue in parallel with the expanded MRET. This will offer a double benefit for small renewable energy plants, which may be able to receive both an increased number of certificates on installation and also an ongoing preferential feed-in tariff.

Historically, one of the interesting outcomes of MRET in Australia has been the emphasis on the development of wind projects. Whereas in Europe and the US such projects have a high degree of public support, the development of wind projects in Australia has not been without its challenges. Initially inadequate planning and environmental codes hampered their development, and in more recent times there has been an increase in public opposition to wind projects, largely as a result of concerns over their siting.

For example, many of the wind sites in Australia are located along the coast, in areas of high residential demand where land is expensive and residents do not want the visual impact of wind farms. In addition, in some communities where one or two individual landholders have received a significant financial benefit from leasing their land to wind project owners, tension has arisen with the surrounding landholders who do not receive a financial benefit but nonetheless must bear the 'cost' of living near a wind farm. There has also been significant litigation against the development of some wind farms, particularly where they have an impact on wildlife species.

An additional point of interest is the fact that the MRET target did not really result in the significant uptake of solar technologies, despite the fact that Australia has vast areas of desert and areas that would be very suitable for siting solar projects. This is partly due to the fact that many of these areas are also very remote, creating transmission issues. In addition, most of the technology to date has not really been targeted at industrial-size solar energy installations, although this is now starting to change with some industrial-scale projects now being developed.

Finally, there has also been significant investment in hot dry rock technology in South Australia, and over the longer term this is expected to play an important role in the make-up of Australia's renewable energy mix.

6. The Voluntary Carbon Market

The trade in carbon rights and carbon permits will, in Australia, be dominated by the Australian emissions trading scheme once established. However, for those many companies not covered by the scheme, the increasing corporate objective of becoming environmentally responsible or carbon neutral has seen a continued growth in the voluntary carbon market over the last year. Currently, the primary source of voluntary carbon credits in the Australian market is the Greenhouse Friendly program, implemented by the previous Liberal Government under John Howard.[13]

6.1 An Australian Offset Standard

The Department of Climate Change is examining the potential for developing an Australian Offset Standard, to replace Greenhouse Friendly accreditation and cover both voluntary and compliance offset credits. Further details on the standard, including the eligibility of offshore projects and projects undertaken at sub-threshold facilities in covered sectors are likely to be revealed later this year.

The announcement by Prime Minister Rudd on 6 June 2007, when he was Opposition Leader, that a Federal Labor Government would introduce a national standard for carbon offsets, may give some clues as to the nature of the Australian Offset Standard. Prime Minister Rudd said that an Australian Offset Standard would be implemented that would:

- require all products on the market to be accredited;
- build on existing standards to avoid duplication;
- provide a nationally consistent approach to offsetting, but take account of international developments;

[13] Greenhouse Friendly provides two different services—certifying abatement certificate providers and certifying carbon neutral products and services. Once certified for a particular sector, abatement certificate providers may carry out projects which can generate Greenhouse Friendly voluntary carbon credits, which can then be sold to companies to offset the embodied emissions in their products and services. Such products and services can then be sold as 'carbon neutral' with a Greenhouse Friendly certification.

- set minimum standards for offsets and include verification and validation protocols;
- require ongoing management where necessary to ensure integrity; and
- incorporate a standard means for calculating carbon neutrality and require credits to be cancelled when used to provide an offset.

This announcement followed a joint communiqué on 9 February 2007 from the Council for the Australian Federation, comprising all Australian Premiers and Chief Ministers, calling for strengthened standards and accreditation in the carbon offset industry, including a possible registry of offset products.

6.2 Australian Competition and Consumer Commission investigation into 'green' claims

The Australian Offset Standard may assist in overcoming one of the difficulties that has faced consumers in Australia's largely unregulated voluntary market—that is, how to determine whether an emissions offset scheme will deliver its claimed benefits. This issue is the subject of a current investigation by Australia's competition watchdog, the Australian Competition and Consumer Commission (ACCC).

The ACCC recently announced that it would be examining carbon neutrality claims more closely. Commissioner John Martin said that a steadily increasing number of inquiries and complaints around 'green' marketing (promoting, for example, 'green flights', 'green cars', or 'green toilet paper') had led the ACCC to examine how the consumer protection provisions of the Trade Practices Act 1974 (Commonwealth) might apply.

The basis for the investigation is the confusion in the market as to the nature and credibility of different offset products and the basis for claims of 'carbon neutrality' in relation to products and services. For example, some carbon offset schemes rely on forestry programs, whereby additional trees are planted to offset a specified quantity of carbon dioxide emissions, while other offset schemes rely on emissions reductions as a result of energy efficiency or utilizing energy from a source that causes fewer emissions (like wind farms or natural gas). The effectiveness of a program used to offset emissions depends heavily on the effectiveness of the methodologies used to measure the emissions reductions or captured carbon and the reliability of the arrangements that the offset provider has put in place to implement and monitor the emissions reductions or ensure the captured carbon will be maintained.

Different forestry schemes may use different methodologies to calculate the amount of carbon dioxide that trees are capable of absorbing and the amount of time for which it will be stored. Likewise, projects that reduce emissions from electricity generation, industrial processes, or waste may also use different methodologies to determine the emissions avoided or reduced. More generally, different schemes utilize different methods for determining the amount of emissions generated by certain practices, such as air travel, leading to significant variations in the number of carbon offsets required to neutralize similar activities. This can create confusion for consumers.

Further, there are a range of subtle differences in the offset products offered, their price in terms of the timing and source of the emissions reductions, other environmental benefits achieved from the underlying projects, and the period over which the emissions reduction will be maintained. For example, some offset products comprise a promise to carry out a project to reduce emissions (for example, to plant trees or reduce energy consumption in the future), whereas others represent emissions reductions that have already been achieved (because the project has already been undertaken and the captured carbon or reduced emissions have already been measured and verified). In some cases the captured carbon underlying the offset product might be retained and monitored over the long-term, whereas in other cases it might not be clear how permanent the emissions sequestrations or how constant the reductions will be.

7. Conclusion

Since the late November 2007 election of the Rudd ALP Government, climate change policy in Australia has undergone a significant reversal. Internationally, Australia's ratification of the Kyoto Protocol sees Australia playing a lead role in international negotiations and domestically, its aggressive policy agenda will see a domestic emissions trading regime in place by 2010.

Annex to Chapter 20

Design feature	White Paper policy
Caps and targets	
Long-term emissions reduction target (WP Chapter 4)	The objects clause sets out the Government's new commitment to reduce Australia's emissions by 25% by 2020 if Australia is a party to a comprehensive international agreement that is capable of stabilizing atmospheric concentrations of greenhouse gases at around 450ppm CO2-e or lower. Up to 5% of this target could potentially coming from the Government purchasing REDD credits. The Government has maintained its long term target to reduce total greenhouse emissions by 60% below 2000 levels by 2050. If a comprehensive global agreement emerges over time involving 'emission commitments by both developed and developing countries that are consistent with long-term stabilization of atmospheric concentrations of greenhouse gases at 450ppm CO2 or lower' then Australia would establish post-2020 targets to ensure that it makes its full contribution to a more ambitious global agreement.
Indicative emissions trajectory and short- and medium-term targets (WP Chapter 4)	A five-year indicative trajectory for Australia-wide emission reductions will be set commencing at the beginning of the Scheme and will be extended by one year, every year, from 2011, so that the trajectory for the current year and four future years are always known. The first indicative national emissions trajectory will be: • In 2011–12, 109% of 2000 levels • In 2012–13, 108% of 2000 levels • In 2013–14, 107% of 2000 levels The Government has stated that the trajectory will start 'soft' and will become steeper in later years. In the first year of the CPRS (2011–12), liable entities will be able to purchase an unlimited number of Australian emission units (AEUs) from the Australian Climate Change Regulatory Authority (Authority) at a fixed price of $10. These AEUs are automatically surrendered upon issue and therefore cannot be banked for future use.

Design feature	White Paper policy
	The price cap will be in place for four years from 2012–13, rising at 5% in real terms, calculated using the CPI rather than an assumed inflation rate of 2.5%. Its value in 2012–13 will be based on the second year price cap being the original price cap of $40 in 2010 adjusted up by $5 and the consumer price index.
Setting medium-term Scheme caps and extending caps (WP Chapter 10)	Scheme caps for the first three full-trading years of the CPRS (2012–13 to 2014–15) will be declared in regulations before 1 July 2010. Caps will be set five years in advance, with the option to extend this period to the end of any existing international commitment period, if this is longer. Scheme caps will be extended by one year, every year, to always maintain a five year cap horizon. Scheme caps for the first two years of the Scheme will be aimed at meeting Australia's Kyoto target, that is, 108% above 1990 emissions for the years 2008–2012. Scheme caps will be set at a level equal to the indicative national emissions trajectory in the relevant year, less the projected emissions from those sources not covered by the Scheme. Where this would lead to Scheme caps that lie outside the bounds of the relevant gateway, the Scheme cap will be set equal to the closest bound of that gateway. When regulations are made declaring scheme caps or gateways, the Minister must table before Parliament her reasons for recommending to the Governor-General that those caps or gateways be set.
Gateways duration and extension (WP Chapter 10)	The Government will use gateways to provide guidance on future Scheme caps beyond the period of fixed Scheme caps. Future caps will be set within the bounds of the gateways. The Government intends to provide up to ten years of gateways beyond the minimum five years of Scheme caps, taking into account progress in international negotiations. Gateways will be extended by five years and existing gateways will be narrowed, every five years, as part of a strategic review of international conditions and Australia's likely future international commitments.

(Continued)

Design feature	White Paper policy
Coverage, liable entities and offsets	
Sectors covered in Scheme (WP Chapter 6)	The following sectors will be covered from Scheme commencement: • stationary energy; • transport; • fugitive emissions; • industrial processes; • synthetic greenhouse gases; and • waste.
Emission sources not included in Scheme (WP Chapter 6)	The following emissions sources will not be included in the Scheme: • deforestation; • emissions from landfill waste sites that closed prior to 30 June 2008, and (until 2018) emissions from waste deposited prior to 1 January 2009; • combustion of biofuels; and • combustion of biomass.
Forestry (WP Chapter 6)	Reforestation is to be included in the Scheme on a voluntary opt-in basis from Scheme commencement. Free AEUs issued in respect of reforestation in 2011–12 will have a vintage year of 2012–13. This means that they will be fully bankable. Only domestic emissions sources and sinks that are recognized in Australia's Kyoto Protocol accounts will be eligible for inclusion in the Scheme. Landowners, certain lease and carbon sequestration right holders would be eligible to opt-in to participate in the Scheme. Scheme legislation will provide further detail as to those leases and carbon property right Schemes that will be considered sufficient for participation in the Scheme. Potential Scheme participants will be required to meet accreditation requirements to reduce the risk of non-compliance, and may be required to provide some form of financial security.

Design feature	White Paper policy
	Scheme obligations will be capped—ie opt-in participants will not be required to surrender more permits than have been issued for an individual forest stand. Scheme liabilities will be enforced for a defined period of time (eg 100 years) following the last issuance of a permit for an individual forest stand.
Agriculture (WP Chapter 6)	The White Paper confirms that agriculture will not be included in the Scheme from Scheme commencement, but that its possible inclusion from 2015 will be reconsidered in 2013.
Points of obligation (WP Chapter 6)	Points of obligation under the Scheme are as follows: • Stationary energy: direct emitters above 25,000t CO_2e are liable, and fuel suppliers are liable in respect of for small emitters • Transport: upstream point of obligation only, via excise system • Industrial process emissions: direct emitters above 25,000t CO_2e threshold are liable • Fugitive emissions: direct emitters above 25,000t CO_2e threshold are liable • Waste: direct emitters are liable if over a threshold of 25,000t CO_2e in rural areas, or 10,000t CO_2e if a landfill facility is operating close to another small landfill facility (within a distance to be determined in the regulations)
	Large users of fossil fuels, coal/coal products and synthetic greenhouse gases (emissions of greater than 25,000t CO_2e from combustion of a single fuel) and other eligible entities will be able to 'net out' (purchase without a carbon price) their fuel purchases and directly manage their Scheme obligations. Upstream liability for natural gas from a wholesale gas market attaches to the entity which re-supplies gas that is supplied to it out of the market. Liability does not arise when gas is supplied into a wholesale gas market because a seller in a wholesale market is not able to determine who purchases its gas.
	Synthetic greenhouse gases: indirect obligations on entities that import or manufacture gases that result in emissions of 25,000t CO_2e a year or more. The Government has undertaken to develop measures to remove the potential for market distortions that may result from the application of this threshold.
	Legacy waste emissions—that is, emissions from waste that was placed in landfill prior to the start of the CPRS—are not covered by the CPRS.

(Continued)

Design feature	White Paper policy
Definition of a liable entity (WP Chapter 7)	In general, entities with operational control over covered facilities or activities will be liable for emission obligations arising from those facilities or activities under the Scheme. Liable entities may be companies, government entities (including local councils), joint ventures, partnerships, trusts or other unincorporated entities. For corporations, obligations will be placed on the controlling corporation of a company group where either the controlling corporation or a member of its group (e.g. a subsidiary) has operational control over a covered facility. A liability transfer certificate can be used to transfer liability to any type of entity with financial control (other than an individual, foreign entity or corporate group member). The test for financial control has been expanded to explicitly cover joint ventures and partnerships. The provisions dealing with quotation of an obligation transfer number (to transfer liability downstream from an upstream importer or producer of fuel) allow for either a one-off quotation in relation to a single supply or a standing quotation in relation to ongoing supply arrangements. The situations in which quotation is mandatory and those in which it is voluntary have also been slightly modified. For corporations, obligations will be placed on the controlling corporation of a company group where either the controlling corporation or a member of its group (eg a subsidiary) has operational control over a covered facility. With the approval of the Scheme regulator and if certain criteria are met, a controlling corporation can transfer Scheme obligations to another legal entity within the same corporate group. If another entity takes on Scheme responsibility for a facility under either of the previous two paragraphs, that entity must also take on emissions reporting obligations for that facility. Where a covered facility is operated by an unincorporated joint venture, a partnership, a trust with multiple trustees or an unincorporated association, the entity with operational control over the facility (usually the operator) will be the liable entity under the Scheme. If a single legal entity does not have operational control over the facility, the relevant parties will need to nominate a single entity to be responsible under the Scheme.
Domestic offsets (WP Chapter 6)	The White Paper confirms there will be no domestic offsets initially. The Scheme's broad initial coverage leaves limited scope for domestic offsets. Where practical, the Government will apply alternative mitigation measures to sources of emissions that cannot be covered or are likely to remain outside the Scheme for an extended period of time.

Design feature	White Paper policy
	Offsets will not be allowed from agriculture emissions in the period prior to coverage of these emissions under the Scheme. The Government will consider the scope for domestic offsets in 2013. Any offset projects accepted under the Scheme in the future would need to comply with world's best standards, and be additional and measurable.
Assistance to industry Mechanisms for emissions-intensive, trade-exposed (EITE) industry assistance (WP Chapter 12)	The Bill states the aim of the EITE assistance program, being to recognize issues relating to the impact of the CPRS on the international competitiveness of EITE activities. Most of the mechanical details of the program will still be contained in regulations. According to the White Paper, the Government has decided to provide the same assistance to both new and existing entities that conduct a given activity on the same basis. Providing assistance on the same basis means that less efficient producers will not be rewarded relative to more efficient producers of the same product. Assistance will be provided to EITE industries with up front (ex-ante) free allocation of permits at the beginning of each compliance period contingent on continuing production. The proposed allocation is based on each entity's level of production in the previous year (a per unit of production basis) to take into account expansion and contraction of EITE activities. This prevents windfall gains flowing to an entity that reduces production in a given year. If an EITE entity ceases to operate in an activity year, it will be required to return carbon pollution permits for production that does not occur. The Government has not included any 'true-up' obligation at the end of the year even though this would result in a closer alignment of allocations and emissions, on the basis that such a true-up would increase the complexity for the regulator and EITE entities. *Review:* Assistance will be reduced by 1.3% per year so that the EITE sector shares the task of meeting the national commitment to reduce emissions. The EITE assistance program will be reviewed at the five year Scheme review or at an earlier date at the request of the Minister. EITE assistance may be withdrawn if 'broadly comparable carbon constraints' apply internationally, or 'an international agreement involving Australia and all major emitting economies is concluded'. Five years' notice will be provided of changes to the program unless required for compliance with Australia's international trade obligations.

(Continued)

Design feature	White Paper policy
Eligibility for and calculation of EITE assistance (WP Chapter 12)	Eligibility: Eligibility for EITE assistance will be determined using industry average estimates of emissions intensity. The Government has decided that for the purposes of determining eligibility, there will be two metrics used: • the weighted average emissions per million dollars of revenue generated by entities conducting the activity; or • the weighted average emissions per million dollars of value added generated by entities conducting the activity. If this test is used, the entity and Government will need to agree on which input costs will be adjusted to calculate the proxy for value-added for the activity. The value-added threshold for eligibility will be set at three times the revenue threshold. *Trade exposed:* The activity must also be demonstrated to be trade exposed, ie, having a trade share greater than 10% in past years or a demonstrated lack of capacity to pass through costs due to the potential for international competition. Emissions intensive activities are those with an emissions intensity above 1000t CO_2/A\$m revenue or 3000t CO_2/A\$m value added. Two assistance rates: There are two assistance rates – 90% and 60% – with the most emissions intensive activities, over 2000t CO_2/A\$m revenue, receiving a 90% assistance rate and those with an emissions intensity above 1000t CO_2/A\$m revenue receiving a 60% assistance rate. Assistance will be provided to new and existing entities at the same rates on the same basis. The Government has not determined the activities that will be eligible but believes that there could be as many as 40 activities that are eligible. At the higher assistance rate of 90%, activities likely to be eligible include: • aluminium smelting; • cement clinker production; • silicon production; • integrated iron and steel manufacturing. • Activities likely to be eligible for assistance at the 60% rate include: • alumina refining; • petroleum refining; • LNG production.

Design feature	White Paper policy
	Additional activities that may be eligible include activities in the pulp and paper manufacturing sector, the iron and steel sector, the plastics and chemical manufacturing sector, other non-ferrous metals sector and the glass manufacturing sector. If the Scheme's coverage is extended to agriculture in 2015 there may be several agricultural sectors, including the production of beef cattle, sheep, dairy cattle, pigs and sugar cane.
Recipients of EITE assistance (WP Chapter 12)	*Who will be the recipients?* The Government has acknowledged that if it provides permits to the entity with operational control over the activity it risks misaligning the recipient of assistance with the party liable for the emissions from that activity. In order to prevent the Government delivering assistance to the 'wrong entity' the policy will be that assistance should be given to the entity or entities that incur costs through the operation of the Scheme (either directly or indirectly). Where more than one entity is liable for direct emissions (because more than one facility is involved) all the parties may make a joint application for assistance. *Public disclosure:* The regulator will be required to publish the recipients and total permits allocated to each recipient.
EITE baselines (WP Chapter 12)	Baselines for allocation of free permits will be based on historic information on the emissions intensity of all entities conducting a given activity. These will be determined using the data from two years prior to the release of the Green and White Paper. Direct and indirect emissions: Assistance will be provided in relation to direct emissions and some indirect emissions. Permit allocations for indirect emissions will be related to the cost increase associated with the use of electricity, steam, and natural gas and its components where these are used as feedstock by an activity.
Assistance to strongly affected industries (WP Chapter 13)	Coal-fired electricity generators are considered to be the only entities satisfying the criteria for 'strongly affected industries': emissions intensive (compared to other entities in the same sector, so carbon costs cannot be passed on), non-trade-exposed, significant sunk capital costs, and no significant cheap abatement opportunities. The Government will provide limited direct assistance to coal-fired electricity generators, in the form of 130.7 million free permits, to the value of A$3.9 billion based on an A$25 carbon price. These permits will be distributed to generators, on a per-output basis (taking into account emissions intensity), in five equal installments over the first five years of the Scheme.

(Continued)

Design feature	White Paper policy
	The provision of assistance will be conditional on the recipient remaining registered with the relevant market operator, with the same actual or planned capacity as at 3 June 2007, unless the relevant market operator assesses that there are likely to be adequate energy reserves in the system to allow the reduction in capacity without breaching the power system reliability standards. In addition, free AEUs issued under the EITE assistance program and to coal-fired electricity generators in the fixed price transition year (2011–12) must be surrendered by 15 December 2012. Free 2011–12 AEUs may also be sold to the Authority during the period from 15 July 2011 to 1 December 2012 for \$10, discounted by a factor specified in regulations. Assisted entities will be subject to a 'windfall gains review'. If a windfall gain is likely, the minister may withhold the last two years of assistance to that entity. Assistance to workers and regions will be delivered through the Climate Change Action Fund. Assistance for the development and deployment of carbon capture and storage technologies will be delivered through existing programs.
Linking to international schemes and acceptance of international permits	
Linking to international Schemes/markets (WP Chapter 11)	In the longer term, the Government has a preference for open linking within the context of an effective global emissions constraint.
Acceptance of international units (WP Chapter 11)	Eligible international emissions units cannot be surrendered in relation to the 2011–12 fixed price transition year. Thereafter, liable entities will be able to meet their obligations by using eligible international units for compliance in the Scheme. The use of eligible international units will not be subject to quantitative restrictions. Liable entities will (initially) be able to surrender certified emission reductions (CERs) (with the exception of long-term and temporary CERs) from CDM projects, emission reduction units (ERUs) from offshore JI projects and removal units (RMUs) from eligible reforestation activities.

Design feature	White Paper policy
	Assigned Amount Units and international non-Kyoto units will not be accepted for compliance in the Scheme. This will be reviewed for the post-2012–13 period in the light of future developments in international negotiations.
Exporting permits (WP Chapter 11)	In the initial years of the Scheme, export of carbon pollution permits will not be allowed. A minimum of five years' notice will be given on a decision to allow the sale and transfer of carbon pollution permits, except where an independent review finds that establishing a bilateral link (most likely with New Zealand) will not have a significant impact on the permit price in the Scheme, and the responsible minister decides to waive or shorten the notice period.
	Australia will not host Joint Implementation projects under the Kyoto Protocol in sectors that are covered by the Scheme. Decisions on Joint Implementation projects in uncovered sectors will be aligned with decisions on domestic offsets. Australia will not host Joint Implementation projects prior to the commencement of the Scheme.
Price cap, banking and borrowing, auctioning	
Price cap (WP Chapter 8)	The Scheme will have a transitional price cap for the period 2010–11 to 2014–15.
	The price cap will be set at $40 commencing in 2010-11. The cap will be in the form of an unlimited store of additional permits available from the Government at A$40 each. However, these permits cannot be transferred or banked.
	The price cap will rise in real terms by 5% per year. There will be no fixed price transitional period.
Banking of permits (WP Chapter 8)	Unlimited banking of permits will be allowed under the Scheme (except those accessed under the price cap arrangements).
Borrowing of permits (WP Chapter 8)	The Scheme will allow liable entities to discharge up to 5% of their obligations by surrendering permits dated from the following year.
Auctioning (WP Chapter 9)	Auctions for AEUs will commence in 2010–11, for AEUs that can be used to meet obligations in 2012–13 and subsequent financial years.

(Continued)

Design feature	White Paper policy
	Participation: There will be no restrictions on who can participate in auctions except for the following: • financial guarantee or cash deposit must be provided; • maximum purchase is limited to 25% of available amount; and • all participants will be required to have a registry account with the regulator.
	Type of auction: Ascending clock auctions will be used for single vintage auctions.
	Uniform pricing: The price paid per permit will be identical for all successful bidders.
	Proxy Bidding: Proxy bidding in a sealed bid format will be permitted.
	Publication of auction results: The results of each auction will be released within 7 days of each auction.
	Reserve Price: The auction will have a reserve price well below the expected market price.
	Internet auction platform: Auctions will be conducted using an internet platform.
	Parcel size: Bidders will be restricted to parcel sizes of no more than 25% of the total number of permits sold at each auction. As there will be sixteen auctions a year this implies bids of no larger than 1.6% of total permits issued for a given vintage.
	Training courses: Mock auctions will be conducted as training in advance of the first auction, in order to allow bidders to familiarize themselves with the auction process.
	Possible auction schedule: Auctions will be held twelve times throughout the financial year.
	At least one auction of the current year's vintage will be held after the end of the financial year in the lead-up to the final settlement date. This will be within one month prior to the final surrender date.
	The first auction will take place as early as is feasible in 2010, before the start of the Scheme.
	Four years of vintages will be advanced auctioned (current vintage plus advance auctions of three future vintages).
	Advance auctions for each future vintage will be held annually, with additional auctions for the first two years to facilitate Scheme start-up.

Design feature	White Paper policy
Compliance, enforcement and reporting	
Compliance (WP Chapter 7)	To comply with their obligations under the Scheme, each financial year liable entities will have to report their emissions (by 31 October) and surrender a number of permits equal to their emissions (by 15 December). The Authority may extend the surrender deadline for an eligible financial year beyond 15 December if at least two liable entities were unable to surrender eligible emission units before 15 December due to a malfunction in the Authority's computer systems. A national registry will be established to track the ownership of eligible compliance permits under the Scheme and to manage their surrender and cancellation (but it will not facilitate payment for transfers of permits). To hold a permit or a CER, ERU or RMU, entities will need to open an account in the registry. Entities will be able to open accounts from mid-2009.
Enforcement (WP Chapter 7)	The Scheme regulator will have a range of compliance, investigative and enforcement powers and a range of mechanisms, including civil penalty and criminal provisions, to respond proportionately to non-compliance with the Scheme. An administrative penalty will be imposed on liable entities if they fail to surrender sufficient permits. The penalty will be: • an amount prescribed in the regulations for the relevant compliance year; and • if no amount is prescribed, or the regulations are disallowed, an amount equal to the benchmark average auction price for permits auctioned in the previous financial year, plus 10% (and this amount will also be the maximum amount that could be prescribed under (i) above). In addition to the administrative penalty, the obligation to surrender permits to meet any shortfall will continue under a 'make-good' requirement, with permits to be surrendered in the next compliance year.
Reporting (WP Chapter 7)	NGERS will provide the framework for monitoring, reporting and assurance under the Scheme, and elements of that system will be strengthened to support the Scheme.

(Continued)

Design feature	White Paper policy
	NGERS will be expanded to impose reporting obligations on all types of legal entities with obligations under the Scheme, including unincorporated entities and entities with upstream obligations. Provisions relating to documentation and record keeping under the Scheme will be those set out under the National Greenhouse and Energy Reporting Act 2007 (NGER Act). Initially, reporting will remain once annually. The Government will consider whether more frequent than annual reporting of emissions is justified following initial experience with the Scheme. Large emitters (emitting 125,000 t CO_2e pa or more) will need to have their NGERS reports audited by independent auditors. The auditors will need to be registered.
Taxation issues	
Income tax issues (WP Chapter 14)	Discrete code will be developed for the income tax treatment of permits. Any proceeds received on the sale of a permit will be assessable income. The cost of acquiring a permit will be deductible when the taxpayer starts to hold the permit. However, when the permit is held at year end, a 'rolling balance' method will be used for the purpose of deferring the timing of the deduction. That is, if a permit is held at year end, the deduction will effectively be reversed by way of the inclusion of an equal amount in assessable income—deferring the deduction until the permit is surrendered or sold. Taxpayers will make an election to use either historical cost or market value to value permits held at the end of an income year. Taxpayers will be able to change valuation methods once during a transitional period of five years from the Scheme's commencement, after which no change will be allowed. *Free permits:* The value of permits allocated for free to coal-fired electricity generators that are on hand at the end of the income year they are received will be included in the taxpayer's assessable income through the rolling balance mechanism referred to above. The value of permits allocated for free to EITE entities will be valued at zero at the end of an income year ending before the last surrender date for the emissions year for which they were issued. If free permits are held at the end of a

Design feature	White Paper policy
	later income year, the permit will be included in the taxpayer's assessable income through the rolling balance mechanism. *Permits created from forestry or destruction of gas*: Created permits surrendered in the year they are issued will have no tax consequences. Created permits held at year end will be included in the entity's taxable income for that year (under the rolling balance method)—with a corresponding reduction in taxable income in the year the permits are finally used (sold or surrendered). Proceeds of selling the permit are assessable income. *Penalties:* A penalty imposed under the Scheme, including one imposed on a liable party for failing to surrender sufficient eligible compliance permits, will not be deductible. *Permits at capped price:* For the first five years of the Scheme, fixed price permits will be available for purchase between the final reporting date and the final surrender date for each emissions year. The cost of these permits will be deductible but they cannot be banked or traded, they can only be surrendered.
GST issues (WP Chapter 14)	The Australian Government will amend the Goods and Services Tax (GST) law to characterize carbon pollution permits and eligible Kyoto units for GST purposes as personal property rights (and not rights within the meaning of real property in the A New Tax System (Goods and Services Tax) Act 1999) to promote certainty. The normal GST rules will apply to Scheme transactions, including the input taxed treatment of supplies of financial derivatives of permits.
Transitional issues	
State Schemes (WP Chapter 15)	*Queensland Gas Scheme:* The Australian Government will continue to work with the Queensland Government to encourage the development of appropriate termination arrangements.

(Continued)

Design feature	White Paper policy
	NSW GGAS: The Australian Government will continue to seek an agreement with the NSW and ACT governments on GGAS termination. However, if agreement is not reached on this approach, the Australian Government will consider providing some limited assistance for the benefit of GGAS participants, with priority given to adversely affected cogeneration (being rewarded for avoided methane creation), landfill gas and waste coal mine methane generators, and, as a lesser priority, to holders of unused NGACs.
	The Australian Government will also allow GGAS forestry projects to opt into the Scheme, provided they meet the Scheme eligibility requirements.
Early action (WP Chapter 15)	A program for allocating early action credits will not be established as companies that have taken action ahead of the introduction of the Scheme may already be benefiting from lower input costs and will continue to benefit under the Scheme through avoidance of higher energy costs or Scheme liabilities.
Cost pass-through (WP Chapter 15)	No cost pass-through legislation will be enacted. Cost pass-through in relation to energy is to be addressed primarily by the Ministerial Council on Energy and the Australian Energy Market Commission. The Government wants to remove unnecessary, inefficient and distortional regulatory barriers to carbon cost pass-through.

21

Canada's Experience in Emissions Trading and Related Legal Issues

Gray E Taylor and Michael R Barrett

1. Introduction

The current Canadian climate change regulatory environment reflects the numerous and often competing forces and issues affecting Canada in the first decade of the 21st century.

In October 2008, the Conservative government was re-elected as another minority national government, but with a larger number of seats and a centre of support in Alberta and its neighbouring western Canadian provinces. This administration's proposed climate change legislation features an emissions intensity reduction mechanism for large industrial emitters, rather than a fixed-cap emission reduction mechanism. As discussed later, the Province of Alberta is the only jurisdiction in Canada that has put in force comprehensive emission reduction legislation for large industrial emitters and it is using a similar emissions intensity reduction mechanism; the Province of Alberta is the largest GHG emitting province in Canada and is the province most likely to see large increases in GHG emissions as the oil sands and coal-fired electricity facilities are further developed. Although the Conservative's plan for Canada has been criticized, the government was faced with the fact that the United States, its largest trading partner by far, and until recently the globe's largest GHG emitter, had not yet implemented any GHG reduction legislation. Moreover, the pre-2000 Liberal's plan had also incorporated an emissions intensity approach. With little incentive to coordinate its efforts with the rest of the world, the Conservative government began moving forward to implement its emission-intensity reduction regime plans.

Within weeks of the Canadian election, Senator Barack Obama was elected as the next president of the United States. President Obama's position on climate change represents a sea change in official United States views. Immediately after the United States election, Canadian national government representatives, including Prime Minister Stephen Harper, were reaching out to the newly elected President to offer to work with his administration on a North

American climate change pact.[1] A much more detailed analysis of President Obama's positions on GHG reductions will be provided in other chapters; however, it would appear that any United States regulatory efforts will include a fixed-cap emission reduction mechanism. The Canadian government's proposal to dovetail Canada's climate change regulatory scheme with that of the United States' has raised many questions, including whether a regulatory regime built on emissions intensity reductions can work with a fixed-cap regime or whether the emissions intensity reduction system will need to be abandoned.

Prior to the United States election, the Canadian government planned to implement its proposals by 1 January 2010; that date now appears much less certain. If it is the Canadian government's current intention to wait for United States legislative efforts to be outlined in some detail, implementation in January 2010 is very unlikely. Further, the entire effort might be abandoned in order to integrate Canada's GHG reduction regime with that of the United States.

Another major influence on Canadian GHG policy is, of course, the current economic climate. The 2008–2009 financial crisis and accompanying global economy contraction has caused many to take a hard look at the costs of implementing comprehensive GHG reduction regimes, and Canada is certainly no exception. In the weeks following the United States election and in the midst of a worsening economic environment, Canada's Environment Minster stated at the Bennett Jones LLP Lake Louise World Cup Business Forum in November 2008:

> We will not—and let me be clear on this—we will not aggravate an already weakening economy in the name of environmental progress.[2]

Notwithstanding the uncertainties described above, the national government's 2008 pre-election and economic crisis plans for GHG reduction exist in reasonable detail and could be implemented reasonably quickly if the government deems it appropriate. Alberta, as mentioned, has already passed comprehensive GHG reduction legislation and other Canadian provinces have also taken significant steps regarding GHG reductions. In December 2002, Canada ratified the Kyoto Protocol and in 2005, the Kyoto Protocol came into force. However, it is very likely that Canada will not meet its target under the Kyoto Protocol of reducing aggregate GHG emissions by 6% below 1990 levels during the 2008–2012 commitment period. The current national government has stated that Canada's Kyoto Protocol goals are unattainable due to the expectation that meeting Canada's these goals would result in unacceptable

[1] D Ljunggren, 'Canada Seeks Climate Pact with United States' *Reuters* (Ottawa 6 November 2008) <http://www.reuters.com/article/environmentNews/idUSTRE4A56KV20081106>, accessed 9 February 2009.

[2] K Cryderman, 'Environment Minister Places Economic Priority over Environment' *Calgary Herald* (Calgary 28 November 2008) <http://www2.canada.com/calgaryherald/news/story.html?id=1c00bef5-02a8-46e8-88bc-dcd7c46f8f2b>, accessed 9 February 2009.

costs and efforts that could plunge Canada into a recession.[3] This view is not shared by the opposition parties that combined to pass the Kyoto Protocol Implementation Act in 2007; the Kyoto Protocol Implementation Act is a piece of legislation that is designed to force the Canadian government to meet its Kyoto Protocol commitments. The national government's view is also not shared by many Canadians, some of whom launched an unsuccessful lawsuit seeking to force the implementation of Canada's Kyoto Protocol commitments using the Kyoto Protocol Implementation Act and international law as the basis for the suit.[4]

This chapter provides an analysis of current and proposed Canadian GHG regimes and issues. Section 2 of this chapter addresses the national government's proposed GHG emission reduction plan, and Section 3 reviews some of the more significant implemented and proposed provincial governments' GHG emission reduction plans. Section 4 discusses issues of significance to Canadian GHG emissions reduction policies, such as trade and trade law considerations and potentially inconsistent US and Canadian GHG emission reduction policies.

2. The Canadian National Government's Proposed GHG Emissions Reduction Plan

2.1 National Plan Regulatory Approach

In April 2007, the national government of Canada released 'Turning the Corner: An Action Plan to Reduce Greenhouse Gases and Air Pollution' (the 2007 Plan).[5] In March 2008, the national government released 'Turning the Corner: Regulatory Framework for Industrial Greenhouse Gas Emissions'[6] updating the 2007 Plan specifically with respect to industrial GHG emissions. The 2007 Plan, as modified by the March 2008 update, will be referred to in this chapter as the 'National Plan'.

[3] Environment Canada, 'The Cost of Bill C-288 to Canadian Families and Business' (Report) (19 April 2007, En4-78/2007E) <http://www.ec.gc.ca/doc/media/m_123/toc_eng.html>, accessed 9 February 2009.

[4] See Friends of the Earth Canada, 'Canada facing legal challenge for breaking federal global warming law' <http://www.foecanada.org/index.php?option=content&task=view&id=340&itemid=2>, accessed 15 June 2009.

[5] Environment Canada, 'Turning the Corner: An Action Plan to Reduce Greenhouse Gases and Air Pollution' (Action Plan) (2007) <http://www.ec.gc.ca/climat-climate/default.asp?lang=EN&n=A3CB096D-1>, accessed 9 February 2009.

[6] Environment Canada, 'Turning the Corner: Regulatory Framework for Industrial Greenhouse Gas Emissions' (2008, En84-60/2008) <http://www.ec.gc.ca/doc/virage-corner/2008-03/541_eng.htm>, accessed 9 February 2009 [hereafter: Turning the Corner].

The National Plan proposes to establish mandatory GHG[7] emission intensity reduction requirements beginning on 1 January 2010 for the following industrial sectors:

> Electricity produced from combustion; oil and gas (including upstream and downstream production, as well as, oil sands and pipelines); pulp and paper; iron and steel; iron ore pelletizing; smelting and refining; cement; lime; potash; and chemicals, which includes fertilizers.

GHG emissions from these sectors constitute approximately 30% of Canadian GHG emissions and 50% of GHG emissions from Canadian businesses.[8]

To achieve GHG emissions intensity reductions, regulated entities must reduce the quantity of GHG emissions per unit of production. Most other jurisdictions currently regulating or proposing to regulate GHG emissions, both in Canada and internationally, favour using a fixed cap on GHG emissions; conversely under the Bush administration's voluntary goals, the United States focussed on emissions intensity reductions, which have been suggested as an appropriate mechanism for developing countries. The National Plan has been criticized due to the possibility that GHG emissions will rise when the production growth of regulated entities exceeds the required intensity reductions. For example, if a regulated entity is required to reduce its GHG emissions per production unit by 5% and it complies exactly, its total GHG emissions will still grow if its total production increases by more than 5%, with all other factors being equal.[9] Under a fixed cap system, a 'hard' cap is placed on total emissions, thereby eliminating any effect of production levels for a complying entity. The GHG emissions reduction targets stipulated in the National Plan are intended to produce an absolute reduction of GHG emissions by 2020 of 20% below 2006 levels,[10] a goal which many critics argue is unlikely to be achieved.

The entities expected to be regulated under the National Plan will be large, final emitters. Each regulated facility must report its emissions to the national government authorities for purposes of establishing a baseline from which its reduction requirements will be based. All established regulated facilities would be required to reduce their GHG emissions intensity by 18% from 2006 levels for the year 2010 and 2% annually thereafter until at least 2020.[11] New facilities in a regulated sector will be required to reduce their GHG emissions intensity by

[7] The GHGs identified for regulation mirror those regulated under the Kyoto Protocol.

[8] Environment Canada, 'Canada's 2006 Greenhouse Gas Inventory' (Table) <http://www.ec.gc.ca/pdb/GHG/inventory_report/2006/tab_eng.cfm>, accessed 9 February 2009.

[9] The purpose of an emissions intensity system is to permit growth in production, while still addressing GHG emissions in some fashion. For Canada, this was particularly attractive for the development of the oil sands and fossil fuel based electricity production in Western Canada. Ironically, in times of declining production like the present, an emissions intensity reduction system would not bring relief from reduction requirements, despite falling absolute emissions.

[10] See Turning the Corner (n 6 above), s 6.

[11] See Turning the Corner (n 6 above), s 4.1.

2% annually starting in the fourth year after commercial operations are commenced.[12] For certain new facilities, a 'cleaner fuel standard' will be used to determine the baseline and the resultant emissions targets. The required reductions will be set using the cleaner fuel identified for that sector, irrespective of whether the new facility actually uses the cleaner fuel. Thus, a fertilizer manufacturer that used oil to provide heat would have its baseline and reduction targets set as if it used the much less GHG emissions intensive 'cleaner' fuel, natural gas. Therefore, a facility that does not actually use the cleaner fuel will have to reduce its emissions intensity to a lower number than would otherwise be the case.

Other notable features of the proposed National Plan's regulatory approach include the following.

2.1.1 No required reductions in uncontrollable 'fixed process' GHG emissions

It does not require reductions in uncontrollable 'fixed process' GHG emissions,[13] as such emissions are tied to production and no alternative exists. Thus, the only way to reduce them is to reduce production. An example of fixed process GHG emissions is the emission of CO_2 from limestone in the production of lime[14] or cement 'clinker'.

2.1.2 Deferral of cleaner fuel standard targets

It contemplates that the 'cleaner fuel standard' based targets will be deferred until 2018 for 'carbon capture ready' new facilities.[15] Facilities prepared to gamble on

[12] See Turning the Corner (n 6 above), s 4.4.
[13] See Turning the Corner (n 6 above), s 4.3.
[14] Note that the GHG emissions from the combustion of fuel to heat the limestone are *not* fixed process emissions. Turning the Corner (s 4.3) defines fixed process emissions as emissions that are:

1. from chemical processes that produce carbon dioxide emissions and are fixed to production; and
2. created in a process where:
    a. carbon that is chemically bound in the raw materials is removed from these materials to produce a carbon-free product (that is, less than 1% carbon by mass); or
    b. carbon is used to remove an undesired component from the raw material and where the raw material is not substitutable; or
    c. unintentional oxidation of hydrocarbon feedstocks results from the catalytic conversation of these feedstocks into products; or
    d. carbon dioxide entrained in ethane gas feedstock is removed and released to the atmosphere in order to process the feedstock.

Fixed process emissions do not include the result of:

a. combustion, where combustion is the exothermic reaction of a fuel with gaseous oxygen; or
b. a process that is for the purpose of reducing emissions of air pollutants from the facility; or
c. the release of formation carbon dioxide from the processing of crude oil or natural gas.

[15] See Turning the Corner (n 6 above), s 4.6.

being able to use carbon capture and storage (CCS) in and after 2018 can thus benefit from using higher GHG emission technology in the interim if the facility is built 'ready' to participate in CCS. CCS is discussed in more detail below.

2.1.3 Encouragement of new co-generation facilities

It intends to establish reduction targets that are designed to encourage new co-generation facilities (ie facilities that produce electricity in addition to useful heat). Emission reduction targets for co-generation facilities will use a baseline that reflects the emissions that would have been created if the heat and electricity had been produced separately.[16] For the emissions deemed to be due to heat production, the reduction target will be the same as other heat producing facilities, ie 18% below the level of 2006 assumed baseline emissions by 2010 with a 2% annual requirement thereafter and parallel requirements for new facilities, while for emissions deemed to be due to electricity generation, there would be no initial or annual reduction requirement.

The Canadian approach raises the issue of the interplay between an emissions intensity system and a fixed-cap system. The most significant difference between these two systems is the manner in which a regulated entity's compliance obligation is established. As outlined above, under an emissions intensity system, a regulated entity's baseline and its compliance obligations are established on a per production unit basis. Once these per unit values are established, the regulated entity's aggregate compliance obligation is determined by multiplying the per unit values by the production levels during the applicable compliance period. Compliance obligations under a fixed-cap system are established by allocating each regulated entity its share of the aggregate cap. While these two systems have different methods of establishing a regulated entity's compliance obligation, they are similar in that they both impose a compliance obligation measured in tonnes of CO_2e which can form the basis for 'demand' in an emissions trading system. However, in an emissions intensity system the atmosphere may not see any reduction in GHG emissions.

Another difference between the two systems is when GHG-based assets can be monetized and how resulting markets develop. In a fixed-cap system, regulated entities receive some or all of their allowances (by grant or by auction) before or early in the compliance period, thereby providing those entities with the ability to trade or otherwise monetize allowances if and when they choose to do so. In comparison, a regulated entity under an emissions intensity system does not crystallize performance credits (ie credits issued equal to the number of GHG tonnes by which the entity's emissions were under the cap imposed by the emissions intensity calculation) until after the period in which the credit was created, thereby delaying the entity's ability to trade or monetize the credit, arguably resulting in a less efficient market.

[16] See Turning the Corner (n 6 above), s 4.11.

Notwithstanding that an emission intensity system and a fixed-cap system have different methods of determining a regulated entity's compliance obligation, once the aggregate obligation is established, compliance under either system may be achieved through the use of offset credits. Both systems measure a regulated entity's compliance obligation in tonnes of CO_2e and therefore, both systems will allow a regulated entity to meet its compliance obligation by purchasing offsets. Consequently, we can see no reason why the Canadian government should not aggressively pursue a comprehensive domestic offset system in the near term, regardless of the type of GHG reduction mechanism ultimately adopted.

The economic downturn provides the backdrop for a final point about these two systems. As noted previously, the National Plan and its Alberta counterpart have been criticized due to the possibility that aggregate GHG emissions could rise, even when regulated entities are meeting their compliance obligations. Under a fixed cap, that outcome is not possible. This would appear to be a valid criticism of an emission intensity system when production levels rise. However if economic output contracts, a regulated entity's aggregate GHG emissions will also very likely fall. In a fixed-cap system, a contraction in production levels will likely mean that a regulated entity will be able to meet its GHG emissions reduction obligations without altering its current business practices. In an emissions intensity system, the regulated entity's per production unit compliance obligation remains, even if production levels fall. Further, one could reasonably argue that as production levels fall, the ability to squeeze GHG reductions out of units of production at acceptable costs becomes more limited.

2.2 National Plan Compliance Options

A regulated entity can comply with its emissions intensity reduction obligations under the National Plan using one or more of the following:

(1) internal abatement;

(2) credits from Technology Fund contributions;

(3) credits:

   (i) generated by regulated entities which have overachieved using internal abatement efforts; or

   (ii) issued as offset credits as a result of projects approved under the proposed Canadian domestic offset system;

(4) Certified Emission Reductions (CERs) from the Clean Development Mechanism (CDM); and

(5) credits issued for early GHG reduction actions in Canada.[17]

[17] See Turning the Corner (n 6 above), s 5.

As previously stated, an emissions intensity reduction mechanism requires related entities to reduce GHG emissions per production unit. Generally speaking, businesses tend to become more efficient at what they do over time; as a result at least some of the annual emissions reductions required under an emissions intensity regime should be achievable under a business-as-usual scenario in a stable economic situation.

Technology Fund credits are proposed to cost $15/tonne until 2012 and then $20/tonne or more (the price will increase at the same rate as nominal GDP). Technology Fund credits will be available for 70% of a facility's reduction requirements in 2010, but decrease to zero by 2018. Thus, the price of offset and performance credits is expected to rapidly escalate. Contributions to the Technology Fund will be invested in GHG reduction technology projects. The Technology Fund is designed to act as an incentive to support the development and deployment of technologies to reduce GHG emissions[18] but, as critics note, it will not result in an immediate reduction in GHG emissions, despite being treated as such for compliance purposes.

Under either the National Plan or under a North American integrated fixed cap regime, there is a clear place for a robust offset system. Under the proposed offset system contemplated in the National Plan (the 'Offset System'), individuals or entities will be permitted to engage in various projects to reduce unregulated GHG emissions and then claim one credit for each tonne of CO_2e reduced or removed by the project. The portion of the National Plan that addresses offsets is supplemented by a companion publication entitled 'Turning the Corner: Canada's Offset System for Greenhouse Gases'.[19] On 9 August 2008, the national government published a draft Guide for Protocol Developers[20] that sets out the parameters for preparing and getting approval for protocols for offset projects in Canada, including a 'fast track' process for protocols substantially similar to existing protocols or methodologies in certain established GHG emission reduction regimes, particularly Alberta and the CDM.[21] To be eligible for offset credits, GHG emission reductions or removals would be required to occur in Canada and the reductions would be required to meet the tests of being real, incremental, quantified, verified and unique[22] to result in offset credits. The project would need to have started on or after 1 January 2000, and credits would only be issued for reductions achieved after 1 January 2008. The process to create offset credits is expected to

[18] Each Technology Fund credit represents the purchaser's right to apply one tonne of CO_2e against its compliance obligation. However, unlike internal abatement efforts or domestic offset credits or CERs, Technology Fund credits are not the result of actual GHG emission reductions.

[19] Environment Canada, 'Turning the Corner: Canada's Offset System for Greenhouse Gases' <http://www.ec.gc.ca/doc/virage-corner/2008-03/526_eng.htm>, accessed 9 February 2009.

[20] Environment Canada, 'Guide for Protocol Developers' (Draft Guide) (9 August 2008) <http://www.ec.gc.ca/creditscompensatoires-offsets/default.asp?lang=En&n=7CAD67C6-1&offset=1&toc=show>, accessed 9 February 2009.

[21] Ibid, s 1.6.

[22] See (n 19 above) [75]–[76].

involve a nine-step project registration process followed by a seven-step reporting, verification and issuance process,[23] similar in many ways to the CDM CER issuance process.

The National Plan contemplates a one-time allocation of credits to firms that took verified 'early action' to reduce GHG emissions during the period 1992–2006. The aggregate of early action credits for all regulated firms will be limited to 15 megatonnes.[24] It is expected that claims for early action credits will vastly exceed the available quantity.

3. Significant Proposed and Implemented Provincial GHG Plans

3.1 Alberta

The Climate Change and Emissions Management Act (Alberta) (CCEMA) provides the statutory framework for North America's first compliance-driven GHG emissions reduction regime. The CCEMA provides that 'the specified gas emission target for Alberta is a reduction by 31 December 2020, of specified gas emissions relative to Gross Domestic Product to an amount that is equal to or less than 50% of 1990 levels'.[25] Alberta implemented its GHG emission reduction regime by enacting a regulation, the Specified Gas Emitters Regulation (SGER), with the initial compliance period beginning from 1 July 2007 to 31 December 2007 and each subsequent calendar year being a further compliance period.

The SGER, like the National Plan, uses an emissions intensity approach that ties targets to emissions per production unit. It seems likely that a powerful factor behind both Canada's and Alberta's adoption of an emission intensity reduction approach is the desire to continue development of fossil fuel resources, particularly oil sands deposits. Alberta, and to a lesser extent its provincial neighbour, Saskatchewan, are home to Canada's oil sands deposits. The extraction of oil from oil sands results in much higher GHG emissions than traditional drilling extraction methods.[26] If Canada were to adopt a fixed-cap approach, its ability to exploit its oil sands deposits might be significantly compromised.[27]

[23] See (n 19 above), 3.

[24] Environment Canada, 'Canada's Credit for Early Action Program' <http://www.ec.gc.ca/doc/virage-corner/2008-03/cea_eng.html>, accessed 9 February 2009.

[25] Climate Change and Emissions Management Act 2007 s 3(1) (Alberta).

[26] The Pembina Institute, 'Oil Sands Fever: The Environmental Implications of Canada's Oil Sands Rush' (Report) (November 2005) <http://pubs.pembina.org/reports/OilSands72.pdf>, accessed 9 February 2009.

[27] The logic applies equally to the use of coal to generate electricity—Alberta has vast resources of high quality coal.

It would appear that any proposed North American integrated GHG reduction regime will have to account for Canadian oil sands deposits in some manner. While some in the United States view oil extracted from oil sands deposits as 'dirty oil' due to its heightened GHG emission properties, the close proximity and relative security of the oil sands presents the United States a means to achieve greater energy security. The Canadian government's proposal to President Obama for a North American climate pact included a request for special positive treatment for oil sands products entering the United States.[28]

Under the SGER, regulated facilities are required to report GHG emissions and reduce GHG emissions intensity relative to a baseline and have a number of options to achieve compliance. In addition to a regulated facility's internal abatement efforts, SGER GHG reduction obligations can be satisfied through the application of any or all of: (i) 'emission offsets'; (ii) 'fund credits'; and (iii) 'emission performance credits'.[29] Emissions offsets would come from reduction or sequestration projects located in Alberta and only from emission reductions achieved since January 2002, while emission performance credits would arise from overachievement of required reductions by regulated entities. Fund credits may be purchased from the Alberta government's Climate Change and Emissions Management Fund at a price of $15/tonne and, unlike contributions to the Technology Fund contemplated in the National Plan which are limited initially to 70%, and then decline to zero by 2018, contributions to the Alberta fund can be used to meet *all* reduction requirements. Monies coming into the fund will be directed to strategic projects or technologies aimed at reducing GHG emissions in the province; an emphasis is expected to be placed on CCS projects and technologies.[30]

During the first compliance period ending 31 December 2007, approximately 100 regulated facilities needed the equivalent of 5.7 million tonnes of CO_2e reductions to meet their aggregate compliance requirements. Of that amount, 2.7 million tonnes were obtained using fund credits, 1.7 million tonnes were achieved through internal abatement, and 1.3 million tonnes were obtained through purchased emission offsets and emission performance credits. It will be interesting to see if fund credits continue to be the favoured compliance option in later compliance periods, particularly as the local offset market matures and in light of potential price increases for fund credits being contemplated by the Alberta government.

[28] S McCarthy and C Clark, 'Ottawa swoops in with climate change offer' *The Globe and Mail* (Toronto, 5 November 2008) <http://www.theglobeandmail.com/servlet/story/RTGAM.20081105.wclimate1106/BNStory/National>, accessed 9 February 2009.

[29] Specified Gas Emitters Regulation 2007 (Alberta).

[30] Climate Change and Emission Management Fund credits, like the national government's Technology Fund credits, represent the purchaser's right to apply one tonne of CO_2e per credit against its compliance obligations.

3.2 Ontario

Canada's largest province, Ontario, announced GHG emissions reduction targets in 2005 that are contained in 'Go Green, Ontario's Action Plan on Climate Change' (the 'Ontario Plan'). The Ontario Plan contains the following general goals:

(1) by 2014, a reduction to 6% below 1990 GHG emission levels;

(2) by 2020, a reduction to 15% below 1990 GHG emission levels; and

(3) by 2050, a reduction to 80% below 1990 GHG emission levels.

According to the Ontario Plan, 50% of the 2014 short-term target is to be achieved by closing Ontario's remaining coal-fired electricity generation plants or converting them to biomass fuel. Practical concerns have been raised, however, that with demand for electricity growing steadily in Ontario, insufficient plans may be in place for new base-load electricity generation (which is intended to include renewable energy and contemplates nuclear energy) to permit the closure of the remaining coal plants by 2014 so that conversion to biomass is being actively considered.

On 2 June 2008, the Premiers of the provinces of Ontario and its neighbour Quebec signed a Memorandum of Understanding (MOU) in which they agreed on behalf of their governments to collaborate on a Provincial and Territorial Greenhouse Gas Cap-and-Trade Initiative. In the MOU, Ontario and Quebec reject intensity-based greenhouse gas emission reduction schemes in favour of hard caps in order to provide 'sufficient certainty of real reductions'. On 18 July 2008, the Province of Ontario joined the Western Climate Initiative ('WCI'), a group consisting of three other Canadian provinces (British Columbia, Manitoba and Quebec) and seven states within the United States led by California, with the intent to create a hard-cap cross-border emissions trading and offset system as soon as 2012.[31] Ontario recently announced that it is considering having a hard cap trading system in operation as early as 2010.

Ontario, British Columbia, Quebec, and Manitoba represent a sizeable majority of the Canadian population and economic base and they find themselves ideologically aligned with the presumed US fixed-cap GHG reduction stance. It will be interesting to follow how the differences between the Canadian fixed cap regimes versus intensity reductions regimes are sorted out, particularly as the next United States administration's plans emerge.

3.3 British Columbia

In conjunction with its involvement with the WCI, British Columbia has also initiated a number of GHG reduction plans. In November 2007, British

[31] See Western Climate Initiative (February 2007) <http://www.westernclimateinitiative.org>, accessed 9 February 2009.

Columbia passed the Greenhouse Gas Reduction Targets Act (British Columbia) (the 'BC Act'), which came into force on 1 January 2008. The BC Act prescribes the following GHG emission reductions:

(1) by 2020, a reduction to 33% below 2007 GHG emission levels; and
(2) by 2050, a reduction to 80% below 2007 GHG emission levels.

The BC Act also requires 'public sector organizations' (including provincial government ministries and agencies) to be carbon neutral from 2010 onwards. Public sector organizations are to achieve carbon neutrality by pursuing internal abatement actions to minimize their GHG emissions and addressing any remaining emissions through the use of offsets.[32] The province's 2008 budget includes $130 million for capital and operating expenses to enable a carbon-neutral public sector by 2010.

In April 2008, the Greenhouse Gas Reduction (Cap and Trade) Act (British Columbia) was introduced by the British Columbia Minister of the Environment. It provides a proposed framework for a market-based cap-and-trade system, many details of which still need to be determined; however proposed compliance options would include government issued allowances, offsets generated in British Columbia and credits from other systems (eg CERs and offsets from other WCI jurisdictions).

On 19 February 2008, British Columbia's provincial government introduced a carbon tax which became effective on 1 July 2008. The tax applies to virtually all hydrocarbons. The tax is intended to be revenue neutral and the legislation requires the provincial government to show each year how the revenue raised will be returned to taxpayers.[33] British Columbia has also introduced legislation regarding a low carbon fuel standard similar to that of California.[34]

4. Specific Issues Related to Canadian GHG Regulation

4.1 Constitutional authority to regulate GHGs

There are differences between Canadian jurisdictions regarding both the magnitude of required GHG emission reductions and the manner in which such reductions are to occur. Given this backdrop, it would not be surprising if some form of challenge is forthcoming regarding a jurisdiction's ability to legislate GHG emission reductions.

[32] The British Columbia government has established the Pacific Carbon Trust as a crown corporation with the mandate to acquire GHG offsets that meet British Columbia requirements. An initial June 2009 requirement of 35,000 tonnes of such offsets has been set.

[33] See Carbon Tax Act 2008 (British Columbia).

[34] Greenhouse Gas Reduction (Renewable and Low Carbon Fuel Requirements) Act 2007 (British Columbia).

In Canada, the ability of the national and provincial bodies to regulate a particular area is dictated by the division of powers contained in the Constitution Act 1867, as amended (the 'Constitution Act'). The Constitution does not specifically assign authority to regulate GHG emissions or other related climate change areas to either the national government or to the provinces.

The distribution of legislative powers in Canada is set out in Sections 91 and 92 of the Constitution Act, which lists the subject matters over which the national government and provinces have legislative jurisdiction. In addition to the enumerated list, the national government has residual power to 'make laws for the Peace, Order and Good Government of Canada'.

Under Section 91 of the Constitution Act, the national government has the power to make laws in relation to 'criminal law' and criminal law can be broadly defined to include areas such as the regulation of GHG emissions.[35] The Canadian courts have defined 'criminal law' as having three elements: (1) a prohibition; (2) a penalty; and (3) a typically criminal purpose.[36] In *R v Hydro-Quebec*,[37] the Supreme Court of Canada dismissed a challenge by the Province of Quebec to the jurisdictional authority of the Canadian Environmental Protection Act, 1999 (CEPA), a federal law that regulates toxic substances. The majority held that the administrative procedure contained in CEPA for assessing the toxicity of substances culminated in a prohibition enforced by a penal sanction, and as a result, the regulatory scheme was sufficiently prohibitive to be a criminal law. Other cases have also upheld the view that the national government's criminal law power may be used to authorize sophisticated regulations in a broad range of areas.[38] In 2005 the national government assessed a number of GHGs under CEPA[39] and concluded they should be added to the list of 'toxic substances' subject to potential regulation under CEPA.

As noted above, the national government has the power to enact legislation that relates to the 'Peace, Order and Good Government of Canada'. The Supreme Court of Canada in *R v Crown Zellerbach*[40] upheld a national law that prohibits dumping at sea in coastal areas inside provincial boundaries. Certain provincial governments had challenged the ability of the national government to regulate in that area. In upholding the law, the court reasoned that the topic of marine pollution was a matter of national concern that is distinct from provincial jurisdiction. Many have observed that a similar argument could be advanced about the control of GHGs. Unlike some other legislative initiatives

35 Peter W Hogg, 'A Question of Parliamentary Power: Criminal Law and the Control of Greenhouse Gas Emissions' (CD Howe Institute, August 2008) <http://www.cdhowe.org/pdf/backgrounder_114.pdf>, accessed 9 February 2009.

36 Ibid.

37 [1997] 3 SCR 213.

38 See *RJR MacDonald v Canada* [1995] 3 SCR 1999 (Tobacco Advertising); *Re Firearms Act* [2000] 1 SCR 283 (Gun Control).

39 The specific GHGs assessed were: carbon dioxide (CO_2), methane (CH_4), nitrous oxide (N_2O), hydrofluorocarbons (HFCs), perfluorocarbons (PFCs), and sulphur hexafluoride (SF_6).

40 [1988] 1 SCR 401.

to regulate emissions which are arguably a local matter (eg smog or other particulates), the manner in which GHGs manifest themselves as a concern is at least national in scope and is realistically a global issue.

Under Section 92 of the Constitution Act, the provinces are granted the power to, among other things, legislate with respect to property and civil rights and the development of natural resources.

In *Friends of the Oldman River Society v Canada (Minister of Transport)*,[41] the Supreme Court of Canada determined that the appropriate approach in determining jurisdictional authority relating to environmental issues is to first look at the catalogue of powers delineated in the Constitution Act, and consider how they might be applied to meet or avoid the environmental concerns. When viewed in this manner, it may be possible that in exercising their respective legislative powers, both levels of government may affect the environment, either by acting or failing to act. For example, one of the heads of power enumerated under Section 92A is the power of a province to legislate with respect to the 'development, conservation and management of sites and facilities contained in the province for the generation and production of electrical energy'. If a province wishes to pass GHG emission reduction legislation concerning electricity generation facilities in that province, this provision may be sufficient authority to do so.

Until some legislative norms are established, industry sectors expected to be regulated will face additional exposure driven by the ambiguous allocation of authority, potentially justifying a dual-level approach. In an attempt to alleviate some of this concern, the national government has indicated its willingness to enter into equivalency agreements with provinces that have established their own GHG regulations provided the provincial GHG emission reduction standards contained in the provincial regime meet or exceed the analogous standards contained in the proposed National Plan. The negotiation and implementation of such equivalency agreements will be important in addressing the systematic uncertainty that the current patchwork approach could present.

A variation of the debate over the power to conduct GHG regulation is with respect to taxation. Section 91(3) of the Constitution Act provides the national government with the broad power to tax. Canadian jurisprudence has consistently stated that the national government's power to tax is tied to its ability to establish that its 'pith and substance' is revenue-raising.[42] Clearly a carbon tax would generate revenue (which could be used in a variety of climate change related ways); however to be constitutionally valid, any proposed national carbon tax would have to articulate revenue-raising, rather than climate change policy, as a primary objective. The second consideration regarding the national government's ability to tax carbon is whether the tax is within an area of provincial jurisdiction. For example, the ownership and taxation of fossil fuels have long

[41] [1992] 1 SCR 3.

[42] Nathalie Chalifour, 'Making Federalism Work for Climate Change: Canada's Division of Powers over Carbon Taxes' (2008) 22 National Journal of Constitutional Law 119, 214.

been established to be a provincial matter. This raises interesting, but difficult issues about the characterization of ownership of carbon. Is the atmospheric release of carbon dioxide, as a result of burning fossil fuels, still considered part of the fossil fuel? Or does it become a separate piece of property subject to other taxation jurisdictions?[43] It is interesting to note the national government has already successfully imposed taxes on fossil fuels on the basis that it is taxing the product after processing (and so it is no longer a natural resource).

4.2 Carbon capture and storage

Carbon capture and storage (sometimes called sequestration, or CCS), involves the process of capturing CO_2 from power plants and other large emitting facilities, compressing the gas to a liquid state, transporting it to suitable locations and injecting it into deep underground geological formations so as to isolate it from the atmosphere.

The 'decarbonization' of fossil fuels is an appealing scenario for many companies involved in Canadian energy-intension industries such as the oil sands, and indeed for Canada as a whole, given its status as both a large producer and consumer of fossil fuels. CCS is also an appealing scenario for many in the United States. CCS may also provide a reasonable basis for the United States to exploit its large coal reserves. The implementation of significant CCS initiatives in the oil sands could allay US concerns about GHGs emitted during the oil extraction phase and allow for a North American integrated GHG reduction regime that provides for stable or even increased oil sands development.

The IPCC estimates that approximately 90% of total CCS costs relate to the capture of CO_2.[44] The costs of transport and storage are much less onerous and can be further minimized by achieving economies of scale and locating close to potential storage sites.

The Alberta government and the national government are both strong proponents of CCS, despite reports that the cost per tonne may be as high as $200 CDN.[45] Notwithstanding its wide-ranging support among Canadian legislative bodies, there are legal uncertainties associated with CCS. The legal issues include the following.

4.2.1 Property rights

Property rights determine who will have access to a project site and are therefore a crucial aspect of any CCS project. The three main areas of property rights are

[43] Ibid, 184.

[44] Intergovernmental Panel on Climate Change, Special Report on Carbon Capture and Storage (CUP: New York, 2005) (Prepared by the Working Group III of the IPCC).

[45] Alberta Environment, 'Accelerating Carbon Capture and Storage in Alberta' (Interim Report) (30 September 2008) <http://www.energy.alberta.ca/org/pdfs/ccsinterimrept.pdf>, accessed 9 February 2009.

surface (injection of the CO_2), sub-surface (reservoir), and the CO_2 itself. Property rights issues may be addressed by contract prior to the initiation of each project and the contracting parties may determine which laws apply and how. However, property rights for CCS is still a relatively new issue, and standards for addressing this issue have not been clearly defined thus making it difficult to determine property rights in the long term. Clear title and transferable rights would facilitate a reliable operating environment and help to establish the chain of liability and responsibility in the event of CO_2 escape, migration, or other related problems.

4.2.2 Monitoring and verification

In order to provide additional project certainty, it is important that standards for the measurement, monitoring, and verification of injected CO_2 be developed so as to establish norms for the collection and reporting of information on containment, reactivity of CO_2 with surrounding materials, seismic activity, escape, and long-term storage, all of which are necessary for establishing liability in the event of leakage or disruption.

4.2.3 Liability

Liability for CCS projects is best understood by differentiating short-term liability and long-term liability concerns.

Short-term liability of CCS projects can be considered an operational liability; operational liability refers to the environmental, health and safety risks associated with the capture, transport, and injection of CO_2. The oil and gas industry has already considered operational liability at its facilities and therefore few new issues should arise when applied to CCS.

Long-term liability is an issue due to the expectation that CCS projects will have a lifespan measured in centuries. The practical ability to tie such liability to the underlying project (or the buyer of resulting credits) becomes untenable. It seems clear that for CCS projects to move forward in Canada, governments must assume some degree of the potential resulting liability.

4.3 International trade issues raised by GHG regulation

As introduced by the national government, the National Plan makes no reference to the possibility of introducing GHG-related trade measures. If the United States initiates comprehensive GHG emission reduction legislation, it is very likely to include requirements on imports from countries that do not implement 'comparable' GHG emission reductions to those of the United States. Adopting some form of GHG trade measures congruent with efforts in the United States may be critical to successfully implementing a North American integrated GHG emission reduction regime.

International trade treaties such as the agreement establishing the World Trade Organization (WTO)[46] and the North American Free Trade Agreement set out rules regarding the imposition of border measures affecting imports from parties to the agreements. Canada's free trade agreements (FTAs) generally incorporate the key WTO rules restricting the imposition of border measures on imports, as well as the main exceptions. Those key rules include:

(1) national treatment—which prohibits discrimination between imported and domestic goods;[47]

(2) tariff binding—which prohibits the imposition of import fees above the duty levels 'bound' by an agreement; [48]

(3) most-favoured nation treatment—which requires treaty parties not to discriminate between goods from different countries;[49] and

(4) prohibition on quantitative restrictions—including quotas on imports and exports.[50]

These rules apply in varying ways to the different policy options for GHG-related trade rules. For example, it is conceivable that a domestic carbon tax might be implemented at the border with respect to imports in a manner consistent with trade rules.[51] However, the challenges of setting tariff rates, taking into account the need to adjust for countries that have adopted their own GHG reduction measures, are fraught with complexity and uncertainty. Imposition of a carbon tariff to support a domestic cap-and-trade regime, however, would raise more fundamental concerns as to its legality under international trade rules since it does not entail implementation of an internal tax.

Another approach to GHG-related trade measures may rely on the specific exception in the international trade treaties relating to conservation. GATT Article XX(g)[52] is of particular significance in this regard since it is incorporated by reference in the North American Free Trade Agreement and Canada's other FTAs. Pursuant to this exception, measures 'relating to the conservation of exhaustible natural resources' are exempt from Canada's principal international trade obligations, provided that they are implemented 'in conjunction with

[46] The Marrakech Agreement Establishing the World Trade Organization (signed 15 April 1994) 1867 UNTS 154, 33 ILM 1144 (WTO Agreement). The WTO Agreement and its various annexes constitute the governing law of the WTO. The 1947 General Agreement on Tariffs and Trade (GATT) is incorporated into an updated GATT 1994 along with other decisions and agreements adopted by the WTO parties; See General Agreement on Tariffs and Trade 1994 (signed 15 April 1994) 1867 UNTS 187, 33 ILM 1153 (GATT 1994).

[47] GATT 1994, Art III. [48] GATT 1994, Art II. [49] GATT 1994, Art I.

[50] GATT 1994, Art XI.

[51] Under WTO law, internal taxes are subject to the national treatment obligation in GATT Article III rather than the 'tariff binding' provisions of Article II that commit WTO members not to increase duties on imports beyond a 'bound' rate. WTO law allows internal taxes to be applied or enforced at the border with respect to imported goods.

[52] It is particularly significant that GATT, Art XX(g) has been incorporated by reference in NAFTA, Art 2101(1) and in Canada's other FTAs.

restrictions on domestic production or consumption.'[53] Since the issue arises in the context of implementation of a Canadian GHG-reduction regime, the 'relating to conservation' criterion is arguably met.[54] However, Article XX imposes additional constraints on the use of this provision, including a requirement that the exempted measures not constitute a 'disguised restriction on international trade'.[55] This has been interpreted by the jurisprudence as requiring the measure to be: (i) 'primarily aimed' at conservation; (ii) related to the conservation of 'exhaustible' natural resources; (iii) implemented in a manner that takes into account local conditions in exporting countries; and (iv) implemented in an 'even-handed' manner as between trading partners.

This conservation-related exception imposes strict parameters on the kinds of trade measures that can be applied to imports from countries that do not implement GHG-reduction commitments. Most significantly, the conservation aim of the measure must be paramount. In other words, the trade measure must be focused primarily on the avoidance of 'leakage' rather than at leveling the competitive playing field. Although it may well have the latter effect, that is not on its own a lawful rationale for introducing restrictions on imports from countries that have not implemented GHG reduction commitments. Moreover, to the extent that a proposed measure is designed to address the competitiveness objective in a manner unrelated to the conservation purpose, it may well be tainted as a 'disguised restriction' on international trade and fail to qualify for the exception.[56]

Regardless of the approach to trade issues ultimately adopted, failure to comply with international trade obligations can have serious ramifications. The WTO Agreements and Canada's FTAs all contain dispute settlement mechanisms that allow Canada's trading partners to seek independent adjudication of allegations that a Canadian measure is inconsistent with trade rules. If a complaining partner is successful, it can impose trade sanctions on Canadian exports. Since many of the countries that have not yet adopted GHG-reduction commitments are among the most dynamic export-driven economies, a failure to comply with international trade rules could have potentially serious economic and political fall-out.

[53] GATT, Art XX(b) relating to measures 'necessary to protect human, animal, or plant life and health' may also be relevant, although the 'necessity' standard contained in this exception makes it more difficult to satisfy than the conservation-based exception in Art XX(g).

[54] There may be some uncertainty with regard to the specific exhaustible natural resource whose conservation is promoted by climate change regulation. However, the mounting evidence of the threat of destruction of natural habitats and extinction posed by climate change suggests that the requirement would likely be met on multiple bases.

[55] See *United States – Import Prohibition of Certain Shrimp and Shrimp Products* (12 October 1998) WT/DS58/AB/R (Shrimp–Turtle case).

[56] See *United States – Standards for Reformulated and Conventional Gasoline* (29 April 1996), WT/DS2/AB/R; *Shrimp–Turtle case* (n 45 above). See also Canada's Landing Requirement for Pacific Coast Salmon and Herring (16 October 1989) CDA-89-1807-01 (CUSFTA Panel Report interpreting GATT, Art XX(g) as incorporated in the 1989 Canada-US Free Trade Agreement). <http://www.worldtradelaw.net/cusfta18/salmon-cusfta18.pdf>, accessed 9 February 2009.

5. Canada's GHG Regulatory Future

As an integral part of the North American market, Canada is motivated to be part of any North American integrated GHG emissions reduction regime. As this chapter has highlighted, there are, however, competing and ideologically diverse positions within the provincial and national governments and as a result, predicting Canada's future is very difficult. Nonetheless, it is safe to assume that Canada, as well as (and perhaps along with), the United States will soon have a national comprehensive GHG emission reduction regime in place.

22

Carbon Law and Practice in China

Christopher Tung

1. Introduction

The pace of economic and social development in China,[1] and its adverse impacts on the environment, are unprecedented. No other country has experienced economic, social and environmental challenges simultaneously on such a massive scale. By many measures, China has succeeded in delivering real improvements to its people but the widespread destruction of the environment means that change is urgently needed to make the country's development sustainable.

China has had considerable success in establishing policy and legal bases for climate change action. A robust industry has grown up in support of Clean Development Mechanism (CDM) projects despite policy hurdles on ownership, pricing, and foreign investment, as well as post-2012 uncertainties. This chapter examines China's sustainable development and climate change policies and laws and their effectiveness in implementing China's international obligations under the United Nations Framework Convention on Climate Change (UNFCCC) and the Kyoto Protocol. It also considers selected carbon law and practice issues in China to date, and finally, considers how China will respond to the challenges, and contribute to the development, of effective climate change solutions after 2012.

2. Climate Change, Sustainable Development Policy, and the Implementation of the UNFCCC and the Kyoto Protocol

China has consistently sought to incorporate sustainable development principles into its policies and laws. While the practical benefits of this approach have not always been apparent, a clear example of success has been in the area of climate change. China has taken a broad range of steps in conservation, environmental

[1] Unless otherwise indicated, references to 'China' mean the People's Republic of China (PRC) and exclude Hong Kong, Macau and Taiwan. Similarly, references to 'Chinese' are to PRC individuals, entities or subjects.

protection, recycling, cleaner energy production and improvements in energy efficiency to meet emissions reduction objectives, so that a proper account of China's performance in combating adverse climate change must refer to these wider initiatives.

2.1 Agenda 21, China's sustainable development policy and law

The dynamic nature of sustainable development; requiring balance between economic, social and environmental considerations, poses considerable challenges to the design and implementation of policies and laws underpinned by it.[2] Laws and policies in China, as in many countries, have traditionally been designed to implement policy objectives in the economic, social and environmental spheres separately. Since 1994, in new laws and policies China's lawmakers have increasingly attempted to take into account the 'triple bottom line' of economic feasibility; environmental capacity; and social equity.[3]

China moved quickly after the Earth Summit held in 1992,[4] where Agenda 21 (A21) established a comprehensive blueprint of action to be taken globally, nationally and locally by the United Nations, its agencies, and governments in every area in which there are human impacts on the environment. In particular, Agenda 21 called on States to develop sustainable development strategies and to participate in the creation of sustainable development law.[5] The White Paper on China's Population, Environment, and Development in the 21st Century (China's A21) was approved by the State Council in 1994.[6] In the same year, the State Council issued a notice calling on government institutions at all levels to consider China's A21 as an overarching strategic guideline for the formulation of economic and social development plans, and particularly to integrate it into the Ninth Five-Year Plan and the Long Term Objectives for the year 2010

[2] The author strongly believes that the parameters of what constitutes sustainable development are now reasonably clear and that the very essence of sustainable development is to reconcile and balance competing economic, social and economic considerations in any form of decision making so as to produce a sustainable outcome or solution for the long term. The real challenge in practice is that the concept of sustainable development brings a new paradigm to the creation, interpretation and practice of the law which requires more fluidity than traditional jurisprudence and practice. Given the criticism that the law and lawyers have on occasions received that law can be inflexible, narrow in scope and even out-of-date, the dynamism that sustainable development and its associated legal principles can bring to legal practice should be welcomed. See n 11 below; see also H C Bugge and C Voigt (eds), *Sustainable Development in International and National Law: What did the Brundtland Report do to legal thinking and legal development, and where can we go from here?* (revised paperback edn, Europa Law Publishing: Groningen, The Netherlands, 2008) 523–9.

[3] Legal Studio, 'Climate Change Law and Sustainable Development in China' (2004) ch 3.

[4] The UN Department of Economic and Social Affairs, Division for Sustainable Development, Agenda 21 (UN Programme on Sustainable Development Rio de Janeiro 1992) <http://www.un.org/esa/sustdev/documents/agenda21/english/agenda21toc.htm>, accessed 31 March 2009.

[5] Agenda 21 chapters 37 and 39.

[6] See China's Agenda 21, 'Chapters and Programme Areas of the White Paper' <http://www.acca21.org.cn/indexe6.html>, accessed 18 March 2009.

relating to environmental protection[7] and into day-to-day management.[8] While China's A21 is the foundation of China's sustainable development strategy and a principal driver of sustainable climate change action, it is a policy document rather than a document with legal force.

2.2 Content of China's A21

China's A21 sets out China's overall strategies for sustainable development; and considers the sustainable development of society, the economy, and resources and the environment. Chapter 3 of China's A21 (Legislation for Sustainable Development and its Enforcement), and chapter 18 (Protection of the Atmosphere) are of particular relevance here.

Chapter 3 provides for the development of Chinese sustainable development laws and the incorporation of sustainable development principles into Chinese legislation.[9] Chapter 3, para 3.1 specifically states:

> Establishing laws related to sustainable development is a mechanism for formalizing and codifying strategies and policies for sustainable development . . .

The implementation of Chapter 3 has therefore given sustainable development and its objectives the force of law within the Chinese legal system.[10] Thus, while China's A21 itself does not have force of law, it has been the main catalyst for new legislation incorporating sustainable development and its objectives. Chinese law does not specifically define sustainable development or its core principles. Nevertheless, it is tolerably clear that the following principles of sustainable development law are accepted under Chinese policy and law:

- equity and eradication of poverty;
- common but differentiated responsibility;
- precaution.[11]

Chapter 18 specifically addresses climate change, with a focus on efforts to protect the atmosphere by developing a national programme for controlling greenhouse gas emissions, research and development, energy conservation and efficiency and follow up activities associated with the UNFCCC and Intergovernmental Panel on Climate Change.[12]

[7] China Development Gateway, 'The National Ninth Five-Year Plan for Environmental Protection and the Long-Term Targets for the Year 2010' (China's Environmental Plan and Targets) <http://en.chinagate.cn/english/536.htm> accessed 18 March 2009.

[8] See The State Council, 'Notice on Implementation of China's Agenda 21—White Paper on China's Population, Environment, and Development in the 21st Century' (4 July 1994) <http://www.qqhrhb.gov.cn/showlaws.aspx?id=14>, accessed 18 March 2009 (Chinese version only).

[9] Bugge and Voigt (n 2 above) 129–138.

[10] Below, Section 3.

[11] See International Law Association, ILA New Delhi Declaration on Principles of International Law Relating to Sustainable Development (ILA, London 2002); See also Cordonier Segger and Khalfan, *Sustainable Development Law, Principles, Practices and Prospect* (OUP, 2004) 99–102.

[12] China's A21, ch18 [18.31]–[.38].

In 1996, China adopted sustainable development as the key guideline and strategic goal for national social and economic development.[13] In 2003, the Chinese government launched the 'Programme for Action for Sustainable Development in China in the Early 21st Century'. The Programme lists the objectives, principles, priority areas and proposed measures for ecological conservation, environmental protection and the 'rational exploitation of resources'. The Programme aims to transform China's A21 into a phased, operational programme of priority projects in order to solve a number of pressing issues.[14]

2.3 Chinese policies on sustainable development

2.3.1 Five-year plans

Since adoption of China's A21, its objectives have been consistently incorporated into successive national five-year plans which direct the strategic course of Chinese policy. The current and key five-year plans in relation to sustainability and climate change priorities are:

(1) Eleventh Five-Year Social and Economic Development Plan (2006–2010)[15]

This is the overall national plan that guides the social and economic development of China from 2006 to 2010. Specifically, the plan outlines the goal of achieving a resource-conserving and environmentally friendly society.

(2) Eleventh Five-Year National Development Programme on Environmental Protection Law and Regulations[16]

This sets out in detail the legislative programme for the amendment of existing legislation as well as the introduction of new environmental and related legislation.

[13] On 4 June 2007, the NDRC issued China's National Climate Change Programme (CNCCP) pt 3 s 3.2. This document reaffirmed the commitment of China to implement sustainable development and to address climate change within the framework of sustainable development.

[14] For further details of progress on China's Agenda 21, see UN Department of Social and Economic Affairs, Division for Sustainable Development, 'China: National Information/ National Reports' <http://www.un.org/esa/agenda21/natlinfo/countr/china/index.htm>, accessed 6 January 2009.

[15] NDRC, 'The Outline of the Eleventh Five-Year Plan for National Economic & Social Development of the PRC' (Profile) <http://en.ndrc.gov.cn/hot/t20060529_71334.htm>, accessed 9 January 2009.

[16] Ministry of Environmental Protection of the People's Republic of China, 'Eleventh Five-Year National Development Programme on Environmental Protection Law and Regulations' (available in Chinese version only) <http://www.mep.gov.cn/plan/hjgh/sywgh/sywgh/200701/P020070110302581257297.pdf>, accessed 5 June 2009.

(3) National Eleventh Five-year Plan for Environmental Protection[17]
This is the detailed strategic plan setting out the policy priorities and focus of the Government in environmental protection between 2006 and 2010.

2.3.2 Major policy shifts

Major policy shifts include:

(1) Economic growth being given equal emphasis to environmental protection
This has been a gradual process. Until very recently, China has driven hardest on economic growth, and while at local or provincial level this is sometimes still the case, there is a move towards requiring that economic development should not be at the expense of the environment.[18]

(2) Environmental problems are intended to be tackled from the outset rather than treated after pollution has occurred
Traditionally, environmental protection involved penalizing polluters rather than banning polluting activities from the start, but there is now a stronger emphasis on vetting projects as potential polluters before environmental problems occur; and

(3) Broad range of measures to deal with environmental issues
There are to be a broad range of methods to deal with environmental issues, including the use of market-based mechanisms.

2.4 National targets

In the Eleventh Five-Year Plan, the Chinese government sets out objectives and targets to be met by 2010 in relation to the reduction of carbon dioxide and sulphur dioxide emissions and the improvement of air and water quality. It also outlines an ambitious objective of reducing energy consumption by 20% per unit of GDP and emissions of major pollutants (chemical oxygen demand and sulphur dioxide) by 10% from 2005 levels by the end of 2010.

In June 2007, the National Development and Reform Commission (NDRC) issued China's National Climate Change Programme (CNCCP).[19] The CNCCP sets out non-binding, but ambitious, targets to be achieved by 2010, including a restatement of the goal of 20% reduction of energy consumption per

[17] State Council of People's Republic of China, 'China's National Environmental Protection Plan in the Eleventh Five-Years (2006–2010)' (Plan) <http://english.mep.gov.cn/down_load/documents/200803/P020080306440313293094.pdf>, accessed 5 June 2009.

[18] Eg the inclusion of the improvement of environmental performance as a criterion in assessments to promote government officials.

[19] See NDRC, 'China's National Climate Change Programme' (June 2007) (Report) <http://www.ccchina.gov.cn/WebSite/CCChina/UpFile/File188.pdf>, accessed 6 January 2009.

unit of GDP and a 10% increase in the share of renewable energy in total primary energy supply.

In August 2007, the NDRC issued the Medium and Long-term Development Plan for Renewable Energy. This Plan outlines the general goals for developing renewable energy in the next 15 years. Its objectives include increasing the proportion of renewable energy in the total energy consumption; addressing the electricity and fuel shortage in rural areas; promoting the utilization of organic waste; and encouraging the industrial development of renewable energy technology. The Plan sets out China's renewable energy targets: an increase in the share of renewable energy in total primary energy consumption to 10% by 2010 and 15% by 2015.

The Plan also sets out quantitative targets for increasing generation from specific renewable sources. By 2020, China aims to increase hydro power generation capacity by 300 million KW; wind power generation capacity by 30 million KW; biomass power generation capacity by 30 million KW; and solar power generation capacity 1.8 million KW.[20] These targets have supported strong growth in the deployment of renewable energy in China and have facilitated the rapid implementation of renewable energy projects as CDM projects.[21]

3. Chinese Laws Supporting Climate Change Action

The CNCCP is China's first comprehensive policy initiative to address climate change. The CNCCP focuses on reducing China's energy consumption per unit of GDP through three key strategies: encouraging the use of renewable energy; improving energy efficiency; and promoting clean technology (such as technologies for the clean and efficient development and use of coal).

The CNCCP[22] states that China will:

Address climate change within the framework of sustainable development;
Follow the principle of 'common but differentiated responsibilities' of UNFCCC;
Place equal emphasis on both mitigation and adaptation;
Integrate climate change policy with other interrelated policies;
Rely on the advancement and innovation of science and technology; and
Participate in international cooperation actively and extensively.

[20] NDRC, 'Medium and Long-term Development Plan for Renewable Energy in China' (September 2007) (Abbreviated English Version, Report) <http://www.cresp.org.cn/uploadfiles/2/967/medium%20and%20long-term%20development%20plan%20for%20re%20in%20china%20eng.pdf>, accessed 31 March 2009.

[21] See 'The Value of Carbon in China: Carbon Finance and China's Sustainable Energy Transition' (July 2008) (Report) <http://wwf.org.hk/eng/pdf/references/pressreleases_hongkong/WWFcarbon_report_FINAL_20080630_ENG.pdf> accessed on 16 March 2009 (commentary on the relationship between national policies, renewable energy targets and the Chinese CDM market).

[22] See CNCCP part 3 s 3.2.

3.1 White Paper

The Chinese State Council Information Office issued a White Paper entitled 'China's Policies and Actions for Addressing Climate Change' (Climate Change White Paper).[23] It is a high level policy statement setting out China's position on combating climate change and summarizing work done thus far. The Climate Change White Paper reiterates the targets previously set out in the CNCCP and Eleventh Five-Year Plan. However, while stating China will place equal emphasis on both mitigation and adaptation, it states that adaptation is a more immediate concern. This suggests a subtle shift of emphasis to more action on adaptation, although there is as yet no specific legislation dealing with adaptation. Rather, Chinese policymakers rely on relevant existing legislation to support adaptation efforts by government agencies around China.[24]

3.2 Chinese laws relevant to action on climate change

A brief outline of key Chinese national laws relevant to combating climate change with comments on support for adaptation where applicable follows.

3.2.1 PRC Law on the Prevention and Control of Atmospheric Pollution

Promulgated by the National People's Congress (NPC) on 29 April 2000, it came into force on 1 September 2000. It aims at preventing and controlling air pollution, protecting and improving the environment, safeguarding human health, and promoting sustainable development.

3.2.2 PRC Cleaner Production Promotion Law

Promulgated by the NPC on 29 June 2002 and came into force on 1 January 2003. The law aims at promoting clean production, increasing resources utilization efficiency, reducing pollutant production, protecting and improving the environment, safeguarding human health, and promoting sustainable development.

3.2.3 PRC Environmental Impact Assessment Law

Promulgated by the NPC on 28 October 2002 and came into force on 1 September 2003. It establishes guidelines for assessing planning and construction projects in order to prevent or mitigate unfavourable environmental impacts.

[23] Information Office of China's State Council, 'Chinese Policies and Actions on Climate Change' (29 October 2008) (White Paper) s VII <http://www.ccchina.gov.cn/WebSite/CCChina/UpFile/File419.pdf>, accessed 6 January 2009.

[24] See Section 3.2 below

3.2.4 PRC Law on Desert Prevention and Transformation

Promulgated by the NPC on 31 August 2001 and came into force on 1 January 2002. The law aims at tackling desertification and promotes sustainable development. Article 3 addresses the transformation of desert land, which arguably provides a legal basis for adaptation measures to eliminate or reduce the impacts of adverse climate change caused by desertification.

3.2.5 PRC Flood Control Law

Promulgated by the NPC on 29 August 1997 and came into force on 1 January 1998. It aims to control and mitigate the damage done by floods. Again, as with measures for the transformation of deserts above, measures to control flooding may support climate change adaptation efforts.

3.2.6 PRC Water Law

Promulgated by the NPC on 29 August 2002 and came into force on 1 October 2002. This law is aimed at the rational development, utilization, preservation, and protection of water, for the prevention and control of water disasters, and for sustainable utilization of water resources in order to meet the needs of national economic and social development. As this law deals with the sustainable utilization of water resources and the prevention and control of water disasters, it provides a legal basis for adaptation action in response to regions with increasingly arid land and constrained water supplies. The South to North Water Transfer Project is a major water project being implemented under this law by the Ministry of Water Resources and can be considered to be one of the single largest projects related to adaptation now underway in China.[25]

3.2.7 PRC Forest Law

Promulgated by the NPC on 20 September 1984 and revised on 29 April 1998. The law aims at protecting and rationally utilizing natural resources in forests.

3.2.8 PRC Environmental Protection Law

Promulgated by the NPC on 26 December 1989 and came into force on the same day. It is the principal law which sets out the basic principles and requirements for environmental protection in China.

[25] See PRC, Ministry of Water Resources, Bureau of South-to-North Water Transfer Planning & Design 'South-to-North Water Transfer Project' <http://www.mwr.gov.cn/english1/20040827/39304.asp>, accessed 14 March 2009. China has unevenly distributed water resources, increasingly severe and long dry seasons in the North of China have strained water resources requiring water diversions from the South.

3.2.9 The Measures on Administration and Operation of Clean Development Mechanism Projects

Promulgated jointly by the Ministry of Foreign Affairs, Ministry of Science and Technology, Ministry of Finance and NDRC on 12 October 2005.

3.2.10 Energy Conservation Law

Promulgated by the NPC on 1 November 1997, revised 28 October 2007 and came into force on 1 April 2008. It aims at promoting energy conservation, enhancing energy utilization efficiency, protecting and improving the environment, and promoting sustainable economic and social development.

3.2.11 Renewable Energy Law

Promulgated on 28 February 2005 and came into force on 1 January 2006. It is China's first law which regulates the use of renewable energy. The key goals of the law include:

- promoting the development and utilization of renewable energy;
- improving China's energy structures and diversifying China's energy supplies;
- safeguarding energy security; and
- protecting the environment and facilitating the sustainable development of China's economy and society.

3.2.12 Circular Economy Promotion Law

Promulgated by the NPC on 29 August 2008 and came into effect on 1 January 2009. It aims to promote the development of the circular economy, improve resource utilization efficiency, protect and improve the environment, and realize sustainable development.

3.3 Draft Energy Law

Following the release of the CNCCP, the Office of the National Energy Leading Group (a high-level entity in charge of energy strategy within the NDRC, set up in 2005) released a draft Energy law for public comment in December 2007. The consultation process closed on 1 February 2008. Although the law was expected to be passed in early 2009, it now appears that this will not occur until 2010.

The proposed Energy Law aims to provide a comprehensive legal framework for China's energy strategy. It will cover all forms of primary energy (including coal, oil, natural gas and renewable) and secondary energy (eg electricity and thermal power). Principles of sustainable development have been incorporated into the proposed Energy Law. Article 20 of the proposed Energy Law provides that 'The national energy strategy is the overall strategy that plans and guides the sustainable development of the nation's energy resources and safeguards its energy security.' The proposed Energy Law also aims at tackling climate change by encouraging the use of renewable and low-carbon energy. The proposed Energy Law is intended (like the Renewable Energy Law) to encourage the development of clean and low carbon energy, but more significantly, it is the first Chinese law to specifically address and define high-carbon and low-carbon energy,[26] and to put renewable energy in context with other energy types.

4. The Effectiveness of China's Implementation of the UNFCCC and the CDM

China is a party to the UNFCCC and the Kyoto Protocol. As a developing country, China has no binding emission reduction targets under the Kyoto Protocol. However, it can be seen from the range of sustainable development and climate change initiatives undertaken by the Chinese government outlined above that China has made substantial efforts to comply with its obligations under the UNFCCC. While China initially proceeded cautiously in the implementation of the CDM, it quickly built institutional capacity and expertise, emerging in 2007 as the largest supplier of Certified Emission Reductions (CERs) from emission abatement projects under the CDM. This dominant position was attained with the help of dedicated and co-ordinated policies and laws, designed to support climate change action and the development of CDM projects in China.[27]

4.1 Policies and laws regulating CDM projects in China

The Instrument 'Measures on Administration and Operation of Clean Development Mechanism Projects', which took effect on 12 October 2005,[28] is the principal law relating to the CDM in China. The CDM Measures aim to facilitate the implementation of CDM projects in China in accordance with the UNFCCC and the Kyoto Protocol. They also provide a specific legal basis for CDM projects in China. Article 4 of the Measures sets out priority areas for CDM projects in China, namely, energy efficiency improvement,

[26] See PRC New Energy Law arts 5, 36, 37 and 139; see also CNCCP pt 1 s 1.3.2 (optimizing the energy mix by developing low carbon and renewable energy).
[27] See Sections 3 and 4.
[28] Superseding the Interim Measures for Operation and Management of Clean Development Projects in China promulgated on 31 May 2004.

development and utilization of new and renewable energy, and methane recovery and utilization.[29]

Articles 6 to 12 set out the conditions for the approval of a CDM project; Articles 13 to 17 detail the institutional arrangements for the management and implementation of a CDM project, including the designation of the NDRC as the national authority for CDM project activity and its main areas of responsibility.

Articles 18 to 20 describe the implementation procedures for the CDM.

4.2 China's CDM Policy

Although Chinese government policy actively supports the CDM,[30] China is unique among host countries in imposing a number of restrictions and requirements on CDM projects. These include a floor price for CERs; a restriction on the sale of post-2012 CERs; a levy of 2% to 65% on CER revenues; a ceiling on the number of CERs that may be transferred to the buyer in any transaction; and a requirement that the CDM project entity be under Chinese control.

4.2.1 Floor price for CERs

From the outset, the NDRC has imposed an unwritten floor price on the sale of CERs, which has been increased with time. The NDRC requires the purchase agreement for the CERs to be submitted with the application for approval of a CDM project, and the floor price is enforced through a refusal to approve projects where the purchase price is unacceptably low. This floor price varies with the technology used in the CDM project. For most projects, the floor price reportedly evolved from €8 to €9 through 2007, dropping back to €8 recently.[31] A separate and higher floor price is applied to wind power CDM projects. While the imposition of a floor price can protect inexperienced Chinese project owners from aggressive intermediaries, it also prevents legitimate discounting of CERs from high-risk projects. This arguably discourages foreign investment in more risky and emerging emission reduction projects, resulting in the use of more

[29] Unlike CDM host countries such as India, Indonesia and Thailand, China makes no reference to any sustainable development criteria in its CDM project approval process. The general understanding is that projects in the priority areas that are consistent with sustainable development strategies (art 6) and policies of China will be approved. The application form for CDM projects requires details of the project's contribution to sustainable development to be set out; see Clean Development Mechanism in China, 'Format for CDM Project Application in China' (updated 20 May 2008) <http://cdm.ccchina.gov.cn/english/NewsInfo.asp?NewsID=2657>, accessed 5 January 2009.

[30] See CNCCP pt 1 s 1.3. See also Section 9 of this chapter for comments on China's position on the CDM post-2012.

[31] Karan Capoor and Philippe Ambrosi, 'State and Trends of the Carbon Market 2008' (The World Bank, May 2008) 31 <http://siteresources.worldbank.org/news/resources/state&trendsformatted06may10pm.pdf>, accessed 9 January 2009.

established technology (a potential reason for the lack of technology transfer in the Chinese CDM market. The control restriction is arguably another deterrence to technology transfer in Chinese CDM projects).

4.2.2 No sale of CERs post-2012

It is NDRC policy to discourage Chinese project owners from selling CERs post 2012 (notwithstanding current uncertainties in international negotiations). Rather, the NDRC approves projects with emission reduction sale and purchase agreements (ERPAs) that contain a right of first refusal over post-2012 CERs or emission reductions.

4.2.3 CER income

Article 24 of the CDM Measures provides that income from the sale of CERs is to be shared by the project owner and the Chinese government. The ratio in which the income is shared is in accordance with the 2% to 65% levy mentioned above. This requirement means that a third party cannot share directly in the proceeds of the sale of CERs. However, by assigning some of the income to a third party, the project implementing entity may share the proceeds indirectly with that third party (which is consistent with Chinese contract law). In other words, if the income is first received by the project implementing entity, it is acceptable for a third party to ultimately benefit from those CERs. The practical effect of Article 24 is to preclude intermediaries from taking a share of CERs or receiving CER income directly upon the issuance of CERs to an Annex B buyer.[32]

4.2.4 The Chinese Control Rule

Article 11 of the CDM Measures requires a company implementing a CDM project to be Chinese funded or owned. In practice, this has been interpreted by the NDRC to mean the CDM project entity must be:

- a Chinese domestic enterprise;[33] or
- an Equity Joint Venture, which is a Foreign Invested Enterprise.[34]

The policy of the NDRC is to require a minimum of 51% of the project entity's shareholding to be owned and controlled by Chinese individuals or entities. As

[32] For a more detailed discussion of the rights to transfer CERs and benefit from CER revenue, see Section 7.4 below .

[33] Limited liability companies (which include wholly state owned enterprises), companies limited by shares, and unlimited liability entities; See PRC Company Law and General Principles of Civil Law of the PRC art 3.

[34] Two other types of Foreign Invested Enterprises are Wholly Foreign Owned Enterprises (WFOEs) and Cooperative Joint Ventures (CJVs); see Law of the PRC on Sino-Foreign Co-operative Enterprises. WFOEs and CJVs currently cannot be used as CDM project implementing entities.

such, foreign shareholding cannot exceed 49%. It is not clear whether this policy applies to both direct and indirect holdings in the project entity.

4.2.5 CDM levy

The levy imposed on CER revenues both allows the State to share the benefits from the implementation of CDM projects and secures a source of funding for sustainable development activities and projects. It also acts as a disincentive to CDM projects that contribute less to sustainable development, as these are subject to the highest levy of 65%. Levies collected are transferred to the CDM Fund—a State-owned fund established in November 2007 to finance the country's efforts to improve energy efficiency and environmental protection. The CDM Fund encourages the use of clean energy through various means, including grants and privileged loans. It is managed by the Ministry of Finance, and funded through grants from financial institutions such as the World Bank and the Asian Development Bank as well as the CDM levies.

Based on figures at the end of November 2007, 885 Chinese CDM projects were expected to supply 1.5 billion metric tons of carbon credits and generate US$15 billion (at US$10 per CER). US$3 billion in levies was estimated to be injected into the China CDM Fund. Although the levies have apparently been collected and transferred to the CDM Fund, there has been very limited application of its funds, as detailed measures on the collection of the CDM levy and the application of funds by the CDM Fund are under preparation.[35]

4.2.6 Enterprise income tax relief

On 23 March 2009, the State Administration of Taxation and the Ministry of Finance issued Cai Shui [2009] No. 30 (Circular 30), which addressed enterprise income tax incentives to Chinese companies engaged in the sale of CERs. Circular 30 confirms that the following portion of revenue, to be paid to the Government from the transfer of CERs, for the following projects is tax deductible:

- the PRC Government's 65% share of revenue for hydro fluorocarbon (HFC) and per fluorocarbon (PFC) projects
- the PRC Government's 30% share of the revenue for nitrous oxide (N_2O) projects, and
- the PRC Government's two per cent share of the revenue for energy efficient improvement, development and utilization of new and renewable energy, recovery and utilization of coal bed methane and CDM reforestation projects.

[35] Enquiry with the China CDM Fund Office (Personal Communication, January 2009).

Also, a three plus three scheme applies from the first year the project generates sale revenue, so that projects in the first two categories above are eligible for a three year tax exemption and a 50% reduction in the tax rate for the fourth to sixth years of the project.

Further, Circular 30 provides for certain types of revenue of the CDM Fund (the Fund) to be exempt from enterprise income tax.

Circular 30 applies retroactively from 1 January 2007 but the exact manner in which this will be carried out is yet to be determined.

5. The Emerging Role of Hong Kong

5.1 Climate change and sustainable development policy in Hong Kong

Hong Kong has no consolidated climate change policy, instead relying on a range of measures in environmental protection, energy efficiency and power generation to partly tackle climate change issues. As an Asia-Pacific Economic Co-operation (APEC) member, Hong Kong has pledged to reduce energy intensity by at least 25% by 2030 (with 2005 as the base year). In September 2007, together with 20 other APEC economies, Hong Kong signed the Sydney Declaration on Climate Change, in which it agreed to address climate change by strengthening co-operation in areas such as energy efficiency, forest protection, energy research and renewable fuels.[36]

Three recent developments show promise for increased carbon market activity in Hong Kong. In January 2008, Hong Kong Exchanges&Clearing Limited announced that it would launch a number of carbon trading initiatives in the near future; in June 2008, the Environmental Protection Department (EPD) announced formal arrangements for the implementation of CDM projects in Hong Kong; and in April 2009, the Hong Kong Secretary for Environment announced qualification criteria for Hong Kong companies to be considered as Chinese enterprises under the CDM Measures.

5.2 Implementation of CDM in Hong Kong

The UNFCCC and the Kyoto framework were extended to Hong Kong by the Chinese government in 2003. As part of a non-Annex 1 country, Hong Kong has no emission reduction targets. On 6 June 2008, the Hong Kong EPD announced formal arrangements for CDM projects in Hong Kong.[37] Under the

[36] The Government of Hong Kong, Environmental Program Development, 'Climate Change: Cooperation with the International Community' <http://www.epd.gov.hk/epd/english/climate_change/coop_int.html>, accessed 5 January 2009.

[37] The Government of Hong Kong, Environmental Program Development, 'Climate Change: Arrangements for the Implementation of Clean Development Mechanism (CDM) Projects in the Hong Kong Special Administrative Region' <http://www.epd.gov.hk/epd/english/climate_change/cdm_project.html> accessed 8 January 2009.

Implementation Arrangements, the EPD is appointed as a liaison agency for CDM projects in Hong Kong. It will collect applications for Letters of Approval for CDM projects implemented in Hong Kong. The applications are then forwarded to the NDRC for processing and approval. Representatives of the EPD will also participate in the work of the National CDM Board, which is responsible for finial review of CDM approval applications.

The Implementation Arrangements are a significant breakthrough in that they pave the way for the implementation of CDM projects in Hong Kong. However, at the time of writing, no project has been presented to the EPD for approval.[38]

5.3 Hong Kong investors in Chinese CDM Projects

Hong Kong investors are currently treated as foreign investors under Chinese law.[39] As such, a Hong Kong investor cannot fully own and control the CDM project entity. Hong Kong investors have to date invested in Chinese CDM projects like any other foreign investor. The Hong Kong EPD and the NDRC have been in discussions for several years to agree arrangements that would give Hong Kong investors the right to wholly own CDM project entities. It is understood that part of the difficulties encountered in negotiations between the Hong Kong EPD and the NDRC is over the qualifications required for a Hong Kong company to control a Chinese CDM project. Issues include the track record of the Hong Kong company, its profitability, level of capital, and extent of Hong Kong management and shareholding. One possible solution to granting Hong Kong investors preferred access to the Chinese CDM market is through an extension of, or a similar approach to, the Closer Economic Partnership Arrangement (CEPA),[40] as CEPA is an existing legal framework that gives Hong Kong service providers and investors the ability to set up wholly-owned enterprises in sectors such as retail and travel in Mainland China.

On 8 April 2009, the Hong Kong Secretary for Environment announced that Hong Kong companies would be recognized as Chinese enterprises under the CDM Measures. Three criteria were outlined for this purpose:

- the company should be registered and established in Hong Kong;

[38] While the author is aware of a number of potential Hong Kong CDM projects, prospective project developers had been expecting the Implementation Arrangement to have been issued for over two years. This delay will mean that only the most determined and experienced CDM developers will be able to implement CDM projects in Hong Kong in the first commitment period (2008–2012).

[39] Investors from Macau and Taiwan are also treated as foreign investors.

[40] The Government of the Hong Kong Special Administrative Region, Trade and Industry Department, 'Mainland and Hong Kong Closer Economic Partnership Arrangement (CEPA)' <http://www.tid.gov.hk/english/cepa/legaltext/cepa_legaltext.html>, accessed 7 January 2009.

- its executive director should be a Chinese national or Hong Kong permanent resident and at least half of the members of the board of directors must be Chinese nationals or Hong Kong permanent residents, and
- in the case of a publicly traded company, at least 50% of its shares must be non-tradable

While implementing arrangements still have to be issued and there are problems in respect of the applicability of non-tradable shares to listed companies under Hong Kong law, this development is a step in the right direction to ramp up Hong Kong investment in China CDM projects and carbon finance. For existing and prospective investors in China CDM projects, real consideration can now be given to using Hong Kong companies to gain full control of a CDM project vehicle.

6. The Practical Effect of China's Climate Change Policy at National, Provincial, and Local Levels

As can be seen from the above analysis, the Chinese government has undoubtedly created a strong foundation for action against adverse climate change within a sustainable development framework. Beyond the clear success of the CDM, with the creation of Chinese carbon law and practice (discussed below) and the diffusion of CDM projects and associated benefits around China, the country has made significant progress in reducing greenhouse gas emissions by controlling pollution and increasing energy efficiency. These improvements have by and large been driven by existing sustainable development policy and environmental laws rather than specific climate change or low carbon initiatives. Indeed, concerted action under the banner of climate change policy really started in earnest in 2007 and 2008 with the issue of the CNCCP and a number of provincial climate change programmes.[41] As such, at provincial and local levels, governments are still formulating detailed climate change action plans and carrying out local research to identify pressing issues such as adaptation to the climate vulnerabilities of their provinces and localities (agriculture, coastal zones, severe natural disasters and water resources). What the CNCCP and provincial programmes are effectively doing is to draw climate change mitigation and adaptation action under a common umbrella across China. Below, we consider China-specific approaches to carbon risk, the development of a low carbon economy, and the emerging concept of establishing low carbon development zones around China. These developments are the real beginning of tangible outcomes from the implementation of national and provincial climate change policies.

[41] Such as Fujian, Guizhou, Jilin, Jiangxi, Heibei, Qinghai, Xinjiang and Yunnan. Guangdong Province is preparing a climate change action plan.

7. Carbon Law and Practice Fundamentals with Chinese Characteristics

7.1 Back to first principles

While Chinese climate change policy and law are in place to facilitate the CDM, practice has shown that there are numerous challenges in the application of those polices and laws. In this part, we focus our discussion on carbon law, which may be described as the body of law that deals with the creation, trade and disposal of emissions reductions or carbon assets. This contrasts with climate change law, which goes beyond the creation and trading of carbon assets and covers the full range of mitigation and adaptation activities as well. Carbon law does not consist of a single discipline of law but constantly draws on principles of international and domestic law, civil law and common law, public and private law, and specific administrative, contract, equity, environmental, and property law concepts. Where the practice of carbon law is particularly unique is the frequent interface between international and domestic laws to implement what are essentially commercial transactions between private entities, and public and private entities.

7.2 A pure legal fiction and creation of policy

Policy and legal schemes at international, regional and national levels aimed at reducing greenhouse gas emissions have introduced market-based mechanisms for combating adverse climate change. These schemes are broadly divided into emissions trading and emissions reduction projects. The introduction of these market-based mechanisms has required the use of a unit of measure for accounting and trading purposes based on the concept of CO_2 equivalence which was already in use scientifically. A tonne of carbon dioxide equivalent (tCO_2e) emerged as the unit of measure for tradable emission reductions. From a legal perspective, the emission reduction unit is generally defined as a verified tCO_2e measured against a baseline (in a baseline-and-credit scheme), or the legal right to emit one tCO_2e of the relevant greenhouse gas under a cap-and-trade scheme. This emission reduction unit is also commonly called a carbon unit or carbon asset, as well as a new commodity.

7.3 The carbon asset: not a commodity and not all equal

There is no definitive statement on the legal status of the carbon asset at international law. The Kyoto Protocol does not define the legal nature of Kyoto units, leaving the issue to national law. Unfortunately, there is also little guidance under national laws (though it is worth noting that the proposed

Australian carbon trading scheme defines the carbon asset as a financial product). Carbon assets have been described variously as intangible assets, permits, property rights, services, securities, commodities, and currency.[42] As the carbon markets grow, the need for certainty over the legal nature of the carbon asset is becoming more urgent.

7.4 The legal nature of CERs in China

While the CDM Measures do not specifically define the legal nature of CERs, they do give an indication of their legal nature under Chinese law.[43] Under Article 24 of the CDM Measures, 'emission reduction resources' are owned by the Chinese Government; emission reductions generated from CDM projects are owned by the enterprise implementing the project;[44] and the right to the income obtained from the transfer of CERs is to be shared between the Government and the enterprise implementing the CDM project (in the form of the CDM levy discussed above).

A recent commentary suggests that the right granted by the Chinese Government to a CDM project owner is analogous to the right to use a natural resource to produce fruits and benefits and hence a 'usufructuary right' under Chinese law.[45] A usufructuary right holder enjoys the right to possess, use and seek proceeds from the real property or movable property owned by another. Under the PRC Property Law, an entity or individual may possess, use and seek proceeds from the natural resources that are state-owned.[46]

The authorization of a project owner to participate in a CDM project by the Chinese Government through a letter of approval confers rights to transfer CERs and to benefit from CER revenue. Although it is not expressly stated in the CDM Measures, the project implementing entity, therefore, receives a right to sell CERs in return for the implementation of the CDM project.

The PRC Security Law classifies property as either 'non-movable property' (sometimes translated as 'real property'); or 'movable property'. Non-movable property is defined as 'land, housing, forestry and other properties being fixed on the ground' and movable property is essentially everything else, as it is simply

[42] Jillian Button, 'Carbon: Commodity of Currency? The Case for an International Carbon Market Based on the Currency Model' (2008) 32 Harvard Environmental Law Review 571.

[43] See also Lin, chapter 6, section 3.1.3 above.

[44] There are inconsistencies between the English and Chinese versions of the CDM Measures and the CDM application form in that the latter and the English version of the CDM Measures refer to a 'project owner', while the definitive Chinese version of the CDM Measures refer to a 'project implementation entity'. While it is generally assumed that 'project owner' and 'project implementation entity' are one and the same, this is not entirely free from doubt.

[45] Charlotte Streck and Xinjun Zhang, 'Implementation of the Clean Development Mechanism in China: Sustainable Development, Benefit Sharing, and Ownership of Certified Emission Reductions' (2005) 16 Yearbook of International Environmental Law 259–84.

[46] Article 118 of the PRC Property Law.

defined as 'properties other than non-movable properties'.[47] If CERs are accepted as a form of 'property', they would be movable property. However, in the absence of an authoritative statement that CERs are property under Chinese law, CERs may be regarded as a bundle of rights. PRC accounting standards define intangible assets as a class of movable property. Intangible assets are 'identifiable non-monetary assets which have no physical shape and are possessed or controlled by enterprises'.[48] To be identifiable, an intangible asset must be capable of being separated or divided from the enterprise, sold, transferred, licensed, rented or exchanged independently or along with other contracts, assets or liabilities; or sourced from any contractual rights or other statutory rights, regardless of whether or not these rights can be transferred or separated from the enterprise or other rights and obligations.[49] On this analysis, CERs are likely to be classified as intangible movable property under Chinese law and accounting standards.

8. Selected Issues in Carbon Procurement and Contracting

8.1 Civil law v common law

Chinese law allows the selection of a foreign governing law for 'foreign related contracts'.[50] An ERPA is a foreign related contract, as a foreign party is a buyer. ERPAs for Chinese CDM projects often stipulate Hong Kong law as the governing law, as a compromise between the selection of a foreign law or Chinese law. China is a civil law jurisdiction and takes a more prescriptive approach to laws and regulations than common law jurisdictions. Great care needs to be taken by both academics and practitioners when analysing legal principles and terms,[51] to ensure they are not confusing a common law interpretation or position with a civil law one, and vice versa.[52]

Generally, English law, Hong Kong law or other common law jurisdictions are preferable for governing carbon contracts, because Chinese contract law is still evolving and has little or no provisions or precedents for many eventualities. But even when the parties can use common law, good Chinese translations of

[47] Article 92 of the PRC Security Law.

[48] PRC, Ministry of Finance, 'Article 3 of the No. 6 Accounting Standards for Enterprises—Intangible Assets' issued 15 February 2006.

[49] Ibid.

[50] PRC Contract Law, art 126.

[51] For example, civil law and common law treat force majeure and liquidated damages differently, and China has recognized principles of fairness and good faith (see PRC Contract Law, arts 5 and 6), which are not generally found in common law jurisdictions.

[52] An interesting observation is that since the return of Hong Kong to China in 1997, the common law of Hong Kong has been affected to some degree by the civil law of China, and vice versa; eg Hong Kong Special Administrative Region Government, 'The Basic Law of the Hong Kong Special Administrative Region of the PRC' (July 2006) <http://www.basiclaw.gov.hk/en/basiclawtext/images/Basic_Law.pdf>, accessed 5 January 2009.

legal documents are important. There may be no equivalent term or word in Chinese for an English term or word,[53] and many unsuitable words have gained currency in Chinese due to the enormous volume of translated English contracts now circulating around China. A common example is the use of the Chinese word? 'ying' which literally means 'ought' or 'should' to translate the English word 'shall', which, unlike 'ying' is imperative rather than exhortative.

8.2 Types of contract

Many forms of ERPAs are used in the Chinese carbon market.[54] Most are heavily adapted after negotiations between the seller and the buyer and are far from standard documents. While parties invariably enter into non-disclosure and exclusivity agreements before entering into discussions about projects and transactions, practice varies between market players as to whether preliminary documents such as term sheets and letters of intent are used before an ERPA is prepared. A general rule of thumb for any contract or document to be used in the Chinese carbon market is to resist complexity and to keep it as clear and simple as possible. Complex concepts or terms may either not be understood by the Chinese party because they are uncommon or no equivalent exists under Chinese law. If such concepts or terms must be included, care should be taken to ensure they are understood by the Chinese party to avoid disputes later on.

8.3 The right of the seller to sell CERs

In the early stages of the Chinese carbon market, ERPAs were concluded between inexperienced sellers and more sophisticated foreign buyers. However in the last two years in particular, Chinese sellers have become more experienced and aware of their commercial and legal positions. Chinese sellers have started to develop and bundle CDM projects and to invite tenders from buyers. In some cases, Chinese sellers have indicated to buyers that they would sell CERs to buyers, when in fact the CDM project that was the subject of the transaction was to be implemented by another related entity. As the project implementing entity is the entity entitled to sell CERs and receive CER income under the CDM Measures, a buyer should not enter into an ERPA with a parent or related company unless there is a clear agency agreement between the parent or related company and the project implementing entity. It is also worth noting the ERPAs are invariably drafted to accommodate a seller that is the project entity, rather than a parent or related entity to the project entity. As such, most ERPAs would need amendment if a parent or related entity is the seller.

[53] A good example of this is the word 'title', which has no direct equivalent in China but is generally translated as 'ownership'.

[54] See Cerspa.org, 'CERSPA: A Carbon Contract Template' (CERSPA Information Note) <http://www.cerspa.org>, accessed 5 January 2009.

8.4 The 'control' rule

As noted above, the CDM Measures require CDM implementing entities to be Chinese controlled. Carbon market players have attempted to find ways to work around this rule legally. Any structure attempting to work around this restriction that is not approved by the NDRC runs the risk of being unwound. Potential ways to overcome or reduce the impact of the control rule are to:

(1) Use a domestic Chinese enterprise as the project implementing entity and appoint foreign directors and managers to run it. Put in place an operation and management agreement with the domestic Chinese enterprise to secure control. The foreign party takes no equity or shareholding in the domestic Chinese enterprise.

(2) Appoint Chinese employees of the foreign party into the domestic Chinese enterprise project implementing entity.

(3) Appoint Chinese employees of the foreign party into the EJV.

(4) Secure veto rights and require unanimity for certain corporate decisions of the EJV (negative control).

(5) Secure indirect control of the Chinese party to the EJV by the foreign party, but ensuring that indirect and direct shareholding by the foreign party does not exceed 49%.

All of the above alternatives rely on contractual arrangements to strengthen the position of the foreign party investing in a Chinese CDM project. None are foolproof and with the exception of the practice of securing negative control in an EJV, they are uncommon and pose some risk. Accordingly, foreign investors that have considered these alternatives have so far invariably adhered to the control rule and invested in up to 49% of an EJV, rather than attempt to secure indirect control or to gain a beneficial stake that exceeds the 49%. A potential breakthrough on this issue would be if the NDRC approves Hong Kong investors to wholly own the project implementing entity.

8.5 Delivery

A buyer will often wish to buy all the CERs from a CDM project. The trend in Chinese carbon transactions has gone from ERPAs with fixed or firm CER volumes to ERPAs with only anticipated or non-firm volumes. As Chinese sellers grew wary of the potential for default under fixed volume ERPAs, they have increasingly sought to water down delivery obligations in ERPAs.

8.6 Payment

The payment obligation of a buyer is usually triggered by the delivery of CERs, which is normally defined as the time when CERs are forwarded into the designated account of the buyer to receive CERs. However, where the buyer is the sole focal point, Chinese sellers have requested payment on the issuance of CERs, on the basis that once CERs are issued, the seller has fulfilled all its obligations. This may not be acceptable to a buyer as there are still Kyoto risks[55] following issuance that could occur to jeopardize the forwarding of CERs into the buyer's account.

8.7 Communications with the CDM Executive Board and focal point

Foreign buyers have generally insisted on being sole focal point of Chinese CDM projects, even if this is confined to the issuance and forwarding of CERs. However, as Chinese sellers have become more experienced, they have quite reasonably sought to become joint focal point with the buyer.

8.8 Dispute avoidance and resolution

In addition to the usual methods of dispute avoidance, disputes in ERPAs with Chinese sellers can be avoided by making sure the parties understand the provisions of the ERPA and how they work; avoiding complex provisions and mechanisms as far as possible; and engaging specialist advisors who understand the CDM and the Chinese legal system.

In terms of dispute resolution, while arbitration is preferred over litigation in Chinese courts and the enforcement of arbitration awards is facilitated by the New York Convention,[56] a risk that is often overlooked is that Chinese courts may refuse the enforcement of a foreign related arbitration award on the ground that to do so would be contrary to the social and public interest of China.[57] This risk is significant in the context of CDM projects, as many requirements and restrictions are a matter of Chinese policy. A dispute involving the floor price policy, for example, would be open to this risk. That said, there is no evidence to date to suggest that Chinese courts would exercise the social and public interest ground to refuse the enforcement of an arbitration award in anything but exceptional circumstances. Indeed, international practice has been to use the public policy ground to refuse enforcement sparingly. While the enforcement of arbitration awards in China has improved in recent times and arbitration is generally a superior option to Chinese court proceedings, foreign parties are still exposed to enforcement

[55] An example would be problems with the International Transaction Log (ITL), although, they would seem to be increasingly unlikely given the operational status of the ITL.

[56] Convention on the Recognition and Enforcement of Foreign Arbitral Awards (adopted 10 June 1958, entered into force 7 June 1959) 330 UNTS 38 (New York Convention).

[57] PRC Civil Procedure Law art 260; See Jerome A Cohen and others, *Arbitration in China: A Practical Guide* (vol 1, Sweet & Maxwell Asia: Hong Kong, 2004) 327.

risks in China, as awards need to be enforced by Chinese courts which vary in quality and experience.[58]

9. A Consideration of Options Post 2012: Evolution or Revolution, a Continuation of the CDM or a New Mechanism?

9.1 Introduction

The development of the CDM and its evolution in China has been well documented. It would be fair to say that the CDM, if not an overwhelming success in China, has certainly contributed significantly to an increase of renewable sources in China's overall energy generation capacity and to renewable energy development. However, as the WWF has stated, 'the CDM market is a time-constrained market and its development cannot continue indefinitely in the absence of a post-Kyoto framework'.[59] The evolution of the existing CDM within a framework which requires only developed countries to commit to emissions caps is unlikely to achieve the level of emissions reductions required to combat climate change and fulfill the aim of the UNFCCC. A compromise must be struck taking into account the notion of 'common but differentiated responsibilities' and the responsibility of some developing nations, like China, to take further mitigation action based on their economic development stages, response capabilities, and contribution to global greenhouse gas emissions. The post-2012 agreement may involve not an evolution, but a revolution of sorts. It may, for example, require developing countries to accept multilateral agreements that deal with intensity-based targets or sectoral targets. However, persuading China and other developing countries to accept targets will be a difficult task without stronger commitments by developed countries and the inclusion of the United States in a new global agreement.

9.2 China's intentions with respect to the CDM

China's Climate Change White Paper recently confirmed the Chinese Government's views in relation to the continuation of the CDM after 2012, stating:

> China holds that CDM, as a comparatively effective and successful co-operation mechanism, should continue to be implemented after 2012. However, efforts should be made to promote fairness, transparency, simplicity, certainty and environmental

[58] This remains the case notwithstanding the issue of the 'Notice of the Supreme People's Court on Several Questions Concerning the People's Court Handling of Issues in Relation to Foreign-related Arbitration and Foreign Arbitration' No 18 (28 August 1995), which established a reporting system by lower courts to higher courts, where the lower court decides to refuse the enforcement of an international arbitration award or a foreign award.

[59] The Value of Carbon in China (n 21 above), 9. See also NDRC 'Implementation of the Bali Roadmap, China's Position on the Copenhagen Climate Change Conference' issued on 20 May 2009 <http://www.ccchina.gov.cn/en/NewsInfo.asp?NewsID=17528>, accessed 6 June 2009.

completeness during the implementing of projects, and to encourage the transfer of advanced technology to developing countries. The host country should play a more important role in developing CDM projects.[60]

9.3 Intensity targets

China and other developing countries view absolute GHG targets, such as those accepted by Annex 1 countries under the Kyoto Protocol, as an unfair restriction on their economic growth and consequently an impediment to eradicating poverty.[61] China has rejected calls to accept caps on its energy use or GHG emissions. Targets for reductions in energy intensity or GHG intensity per unit of economic output, which do not directly require a reduction in economic production, may be more acceptable to China.[62]

In order to meet intensity-based targets, China would need to reduce emissions levels while maintaining economic growth—encouraging signs of change are occurring through the increasing deployment of renewable energy, high efficiency supercritical coal power technology, and low carbon transport. This may enable China to pursue what it sees as its right to sustainable economic development in order to address poverty, and at the same time assist it to shoulder further responsibility and contribute to the international efforts to combat global warming.

9.4 Sectoral targets

Sectoral emission targets provide an alternative to economy-wide emission targets. With the focus on energy-intensive sectors, rather than the whole economy, the sectoral approach is likely to be easier for developing countries to implement. Estimates of national emissions reductions are typically subject to high levels of uncertainty in developing countries. This problem would decrease if a smaller number of sources and baseline emissions across sectors was considered,[63] since sources of emissions across a sector are more likely to be known and well documented. By developing an emissions baseline across an entire sector, either internationally or nationally, standards can be benchmarked and applied across emission-intensive sectors, such as electricity or cement production.[64] Subject to

[60] White Paper (n 23 above), s VII; See PRC, 'China's Policies and Actions for Addressing Climate Change' <http://english.gov.cn/2008-10/29/content_1134544_9.htm>, accessed 8 January 2009.

[61] White Paper (n 23) s VII; See PRC, 'China's Policies and Actions for Addressing Climate Change' <http://english.gov.cn/2008-10/29/content_1134544_5.htm>, accessed 8 January 2009.

[62] Joanna I Lewis, 'China's Strategic Priorities in International Climate Change Negotiations' (Winter 2007–08) *The Washington Quarterly* 166 <http://www.twq.com/08winter/docs/08winter_lewis.pdf>, accessed 21 May 2009.

[63] Ibid, 166–7.

[64] The Value of Carbon in China (n 21 above), 43; see also Centre for Clean Air Policy 'Sectoral Approaches: A Pathway to Nationally Appropriate Mitigation Actions' (December 2008) (Interim Report) <http://www.ccap.org/docs/resources/560/CCAP-sectoral%20interim%20report%20final%20012209.pdf>, accessed 5 June 2009.

improvement in data collection and confidence in emissions estimates within sectors, China may be persuaded to accept binding sectoral emission reduction targets.[65]

Resistance to sectoral agreements from China and other developing countries is based on the fact that developing countries typically use less efficient technology, and would face higher costs in upgrading facilities to meet sector-wide standards.[66] Further, there might be concerns over emissions leakage between sectors if some sectors are targeted for mitigation and others are left unregulated.[67]

9.5 Revolution?

A more ambitious modification to the climate change framework would be to re-categorize State Parties.[68] The current UNFCCC and Kyoto Protocol essentially only accommodate States in two categories, Annex 1 or Non-Annex 1. Adding categories would be consistent with the principle of common but differentiated responsibilities, and would allow different compliance obligations to be applied to advanced developing countries such as China and South Korea. Unfortunately, absent a major change in China's current climate change negotiation position, a proposal to support a recategorization is unlikely.

10. The Potential for the Emergence of a Carbon Emissions Trading Scheme in China

While the introduction of a domestic carbon emissions trading scheme (ETS) in China has been debated by academia and the carbon industry, some commentators have suggested that the Chinese government is unwilling to introduce a domestic ETS as this would require the adoption of a nation-wide mandatory cap on GHG emissions, which would run counter to China's current resistance to accepting mandatory emissions caps as part of a global agreement. As noted in the Climate Change White Paper:

> China's ongoing industrialization process and its coal-dominated energy mix determine that its task of controlling greenhouse gas emissions is a tough one . . . its coal-dominated energy mix cannot be substantially changed in the near future, thus making the control of greenhouse gas emissions rather difficult.[69]

This suggests that China will continue to be unwilling to accept restrictions on its emissions or to introduce a mandatory national carbon emissions trading scheme in the short to medium term.

[65] Lewis (n 62 above), 167.
[66] Lewis (n 62 above), 167.
[67] Lewis (n 62 above), 167.
[68] A move that China would currently appear to reject.
[69] White Paper (n 23 above), Foreword: <http://english.gov.cn/2008-10/29/content_1134544_2.htm> (accessed 8 January 2009).

China's recent experience in sulphur dioxide emissions trading does not bode well for a carbon emissions trading scheme. To date, despite a number of pilot schemes[70] and announcements that a national sulphur dioxide ETS will be launched, no meaningful progress towards a national ETS has been made.

11. China's Specific Responses to Carbon Risk and a Low Carbon Economy

The Climate Change White Paper emphasizes that China's priority is to adapt to inevitable climate change, noting that '[D]eceleration [mitigation] is a long and arduous challenge, while adaptation is a more present and imminent task.'[71]

The theme that China, as a developing country, should not be held to account for mitigating climate change in the same way as developed nations runs throughout the Climate Change White Paper. It is clear from the White Paper that the Chinese government will continue to rely on the principle of 'common but differentiated responsibilities', which it reiterates 'is a core principle of the UNFCCC'. The White Paper stresses that:

> [D]eveloped countries should be responsible for their accumulative emissions and current high per-capita emissions, and take the lead in reducing emissions in addition to providing financial support and transferring technologies to developing countries.[72]

China's focus remains on reducing poverty, and the White Paper claims that China's per-capita emissions are relatively low. However, there is considerable scope for China to take on binding obligations that take into account development priorities on a sustainable basis.[73]

Despite this public stance, China has been making progress in a wide range of low carbon initiatives and innovations in recent years, and has become a leader in a number of critical renewable energy markets.[74] China argues in the Climate Change White Paper that it is taking broad-based action to reduce greenhouse gas emissions, and that economy wide absolute emission reduction targets are not suited to a country of its size at this stage of its development. China's intention to rely on advancement and innovation in science and technology to respond to carbon risk is another common theme throughout the White Paper.[75]

[70] See The Government of Hong Kong, Environmental Program Development, 'Implementation Framework of the Emissions Trading Pilot Scheme for Thermal Power Plants in the Pearl River Delta' (Pilot Scheme) <http://www.epd.gov.hk/epd/english/action_blue_sky/files/PRD_emission_trading_eng.pdf>, accessed 8 January 2009.

[71] White Paper (n 23 above), s VII.

[72] White Paper (n 23 above), s VII.

[73] Well designed and reasonable 'no lose' sectoral targets would be one.

[74] Steve Howard and Changhua Wu, 'China's Clean Revolution' (The Climate Group Report, August 2008) <http://www.theclimategroup.org/assets/resources/Chinas_Clean_Revolution.pdf>, accessed 9 January 2009.

[75] White Paper (n 23 above), s VII.

11.1 Low carbon development zones

An approach which has recently gained traction with the NDRC that ties in with this push for innovation is the concept of 'low carbon development zones' (LCDZs) around China.[76] In essence, the concept advocates the use of a similar development model to Special Economic Zones (SEZs) [77] that were introduced by the Chinese government to encourage foreign investment when China was opening up its economy to foreign investors in the early 1980s. The legal bases for the establishment of SEZs are straightforward, well tried and have worked successfully for many years.[78] Foreign invested businesses established in SEZs benefit from tax concessions, preferential land use grants, import duties, superior infrastructure and streamlined investment procedures. Where LCDZs might differ from SEZs in approach is that LCDZs would seek to create an investment environment that attracts both foreign and domestic investment in low carbon businesses and projects in order to create low carbon goods, services and technologies. LCDZs would also create scaleable opportunities for carbon finance, from individual projects to integrated developments and low carbon business parks. Indeed, LCDZs could form part of sectoral commitments under a new international climate change agreement.

12. Conclusions

It may be argued that China is making progress towards a sustainable future and towards a move away from a carbon-intensive economy.[79] However, the Chinese government has also made it clear that it does not intend to accept a binding emissions target in the near future. Accordingly, China's specific responses to carbon risk in the post-2012 period is likely to be a continuation of progress towards a low-carbon economy and adopting adaptation measures to cope with climate change, so as to ensure continued economic growth and development. However, there is still some possibility that 2009 may be a turning point in climate change negotiations. It is difficult to predict entirely what actions China will take post-2012, as much will no doubt depend on proposals put forward by developed countries to incentivize greater action by developing countries.

[76] Chatham House and E3G, 'Low Carbon Zones: A Transformational Agenda for China and Europe' (4 December 2008) <http://www.chathamhouse.org.uk/files/13470_1208_lczs.pdf>, accessed 9 January 2009.

[77] SEZs were established in Shenzhen, Zhuhai and Shantou in Guangdong Province, and Xiamen, Fujian Province and Hainan Island, Hainan Province.

[78] Once authority is granted at the national level by the State Council and the National People's Congress, considerable autonomy has been given to the provincial authorities to set up the SEZs.

[79] Howard and Wu (n 74 above), 28.

PART VI

VOLUNTARY MARKETS

23

The Voluntary Carbon Market: Its Contributions and Potential Legal and Policy Issues

*Michelle Passero**

1. Introduction

Over the past several years, transactions for the voluntary purchase and sale of greenhouse gas (GHG) emissions reductions have grown significantly. Also known as the 'voluntary carbon market', this market has provided a benefit to the climate by building public awareness and infrastructure for transactions and informing government climate policy. There is opportunity for the voluntary market to continue and expand. To facilitate this expansion, consideration should be given to the continued development and refinement of related legal issues, such as the harmonization of GHG accounting standards and registries and the clarification of the legal nature of voluntary GHG reductions.

This chapter is divided into five sections: an introduction is provided in Section 1; Section 2 gives background information on the voluntary carbon market, what it is and reasons for its growth; Section 3 elaborates on the contribution of the voluntary market including the development of GHG reduction accounting standards, registries and policy; Section 4 discusses the additional action that could be taken in this respect and identifies the potential need to clarify the legal nature of reductions in the voluntary market; and finally, Section 5 provides a summary and conclusions.

2. Background

For the purposes of this chapter, the voluntary GHG emissions reduction market (hereinafter 'voluntary market') refers to the voluntary creation of GHG

* Senior Climate Policy Advisor for The Nature Conservancy. The views expressed are those of the author. The author thanks Linda Sayler, Jonathan Rotter, Professor Eldon Reilly, Ira Chin, and Ricardo Bayon for providing helpful insights and expertise.

emission reductions[1] by one party and subsequent sale (or transfer) of these reductions to another party. It also includes the infrastructure that facilitates these transactions. In exchange for this transfer of reductions, the receiving party generally provides financial compensation and obtains the right to claim and use their benefits or transfer the reductions to another party. These reductions are generally intended to compensate for GHG emissions that are distinct from the voluntary reduction and are often called 'offsets'.

Unlike the voluntary market, the regulatory GHG emissions reduction market refers to the creation and subsequent trade of reductions pursuant to treaties and laws that place a legally binding cap, or limit, on specified sources of GHG emissions. Examples of regulatory markets include the Kyoto Protocol, the European Emission Allowances Trading Scheme (EU ETS), the Australian New South Wales GHG Abatement Scheme, and the Regional GHG Initiative in the United States, among others. Regulatory systems may permit the trading of emissions allowances[2] as a means to meet the regulatory cap and may also allow the use of reductions attained outside of the cap to meet the required emission limitations.

These 'uncapped' reductions are often called regulatory offsets or credits. The use of the term 'credit' has been used interchangeably with offsets in the voluntary and regulatory markets. Also, the term 'verified emission reduction,' or VER, is often used in lieu of the term offset or credit and is used throughout this chapter to refer to offsets in the voluntary market except where noted.

Voluntary markets may be organized into two categories. 'Over the counter' (OTC) transactions are generally transactions that occur on a deal-by-deal basis at a retail level. The other category includes trades facilitated by exchanges, such as the Chicago Climate Exchange.[3] To function smoothly, the voluntary market is also reliant upon the development of a supporting legal and institutional infrastructure, including greenhouse gas accounting standards and contracts that facilitate transactions in the market.

Over the past decade, there has been significant growth in the voluntary carbon market. While there has been some speculation that the growth has largely been driven by the anticipation of regulation and therefore, the occurrence of regulation would lead to a decline in the popularity of the voluntary market, this has yet to occur. To the contrary, the voluntary market has continued to expand and some speculate that the voluntary GHG market will continue to grow for the foreseeable future in spite of parallel growth in the regulatory market. Such anticipated growth emphasizes the importance of the

[1] For the purposes of this chapter, emissions reductions refer to a decrease or avoidance of GHG emissions or the increased removal and storage of GHG emissions from the atmosphere.

[2] Allowances represent an authorization to emit a defined amount (ie one metric ton) of GHG emissions pursuant to a cap, or emissions limit.

[3] Chicago Climate Exchange, 'Membership Categories' <http://www.chicagoclimateexchange.com/content.jsf?id=65>, accessed 30 December 2008.

voluntary market and the potential need for the market to maintain and strengthen its infrastructure.

According to one of the leading annual reports on the voluntary market, the number of organizations supplying greenhouse gas reductions to the marketplace has grown by 200% since 2002.[4] It is estimated that in 2007 a total volume of 65 million metric tonnes (value of 331 US$ million) of carbon dioxide equivalents (MtCO2e) were traded in the voluntary marketplace, up from 24.6 MtCO2e in 2006 (96.7 US$ million).[5] It is predicted that the voluntary market will continue to grow, at least in the near-term, with initial indications at the time of drafting that 2008 transactions could reach as much as 150 MtCO2e.[6] While the absolute volume of reductions is relatively small compared to the transactions in the regulatory market (nearly 3 billion MtCo2e in 2007), the rate of growth in the voluntary market is 165% compared to 71% in the regulated markets.[7]

The types of voluntary GHG reductions entering the marketplace are varied. They range from geologic sequestration to industrial gas destruction to agricultural soil carbon sequestration. However, several types of GHG reductions appear to be dominating the more recent transactions.[8] These reductions include renewable energy, energy efficiency, forestry, and methane destruction.[9] In general, GHG reductions that are associated with other sustainability benefits have greater public appeal.

A variety of reasons appear to be driving the growth of the voluntary market. One could speculate that a primary motivation is the anticipation of a broader regulatory market, with industries anticipating future regulatory compliance periods under the Kyoto Protocol or the establishment of a regulatory cap and trade system in the United States. These types of GHG reductions or reasons for purchase are often considered 'pre-compliance' reductions. However, some analysis indicates that while this pre-compliance obligation may be a motivating factor, it is not necessarily the primary driver behind the voluntary market.[10] Rather, a host of reasons extending beyond regulation appears to be the motivating factor, including corporate responsibility and environmental ethics, public relations, climate change oriented business models, sales of carbon neutral products and investment.[11] This broad

[4] K Hamilton and others, 'State of the Voluntary Carbon Markets 2007: Picking Up Steam' (Report) (18 July 2007). <http://portal.conservation.org/portal/server.pt/gateway/PTARGS_0_2_144235_0_0_18/StateoftheVoluntaryCarbonMarket18July_Final.pdf>, accessed 18 December 2008.

[5] K Hamilton and others 'Forging a Frontier: State of the Voluntary Carbon Markets 2008' (2008) 6. <http://www.ecosystemmarketplace.com/documents/cms_documents/2008_StateofVoluntaryCarbonMarket2.pdf>, accessed 10 December 2008.

[6] Ibid, 10; see also Karan Capoor and Philippe Ambrosi, 'State and Trends of the Carbon Market, 2008' (World Bank Report) (May 2008). <http://siteresources.worldbank.org/NEWS/Resources/State&TrendsformattedO6May10pm.pdf>, accessed 24 December 2008.

[7] Ibid, 7. [8] Recent transactions are those that have taken place in 2007.

[9] Hamilton (n 5 above), 7. [10] Hamilton (n 5 above), 68. [11] Ibid, 68.

range of motivations suggests that the voluntary market may thrive alongside a regulatory market, as indicated by its popularity in Europe despite the growth of the regulatory EU ETS.[12]

3. Contributions by the Voluntary Market

The voluntary market has been valuable for the climate and climate policy for many reasons, legal and otherwise. At a minimum, it has helped raise public awareness regarding how individuals or entities contribute to global climate change, how to measure GHG emissions, and how these emissions may be avoided or mitigated. The growth of OTC VER retailers and news stories on this issue are evidence of this increased awareness.[13]

The voluntary market can also empower and motivate individuals and companies to address global climate change. Individuals and certain businesses or industries may not be subject to GHG regulations or a regulatory cap-and-trade system,[14] so the voluntary market provides them with an opportunity to reduce emissions through voluntary actions. These voluntary actions can, in turn, lead to greater GHG emission reductions than might otherwise happen, notwithstanding a regulatory market or other GHG regulations.

As mentioned earlier, it is estimated that the voluntary market resulted in the sales of 65 MtCO2e in 2007 and the sale of GHG reductions could be as great as 150 MtCO2e in 2008. These reductions are roughly equivalent to the reductions of the annual emissions of 12 million and 27 million cars respectively.[15] Thus, the voluntary market does present an opportunity to achieve additional GHG reductions by appealing to individuals, industries or sectors not otherwise included in a cap or GHG regulation and by empowering and inspiring action to address global climate change.

The voluntary market has also fostered the development of GHG accounting standards and registries. This kind of infrastructure is essential to promote accurate, standardized and transparent approaches to quantify GHG emissions

[12] Roughly 47% of the purchases of GHG reductions in the voluntary market were from Europe in 2007, representing the largest share of purchasers; See Hamilton (n 4 above), 67.

[13] David Biello, 'A Drive to Offset Emissions' Ecosystem MarketPlace (Washington DC, 21 February 2006). <http://ecosystemmarketplace.com/pages/article.news.php?component_id=4192&component_version_id=6012&language_id=12>, accessed 25 December 2008.

[14] California Air Resources Board, 'Climate Change Proposed Scoping Plan: A Framework for Change' (Plan) (October 2008) <http://www.arb.ca.gov/cc/scopingplan/document/psp.pdf>, accessed 30 December 2008.

[15] The equivalent is based on the United States Environmental Protection Agency's estimates of annual GHG emissions from standard passenger vehicles; United States Environmental protection Agency, 'Emissions Facts: Greenhouse Gas Emissions from a Typical Passenger Vehicle' (February 2005, EPA420-F-05-004). <http://www.epa.gov/OMS/climate/420f05004.htm>, accessed 20 December 2008 (EPA Fact Sheet).

and reductions. The registries centralize the accounting data, provide transparent monitoring of reductions, and track transactions to avoid the double counting of reductions. From a legal and policy perspective, they inform the construction of purchase and sale contracts of GHG reductions, otherwise known as verified emission reduction purchase agreements (VERPAs), and serve as models for regulatory measures and government policies.

Greenhouse gas programs in the voluntary market generally include, but are not limited to, the following accounting standards:

(1) *Baselines and additionality*: Measurable reductions should be in excess of what would have happened in the absence of the GHG reduction activity. This assessment requires the establishment of a baseline that reflects a 'business as usual' or 'without project' scenario and actions that lead to GHG reductions that exceed this scenario. Certain programs also require a demonstration of financial additionality.[16]

(2) *Permanence*: Permanence refers to the duration of greenhouse gas reductions and the ability to address any reversals of emission reductions that may have already been verified as a reduction. The issue of permanence is particularly relevant for VERs that correspond to land-based reductions (eg forests, soils, grasslands, etc), which can be released back into the atmosphere due to disturbance and the loss of biomass. GHG programs and buyers often require forest-based reductions to remain stored in forests for long durations, sometimes up to 100 years or more, to be equivalent to emissions reductions in other sectors.[17]

(3) *Verifiability*: The credibility and enforceability of reductions rely on third-party verification. Third parties independently verify the accrual, permanence and ownership of the reductions over time.

(4) *Leakage*: GHG reductions should avoid, minimize or account for (ie deduct) any displacement of GHG emissions that are caused directly by the GHG reduction offset.

These GHG accounting standards and their corresponding measurement methodologies (ie protocols) have been developed and refined in the voluntary market and inform the construction of VERPAs. They help identify the 'product', or GHG reduction, being sold and can influence how the reductions are monitored and delivered over time, as well as the roles and responsibilities of the parties to the VERPA. For instance, an accounting standard may require

[16] Financial additionality requires, in general, the demonstration from a GHG reduction project that it would not have happened 'but for' financial payments or investments made specifically to the project for the production and acquisition of GHG emission reductions. For a discussion of 'additionality' in the context of the CDM, see chapter 12 above.

[17] The permanence of emissions reductions in sectors that are only emissions sources (and not a 'sink') is often judged on an annual basis. If emissions from a source are not emitted (or fewer are emitted than the previous year), the reductions are generally considered permanent.

GHG reductions from forest activities to be maintained for 100 years in order to be deemed permanent.[18] This 100-year obligation would be an important factor for the construction of a VERPA, as it could impact the length of the contract. It could also have implications for the associated long-term liability and remedies that are typically identified in contracts. Given the duration of the reduction obligation, the seller could seek to bind subsequent landowners of the forest property through the use of a conservation easement or covenant and incorporate this instrument in the VERPA.

Greenhouse gas emission reduction initiatives in California illustrate the contributions of the voluntary market. The California Climate Action Registry (CCAR), a voluntary greenhouse gas emissions reductions registry, adopted accounting standards and developed detailed protocols to quantify GHG reductions in the forest, landfill and agricultural sectors, among others. The CCAR's accounting standards and protocols have facilitated voluntary GHG transactions,[19] and VERPAs have incorporated and relied upon these accounting standards. The CCAR has also served as a centralized system to compile GHG emissions and reduction information and recently adopted a system to serialize reductions and track transactions to provide greater transparency for transactions and avoid the double counting or double selling of VERs.[20]

With the establishment of the CCAR Forest Protocols, the California-based Pacific Gas and Electric (PG&E) Utility Company launched its voluntary ClimateSmart program in 2007. This program provides an opportunity for utility ratepayers to offset the GHG emissions associated with their energy use by paying a monthly fee of US$ 5.00 to PG&E.[21] In exchange, PG&E enters into contracts with entities that develop and verify GHG reductions pursuant to the Registry's GHG accounting standards. PG&E then purchases the reductions and retires them on behalf of ratepayers.[22] The accounting standards of the CCAR are incorporated into these contracts, helping to define what is being purchased by the program (ie a real GHG reduction). Its registry also provides transparent recording and tracking of the purchased reductions to avoid double counting and selling and

[18] The California Climate Action Registry, 'Draft Forest Project Protocol, Version 3.0' <http://www.climateactionreserve.org/how-it-works/protocols/adopted-protocols/forest/forest-project-protocol-update/>, accessed 15 February 2009.

[19] Since the adoption of GHG accounting standards, several sales of VERs have relied on these standards for contractual agreements, including The Conservation Fund, The Pacific Forest Trust and the Sempervirens Fund. Climate Action Reserve <https://thereserve1.apx.com/myModule/rpt/myrpt.asp?r=111>, accessed 31 December 2008.

[20] The California Climate Action Registry, 'The Carbon Reserve' <http://www.climateregistry.org/reserve.html>, accessed 16 February 2009.

[21] Pacific Gas and Electric, '2008 Request for Proposals: ClimateSmart Greenhouse Gas Emission Reduction Projects' <www.pge.com/myhome/environment/whatyoucando/climatesmart/climatesmartrequests>, accessed 30 December 2008.

[22] Pacific Gas and Electric, 'What is the ClimateSmart Program?' <http://www.pge.com/myhome/environment/whatyoucando/climatesmart/climatesmartabout/index.shtml> accessed 30 December 2008 (PG&E Information Note on the ClimateSmart Program).

assures the reductions are permanently retired, which is a requirement of the ClimateSmart program. As of 2008, over 17,500 ratepayers had signed up for the program[23] to offset their emissions through VER projects verified by CCAR.

Voluntary GHG accounting standards and methodologies also inform and support government climate policies, including regulatory markets. In 2007, and subsequently in 2008, the California Air Resources Board adopted the CCAR GHG reduction protocols for the forests, urban forestry and agricultural sectors as part of its statutory responsibility under the California Global Warming Solutions Act of 2006 to promote voluntary early action GHG reductions to meet California's reduction goals.[24] The California Air Resources Board has indicated that it will review the existing GHG accounting rules that are available in the voluntary market as a basis for developing accounting rules in a regulatory cap-and-trade system.[25]

The voluntary GHG emission reduction market has made significant gains to develop GHG accounting standards, methodologies and registries to quantify, verify and track real emissions reductions. These developments are necessary from a legal and policy perspective, as well as from an environmental perspective, so that parties that are buying and selling GHG reductions are able to identify clearly what is being transacted in a VERPA, as well as associated responsibilities and liabilities over time. Registries also provide assurances that reductions are maintained over time and are not sold more than once through transparent reporting and tracking of reductions. The development of this infrastructure not only strengthens and builds confidence in the voluntary market to perpetuate its success, but also informs the regulatory market as seen in California.

4. Additional Actions to Support the Voluntary Market

4.1 Harmonization of GHG accounting standards

While the development of accounting standards and protocols by a number of GHG programs has led to significant gains in the voluntary market, their proliferation has also caused some challenges. At the time of drafting, there are at least 13 different programs offering voluntary GHG accounting standards and methodologies in the marketplace;[26] it is likely that even more will have been

[23] Pacific Gas and Electric, 'Pacific Gas and Electric Company's ClimateSmart Program Makes Largest Purchase of Greenhouse Gas Emissions Reductions in California' <http://www.pge.com/about/news/mediarelations/newsreleases/q1_2008/080226.shtml>, accessed 15 June 2009.

[24] California Air Resources Board (n 14 above), 69. [25] Ibid.

[26] As of December 2008, available voluntary GHG accounting standards include, but are not limited to: Gold Standard, The Voluntary Carbon Standard, Greene- Climate, Climate Community and Biodiversity Standard, Chicago Climate Exchange, The California Climate Action Registry, US '1605b' program, Greenhouse Friendly, VER+ (Verified Emission Reduction Standard), Social Carbon, Plan Vivo, World Resources Institute, and ISO 14064.

established by the time this chapter is in print.[27] The sheer number of programs and their different interpretations and level of guidance with respect to the basic accounting standards have caused some confusion among buyers regarding the most appropriate standard to use.[28]

While some program standards are comparable, many address different types of GHG projects (eg renewable energy, forestry, landfill, etc). For instance, The Gold Standard, a certification scheme for regulatory and voluntary offsets, provides accounting standards and methodologies for only renewable energy and energy efficiency projects. On the other hand, the Voluntary Carbon Standard, which is an international GHG accounting and verification program, provides standards and accounting methodologies for a spectrum of voluntary GHG reduction projects, from landfill methane capture to forest carbon sequestration. This discrepancy in coverage could cause confusion about the most appropriate standard to use, as well as raise doubts about the legitimacy of certain sources of VERs (eg energy projects are better that forest projects).

Moreover, some GHG programs interpret accounting standards and methodologies differently, providing varying levels of detail. For instance, the Chicago Climate Exchange effectively defines a baseline for forestry VERs as a measurement of GHG emissions at one point in time (also known as base-year accounting), while other VER programs, like the California Climate Action Registry, interpret the baseline to be a projection of emissions over time. The differences in baseline approaches could ultimately result in different calculations of GHG reductions for the same project and raise doubts about the additionality of certain VERs. The permanence of VERs, which is an issue associated with forestry GHG reductions, is also handled differently among some programs, whereby some require reductions to be stored in perpetuity or at least 100 years, while others allow shorter-term storage of reductions so long as the VERs are replaced with others that are equivalent.

While some differences among the GHG programs may be beneficial by fostering competition and innovation, they may also foster uncertainty in the market. From a legal standpoint, disparities among programs could make transaction negotiations and VERPAs more complicated, as parties must negotiate and choose among different standards. Contracts may need to be tailored to reflect the differences in accounting standards and methodologies and corresponding liabilities. These differences could make it more difficult to standardize contracts and transactions, thereby hindering the liquidity of the market. They could ultimately lead to public and market uncertainties about the legitimacy of GHG reductions and transactions.

[27] The American National Standards Institute is in the process of establishing VER standards for forest activities.

[28] Beth Daley, 'Carbon Confusion: Buying Emission Offsets is a Challenge for Consumers' The Boston Globe (Boston, 12 March 2007) <http://www.boston.com/news/nation/articles/2007/03/13/carbon_confusion>, accessed 30 December 2008.

In recognition of these potential issues, efforts have been made to foster greater consistency and transparency among voluntary GHG program accounting standards. In the voluntary market, The Voluntary Carbon Standard (VCS) has been established, at least in part, as a 'meta-standard' to provide general accounting guidance for the development of voluntary GHG reductions.[29] This standard can approve other registries and standards as consistent with its general approach and has done so with the California Climate Action Registry. Thus, the VCS or a similar program could help promote harmonization and transparency among the voluntary standards.

Governments may also promote greater consistency and transparency across VER standards. For instance, in 2008, the US Federal Trade Commission, pursuant to its authority under the Federal Trade Commission Act to prohibit unfair or deceptive practices, undertook a public process to revise its Guides for the Use of Environmental Marketing Claims (Green Guides) to include GHG emissions offsets. The purpose of these Green Guides is to assist marketers so that they will avoid making deceptive or unfair claims with respect to offsets or GHG emission reductions.[30] The effect of the Green Guides is intended to be greater consistency and transparency with respect to GHG reductions in the voluntary marketplace, which would also be supported by a right of enforcement by the US Federal Government.[31]

California has also undertaken efforts to promote greater consistency and transparency across VER accounting standards in the voluntary market. As mentioned earlier, the California Air Resources Board adopted the California Climate Action Registry Forest Protocols as an accounting standard and methodology to promote voluntary GHG reductions in the forest sector. It subsequently adopted GHG protocols for urban forestry and methane digesters to promote voluntary reductions in a standardized manner.[32] The California Legislature has also made recent efforts to adopt legislation to ensure transparency and consistency of VER standards.[33]

Likewise, the United Kingdom has also undertaken similar efforts through the Department of Environment, Food and Rural Affairs (DEFRA).[34] Beginning in February of 2008, the Department undertook an effort to develop a Code of

[29] Voluntary Carbon Standard Association, 'Voluntary Carbon Standard 2007.1' (VCS Program) (18 November 2008) <http://www.v-c-s.org/docs/voluntary%20Carbon%20Standard%202007_1.pdf>, accessed 10 December 2008.

[30] Federal Trade Commission, 'Reporter Resources: The FTC's Green Guides' <http://www.ftc.gov/opa/reporter/greengds.shtm>, accessed 20 December 2008.

[31] At the time of drafting, these Green Guides were not yet completed.

[32] California Air Resources Board (n 14 above), 69.

[33] California Assembly Bill (2007–08)1851 <http://www.leginfo.ca.gov/pub/07-08/bill/asm/ab_1851-1900/ab_1851_bill_20080822_amended_sen_v90.pdf>, accessed 2 February 2009.

[34] As of December 2008, this effort is now being undertaken by the Department of Energy and Climate Change; Department for Environment Food and Rural Affairs, 'Climate Change: Carbon Offsetting' (2008) <http://www.defra.gov.uk/environment/climatechange/uk/carbonoffset/index.htm>, accessed 24 December 2008 (DEFRA's Carbon Offsetting Link).

Best Practice for greenhouse gas offsets. Upon approval of the Code, offset providers would become accredited and would receive a quality mark that would provide consumers with some certainty and transparency that the offsets have met a particular Government-approved standard.[35]

It is not unique for governments or the private sector to undertake efforts to harmonize standards and create greater transparency in evolving markets. For instance, the Uniform Commercial Code was developed in the United States (US) to foster greater standardization of transactions in the commodities market.[36] The Code was developed largely by private sector interests, but was ultimately adopted as law by many states. Similarly, the globalization of the securities market has highlighted transaction challenges due to the differences among financial accounting standards of countries.[37] As a consequence, efforts have been made to reach convergence among the different standards to create greater liquidity in the securities market.[38]

In spite of these analogous issues and efforts in other markets, there is some concern regarding government involvement in or regulation of the voluntary carbon market. Some of this resistance is based on the concern that government interference, or too much interference, at this stage would diminish the many positive aspects of the voluntary market, namely its flexibility and innovation.[39] Therefore, if governments do ultimately assume the role of harmonizing standards and protocols for VERs or develop additional requirements for the voluntary market, consideration should be given to the potential downside of intervention, particularly with respect to the timing and extent of such interventions.

The voluntary market has been quite effective in developing a number of GHG accounting standards and protocols to quantify GHG emission reductions. These standards are critical for the climate, public confidence, and to facilitate transactions. Due to the number of standards that have evolved over the years, it may become necessary for the growth of the market to reconcile core differences in standards. Efforts in the voluntary market and among governments are underway to create a more unified foundation for the growth of the voluntary market. However, the timing and extent to which government should intervene may be important considerations, as indicated by some

[35] As of December 2008, the Best Practice Code has yet to be adopted; ibid.

[36] Uniform Commercial Code 2005 <http://www.law.cornell.edu/ucc>, accessed 10 December 2008.

[37] Bo Harvey, 'Exchange Consolidation and Models of International Securities Regulation' (2007) 18 Duke Journal of Comparative and International Law 151.

[38] The Committee of European Securities Regulators, 'CESR Advice on the Equivalence of Chinese, Japanese, and US GAAPs' (March 2008) <http://www.cesr-eu.org/data/document/08_179.pdf>, accessed 10 December 2008.

[39] Terra Pass, 'Is It Time to Regulate The Voluntary Carbon Market?' (May 29, 2007) <http://www.terrapass.com/blog/posts/is-it-time-to-r>, accessed 31 January 2009; Standard Carbon, 'California Poised To Regulate Voluntary Carbon Market' (July 2, 2008) <http://www.standardcarbon.com/resources/?p=68>, accessed 31 January 2009.

market participants, to ensure that key attributes of the voluntary market are not diminished.

4.2 Determining the legal nature of VERs

As the voluntary carbon market grows, it may ultimately need greater clarity regarding the legal nature of VERs. In both the regulatory and voluntary markets, there has been ambiguity regarding the legal nature of emissions reductions (ERs)[40] and debate in the regulatory market with respect to this issue. In particular, it has been unclear whether these reductions should be treated as commodities, securities or financial instruments, real property rights, permits, or some other type of legal classification. In the absence of clear and consistent guidance on this legal issue, it may be difficult for market participants to understand fully the range of laws, rights and liabilities to which they may be entitled or exposed, which could ultimately become a barrier in the market. Just as resolution of this issue in the regulatory market has been considered essential, resolution of this issue in the voluntary market could be just as useful.

To understand the potential issues that may arise in the voluntary carbon market with respect to the legal nature of VERs, it is helpful to examine briefly this issue in the regulatory market. The Kyoto Protocol defines the measurable unit of a tradable ER as one tonne of carbon dioxide equivalent and acknowledges that it does not create or grant 'any right, title or entitlement to emissions of any kind'.[41] However, it is vague with respect to its precise legal nature. As a consequence, Parties (ie countries) to the Protocol were left to make this determination themselves.

Likewise, the EU ETS Directives issued in accordance with the Kyoto Protocol have deliberately avoided prescribing the legal nature of ERs.[42] By default, each EU Member State could decide whether these tradable reductions are permits, goods or services, securities or some other type of legal distinction.[43] The decision by policy-makers to effectively delegate this legal clarification has fostered some uncertainty and risk in the market by: (1) leaving the issue unanswered until countries make such a determination; and (2) allowing

[40] For the purposes of this Part, emissions reductions or ERs refer to the suite of emission reduction units (eg certified emission reductions, assigned amount unit, removal unit, verified emission reductions, etc) that are traded in the regulatory and voluntary markets.

[41] FCCC/CP/2001/13/Add.2, Decision 15/CP.7 Principles, nature and scope of the mechanisms pursuant to Arts 6, 12 and 17 of the Kyoto Protocol, preamble; draft Decision -/CMP.1 (Mechanisms), preamble <http://unfccc.int/resource/docs/cop7/13a02.pdf>, accessed 15 February 2009.

[42] See Matthieu Wemaere and Charlotte Streck, 'Legal Ownership and Nature of Kyoto Units and EU Allowances' in David Freestone and Charlotte Streck (eds), *Legal Aspects of Implementing the Kyoto Protocol Mechanisms: Making Kyoto Work* (Oxford University Press, 2005), 48–9. For a contemporary discussion see also chapter 2 above.

[43] Ibid.

countries to make different determinations with respect to the legal nature of ERs.[44]

Participants in both the regulatory and voluntary markets should be able to readily identify the range of laws that would apply to an ER transaction so corresponding rights and liabilities can be clearly identified and addressed in transactions. Knowing the legal nature of these reductions is a key aspect of being able to identify these laws. For instance, if it were determined through the Kyoto Protocol or the EU ETS that ERs (and related transactions) were securities, private parties transacting ERs in Europe could identify the securities laws that would likely apply to the transactions, such as those issued pursuant to Investment Services or Market Abuse Directives.[45]

Likewise, a determination that these reductions are commodities or services could provide more certainty and direction for parties to focus on whether transactions are subject to trade agreements of the World Trade Organization,[46] the Convention on the International Sale of Goods, or other laws governing services and commodities. If the tradable ERs were identified as permits, similar to the sulfur dioxide emissions trading program of the United States Clean Air Act,[47] the rights associated with the reductions could be clearly identified and limited by the permit.

Regardless of whether the ultimate legal classification of reductions should be one of those identified in the preceding paragraph or another legal distinction, the initial ambiguity associated with no determination by either the Kyoto Protocol or the EU ETS has fostered a level of uncertainty in the regulatory market.[48] This void has left market participants to debate or assume which laws and corresponding legal rights and responsibilities could apply to a transaction.[49] Also, such uncertainty has likely increased the time dedicated to transactions and as a consequence, increased the transaction costs.

The opportunity for individual countries to classify the legal nature of ERs differently could also foster complications and market barriers for international transactions. For instance, if one country determines ERs to be a form of permit

[44] Ibid; Foundation for International Environmental Law and Development, 'Background Paper: International Workshop on the Legal Nature of Emissions Reductions 05/04' (2004), 15–16; John Palmisano, 'Property Rights and Risk Management' (Emissions Marketing Association website) <http://www.emissions.org>, accessed 5 January 2009. [45] Ibid.

[46] United Nations Conference on Trade and Development, *A Pilot Greenhouse Gas Trading System: The Legal Issues* (United Nations:, Geneva, 1996, UNCTAD/GDS/GFSB/Misc. 2) 11.

[47] 42 USC § 7651b (f) (1999).

[48] While a number of EU Member States have undertaken efforts to clearly identify the legal nature of regulatory ERs, such as Sweden and France, others, like Germany, have yet to clearly articulate this issue. K Upston-Hooper, K Anttonen, M Mehling and A Perrells, 'Overview Report: Legal Frameworks for Emissions Trading in the European Union' (2006) 7. <akseli.tekes.fi/opencms/opencms/OhjelmaPortaali/ohjelmat/ClimBus/fi/DoKumenttiarkisto/Viestinta_ja.../Legal_Final_report.pdf>, accessed 10 February 2009.

[49] Wemaere and Streck (n 42 above); Foundation for International Environmental Law and Development (n 44 above); United Nations Conference on Trade and Development (n 46 above), 11.

and not a type of security or commodity, yet another country classifies them as financial instruments, there could be conflict regarding which laws should be applicable to the transaction. In fact, this has been the case among certain EU Member States. For instance, Sweden defines allowance ERs as financial instruments[50] and France classifies allowance ERs as permits.[51] Transactions between these countries raise uncertainty regarding the applicability of securities laws since France does not consider ERs to be securities.

Such differences in legal classifications could also foster market imbalances, whereby transactions become favored in certain countries because their legal classifications are more comparable (thereby reducing uncertainty and related costs). Different legal distinctions in countries could also result in more favorable tax treatment.[52] For example, the decision by Sweden to treat allowance ERs as financial instruments created the presumption that the Swedish Value Added Tax (VAT) Act would not be applicable.[53] In contrast, the United Kingdom views the transfer of ERs in a market as a 'supply of services' and therefore subject to VAT at a standard rate.[54] While it is likely not feasible to remove all discrepancies among countries with respect to ERs, it could be possible to minimize them by at least clarifying and harmonizing their legal nature.

The voluntary carbon market could share legal uncertainties that are similar to those in the regulatory market. As with the regulatory market, there is no global legal agreement that specifies or seeks to harmonize the legal nature of VERs in the voluntary market.[55] While VER transactions are generally administered through private contract law, questions could still arise regarding the appropriate legal classification of VERs. That is, there could still be legal uncertainty or conflicts over whether they are a type of commodity or service, security, right in real property, or some other legal distinction. As a consequence, the voluntary market could experience the same kind of legal uncertainties and market disparities as those identified in the regulatory market.

For instance, the legal uncertainty of whether a VER is a right in real property could be prevalent in the voluntary market given the popularity of VERs from forest activities.[56] The GHG reductions from forest activities, such as reforestation, changes in forest management and avoided deforestation entail the protection and storage of biological carbon in trees and soil, aspects that are generally deemed a part of real property. As a consequence, legal questions and uncertainties could arise for some VERs and related transactions regarding whether they are a real property right and therefore affected by the suite of laws

[50] Upston-Hooper (n 48 above) 7. [51] Wemaere and Streck (n 42 above) 39.

[52] Foundation for International Environmental Law and Development, 'Background Paper: International Workshop on the Legal Nature of Emissions Reductions,' (May 2004, London), 24–25; Wemaere and Streck (n 42 above), 49, and chapter 2, above.

[53] Upston-Hooper (n 48 above), 9.

[54] Foundation for International Environmental Law and Development (n 52 above), 25.

[55] At the time of writing (February 2009), there were no legal structures in place to define the legal nature of VERs in the voluntary market. [56] Hamilton (n 5 above), 7.

and legal protections related to real property. These could include property taxes, Constitutional 'takings' protections (in the United States),[57] and conservation easement laws, among others.

A lawsuit filed in 2005 in the United States illustrates the potential for VERs to be deemed rights in real property. In Louisiana, a legal case was filed in federal court by a private forest landowner against the United States Fish and Wildlife Service (USFWS). The landowner sought a legal determination that the retention of a 'conservation servitude' by the USFWS on the landowner's property did not mean that the carbon credits contained in the trees were also retained by the federal agency.[58] The court's dicta indicated that these credits were viewed as the bundle of rights associated with real property.[59] While this case was ultimately dismissed, it demonstrates the potential for at least some VERs to be interpreted as rights in real property, which could ultimately expose these VER transactions to laws related to real property. However, in the absence of any legal clarification that makes this issue clear, it is open to debate and could foster uncertainty in the market.

Likewise, there may be some uncertainty whether VERs could be commodities. A number of business authorities refer to VERs as commodities, including publications from Businesswire, a business news service, and APX, a financial service provider.[60] Some policy-makers may view VERs as a commodity, as demonstrated by efforts in the US legislature in 2009 to clarify this legal distinction through draft legislation.[61] If VERs were deemed to be commodities, their transactions could be subject to a host of related domestic and international laws that govern commodity transactions, such as the United Nations Convention on Contracts for the International Sale of Goods (CISG), World Trade Organization agreements, and in the United States, the Uniform Commercial Code (UCC) and Commodities Exchange Act.

However, it is debatable whether VERs are truly commodities as traditionally defined. Legal interpretations of commodities, whether they are related to the

[57] The Fifth Amendment of the United States Constitution prohibits the federal government from taking property from citizens without due process and just compensation (US CONST amend V). Over the years, the adoption of environmental laws and land use regulations has led to regulatory 'takings claims' in the US as regulations have, in some instances diminished the economic viability or value of property. To the extent that VERs are deemed to be a part of real property, land use regulations that diminish the economic value of VERs could potentially result in additional takings claims.

[58] In this case, the court referred to the forest VERs as carbon credits.

[59] *Roseland Plantation, LLC v United States Fish and Wildlife Service* (No 05-0793) 2006 US Dist LEXIS 29334, 3 (W D Louisiana 5 April 2006).

[60] 'Global Market Grows 80 Percent in 2007' *Business Wire* (San Francisco 18 January 2008) <http://findarticles.com/p/articles/mi_m0EIN/is_2008_jan_18/ai_n24236514>, accessed 20 December 2008; APX, 'Carbon Market Infrastructure' (Information Note) <http://www.apx.com/environmental/carbon-market-infrastructure.asp>, accessed 30 December 2008.

[61] Staff of House Agriculture Committee, 111th Cong, Draft Bill to Amend Commodities Futures Trading Act of 1974. (As of 10 February 2009, the draft bill language sought to include carbon credits transactions, without voluntary or regulatory distinction, as part of CFTA).

CISG or the UCC, generally refer to commodities as tangible, moveable goods, such as grains, metals and foods.[62] In contrast, VERs are not tangible goods, as they represent removals, reductions or avoidances of GHG emissions, an intangible substance. The potential conflicting interpretations of whether a VER is a commodity, combined with the absence of policy to make this determination, could be viewed as a risk in the voluntary market.

Regardless of whether a VER qualifies as a commodity, a real property right, or some other legal distinction (eg a financial instrument), the absence of policy clarifying this issue could ultimately become a market barrier. Market participants could determine that it is too risky to enter into transactions without understanding the full range of laws that could or should apply to a transaction, a similar concern raised in the regulatory market.

Furthermore, countries could also arrive at different conclusions with respect to the legal nature of VERs in the voluntary market. Thus, the same issue of potential market disparities, such as differential tax treatment, that is present in the regulatory market could also be applicable to the voluntary market.

To a certain extent, the legal nature of VERs could be even more ambiguous in the voluntary market compared to ERs in the regulatory market as there is no legal agreement like the Kyoto Protocol to clarify that the creation and sale of VERs do not create or grant 'any right, title or entitlement to emissions'.[63] This issue raises a larger public policy question regarding the potential need to clearly articulate that the creation of VERs and their transactions, like the issuance and trade of allowances in a regulatory market, do not lead to the 'privatization of the atmosphere',[64] or create a right to the emissions. This has been a public policy concern raised in the development of regulatory emissions trading programs for sulfur dioxide that extends to the regulatory GHG emissions trading programs[65] and could extend to the voluntary market.

Arguably, one of the reasons behind the US Clean Air Act articulation of regulatory allowance as permits was to clarify that those allowances were only limited rights that did not create a private property right in the atmosphere or the right to pollute.[66] While it may not be necessary for laws to specify that

[62] Existing case law suggests that the CISG treaty will probably not apply to GHG reduction transactions in the voluntary market as the CISG applies to goods (ie commodities), which are defined as 'corporeal moveable goods'. Franco Ferrari, 'Specific Topics of the CISG in the Light of Judicial Application and Scholarly Writing' (Excerpts) [1995] 15 Journal of Law and Commerce 1–126 <http://www.cisg.law.pace.edu/cisg/wais/db/editorial/ferrari931220h1.html>, accessed 15 June 2009; the Uniform Commercial Code, which has been adopted by a majority of states in the US, defines goods (commodities) as 'all things that are moveable at the time of identification to a contract for sale.' Uniform Commercial Code 2005 s 2-103 (1)(k) <http://www.law.cornell.edu/ucc/2/article2.htm#s2-103>, accessed 10 December 2008.

[63] FCCC/CP/2001/13/Add.2, Decision 15/CP.7 <http://unfccc.int/resource/docs/cop7/13a02.pdf>, accessed 15 February 2009.

[64] Wemaere and Streck (n 42 above), 39.

[65] Tom Tietenberg, 'Tradable Permit Approaches to Pollution Control: Faustian Bargain or Paradise Regained?' <http://www.colby.edu/personal/thtieten>, accessed 1 December 2008.

[66] Ibid.

VERs are permits, it may be useful to have some legal articulation that VERs do not create a private property right in the atmosphere or a right to the emissions so that the voluntary market is more consistent with the regulatory market and does not somehow undermine the public policy intent of the Kyoto Protocol, or other GHG emissions trading laws that try to make this policy and legal issue clear.[67]

Similar legal uncertainties that have been raised the regulatory market could also arise the voluntary market. In the voluntary market, there is no legal determination that VERs should be treated as a commodity, right in real property or some other legal classification and no legal structure to mandate such determinations have been made. If countries ultimately choose to make these legal distinctions, they may be different. If so, disparities in the voluntary market could be created. There may even be some additional legal uncertainty regarding VERs, as to whether their creation and sale could somehow implicitly confer a legal right to emit or a private right to the atmosphere. In the voluntary market, these uncertainties have generally been addressed on a contract by contract basis, but these issues could become more significant over time if the market persists and continues to grow.

5. Summary and Conclusions

The voluntary GHG market has contributed significantly to climate change mitigation. It has attained additional GHG reductions beyond those achieved by the regulatory market by raising awareness and engaging individuals and businesses to reduce emissions. It has also served as an innovative forum for developing market infrastructure, such as GHG accounting standards, protocols and registries, which are essential for transactions and a functioning market. The voluntary market has also informed the regulatory market.

As the voluntary market evolves, it may also need to address other legal issues to minimize any market uncertainties or barriers. The issues include greater harmonization of GHG accounting standards and protocols. Both the private sector and government have taken steps to address this issue, which could be effective. If governments ultimately assume this responsibility, consideration should be given to concerns raised that too much government

[67] Given the precedent established by the US CAA, it is possible that a regulatory system for GHG emissions trading in the United States would establish ER allowances and associated rights in the same fashion. For instance, one of the more prominent climate bills introduced in US Congress in 2007, the Lieberman-Warner Climate Security Act of 2008, followed the CAA precedent by defining GHG allowances as limited authorizations to permit one ton of carbon dioxide and states that emissions allowances 'shall not be a property right' and includes emission reductions offsets as 'offset *allowances*.' While this bill did not become law, it will influence future climate bills introduced in US Congress. Lieberman-Warner Climate Security Act of 2008 (S 3036 110th Congress) s 2403(a) (2008).

intervention or intervention at an early stage could have unintended consequences for the market.

The uncertain legal nature of VERs in the voluntary market may also need to be addressed by governments. Like the regulatory market, the question of whether VERs is a commodity, real property right, or other legal distinction could arise. The lack of a broad legal framework to address these questions and create some consistency of treatment among countries could ultimately cause uncertainty in the market and market disparities, similar to the concerns expressed in the regulatory market. Countries could reach different conclusions on this issue, if they do at all, exposing transactions to different or conflicting laws and legal treatment, thereby increasing risk. Furthermore, the legal nature of VERs may even be more uncertain due to the lack of policy to clarify that VERs and their transactions to do not create private property rights in the atmosphere, an issue that has been clarified in the regulatory market by the Kyoto Protocol.

For the voluntary market to maintain its momentum and build upon its success, these policy and legal issues may need to be addressed. Ultimately, these clarifications could reduce legal conflict and uncertainty that could arise, similar to the regulatory market. Addressing these issues could be part of a natural evolution that other markets have experienced as a part of growth and have done so successfully.

PART VII

POST KYOTO: MOVING TOWARDS COPENHAGEN

24

What Might a Future Global Climate Change Deal Look Like?

Murray Ward

1. Introduction

2009 is the year when global leaders are expected to come together and agree a new multilateral framework to protect the world's climate system. Under the United Nations Framework Convention on Climate Change (UNFCCC) negotiations have been underway for some years already on an update and enhancement of the Kyoto Protocol. Its first commitment period, under which most developed countries have taken on binding emissions limitation and reduction targets, ends in 2012.[1]

The current negotiations are framed by the 'Bali Action Plan' agreed at the 13th session of the UNFCCC Conference of the Parties (COP13) meeting in December 2007.[2] This set in place a two-year process expected to result in a global agreement at COP15 in Copenhagen in December 2009.

This paper sets out ideas for what the architecture of a global deal might look like.[3] It describes a concept involving enhanced commitments to constrain emissions in a quantitative and legal manner, as well as a broad set of other elements that are expected to be needed to bring a package together that can win the support of all countries. One issue raised, but left as an open question, is the extent to which all the elements needed to bring together a global deal are part of the UNFCCC Copenhagen Agreement package, and

[1] The US agreed to but did not subsequently ratify the Kyoto Protocol.

[2] See <www.unfccc.int/files/meetings/cop_13/application/pdf/cp_bali_action.pdf>, accessed 31 March 2009.

[3] This article draws, in particular, from a Briefing Paper 'The Architecture of a Global Climate Change Agreement' prepared as part of the Tony Blair/The Climate Group 'Breaking the Climate Deadlock' initiative in June 2008. (See <http://www.theclimategroup.org/special_projects/breaking_the_climate_deadlock>, accessed 31 March 2009.)

which may support the deal coming together, but might better remain outside the package per se.[4]

2. The Scale of the Task

For the world to have a reasonable chance of avoiding the worst effects of climate change, the challenge is have global emissions of greenhouses gases peak by 2020 and then to reduce them to 50% of 1990 levels by 2050. These goals have been articulated by international science and policy leaders—the second, even by the heads of State of the G8 countries at their 2008 meeting in Japan.[5]

For developed countries, the challenge is much greater than the global percentage of required reductions suggests. It is not expected that the goal of a 50% reduction will be equally shared by all countries. Figure 24.1, taken from the 2008 report by Tony Blair and the Climate Group,[6] provides an illustrative set

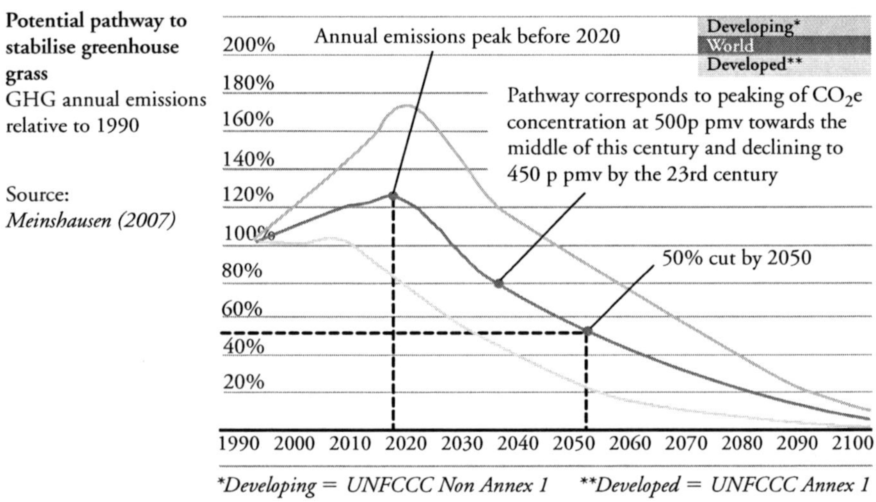

Figure 24.1. Emission pathways towards 450 ppm and 2°C

[4] UNFCCC negotiations work under a consensus rule of procedure. This requires the agreement, or at least non-blocking, of the over 180 countries engaged in the negotiations. It might therefore be easier to develop elements outside the UNFCCC process per se, such as sectoral technology cooperation agreements that might be of particular importance to a smaller group of key parties. Moreover, the reach of climate change, and climate change policy, is very wide, so cross-cuts with other key global agendas such as development, disaster relief and multilateral trade and investment.

[5] For example, see the news report of the G8 meeting at <http://search.japantimes.co.jp/cgi-bin/nn20080709a1.html>, accessed 9 April 2009.

[6] 'Breaking the Climate Deadlock—A Global Deal for Our Low-Carbon Future', The Climate Group, June 2008.

of pathways for global emissions and respective shares of emissions for developed and developing countries.

Lord Nicholas Stern points out in his 2008 report on the global deal,[7] that what the global 50% by 2050 goal means 'as a matter of arithmetic' is average global *per capita* emissions of about 2 tonnes carbon dioxide equivalent (CO_2e) by 2050—and that if there are any large economies greater than this, there need to be other equally large economies commensurately less than this.[8]

3. Meeting the Needs

The political challenge in front of the global community, then, is to craft, agree and *ratify* a multilateral agreement that adequately addresses the scale of the climate change problem. The agreement must therefore be able to meet the needs and expectations of all countries and relevant constituencies. This also must be done in the context of other imperatives that are bearing on country leaders, in particular energy security, water security, food security and sustainable development—and now, most recently, the global financial crisis.

To achieve, and be mindful of, these many objectives, of necessity, any agreement will have to be both very comprehensive and very flexible. For emissions to peak by about 2020 (in the face of current emissions growth trends), the agreement needs also to build on the international climate change policy that exists and is working. There is not enough time to start with a new clean page.

In particular, the Kyoto Protocol set in place a number of market mechanisms that have led to the emergence of an international carbon market. Of particular note, under the Clean Development Mechanism (CDM) the public and private sectors in developed countries have financed projects in developing countries that reduce emissions of greenhouse gases and contribute to their sustainable development. To support the various mechanisms, new legal structures have been established based on the monitoring, reporting, and review (and ultimately compliance) of emission inventories across developed countries' economies, and for projects undertaken in developing countries.

While the infrastructure of a workable global international treaty has been put in place by the Kyoto Protocol, the sheer complexity of the challenge ahead means that the new global agreement must additionally be open to further innovation and diplomacy at the highest levels. Put simply, the Kyoto Protocol was mostly about agreeing binding emission targets for developed countries and a set of mechanisms for how these might be met at least cost. The new global deal must also deal with this, but among a much broader set of objectives. It must additionally cover the nature

[7] 'Elements of a Global Deal on Climate Change', London School of Economics, April 2008.

[8] For reference, current per capita emissions of some large economies (and current average annual growth rates) are: United States 22T (0.4%); Russia 16T; (1.2%); Japan 10T (0.7%); EU-27 10T (0.4%); Brazil 15T (0.2%); China 6T (2.7%); India 2T (2.8%).

and scale of commitments that rapidly industrializing major developing country economies might now take on; the challenge of deforestation and forest degradation in tropical countries (which accounts for some 18% of global greenhouse gas emissions); the needs of countries, especially the most vulnerable, to adapt to the current and future effects of climate change related to historical and future emissions; and, in the light of today's global challenges, operationalize commitments made in the UNFCCC by developed countries to developing countries for technology transfer, capacity building and financing.[9]

4. Key Elements of a Global Agreement

Taken together, these above points clearly signal the need for an agreement that has at its core both some quantitative elements and some 'bigger picture'—new and innovative—elements. Quantitative elements of a Copenhagen Agreement are those that constrain emissions in a predictable way. This means a cap on the aggregate emissions of developed countries. It can also mean other forms of targets or commitments—for example of a sectoral nature in some key developing countries, or covering international aviation and marine bunker fuels. Importantly, it is these quantitative elements that create the basis for the emergence of a more robust and broad-based international carbon market. Emission limitation targets are a fundamental element of a future climate change agreement, among others, because they help to put a price on greenhouse gas emissions. They therefore are able to engage the world's private sector and mobilize the levels of investment in low carbon technologies and practices needed worldwide. Other carbon price signalling policy instruments such as carbon taxes have neither the dynamic engagement of the private sector, nor are they politically feasible as international policy instruments.

But these quantitative elements can be seen as threatening in some circles. In developed countries they can appear to exacerbate competitiveness concerns in key industry sectors, especially at a time of economic slow-down and job losses. And for developing countries, any suggestion that emission constraints might place a cap on their right to industrialize and address their priority development concerns is an anathema. They can rightly point to the fact that current levels of greenhouse gas concentrations have been caused by the emissions of the developed world, that emissions in developing countries are typically just a fraction of those in developed countries on a per capita basis and that growth in emissions in developing countries over the last few decades is frequently tied to the fact that emissions intensive manufacturing of products consumed in developed countries is now being done in their countries.

It is therefore unrealistic to expect agreement to be reached on a framework which, in its nature, is just a set of quantitative emissions constraints. Moreover,

[9] The 'Bali Action Plan' calls, inter alia, for... 'Nationally appropriate mitigation actions by developing country Parties in the context of sustainable development, supported and enabled by technology, financing and capacity-building, in a measurable, reportable and verifiable manner.'

such quantitative elements and the carbon markets they engender do not, in themselves, ensure an adequate mitigation response. And dealing with the climate change challenge is not just about mitigation. As set out above, there is more that the new global deal needs to address. Importantly, it is this new and innovative side of the agreement that provides the greatest scope for the needed innovations, leadership, and diplomacy.

Some of the bigger picture elements will be important to help enable acceptance of the elements on the quantitative side of the agreement. They represent the compensation or additional action that countries expect in return for committing to undertake greenhouse gas emission reductions—whether this be developed or developing countries. But there is more needed than just the facilitation of the quantitative side of the agreement and the establishment of the emission reductions ambition. By their nature, quantitative elements, especially those like emission caps that involve international carbon prices as a key driver for action, cannot be expected either to cover all sectors in all countries, nor necessarily achieve the full potential of emissions reductions in those sectors that are covered. In some circumstances, it will be additional measures that may play the greatest role in achieving the potential mitigation outcomes.

And it is not all about mitigation. The effects of climate change are large and increasingly looming, especially for some of the world's most vulnerable populations and ecosystems. An effective global agreement must now also take a firm stance on means to address adaptation needs.

To give some substance to the concepts just set out, Figure 24.2 provides a depiction of the two-sided architecture that is proposed. However, especially on the 'bigger picture' side, this is not intended to be exhaustive of the possible innovative ideas for needed and helpful elements. It should be seen as high-level and generally illustrative.

On the quantitative side, the emissions circles represent aggregate emission totals under the various forms of management. The difference between the solid line around the largest circle, ie the large 'FBTs in ICs' (Fixed and Binding Targets in industrialized countries) circle compared with the dashed lines surrounding the smaller three circles, is intended to denote the fixed nature of the former and the likely rate-based (intensity) nature of the latter. In an emissions trading context, the trading unit associated with 'FBTs in ICs' would be 'allowances'[10] and for the other three it would be 'emission credits' issued where performance was better than the crediting baselines that these targets and commitments represent.

[10] The technical term for 'allowances' used in the Kyoto Protocol is 'assigned amount units'. A target can be seen as a level of allowed emissions. Where a policy framework allows emissions trading by parties with targets, this is put into effect by issuing the target tonnes as units that can be traded by parties. The compliance requirement, then, is not for parties to meet their target emissions; rather it is for them to have sufficient units to match their emissions. The Kyoto Protocol also had mechanisms that allowed an increase in the supply of total units. Compliance-grade units could also be issued into the overall pool of units from some forest and land-use activities that sequester 'carbon' and from emission reduction (and sinks enhancement) projects in developing countries under the CDM.

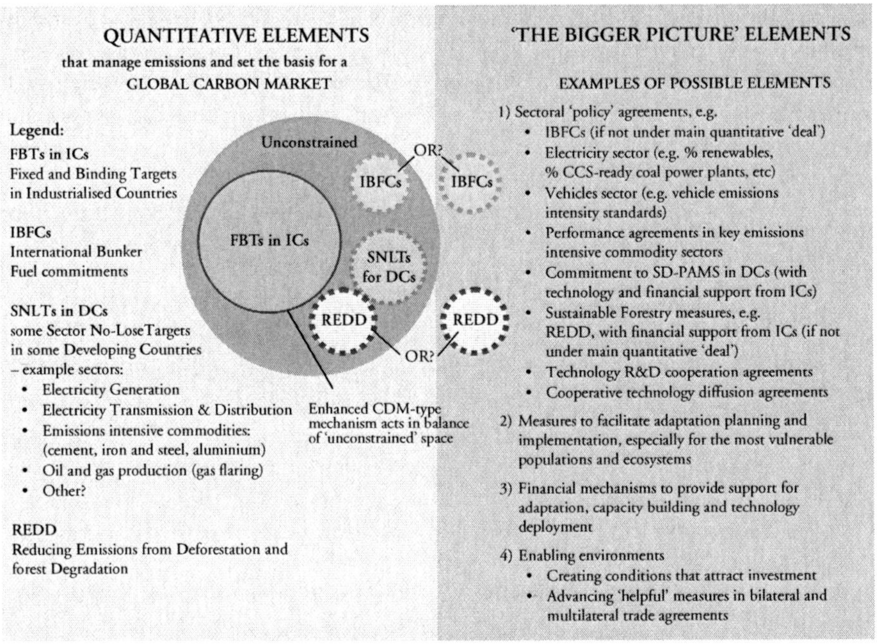

Figure 24.2. The two-sided architecture of a comprehensive and flexible agreement

In addition, as noted in the depiction the remaining 'unconstrained' space'[11] is where current and future enhanced CDM-type activities could be undertaken and provide a supply of credits into the 'FBTs in ICs' circle to enable those countries to meet their targets at lower cost.

Further with respect to the 'FBTs in ICs' circle, such a circle is made up of the aggregate of the allowed emissions represented by the developed countries' targets—eg the targets that Annex B countries took under the Kyoto Protocol. This is shown as a single circle, and this may result from a single agreement that all developed countries are party to. It may also be the sum of targets collectively negotiated by a group of countries under such an agreement, plus targets applying to countries or specific sectors that may sit outside the multilateral agreement but, nevertheless, represent self-imposed fixed and binding targets—for all or parts of their economy.[12] This situation may necessarily result in the

[11] The 'unconstrained space' represents any sources or sectors that are not covered under one of the policy forms represented by the quantified emission circles. There may of course be other policies that bear on these sources or sectors, just not ones that have quantitatively bound their emissions in the international agreement.

[12] A reality that faces negotiators is that the United States may, as with the Kyoto Protocol, struggle at home to ratify an international UN agreement. But this time round, the signs are that, under its new Administration, the US may agree, at a minimum, to establish domestic targets and a binding internal emissions trading scheme.

need for some rules that link the different emissions trading schemes, ie between those in the collective and those outside. However, the overall point is that there is a sum of allowed emissions stemming from the targets of all these developed countries, and these form the basis for international emissions trading of the cap-and-trade variety among them.

Sector No-Lose Targets (SNLTs) for developing countries are expected to be expressed in intensity terms (eg carbon dioxide (equivalent) emissions per Megawatt-hour of electricity (CO_2e/MWh), or per tonne of cement). The no-lose nature of these targets simply means that there is no compliance penalty if the targets are not met. However, because the purpose of such a mechanism is to scale up significantly the inward flows of carbon finance-supported low carbon technology (compared with the current CDM), these targets would be something that can reasonably be expected to be met and beaten. Any emission reductions below the SNLT would be eligible for the issuance of emission reduction credits. SNLTs should be realistic and not be seen as overly soft targets, opening the door to large credit generation for actions that are likely to be done anyway. Given that these targets will be negotiated as part of the package of the quantitative agreement, subsequent additionality assessments would not be required. This is one of the means by which this mechanism is different from any form of CDM.[13]

The process of negotiating an acceptable intensity baseline should be educated by all relevant factors by which a given developing country could improve its intensity in the sector through policy measures. Additional actions should be incentivized by carbon finance and through the prospect of generating carbon credits. In the jargon of the negotiations, this information is expected to be set out in 'NAMAs' (the developing country plans for 'nationally appropriate mitigation actions').[14] In this way, there will be an expectation on the developed countries' side that major developing countries will commit to some level of self-funded mitigations efforts, rather than just have all their efforts positively incentivized by developed countries through carbon credits.

The nature of possible international aviation and marine bunker fuel commitments (IBFCs) is somewhat less discernible. It is for this reason that they are shown as possibly occurring on either side of the quantitative and 'bigger picture' divide line. If on the quantitative side, this would imply that these sectors accept (and be held to) some form of a binding emissions target (whether of a

[13] Fundamental to the CDM, is a requirement that projects show they would not just have happened anyway. This is the principle of *additionality*. This is necessary because CDM projects take place outside the previously set quantitative constraint set by developed country targets under the Kyoto Protocol. CDM credits add to this Kyoto emissions cap, so allow additional emissions. If the projects would have been undertaken even without the CDM, the atmosphere will see greater emissions. See further above, chapter 12.

[14] A specific methodological tool, Sectoral Proposal Templates, has been developed by Ecofys and GtripleC to help developing countries prepare and present proposals for sector no-lose targets. See <www.sectoral.org>, accessed 9 April 2009.

fixed or intensity nature). This would set the stage for these two sectors to be sellers into the international carbon market if they met and beat their targets, and be buyers from it if they did not.

A possible scenario for their being on the bigger picture side would be if they struck a deal with the international community to voluntarily reduce or offset a certain percentage of their emissions. Their activity might therefore occur in the voluntary carbon market, not the compliance market that occurs through the elements on the compliance side of this proposed agreement.

Similarly, on the issue of Reducing Emissions from Deforestation and forest Degradation (REDD) opinions currently are still quite divided as to whether this sector in developing countries should be one which receives its needed financial support through public sector (or even voluntary carbon market) funds, hence on the bigger picture side. Alternatively it may be feasible to have its potential supply of credits incorporated into the compliance carbon market—with the risk of perhaps swamping the market and severely lowering the cost of carbon.

The issue of balance in supply and demand in carbon markets is not just relevant as to whether REDD is financed through public sector funds or the carbon market. It applies to any mechanism that may significantly increase the scale of the supply of credits coming from mitigation actions in developing countries, so it applies for SNLTs as well. Principally, this comes back to the level of ambition of the emission reduction targets of developed countries. But there is also a question as to whether compliance carbon markets should be seen as a source of financing for negative, zero, and low cost mitigation actions in developing countries. Or is enhanced public policy (and public funding, as applicable) the right vehicle to go after these reduction opportunities?

5. A Sectoral Perspective

When considering the proposed two-sided architecture from a sectoral perspective, a number of groupings and linkages become apparent. This has usefully been elaborated in work of the Pew Centre on Global Climate Change on what they refer to as an integrated multi-track climate framework.[15] In a subsequent 'Background Note' on sectoral approaches[16] (from which Figure 24.3 below is taken), they note that

> ... Sectoral approaches could sit alongside other types of action/commitments as elements of a comprehensive post-2012 framework. For example, the framework could

[15] D Bodansky and E Diringer, 'Towards an Integrated Multi-track Climate Framework,' Pew Centre on Global Climate Change, December 2007. <http://www.pewclimate.org/docUploads/Multi-Track-Report.pdf>, accessed 9 April 2009.

[16] Pew Centre 'Background Note: Sectoral Approaches in a Post-2012 Climate Framework,' Pew Centre on Global Climate Change, 2008. <http://www.pewclimate.org/docUploads/Sectoral-Background.pdf>, accessed 9 April 2009.

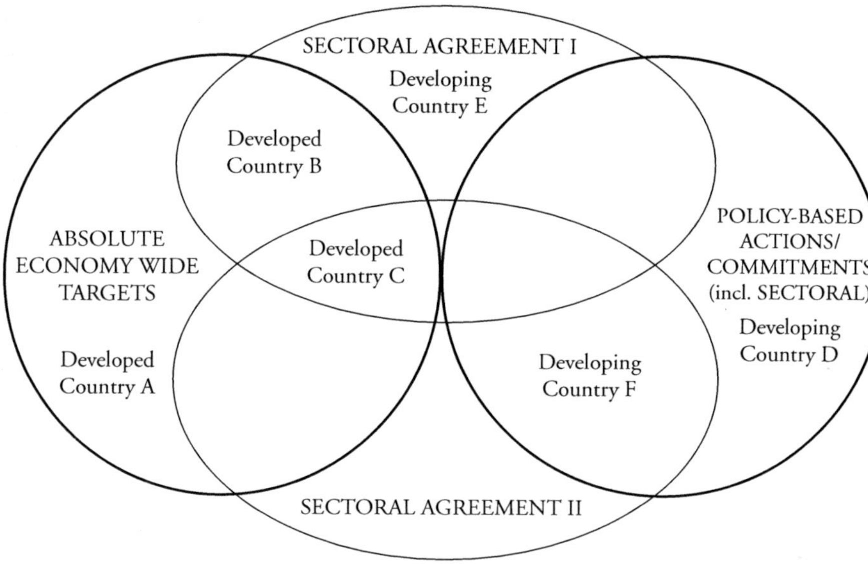

Figure 24.3. Possible means of engagement at a sector level[17]

Source: Pew Centre 'Background Note: Sectoral Approaches in a Post-2012 Climate Framework,' Pew Centre on Global Climate Change, 2008. <http://www.pewclimate.org/docUploads/Sectoral-Background.pdf>, accessed 9 April 2009

include absolute economy-wide targets for some countries; policy-based actions/commitments (sectoral or economy-wide) for other countries; and one or more overlapping sectoral agreements (with different country groupings in each).

In the illustration in Figure 24.3 above:

- Developed Country A takes an absolute economy-wide target;
- Developed Country B takes a target and participates in one sectoral agreement;
- Developed Country C takes a target and participates in two sectoral agreements;
- Developing Country D takes a policy-based (possibly sectoral) action/commitment;
- Developing Country E participates in a sectoral agreement; and
- Developing Country F takes a policy-based (possibly sectoral) action/commitment and participates in a sectoral agreement.
- The two-sided architecture depicted in Figure 24.2 is fully consistent with this sectoral concept helpfully set out here by the Pew Centre.

[17] <http://www.pewclimate.org/docUploads/Sectoral-Background.pdf>, accessed 9 April 2009.

6. Negotiation Process Issues

Leaders have affirmed a number of times that the UNFCCC is the proper forum within which to conclude a comprehensive climate agreement. As noted in the introduction section, this process is looking to achieve its primary outcomes at the 15th Session of the Conference of the Parties to the UNFCCC to be held in Copenhagen (COP15) at the end of 2009. In particular, it would be expected that all of the quantitative elements of the agreement shown in Figure 24.2 would be negotiated under the UNFCCC process.

However, it is quite feasible (indeed probably necessary) that some of the agreements on the 'bigger picture' side might better be struck outside of the UNFCCC process per se, but then be recognized as existing as the overall package of the UNFCCC deal is coming together. These, for example, might be cooperative financing or technology sharing elements that complement those in the UNFCCC agreement, for example agreed bilaterally between key countries or among smaller groups of countries, or even among key industries operating in some countries. It can be expected that these would be set out for given sectors and may result from sector-focussed efforts such as those that have been undertaken, for example, through task forces of the Asia Pacific Partnership on Clean Development and Climate[18] or similar EU technology assistance efforts on renewable energy. There might also be elements of the agreement, for example related to sustainable forest management or international trade, which might be struck in other multilateral fora.

While this model of including elements inside and outside the UNFCCC suggests a more complex negotiating process, it has the great advantage of providing significant flexibility for the necessary diplomacy and leadership to bring an overall global deal together. In particular, it gets past the problem that the UNFCCC is a forum where it is only national governments at the negotiating table. A global climate change deal will need leadership initiatives also from international business, local government and civil society non-government groups at large. This is difficult when they can only be observers—and are out of the room completely when the final deals are getting done.

7. Legal Issues—Beginning Reflections and Questions

The nature of this two-sided architecture and 'inside/outside the UNFCCC' deal raises significant questions about the associated legal framework required.

[18] The Asia-Pacific Partnership on Clean Development and Climate is an effort to accelerate the development and deployment of clean energy technologies. <http://www.asiapacificpartnership.org/english/default.aspx>, accessed 9 April 2009.

For the quantitative side, the legal framework of the Kyoto Protocol and Marrakech Accords can serve as a significant base—in particular the elaborated rules for Article 6 (joint implementation), Article 12 (clean development mechanism), Article 17 (international emission trading), Article 7(4) (accounting of assigned amounts) and Articles 5, 7 and 8 (monitoring, reporting, accounting and expert review). However, any elaborations or enhancements of existing mechanisms (such as the CDM) or new mechanisms (such as SNLTs) will require appropriate legal rules to be developed.

Some of the legal framework needs to be included in a Copenhagen Agreement. But if the Kyoto Protocol and Marrakech Accords[19] example is a model for what may also happen in the next years, much of the legal detail will be developed in the following year(s) prior to the ratification of the treaty by Parties. However, given the urgency to ensure that new commitments for developed countries start on 1 January 2013, the details on these commitments will be needed to be added quickly, not four years later as with the Marrakech Accord.

A larger set of legal questions needs to be considered with respect to the 'bigger picture' side, especially where such elements are brought together outside the UNFCCC process. How much of this, in fact, requires a clear legal framework? Put another way, how much can be done by voluntary, 'in good faith' declarations or memoranda of understanding by sovereign states and non-government entities? Where critical parts of the deal have come together on the basis of quid pro quo or mutual consideration, how can some of the elements that form 'part of the deal' be covered by clear and binding legal rules and agreements, and others not? These questions appear to be highly salient and provide plenty of challenge to innovative legal thinkers.

8. Concluding Thoughts

How likely is it that things will play out along the lines suggested here? This, of course, is also for others to judge. But, in many ways, the two-sided architecture, and the notion of elements coming together inside and outside of the UNFCCC process, is just a reflection of what appears to be occurring in practice. This paper attempts to describe this and give a sense of possible coherency and order to what, otherwise, may seem to lack these desirable features.

[19] See Decisions 15–19/CP.7 and Annexes.

25

The Role of Project-Based Mechanisms in the Future Carbon Market

Jos Cozijnsen and Michael J Coren

1. Introduction

We have learned much during the last few decades of managing environmental markets. Beginning with US experimentation with sulfur dioxide markets during the 1990's and now Europe's multi-billion dollar market for greenhouse gas emissions (GHG), trading mechanisms have been shown to dramatically cut the cost of pollution mitigation. From measures to reduce acid rain to ozone-depleting chlorofluorocarbons, market competition has spurred technological innovation and decreased costs. Yet we still have much to learn if a global emissions market is to support both industrial economies and a climate where concentrations of CO_2 stabilize below 450 parts per million by next century.[1]

Market mechanisms are central to this approach. Without them, the flexibility and time to develop crucial technologies and implement policies will be lost, along with almost a decade of learning from the cap-and-trade and project-offset provisions under the Kyoto Protocol. Offset mechanisms are influential initial tools to involve countries without binding emission reduction commitments, lower costs of compliance in countries with such commitments, and mobilize resources for sustainable development. The Kyoto Protocol defined a model that prepares countries for low-carbon economic growth (the Clean Development Mechanism or the CDM) and a project mechanism that mobilized net GHG emission reductions in countries with an emissions cap (Joint Implementation or JI). We should not discard it. In fact, the potential to refine and expand these concepts during the upcoming negotiations on a post-2012 legal framework scheduled for late 2009 in Copenhagen will strengthen the global architecture of any climate agreement.

This chapter asks, 'Can offset mechanisms help promote a low-carbon economy at a reasonable cost?' The answer, we believe, is yes. They are not the

[1] Tony Blair, 'Breaking the Climate Deadlock: A Global Deal for Our Low-Carbon Future'. Report submitted to the G8 Hokkaido Toyako Summit. June 2008.

entirety of the solution—regulation and economy-wide incentives are needed—but as an addition to cap-and-trade, an essential component. Reforms will go a long way toward making offset mechanisms an integral and effective part of the global climate architecture. Anticipated benefits are:

- offsets will link domestic emission trading systems (ETS) in advance of formal bilateral and multilateral agreements leading to more efficient global architecture;
- offset mechanisms can move advanced developing countries toward more ambitious market-based schemes, and rapidly direct investment to countries where emissions are rising and mitigation is needed most.

The first section of this chapter covers existing cap-and-trade systems and progress toward a global system. The second section identifies the building blocks needed to design an effective globally linked emissions trading system. Finally, a vision for a post-2012 climate treaty incorporating project mechanisms is outlined and the policies needed to increase global emission reductions and achieve the target of limiting temperature increase to 2°C are discussed.

1.1 Where we are: emission reductions and carbon markets

The Intergovernmental Panel on Climate Change (IPCC) concluded in 2007 that 'warming of the climate system is unequivocal' and humans are 'very likely' the cause.[2] The prescription was clear: to contain temperatures rising above 2°C, global GHG emissions should peak within 10 to 15 years[3] and fall to less than half of 1990 levels by 2050.[4] The target is unambiguous, and formidable. Industrialized countries will need to cut their emissions by 80 to 95% by 2050—a target now under discussion within the context of the UNFCCC process.[5]

Developing countries, with over half of global emissions (and growing), face a no less daunting task. To prevent carbon-polluting economic growth from swamping emission reductions in the developed world, these countries will need to chart a new path toward economic growth. Today, annual global GHG emissions stand at 8 tons per person: 6 tons in China, 2 tons in India, 10 in Europe and Japan, and 20 tons in the US.[6] This must drop to 2 tons per person on average by 2050 if we are to meet the IPCC targets, even as hundreds of

[2] IPCC, Climate Change 2007: Synthesis Report of the Fourth Assessment Report. 12–17 November 2007.

[3] Apart from land use and forestry

[4] Ibid.

[5] R Maas, 'Aspirational targets for 2050 Workshop organized by TFIAM and ACCENT', Utrecht 5–6 March 2009; 12 March 2009.

[6] IPCC, 'Climate Change 2007: Synthesis Report of the Fourth Assessment Report'. 12–17 November 2007.

millions of people strive to adopt the energy-intensive lifestyles prevalent in the industrialized world.[7]

But reaching these emission reduction goals will be impossible without rapid and sweeping changes to the growth model of the last four decades. It need not be impossibly expensive. While gross expenditures are large, net costs may not significantly curtail economic growth. An IPCC review of mitigation studies found that the projected cost of addressing climate change ranged from a 5.5% drop to a 1% increase in global GDP by 2050.[8] Studies have also identified many low-cost emission reductions opportunities in the developing world.[9] This includes 68% of emission reductions for stabilization and 46% of the total global mitigation, according to the UNFCCC.[10]

Simply making better use of the current market structures like the CDM in developing countries will not be enough to ensure required global reductions are achieved. For one thing, it is still too small. The CDM has leveraged about US $45.9 billion to date, but the sum still does not translate into the resources needed to move the industrializing world toward a low-carbon development path.[11] Cost estimates to address climate change are about US$175 billion per year through 2020, and US$23–54 billion in following years.[12] This is far more than official development assistance (ODA) for climate activities.[13] This implies that neither direct contributions from governments nor private sector investment alone is sufficient.[14] The key, therefore, is to make efficient use of public funds and leverage them to reduce emissions as quickly as possible through private capital and investments. In general, this will require three strategies:[15]

- Optimize deployment of funds by spreading risk across the private and public sector (e.g. subsidies, loan guarantees).
- Shift routine infrastructure investments toward more low-carbon systems and technologies (eg energy efficiencies, smart grids)
- Grow the pool of public and private investors willing to fund the infrastructure for a more sustainable growth model

[7] Ibid. [8] Ibid. [9] Ibid.

[10] UNFCCC, 'Investment and Financial Flows Relevant to the Development of an Effective and Appropriate International Response to Climate Change', 2007.

[11] UNFCCC, 'Update on Investment and Financial Flows', 2008.

[12] Jay S Constantine, IL Azevedo, LB Lave, MG Morgan and ES Rubin, 'Cap and Trade is Not Enough: Improving US Climate Policy'. Briefing Note from the Department of Engineering and Public Policy Carnegie Mellon University. March 2009.

[13] T Blair, 'Breaking the Climate Deadlock: A Global Deal for Our Low-Carbon Future'. Report submitted to the G8 Hokkaido Toyako Summit, June 2008.

[14] Stephen Groff, OECD. 'Presentation. International Policy Dialogue on Financing the Climate Agenda: The Development Perspective', 19–20 March 2009.

[15] BS Gentry, 'Investing in a Lower Carbon, More Climate Proof Future: Options, Tools, Mechanisms'. Background Paper on Investment and Finance to Address Climate Change. Center for Business and the Environment at Yale, July 2007.

Table 25.1. International investment pools for climate mitigation[16]

		Direct public investment	Grants		Debt		Equity	
			Public	*Private*	*Public*	*Private*	*Public*	*Private*
Value (total pools)		Large	Small	Small	Medium	Large	Small	Large
Returns sought	*Social*	High	High	High	High	Low	High	Low
	Financial	None	None	None	Low	Medium	Medium	High
Risks	*Project*	Yes	Yes	Yes	Some	Little	Some	Yes
	Technology	Little	Yes	Yes	Some	No	Yes	Yes
	Country	Yes	Yes	Yes	Some	Some	Yes	Some

At least one way to advance these aims is through project-generated finance. As the table by Yale University's Brad Gentry above describes (see Table 25.1), private equity and debt—perhaps the largest capital pools for financing climate mitigation—are also among the most amenable to project approaches with significant implementation, country and technology risks. While a mix of funding sources will be necessary, particularly at first, the largest pool of private investments will probably need to be mobilized by the medium to small pool of public financing.

On such a tight schedule, with such massive political and implementation challenges ahead, failure to use all the potential of tools at our disposal will put global emission reduction targets further out of reach. Mechanisms that support carbon finance in projects and programs in developing countries need to scale beyond today's structures to stimulate clean investments, reduce GHG emissions abatement costs, and spur technological innovation. This chapter derives lessons from the last decade of experimentation with regulated carbon markets, and outlines how these lessons might be applied to achieve goals in the context of a post-2012 regime.

1.2 Emerging national emission markets

First proposed by the United States in the negotiations preceding the Kyoto Protocol's adoption in 1997, 'cap-and-trade' systems are an essential instrument by which industrialized countries have chosen to reduce emissions.[17] Experience suggests costs will be lower—possibly much lower—than projected if the private sector is given proper incentives to reduce emissions. Two examples of early incentive mechanisms, each imposing a cost on business as usual practices,

[16] Ibid.

[17] RE Stavins and JE Aldy, (eds). 'Designing the Post-Kyoto Climate Regime: Lessons from the Harvard Project on International Climate Agreements. An Interim Progress Report for the 14th Conference of the Parties, Framework Convention on Climate Change'. Poznan, Poland, December 2008.

exemplify this phenomenon. The first occurred in the late 1980's as the United Sates began to comply with the phase-out of chlorofluorocarbon (CFCs) after signing the Montreal Protocol on Substances that Deplete the Ozone Layer in 1987.[18] Compliance costs were ultimately US$2.7 billion, just a fraction of the expected US$21 billion while yielding twice the expected CFC reductions.[19] A decade later, the US imposed an SO_2 (acid rain) cap-and-trade system on emitters and achieved its targeted reductions for USD1 billion to USD1.4 billion, a fraction of the expected USD3 billion to USD25 billion.[20]

Why the savings? A price on marginal emissions—that is, a price on every additional unit of GHG emitted into the atmosphere—does two things: (1) encourages future investment in efficient technologies (eg low-carbon power plants, insulation); and (2) changes current behaviour (eg reducing power usage).[21] Regulators in cap-and-trade systems impose a limit on emissions of GHG among many economic players, usually large industrial or energy emitters. Credits representing the right to release GHG (usually 1 ton) are traded among these emitters. A planned shortage of allowances or credits forces market actors to either reduce emissions internally, or buy and sell rights to release GHG gasses.

This creates economic efficiencies since firms with the lowest cost of abatement will reduce their emissions and sell the excess credits. Those with highest costs will buy the resulting surplus. This lowers total emissions, along with the cost of complying with the cap compared to fixed targets for each emitter. The larger the pool, the more efficient the system becomes by identifying the lowest-cost options. As a result, a truly global carbon market could cut abatement costs by 50%.[22] Shortages (lowering the cap) create ongoing incentives for emission reductions.

One notable addition to this cap-and-trade system is emission-reduction crediting. Although similar to cap-and-trade, it differs in important ways as it is (a) voluntary; (b) acts as credit sources for entities facing compliance obligations; (c) does not impose obligations on suppliers to surrender credits; and (d) grants credits for projects or activities based on a business as usual (BAU) baseline.[23] This may apply to 'no lose' sector targets (ie steel or glass), where a BAU baseline (or better than BAU) is determined and reductions below this baseline are creditable as offsets but carry no penalty if the targets are not met. It can also apply to project-level offsets through crediting of discrete projects or aggregation

[18] IH Rowlands, 'The fourth meeting of the parties to the Montreal Protocol: Report and reflection' (1993) 35(6) Environment 25–34.

[19] T Blair, 'Breaking the Climate Deadlock: A Global Deal for Our Low-Carbon Future'. Report submitted to the G8 Hokkaido Toyako Summit, June 2008.

[20] Ibid.

[21] S Constantine, A Jay, IL Azevedo, LB Lave, MG Morgan and ES Rubin, 'Cap and Trade is Not Enough: Improving US Climate Policy'. Briefing Note from the Department of Engineering and Public Policy Carneigie Mellon University, March 2009.

[22] IPCC, AR4 Synthesis Report: Topic 3, 2007.

[23] J Jaffe and RN Stavins, 'Linkage of Tradable Permit Systems in International Climate Policy Architecture'. Discussion Paper 2008–07, Cambridge, Mass: Harvard Project on International Climate Agreements, September 2008.

of project activities. The most significant of these programs are the so-called flexibility mechanisms under the Kyoto Protocol including the Clean Development Mechanism and Joint Implementation. These two mechanisms create offsets, or project-based emission reductions, that are sold into a cap-and-trade system for compliance targets under the Kyoto Protocol.

CDM projects are hosted by developing countries without emission caps that ratified the Kyoto Protocol. Offsets, known as certified emission reductions (CERs), satisfy binding emission reductions in Annex I countries. As at February 2009, more than 4,474 projects were in the pipeline, producing 2.9 billion tons of CO_2 reductions. The 1,370 registered CDM projects will generate 1.5 billion tons of CO_2 reductions up to 2012, and 465 CDM projects have already issued CERs.[24] Offsets created in this manner do not result in absolute emission reductions to the atmosphere. Instead, they allow Kyoto Protocol parties to use low-cost emission reductions achieved in developing countries to meet their targets. CDM-type project mechanisms are nevertheless an important step to involve developing countries into the mitigation and compliance system of the Kyoto Protocol.

Joint Implementation, or JI, operates in Kyoto parties that have accepted an emission limitation target. By allowing Annex I countries to exchange emission reductions, JI gives project-based flexibility for capped countries under the Kyoto Protocol and also for companies under a cap. By trading among countries, the environmental integrity of the systems is preserved and no emissions above the pre-determined cap are emitted without a penalty. This is the essential difference between JI and the CDM—JI credits convert one type of unit into another type of unit (under a cap) whereas CDM credits increase the total cap. At the moment, JI projects produce far fewer credits than the CDM. The Joint Implementation Supervisory Committee recorded only 160 projects at determination in March 2009, with 33 registered projects, accounting for an estimated 322 million tons of CO_2-equivalent emission reductions over the Kyoto Protocol's commitment period until 2013.

National emission trading systems to help Annex I countries reduce domestic emissions have already been adopted by the European Union, while others such as Australia, Canada, and the United States, have announced plans to adopt similar systems. Their scope and implementation strategies differ, but they share basic structures. Most try to cover a large proportion of economy-wide emissions, allow flexible compliance across time and space, and set the volume of both allowances and project-based offsets. Existing cap-and-trade systems include the European Union's ETS (2005), Japan's Voluntary Emissions Trading System (2006), the Norwegian Emissions Trading System (2005) (which became part of the EU ETS as of 2008), and the US Regional

[24] UNEP Risoe, 'CDM/JI Pipeline Analysis and Database', UNEP Risoe Centre, Roskilde, 1 February 2009.

Greenhouse Gas Initiative including 10 states (2009). Their scope of emissions differs dramatically; leading the way is the EU ETS with at least 11,000 industrial and power facilities in 25 European countries emitting approximately two billion metric tons of CO_2 (2005), or 45% of the EU's GHG.[25]

There is a perception that the EU ETS pilot phase, 2005–2007, did not provide additional overall emissions reductions due to surplus allowance allocation.[26] This was not entirely surprising as the process to determine a cap relied on EU Member States to submit National Allocation Plans (NAPs) based on projections, not verified, historical emission records as is now the case. Consequently, allowances exceeded emissions by about 4.5% during this period.[27] However, new research suggests the EU ETS is beginning to spur carbon solutions across large, medium and small enterprises in Europe, as well as within companies that operate outside the EU ETS.[28] Emissions were lower than expected during the first phase of the EU ETS according to a 2008 study that found annual emissions 2 to 5% below BAU projections—about 50 to 100 million tons of CO_2—between 2005 and 2006.[29] In the most recent phase, 2008–2012, preliminary evidence suggests the EU has experienced a 3% drop in industrial carbon emissions in 2008 relative to 2007. Analysts attributed 40% of this fall to the ETS and 30% to the economic recession. [30] They did not attribute the remaining 30%—other factors may be, for example, closures or temporary shutdowns of installations

Businesses also report that the EU ETS is having an impact on corporate behavior and planning. Of the 167 industrial companies that responded to a 2005 McKinsey&Company survey, about half claim that the EU-ETS has a strong or medium impact on decisions to develop innovative technology.[31] Quantifying the impact of these decisions is difficult, but anecdotal evidence suggests that they may be resulting in real emission reductions. A Deutsche Bank manager reported in 2009 that some of the Bank's clients, especially large

[25] J Jaffe, and RN Stavins, 'Linkage of Tradable Permit Systems in International Climate Policy Architecture'. Discussion Paper 2008–07, Cambridge, Mass: Harvard Project on International Climate Agreements, September 2008.

[26] Fraunhofer Institute, CEEM and Jos Cozijnsen. 'Increasing the Ambition of EU Emissions Trading, An Assessment of the Draft Second Allocation Plans and Verified Emission Reports of Germany, the United Kingdom and the Netherlands'. A report to Greenpeace International, June 2006, text available at: <http://www.isi.fraunhofer.de/publ/downloads/isi06b30/Ambition-eu-emissions-trading.pdf>, accessed 21 May 2009.

[27] Optimistic economic growth and unusually warm winters may have also contributed to this.

[28] Annie Petsonk and Jos Cozijnsen, 'Harvesting the Low-Carbon Cornucopia: How the European Union Emissions Trading System is Spurring Innovation and scoring results', Environmental Defense Fund, Washington DC (March 14, 2007).

[29] A Ellerman and B Buchner, 'Over-allocation or Abatement: A Preliminary Analysis of the EU ETS', based on the 2005–06 Emissions Data, Environmental and Resource Economics, V.39, 2008.

[30] Press release New Carbon Finance: 'Emissions from the EU ETS down 3% in 2008', 16 February 2009.

[31] McKinsey & Company, 'Review Of EU Emissions Trading Scheme–Survey Highlights', November 2005.

utilities in Europe, 'look at emissions trading and factor a long term price signal into their investment decision making' and have taken actions such as increasing the efficiency of Eastern European industrial assets, expanding natural gas capacity before the third phase of the EU ETS compliance period (2012–2020), and refurbishing facilities in cement and glass manufacturing.[32]

2. Project-Mechanisms as a Transitional Tool

Offsets can serve two critical functions in the emerging global climate regime. First, they will act as the preliminary links among domestic trading systems. Secondly, they will inject critical public and private sector funds into developing countries, some of which may adopt emission targets in the future. The transitional function of project-based mechanisms is often overlooked amid criticism of the CDM and related project mechanisms. These criticisms, roughly divided into concerns about environmental integrity, inefficient subsidies or incentives, low sustainable development benefits, and poor governance, do not necessarily detract from benefits that arise from their ability to serve as linking mechanism among developed and developing nations.

Linking occurs when 'a tradable permit system's regulatory authority allows regulated entities to use emission allowances or emission reduction credits from another system in order to meet their domestic compliance obligations'.[33] Linkages can be one-way (unilateral) or two-way (bilateral or multilateral), or even occur without explicit decisions to link them where systems allow 'internationally recognized' offsets to enter the trading system.[34] These linkages enhance the ability of emission trading systems to accommodate a diversity of abatement technologies and act as cost containment mechanisms (assuming the environmental integrity of the project-mechanism). Mutual recognition of offsets is a way to gain the benefits of linkage without harmonizing systems.[35]

While project-based mechanisms have significant limitations, this typically reflects the flip side of their advantages: efficiency, flexibility, and low cost. Project mechanism will not naturally lead to equal sectoral (or regional) changes. By definition, they seek out low-cost, low-risk, high-volume projects (giving priority to potent non-CO_2 GHG gases) while excluding some sectors where fund-based or public interventions may be necessary to mobilize emission reductions. As a result, CDM or JI mechanisms do not typically incentivize

[32] CarbonTradeEx 2009. 'Session P-7: 2013: Architectures for a Global Carbon Market,' public communication.

[33] RE Stavins and JE Aldy (eds). Harvard. 'Designing the Post-Kyoto Climate Regime: Lessons from the Harvard Project on International Climate Agreements. An Interim Progress Report for the 14th Conference of the Parties, Framework Convention on Climate Change'. Poznan, Poland, December 2008.

[34] Ibid.

[35] Ibid.

large-scale technology change or accomplish sectoral emission reductions when operating alone. Project-based mechanisms, therefore, should be considered a complementary tool to be used alongside policies that encourage more systematic changes.

Reinventing or abandoning project-based mechanisms would however discard an invaluable tool in financing future mitigation activities. Retaining project-based mechanisms in the post-2012 climate agreement depends on the ability of negotiators to address and implement technical fixes in project-based policies, and to communicate the benefits these bring to a global architecture.

2.1 Linkages among markets

Advantages of offset mechanism for markets are clear: the more low-cost offsets are allowed to be imported into a cap-and-trade environment, the more ambitious emissions trading participants can be in cutting emissions at an acceptable cost. Project-mechanisms act as indirect linkages between markets offer a price safety valve and private sector incentives to reduce emissions. Wider economic benefits for the functioning of the market include:[36]

- improved market liquidity;
- reduced price volatility;
- reduced market power and manipulation concerns; and
- more efficient market for allowances and credits.

These benefits have already led to a more integrated global market. The most comprehensive model for the operation of a global emissions market is the EU ETS. The European GHG markets are indirectly linked through the European Commission's 'Linking Directive' (2004/101/EC) permitting ETS participants to use CERs (CDM) and Emission Reduction Units (JI) to meet compliance obligations giving companies under the EU ETS more flexibility to reach emissions reductions and scope to cut overall costs.[37] The EU ETS may also move toward a more internationally compatible system: Article 25 of the EU ETS Directive foresees linking to other cap-and-trade systems by mutual recognition of ETS's and allowances. Although no linking has occurred through this provision to date, the EU is promoting a 'robust OECD-wide carbon market by 2015', and inclusion of economically advanced developing countries by 2020. The EU Commission has already explicitly called for 'a global carbon

[36] J Jaffe and RN Stavins. 'Linkage of Tradable Permit Systems in International Climate Policy Architecture'. Discussion Paper 2008–07, Cambridge, Mass: Harvard Project on International Climate Agreements, September 2008.

[37] Directive 2009/29/EC of the European Parliament and of the Council of 23 April 2009 amending Directive 2003/87/EC so as to improve and extend the greenhouse gas emission allowance trading scheme of the Community: <http://eur-lex.europa.eu/LexUriServ/LexUriServ.do?uri=OJ:L:2009:140:0063:0087:EN:PDF>, accessed 15 June 2009.

market [that] can and should be built by linking comparable domestic emission trading systems'.[38]

Other nations appear to be following the lead. The new Australian government's scheme has been designed to link with international carbon markets. It would start with an unlimited access to the Kyoto market (AAU, ERUs and CERs) eventually link with other ETS.[39] The United States' 10-state RGGI program has included a similar policy in the Model Rule allowing several one-way links where up to 10% of a source's emissions can be covered by offsets. The recently proposed Waxman-Markey bill (2009),[40] outlining a national ETS for the US, also allows use of international allowances and international offsets under strict conditions, such as a 20% discount. An analysis of the bill by Point Carbon estimates that 30% of the overall cap can be met by offsets between 2012 to 2030, rising to 65% by 2050 (equivalent to 1,409 million tons of CO_2e in 2012) with offsets equally split between domestic and international projects.[41] The actual number of permitted offsets is determined by through executive authority of the ETS administrator based on environmental integrity and the price of allowances in the system.

2.2 Investment mechanism for developing countries

Flexibility mechanisms are also seen as a pioneer of crediting mechanisms for emerging economies to adopt low-carbon technologies and market-based emission strategies. They are designed to jump-start emission reduction activities in the private sector as well as facilitating the flow of technology and know-how to, between, and within developing countries. This is an area where the CDM can claim some success: CDM was adopted more rapidly than many expected once the structures were in place to support it. Although the majority of projects have been concentrated in a few nations—those where emissions were growing the fastest, such as China and India—its diffusion in recent years has accelerated, despite room for improvement, especially in Africa.

Researchers critical of the CDM believe a far more efficient and environmentally effective trading system is possible, but some including Stanford University researcher Michael Wara recognize that the mechanism has at least 'achieved remarkable participation in the developing world'.[42] Wara outlines

[38] EC Communication, 'Towards a comprehensive climate change agreement in Copenhagen', COM(2009) 39, page 2, January 29, 2009.

[39] Australian Government, Department of Climate Change, 'Carbon Pollution Reduction Scheme: Australia's Low Pollution Future: White Paper', <http://www.climatechange.gov.au/whitepaper/summary/index.html>, accessed 15 December 2008.

[40] <http://markey.house.gov/index.php?option=content&task=view&id=3583&itemid=125>, accessed 15 April 2009.

[41] PointCarbon, 'Waxman-Markey Bill: Focus on Offsets'. 10 April 2009.

[42] MW Wara, 'Measuring the Clean Development Mechanism's Performance and Potential'. Working Paper #56. Program on Energy and Sustainable Development, Stanford University, July 2006.

measures like discount rates, regulatory oversight and improved verification systems as critical reforms if the CDM can realize ambitious expectations of global emission reductions.[43]

Whether the CDM remains the primary offset mechanism, or informs the evolution of a new generation of offset policies, it is likely to remain a support and transition tool for developing countries. In Copenhagen, UNFCCC Parties will be considering an array of measures outlined by the Ad Hoc Working Group on Further Commitments for Annex I Parties under the Kyoto Protocol in April 2009 to assess what direction the CDM—and other project mechanisms—will move.

2.2.1 Reformed CDM

Reforms will be a prerequisite to winning acceptance of an expanded CDM by Annex I countries. CDM reform may range from relatively modest changes, to a new suite of mechanisms that preserve only the core structure of the original. Procedural reforms being considered include standardization of CDM project baselines, expanded access for least-developed countries, adoption of rules for sectoral CDM approaches, limiting project-based CDM to least developed countries, and phase-out for advanced developing countries. New activities that may come under the CDM also include agriculture, forestry and other land use (AFOLU), as well as carbon capture and storage.

Efforts at CDM reform will meet entrenched resistance from countries which benefit from the current system. The recent push by the EU for rules to enhance objectivity in baselines and emission removal benchmarks in the CDM was strongly resisted by Brazil, China, Columbia, India, and South Africa during meetings in 2008. Despite support among developed Annex I nations, developing countries are generally seeking to preserve the project-based nature of the CDM and prevent sectoral targets for developing countries.[44] The debate over the CDM indicates the momentum behind the preservation of project mechanisms in a future framework, but its long-term survival will likely depend on the dramatic reforms that regard such policies 'as a reward mechanism to stimulate investments, not as an accounting tool for perfect offsetting'.[45]

2.2.2 Crediting for NAMAs

Nationally appropriate mitigation actions (NAMAs) are an attempt to stimulate further GHG reduction measures in developing countries. NAMAs (decision 1/CP.13 in the Bali Action Plan) can turn part of the voluntary mitigation actions

[43] MW Wara. and DG Victor, 'A Realistic Policy on International Carbon Offsets'. Working Paper #74, Stanford University, April 2008.

[44] iisd Reporting Services, COP 14 Highlights: Wednesday, 10 December 2008. International Institute for Sustainable Development (IISD) Vol. 12 No. 393, 11 December 2008.

[45] AA Niederberger, 'Scaling up energy efficiency under the CDM' in A Reformed CDM, UNEP Riso Centre, Roskilde, December 2008.

by developing countries into international credits. Emissions reductions from NAMA—established against a reasonable baseline in the sector where national actions are undertaken—could fully or partly be credited for countries with emission targets, provided they lead to an overall contribution to global reductions. These actions include but are not limited to: (a) sustainable development policies and measures; (b) low-carbon development plans; (c) sector-based mitigation actions and standards; (d) reduction of emissions from deforestation and forest degradation; (e) technology deployment programmes; (f) relevant standards, laws, regulations and targets at a national or sectoral level; and (g) voluntary cap-and-trade schemes in non-Annex I Parties..

2.2.3 Reduced Emissions from Deforestation and Degradation (REDD)

The Bali Action Plan committed the parties of the UNFCCC to explore the feasibility of REDD as a global GHG mitigation activity. Quantitative targets were not agreed upon at the meeting, but parties such as the EU have called for gross[46] tropical deforestation reductions of at least 50% below current levels and a full stop to global forest cover loss by 2030.[47]

REDD is being discussed as mechanism under the UNFCCC that generates credits for compliance purposes through the carbon market and or as part of an international fund. Crediting may be preferably at the national or subnational level. International opinion has stressed the importance of sub national accounting allowing for demonstration projects in the initial stages. Ultimately, REDD credits could be generated below a national, conservative historic emission baseline creating emission reductions over time relative to the BAU scenario, a position advocated by the Coalition of Rainforest Nations, led by Papua New Guinea.[48] The proposal has the potential to direct billions of dollars annually into the land use and agricultural sectors providing early and large emission reduction opportunities, especially by halting deforestation and degradation responsible for about 15 to 20% of human-induced greenhouse gas emissions.[49]

Outside the UNFCCC negotiations, NGOs have advocated ways to support immediate investments in developing countries while securing long-term reductions. An example is the CLEAR Path (Carbon Limits + Early Action = Rewards) advocated by the Environmental Defense Fund (EDF) will 'award carbon allowances to emerging nations who voluntarily take on early carbon caps, thereby helping to finance their shift to a low-carbon future'.[50] This

[46] The stricter 'gross' language is more demanding than a net target which would allow plantations and other such cultivated forests to replace natural forests from a climate perspective.

[47] EC Communication, 'Towards a comprehensive climate change agreement in Copenhagen', COM(2009) 39, page 2, January 29, 2009.

[48] Submission of several countries gathered in the Coalition of Rainforest Nations to the UNFCCC, in: FCCC/AWGLCA/2009/MISC.1/Add.4, April 7 February 2009.

[49] R Pirard, 'Reducing Emissions from Deforestation and Degradation in non Annex 1 Countries: Breaking the Climate Deadlock'. Briefing Paper. The Climate Group.

[50] Environmental Defense Fund, 'Carbon Limits + Early Action = Rewards'. 3 December 2008.

voluntary measure rewards developing countries for assuming early caps by creating a multi-year national cap, initially set above current emissions levels. The difference between the cap and actual emissions represent fully tradable global carbon credits for additional financing investments in clean technologies and mitigation. This idea can be an elaboration of current proposals by Parties in the framework of the Bali Action Plan, for example the proposal of setting aside a percentage of post 2012 Annex I Parties AAUs to compensate for mitigation actions by developing countries.[51]

3. Getting There From Here: Mechanisms and Structure For a Future Global Market

There is no easy road to guide reforms of project-mechanisms acceptable for a global climate agreement. The multilateral process required to reach an agreement through the UN involves elaborate negotiations among more than a hundred countries to reach a consensus on the details of emission reduction mechanisms. Linking mechanisms, some tied directly to project offsets, can play a central role developing consensus in the coming negotiations. If climate negotiators succeed in securing a global scheme, project mechanisms can offer a primary conduit by which the global climate architecture is connected to the developing world. If a deal fails to emerge from Copenhagen, then an interim solution may consist of internationally arbitrated or bilateral linking mechanisms (CDM, sectoral agreements) with domestic cap-and-trade systems. Excluding project-based mechanisms from either one of this approaches will make global reductions goals that much more difficult to achieve.

This chapter has sought to outline the role of project-based mechanisms in stimulating emission reductions in a future climate regime. Regardless of the outcome of the talks, unilateral or bilateral linkages will almost certainly lead to a stand-alone climate architecture composed of linkage agreements and unilateral emissions reduction commitments. Ultimately, if an agreement is not immediately reached, this could evolve into a coordinated, international system to ensure the environmental integrity of various trading schemes. Elements of this approach would need to be elaborated as the world shifted from national or regional ETS to a more globally harmonized system. The specifics of such a system will be a source of future debate from harmonized emissions cap trajectories to common registries.

[51] UNFCCC, Ideas and Proposals on paragraph 1 of the Bali Action Plan, FCCC/AWGLCA/2008/16/Rev.1.

This transitional approach is important as full multilateral carbon markets are probably a decade or more away.[52] As the CDM and JI mechanisms have shown, technical barriers to linking are low. The crucial design aspect of linking domestic emission trading systems is not technical compatibility, but regulatory control over the trading schemes and the contribution of the mechanisms to global reductions. This is evident in the proposed US schemes that have taken a cautious stance toward international offsets, imposing both qualitative and quantitative screens. This comes at a cost. Trade improves flexibility, lowers costs, and enhances diversity; restrictions on ETS linkages will initially diminish these benefits in favor of domestic price and quality control. But as with trade, linked compliance systems will be a two-way street. Linkages depend on reciprocity and movement toward compatibility before full fungibility is possible. Markets allow for interim solutions such as arbitrage and discounting to compensate for gaps. In the near future, the CDM and other offset provisions should be tools enabling developing countries to benefit from market-based mechanisms and developed countries to make the transition to linked cap-and-trade systems within a global architecture.

Delegates arriving in Copenhagen at the end of 2009 will arrive with political winds at their back encouraging them to leave with a robust climate framework. Yet there is still the need for strong and explicit mandates to be given before the multilateral negotiations begin in Copenhagen this year. Former UK prime minister Tony Blair argued in a statement before the G8 Major Economies Meeting in 2008 that 'without a clear political direction' for the UN talks, they will be a 'negotiator's nightmare'.[53] Early indications hold some promise. UNFCCC Executive Secretary Yvo de Boer stated at the close of the 2009 Bonn talks, the first of five negotiating sessions before COP 15 in Copenhagen, that the countries had 'narrowed gaps in many practical areas,' and 'have the necessary clarity to move into intensified negotiations based on texts'.[54] Many pitfalls remain. Project-based mechanisms that link national and regional cap-and-trade systems with the developing world, as well as each other, should be a central principle of any treaty that emerges from these talks if it is to have a realistic hope of achieving swift, globally significant emission reductions.

[52] A Tuerk et al. 'Linking of emissions trading schemes Synthesis Report Draft. Climate Strategies. 28 February 2009.

[53] T Blair, 'Breaking the Climate Deadlock: A Global Deal for Our Low-Carbon Future'. Report submitted to the G8 Hokkaido Toyako Summit. June 2008.

[54] Climate-L.org, 'UNFCCC Bonn Talks Conclude'. <http://climate-l.org/2009/04/09/unfccc-bonn-talks-conclude>, accessed 8 April 2009.

26

A Post-2012 Vision for the Clean Development Mechanism

*Christiana Figueres and Charlotte Streck**

1. Introduction

Despite the many calls to reform the flexibility mechanisms of the Kyoto Protocol, their conceptual underpinnings are strong and it is likely that the idea of the mechanisms, in fact the mechanisms themselves, will survive in a post-2012 climate regime. It is therefore timely to review the strengths and weaknesses of the architecture of the mechanisms, in particular of the Clean Development Mechanism (CDM), in the light of a post-2012 international climate treaty. Whether and how the CDM should continue serving the UN Framework Convention on Climate Change (UNFCCC), and which additional mechanisms could complement its mandate is the subject of this paper.

The core of this chapter is dedicated to investigating how a post-2012 climate treaty could provide incentives for greater emission reductions in developing countries. We start by summarizing the current frustrations with the CDM and review incremental modifications that could be considered to strengthen the mechanism without requiring a decision of the Parties to the Kyoto Protocol or changes to the Marrakech Accords.[1] We then explore a potential post-2012 scenario that combines three possible financial mechanisms to promote large-scale mitigation in developing countries: the activity-based CDM as we know it albeit improved, a second market mechanism which would seek to improve the long-term emission trends of developing countries, and a third, non-market based financial mechanism which would follow a scheme of payment that is not commoditized and does not create international offsets.

* This chapter is based on an article by the same authors, 'The Future of the CDM in a Post-2012 Climate Agreement' (2009) 3 Journal of Environment and Development.

[1] The Marrakech Accords were adopted at the 7th Conference of the Parties to the UNFCCC and confirmed by the 1st Meeting of the Parties of the Kyoto Protocol and contain the implementation guidelines for the Kyoto Protocol. Decisions 1-19/CP7 reflected in decisions 2-13/CMP.1

2. The Clean Development Mechanism: the Expectations of Success

The Intergovernmental Panel on Climate Change (IPCC) has left no doubt that there is an urgent need to scale up mitigation to levels that go far beyond that which was intended by the Kyoto Protocol and its flexibility mechanisms. We now know that the CDM falls short of delivering the emission reductions needed to lower the emission trajectories of developing countries in the longer term. But the CDM must be recognized as a crucial starting point in developing country efforts to contribute to global emission reductions. So much has been accomplished over the past decade that we have fallen into a permanent inflation of expectations, which obscures the beginnings of the instrument. The CDM was created in 1997 by Article 12 of the Kyoto Protocol, that is nothing more than a short 10-paragaph text that defines the mechanism and its two goals. These emphasize that eligible 'emission reductions resulting from each project activity' must be certified on the basis of 'real, measurable, and long term benefits related to climate change', and that reductions must be 'additional to any that would occur in the absence of the certified project activity'.[2] Parties to the Protocol have since then elaborated the full modalities and procedures for project activities and established the Executive Board (EB) as the operational decision making body, assisted by various expert panels and working groups.[3] With the linking into the International Transaction Log[4] and the connection to the European Union's Community Independent Transaction Log,[5] the international operational structure for the functioning of the CDM is complete.[6] The volume of transactions under the CDM has more than tripled from US$5.25 billion in 2006 to US$18 billion in 2008.[7] The achievements of the past 10 years cannot be underestimated: the CDM has established a structure that facilitates a functional carbon market by defining the standards and processes for creating tradable emission reductions, consolidating methodologies, streamlining procedures, and reducing global mitigation costs.

[2] Article 12.5(b) and (c) of the Kyoto Protocol.

[3] For a description of all Panels, Working Groups and Teams see <http://cdm.unfccc.int/Panels/index.html>, accessed 21 May 2009. For a list of operational entities see <http://cdm.unfccc.int/DOE/list/index.html>, accessed 21 May 2009.

[4] The International Transaction Log (ITL) is the system that verifies emission reduction transactions among the industrialized countries that have emission reductions obligations under the Kyoto Protocol. The ITL ensures the transactions are consistent with all established rules before authorizing the transfer of reductions from one registry account to another.

[5] The European Union established its own transaction log to verify the transactions among members of the EU and this transaction log is now electronically linked to the International Transaction Log.

[6] As a result the price spread between EU allowances and secondary CERs decreased substantially: Axel Michaelowa and A Vasa, 'Uncertainty in climate policy-impacts on market mechanisms' in G Gramelsberger and J Feichter (eds) *Dealing with Uncertainty in Climate Research and Policy* (Springer, forthcoming 2009).

[7] World Bank, 'State and Trends of the Carbon Market' (2009) <http://carbonfinance.org/docs/Carbon_Trends_2009>, accessed 27 July 2009.

Ironically, however, success breeds rising expectations. As the supply of CDM projects has grown, and as the need for increased mitigation in developing countries has become clearer, stakeholders clamor for 'more and better' from the CDM. The past two years in particular have seen a plethora of publications expressing frustration with the CDM. Given the breadth and depth of those writings we simply summarize here the main areas of discontent.

2.1 Shortcomings

The long-term success of the CDM can be best measured in respect to its capacity to measurably mitigate the emissions of greenhouse gases (GHG). However, this evaluation is not easy as the CDM creates emission reductions that can be used to offset emissions in industrialized countries. A crucial aspect of CDM emission reductions is that they be 'additional' to what would have occurred otherwise. The additionality concept of the CDM has been debated vigorously, with some authors claiming that many registered projects would have occurred anyway[8] while practitioners in the field and business associations complain that the EB is being excessively stringent in its assessment of additionality.[9] In between both arguments remains the fact that the additionality of individual projects is difficult to prove and even more difficult to validate.

2.1.1 Inefficient operation

After a slow start, the CDM has seen an explosion of projects that was unimaginable in 1997 or even in 2004 when the first project was registered. Over the past four years the CDM has registered 1,500 projects and is gearing up for an additional 2,800 projects which are now in the pipeline.[10] After long delays the CDM system is now fully funded and the technical support provided by the Secretariat has grown to meet the demand. There have been copious complaints

[8] Axel Michaelowa and Pallav Puohit, 'Additionality determination of Indian CDM projects. Can Indian CDM project developers outwit the CDM Executive Board?' (Discussion Paper CDM-1, Climate Strategies, London, 2007) <www.noe21.org/docs/Michaelowa-teripress-2007>, accessed 21 February 2009; Lambert Schneider, 'Is the CDM fulfilling its environmental and sustainable development objective? An evaluation of the CDM and options for improvement' (Report prepared for WWF by Oko Institut, Berlin 2007) <www.oekoinstitut.de/oekodoc/622/2007-162-en.pdf>, accessed 21 February 2009; Rie Watanabe et al, *The Bali Roadmap for Global Climate Policy–New Horizons and Old Pitfalls* (Wuppertal Institute for Climate, Environment and Energy, 2008).

[9] In particular, for its use of investment analysis as a litmus test. International Emissions Trading Association (IETA), 'State of the CDM 2008' (2008) <http://www.ieta.org/ieta/www/pages/getfile.php?docID=3111>, accessed 21 February 2009; UNFCCC, 'Call for Input on Non-Binding Best-Practice Examples on the Demonstration of Additionality to Assist the Development of PDDs, Particularly for SSC Project Activities' (2007) <http://cdm.unfccc.int/public_inputs/dev_PDDs/cfi/R5142ZGKFD6FF6IW5MZEID3DJJKYH>, accessed 12 March 2009. Some of the comments prove the dissatisfaction with the additionality tool.

[10] It has been shown that not all pipeline projects move forward to registration due to barriers including financing, regulatory risks, transaction losses, etc.

about year-long delays in the approval of methodologies,[11] about the one- to two-year time lag in the assessment of projects,[12] and, recently, criticisms about the ineffectual operation of Designated Operational Entities (DOEs), which have become new bottlenecks in the functioning of the CDM.[13] It seems that one of the consequences of demand-driven growth is an inevitable delay between appearance of the demand and response of the system.

2.1.2 Insufficient contribution to sustainable development

The CDM was created with two objectives: lowering the cost of global climate change mitigation and contributing to the sustainable development of developing countries.[14] However, as a market mechanism that gives monetary value only to emission reductions and searches for the highest volumes at the lowest price, the CDM has been more effective in reducing mitigation costs than in broad contribution to sustainability.[15] In this criticism most authors assume a definition of sustainable development that focuses on local stakeholder participation, local job creation, and small scale renewable energy supply.[16] From a climate change perspective, however, one could argue that it is much more worrisome that the CDM has not promoted sustainable development writ large: it has not moved developing countries toward low carbon development paths[17] based on more sustainable energy production and consumption patterns and more sustainable forest management practices.

[11] IETA (n 9 above). [12] Michaelowa and Purohit (n 8 above); IETA (n 9 above).

[13] J Hoogzaad, A Korthius, C Streck, 'A Call to Reform' (October 2008) 5 Carbon Finance, 16, 17.

[14] Kyoto Protocol to the United Nations Framework Convention on Climate Change (adopted 11 December 1997, entered into force 16 February 2005) 37 ILM 22, art 12.

[15] Karen Holm Olsen, 'The Clean Development Mechanism's Contribution to Sustainable Development: A Review of the Literature' (2007) 84 Climatic Change 1; C Sutter and JC Parreño, 'Does the Current Clean Development Mechanism (CDM) Deliver its Sustainable Development Claim? An Analysis of Officially Registered CDM Projects' (2007) 84 Climatic Change 75; Karen Holm Olsen and Jørgen Fenhann, 'Sustainable Development Benefits of Clean Development Mechanism Projects. A New Methology for Sustainability Assessment Based on Text Analysis of the Project Design Documents Submitted for Validation' (2008) 36 Energy Policy 8; Patrick Nussbaumer, 'On the Contribution of Labelled Certified Emission Reductions to Sustainable Development: A Multi-Criteria Evaluation of CDM Projects' 37 Energy Policy 91 (2008).

[16] A Cosbey and others, 'Realizing the Development Dividend: Making the CDM Work for Developing Countries' (International Institute for Sustainable Development Research Paper 2005) <www.iisd.org/pdf/2005/climate_realizing_dividend.pdf>, accessed 21 February 2009; K Brown and others, 'How do CDM projects contribute to sustainable development?' (Tyndall Centre for Climate Change Research Technical Report 16, 2004) <http://www.tyndall.ac.uk/research/theme2/final_reports/it1_13.pdf>, accessed 21 February 2009; Axel Michaelowa and K Umamaheswaran, 'Additionality and Sustainable Development Issues Regarding CDM Projects in Energy Efficiency Sector' (HWWA Discussion Paper No. 346 2006) <http://ssrn.com/abstract=908824> accessed 21 February 2009.

[17] Figueres and others, 'Programmatic CDM Project Activities: Eligibility, Methodological Requirements and Implementation' (Study for the Carbon Finance Business Unit of the World Bank, 29 November 2005) <http://figueresonline.com/publications/Programmatic_CDM.pdf>, accessed 26 February 2009; Michael Wara, 'Is the Global Carbon Market Working?' (2007) 445 Nature 595; Michael Wara and David Victor, 'A Realistic Policy on International Carbon Offsets' (April 2008) PESD Working Paper, Stanford University <http://pesd.stanford.edu/publications/a_realistic_policy_on_international_carbon_offsets>, accessed 21 February 2009.

2.1.3 Weakness of the incentive

The incentive of the CDM has been too weak to foster the necessary type of transformation in the economy, without which emission paths in developing countries will continue to increase.[18] The CDM was effective in quickly eliminating a substantial portion of HFC_{23} and N_2O industrial gases which gave an early spur to the market although contributing little to sustainable development.[19] More recently, the CDM has shown that it can catalyze uptake of commercially proven technologies to capture waste heat and waste gases in carbon-intensive manufacturing industries (iron and steel, cement, chemicals). Likewise, CDM has begun to support a wave of renewable energy projects, methane capture and use projects, efficiency activities in coal mining, and oil and gas exploration and distribution. However, NGO concerns, and even more importantly the relatively low price of carbon, have impeded the CDM from appropriately addressing the largest source of greenhouse gas emissions globally coal fired power plants.[20] Up until recently the CDM was practiced on the basis of single site projects (eg, a wind farm, a hydroelectric dam, a biodigester, etc), which were submitted and registered as CDM project activities. The 2005 decision of the Conference of the Parties to the UNFCCC serving as the Meeting of the Parties to the Kyoto Protocol COP/MOP to also open the option for submitting and registering a potentially unlimited number of mitigation activities that are widely dispersed (eg, efficient cooking stoves, compact fluorescent lights, etc) under a single CDM program, may be a first step to including demand side energy efficiency and distributed clean energy generation in the CDM. However, programmatic CDM is still in its infancy and the fact remains that the CDM has not supported any increased efficiencies in households, buildings, or transportation systems—which together comprise more than half global carbon emissions[21] and are the fastest growing sources of carbon emissions in the emerging markets.[22]

[18] Wolfgang Sterk, 'From Clean Development Mechanism to Sectoral Crediting Approaches-Way Forward or Wrong Turn?' (JIKO Policy Paper Wuppertal Institute for Climate, Environment and Energy January 2008) <http://www.wupperinst.org/en/publications/entwd/index.html?&beitrag_id=790&bid=86>, accessed 21 February 2009.

[19] 'UNEP Risoe CDM/JI Pipeline Analysis and Database' <http://www.cdmpipeline.org/overview.htm>, accessed 26 February 2009.

[20] As energy demand growth in some developing countries is currently satisfied with coal-fired plants, the introduction of ultra-critical coal-fired plants is both a promising and contentious issue. The first methodology for potential power plant energy efficiency and supercritical coal-fired CDM projects, AMCM13 was approved in 2007 (EB39). See UNFCCC, CDM Meeting Archive <http://cdm.unfccc.int/EB/archives/meetings_08.html#039>, accessed 26 February 2009.

[21] Kahn S Ribeiro and others, 'Transport and its infrastructure' in B Metz, OR Davidson, PR Bosch, R Dave, LA Meyer (eds) 'Climate Change 2007: Mitigation'. Contribution of Working Group III to the Fourth Assessment Report of the Intergovernmental Panel on Climate Change (2007) (Cambridge University Press: Cambridge) (IPCC, 2007).

[22] D Sperling and D Salon, 'Transportation in Developing Countries: An Overview of Greenhouse Gas Reduction Strategies' (Paper Prepared for Pew Center on Global Climate Change 2002) <http://www.pewclimate.org/doCUPloads/transportation_overview.pdf>, accessed 26 February 2009.

2.1.4 Weak governance

The CDM has failed to develop a due process to guarantee fundamental fairness, justice, and respect for property rights. The overwhelming majority of the entities trading in the CDM market belong to the private sector. The CDM is unique in regulating a market dominated by private players that depend on a United Nations committee (the EB) to approve the calculation methods and projects that create the market's underlying asset. The credibility of the CDM market depends largely on the robustness of its regulatory framework and private sector's confidence in the opportunities provided by the mechanism.[23] There are mounting complaints about the continued lack of transparency in the Board's decision-making and the lack of predictability.[24] The governance structure of the CDM would have to be reviewed taking into account the need to provide in particular the private sector participants that are not represented in the Conference of the Parties with due process, ensuring the conditions for fair and predictable decisions.

Not all of the above issues will ever be solved by the CDM, but there is ample room for improvement toward 2012 and evolution beyond 2013, and there has certainly been no lack of suggestions on how to change and enhance the CDM. Here we do not undertake an in-depth explanation of these proposals instead evaluate what may be politically acceptable to all Parties, given the necessary scale of mitigation post-2012 and particularly the short time-frame available to debate a 2009 Copenhagen agreement.

3. Political Realities

There are three main political constellations that shape the potential consensus around a future climate deal. First, the US (until recently the largest national source of CO_2 emissions[25] and by all counts the most historically responsible) has made little attempt to reduce emissions. The administration of President Obama has initiated the development of domestic legislation, and the

[23] Charlotte Streck and Thiago Chagas, 'The Future of the CDM in a Post-Kyoto World' (2007) 1 Carbon & Climate Law Review 53; Ernestine Meijer, 'The International Institutions of the Clean Development Mechanism Brought Before National Courts: Limiting Jurisdictional Immunity to Achieve Access to Justice' (2007) 39 NYU JILP 877; Charlotte Streck and Jolene Lin, 'Making Markets Work: A Review of CDM Performance and the Need for Reform' (2008) 19 EJIL 409.

[24] International Emissions Trading Association, 'Strengthening the CDM' (IETA Position Paper for COP 11 and COP/MOP 1 2005) <http://www.ieta.org/ieta/www/pages/getfile.php?docID=1132>, accessed 26 February 2009.

[25] China is believed to have overtaken the US in absolute national terms during 2006, see JS Gregg and others, 'China: Emissions Pattern of the World Leader in CO_2 emissions from fossil fuel consumption and cement production' (2008) Geophys Res Lett 35 <http://www.agu.org/pubs/crossref/2008/2007GL032887.shtml>, accessed 26 February 2009.

expectation is that the US will undertake efforts that are comparable with other industrialized nations. The political will has been established by the new leadership, but the procedural necessities will take time. Even with all good will, the US will first develop its domestic emissions regulations before considering entry into a multilateral agreement and this is unlikely to occur by December 2009.[26] The consequence for the international agreement could be either that the 2009 Conference of the Parties (COP) to the UNFCCC may be suspended in December to be continued in June 2010 (as was done with COP 6), or that the Copenhagen agreement, in close consultation with the US team, may have to be crafted as an architectural 'docking station' where the major architectural design is set for the US to later 'dock in'.

The second political reality is that among the industrialized nations the EU has announced the strongest post-2012 reduction commitments, and even those are insufficient and tenuous. Unlike the US where international commitments tend to follow domestic policies, the EU has had a practice of defining international targets first before deciding on how to attain those targets. Although the EU has been careful to differentiate its unilateral mitigation commitment (20% below 1990 levels by 2020) from an additional 10% reduction in case of a satisfactory—from the point of view of the EU—international agreement,[27] it is not guaranteed that Member States would be willing to uphold the unilateral reduction level in the event of a failed or unsatisfactory international agreement.

The third political constellation is that of developing countries. These countries understand that the reduction in GHG emissions needed to avoid catastrophic climate disruptions cannot be attained by industrialized countries alone, even if those reductions reach 80 or 90% below 1990 levels. However, developing countries' contribution to global mitigation must take into account that:

- Developing countries will consider emission reductions only once all industrialized countries (including USA, Japan, Canada and Australia) have demonstrably taken the lead.
- Developing countries include a broad spectrum of economy sizes, which has a direct bearing on both responsibility,[28] as well as capability,[29] to mitigate. A few emerging economies have felt the pressure to contribute to mitigation efforts in the near term due to their current rapidly increasing

[26] The Waxman-Markey climate bill which sets short-term greenhouse gas emission reduction targets, will likely face substantial resistance in the Senate.

[27] The EU climate goal forms part of an energy and climate strategic package presented by the EU Commission on 23 January 2008, and endorsed by the Council on 4 December and the European Parliament on 13 December 2008. The reference line for reductions has been defined as 2005, as opposed to the previous 1990 levels, due to resistance of some Member States. See European Commission, 'EU ETS post 2012' <http://ec.europa.eu/environment/climat/emission/ets_post2012_en.htm>, accessed 26 February 2009.

[28] Responsibility could be proxied as either cumulative emissions or annual emissions, and either could be measured on a per capita basis.

[29] Capability could be proxied as either GDP or GDP per capita.

emission levels and growing economic development levels. China, India, South Korea, Mexico, Brazil, and South Africa have all come forward with first estimates of their mitigation potential. In addition, there is a group of middle-income developing countries that are not currently being 'targeted', but whose growth patterns could lead them, under a business as usual scenario, to relatively high GDP and emission levels over the next 20 years (eg, Chile, Argentina, Iran, Saudi Arabia, etc). The largest number of developing countries is however comprised of small economies whose emissions are negligible now and in the future. They may continue to contribute to mitigation efforts, but most likely under little or no international pressure to curb their emission growth.

- Developing countries will not immediately enter into absolute reduction commitments but rather may sequence their nationally appropriate mitigation actions[30] to gradually move up the stringency ladder. The larger developing countries could start with a focus on climate-friendly development policies without explicit mitigation commitments, and transit over time, based on demonstrated responsibility and capability, to limiting emission growth and finally at some point in time, to adopting emission reduction or at least emission intensity targets (Figure 26.1). In order to uphold the integrity of the system, all mitigation efforts would have to be domestically measured and reported, and independently verified.

- As two-thirds of future emissions will come from developing countries, they could potentially provide most of the mitigation needed for stabilization, but they cannot be expected to pay for it. Financial support for substantial mitigation in developing countries is made more palatable to industrialized nations by the fact that these are the most cost effective mitigation efforts. The UNFCCC Secretariat estimates that 68% of the mitigation necessary for stabilization is achievable in developing countries and would cost 46% of the total global mitigation.[31]

The conditional transfer of resources from the industrialized countries for purposes of underwriting part of the mitigation costs in the developing world is what is at the heart of the CDM and is likely to continue to be the *modus operandi* of the financial mechanism(s) that could be devised for post-2012 mitigation.

[30] The Bali Action Plan, Decision 1/CP13 envisages '[n]ationally appropriate mitigation actions by developing country Parties in the context of sustainable development, supported and enabled by technology, financing and capacity-building in measurable, reportable and verifiable manner.'

[31] UNFCCC, 'Investment and Financial Flows Relevant to the Development of an Effective and Appropriate International Response to Climate Change' (2007) <http://unfccc.int/cooperation_and_support/financial_mechanism/items/4053.php>, accessed 26 February 2009.

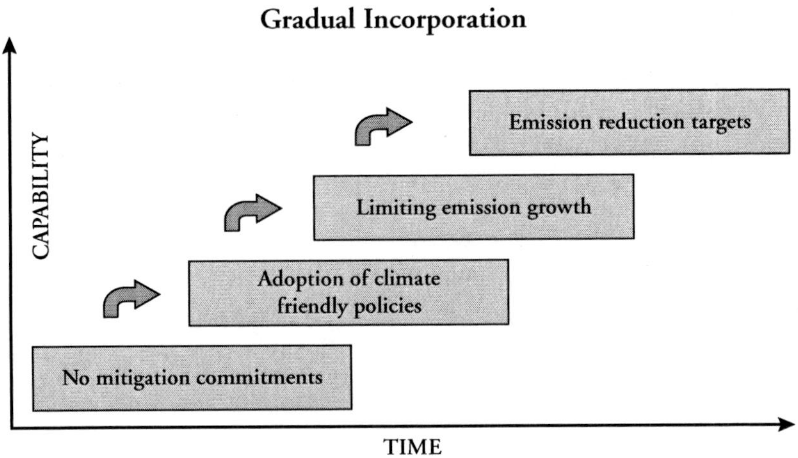

Figure 26.1. A possible scheme for gradual incorporation of developing countries
Source: Christiana Figueres, 'The Bali Batik: Design Options for the post 2012 Climate Regime' (Paper Written for Economic Commission of Latin America and the Caribbean, Santiago de Chile, July 2007) <http://figueresonline.com/publications/balibatikenglish.pdf>, accessed 26 February 2009.

4. Scaling up Emission Reductions in Developing Countries

One of the main challenges of a post-2012 agreement is thus to define a framework which creates incentives for developing countries to reduce emissions beyond those currently mobilized by the CDM. The following paragraphs summarize the challenges of scaling up the current CDM to access broader emission reductions.

4.1 Environmental integrity

Environmental credibility is particularly important in the CDM where emission reductions are credited as offsets, thus not leading to an overall reduction but only to a displacement of the emission reduction to a source of more cost efficient GHG emissions abatement. The current CDM is based on project-based crediting, supported by rigorous checks and balances to ensure that emission reductions are additional, and that any emission reductions are correctly monitored and verified. The environmental integrity of any and all future market mechanisms must be upheld.

4.2 Uncertainty over demand

As with any market, the carbon market is based on the notion of scarcity. Incentives to innovate, seek low-cost emission reduction options and invest in relevant technology are dependent on an environment where the circulating number of allowances and credits are below the actual level of emissions. In order to ensure that credits remain scarce and prices high, the design of emission trading schemes normally involve limitations for the import of credits. In order to protect the EU Emission Trading Scheme from the flooding of CDM credits, for example, the market regulator has capped the number of Certified Emission Reductions (CERs) from the CDM authorized for transfer.[32] Any future market mechanism must have appropriate demand as a key design feature.

4.3 Market and non-market

The post-2012 international climate agreement will probably have to rely on a mix of market and non-market based mechanisms. The carbon market so far has proven to be a successful way to involve private entities in treaty compliance, but has yet to be tested as a means to generate public finance. Carbon finance and the private sector contributions to a future climate regime are essential to mobilize the required scale of funding, but governments need to have access to stable resources in order to adopt and implement policies and programs that lead to GHG emission reductions.

Without minimizing the need for more effectiveness, it is probably not realistic to attain anything more than incremental changes to the CDM during the current commitment period. While practitioners and some parties clamor for significant improvements, many developing countries are still learning the ropes of the instrument, and others are focused on the new agreement and tools for scaling up post-2012 mitigation.

4.4 Three possible levels of improvement

In order to organize the steps in a possible evolution, it may be helpful to distinguish three levels of improvements/changes to the current CDM, which could ultimately result in two or three financial mechanisms, functioning in parallel and in a complementary manner to one another.

4.4.1 The activity-based CDM

Both projects and programs can be submitted to the CDM for registration, however they must be activity-based. By 'activity-based' we mean that the emission reductions must be directly traceable to a specific and concrete mitigation activity. Assuming the CDM will continue to operate within this restriction imposed by the Marrakech Accords,[33] at least until the end of the current commitment period, there are incremental improvements that could be

[32] EU ETS post-2012 (n 27 above).
[33] Decision 17/CP 7 as adopted by COP/MOP 1 as Decision 3 CMP1.

undertaken by the EB itself, supported by a decision of the COP/MOP if it occurs, but not dependent on that mandate.

4.4.2 A trend-changing market mechanism

Given the need to scale up mitigation in developing countries beyond what can be delivered by the CDM, it may be necessary to create a second market mechanism, inside or outside of the current CDM, inside or outside of the Kyoto Protocol, which focuses on impacting long-term emission trends particularly in large emerging economies.

4.4.3 A non-market mechanism

In the post-2012 regime it may be necessary to limit the scope of the offset market and create a complementary financial mechanism to promote certain mitigation efforts in developing countries without creating international offsets in order to keep the global reductions and timelines within the ranges demanded by science.

We discuss each of these possible mechanisms below.

5. The Activity-based Clean Development Mechanism

Today's wisdom demands higher emission reduction levels than those made possible through a project-based CDM, and many voices have understandably joined the chorus to expand the CDM (or create another mechanism) toward sector-wide mitigation efforts. However, it is likely that the CDM (or at least part of it) may remain an activity-based market mechanism, now and beyond 2012. The logic of a system based on the measurement of concrete mitigation activities is to measure tons reduced by individual projects rather than encourage a shift in emission trends.[34] In this system, baselines and additionality are determined at the individual project activity level.

The Marrakech Accords flow from an activity-based logic. To change that logic the CDM Executive Board would need a decision of the COP/MOP, and there is little political space to make such a shift before 2012.[35] Over the next two years negotiators are not likely to consent to material changes, particularly if they could lead to higher CERs volumes. Industrial and developing countries

[34] Christiana Figueres, 'From Tons to Trends' in Lydia Swart and Estelle Perry (eds), *Global Environmental Governance, Perspectives on the Current Debate* (Center for UN Reform Education: New York, 2007).

[35] Substantial changes to the CDM, which were discussed under Art 9 negotiations at COP/MOP 4 in Poznan, did not prosper and were deleted from the text even before the entire Art 9 was dropped.

alike fear that a substantial change in rules could open a floodgate to supply which would overwhelm demand, depress prices, and cause environmental damage due to the offsetting nature of the CDM. Thus it appears unlikely that the CDM will be subjected to much serious revamping during the first commitment period.

This is not to say that the CDM is a finished product. Importantly, it should incorporate several key administrative improvements and the sooner the better. In the following we list: (i) measures that could be undertaken directly by the EB without changing the Marrakech Accords and without the need for a COP/MOP decision; and (ii) improvements of the CDM requiring a COP/MOP decision that revises the Marrakech Accords but without touching the fundamental principles of the CDM. Any changes undertaken by the EB for the first commitment period would presumably stay in place after 2012, while reforms that require a more profound reform of the CDM would likely only take effect in 2013.

5.1 Improvements that can be enacted by the Executive Board itself

5.1.1 Increased delegation to the Secretariat

Assuming that the level of CDM submissions continues to rise and that nominations to the Board will, at least for the next three years, continue the current practice of part-time voluntary Board members who have limited technical knowledge or regulatory experience, it is inevitable that the Board will have to delegate even more responsibility to the Secretariat than it has in the past. Over the past two to three years the Secretariat has been recruiting and training staff to support the various functions of the EB. This trend will probably continue. The ultimate responsibility will certainly continue to reside with the Board, but project review could be shifted even further to the Secretariat, avoiding unnecessary delays and allowing the EB to focus more on policy decisions.

5.1.2 Review of the EB's administrative rules

Currently, there are only a few formalized provisions governing the interaction between project proponents, the EB, and its panels. Insecurities regarding communications, hearings and time lines often make processes cumbersome and opaque. From the perspective of project participants, there is a perception of insufficient and circuitous communication. As a result, there is an undefined period of legal and planning insecurity during which project participants: (i) must retain resources to answer an undefined and unlimited number of new questions; and (ii) have no indication on whether they can move ahead with developing the corresponding CDM project activities. Parties could consider the adoption of administrative due process rules governing communication amongst the various CDM actors.[36]

5.1.3 Role of Designated Operational Entities

The functional relationship between the EB and the Designated Operational Entities (DOEs) requires urgent improvement. DOEs are charged with ensuring that any project submitted for registration meets all CDM requirements. However, the reality is that DOEs have not been performing to expectations. DOEs argue that they are understaffed and that it is difficult to find suitably qualified people. The EB argues that DOEs should learn from the published EB reports. Either way, DOEs cannot stand outside EB decisions—they need to be incorporated more organically into the EB decision-making process. The recent approval of a Validation and Verification Manual should be of help. However, the EB reaches decisions with direct regulatory implications at every meeting and the DOEs should be the first to know of, and understand, these decisions. Were the EB to clearly state the rationale for decisions on registration and issuance, DOEs and project proponents would be able to more effectively derive the lessons learned and then apply these in the preparation of future projects, over time improving the performance of DOEs.

5.2 Improvements requiring a COP/MOP decision

The above issues could be considered directly by the EB or decided by Parties in a COP/MOP decision for the current commitment period. In addition, there are other elements which represent more fundamental changes to the CDM which are not likely to be considered for enactment before 2013 and would have to stem from a decision of the Parties.

5.2.1 Professionalizing the EB

The current EB has been established as a United Nations committee, rather than as a professional regulatory authority overseeing the carbon market. However, the EB is in the position of a *de facto* market regulator. In order to effectively fulfill this role, a first step in this direction would be to professionalize the EB.[37] Presently, the majority of its members have a background in international environmental negotiations, not in market regulatory work (for example, work experience in financial or energy regulatory authorities). As a result, the considerations of the EB tend to be oriented towards agendas raised during international negotiations rather than to the sort of issues related to the creation and maintenance of an efficient international market. The professionalization of the

[36] Charlotte Streck and Jolene Lin, 'Making Markets Work: A Review of CDM Performance and the Need for Reform' (2008) 19 European Journal of International Law 409. [37] Ibid.

EB would require the recruitment of full-time salaried individuals whose collective experience spans the entire range of responsibilities (including project finance, law, business management, science) and is grounded in practical, project-level experience and knowledge of the CDM. Staffing the EB with professional staff would also help avoid conflict of interests since individuals would no longer have to serve several agendas and interests in parallel, and could devote themselves full time to the EB. The current regional UN representation could be maintained by ensuring that the professionals stem from the respective regions of the world.

5.2.2 Review mechanism

In addition, a review mechanism of the decisions of the EB could be put in place. The COP/MOP decisions foresee a review procedure of some contested decisions when a decision improperly affects a Party's interest. The review is conducted by the enforcement branch of the Protocol's Compliance Committee.[38] These procedures, however, do not extend to non-Party participants in CDM projects. Under the existing guidelines, procedures and rules, the procedural rights of private parties are very limited. Affected project participants are afforded no opportunity for review of EB decisions. In order to strengthen the quality and legitimacy of the EB's decisions, the COP/MOP could establish an appeal mechanism which gives standing to individuals that are granted rights and obligations under the CDM and guarantees a full review of EB decisions.[39]

6. A Trend-changing Market Mechanism

The reductions achieved under the Kyoto Protocol will not make a dent in global emissions, and so far has had even less of an effect on emission trajectories. The decarbonization that the current CDM has not been able to achieve in developing countries needs to be aggressively pursued in the post-2012 period. While economic growth must continue, particularly in developing countries, the efficiency of energy consumption must improve and the carbon intensity of production must swiftly decrease. Decoupling growth from emissions is the only way to

[38] Decision 27/CMP.1, Annex, Procedures and mechanisms relating to compliance under the Kyoto Protocol, sections IX and X.

[39] Charlotte Streck, 'The governance of the Clean Development Mechanism—the case for strength and stability' in David Freestone and Charlotte Streck (eds), *The Kyoto Protocol and Carbon Finance: Current legal status of Carbon Finance and the Flexible Mechanisms (CDM and JI)* [2007] Environmental Liability 91; Charlotte Streck and Thiago Chagas, 'The Future of the CDM in a Post-Kyoto World' (2007) 1 Climate and Carbon Law Review 53; also chapter 15 above. An alternative remedy has been proposed by Ernestine Meijer under which decisions of the EB would be open for review by national courts E Meijer, 'The International Institutions of the Clean Development Mechanism Brought Before National Courts: Limiting Jurisdictional Immunity to Achieve Access to Justice' (2007) 39 New York University Journal of International Law and Policy 877.

pursue economic development and climate protection simultaneously, and this requires a radical shift in the policies that regulate the productive sectors.

Unlike the current CDM, the next chapter of the carbon market must play an important role in introducing and implementing the policy changes needed to put developing countries onto a low carbon path. It could be argued that programmatic CDM is a first step in the direction of decarbonization by enacting climate friendly policies. Indeed, that was the original intent. Stemming from the firm conviction that the market needs to promote climate friendly policies and not just isolated projects, proponents tried to introduce the eligibility of policies and standards in the CDM during the COP/MOP negotiations held in Montreal in November 2005. The compromise that was reached is encapsulated in the COP/MOP-1 decision that gives rise to programmatic CDM:[40]

> A local/regional/national policy or standard cannot be considered as a clean development mechanism project activity, but project activities under a programme of activities can be registered as a single clean development mechanism project activity.

The decision differentiates between the existence of a policy (not eligible) and its implementation through specific activities (eligible). The adoption of rules governing programmatic CDM was important for opening the CDM to projects in sectors that are highly dispersed over space and time. However, programmatic CDM continues to fall short of triggering the needed level of GHG emission reductions at the scale of whole economies and does not allow the crediting of GHG emission reductions at the policy level.

The next chapter of the climate regime must be built around the active and deliberate promotion of climate friendly policies throughout the developing world. Over the past few years a plethora of proposals has emerged suggesting options for making this shift.[41] Here we highlight the subset of proposals that revolve around sectoral concepts, as they seem to have most political attention. While authors take various perspectives on what has become the 'magical' and overly used term 'sectoral', there are basically two groups of proposals with varying degrees of elaboration: those that stem from an agreement among industries that operate in the same sector but are located across different countries, and those that evolve from a national government decision to implement a specific policy or measure within a sector.

6.1 Industry agreements

Arising from concerns over leakages and negative competitiveness effects associated with country-specific mitigation commitments,[42] industry has taken the

[40] Paragraph 20, Decision 2/CMP1.

[41] Cosbey et al describe 44 proposals which have been made within and outside of formal UNFCCC processes, and additional ones have emerged since then. Cosbey (n 16 above).

[42] International Energy Agency, 'Issues Behind Competitiveness and Carbon Leakage' (IEA Information Paper, 2008) <http://www.iea.org/Textbase/papers/2008/Competitiveness_and_Carbon_Leakage.pdf>, accessed 27 February 2009.

initiative to establish transnational, industry-led networks that promote climate change mitigation policies involving different sectors in developed and emerging countries alike. A key purpose is to avoid that competitiveness gains could be obtained through regulatory arbitrage, a particular concern for trade-exposed industries such as cement, aluminum, and steel, which are so energy intensive that they alone represent a significant share of emissions.[43] Transnational voluntary policy co-operations exist for the aluminum sector (under the auspices of the International Aluminum Institute), for the cement sector (Cement Sustainability Initiative, under the auspices of World Business Council for Sustainable Development), for the iron and steel sector (administered by the Iron and Steel Institute) and in the form of multi-sectoral, public/private partnerships (Asia-Pacific Partnership on Clean Development and Climate).

The private sector approach to sectoral crediting has three major flaws. First, it would need to be based on firm commitments, and yet the participation in private sector agreements is voluntary by definition. Second, even though industry today confesses wholeheartedly to the objectives of sustainable development, the level and quality of commitments taken are meager. Thirdly, voluntary commitments by private stakeholders are by definition outside the purview of the UNFCCC process where only states can enter into international agreements. Therefore, agreements that are reached in the realm of the private sector could be complementary to, but cannot substitute, agreements reached among governments within the UNFCCC process.

6.2 Government actions

Referred to by some authors as the 'policy-based approach',[44] but by others as a form of a 'sectoral' approach,[45] this approach centers on the generation of emission reductions by developing countries that adopt binding or non-binding policies, voluntary or mandatory standards that measurably reduce GHG

[43] C Egenhofer and others, 'Global Sectoral Industry Approaches to Climate Change: The Way Forward' (Center for European Policy Studies: Brussels 2008); N Höhne, and C Ellermann, 'A sectoral approach and technology transfer for the cement sector' (Paper Commissioned by Federal Office for the Environment FOEN: Switzerland, 20 August 2008) <http://www.bafu.admin.ch/klima/index.html?lang=en&download=NHzLpZeg7t,lnp6I0NTU042l2Z6ln1ad1iZn4Z2qZpnO2Yuq2Z6gpJCFeYR6fGym162epYbg2c_JjKbNoKSn6A–.pdf>, accessed 27 February 2009.

[44] W Sterk and B Wittneben, 'Enhancing the clean development mechanism through sectoral approaches: definitions, applications and ways forward' (2006) 6 International Environmental Agreements: Politics, Law and Economics 271-87; Bosi and Ellis, 'Exploring Options for Sectoral Crediting Mechanisms' (Organization for Economic Cooperation and Development 2005), 6 <www.oecd.org/dataoecd/55/61/34902644.pdf>, accessed 12 March 2009.

[45] Dan Bodanksy, 'International Sectoral Agreements in a Post-2012 Climate Framework' (Pew Center 2007) <http://www.pewclimate.com/docUploads/International%20Sectoral%20Aggreements%20in%20a%20Post-2012%20Climate%20Framework.pdf>, accessed 12 March 2009; Bosi and Ellis (n 44 above).

emissions. Under the approach originally proposed by Samaniego and Figueres,[46] developing countries would develop regional, sectoral, sub-sectoral, or cross-sectoral mitigation efforts, which would be the result of specific sustainable development policies, measure the attained reductions against a sector-wide baseline, and sell those on the international emission reduction market. The mechanism would be comparable to the CDM, but covering a whole sector rather than a particular activity.

In order to assign some of the cost to developing countries and increase the net gain to the climate by not converting all reductions into offsets, some authors have evolved the above concept toward Sectoral No-Lose Targets (SNLTs). Under SNLTs developing countries would voluntarily propose a domestic interest crediting baseline over a commitment or 'management' period of time which would be below the business as usual projection and be negotiated internationally. The country would reach the crediting baseline through domestic efforts, and would then be allowed to sell any surplus emission reductions which are achieved beyond the crediting baseline, but would have no penalty for not achieving that baseline.

Proponents of the SNLT mechanism propose that crediting baselines be negotiated at the same time as Annex I country targets for post-2012 in order to avoid the need for proving additionality. However, benchmarking has its own set of enemies, is very data-intensive and may not be realistic in some countries or some sectors,[47] particularly if the target date of agreement is COP15 at the end of 2009. Furthermore, SNLTs make sense for larger developing countries with a stable investment climate that seeks to significantly scale up private sector investment according to their sustainable development priorities, and where current carbon market policy tools, such as the various forms of CDM, are not considered adequate to the task.[48] The aggregation of revenue potential could provide financial leverage sufficient to transform the sector over a 10–20 year period. Developing countries, however, will be careful to not accept any target that would operate as a cap on development. They are, therefore, likely to argue for setting baseline emissions, if at all, on the basis of the national emission intensity of the sector in question. Developing countries will also avoid any

[46] J Samaniego and C Figueres, 'Evolving to a Sector-Based Clean Development Mechanism' in K Baumert, O Blanchard, S Llosa and J Perkaus (eds) *Building on the Kyoto Protocol: Options for Protecting the Climate* (World Resource Institute: Washington DC, 2002).

[47] C Egenhofer and N Fujiwara, 'Global Sectoral Industry Approaches to Climate Change. The Way Forward' (Center for European Policy Studies: Brussels, 2008) <http://se1.isn.ch/serviceengine/FileContent?serviceID=ISN&fileid=E66EBE8F-1AF5-0C10-8914-FB462048AB24&lng=en> accessed 12 March 2009; Ward and others, 'The Role of Sector No-Lose Targets in Scaling up Finance for Climate Change Mitigation Activities in Developing Countries' (Paper prepared for International Climate Division, Department for Environment, Food and Rural Affairs 2008) <http://www.sectoral.org/images/presentations/defra_paper%20on%20sector%20no%20lose%20targets_final.pdf>, accessed 12 March 2009.

[48] Ward and others (n 47 above).

mechanism that is perceived as a back door strategy to push them into binding national targets.

The policy-based or sectoral crediting approaches open an avenue for a financial mechanism that can complement current CDM practices and help to take mitigation efforts to scale. In order to be successfully implemented, the expanded mechanism would require major alterations to the carbon market as we know it, if only to safeguard the integrity of a system that will trigger mitigation efforts at a scale heretofore unknown. Taking into account the risk of driving emission reduction prices down by enabling high supply levels, the comparative advantages of market based against non-market based mechanism will have to be carefully assessed. Parties would decide whether this new market channel is created inside the CDM under the guidance of the EB (but with modalities different from those of Marrakech), or whether it should have a separate structure and regulating body.

In any case, the following issues would need to be addressed.

6.2.1 At the mitigation level

The logic of the system needs to evolve from focusing on the project activity to focusing on the policy that spurs the emission reductions. Decarbonization will simply not occur without the necessary regulatory framework. Only regulatory certainty will stimulate an adequate and reliable new source of risk capital to finance technology shifts on the scale of whole economies. In addition to promoting activity-based emission reductions as in the traditional CDM, the next (or expanded) emission reduction mechanism must promote the necessary sector-wide transformation, attained by cost effectively channelling capital and know-how to decarbonize carbon intensive sectors such as energy, transport, and infrastructure.[49] As discussed, some are interested in transforming some of these sectors at the international level, but it would seem that developing countries will claim their sovereignty and consider only national level policies which lead to decarbonization.

6.2.2 At the financing level

Mitigation action needs to be appropriately rewarded, reflecting the different strengths and constraints of private versus public financing. The conditions under which the private sector accepted the CDM as an international incentive mechanism include: (i) low exposure to host country risk, which in the CDM is limited to the issuance of a letter of approval; (ii) ability to control project risk and independence of carbon credit allocation to an individual project from

[49] Figueres and Newcombe, 'Evolution of the CDM, 2012 and Beyond' (United Nations Foundation Document 2007), <figueresonline.com/publications/Post_2012_CDM.pdf>, accessed 12 March 2009.

broader policy failure; and (iii) despite all flaws, trust that the international governance structure will reward emission reductions with tradable carbon credits. Governments on the other hand rarely act as carbon speculators. They are unlikely to create budget lines on the basis of a future promise of carbon credits unless they receive a price guarantee and assurance that credits will actually be issued. Experiences with the establishment of Assigned Amount Unit (AAU)-backed Green Investment Programs in countries with an over-allocation of AAUs in Eastern Europe have shown that trading of carbon credits by governments, even if they come in the form of allocated allowances as in the case of AAUs, raises issues related to state budget rules, sale of state assets, ownership of emission rights, constitutional limitations, predictability of funding, and allocation of proceeds.[50]

6.2.3 At the crediting level

Governments could have the right to propose sectoral or policy-based crediting schemes involving tradable carbon credits or opt for other negotiated and determined incentives (eg, cash, loans, guarantees). Performance could be measured against an agreed and adopted baseline, an SNLT, or any other performance indicator. The mechanism could foresee the allocation of tradable carbon credits based on a reduction of emissions below a certain baseline. In order to reduce the price risk, governments could negotiate the sale of the credits in advance against a fixed price per ton of CO_2e equivalent reduced. Annex I governments would have to decide whether they will open private carbon markets to these credits; if not, other agreements would be needed to give developing countries the assurance that there is a real demand for emission reductions. Alternatively, governments could also opt not to receive tradable carbon credits, instead receiving cash for emission reductions eg, via an abatement fund. This would spare governments the additional complication of managing and selling carbon credits, while at the same time reduce the supply of credits to international carbon markets. A private sector crediting scheme, a CDM-like mechanism, could be integrated into the system to attract private sector financing against the reward of tradable carbon credits. Double counting would have to be eliminated by deducting emission reductions that form part of an activity based crediting mechanism from government achievements.

6.2.4 At the administrative level

An international regulatory body would have to administer the mechanism and any agreements concluded or programs approved. To ensure consistency among various mechanisms, the mandate of such regulatory body—either a reformed Executive Board or a newly constituted body—could include the management

[50] See further Simonetti and de Witt Wijnen, chapter 7 above.

and supervision of an expanded crediting mechanism. This body would have to be composed of professional regulators who understand and have expertise in the relevant sector. Technical experience should therefore be the governing criterion for the selection of relevant experts.

The participation of private sector entities in an expanded crediting mechanism would require the establishment of administrative procedures that ensure a transparent, legitimate and fair process. This would imply the establishment of due process based on administrative procedures on the international level and relevant laws on the national level. Where carbon crediting takes place on the government level, laws would have to ensure that rights to emission reductions from a particular actor or activity be transferred to governments and that the initial holders of emission rights be duly compensated.

7. A Non-market Mechanism

While the CDM has proven to be an effective vehicle for stimulating investment into emission reductions in developing countries it encapsulates the major flaw of creating offsets which are used to cover emissions elsewhere, and hence do not contribute to limiting or reducing overall global emissions. A non-market mechanism could be used to scale up mitigation efforts.

While during the first commitment period demand and supply seem to be approximately in balance, in the post-2012 period supply may overwhelmingly dwarf demand. Leaving the environmental integrity of offsets aside for the moment, the limitations in demand for offset credits dictates a necessary restriction to the supply of carbon credits.

A non-market financial mechanism could be created to reward policy efforts and emission reductions without creating tradable carbon credits. In principle, the concept is not much different from the Sustainable Development Policies and Measures (SD-PAM) proposal originally suggested by Baumert and Winkler,[51] with the difference that emission reductions resulting from the policies would have to be quantified. In contrast to the current CDM, the focus would be on large-scale policies and measures, not individual projects. Furthermore, developed countries would support the voluntary efforts of developing countries, both financially and through technology transfers, but not on the basis of the purchase of offsets.

Many questions remain unanswered by this non-market financial mechanism, but perhaps the most important is the actual appetite that industrialized countries might have for such funding, considering there would be no emission

[51] K Baumert and H Winkler, 'SD-PAMs and International Climate Agreements' in R Bradley, K Baumert and J Pershing (eds) *Growing in the Greenhouse: Protecting the Climate by Putting Development First* (World Resource Institute: Washington, DC, 2005).

crediting. Another question relates to the predictability and stability of any particular funding source. The details of the mechanism will have to be answered by the negotiations. However, industrialized countries could be reassured by the fact that any distribution of funds would be success based, rewarding only those emission reductions that have actually occurred, as confirmed by measurement, reporting, and verification.

8. Conclusion

The gradual and incremental approach that has been outlined above is the 'path of least resistance' to the next iteration of financial mechanisms for long-term (2013 and beyond) mitigation. As we continue with the international negotiations for the future climate regime, options that can be agreed to by all nations are constrained by the current political constellation. At a time in which rapidly emerging countries are focused on economic growth, climate science requires them to initiate efforts to deviate from uncontrolled emissions, and to soon impose some type of restriction on emission growth. However, neither the North nor the South are racing to meet the requirements of science. Entrenched in their traditional defensive positions that reflect a deep lack of trust of the other side, both sides are currently only willing to contribute to a solution that represents the minimum common denominator, and is at best gradual and incremental.

There is however one factor that could substantially alter the path forward and that is the global financial crisis. A traditional interpretation of the crisis would foresee even less engagement on climate given the dwindling availability of capital. Ironically, it is also possible that this is precisely the pressure that is needed to radically shift the course of events.

The low carbon agenda can act as an engine for job creation and economic recovery while at the same time increasing energy and climate security. Governments and private sector could make strategic investments to dramatically improve efficiency in buildings and power, and to replace 19th century technologies that depend on carbon-based fuels with 21st-century technologies that use renewable fuels, all as part of a concerted effort to revive and redirect the economy. Contrary to historical behavior, the challenge of solving the climate crisis could move us into the realization that we now live in a multi-polar world where solutions cannot be implemented only by a very few. The recent G20 meetings are a first evidence of what could be a profound shift in global power and influence. A new world order that incorporates emerging economies into the solutions group, albeit gradually, is a sound harbinger of a durable development framework.

27

International Market Solutions to Protect Tropical Rainforests

Robert O'Sullivan and Rick Saines[*]

1. Introduction

Tropical forests play a crucial role in the global carbon cycle and have consequently moved to the forefront of the discussion about how to effectively reduce global greenhouse gas (GHG) emissions and mitigate climate change. Tropical forests encompass a wide variety of eco-systems on several continents. Each of these forests, while immensely important for their biodiversity and unique characteristics, are also an enormous store of carbon dioxide. The amount of carbon currently locked up in forest ecosystems is greater than the total amount of carbon already in the atmosphere.[1] When forest areas are cleared or degraded, however, the stored carbon is released back into the atmosphere. Estimates suggest that annual GHG emissions from deforestation and degradation of forests account for 20% of global GHG emissions, a figure roughly equivalent to the total emissions of the United States, and greater than emissions from the entire global transport sector.[2]

There is consensus that in order to achieve meaningful global GHG emission reductions, the international community must address emissions from tropical deforestation.

At the United Nations Framework Convention on Climate Change (UNFCCC)[3] 13th meeting of the Conference of the Parties (COP) in Bali in

[*] The authors wish to thank Daniel DeDeo, Associate, Baker and McKenzie, for his significant contributions to this chapter.

[1] Nicholas Stern, *Stern Review: The Economics of Climate Change* (Office of Climate Change: London, 2007) 537; Food and Agriculture Organization (FAO), 'Global Forest Resources Assessment 2005: Progress towards Sustainable Forest Management' (2006) Forestry Paper 147, 14 <ftp://ftp.fao.org/docrep/fao/008/A0400E/A0400E00.pdf>, accessed 22 March 2009.

[2] Climate Action Network, 'Reducing Emissions from Deforestation and Forest Degradation (REDD)' (November 2007) UNFCCC Bali Discussion Paper, 2 <http://unfccc.int/files/meetings/intersessional/awg_4_and_dialogue_4/press/application/pdf/awg4_com_plus_red.pdf>, accessed 22 March 2009; Stern (n 1 above) 537–8.

[3] United Nations Framework Convention on Climate Change (adopted 9 May 1990, entered into force 21 March 1994) 1771 UNTS 107 (UNFCCC).

December 2007, governments from around the world agreed to consider, over the next two years, 'policy approaches and positive incentives' for reducing emissions from tropical deforestation that might become part of an agreement that would follow on after the current commitments under the Kyoto Protocol[4] expire.[5] UNFCCC member states are currently undertaking a work program to consider different policy approaches to address emissions from tropical deforestation in the post-2012 agreement, to be finalized at COP 15 in Copenhagen in 2009. One of the policy options under consideration is a mechanism that would create positive incentives to reduce emissions from deforestation and forest degradation (REDD). This chapter will provide background on international forest carbon policies, making the case that market-based mechanisms are the most appropriate ways to provide the bulk of finance needed to address REDD. In this context it assesses challenges related to REDD and the recognition of credits for REDD in global carbon markets.

The essential policy problem created by tropical forests is that while the forests provide critical environmental benefits in terms of wildlife habitat, biodiversity and carbon sequestration, they also represent a highly exploitable resource, providing reservoirs of timber, fuel, and potential for agriculture.

Over the last few decades, international organizations have increasingly sought to align both environmental and economic interests with principles of sustainable forest management. The first International Tropical Timber Agreement was signed in 1983 and subsequently revised in 1994 and 2006.[6] These agreements establish soft guidelines for forest management, encourage sustainable practices in the global timber trade, and provide an administrative framework for the International Tropical Timber Organization (ITTO). The 1992 UN Conference on Environment and Development (UNCED) also initiated a series of agreements that have impacted the development of international forest policy. In particular, the Convention on Biological Diversity (CBD), and the 'Statement on Forest Principles' have provided significant guidance on sustainable forestry practices.[7] The CBD, for example, has maintained a work program on forest biodiversity for over a decade, consistently updating and expanding the program based on scientific and technical advice from its subsidiary bodies and other forest-related international organizations.[8]

In the years following UNCED combined UN efforts to promote forest management and conservation led to the formation of an Intergovernmental Panel on Forests (IPF) in 1995, and an Intergovernmental Forum on Forests in

[4] Kyoto Protocol (adopted 11 December 1997, entered into force 16 February 2005) 37 ILM 22.

[5] UNFCCC COP 13 Decision 1/CP.13 *Bali Action Plan*, para 1(b)(ii).

[6] See International Tropical Timber Organization, 'About ITTO' <http://www.itto.int/en/about_itto>, accessed 22 March 2009.

[7] John O Niles, 'Tropical Forests and Climate Change' in Stephen Schneider, Armin Rosencranz, John O Niles (eds), *Climate Change Policy* (Island Press, 2002) 337.

[8] See Convention on Biological Diversity (CBD), 'Forest Biodiversity Programme, Background' <http://www.cbd.int/forest/background.shtml>, accessed 24 March 2009.

1997. The work of these two groups culminated in a set of proposals for action in forest management. These proposals are collectively known as the IPF/IFF Proposals for Action. The proposals are non-binding, but countries are expected to conduct national assessments using the proposals and develop a plan for their implementation.[9] The United Nations Forum on Forests (UNFF) was established in 2000 to carry on the work of the IPF and the IFF. The UNFF has adopted its own set of non-binding principles of forest stewardship and oversees collaboration between different forest-related international organizations, such as the ITTO and the CBD.[10]

Despite these efforts, deforestation in the tropics continues at an alarming rate.[11] The high opportunity costs of protecting standing forest, especially for populations living in or around the forests, have, in large part, undercut attempts to control deforestation through remote international agreements and soft guidance.

At COP 15 in Copenhagen the Parties to the UNFCCC will consider policy options and positive incentives to reduce deforestation that will in effect put a price on forest carbon, paid for largely by the developed world. This will capture the previously 'externalized' climate costs of deforestation and carbon pollution in the developed countries, thereby creating an incentive to invest in developing countries to help protect the world's tropical forests.

However, there are legitimate legal, policy, and methodological challenges related to the monetization of forest carbon. In particular, concerns about uncertain land tenure and governance in forest areas, as well as concerns about effective monitoring and enforcement, will require careful consideration in the development of any incentive mechanism for REDD. Other over-arching technical challenges that arise in relation to the creation of a REDD mechanism include avoiding the danger that preservation in one area could merely lead to increased deforestation in another (leakage), and ensuring that credited emissions reductions from avoided deforestation are sustained over time (permanence).

International policymakers and stakeholders have already begun to consider methodological implications of certain REDD mechanisms. The COP started discussing REDD at its 11th session in 2005, and it has been high on the agenda since. There are also efforts underway to address REDD in domestic legislation, such as the recognition of international forestry credits in the recent versions of US federal legislation in both the Senate and the House of Representatives.[12]

[9] United Nations Forum on Forests (UNFF), 'History and Milestones of International Forest Policy' <http://www.un.org/esa/forests/about-history.html>, accessed 24 March 2009.

[10] See UNFF, 'Non-legally binding instrument on all types of forests' E/CN.18/2007/8, Apr. 27, 2007.

[11] Niles (n 7 above) 337. Estimates suggest that deforestation is occurring in the tropics at the rate of 40 million acres per year. Ibid.

[12] See Lieberman-Warner Climate Security Act of 2008, S 3036, 110th Cong (2008); American Clean Energy and Security Act of 2009, H.R. 2454, 11th Cong (2009).

Developing a politically acceptable and technically sound plan for confronting the problem is essential. This chapter will review the development of REDD under the UNFCCC process to provide some context to a number of cross cutting issues that are key to making REDD a success. It concludes with a suggestion for developing a REDD market mechanism that would sit alongside non-market funding.

2. Consideration Under the UNFCCC

As with most aspects of international climate policy, the treatment of deforestation has been dictated by a mixture of science and politics and their evolution over time. Forest conservation was first referenced in the UNFCCC in 1992, but as is the case generally under the UNFCCC, consideration is limited to somewhat vague commitments.[13] Deforestation was considered in detail under the Kyoto Protocol, where Annex I countries are required to account for emissions from deforestation under Art. 3.3 which can affect an Annex I country's[14] ability to meet its 'Quantified Emission Limitations or Reduction Commitments' (QELRCs). The agreement on the land use sector in general was agreed after the QELRCs were negotiated, and scientific uncertainty surrounding the role of forests in the climate system and how to monitor emissions in this sector led to intense negotiations and restricted use of credits from the forest sector. The negotiations around the use of forest-related credits to meet QELRCs dragged on from Kyoto in 1997 and many issues were not resolved until the Milan round of negotiations in 2003,[15] with some interpretations on forestry credits in Annex I countries still being made through 2006.[16] In the end, reducing emissions from deforestation in Annex I countries was creditable under a country's national accounting, but not clearly eligible for project-based credits under Joint Implementation—the project-based flexible mechanism for Annex I countries.[17] Forestry activities under the Clean Development

[13] See Article 4(1)(d) of the UNFCCC which refers to the sustainable management, conservation and enhancement of forests.

[14] Annex I countries are the industrialized countries listed under Annex I to the UNFCCC.

[15] For a more detailed account see E Trines, 'History and Context of LULUCF in the Climate Regime' in Streck, O'Sullivan, Janson-Smith, and Tarasofsky (eds) *Climate Change and Forests, Emerging Policy and Market Opportunities* (Brookings/Chatham House, 2008).

[16] For example the decision to require Joint Implementation forestry projects to only convert RMUs (as opposed to RMUs or AAUs) into ERUs.

[17] This is because under Art 3.3 and 3.4 RMUs can only be issued for the net sequestration of carbon. As reducing deforestation does not result in net sequestration of carbon, REDD activities will not generate RMU. As JI LULUCF projects can only convert RMUs into ERUs, REDD projects in Annex I countries will not be able to generate ERUs. As a result, REDD activities in Annex I countries can only be awarded AAUs directly by the host country. See M Ward, B Schlamadinger, R O'Sullivan, and M Burian, 'LULUCF Projects under JI: Will they be impossible?' (SB26 Side Event Discussion organized by Joanneum Research) (2007) <http://unfccc.meta-fusion.com/kongresse/SB26/templ/ply_sideevent.php?id_kongresssession=531&player_mode=isdn_real>, accessed 24 March 2009.

Mechanism (CDM)—the project based mechanism for non-Annex I countries—was limited to afforestation and reforestation only. Deforestation was explicitly excluded from the CDM for a number of reasons, including concerns over leakage, permanence, potential for market flooding, and general opposition from a number of nongovernmental organizations (NGOs) to the use of forest-based credits to offset emissions from fossil fuels.[18]

Deforestation was brought back to the international climate agenda at the 11th COP in 2005 by the governments of Papua New Guinea and Costa Rica.[19] These countries were able to put deforestation back on the agenda because the science around forestry had improved and the political landscape had changed. In 2005 there was a greater understanding of the role forests play in climate including better satellite data and monitoring capabilities. This allayed some of the fears around uncertainty and the ability to monitor emissions that were raised in the past. The better science was also coupled with a new geo-political landscape. The next round of negotiations were starting on what to do when the QELRCs negotiated in Kyoto expire in 2012, and Annex I countries were particularly keen to solicit greater engagement from developing countries. This desire was met by a number of developing countries who agreed that incentives for reducing deforestation should be assessed based on national performance measured against a national reference level. The offer to engage in emission reductions activities and to account for results at the national level satisfied both the political needs of Annex I countries and the technical concern associated with domestic leakage—one of the reasons why deforestation was left out of the CDM.

The 2005 submission started a two-year learning process for the UNFCCC's Subsidiary Body for Scientific and Technical Advice (SBSTA),[20] during which time the SBSTA focused on 'relevant scientific, technical and methodological issues, and the exchange of relevant information and experiences, including policy approaches and positive incentives'. The SBSTA held two technical workshops, one from 30 August to 1 September 2006, in Rome, Italy, and the other from 7–9 March 2007, in Cairns, Australia. A report summarizing the two-year process and any recommendations was submitted to the thirteenth COP in Bali in December 2007. The official UNFCCC events were augmented by a number of additional initiatives, conferences, workshops, and meetings organized by governments, NGOs, the World Bank, research institutions, and the private sector.

[18] Some of these issues do not arise for industrialized countries that have accepted caps under the Kyoto Protocol, because these emissions can be accounted for as part of the country's overall emissions accounting.

[19] See UNFCCC, 'Reducing Emissions from Deforestation in Developing Countries: Approaches to Stimulate Action' FCCC/CP/2005/MISC.1 <http://unfccc.int/resource/docs/2005/cop11/eng/misc01.pdf>, accessed 24 March 2009. Official support for including the issue on the COP's agenda was sent by Bolivia, the Central African Republic, Chile, Congo, the Democratic Republic of Congo, the Dominican Republic, and Nicaragua.

[20] The SBSTA is one of two permanent subsidiary bodies established by the UNFCCC. It provides advice to the COP.

Through the course of this two-year process, culminating in Bali, the negotiations around deforestation evolved to include degradation, conservation, sustainable management of forests, and enhancement of forest carbon stocks, and also consideration of national and subnational demonstration activities.

After the negotiations in Bali, an additional SBSTA workshop was organized in Tokyo in June 2008 and a workshop was organized in Accra in August 2008 under the AWG-LCA[21] in advance of COP 14 in Poznan. The Tokyo workshop focused on a set of methodological issues raised in Bali aimed at assessing emissions, degradation, sustainable management of forests, reference scenarios, the implications of national and subnational approaches, and how to evaluate actions to address REDD. The Accra workshop focused on mitigation activities and policy approaches, positive incentives/and or financing options, reference emission levels, national and subnational approaches, capacity building and other means of support, and early action and demonstration activities. There was also 'a common understanding that the current knowledge of methodological issues was sufficient to initiate discussions on policy approaches and positive incentives'.[22] Despite this common understanding, the negotiations at COP 14 in Poznan remained stuck on methodological issues, and did not make progress on policy approaches and positive incentives.

3. Analysis and Evolution of Issues

3.1 Scope

The scope of REDD has expanded beyond the initial consideration of reducing emissions from deforestation only to the final Bali text, which refers to 'policy approaches and positive incentives on issues relating to reducing emissions from deforestation and forest degradation in developing countries; and the role of conservation, sustainable management of forests, and enhancement of forest carbon stocks'.[23]

The inclusion of degradation was initiated by Vanuatu and the Coalition for Rainforest Nations early in the process. They argued that its inclusion would enable a more comprehensive coverage of emissions from the forest sector as

[21] The Ad Hoc Working Group on Long-term Cooperative Action under the Convention (AWG-LCA) was established under the Bali Action Plan in COP 13 (see Decision 1/CP.13 paragraph 1) to set out the scope of negotiations under the UNFCCC in order to come to an agreement on a future climate change agreement in Copenhagen at COP 15.

[22] See 'Report of the workshop on policy approaches and positive incentives on issues relating to reducing emissions from deforestation and forest degradation in developing countries; and the role of conservation, sustainable management of forests and enhancement of forest carbon stocks in developing countries' FCCC/AWGLCA/2008/CRP.5 para 7 <http://unfccc.int/resource/docs/2008/awglca3/eng/crp05.pdf>, accessed 24 March 2009.

[23] See Decision 2/CP 13, 'Reducing emissions from deforestation in developing countries: approaches to stimulate action' para 11 <http://unfccc.int/resource/docs/2005/cop11/eng/misc01.pdf>, accessed 26 March 2009; Bali Action Plan (n 6) para 1(b)(iii).

degradation is more significant that deforestation is some countries, can lead to biodiversity loss,[24] and reduces a forest's ability to adopt to climate change. However, during the Bali negotiations the Congo Basin countries were the most outspoken and fought successfully for the inclusion of degradation. This changed RED into REDD (with the second D corresponding to degradation).

The inclusion of conservation, sustainable management of forests, and the enhancement of carbon stocks was included late in the Bali negotiations by India, which has already gone through a period of deforestation and is now increasing its total forest cover. It can also apply to countries with high forest cover and low rates of deforestation (HFLD).[25] When first introduced this expansion threatened to derail the entire REDD negotiations as a number of countries did not want to create an emissions trading system based on standing carbon stocks not under threat of deforestation. This was overcome by last minute compromises and agreement by the US and other countries. India pushed the issue further in Poznan, replacing with a comma the semicolon that divided the text between deforestation and degradation and the role of conservation, sustainable management of forests, and enhancement of forest carbon stocks.[26]

The types of policy approaches or positive incentives that may apply to countries such as India with low forest cover and low deforestation rates, and HFLD countries such as Guyana has not been the subject of negotiations to date. However, there are two main options that can be considered together or separately. One option is to compensate these countries via a fund financed through eg, a levy on AAUs or other carbon credits, or direct sovereign contributions.[27] The other option is to adjust a country's baseline or reference scenario to take future deforestation into account. This may be more appropriate for HFLD countries than the low forest cover/low deforestation countries which may not have significant deforestation in the future. Other market-based mechanisms associated with sequestration of carbon via forest management or

[24] For example, in Vanuatu the invasive vine *Merremia peltata* becomes established if forest degradation by selective logging opens up too much of the forest canopy. The vine rapidly covers the trees surrounding the break, killing parts of the forest, which leads to further degradation and emissions from loss of biomass.

[25] There are 11 HFLD developing countries that are estimated to store 18% of the world's tropical forest carbon. See Fonseca and others, 'No forest left behind' (2007) 5(8) PLoS Biology 216. doi:10.1371/journal.pbio.0050216 <http://www.plosbiology.org/article/info:doi/10.1371/journal.pbio.0050216>, accessed 15 June 2009.

[26] The text therefore changed from 'policy approaches and positive incentives on issues relating to reducing emissions from deforestation and forest degradation in developing countries; and the role of conservation, sustainable management of forests, and enhancement of forest carbon stocks' to 'policy approaches and positive incentives on issues relating to reducing emissions from deforestation and forest degradation in developing countries, and the role of conservation, sustainable management of forests, and enhancement of forest carbon stocks'. See FCCC/SBSTA/2008/L.23, 'Reducing emissions from deforestation in developing countries: approaches to stimulate action', para 4.

[27] See Decision 10/CP.7 'Funding under the Kyoto Protocol' para 2 and 3, for an example of how the Adaptation Fund was created and financed under the Clean Development Mechanism.

enhancement of carbon stocks may be more appropriate. See section 3.4 below for additional discussion on market and non-market approaches.

3.2 Scale

The scale for addressing REDD moved from purely national approaches to also include consideration of 'subnational'[28] activities where they 'constitute a step towards the development of national approaches, reference levels and estimates'.[29] However, to date this inclusion is limited to demonstration activities and methodological issues and is being considered 'without prejudice' to any future COP decision on REDD. As discussed above, national monitoring and accounting was the favored approach when deforestation was first reintroduced onto the agenda in 2005. However, there are a number of different interpretations of what 'national monitoring and accounting' means. The most common view involved national governments: (1) determining and negotiating national deforestation reference levels; (2) implementing national policies and measures to ensure that deforestation is reduced nationally; (3) monitoring to check actual rates of deforestation against the reference level; and (4) receiving credits for national reductions. This type of national monitoring and accounting was favored by some because it centralizes management and control over land-use choices with the government. It is argued that a top-down approach in which the central government is responsible for reforming land and forest policy and laws, addressing corruption, and law enforcement is the most appropriate approach to reduce deforestation and effectively address concerns over leakage and permanence. This also centralizes income generated from the sale of credits with the federal government, which is appealing to some governments.

Criticisms of national-level monitoring and accounting include the potential for reduced accuracy and increased uncertainty associated with national monitoring and accounting, increased sovereign responsibilities and potential for associated corruption, and uncertainty over how the private sector might participate in this type of system, which is of critical importance given the private sector driven carbon markets are often expected to drive funding for REDD. These issues can be addressed by creating the flexibility to implement (and account) for REDD activities via subnational efforts.

Subnational and project-based approaches focus primarily on monitoring and accounting for emissions from discrete areas within a country. Such approaches are preferred if the obligations of the national accounting method are viewed as potential risks rather than advantages. A lack of capacity or challenges regarding governance in some developing countries may make it difficult to reliably and

[28] 'Subnational' is seen by many as a euphemism for 'projects', a term that historically generated significant opposition in REDD due to concerns over the potential for leakage. It can also refer to regions within a country.

[29] See Decision 2/CP 13, 'Reducing emissions from deforestation in developing countries: approaches to stimulate action' Annex, para 7.

accurately account for emissions and effectively reform current practices nationally to reduce deforestation. The private sector has also expressed concern that the national approach creates too much risk to warrant up-front or on-the-ground private sector participation.[30] This is because if the private sector invests in a project in a country that has national accounting and crediting, then investors will earn credits only if the project performs as expected *and* the government effectively protects other forests outside the project's boundary. This is true even if an increase in deforestation outside the project boundary is not due to leakage. A number of private sector investors have indicated a preference for project-based accounting and crediting because it allows private sector investment in individual projects without exposure to sovereign performance risk.[31] Project-based accounting may also enable funding to flow more readily to local communities and landowners, because it promotes the direct participation of these stakeholders. It also allows for more accurate monitoring of emissions and emission reductions in a distinct area. As noted earlier, issues of leakage have traditionally been the main stumbling blocks of project-level accounting.[32]

3.3 Permanence

Permanence refers to the likelihood that emission reductions from reducing deforestation will not be reversed in the future. Concerns about the permanence of REDD are similar to those for afforestation and reforestation, and relate to the vulnerability of forest carbon to fire, pests, changes in ownership or management of a forest area, and other natural or anthropogenic disruptions.[33]

Several approaches to addressing permanence have been proposed.[34] At the national level, one alternative is to make the government participating in REDD

[30] See P Moura Costa and J Ebeling, 'EcoSecurities Position Statement Reducing Emissions from Deforestation and Degradation', EcoSecurities 2008 (unpublished, on file with authors). This was also expressed at the workshop 'International Roundtable on Reducing Emissions from Deforestation and Forest Degradation in Developing Countries,' organized by Climate Focus, the Tropical Agricultural Research and Higher Education Center (CATIE), and Avoided Deforestation Partners, held in Brussels, 24–25 October 2007. For a summary of the workshop see Avoided Deforestation Partners.org 'Putting Forests on the Agenda' (News and Events) <http://www.adpartners.org/news_putting_agenda.html>, accessed 26 March 2009; see also 'International Roundtable on Reducing Emissions from Deforestation and Forest Degradation in Developing Countries' (Summary) <http://www.climatefocus.com/downloads/SummaryofBrusselsInternationalRoundtableonREDD.pdf> accessed 26 March 2009 (International Roundtable Summary).

[31] Ibid.

[32] See Moura Costa (n 30 above).

[33] Erin C Myers, 'Policies to Reduce Emissions from Deforestation and Degradation (REDD) in Tropical Forests: An Examination of the Issues Facing the Incorporation of REDD into Market-Based Climate Policies,' Resources for the Future (2007) 48. However, it should be noted that there is no consensus on the relevance of permanence in REDD. REDD, unlike AR, involves the reduction of emissions as opposed to the sequestration of CO_2. It can be argued that REDD emission reductions are intrinsically permanent in the same way that the reduction in fossil fuel emissions are. See ibid.

[34] For a more complete review of options see M Deutsche and A Arild, 'How do we ensure permanence and assign liability?' in A Arild (ed) *Moving Ahead with REDD Issues, Options and Implications* (CIFOR, 2008).

responsible for ensuring permanence by accounting for any reversal over time against the national reference level. This in effect passes liability for permanence onto the country.

Another option is to create a buffer or reserve account of credits to counter the risk of future increases in emissions from deforestation. This would involve setting aside a certain percentage of all credits earned by a developing country and holding that amount in reserve to guard against the reversal of any compensated reductions.[35] For example, a crediting mechanism might automatically set aside 10% of the credits earned by a country in a given year. If at any point in the future deforestation rates rose above a country's baseline, that country could dip into its reserve account to offset the surplus emissions. The reserve amount could vary based on the risk level and performance indicators associated with the country. Reserve accounts would limit the amount of capital available to forest countries in the near-term, but they would also prevent forest countries from going into carbon debt over time and insure the environmental benefit of reductions credited by a REDD mechanism. This accounting partially removes the permanence liability from the country by shifting it to the reserve.

Irrespective of the accounting and liability approach taken, permanence issues should also be addressed in the countries themselves. Monitoring systems and satellite data should be used to track land cover and provide real time updates on land use changes. Such a system would have to be coupled with effective enforcement of laws against clearing forest. Forested countries would greatly benefit from a multilateral deployment of technology and training to improve monitoring and enforcement capabilities. Such a deployment could be used as a bargaining chip with developing countries in exchange for greater liability for the permanence of forest carbon credits.

At the subnational or project level, the permanence analysis becomes linked to the length of an individual project and the health of a particular forest area is more relevant than in a national baseline scenario. What are the guarantees that a lowered rate of deforestation in the project area will be sustained over time and especially after the formal project activity ends? While the CDM has used temporary credits to account for project-level reductions over time, these credits have proven more difficult to market than full-value carbon credits. Commentators have suggested that it may be appropriate to apply a required minimum length to forestry project activities in order to ensure that reductions are sustained for a long enough period of time as to allow for a positive adjustment in carbon emissions and forest activities in the project activity.[36] The Voluntary Carbon Standard (VCS) takes such an approach to permanence by recognizing that in order to be successful in the long term, a REDD project must address the

[35] Ibid, 55.

[36] John O Niles, 'Tropical Forests and Climate Change' in S Schneider, A Rosencranz, and J Niles (eds) *Climate Change Policy* (Island Press, 2002), 351.

drivers of deforestation and provide sustainable alternatives for local agents of deforestation.[37] The VCS applies a minimum project duration of 20 years with the intention that after at least 20 years, a successful REDD project 'should have induced structural changes that make a return of pre-project levels of deforestation unlikely'.[38]

3.4 Market and non-market funding

Total official development assistance for the entire forestry industry was estimated at approximately US$1.1 billion in 2004—and only a portion of this has been dedicated to forest protection.[39] The Stern Review, produced for the British government in October 2006, determined that reducing emissions from deforestation was one of the most cost-effective ways of reducing GHG emissions globally and estimated that the opportunity cost of stopping deforestation in the eight countries responsible for 70% of global deforestation would be between US$5 billion and US$10 billion a year. Other estimates put the cost of reducing 50% of CO_2 emissions from deforestation over the next 20 years at US $33 billion a year.[40] There has been a significant shortfall in public funding to address deforestation to date. Most groups recognize this and doubt whether public sector funding can be sufficiently scaled up and maintained at a high enough scale over time to address REDD.

Parties' positions on funding REDD can be broken into two broad groups. One group supports the use of both market and non-market funding, and the other advocates for non-market funding only.[41]

Non-market approaches that have been proposed focus on establishing one or more international funds to pay for REDD. To overcome the likely shortfall in public sector funding from developed countries, a number of options for financing these funds have been proposed including (i) through donations from developed countries; (ii) a tax or auction proceeds from developed countries'

[37] Voluntary Carbon Standard (VCS), 'Guidance for Agriculture, Forestry and Other Land Use Projects' (November 2007), 35 <http://www.v-c-s.org/docs/Guidance%20for%20AFOLU%20projects.pdf>, accessed 24 March 2009. [38] Ibid.

[39] I Tomaselli, 'Brief Study on Funding and Finance for Forestry and Forest-Based Sector' (Report prepared for the United Nations Forum on Forests) (January 2006) <http://www.un.org/esa/forests/pdf/publications/brief_funding_tomaselli.pdf>, accessed 24 March 2009.

[40] Stern (n 1 above), 217. The passage cites a study carried out for the report by M Grieg-Gran, 'The Cost of Avoiding Deforestation' (2006) International Institute for Environment and Development; M Obersteiner and others, 'Economics of Avoided Deforestation' (Paper presented at the conference 'Climate Change Mitigation Measures in the Agro-Forestry Sector and Biodiversity Futures,' International Centre for Theoretical Physics, Trieste, Italy, 16–17 October 2006).

[41] See the summary from the AWG-LCA Accra meeting: FCCC/AWGLCA/2008/CRP.5 'Report of the workshop on policy approaches and positive incentives on issues relating to reducing emissions from deforestation and forest degradation in developing countries; and the role of conservation, sustainable management of forests and enhancement of forest carbon stocks in developing countries' at para 16.

assigned amount units (AAUs);[42] and/or (iii) proceeds from auctioning emission allowances under domestic emissions trading schemes. Funding would be used to help countries develop the capacity to monitor and account for REDD, and then also compensate them for any emissions reductions they achieve.

Market approaches focus on integrating emission offset credits generated under a REDD trading mechanism into the carbon market.[43] While creating a robust REDD credit trading market has been raised in the international discussions, advocates of market approaches also recognize that additional, non-market based funds will need to be mobilized as part of the broader REDD policy.[44] For robust REDD trading markets to develop, a certain threshold level of readiness will need to be established among the key forest jurisdictions. This traditional fund-based financing would help develop the capacity needed to engage the market. Funding will be needed to implement this capacity building to jump start market activities as investors will refrain from investing in jurisdictions that do not have adequate legal, technical, and political frameworks in place.

4. Cross-cutting issues to trading REDD

The scale of the challenge to dramatically reduce GHG emissions from global deforestation and the estimated costs associated is significant. We suggest it will require a concerted effort of both governments and the private sector, as history tells us public sector funding is not sufficient on its own. Despite the current global economic crisis, substantial participation by the private capital markets are likely to be needed over the coming decades to achieve the necessary sustained level of investment in REDD. The best way to engage the private sector is to allow the use of fully fungible REDD units that are integrated into the carbon market. The policy measures currently being debated at the UNFCCC COP level and at the various national levels need to strike the right balance between sovereign and private sector participation.

The recent evolution of REDD under the UNFCCC and its potential inclusion in the UN regulated segment of the carbon market (along with potential inclusion

[42] AAUs are the units issued by Parties who agreed to cap their emissions under Kyoto. The amount of AAUs each Party issues is equal to the number of emissions they agreed to release during the period 2008–2012.

[43] There are a number of variations of this, such as creating parallel markets or dual markets for REDD units. See, eg M Ogonowski and others, 'Reducing Emissions from Deforestation and Degradation: The Dual Markets Approach' (2007) Centre for Clean Air Policy; (2008) 17 Review of European Community and International Environmental Law. ISSN 0962 8797 <http://www.fern.org/media/documents/document_4263_4264.PDF>, accessed 26 March 2009; and B Hare, and K Macey, 'Tropical Deforestation Emission Reduction Mechanism A Discussion Paper' (Greenpeace International Report) (2008) <http://www.greenpeace.org/raw/content/international/press/reports/TDERM.pdf>, accessed 26 March 2009.

[44] See the summary from the AWG-LCA Accra meeting: FCCC/AWGLCA/2008/CRP.5 'Report of the workshop on policy approaches and positive incentives on issues relating to reducing emissions from deforestation and forest degradation in developing countries; and the role of conservation, sustainable management of forests and enhancement of forest carbon stocks in developing countries' at para 16.

in a US regulatory system) has created growing interest in developing REDD projects in the voluntary market. This interest has mainly come from NGOs who see REDD as a financing option for their conservation aims (which represents a change of position of some NGOs that opposed forestry projects during the Kyoto round of negotiations), large emitters that are likely to be regulated and seek cost effective compliance, and some private sector companies that see REDD as a new investment opportunity. In 2007, REDD deals accounted for 5% of the volume of all over-the-counter voluntary market credits transacted on the voluntary market, and 28% of all biological carbon sequestration project deals.[45]

For REDD to move from these isolated speculative investments to a massive capital markets shift, international and national climate policy must create a structure to attract that investment. Among the most critical aspects of creating a successful market for REDD are addressing governance and rule of law in certain jurisdictions where tropical deforestation is occurring, and understanding the underlying legal interest that is to be created by a REDD credit. The next section discusses these issues.

4.1 Governance

The importance of governance in improving the forest sector has been widely accepted and included in the EU's Forest Law Enforcement and Governance Programme, the UN Forum on Forests' Non Legally Binding Instrument, the FAO, the ITTO, and the G8.[46] As two forest governance experts have put it 'tackling systematic poor governance is a prerequisite for achieving investment in long-term forest management or any broader environment or development aims for the sector'.[47]

Discussions around governance in REDD were informally raised and discussed by a number of parties in the lead up to the Bali COP. Some developing countries were opposed to any discussion of governance in REDD, arguing essentially sovereign issues such as this should not confuse the focus on international climate change policy. However, the national level approaches favored by many countries require strong and effective governments that can develop, monitor and enforce national policies to address REDD. Most of the eight countries responsible for 70% of the total land use change emissions[48] have

[45] Hamilton, Sjardin, Marcello, and Xu, 'Forging a Frontier: State of the Voluntary Carbon Markets' (Report by Ecosystem Marketplace & New Carbon Finance) (2008) <http://www.newcarbonfinance.com/download.php?n=2008_StateofVoluntaryCarbonMarket.pdf&f=fileName&t=NCF_downloads>, accessed 26 March 2009. [46] Ibid.

[47] J Saunders and R Nussbaum, 'Forest Governance and Reduced Emissions from Deforestation and Degradation (REDD)' Chatham House Briefing Paper, Energy Environment and Development 03 (December 2007), 3 /03 (December 2007), 3 <http://www.illegal-logging.info/uploads/REDDupdate.pdf> accessed 26 March 2009.

[48] See M Grieg-Gran, 'The Cost of Avoiding Deforestation: Report prepared for the Stern Review of the Economics of Climate Change' (2006) London: International Institute for Environment and Development.

problems with governance (see table 27.1) and the private sector raised questions about whether they would invest in REDD if it required national level programs that were essentially dependent on good governance for success.[49]

Table 27.1[50] shows a country's rank for each governance indicator against other countries around the world. The rank is a percentile rank that indicates the percentage of countries that ranks below the particular country. Each governance indicator noted above is defined[51] as follows.

(1) Voice and accountability measures

Perceptions of the extent to which a country's citizens are able to participate in selecting their government, as well as freedom of expression, freedom of association, and a free media.

(2) Political stability and absence of violence measures

Perceptions of the likelihood that the government will be destabilized or overthrown by unconstitutional or violent means, including politically-motivated violence and terrorism.

(3) Government effectiveness measures

Perceptions of the quality of public services, the quality of the civil service and the degree of its independence from political pressures, the quality of policy formulation and implementation, and the credibility of the government's commitment to such policies.

(4) Regulatory quality measures

Perceptions of the ability of the government to formulate and implement sound policies and regulations that permit and promote private sector development.

[49] See M Estrada, E Corbera, and K Brown, 'Reducing Greenhouse Gas Emissions From Deforestation In Developing Countries: Revisiting The Assumptions' (December 2007) Tyndall Centre for Climate Change Research, Working Paper 115 <http://www.scribd.com/doc/5283247/REDUCING-GREENHOUSE-GAS-EMISSIONS-FROM-DEFORESTATION-IN-DEVELOPING-COUNTRIES-REVISITING-THE-ASSUMPTIONS#>, accessed 26 March 2009. This was expressed at the workshop 'International Roundtable on Reducing Emissions from Deforestation and Forest Degradation in Developing Countries.' See International Roundtable Summary (n 30 above).

[50] This table is updated from Estrada, Corbera, and Brown (n 49 above), 30. The update prepared by the authors includes all governance indicators analysed by the World Bank and is based on its most recent assessment of governance in D Kaufmann, A Kraay, and M Mastruzzi, 'Governance Matters VII: Aggregate and Individual Governance Indicators 1996–2007' (2008) The World Bank Development Research Group Macroeconomics and Growth Team & World Bank Institute Global Governance Program, Policy Research Working Paper 4654 <http://papers.ssrn.com/sol3/papers.cfm?abstract_id=1148386>, accessed 26 March 2009.

[51] Kaufmann, Kraay, and Mastruzzi (n 50 above) 10, 11.

Table 27.1. Governance indicators and trends in the eight countries responsible for 70% of emissions from land use change

Governance Indicators	Year	Bolivia	Brazil	Cameroon	Congo DR	Ghana	Indonesia	Malaysia	PNG	Average
Voice and Accountability	2007	49.5	59.1	21.2	8.7	62	42.8	31.3	51.9	40.8↑
	2003	47.1	62.5	16.8	5.8	53.4	38.5	38	45.7	38.5↑
	2000	51.9	53.4	16.3	1.9	47.6	36.5	40.4	48.6	37.1
Political Stability	2007	18.3	36.5	31.3	2.4	53.4	14.9	52.4	22.6	29.0↓
	2003	26.9	47.6	28.4	1.9	46.6	3.4	55.8	25	29.5↓
	2000	35.6	50	28.4	0.5	37.5	7.7	55.3	31.3	30.8
Governmen Effectiveness	2007	19.9	52.6	17.1	1.4	55	41.7	82.9	25.1	37.0↓
	2003	41.2	61.1	28	3.3	47.4	33.2	78.2	26.5	39.9↓
	2000	47.9	58.8	23.7	2.4	57.8	35.5	78.7	34.1	42.4
Regulatory Quality	2007	11.7	53.4	24.3	8.3	53.9	43.7	67	30.1	36.6↓
	2003	51.7	62.9	19.5	3.9	41.5	26.3	70.2	24.9	37.6↓
	2000	56.1	62.9	31.2	1.5	51.7	36.6	64.9	30.7	42.0
Rule of Law	2007	17.6	43.3	12.9	1.4	52.4	27.1	65.2	21	30.1↓
	2003	39.5	44.8	16.2	0.5	47.1	18.6	63.3	15.2	30.7↓
	2000	41.9	45.2	11	1.4	51.9	24.8	61.9	23.3	32.7
Control of Corruption	2007	38.6	52.2	15.9	3.9	56	27.1	62.3	9.2	33.2↑
	2003	24.3	56.3	25.2	2.4	48.1	14.1	62.6	18.9	31.5↓
	2000	36.4	58.7	9.2	2.4	48.5	13.6	66	24.8	32.5

(5) Rule of law measures
Perceptions of the extent to which agents have confidence in and abide by the rules of society, and in particular the quality of contract enforcement, property rights, the police, and the courts, as well as the likelihood of crime and violence.

(6) Control of corruption measures
Perceptions of the extent to which public power is exercised for private gain, including both petty and grand forms of corruption, as well as 'capture' of the state by elites and private interests.

Each of the six governance criteria set out in Table 27.1 are important for national level REDD initiatives to succeed. This importance is heightened when national REDD initiatives extend beyond monitoring and accounting only, to include implementation and benefit sharing or distribution of any financial incentives the government may receive from implementing REDD policies and reducing emissions. As a result, government failure in any one of these indicators could undermine socially and environmentally effective REDD implementation.

Of these six indicators political stability, regulatory quality, and rule of law are most relevant for project level REDD activities. Poor voice and accountability, and ineffective national policies to preserve forests can be overcome by well designed project level activities. Corruption will be a factor in project level activities, but it will not play as significant a role as it would in national level activities where financial flows were directed through the government.

Recognition of subnational activities as 'a step towards the development of national approaches, reference levels and estimates'[52] would create sufficient flexibility to allow countries to respond to REDD as they deem most appropriate. It would allow those countries with the capacity to test national level initiatives do so, and those that preferred a step wise approach as they developed national capacity to initiate subnational demonstration activities first.

4.2 Defining the REDD credit

Under international public law an internationally assigned REDD credit is most likely defined as a sovereign right.[53] How the REDD credit is defined within the various sovereigns, therefore, will directly impact the potential future market for REDD credits as it will influence where and how much private capital flow for

[52] See Decision 2/CP 13 'Reducing emissions from deforestation in developing countries: approaches to stimulate action', Annex, para 7.

[53] For a more complete legal analysis of carbon credits see M Wemaere and C Streck, 'Legal Ownership and Nature of Kyoto Units and EU Allowances' in D Freestone and C Streck (eds), *Legal Aspects of Implementing the Kyoto Protocol, Making Kyoto Work* (OUP: 2005), and chapter 2 above.

REDD is applied. In general, tradable carbon rights are intangible property interests that are created either by legislation or by contract under general property law principles within a given jurisdiction. While examples of legislation defining the REDD credit exist,[54] developing countries have not yet established the necessary legal frameworks to specifically define the legal nature of the REDD credit.

Perhaps the most important aspect of defining the REDD credit domestically is whether REDD credits will be owned by the sovereign only or subject to private ownership. If REDD credits are owned by the sovereign (and not subject to private ownership), private sector involvement and investment towards REDD will be reduced substantially. One the other hand, if REDD credits are recognized as private property that can be traded among private parties, the market will likely be underpinned by private sector investment. Where private ownership of REDD credits will be allowed, project participants must also consider the requisite coordination of approvals at various levels of government within a jurisdiction. Early REDD projects have suffered as a result of inconsistent and overlapping governmental involvement at the local, state and federal levels.[55] The most likely scenario for a broader global market is a hybrid system incorporating a mix of sovereign and private ownership of the REDD credit across the numerous jurisdictions involved.

The lack of clear land tenure regimes and inconsistent enforcement where such regimes exist is another hurdle to scaled investment in REDD. REDD projects require long-term land use preservation, which is made more challenging in countries that do not have adequate land title recording systems and clear property transfer laws. The lack in certain jurisdictions of adequate legal systems to address land tenure will, at the least, make undertaking REDD projects in those jurisdictions more costly and less competitive as a result of the need to

[54] Each of the States in Australia has passed legislation to create a form of carbon sequestration right which landowners may register on title in favor of another party. This right is similar in law to an easement or a covenant, transferring the carbon rights and responsibilities associated with the land to a third party. In Indonesia, Government Regulation No 6/2007 has been enacted which authorizes provincial and district governments to issue permits for the utilization of environmental services, called *Izin Usaha Pemanfaatan Jasa Lingkungan* (*IUPJL*). IUPJLs entitle their holder to store and absorb carbon and other environmental services in certain government-owned forests, and are granted for a term of at least 30 years. See M Wilder and J Crittenden, 'Bringing the Forests to Market: Structuring Avoided Deforestation Projects' in *Climate Change and Forests, Emerging Policy and Market Opportunities* (n 15 above).

[55] As Martijn Wilder notes, '[p]rojects have failed in the past when, following significant time and capital investment, and lengthy negotiations with state governments, the federal government has stepped in to prohibit sale of the Environmental Benefits. In Brazil, some sections of the Federal Government have argued that the country cannot allow rights to carbon to be sold internationally, as this will simply result in a lower national baseline, which is a concern if Brazil adopts a post-2012 target. In Papua New Guinea, different levels of government have issued different levels of support to projects and articulated different levels of rights to deal in carbon, resulting in a range of conflicting claims. Project developers must be aware of such risks, and should engage in active dialogue with all relevant levels of government to prevent such outcomes to the extent possible.' Ibid.

perform increased levels of due diligence and to identify and secure approvals of all parties that could have potential claims to rights in the land, including claims from indigenous communities.[56]

Under either a sovereign right or a private property model, a key aspect of ensuring the success of REDD credit trading is the degree to which the newly defined economic benefits accrue to the communities on the ground. If a disproportionate amount of the economic benefits of REDD remain with central governments or private companies, and do not benefit the people at the forest's edge, there will be little chance of sustaining the preservation of a forest. To date, the initial voluntarily undertaken REDD projects have largely recognized the importance of ensuring local benefits from REDD projects, and third party certification programs such as the Climate, Community and Biodiversity Standard[57] are widely adopted as standard industry practice. One of the policy challenges to be sorted out is how to balance the need to create economic returns for the private capital market investors while ensuring proper levels of economic returns for the forest communities.

4.3 Market risks

The discussion on transaction structures below touches on market risks governments or project developers are exposed to. These market risks centre on the price of credits, which may be affected by the market being flooded by REDD credits, restrictions on the use of REDD credits by capped actors to meet commitments, or other market fluctuations. These market risks are important for anyone engaging in REDD transactions, be it sovereign governments, local communities, or private investors. The level of engagement of each of these actors in REDD transactions will be affected by their appetite to take on these types of market risks. Sovereign governments, for example, may find it more difficult to justify (and pass through parliament) far reaching and long term policy reform where the pay-off is subject to significant market risks. Many governments will need more stable sources of non-market funding to support REDD. The private sector, on the other hand, is expected to be able to manage their exposure to these market risks by engaging in hedging strategies and having a diversified portfolio of investments.

[56] For example, in Papua New Guinea, approximately 97% of all land is held as communal or clan commons, a status guaranteed by the Papua New Guinean Constitution. See International Tropical Timber Organization, 'Status of Tropical Forest Management 2005-Papua New Guinea' (2005) 177. <http://www.itto.int/direct/topics/topics_pdf_download/topics_id=12430000&no=1>, accessed 26 March 2009.

[57] The Climate, Community and Biodiversity Standard is administered by the Climate, Community and Biodiversity Alliance. See 'CCB Standards' <www.climate-standards.org>, accessed 26 March 2009.

4.3.1 Flooding

Opening existing carbon markets to REDD credits has caused concerns that the market will be flooded by an influx of large numbers of 'cheap' credits, which would decrease the price of carbon credits. This would in turn decrease the incentive to invest in low carbon technologies, reduce the transfer of clean technology to developing countries, and reduce the amount of funding available for REDD.[58]

However, modeling of price affects of including forest carbon in the EU ETS conducted for the Eliash Review found this was not necessarily the case. Increasing the number of international credits that could be imported into the EU will lower the carbon price and create short term efficiencies. The inclusion of forest carbon may provide the political cover to seek steeper reductions both at the international treaty level and within key developed countries. As a result, prices can be maintained or even increased by committing to such steeper reductions. This combination of increasing domestic emission reduction commitments combined with increasing the number of international credits into the system will therefore lead to greater emission reductions globally *without* affecting the price of carbon.

4.3.2 Use restrictions

The EU ETS is currently the main source of demand for carbon credits but prohibits all forest based carbon credits being used to meet the emission reduction commitments it sets. This ban was justified by a mix of ideological opposition to offsets from forests and registry complexities associated with CDM forest projects generating 'temporary' CERs that pass certain liabilities onto the entity that buys them.[59]

As a result, the market for forest carbon offsets is limited to the lower priced voluntary market and a small amount of demand from some Annex I governments. This lack of market acceptance and lack of fungibility effectively killed the market for CDM forest projects. If similar restrictions were placed on REDD credits in the EU ETS or other domestic emissions trading schemes REDD is unlikely to attract significant investment from the private sector. It is also unclear how large or sustainable public sector funding would be to compensate REDD.

[58] Johan Eliash, 'Climate Change: Financing Global Forests' *The Eliash Review* (2008) 188 <http://www.occ.gov.uk/activities/eliasch/full_report_eliasch_review(1).pdf>, accessed 26 March 2009.

[59] A country that uses a tCER or lCER (the CERs created by CDM A/R projects) for compliance it is responsible for replacing these credits when they inevitably expire or are cancelled. For a more detailed discussion of this see S Scholtz and I Noble, 'Generation of Sequestration Credits under the CDM' in D Freestone and C Streck (eds), *Legal Aspects of Implementing the Kyoto Protocol Mechanisms, Making Kyoto Work* (OUP: 2005); B Bosquet, 'Specific Features of Land Use, Land-Use Change, and Forestry Transactions' in D Freestone and C Streck (eds), *Legal Aspects of Implementing the Kyoto Protocol Mechanisms, Making Kyoto Work* (OUP: 2005).

4.3.3 Other market risks

Assuming the policy-related market risks discussed above are resolved, price fluctuations can still be expected—the carbon market is still a market exposed to a plethora of factors that affect prices. Pricing factors that affect current carbon prices such as weather, oil prices, gas prices, and coal prices will still exist and influence the price paid for carbon credits in a liquid carbon market. One of the key risks for REDD inclusion is the overall economic decline, which has depressed the carbon prices across all of the markets. A depressed carbon price reduces modelled return on investment and makes any carbon project more difficult to finance. On the other hand, the need to demonstrate that any new carbon mandates emerging out of the United States and in the post-2012 international policy will be cost effective is at an all time high. In that sense, REDD may be viewed as a means to achieve real commitments that are cost effective in light of the global recession.

Another risk for REDD is the need for long-term, stable policy that provides a clear price signal over decades to enable the carbon return on investment in REDD projects to be modeled over the life of the projects. Currently, there still exists significant uncertainty surrounding the outcome of the COP meeting in Copenhagen, including how REDD policies will be shaped. REDD projects typically have long crediting periods and there is no 'multiplier effect' as with other industrial gas or methane projects, so the policy for REDD needs to be clear and long term to enable proper modeling of returns for investors. If the negotiations in Copenhagen result in another five-year commitment from the international community, thereby leaving uncertainty after 2018, market-based financing for REDD will likely suffer. One potential counterbalance to these international policy risks is the possibility of a long-term policy signal from the United States. A US mandatory program that recognized REDD for its own compliance needs would operate as part of its re-engagement with the international community, but would not be limited by any failures of the COP to achieve a long-term policy signal.

4.4 Transactions

How REDD transactions are structured, including who bears what risks associated with the trade of any REDD credits and whether or not it will attract private capital, will be significantly affected by international REDD policy.

Transaction structures will be examined in three different scenarios of possible REDD policy: (1) national-level monitoring, accounting, implementation, and awarding of credits; (2) project-level monitoring, accounting, implementation, and awarding of credits; and (3) national-level monitoring and accounting and project-level implementation and awarding of credits. All three scenarios assume REDD policy allows for the creation and trading of REDD credits.

Scenario 1: National

Under this scenario the central government would be intimately involved in all REDD transactions as they would be the initial recipient and owner of all REDD credits. REDD credits—as a sovereign right/asset under public international law—would be likely to be treated as State assets, and decisions regarding their sale and use of proceeds may need to go through parliament. Proceeds from the sale of credits could form part of the government's general budget or be separately allocated to specific programs or policies. Where the government is not the owner of the forest, it would need to determine under domestic law how their sovereign rights to the REDD credits related to any rights local people have to the land or forest. Where the government restricted the rights of these groups in the implementation of its REDD agenda, there should be consideration given to concepts of 'just compensation' under local law for these groups as well as appropriate benefit sharing.

The central government would be liable for any under delivery of credits under a forward contract. It would also assume liability associated with any funding it received to implement REDD initiatives that were repaid with REDD credits or proceeds from the sale of REDD credits. Similarly, the government would also be exposed to the uncertainty surrounding (i) how many (if any) REDD credits are generated and (ii) the market price for those credits at the time of sale. Market risk may be reduced where forward sales are entered into at pre-determined prices, but these prices are typically discounted and such contracts often contain liability associated with under delivery.

The allocation of sovereign REDD credits to local communities or other project developers that may have been involved in implementation would be the responsibility of the government. These groups would therefore be exposed to sovereign risks associated with (i) the country as a whole not generating sufficient REDD credits and (ii) the government failing to honour any previous agreement to distribute a certain number of credits to the project. These risks may be reduced where the government knows in advance that it will have surplus credits.[60] Leakage is captured within national monitoring, however any leakage associated with subnational implementation will still need to be tracked if credits or performance based incentives are passed down to subnational actors. This creates additional government performance and governance related transaction risks for communities or other project developers. It is likely that local communities or other right holders to forests or land affected by national REDD policy would want any such allocation to be separated from any consideration of just compensation associated with any loss or restriction of their rights discussed above.

[60] This is the case in green investment schemes regarding the sale of AAUs in Eastern European countries. It could also be the case if, for example, a HFLD country negotiates a national reference scenario for deforestation and degradation emissions that is greater than its historic emissions to take into account likely future deforestation.

If permanent REDD credits are issued, and policy does not address permanence via a buffer, the government would be responsible for ensuring permanence of its forests into the future. A permanence failure may result in a lack of new credits being issued in future crediting periods. If permanence is addressed via the temporary crediting approach currently adopted in the CDM for afforestation and reforestation projects[61] the government would only be liable for permanence if they contracted to deliver a stream of credits into the future and were unable to because of a permanence failure. However, it must be emphasized that experience to date shows that the carbon markets discount temporary credits heavily, valuing them at a fraction of permanent credits, and is not recommended by the authors.

Scenario 2: Project

Under the project scenario involvement of the host country's government is kept to a minimum, which eliminates all government risks and potential liability associated with participating in REDD. It also reduces governance and REDD implementation risks for projects found in Scenario 1. Local communities and other project developers are able to receive direct benefits from REDD if their efforts are successful. They are also not exposed to the risk in national approaches of not receiving credits if deforestation increases in other areas of the country for reasons not related to their project (ie not associated with leakage).

As with national transactions, the project developer or community is exposed to uncertainty regarding generating credits, market prices, and any credit generation and delivery liability assumed under project finance or credit sale agreements. The same considerations regarding permanence discussed under Scenario 1 also apply.

Concern over leakage is typically the biggest criticism of project approaches. To be credible, project activities will need to monitor and account for domestic leakage associated with their project. This risk of leakage creates additional uncertainty and delivery risk for projects. However, this risk may be less than in Scenario 1 where the government would be involved in assessing leakage and discounting project level credits.

Scenario 3: National/Project

The national/project scenario is, as its name suggests, a hybrid between the national and the project scenarios. There are a number of different options associated with this concept. In one variation, the government's role is limited to monitoring and accounting, along with any implementation they choose to undertake. As they are not primarily responsible for national level implementation they are not exposed to risks associated with uncertain credit

[61] In this approach new credits are issued every five years in the case of tCER (with previously issued credits expiring) or the credits are re-verified every five years in the case of lCERs (with unverified credits being cancelled).

generation or market prices or potential liability associated with under delivery.[62] These liabilities are passed onto the communities or project developers implementing any REDD projects. Domestic leakage is captured under the national monitoring. Project level and any national level monitoring and accounting for leakage would need to be reconciled, but this could be done eg through independent third party assessments of project emission reductions that would also assess leakage.

5. Conclusions

If the worst consequences of the climate crisis are to be averted, the international community needs to effectively halt and reverse tropical deforestation. Include a REDD mechanism in a future, broad agreement on climate change would be a major step towards that goal. For this inclusion to result in significant reductions in deforestation in developing countries, several steps still need to be taken. First, in this chapter we have argued that tapping into private sector capital via the inclusion of REDD within the carbon market is needed to generate sustained funding at sufficient scale. This integration would allow for cost effective steeper global reductions of GHG emissions.

However, for this integration to produce meaningful results, policymakers will need to create a clear, long-term demand for REDD credits through express inclusion of the credits in a mandatory GHG reduction regime. This needs to be in international and domestic climate change law. In addition, while REDD may ultimately be addressed at the national level, sub-national REDD projects should not be discounted entirely due to concerns about leakage and permanence. There are a number of ways to address these issues at the project level and project level efforts to address REDD have a number of advantages over national level implementation. Project-level REDD will help mitigate some of the challenges of the transition to national level REDD programs.

Private capital and the carbon markets are not the complete solution. Additional funding is needed to support government action on REDD and the conservation of forest not under threat of deforestation. Increased multilateral investment in capacity building and technology for developing countries combined with policy reform to shore up the governance institutions that relate to land tenure, monitoring and enforcement are all essential elements of a successful strategy.

[62] It should be noted that this does not prevent the government from implementing subnational projects or programs that reduce emissions. In this instance the government's activities would be treated the same as any local community or other project developer.

28

Aviation and Climate Change Regulation

Claybourne Fox Clarke and Thiago Chagas

'I think everybody agrees that we have to find some way of addressing emissions from aviation' Yvo de Boer, UNFCCC Executive Secretary*

1. Introduction

Aviation emissions are the fastest growing source of anthropogenic greenhouse gas emissions. Aviation emissions from Annex I Countries increased by 67% between 1990 and 2005 and, according to some estimates, by as much as 90% when aviation emissions from non-Annex I Countries are included for this period.[1] At present, roughly 3% of both the United States' and European Union's (EU) total greenhouse gas (GHG) emissions are from aviation. The EU projects that by 2012, with current growth trends, emissions from international flights using EU airports will increase by 150% over 1990 levels.[2] Growth at this rate, the EU determines, would negate over 25% of the reductions the EU projects achieving in other areas.[3]

To date, despite heated debate on the issue, no international measures have been adopted to address these emissions. While there is a general consensus that steps will need to be taken to address the growing emissions from international air travel in the near future, two over-arching questions arise in the ongoing

* Addressing UNFCCC delegates at the climate change talks in Bangkok, April 2008.

1 National greenhouse gas inventory data from Annex-I Parties for the period 1990–2005, Note by the Secretariat, Subsidiary Body for Implementation, 24 October 2007, FCCC/SBI/2007/30; see also Institute for European Environmental Policy, J Chiavari et al, *The Role of the EU in Attempting to 'Green' the ICAO* citing European Federation for Transport and Environment report '*Clearing the Air: The Myth and Reality of Aviation and Climate Change* (2006).

2 See 'Communication from the Commission to the Council, the European Parliament, the European Economic and Social committee and the Committee of the Regions—Reducing the Climate Change Impact of Aviation' {SEC(2005) 1184}, COM/2005/0459 final [hereafter Communication from the Commission—Reducing the Climate Change Impact of Aviation].

3 Ibid.

debate: which entity or entities should lead the development of such measures; and, what is the most effective method for carrying out such measures?

The International Civil Aviation Organization (ICAO), an agency within the United Nations established by the 1944 Chicago Convention, is charged with coordinating international measures in the aviation sector.[4] While ICAO has asserted itself as the appropriate body to lead international aviation's efforts to reduce GHG emissions, its failure thus far to achieve substantive results has led to criticism and for the EU and others to push for the sector's inclusion under the United Nations Framework Convention on Climate Change (UNFCCC) in a post-2012 successor-agreement to the Kyoto Protocol.

Because ICAO and proponents of UNFCCC inclusion both support international aviation taking part in some form of emission trading scheme to achieve emission reductions, it is likely the sector will eventually utilize this approach. However, debate again arises over the form. Should the sector's emissions be included in national totals and reduction commitments under a post-2012 climate agreement? Or, should a sectoral approach be utilized, possibly with aircraft operators directly responsible for meeting reduction targets?

The following chapter discusses these and related issues in the context of examining the current status of international aviation under the UNFCCC, the ongoing debate over developing measures under ICAO or including the sector in a post-2012 agreement, the moves by the EU to bring international aviation under the European Union Emission Trading Scheme (EU ETS), and issues that will need to be considered when implementing measures at the international level to reduce the sector's emissions.

2. Current Efforts at International Regulation

2.1 Aviation under the UNFCCC and the Kyoto Protocol

Domestic aviation emissions are accounted for within the GHG inventory of Parties and required to be included in the calculation of the total national GHG emissions (also referred to as 'national totals') when Parties report to the UNFCCC. The Intergovernmental Panel on Climate Change (IPCC) Guidelines for the preparation of GHG inventories and the UNFCCC reporting guidelines on annual inventories call for emissions from international aviation to be calculated as part of the national GHG inventories of Parties, but to be

[4] ICAO now has 190 contracting states which comprise ICAO's Assembly. The constitution of ICAO is the Convention on International Civil Aviation, drafted in Chicago in 1944, to which each ICAO Contracting State is a party. The Assembly meets every three years and elects the organization's governing body, the Council. As stated in the Convention, the purpose of ICAO is to promote co-operation between nations 'in order that international civil aviation may be developed in a safe and orderly manner and that international air transport services may be established on the basis of equality of opportunity and operated soundly and economically'.

excluded from national totals and reported separately.[5] Therefore, national inventories of GHG emissions only include aviation emissions occurring within the national territory, with the emissions associated with fuel sold for use in international aviation reported separately as what have become known as 'memo items'.

Article 2(2) of the Kyoto Protocol specifically excludes emissions from international aviation and maritime transport (also known as international bunker fuel emissions) from the limitation and reduction commitments of Annex I Parties. Instead, the Kyoto Protocol calls for Annex I Parties to pursue reductions in emissions from these sources through ICAO and the International Maritime Organization.[6] The exclusion for aviation was based upon recognizing ICAO's pre-existing authority, but also in large part due to the difficulty of determining national responsibility for international emissions and developing an effective international instrument for the sector.

The final version of the 'Bali Road Map' provides for '(c)ooperative sectoral approaches and sector-specific actions, in order to enhance implementation of Article 4(1)(c) of the Convention.'[7] Inclusion of this language in the path forward agreed upon at the 13th Session of the UNFCCC Conference of the Parties (COP13) in Bali allows for specific measures regarding international aviation, along with shipping, to be included in the negotiations toward an agreement that will replace the emission limitations set in the first commitment period of the Kyoto Protocol. Even with this language, including the regulation of international aviation in a post-2012 UNFCCC agreement may prove difficult. As discussed below, the EU, with little support from other Parties, has been the only major proponent of change, with its move to address international aviation under the UNFCCC.

2.2 Developments under ICAO

Since the Kyoto Protocol's adoption, ICAO has asserted itself as the recognized 'global instrument for developed countries to pursue the limitation or reduction of greenhouse gas emissions from international aviation'.[8] ICAO has publicly discussed voluntary mechanisms and, in particular, emissions trading as a means

[5] See 2006 IPCC Guidelines for National Greenhouse Gas Inventories.

[6] Kyoto Protocol, Art 2(2).

[7] See Para 1(b)(iv), Decision 1/CP.13 (Bali Action Plan); UNFCCC, Art 4(1)(c) commits all Parties to ... 'Promote and cooperate in the development, application and diffusion, including transfer, of technologies, practices and processes that control, reduce or prevent anthropogenic emissions of greenhouse gases not controlled by the Montreal Protocol in all relevant sectors, including the energy, transport, industry, agriculture, forestry and waste management sectors.'

[8] ICAO, 'Kyoto Protocol Emphasizes ICAO's Role in Addressing Greenhouse Gas Emissions from International Aviation', International Civil Aviation Organization website, Media Release, December 12, 1997; available online at: <http://www.icao.int/icao/en/nr/1997/pio199725_e.pdf>, accessed 31 March 2009.

of reducing the sector's emissions. However, to date, no definitive measures have been proposed. ICAO has recently developed draft guidance intended for its Contracting States (member states to the Chicago Convention) to use in voluntarily implementing emission trading measures.[9] However, since being designated under the Kyoto Protocol as the forum for addressing international aviation's GHG emissions, ICAO has served as much, if not more, as a forum for championing causes to preclude the sector from mandatory measures aimed at reducing GHG emissions as it has for developing such measures.[10] For this reason, ICAO has been accused of failing to be sufficiently proactive and of, in effect, holding up the development of substantive GHG reduction measures for the sector.[11]

It has been asserted that close ties with and the influence of the aviation industry has affected ICAO's stance on GHG measures.[12] Regardless of this assertion, ICAO's approach thus far may be more fundamental, resulting from its initial purpose and the Chicago Convention's primary aim of assuring 'safe, regular, efficient and economical air transport.'[13] While environmental protection is now recognized among the 'strategic objectives' of ICAO, this objective does not relate back to specific language in the Chicago Convention.[14] The

[9] CAEP, 'Report on Voluntary Emissions Trading for Aviation (Draft Guidance on the Use of Emissions Trading for Aviation,' DOC 9885, adopted at the Seventh Meeting of CAEP—CAEP/7), 2007, 15/04/2007, <http://www.icao.int/env/vets_report.pdf>, accessed 31 March 2009. For a list of Contracting States to ICAO see <http://www.icao.int/cgi/goto_m.pl?cgi/statesdb4.pl?en>, accessed 31 March 2009.

[10] See, eg, ICAO, Council Resolution A33/7, Consolidated statement of continuing ICAO policies and practices related to environmental protection (setting forth the 33rd Assembly's resolution in opposition to fuel taxes and preference for GHG charges, which it later requested not to be applied until 2007, at the earliest), 2001; CAEP, Report of the Fifth Meeting of ICAO's Committee on Aviation Environmental Protection (CAEP/5) (concluding CAEP should not pursue developing a carbon dioxide standard), 2001; CAEP, 'Market-Based Measures'Report from Working Group 5 to the fifth meeting of the Committee on Aviation Environmental Protection, CAEP/5-IP/22, 5/01/2001 (concluding, based upon cost benefit analysis, a closed emission trading scheme for aviation does not warrant further consideration; thereafter, ICAO's 33rd Assembly recommended an open emission trading scheme) 2001; and ICAO, Council Resolution A35/5, Consolidated statement of continuing ICAO policies and practices related to environmental protection (setting forth the 35th Assembly's dismissal of developing an open emission trading scheme, instead recommending voluntary inclusion in existing emission trading schemes), 2004.

[11] See eg, DG Internal Policies of the Union (IP/A/ENVI/NT/2006–08), 'Reducing the Impact of Aviation on Climate Change,' page 1, (2006); and K Ribeiro et al, 2007: 'Transport and its infrastructure, Climate Change 2007: Mitigation, Contribution of Working Group III to the Fourth Assessment Report of the IPCC,' page 375, (2007). Available online at: <http://www.ipcc.ch/pdf/assessment-report/ar4/wg3/ar4-wg3-chapter5.pdf>, accessed 31 March 2009.

[12] Institute for European Environmental Policy, J Chiavari et al, 'The Role of the EU in Attempting to "Green" the ICAO' (n 1 above), 12.

[13] The Chicago Convention establishes, in its preamble, that Contracting States have agreed on principles related to international civil aviation so that 'it may be developed in a safe and orderly manner and that international air transport services may be established de on the basis of equality of opportunity and operated soundly transport and economically'. The Chicago Convention's Article 44(d) establishes ICAO and its objectives.

[14] ICAO (2005), Strategic Objectives of ICAO for 2005–2010—Consolidated Vision and Mission Statement, <http://www.icao.int/icao/en/strategic_objectives_2005_2010_en.pdf>, accessed 21 May 2009.

Chicago Convention's mandate of equal treatment and uniform application of measures to international aviation has also proven difficult to reconcile with the UNFCCC principle of common but differentiated responsibilities. The distinction between Annex I countries and developing countries under the Kyoto Protocol has made it difficult for ICAO Contracting States to agree on measures that can be implemented uniformly by all its Contracting States, which include both developed and developing countries.

2.3 The EU approach

The EU has been the most vocal critic of ICAO's delay in achieving measures to reduce international aviation's GHG emissions. Based upon this delay and ICAO's 2004 endorsement of including international aviation under Contracting State's emission trading schemes,[15] the EU decided to move towards the inclusion of aviation in the EU ETS.

The EU first expressed an intention to address aviation emissions in 2002, when the European Parliament and the Council established as an environmental priority the need to identify and undertake specific actions to reduce GHG emissions from aviation if no such actions were agreed upon within ICAO.[16] In 2005, the European Commission commissioned a consultancy study—'Giving Wings to Emissions Trading'—in order to assess viable policy options to include aviation emissions within the EU ETS.[17] Supported by the findings of this study, in September 2005 the Commission adopted a Communication outlining plans to reduce the impact of aviation on climate change, in which it re-affirmed its concerns with the rapid growth of national and international air travel emissions and pointed to the inclusion of aviation emissions under the EU ETS as the 'most promising way forward'.[18] Through the September 2005 Communication, the Commission also established an Aviation Working Group under the European Climate Change Programme to consider ways of including aviation in the EU ETS and to inform the Commission's legislative proposal on the subject.[19]

[15] COM (2006) 818 Final, explanatory note. <http://eurlex.europa.eu/LexUriServ/site/en/com/2006/com2006_0818en01.pdf>, accessed 31 March 2009.

[16] Decision No 1600/2002/EC of the European Parliament and of the Council of 22 July 2002 laying down the Sixth Community Environment Action Programme of 10 September 2002, OJ L 242, 10.9.2002.

[17] See RCN Wit, BH Boon and A van Velzen (CE Delft); M Cames and O Deuber (Oeko-Institut, Berlin); and DS Lee (Manchester Metropolitan University), 'Giving wings to emissions trading – Inclusion of aviation under the European emissions trading scheme (ETS): design and impacts', July 2005, available at <http://ec.europa.eu/environment/climat/aviation_en.htm>, accessed 4 February 2009).

[18] See 'Communication from the Commission—Reducing the Climate Change Impact of Aviation,' (n 2 above).

[19] Ibid, Annex.

After obtaining the endorsement of the Council[20] and of the Parliament,[21] the Commission adopted in December 2006 a proposal for amendments to the EU ETS for the inclusion of aviation.[22] In September 2008, the European Parliament, Council and Commission reached a final agreement on the text and amendments to the proposal.[23] Directive 2008/101/EC to include aviation in the EU ETS was finally adopted in November 2008 (the 'Aviation Directive').[24]

The Aviation Directive establishes an EU-wide cap for GHG emissions in the aviation sector, effective as of 2012. The cap equates to 97% in the first year (2012) and 95% for each year thereafter (during the period comprising the third phase of the EU ETS) of the average annual aviation emissions in the EU from 2004 to 2006. Under the EU approach, with the exception of small aircraft operators and other limited exemptions,[25] airlines flying in and out of the EU[26] will be required to comply with the EU ETS obligation of surrendering allowances or offset credits[27] to match their actual emissions starting from 2012. This includes airline operators based outside the EU.

Due to the fact that reductions of emissions from international aviation are not part of the commitments made by parties under the Kyoto Protocol, aviation allowances allocated to aircraft operators can only be used to meet the obligations imposed under the Aviation Directive. Aircraft operators are allowed to use allowances initially allocated to operators of stationary installations for

[20] See Council of the European Union, Presidency Conclusions, Brussels 15/16 December 2005, 30 January 2006, 15914/1/05 REV 1, available at <http://ec.europa.eu/environment/climat/aviation_en.htm>, accessed 4 February 2009.

[21] See European Parliament resolution on reducing the climate change impact of aviation (2005/2249(INI)), 4 July 2006, available at <http://ec.europa.eu/environment/climat/aviation_en.htm>, accessed 4 February 2009.

[22] See Proposal for a Directive of the European Parliament and of the Council amending Directive 2003/87/EC so as to include aviation activities in the scheme for greenhouse gas emission allowance trading within the Community {SEC(2006) 1684} {SEC(2006) 1685}, COM/2006/0818 final.

[23] See Opinion of the Commission pursuant to Article 251(2), third subparagraph, point (c) of the EC Treaty, on the European Parliament's amendments to the Council's Common Position regarding the proposal for a Directive of the European Parliament and of the Council amending Directive 2003/87/EC so as to include aviation activities in the scheme for greenhouse gas emission allowance trading within the Community amending the proposal of the Commission pursuant to Article 250(2) of the EC Treaty, COM/2008/0548 final.

[24] See Directive 2008/101/EC of the European Parliament and of the Council of 19 November 2008 amending Directive 2003/87/EC so as to include aviation activities in the scheme for greenhouse gas emission allowance trading within the Community, OJ L8, 13.1.2009.

[25] Small aircraft operators are considered those operators flying on an average of less than two flights per day or whose annual emissions are below 10,000 tonnes.

[26] Flights in and out of Norway, Iceland and Lichtenstein will also be included.

[27] Under the Aviation Directive, aircraft operators are able to use offset credits approved under the Kyoto Protocol to meet a percentage of their obligations to surrender allowances. For the period running from 1 January 2012 to 31 January 2012, the limit on the use of UN offset credits has been set at 15%. For the period from 2013 to 2020, the review of the EU ETS for a third trading period established that the aviation sector will be able to use offset credits up to an amount corresponding to a certain percentage of their verified emissions during the third phase, such amount not being set below 1.5%.

compliance under the Aviation Directive, but stationary installations cannot use aviation allowances for compliance purposes under the EU ETS.

3. International Response to the EU Approach and the Ongoing Debate over the Development of Measures through ICAO

The extension of obligations to monitor and verify emissions and surrender allowances to all air operators arriving at or departing from an airport in the EU has sparked fierce opposition from both developed and developing countries. In a letter addressed to the European Parliament and to the German Presidency of the EU in April 2007, the US, Canada, Australia, Japan, Korea and China have jointly voiced their dissatisfaction with the inclusion of their domestic airlines under the EU ETS without their prior consent. The letter asserts that a unilateral regulatory measure of this type amounts to a breach of international duties the EU assumed under the Chicago Convention and other bilateral international agreements governing aviation.[28]

The authority of Contracting States to mandate foreign airlines' participation in an emission trading scheme had also been heavily debated in 2007 at ICAO's Committee on Aviation Environmental Protection (CAEP).[29] This debate is ongoing, with Contracting States' positions divided between two approaches:

(1) *The Mutual Agreement Approach* under which the State implementing the emission trading scheme would need to have the consent of the airline's home country prior to including the airline.

(2) *The Unilateral or Alternative to Mutual Agreement Approach*,[30] under which the State (or group of States) implementing the emission trading scheme could include an airline without the consent of the airline's home country.

At the 36th Assembly of ICAO held in September 2007 in Montreal, the Assembly called on its Council to develop substantive proposals for reductions in aviation emissions, through both technical solutions and a market-based approach, by the end of 2009 (taking into consideration the date of the December 2009 Conference of the Parties to the UNFCCC).[31] However, in

[28] Official letter from the governments of the United States, Australia, Canada, China, Japan, and Korea available at <http://ec.europa.eu/environment/climat/poznan_presentations_08.htm>, accessed 30 December 2008.

[29] CAEP/7 Information Paper 20: Guidance on Emissions Trading for Aviation. Presented by the Rapporteurs of the ICAO Emissions Trading Task Force.

[30] Some EU representatives have objected to the title of 'unilateral' since multiple Member States have determined to carry out the requirement.

[31] The language of the Resolutions only recognizes the date of the December 2009 UNFCCC COP 15. It does not state that the program developed will be completed in time for consideration at COP 15.

response to the EU intent to unilaterally include foreign airlines under the EU ETS, Resolution A36/22 was also passed. The Resolution calls for Contracting States to utilize the mutual agreement approach before applying market-based measures, such as emissions trading, to a foreign airline. Contrary to normal practice, Resolution A36/22 was not adopted by consensus. Forty-two European States registered a formal reservation, stating their intent not to be bound by this part of the Assembly's conclusions.[32]

All non-European Contracting States of ICAO expressed support for the mutual agreement approach at the 36th Assembly of ICAO. Australia, Canada, the US, China, Argentina, Brazil, Japan, South Korea and Nigeria led support for this approach. These countries, and in particular the US, have argued that the EU imposition of measures upon international flights may have little benefit since airlines may shift their older, less efficient planes to domestic and international routes not subject to the EU ETS, and that higher ticket prices in the EU will cause passengers on international transit flights to chose alternative routes not connecting through the EU.[33] The US has also been critical of the requirement that non-EU airlines would be required to register with and be regulated by the EU Member State in which they have the most flights. A Member State that regulates its own domestic airlines could be prone, the US has asserted, to favouring its own airlines and discriminating against the foreign airlines it regulates.[34] Developing countries, such as Nigeria, have urged also the development of measures that do not impact the growth of aviation in developing countries.[35]

The US will undoubtedly refine and expand its arguments under the provisions of the Chicago Convention if it proceeds to take action against the EU. Thus far, the US has asserted before the ICAO that the requirement that non-EU airlines participate in the EU ETS is a violation of the Convention Article 15 which states that 'no fees, dues or other charges shall be imposed by any contracting State in respect solely of the right of transit over or entry into or exit from its territory of any aircraft of a contracting State or persons or property thereon'.[36]

[32] ICAO, Minutes of the Eleventh and Tenth Meetings of the Executive Committee, Reservations made to Assembly Resolutions A36–22 and A36–28, 25/02/08, <http://www.icao.int/icao/en/assembl/a36/docs/A36_MIN_EX_10_11_en.pdf>, accessed 31 March 2009.

[33] US Comments in the Executive Committee of the Council of ICAO on Agenda Item 17 Concerning Aviation Emissions, by Andrew B. Steinberg, Assistant US Secretary of Transportation for Aviation and International Affairs, September 22, 2007. Available online at: <http://ostpxweb.dot.gov/aviation/speeches/2007_9_22_US_STATEMENT_ON_EMISSIONS.pdf>, accessed 31 March 2009.

[34] Ibid.

[35] Ibid, citing Comments of Her Excellency, Mrs. Alison-Madueke of Nigeria in her comments before the Executive Committee of the Council of ICAO.

[36] Ibid; For further analysis of potential arguments against the EU's imposition of the EU ETS upon non-EU carriers pursuant to the Chicago Convention see Daniel Reagan, 'Putting International Aviation into the European Union Emission Trading Scheme: Can Europe do it flying solo?' (2008) 35 Boston College Environmental Affairs Law Review 349.

Similarly, the International Air Travel Association (IATA), aviation's largest trade group, has been particularly critical of the EU's unilateral approach, calling for measures to be taken through ICAO and voluntarily.[37] IATA has also proposed a voluntary target that airlines should operate as 'carbon neutral' by 2057.[38] The feasibility of this voluntary target, without relying purely on the purchase of offset credits from outside the sector, has been questioned given current fuel technologies. However, the feasibility of achieving reductions within the sector, at least in part, has received a boost with Rolls Royce's recent certification for use of a jatropha-based fuel for one of its jet engines. Air New Zealand is to begin using the new fuel in test flights. IATA and the US-based Air Transit Authority (ATA) and the Association of European Airlines (AEA) have all pushed for measures to be taken to increase efficiencies in air travel rather than mandatory reductions in GHG emissions being imposed upon the sector. The industry groups support implementing the Single European Sky and US NextGen Air Transport System, satellite-based air traffic management systems allowing airlines to fly routes that are more direct and less dictated by national borders.[39]

In its statement of reservation to Resolution A36/22, the EU, in turn, sets forth the basis for its objections to the mutual agreement approach and the right to proceed unilaterally with mandating foreign airlines to comply with the EU ETS. The EU reservation asserts:

> there is no provision in the Chicago Convention which may be construed as imposing upon the Contracting Parties the obligation to obtain the consent of other Contracting Parties before applying market-based measures referred to in Annex L to operators of other states in respect of air services to, from or within their territory. On the contrary, the Chicago Convention recognises expressly the right of each Contracting Party to apply on a non-discriminatory basis its own air laws and regulations to the aircraft of all States.[40]

In a subsequent legal opinion that the EU has made public, further basis is provided for assertion that extension of the EU ETS to include international aviation emissions does not breach any provisions of the Chicago Convention.[41]

[37] IATA publication, *Building a Greener Future*, 3rd Edition, 2008. Available online at: <http://www.iata.org/NR/rdonlyres/C5840ACD-71AC-4FAA-8FEE-00B21E9961B3/0/building_greener_future_oct08.pdf>, accessed 31 March 2009; see also IATA press release on EU ETS inclusion of aviation at: <http://www.iata.org/pressroom/pr/2008-10-24-02.htm>, accessed 31 March 2009.

[38] Ibid.

[39] Ibid; see also speech of James May, ATA President and CEO, addressing the European Aviation Club in October 2008, <http://www.airlines.org/news/speeches/speech_10-30-08.htm>, accessed 31 March 2009.

[40] Written statement of reservation on behalf of the member states of the European Community (EC) and the other states members of the European Civil Aviation (ECAC) [made at the 36th Assembly of the International Civil Aviation Organization in Montreal, 18–28 September 2007] <http://europa.eu/rapid/pressReleasesAction.do?reference=MEMO/07/391&type=HTM>, accessed 31 March 2009.

[41] See legal memo of John Balfour, Clyde & Co, of 19 November, requested by the European Parliament: <http://ec.europa.eu/environment/climat/poznan_presentations_08.htm>, accessed 4 February 2009.

The main arguments put forward in favour of the EU approach are that: (i) the EU ETS is devised as a non-discriminatory measure that is applicable to all flights in and out of the EU and thus compliant to Article 11 of the Convention;[42] and (ii) an emissions trading scheme creating obligations to monitor emissions and surrender allowances is an administrative measure that is in nature different from any 'fees, dues or other charges' imposed with respect to the right of transit over, entry into or exit from the territory of a Contracting State by an aircraft of another Contracting State and would therefore fall outside the scope of Article 15 of the Convention. In distinguishing between the meanings of taxes and charges, ICAO has defined charges as 'levies to defray the costs of providing facilities and services for civil aviation'.[43]

Whether these arguments would indeed hold up in the event of a challenge under the ICAO dispute proceedings is still uncertain. While the EU ETS is designed to apply without distinction as to the nationality of carriers, it remains to be seen how the EU Member States conduct and implement the scheme in practice. In addition, outside the context of the Chicago Convention, there is at least some authority to consider payments for allowances auctioned to regulated entities as a form of tax or unrequited charge under international law.[44]

4. Post-2012 Measures to Reduce Emissions

Citing ICAO's failure to move forward and the sector's growing emissions, the EU has continued to call for international aviation to be included under the UNFCCC in a post-2012 agreement. At the UNFCCC climate change talks since Bali, the EU has continued to urge Parties under the Kyoto Protocol to establish 'clear, meaningful targets' for the sector within the framework of an agreement for the post-2012 period.[45] Norway and New Zealand have also

[42] Article 11 of the Chicago Convention states that '[...] the laws and regulations of a contracting State relating to the admission to or departure from its territory of aircraft engaged in international air aviation, or to the operation and navigation of such aircraft while within its territory, shall be applied to the aircraft of all contracting States without distinction as to nationality, and shall be complied with by such aircraft upon entering or departing from or while within the territory of that State'.

[43] See ICAO Council Resolution On Environmental Charges and Taxes. Available online at: <http://www.icao.int/HyperDocs/Display.cfm?name=at%2dwp%2f1900&lang=E>.

[44] J De Cendra, using as a basis the definitions of taxes and charges provided by the OECD, observes that payment for allowances under an emissions trading scheme (when these are auctioned by governments) could qualify as a tax or unrequited charge because: (i) they constitute a payment to the government; and (ii) the government does not provide any benefit in proportion to the payment. See J De Cendra, 'Can Emissions Trading Schemes be Coupled with Border Tax Adjustments? An Analysis vis-à-vis WTO Law' (2006) 15 Review of European Community and International Environmental Law 135. See also Ismer and Neuhoff, 'Border Tax Adjustments: A Feasible Way to Address Nonparticipation in Emissions Trading', Cambridge-MIT Institute Working Paper 2004.

[45] EU submission to the UNFCCC's Ad Hoc Working Group on Further Commitments for Annex I Parties under the Kyoto Protocol (AWG KP 5).

supported the inclusion of aviation in a post-2012 agreement and the use of emissions trading under such an agreement to achieve reductions for the sector. Opposition to the EU has come from countries such as Thailand, Australia and China. These countries maintain the position that ICAO is the appropriate forum for developing measures to reduce the sector's emissions.

It has also been argued that those developing countries with economies relying more heavily on tourism would be disproportionately impacted. In response to this latter concern, the EU asserts that including aviation under the EU ETS will increase demand for projects under the CDM, and that the economic benefits of this to developing countries will outweigh potential impacts on tourism.[46]

Before the UNFCCC, ICAO has maintained that it is still the rightful forum for continuing to address aviation's emissions post-2012. In a statement to the third session of the Ad hoc Working Group on Long-term Cooperative Action under the UNFCCC in Accra, Ghana, in August 2008, representatives of ICAO asserted that the bases for the initial designation of ICAO under the Kyoto Protocol 'are equally relevant for a post-Kyoto agreement'.[47] The statement further pledged that the ICAO Group on International Aviation and Climate Change (GIACC) is developing an 'aggressive programme of action on international aviation and climate change' that will be completed 'at a time which would take into account COP 15, in December 2009'.[48]

4.1 The appropriate post-2012 authority

Future measures to reduce GHG emissions from international aviation can be achieved under three potential authorities: ICAO, the UNFCCC, or by handing authority over to regional bodies. If the ICAO promised 'aggressive programme of action' is acceptable to the EU, it is likely that ICAO will remain the primary authority. In Accra, ICAO emphasized that 'effective solution for aviation emissions will involve a global framework encompassing a basket of measures of a technological, operational and market-based nature to deal with the specific realities of States and regions'.[49] However, given the fact that to date ICAO has had difficulty in achieving agreement among its Contracting States and in reconciling the principles of the Chicago Convention with significant GHG reductions, the EU and its supporters will likely find it prudent to simultaneously pursue action under the UNFCCC and at the regional level until ICAO delivers.

[46] EU Commission presentation at Third Session of the Ad hoc Working Group on Long-term Cooperative Action under the Convention (AWG-LCA3), Accra, Ghana, (2008) available at: <http://unfccc.meta-fusion.com/kongresse/AWG_08_Ghana/templ/ply_page.php?id_kongresssession=1358&player_mode=isdn_real>.

[47] Statement by ICAO to AWG-LCA, Accra, Ghana, (2008) available at: <http://www.icao.int/icao/en/env/awg_lca.pdf>.

[48] Id.

[49] Ibid.

4.2 Potential post-2012 options

A post-2012 agreement could, expanding upon the current UNFCCC approach, allocate emissions from international aviation to Parties for inclusion under national totals, or adopt a sectoral approach requiring aircraft operators to be directly responsible for achieving reductions.[50] Without focusing upon which international body that will serve as the primary authority over international aviation emissions, the following sub-sections discuss two distinct options for post-2012 regulation of these emissions.

4.2.1 Inclusion under national totals

Emissions from international aviation could be accounted for by countries as part of national GHG inventories and national emission reductions or limitation targets. Each country would then determine how best to address the emissions. Under this approach, countries could either adopt measures to reduce their respective share of international aviation emissions or increase mitigation efforts in other sectors.

A primary hurdle to this approach is determining and assigning the emissions from international aviation to specific countries. Because the emissions are occurring outside the borders of the responsible country, reaching agreement on how to make this determination and on taking responsibility for these emissions could prove difficult. Potential alternatives have been proposed, such as assigning emissions to states according to the emissions generated in their national air space or distributing emissions in proportion to the state's total national emissions, but at this point no alternative has gained broad support.[51]

International aviation emissions attributable to developing countries that do not have national reduction targets under the UNFCCC would raise further issues. Developing countries generally oppose the imposition of national reduction targets. If emissions attributable to developing nations' airlines went unregulated, this could result in market distortions, significant international leakage, and a competitive advantage for airlines based in these countries.

4.2.2 A sectoral approach

With a sectoral approach, emissions from international aviation would not be included in national emission reduction or limitation targets. Instead, the Parties would be tasked with enforcing a sectoral scheme and aircraft operators would be directly responsible for surrendering allowances or offsets covering their

[50] See EU presentation in thematic session 4 before the AWG-KP 5, Bangkok, 2008. AWG-KP 5, <http://unfccc.int/kyoto_protocol/items/4336.php>, accessed 31 March 2009.

[51] See M Grubb et al, 'Energy and Climate: Opportunities for the G8' (2008), Climate Strategies and Centre for Energy Studies, available at <www.climatestrategies.org>, accessed 26 March 2009.

emissions. Under this approach, a baseline would be determined, a cap established, and global reduction target set for the sector.

Realizing a sectoral approach would require that the Parties first agree upon a sectoral scheme, and that this scheme then be equitably enforced by each of the Parties on the airlines under their jurisdiction. To assure a harmonized application of the scheme by each of the Parties, many of its requirements would need to be agreed upon in advance. Key elements, such as allocation methods and penalties for non-compliance, among others, would need to be determined to assure a uniform application across the sector in different countries. The relatively low number of airlines, at least compared with emitters in other sectors, should help make the debate over terms of an agreement between the Parties less complicated and more achievable.

While the Parties would be responsible for the implementation and enforcement of the scheme at the country level, an international body would be needed for development and for oversight of a sectoral scheme. This international body could also serve other functions, ranging from hearing disputes to managing a single registry system for allocation, trade, and surrender of allowances for the sector.

Under this approach, complications over attributing a share of international aviation emissions to specific countries can be avoided. The participation of States in the actual scheme would likely be limited to an administrative role (eg collecting data and enforcing rules upon airlines). Notwithstanding these advantages, the sectoral approach also presents complexities of its own. A sectoral scheme could follow a baseline-and-credit system based on international benchmarks or apply a cap-and-trade approach. Developing countries are likely to reject international benchmarks, but a cap-and-trade system for the airline industry faces its own set of challenges. Determining the number of allowances to be allocated to airlines in each country and the method of such allocation, establishing the rules for new entrants, and deciding the end-use of revenues for auctioned allowances, are all issues that are prone to raise competitiveness and distributional concerns. Equity considerations and differential treatment of developing countries, in particular those with economies heavily dependent on tourism, would also need to play a central role in the design if it is to gain acceptance among these countries.

Another key determination would be if the scheme is open or closed. Under an open scheme, the use of allowances from other sectors or linkages to other regimes would be allowed. Under a closed scheme they would not. Variations are also possible, whereby differing value or limited eligibility is given to allowances from other sectors. Both ICAO and IATA have emphasized that aviation should only be included under an open emission trading scheme. However, the differences between aviation and other sectors would need to be reconciled in an open scheme. Issues such as the differing effect of emissions at high altitude and the impact upon other sectors of the fact that, given current aviation technology, aviation will need to be a significant net purchaser of allowances and CERs (or other offsets), will need to be better explored. This latter issue could pose a negative effect on other industrial sectors. Under an open emission trading model, if the least costly abatement options exist outside the aviation sector, then these are the ones that will be pursued. This could cause mitigation costs for industrial sectors to significantly rise, and may result in little in the way of actual emission reductions in the aviation sector. A closed sectoral approach for aviation would have the advantage of posing less impact on mitigation costs in other sectors and potentially drive greater actual emission reductions in the aviation sector. These advantages would need to be weighed against the higher compliance and mitigation costs for airlines.

A sectoral approach also has the advantage of being able to be implemented at international, regional or national levels. Regional authorities, as discussed next, would benefit from a sectoral approach to regulate emissions from international aviation. While reaching international agreement on the terms may not prove easy, the sectoral approach could present the most viable near-term solution for achieving reductions in the sector.

4.3 Regional authorities

Another approach for addressing GHG emissions from international aviation would be to place this authority with regional bodies. It has been noted that not only the EU, but also other regional authorities, such as the CARICOM in the Caribbean region, the Yamoussoukro process in Africa and the ASEAN in Asia could provide appropriate forums for this approach.[52]

Even if there is no agreement under ICAO or UNFCCC processes for turning over greater control to regional authorities, it is likely, as the EU has demonstrated, that regional and national bodies supporting the development of reduction measures for the sector will continue moving forward until there is agreement at the international level. The legality of this approach has been challenged. As discussed above, the EU stands alone in its position that ICAO Contracting States can unilaterally mandate foreign airlines to comply with regional measures to reduce GHG emissions. This is not the first time the EU has sought to impose environmental measures upon foreign aircraft, and not the first time such efforts have been challenged. In the period between 1999 and 2001, the EU and the US were embroiled in a dispute over the EU adoption of noise restrictions that would result in the prohibition of 'hush-kitted' aircraft. In the negotiations, the EU maintained that it would not go forward with implementing the restriction if measures could be reached through ICAO. In the end,

[52] See J Chiavari (n 1 above), at 4. It has to be said however that none of the listed regional authorities has the standing and legislative power comparable to the EU.

compromise measures were adopted through ICAO and the US effectively withdrew the complaint it had filed under the ICAO dispute procedures.

Analogous to its approach with imposing noise restrictions, in addressing aviation's GHG emissions the EU has proceeded unilaterally while maintaining its support for measures at the international level and providing potentially sufficient time for such measures to be developed. The inclusion of aviation under the EU ETS does not become effective until 2012. The amendments to the EU ETS also specifically carve out an exception for foreign airlines from countries adopting comparable measures to reduce GHG emissions from aviation. The new US administration will likely be more supportive of addressing climate change both at the international level and domestically. It is foreseeable that with a more supportive US administration, international measures could be more easily achieved and that the US could have equivalent domestic measures in place by 2012. This, of course, is yet to be known. However, the fact that the EU is pushing ahead unilaterally does ensure that the issues regarding international aviation's GHG emissions will be addressed—amicably or otherwise.

5. Conclusions

The Kyoto Protocol exemption of international bunker fuels and the designation of ICAO as the forum for developing measures to reduce international aviation GHG emissions has resulted in the sector not being subject to emission reduction requirements during the Kyoto commitment period. ICAO has pledged to provide a comprehensive plan for achieving reductions for the sector by 2009. However, the difficulties that ICAO has had, thus far, in achieving agreement among its Contracting States and reconciling its fundamental objectives with those of the UNFCCC, means that for proponents of such measures pursuing the inclusion of international aviation under the UNFCCC post-2012 is necessary until the ICAO programme of action is delivered. If ICAO fails to deliver a complete programme of action, the collective strengths of both ICAO and the UNFCCC could be utilized. Even if ICAO does undertake an aggressive plan of action, greater UNFCCC involvement in developing measures for the sector would be advantageous. Until meaningful international measures are developed, regional and national bodies supporting reductions in the sector will most likely proceed unilaterally, as demonstrated vividly by the EU.

Seeing the writing on the wall, ICAO may chose to act rather than risk losing control over the process of developing measures to address the sector's GHG emissions. The new US administration seems more likely be more supportive of reduction targets internationally and of the development of domestic measures for the sector. Either of these approaches, if achieved, allows the new US administration to demonstrate a commitment in leading climate action as well as avoiding the dispute it has threatened over the projected EU regulation of US aircraft emissions.

For the development of post-2012 measures, utilizing a sectoral approach requiring aircraft operators to be directly responsible for emissions should be explored and may prove preferable. Reaching international agreement over the terms of a sectoral approach, while likely not easy, may prove more workable than overcoming the difficulties of assigning emissions from international aviation to individual countries. Under a sectoral approach, the potential impacts of including aviation in an open or closed scheme will still require further analysis. Such analysis will need to consider the impacts not only on the aviation sector, but on other sectors as well. The likelihood of achieving actual reductions within the sector, and the greater control over this that a closed scheme may provide, will need to be weighed against the more cost effective approach of allowing aviation to utilize less costly reductions in other sectors.

PART VIII

SUMMARY AND OUTLOOK

Summary and Outlook

Charlotte Streck and David Freestone

In early 2010 it will be five years since the Kyoto Protocol came into force. This short period has been truly astonishing for the new world of carbon trading. The market for trading emission rights and carbon credits has exploded—more than doubling each year. In 2005 the market was worth US$10 billion, in 2006 US $30 billion, in 2007 US$67 billion, and in 2008 more than US$ 120 billion.[1] At the same time as the amounts of money involved have increased so has the level of sophistication of those involved—investors, project proponents, host countries, and the international bodies responsible for overseeing these developments. However, rather than moving into a settled era of maturity and certainty, the future of the carbon market is still wracked with uncertainty.

Uncertainty is in fact one of the key themes of this book. But so also is the spirit of experimentation and enterprise—of *Innovation* and of *Learning by Doing*. Each of these three themes will be explored in this summary.

1. Originality and Innovation

The challenge of implementing the obligations of the Kyoto Protocol has had an energizing impact on world carbon markets. Not only is the Kyoto Protocol the first international instruments to envisage market mechanisms, but the scale of the challenge facing countries in moving to a lower and ultimately to a no-carbon economy is staggering. To achieve such an objective requires a paradigm shift by the major industrialized economies. The legally binding emission reduction targets established for Annex I countries by the Kyoto Protocol constitute such a shift[2]—however, these targets can only represent the very modest beginning of a much more ambitious effort to curb global climate change. The Kyoto Protocol by itself falls short of achieving the sort of reductions that would even arrest the increases in greenhouse gas emissions let alone achieve stabilization. At the time of writing these conclusions, the negotiations on a post-2012 agreement have entered a critical phase with only a few months until the next

[1] See K Capoor and P Ambrosi, 'State and Trends of the Carbon Market 2009' (The World Bank: Washington DC, 2009) <http://siteresources.worldbank.org/INTCARBONFINANCE/Resources/State___Trends_of_the_Carbon_Market_2009-FINAL_26_May09.pdf> accessed 4 June 2009.

[2] For more detail, see Freestone in the introduction to this volume, chapter 1.

session of the conference of the parties to the UNFCCC and the deadline established in the Bali Action Plan for an agreement. While the Kyoto Protocol represents an important first step towards a comprehensive climate agreement it falls short in ambition and scope. A Copenhagen Agreement will have to go beyond establishing stricter absolute targets for industrialized countries. Such emission reductions, however ambitious they are, will not by themselves be enough to achieve the IPCC goal of limiting global climate change to an average temperature increase of 2°C. The challenge for negotiators and policy makers consists in formulating incentives for developing countries to achieve economic growth while avoiding the creation of further more serious environmental problems for future generations.[3] It has already become obvious in the negotiations that emission reduction targets are not going to be the appropriate tool to stimulate action by developing countries. Instead the international community will have to engage in the much more ambitious and complicated exercise of creating incentives and providing assistance in defining policies that will allow countries to develop while avoiding further climate damage. Not only must carbon reductions be achieved at a much more significant scale, but trade and investment in sustainable low-carbon alternatives will need to increase substantially. Of equal importance is the improvement of technology transfer to ensure developing countries have the capacity to be able to develop more sustainable systems for transportation, energy, housing, agriculture, forestry and other sectors.[4] Developing countries have also to find innovative and unprecedented solutions to development, which do not involve the unsustainable logging of their forests[5] or following the pathway of high carbon development that the industrialized countries have pursued in the past. This obviously poses a challenging intergenerational equity issue.

A key aspect of the Kyoto Protocol is the introduction of emission allowance trading into international law. Article 17 of the Kyoto Protocol authorizes countries to transfer Assigned Amount Units (AAUs), the base unit of the Kyoto Protocol, among national registries. To ensure the environmental integrity of AAU transfers, many sellers and buyers of AAUs require the linking of proceeds from the sale of AAUs to investments in emission reductions or other climate-relevant investments—so-called 'green investment schemes.'[6] More important however than these government-to-government transactions, is the facility that the Kyoto Protocol provides to harness the investment power and the energy of the private sector by defining a set of market mechanisms that involve the private sector actively in trading and investment activities that contribute substantially

[3] For the potential elements of a global climate deal see Ward, chapter 24.

[4] See Seggers and Gehring, chapter 4.

[5] For potential incentives to reduce emissions from deforestation see O'Sullivan and Saines, chapter 27.

[6] Simonetti and De Witt Wijnen describe details of Article 17 emission trading and green investment schemes in chapter 7.

to compliance with this multilateral environmental treaty.[7] This facility is unprecedented for an international agreement. Domestic emission trading schemes have followed suit and transposed the polluter-pays principle from the international to the domestic level: from Annex I countries to private legal entities operating in these countries.[8] Faced with the reality or possibility of regulatory caps on their carbon intensive activities the private sectors in Europe, Canada, and Japan—and some states of the US—have started to take action. Companies have begun to invest in reducing the emissions from their operations or—to the extent permissible under the various emission trading schemes—invest in the acquisition of offsets. In doing so they can rely on the experiences accumulated in the voluntary market as well as in the implementation of the Kyoto Protocol's project based mechanisms: the Clean Development Mechanism (CDM) and Joint Implementation (JI). While the voluntary market remains largely unregulated, a trend to standardization of emission offsets helps to enhance credibility in the environmental robustness of voluntary offsets.[9] As markets grow legislation may emerge that regulates the voluntary market at a domestic level but there will inevitably be differences and perhaps conflicts between national perceptions of the proper role for the voluntary sector. The CDM and JI in contrast provide a regulatory framework for investments abroad, utilizing the differences in marginal abatement costs between developed economies and those of transition economies and developing countries to finance greenhouse gas reduction activities off-shore. While falling short of triggering the policy changes needed to decarbonize developing economies, project-based mechanisms constitute an important first step towards the greater involvement of developing countries in climate mitigation investments.[10]

The establishment of the EU Emissions Trading Scheme has provided a major boost, encouraging industrialized economies to begin to tap into the power of the markets to reduce greenhouse gas emissions. The link between the EU trading scheme and the Kyoto Protocol CDM and JI has created powerful demand for offsets from developing and transition economies. Similarly it is expected that the US climate legislation will provide the possibility for major emitters to offset part of their emissions with domestic and international carbon credits.[11] The US will however be more reluctant to embrace a mechanism that channels the greatest portion of foreign investments to China, which holds the

[7] Lin elaborates on the opportunities and challenges that come with such an involved relationship between international organizations and private entities in chapter 6.

[8] See Pohlmann, chapter 16, for the EU Emission Trading Scheme; Wilder and Fitz-Gerald, chapter 20, for developments on the establishment of a national emission trading scheme in Australia; Gray and Barrett, chapter 21, describe recent developments in Canada and LaMotte, Williamson and Hopkins, chapter 18, as well as Danish, chapter 19, provide an update on emerging emission trading schemes in the US.

[9] For a detailed discussion see Passero, chapter 23.

[10] The role of carbon offset mechanisms for countries and sectors not involved in a cap-and-trade scheme is described by Cozijnsen and Coren in more detail in chapter 25.

[11] For the role of offsets in the discussion on a US emission trading scheme see Danish, chapter 19.

lion's share of the CDM market.[12] The EU has confirmed on several occasions its long-term vision of a global carbon market and has included in the EU emission trading legislation a provision that authorizes the exploration of possible linkages to other emission trading schemes. Such linking is however burdened by a number of legislative and economic challenges that will take time to overcome. For the time being the recognition of the same offset standards may provide the simplest way of connecting various carbon markets.[13]

Newcomers will doubtless find it difficult to find their way through the jungle of terms and acronyms describing different emission allowances and carbon credits. Assigned Amount Units, Certified Emission Reductions, Emission Reduction Units, and EU Allowances have been supplemented by others such as RMUs, tCERs and lCERs as well as VERs. These are all new creatures entering the legal language, describing different type of international or domestic, private or regulatory rights. The creation of the various rights calls for definition of the legal nature of such rights, leading to a classification that is relevant for contractual interpretation as well as fiscal and accounting implications of carbon transactions.[14]

2. Learning by Doing

Despite the great success that the CDM has become, it has proved difficult to marry the various objectives of the mechanism: Providing greenhouse gas emissions at the scale necessary and at the lowest costs does not necessarily maximize the CDM's contribution to local sustainable development.[15] Academics and NGOs have voiced numerous concerns about the potential detrimental impacts that some projects may have.[16] Even more relevant are concerns related to the issue of the required additionality of some projects that have been registered by the CDM Executive Board. Environmental concerns are complemented by strident complaints about what are perceived to be the inefficient operating practices of the Executive Board, a shortage of qualified verifiers of emission reductions, unpredictable Board decisions and a weak governance structure which allows neither administrative recourse against, nor judicial review of, any decisions of the Board.[17] A number of key institutional changes, such as the establishment of Registration and Issuance Teams and enhanced allocation of responsibilities to the UNFCCC professional secretariat, have not

[12] Tung describes the implementation of the CDM in China in chapter 22.

[13] For a discussion on linking of carbon markets see Mehling, chapter 5.

[14] Wemaere, Streck and Chagas shed light on the different classes of emission rights and their ambiguous legal nature in chapter 2. In chapter 3, Cook outlines the discussion surrounding the classification of emission rights in the context of international accounting standards.

[15] Barani-Schmidt and Netto introduce the reader to the CDM project cycle and the various bodies and panels regulating the implementation of the mechanism in chapter 10.

[16] See Voigt's concerns as expressed in chapter 13.

[17] For suggestions on how project developers can successfully manoeuvre through the complicated and partly dysfunctional CDM operations, see Krey and Santen, chapter 11.

yet had the impacts necessary to address these various concerns. In order to fulfil a meaningful and enhanced role within a post-2012 legal framework, the CDM needs to be reformed. Its environmental credentials need to be further strengthened,[18] the rule of law and good governance firmly established,[19] and its scope expanded to provide incentives not only for project investment but also for policy change.[20] Provided that it proves able to reform, the CDM can remain a vibrant mechanism with a firm place in the implementation of a post-2012 climate agreement. While it is certain that the CDM can no longer, single handedly, provide the sole mechanism for mobilizing emission reductions in developing countries, it has the potential to be much more than a vehicle, as proposed by the European Commission, that is exclusively for inward investment in the smallest and the poorest states. Such a proposal seems unlikely in any event to be accepted: the CDM is dependent on the private sector appetite for foreign direct investment, which is available at only a very limited scale in the least developed countries. The future of the CDM will thus depend on its ability to learn and improve. It is also likely to be complemented by other mechanisms such as a wider use of caps or 'no-lose' targets.[21] Such mechanisms would weaken, or even eliminate, the significance of the 'additionality' concept. The discussion of the environmental robustness of emission reductions would then centre on how to set targets, monitor emission reductions, and verify reported results.

But not only has the CDM to digest some serious lessons learned. Domestic emission trading schemes remain in their infancy and have to be adjusted carefully to live up to their potential. The European lawmakers have just concluded a major review of the EU Emissions Trading Scheme and the reformed system will in 2012 enter its third phase of operation. This phase covers the period from 2013–2020 and will be characterized by a further centralization of the scheme. Whereas today national governments establish their own overall national emission limitation target, the EU Commission will take on that role from 2012 onwards. Free allocation of EU allowances will also be gradually phased out and replaced by the auctioning of allowances. It remains to be seen whether these measures will address the most serious problem in the EU emission allowance market. This is the volatility in the carbon market that results from the elastic supply of emission reductions, compounded by the fact that the EU allowance scheme does not currently allow adjustments in demand. It may well be that the EU lawmakers will have to consider further measures to address this problem. Adaption and change may not only be triggered by economic

[18] See Michaelowa, chapter 12, for a discussion of various proposals of how to reform the additionality testing of CDM projects.

[19] For various proposals see Voigt, Lin or Figueres and Streck, chapters 13, 6 and 26 respectively.

[20] For a discussion of CDM reform see Figueres and Streck, chapter 26.

[21] As discussed by Ward, chapter 24.

realities and the need to facilitate rigorous reductions in emissions, but also by a growing body of emission trading and transaction-specific jurisprudence.[22]

Similarly, the success of emerging US and Australian emissions trading schemes will depend on their ability to adapt and learn. Trading schemes that cover wide parts of a country's economic activity are inherently complex and can only be effective if they are able to respond to changes and disruptions in the economic sectors that they cover.

3. Uncertainty

The early movers in the carbon market were accustomed to dealing with uncertainty. It took seven years for the Kyoto Protocol to come into force and four years to negotiate the Marrakech Accords. During that time a large number of investments were already being made, on the basis of the future value of emission reductions. In 2000, The World Bank launched its Prototype Carbon Fund and soon afterwards the Dutch Government implemented the first emission reduction purchase tenders; this was all undertaken before the Executive Board of the Clean Development Mechanism began to meet. Much of the regulatory uncertainty was reflected in a move towards contractual certainty and risk sharing, with higher levels of due diligence and careful contractual drafting at an individual project level. Contracts in the primary, project-based, carbon market still remain project-specific investment agreements that reflect the particularities of the specific project. Although there have been some moves towards standardization, these contracts address uncertainties by means of carefully negotiated allocation of risks and responsibilities to the various contractual partners.[23] While the CDM specific risks relate mostly to the slow and unpredictable performance of verifiers and the Executive Board, the success of JI transactions depends largely on the reliability of the host country's undertaking to convert AAUs into emission reduction units than can be transferred to project proponents and carbon buyers.[24] Secondary trading contracts tend to deal less with project risk and are therefore easier to standardize.[25] Indeed, while secondary and more sophisticated emission-based transactions are still special in the carbon market because of the specific risks that they have to address, their mere existence evidences the growing confidence of more mainstream financial players in the future of carbon markets.

However the most important uncertainty is the international climate change negotiation process itself. Not only are outcomes of the climate negotiations in

22 The relevant jurisprudence of the European Court of Justice is discussed by Singh Ghaleigh, chapter 17.

23 On the experience in carbon contracting see Wilder and Fitz-Gerald, chapter 14.

24 For details of JI and JI transactions, see Hobley and Roberts, chapter 9; for a general overview of JI see Hoogzaad and Streck, chapter 8.

25 Hedges provides details of secondary trading transactions in chapter 15.

Copenhagen in late 2009 still unclear but the world is now in the midst of a major economic recession. Some, such as Lord Stern in his submission to the 2009 G20 Meeting,[26] see massive investment in green low carbon technologies as a way out of the recession. Others, including some powerful policy makers, see economic recovery as a prior condition for further action on climate change. At the same time, the world's largest economy, under the new Obama administration, is beginning to play a more active role in the climate negotiations and is supporting the conclusion and implementation of major federal legislation to reduce radically its national carbon emissions.

The failure, indeed the refusal, of the US to ratify the Kyoto Protocol has meant that what was until recently at least the world largest emitter of greenhouse gases[27] is some five to eight years behind the other major developed economies in implementing GHG emission restrictions. The unique circumstances of the US federal system suggest that implementing an economy wide system in the US will not be an easy, trouble-free, or even particularly swift, process. It is significant that states and regions of the US, frustrated by the inaction of the previous administration under GW Bush, have already started to implement emission reduction schemes and cap and trade systems at a sub national level. These schemes face a number of legal hurdles, although nearly thirty states are actively participating, with a further ten states acting as observers in collaboration with a number of Canadian provinces and Mexican states. [28] Although critics have complained of the modest targets, the exposure to reduction targets, and carbon trading at a sub-national level, is also experience that is bound to be useful to the federal authorities. It may also be difficult for the US to make meaningful international commitments until it is able to witness the effects of its domestic actions. A lot will depend upon the vision, farsightedness and courage of its politicians as well as on their diplomatic skill to broker an ambitious, but acceptable international climate deal.

Whether such a deal will come as an amendment to the Kyoto Protocol or the UNFCCC; whether it will be formulated in one or several legal agreements; whether it will provide operational detail or delegate all technical decision-making to further sessions of the conference of the parties and subsidiary bodies remains to be seen. Elements of a future international climate architecture are only slowly emerging. The architecture seems likely to include emission reduction targets for industrialized countries and incentives for low carbon development measures for developing countries. Broader policy change in developing countries may have to be voluntary and be supported by all relevant

[26] Ottmar Edenhofer and Lord Nicholas Stern, 'Towards a Green Recovery: Recommendations for Immediate G20 Action.' Grantham Research Institute on Climate Change and the Environment and the Potsdam Institute for Climatic Change Research, March 2009.

[27] Whether indeed China has surpassed the US in absolute greenhouse gas emissions is disputed; complete data sets are missing as China has yet to submit its recent emission inventories to the UNFCCC.

[28] See LaMotte, Williamson and Hopkins, chapter 18.

constituencies at the national level. It will not be an easy exercise as it will require a shift in the current development paradigm; the accumulation of wealth needs to be decoupled from the accumulation of emissions. The international negotiations on creating incentives for reducing emissions from deforestation and forest degradation may serve as an example of a constructive dialogue between developing and industrial countries. This dialogue was triggered by a 2005 submission to the UNFCCC process from Papua New Guinea and Costa Rica, developing countries that have shown unprecedented leadership in tabling the topic and proposing potential solutions.[29] Industry has been less eager to engage and the inclusion of maritime and aviation emissions into a post-2012 climate agreement still remain important but uncertain. Because the International Civil Aviation Authority has been slow to propose meaningful action to reduce greenhouse gas emissions of international aviation, the EU has unilaterally moved forward and brought aviation into the third phase of its emission trading scheme.[30]

Uncertainty also hangs over the future of the international carbon markets themselves. The liquidity of these markets will depend on a number of factors, including the rigour with which emission reduction targets are set for industrialized countries, the creation of broader market incentives and a supporting infrastructure, the reform of the institutions regulating international carbon markets, and the re-definition and reform of the mechanisms that stimulate private sector participation in carbon markets.

The overarching objective of this book has been to explore the key legal issues surrounding the twenty-first century phenomenon of carbon trading. As the international community works to develop a post-2012 regime the chapters in this volume have demonstrated that although there are a host of issues to be negotiated, there are also a host of opportunities. The experience and the issues that are collected in this volume demonstrate the vital role that legal issues and lawyers need to play in this venture. As an international law instrument the Kyoto Protocol is itself highly innovative both in the way that it imposes emission reduction commitments on industrial countries and in the way that it envisages market mechanisms operating to reduce the overall costs to these countries in meeting these commitments. To move this debate to the next level, to engage developing countries in an equitable and meaningful way will require mobilizing even more innovative and adventurous options. It is clear that a simple extension of the Kyoto regime to a new range of countries will not work. But there is a lot of value still in the systems established by the Protocol. As this volume has amply demonstrated, market mechanisms have been shown to work. The scale of the financing mobilized by the private sector has dwarfed the efforts

[29] For a detailed account of the REDD debate and status of negotiations, see chapter 27 by O'Sullivan and Saines.

[30] Chagas and Clarke discuss the inclusion of greenhouse gas emissions from aviation in chapter 28.

of the public sector. But as we have learned too vividly in the recent past, private markets do not function through altruism. The appropriate incentives and regulatory frameworks do need to be put in place to harness them effectively. Moving to a post-2012 regime does not mean—or should not mean—discarding the lessons and the successes of the Kyoto regime. Bringing new agendas into the negotiations also opens up new opportunities to harness the particular talents of the private sector to use innovative carbon finance instruments to contribute to sustainable development in a range of ways—from the financing of clean energy production to avoided deforestation and sustainable land use techniques. As was said in the Introduction, the objective of this volume has been to explain in detail the way that these markets can be, and have already been, put to work in the global effort to lower greenhouse gas emissions, to decarbonize our economies, and to achieve truly sustainable development.

Glossary

Additionality: Measurable, long-term GHG emission reductions and/or removal enhancements that would not have occurred in the absence of a particular project, policy, or activity.

Afforestation: Planting of new forests on lands that historically have not contained forests.

Annex B Parties: Countries listed in Annex B to the Kyoto Protocol, which is based upon countries listed in Annex I to the UNFCCC and sets out the Quantified Emission Reduction and Limitation Commitment of each country.

Annex I Parties: The industrialized countries listed in Annex I to the UNFCCC, which are committed to adopt policies that contribute to limiting their greenhouse gas emissions to 1990 levels by the year 2000 as per Article 4.2(a) and (b). Annex I Parties have also accepted emissions targets for the period 2008–12 as per Article 3 and Annex B of the Kyoto Protocol.

Assigned Amount: The quantity of greenhouse gases that an Annex I Party is allowed to release in accordance with Article 3 of the Kyoto Protocol, during the first commitment period of that protocol (2008–12).

Assigned Amount Unit: The unit quantifying Assigned Amounts, equivalent to one metric tonne of carbon dioxide emissions.

Bali Action Plan: In December 2007 in Bali, the 13th Conference of the Parties to the UNFCCC adopted the Bali Action Plan describing a two-year process to finalizing an agreed outcome in 2009 in Denmark (UNFCCC Decision 1/CP.13).

Baseline: The emissions of greenhouse gases that would occur without the contemplated policy intervention or project activity.

Cap and Trade: Cap and trade refers to an emission trading system where an international or national regulator establishes an overall cap on emissions, issues emission units or rights, and allows the transfer and acquisition of such rights.

Carbon dioxide equivalent (CO_2e): The universal unit of measurement used to indicate the global warming potential of each of greenhouse gases. Carbon dioxide—a naturally occurring gas which is a by product of burning fossil fuels and biomass, land-use changes, and other industrial processes—is the reference gas against which the other greenhouse gases are measured.

Carbon Market: A market that creates and transfers greenhouse gas emission units, allowances or rights.

Certification: is the written assurance by an independent third party that during the relevant period a project, program, activity, sector, region or country has achieved the GHG Reductions as reported in the verification report.

Certified Emission Reduction: A unit issued pursuant to Article 12 of the Kyoto Protocol for the purposes of the Clean Development Mechanism that is equal to one metric tonne of carbon dioxide equivalent.

Clean Development Mechanism: The mechanism provided by Article 12 of the Kyoto Protocol to assist developing countries in achieving sustainable development by permitting industrialized countries to finance climate mitigation projects in developing countries and receive credit for doing so.

Clean Development Mechanism Executive Board: The committee supervising and monitoring the implementation of Article 12 of the Kyoto Protocol.

Conference of the Parties (COP): The supreme body of the Convention. It currently meets once a year to review the Convention's progress. The word 'conference' is not used here in the sense of 'meeting' but rather of 'association', which explains the seemingly redundant expression '4th session of the Conference of the Parties'.

Conference of the Parties serving as the Meeting of the Parties (CMP): The Convention's supreme body is the COP, which serves as the meeting of the Parties to the Kyoto Protocol. The sessions of the COP and the CMP are held during the same period to reduce costs and improve coordination between the Convention and the Protocol.

Deforestation: As defined in Marrakech Accords, the direct human-induced conversion of forested land to non-forested land.

Designated National Authority: The national authority countries participating in the CDM are required to establish pursuant to the Marrakesh Accords.

Designated Operational Entity: An independent third party accredited by the CMP to validate CDM projects, and verify and certify emission reductions.

Emission Reduction Unit: Emission reduction credits issued under Joint Implementation measured in metric tonnes of carbon dioxide equivalent.

Emission Right: A (tradable and transferrable) right to emit into the atmosphere a certain quantity of a specified substance during a defined period of time.

European Emission Trading Scheme: The Scheme envisaged by Directive 2003/87/EC of the European Parliament and the Council of 13 October 2003 for imposing carbon dioxide emissions caps on installations in specified sectors and permitting the use and transfer of allowances for the purposes of achieving these caps as implemented in each EU Member State by national laws.

Executive Board of the Clean Development Mechanism: A 10-member panel elected at COP-7 which supervises the CDM and has begun operation in advance of the Protocol's entry into force.

GHG Reduction: The removal or mitigation of greenhouse gases by a project, program, activity, sector, region or country below the baseline emissions and measured in tCO2e as monitored in accordance with a Monitoring Plan.

Greenhouse Gases: The six gases listed in Annex A of the Kyoto Protocol which are carbon dioxide (CO_2), methane (CH_4), and nitrous oxide (N_2O), as well as hydrofluorocarbons (HFCs), perfluorocarbons (PFCs), and sulphur hexafluoride (SF_6).

Green Investment Scheme: A contractual arrangement between a buyer and a seller of emission rights whereby the seller is obliged to 'green' the sold emission rights by reinvesting the sale proceeds in greenhouse gas emission reduction projects or other projects beneficial to the environment.

Host Country: The country where an emission reduction project is physically located.

Intergovernmental Panel on Climate Change (IPCC): Established in 1988 by the World Meteorological Organization and the UN Environment Programme, the IPCC surveys world-wide scientific and technical literature and publishes assessment reports that are widely recognized as the most credible existing sources of information on climate change. The IPCC also works on methodologies and responds to specific requests from the Convention's subsidiary bodies. The IPCC is independent of the Convention.

International Emissions Trading: Mechanism provided by Article 17 of the Kyoto Protocol, which allows for the international transfer and acquiescence of kyoto emission rights (AAUs, CERs, ERUs and RMUs).

Joint Implementation: Mechanism provided by Article 6 of the Kyoto Protocol, whereby an Annex I country may acquire emission reduction units from projects that reduce net emissions in another industrialized country (including countries with economies in transition (EITs).

Joint Implementation Supervisory Committee: A committee providing international oversight of 'track-two' Joint Implementation projects. Joint implementation projects are carried out by sponsoring and recipient Annex B Parties under Article 6 of the Kyoto Protocol. Track-two is used if one or both of the countries does not meet requirements for the standard ('track-one') joint implementation programme.

Kyoto Mechanisms: Three procedures established under the Kyoto Protocol to increase the flexibility and reduce the costs of making greenhouse-gas emissions cuts: Joint Implementation in Article 6, the Clean Development Mechanism in Article 12, and International Emissions Trading in Article 17.

Kyoto Protocol: The Protocol to the UNFCCC setting out Annex B Party Quantified Emission Limitation and Reduction Commitments, and establishing the three 'Kyoto Mechanisms', Joint Implementation in Article 6, the Clean Development Mechanism in Article 12, and International Emissions Trading in Article 17. Adopted at the third COP held in Kyoto, Japan in December 1997, the Kyoto Protocol commits industrialized country signatories to reduce their greenhouse gas emissions compared with 1990 emissions, in the period 2008–12.

Land use, land-use change, and forestry (LULUCF): A greenhouse gas inventory sector that covers emissions and removals of greenhouse gases resulting from direct human-induced land use, land-use change and forestry activities.

Least Developed Countries (LDCs): The World's poorest countries. The criteria currently used by the Economic and Social Council (ECOSOC) for designation as an LDC include low income, human resource weakness, and economic vulnerability. Currently 50 countries have been designated by the UN General Assembly as LDCs.

Least Developed Country Fund (LDCF): A fund established to support a work programme to assist Least Developed Country Parties to carry out, inter alia, the preparation and implementation of national adaptation programmes of action (NAPAs). The Global Environment Facility, as the entity that operates the financial mechanism of the Convention, has been entrusted to operate this fund.

Marrakech Accords: Agreements reached at COP-7 which set various rules for operationalizing the more complex provisions of the Kyoto Protocol. Among other things, the accords include details for establishing an AAU based trading system; implementing and monitoring the Protocol's Clean Development Mechanism and Joint Implementation; and setting up and operating three funds to support efforts to adapt to climate change.

Monitoring Plan: A set of requirements for monitoring and verification of GHG Reductions achieved by a project.

No-Lose Targets: Best-effort emission reduction targets that do not trigger any penalty or other liability in case the target is not being achieved.

Project Design Document: A project-specific document required under CDM which will enable the DOE to determine whether the project (i) has been approved by the parties involved in a project, (ii) would result in reductions of greenhouse gas emissions that are additional, (iii) has an appropriate Monitoring Plan.

Quantified Emission Reduction and Limitation Commitment: These are the target GHG emission reduction limitations that Annex B Parties agreed to be bound to under the Kyoto Protocol.

Reforestation: Replanting of forests on lands that previously contained forests but that have been converted to some other use.

Registration: The formal acceptance by the CDM Executive Board of a validated project as a CDM project activity.

Removal unit (RMU): A Kyoto Protocol unit that is generated in Annex I Parties by LULUCF activities that absorb carbon dioxide.

Sectoral Approaches: This term describes the various models and proposals for creating positive incentives for the reduction of emissions from particular sectors of a country's economy. Sectoral approaches can be based on sectoral CDM, policy crediting, or No-Loose Targets.

Small-scale Projects: Projects which are compatible with the definition of 'Small-scale CDM Project activities' set out in decision 17/CP.7 of the Conference of Parties to the UNFCCC.

Special Climate Change Fund (SCCF): The SCCF was established to finance projects relating to adaptation; technology transfer and capacity building; energy, transport, industry, agriculture, forestry and waste management; and economic diversification. This fund should complement other funding mechanisms for the implementation of the Convention. The Global Environment Facility (GEF), as the entity that operates the financial mechanism of the Convention, has been entrusted to operate this fund

Subsidiary body: A committee that assists the Conference of the Parties. Two permanent subsidiary bodies are created by the Convention: the Subsidiary Body for Implementation (SBI) and the Subsidiary Body for Scientific and Technological Advice (SBSTA). COP-1 also established two temporary bodies: the Ad hoc Group on the Berlin Mandate, which concluded its work on 30 November 1997, and the Ad hoc group on Article 13. Additional subsidiary bodies may be established as needed.

Subsidiary Body for Implementation (SBI): The SBI makes recommendations on policy and implementation issues to the COP and, if requested, to other bodies.

Subsidiary Body for Scientific and Technological Advice (SBSTA): The SBSTA serves as a link between information and assessments provided by expert sources (such as the IPCC) and the COP, which focuses on setting policy.

Track-two JI: One of two approaches for verifying emission reductions or removals under joint implementation, whereby each JI project is subject to verification procedures established under the supervision of the Joint Implementation Supervisory Committee. Track-two procedures require that each project by reviewed by an accredited independent entity.

United Nations Framework Convention on Climate Change: The multilateral Treaty adopted in June 1992 that commits Parties to stabilize anthropogenic (ie human-induced) GHG emissions to levels that would prevent dangerous anthropogenic interference with the climate system. In December 1997, the Parties to the UNFCCC adopted the Kyoto Protocol.

UNFCCC Secretariat: The office staffed by international civil servants responsible for 'servicing' the UNFCCC Convention and ensuring its smooth operation. The Secretariat makes arrangements for meetings, compiles and prepares reports, and coordinates with other relevant international bodies. The UNFCCC Secretariat, which is based in Bonn, Germany, is institutionally linked to the United Nations.

Validation: The assessment of a project's design, including its baseline, before the implementation of the project against the requirements of the CDM, by an independent third party.

Verification: The periodic independent review and ex post determination by an independent third party of GHG Reductions that have occurred during the relevant period as a result of a project activity.

Verification Report: A report prepared by an independent third party, pursuant to a Verification, which reports the findings of the Verification process, including the amount of GHG Reductions that have been found to have been generated.

Voluntary Carbon Standard: Certification schemes for emission credits not regulated under the Kyoto Protocol.

Index